Cancer
Epigenetics

Cancer Epigenetics

Edited by
Trygve Tollefsbol

CRC Press
Taylor & Francis Group
Boca Raton London New York

CRC Press is an imprint of the
Taylor & Francis Group, an **informa** business

CRC Press
Taylor & Francis Group
6000 Broken Sound Parkway NW, Suite 300
Boca Raton, FL 33487-2742

First issued in paperback 2019

ISBN-13: 978-1-4200-4579-6 (hbk)
ISBN-13: 978-0-367-38686-3 (pbk)

Library of Congress Cataloging-in-Publication Data

Cancer epigenetics / edited by Trygve Tollefsbol.
    p. ; cm.
  Includes bibliographical references and index.
  ISBN-13: 978-1-4200-4579-6 (hardcover : alk. paper)
  ISBN-10: 1-4200-4579-2 (hardcover : alk. paper)
  1. Cancer--Genetic aspects. 2. Epigenesis. 3. DNA--Methylation. 4. Post-translational modification.
I. Tollefsbol, Trygve O.
  [DNLM: 1. Neoplasms--genetics. 2. DNA Methylation. 3. Epigenesis, Genetic. 4.
Histones--metabolism. QZ 202 C21523 2009]

RC268.4C3493 2009
616.99'4042--dc22   2008012865

Visit the Taylor & Francis Web site at
http://www.taylorandfrancis.com

and the CRC Press Web site at
http://www.crcpress.com

Contents

PART III   Other Epigenetic Aspects of Cancer

PART IV   Epigenetics in the Diagnosis, Prognosis and Therapy of Cancer

PART V   Future Directions

Preface

Since the field of cancer epigenetics is relatively new, the number of books focusing on this topic is currently rather small. However, given the international interest in this field and its recent expansion, a new book on cancer epigenetics seems appropriate. This book covers aspects of cancer epigenetics ranging from the role of epigenetics in the basic mechanisms of tumorigenesis to the newest epigenetic drugs being developed or used for cancer therapy. Five key aspects of cancer epigenetics are highlighted: (1) DNA methylation and cancer, (2) histone modifications in cancer, (3) other epigenetic aspects of cancer, (4) epigenetics in the diagnosis, prognosis, and therapy of cancer, and (5) future directions in epigenetic cancer research.

The field of cancer epigenetics is now the focus of many of the most exciting and significant advances in cancer research. DNA methylation has long been associated with cancer and it is now apparent that alterations in DNA methylation are not only intimately involved in tumorigenesis but also diagnosis, prognosis, and therapeutic approaches relating to DNA methylation are on the horizon. Histone modifications such as changes in acetylation and methylation are also rapidly emerging as major epigenetic factors in tumorigenesis. In addition, modifications of histones are important in the diagnosis, prognosis, and treatment of cancer.

Many dietary and environmental factors have been shown to modulate epigenetic processes such as DNA methylation and histone modification, and may have important implications not only in cancer prevention but also in slowing cancer progression and in cancer therapy. Epigenetic processes such as key RNA modifications and chromosome position effects as well as changes mediated by the polycomb group of proteins are gaining increasing importance in cancer. Major aspects of cancer such as its age relatedness appear to be influenced by epigenetic processes.

This book is intended for those with interests ranging from the basic mechanisms of tumor biology to cancer therapy. Undergraduate and graduate students, postdoctoral fellows, university researchers, pharmaceutical companies interested in anticancer drug development, and biotechnology companies should find this book useful in illuminating the most recent advances in epigenetic cancer research.

It is clear that epigenetics is rapidly moving to the forefront of cancer research and the purpose of this book is to provide the most up-to-date and relevant information pertaining to the role and potential of epigenetics in cancer.

Trygve O. Tollefsbol

Editor

Dr. Trygve O. Tollefsbol is a tenured professor in the Department of Biology at the University of Alabama at Birmingham. His graduate work at the University of North Texas Health Sciences Center led to the discovery that induction of glycolysis is impaired during transformation of aging lymphocytes, which contributes to the increased incidence of cancer in aging. He continued this research as a postdoctoral fellow and junior faculty member in the Department of Medicine at Duke University. Dr. Tollefsbol later sought additional training at the University of North Carolina in epigenetics and was joint first author on a *PNAS* paper reporting the first use of reversal of 5-methylcytosine deamination to resurrect an extinct ancestral gene. Additional work led to the first cloning and characterization of the mammalian maintenance DNA methyltransferase in *E. coli*. At least 14 of his publications on DNA methylation and telomerase in cancer genetics have received international recognition as high impact articles. His laboratory has discovered that the telomerase gene is induced as normal cells are exposed to defined genetic elements and undergo cancer formation. Dr. Tollefsbol has published about 80 scientific papers dealing primarily with the epigenetics of cancer and he has edited several books including *Epigenetics Protocols*; *Telomerase Inhibition: Strategies and Protocols*; and *Biological Aging: Methods and Protocols*. He also serves on the editorial board of *Molecular Biotechnology* and *The Open Aging Journal*.

Contributors

Lucy G. Andrews
Department of Biology
University of Alabama at Birmingham
Birmingham, Alabama

Ada Ao
Department of Biochemistry
  and Molecular Biology
University of Florida College
  of Medicine
Gainesville, Florida

Wenlong Bai
Department of Pathology and Cell Biology
University of South Florida College
  of Medicine and H. Lee Mottiff
  Cancer Center
Tampa, Florida

Jan Barciszewski
Institute of Bioorganic Chemistry
Polish Academy of Sciences
Poznan, Poland

Joel B. Berletch
Department of Biology
University of Alabama at Birmingham
Birmingham, Alabama

Jeremiah Bernier-Latmani
Institute of Pathology
University of Lausanne
Lausanne, Switzerland

Ching-Shih Chen
Division of Medicinal Chemistry
  and Pharmacognosy
College of Pharmacy
The Ohio State University
Columbus, Ohio

Ann-Lii Cheng
Department of Internal Medicine and Oncology
National Taiwan University Hospital
Taipei, Taiwan

Sang-Woon Choi
Vitamins and Carcinogenesis Laboratory
Jean Mayer USDA Human Nutrition Research
  Center at Tufts University
Boston, Massachusetts

Craig A. Cooney
Department of Biochemistry and Molecular
  Biology
University of Arkansas for Medical Sciences
Little Rock, Arkansas

Tim Crook
Cancer Genetics and Epigenetics Laboratory
The Toby Robins Breakthrough Breast Cancer
  Centre
Institute for Cancer Research
London, England

J. Tyson DeAngelis
Department of Biology
University of Alabama at Birmingham
Birmingham, Alabama

Melanie Ehrlich
Hayward Human Genetics Center
Department of Biochemistry
Tulane Medical School
New Orleans, Louisiana

and

Tulane Cancer Center
Tulane Medical School
New Orleans, Louisiana

Ada Elgavish
Department of Genetics
University of Alabama at Birmingham
Birmingham, Alabama

Simonetta Friso
Department of Clinical and
  Experimental Medicine
University of Verona School of Medicine
Verona, Italy

Edna Gordian
Miami Veterans Affairs Medical Center
Miami, Florida

Eleftheria Hatzimichael
Department of Hematology
University Hospital of Ioannina
Ioannina, Greece

Debby M.E.I. Hellebrekers
Department of Pathology
GROW School for Oncology
  and Developmental Biology
Maastricht University
Maastricht, The Netherlands

Tung I. Hsieh
Department of Pathology
  and Cell Biology
University of South Florida College
  of Medicine
Tampa, Florida

Shi Huang
The Burnham Institute
  for Medical Research
La Jolla, California

Samuel K. Kulp
Division of Medicinal Chemistry
  and Pharmacognosy
College of Pharmacy
The Ohio State University
Columbus, Ohio

Deepak Kumar
Department of Biological and
  Environmental Sciences
University of the District
  of Columbia
Washington, DC

Siavash K. Kurdistani
Biological Chemistry
David Geffen School of Medicine
University of California
  at Los Angeles
Los Angeles, California

Jianrong Lu
Department of Biochemistry and Molecular
  Biology
University of Florida College of Medicine
Gainesville, Florida

Yen-Shen Lu
Department of Oncology
National Taiwan University Hospital
Taipei, Taiwan

Matthew A. McBrian
Department of Biological Chemistry
David Geffen School of Medicine
University of California at Los Angeles
Los Angeles, California

Richard R. Meehan
Human Genetics Unit
Medical Research Council
Edinburgh, United Kingdom

Peter L. Molloy
Commonwealth Scientific and Industrial
  Research Organization
Molecular and Health Technologies
North Ryde, New South Wales, Australia

Vijayalakshmi Nandakumar
Department of Biology
University of Alabama at Birmingham
Birmingham, Alabama

Santo V. Nicosia
Department of Pathology and Cell Biology
University of South Florida College
  of Medicine and H. Lee Mottiff
  Cancer Center
Tampa, Florida

Hanneke E.C. Niessen
Department of Molecular Genetics
GROW School for Oncology and
  Developmental Biology
Maastricht University
Maastricht, The Netherlands

Panagiotis Papageorgis
Department of Medicine
Boston University School
  of Medicine
Boston, Massachusetts

and

Department of Genetics and Genomics
Boston University School of Medicine
Boston, Massachusetts

Jacob Peedicayil
Department of Pharmacology
  and Clinical Pharmacology
Christian Medical College
Vellore, India

Sari Pennings
Queen's Medical Research Institute
University of Edinburgh
Edinburgh, United Kingdom

Sharla M.O. Phipps
Department of Biology
University of Alabama at Birmingham
Birmingham, Alabama

Kavitha Ramachandran
Sylvester Comprehensive Cancer Center
University of Miami
Miami, Florida

Sabita N. Saldanha
Department of Biology
University of Alabama at Birmingham
Birmingham, Alabama

Aaron M. Sargeant
Division of Medicinal Chemistry
  and Pharmacognosy
College of Pharmacy
The Ohio State University
Columbus, Ohio

and

Department of Veterinary Biosciences
College of Veterinary Medicine
The Ohio State University
Columbus, Ohio

David B. Seligson
Pathology and Laboratory Medicine
David Geffen School of Medicine
University of California at Los Angeles
Los Angeles, California

Rajeshwar Nath Sharan
Department of Biochemistry
North-Eastern Hill University
Shillong, India

Phillip Shaw
Institute of Pathology
University of Lausanne
Lausanne, Switzerland

Rakesh Singal
Sylvester Comprehensive Cancer Center
University of Miami
Miami, Florida

Justin Stebbing
Department of Medical Oncology
Imperial College School of Science,
  Medicine, and Technology
London, England

Maciej Szymański
Institute of Bioorganic Chemistry
Polish Academy of Sciences
Poznan, Poland

Sam Thiagalingam
Department of Medicine
Boston University School of Medicine
Boston, Massachusetts

and

Department of Genetics and Genomics
Boston University School of Medicine
Boston, Massachusetts

and

Department of Pathology and Laboratory
  Medicine
Boston University School of Medicine
Boston, Massachusetts

Trygve O. Tollefsbol
Department of Biology
University of Alabama at Birmingham
Birmingham, Alabama

Manon van Engeland
Department of Pathology
GROW School for Oncology and
  Developmental Biology
Maastricht University
Maastricht, The Netherlands

Mukesh Verma
Analytical Epidemiology Research Branch
National Cancer Institute
Bethesda, Maryland

Jan Willem Voncken
Department of Molecular Genetics
GROW School for Oncology and
  Developmental Biology
Maastricht University
Maastricht, The Netherlands

Sabrina L. Walthall
Department of Biology
University of Alabama
  at Birmingham
Birmingham, Alabama

Xiaohong Zhang
Department of Pathology and Cell Biology
University of South Florida College
  of Medicine and H. Lee Mottiff
  Cancer Center
Tampa, Florida

1 Role of Epigenetics in Cancer

Trygve O. Tollefsbol

CONTENTS

1.1   INTRODUCTION

Epigenetics is the study of heritable changes in gene regulation that do not involve a change in the DNA sequence or the sequence of the proteins associated with DNA. The most studied of epigenetic changes is that of DNA methylation, which occurs primarily in CpG dinucleotides and is often altered in cancer cells. Both hypermethylation and hypomethylation contribute to tumorigenesis. Gene-specific hypermethylation can affect many different types of cancers and is often mediated through the silencing of tumor suppressors [1]. Hypomethylation can contribute to genomic instability that is frequently observed in cancer [2], activation of oncogenes, or loss of imprinting which are also causes of oncogenesis [3].

Histone modifications are also important in cancer and result in dramatic changes in chromatin structure as well as the accessibility of DNA to transcription factors that mediate gene expression. For example, histone acetylation has been associated with an increase in gene activity whereas histone deacetylation normally prevents transcription [4]. Other histone modifications such as histone methylation can have varied effects and taken together the many types of histone alterations seen in cancer have a major impact not only on the initiation of cancer processes but also on its progression potentially to malignant cells [5].

1.2   CONNECTION BETWEEN DNA METHYLATION AND CANCER

Hypomethylation of the genome in cancer impacts not only single-copy genes such as oncogenes, but also repeat sequences. As outlined in Chapter 2 of this book, there is great variability of hypomethylation which applies not only to different types of cancers, but also within the same type of cancer. Genomic hypomethylation increases the probability of genomic rearrangements through chromosomal instability and also contributes to oncogene activation that can lead to the development of cancer. Clinical advances are currently being made toward the classification of tumors based on genomic hypomethylation as well as the diagnosis and prognosis of cancer. Silencing of genes through hypermethylation can contribute to many of the hallmarks of cancer including evading apoptosis, insensitivity to antigrowth signals, sustained angiogenesis, limitless replicative potential, and tissue evasion and metastasis (Chapter 3). Hypermethylation may provide one of the most promising targets of cancer therapy and major advances are currently underway to further develop drugs that target hypermethylation abnormalities in cancer.

The maintenance of proper parent of origin gene expression is often aberrant in cancer leading to loss of imprinting (Chapter 4). This alteration is observed in many different types of cancer cells and frequently influences the timing of key developmental events or potentially the loss of imprinting in stem cells leading to the genesis of cancer. Of course DNA methylation aberrations in cancer are modulated through many different types of proteins such as the DNA methyltransferases and methyl-CpG-binding proteins. The imbalance in these proteins and the role that they have in the epigenetic events that contribute to cancer are detailed in Chapter 5. The progression of cancer is also greatly impacted by DNA methylation (Chapter 6) which is involved in gene silencing, chromosomal instability, and differentiation of cancer stem cells.

1.3 HISTONE ALTERATIONS IMPACT CANCER

Many genes that are altered in cancer cells are the result of histone modifications. In Chapter 7, the alterations in histone modifications that influence genes such as hTERT, c-Myc, p21, p53, Sp1, Mad, and Mnt that are involved in tumorigenesis are discussed. The role of histone methylation in the initiation of cancer (Chapter 8) may be controlled through dietary influences. For instance, loss of histone methyltransferase function is frequently associated with carcinogenesis and chronic changes in the functions of these enzymes may contribute to heritable changes in cell phenotypes leading to sporadic cancers. The sirtuins are a member of the SIR2 family of histone deacetylases (HDAC) that have been associated with aging. However, SIRT1 is a mammalian histone deacetylase that undergoes an increase in expression in cancer cells due to loss of tumor suppressors that normally suppress SIRT1 expression (Chapter 9). The increase in SIRT1 may promote tumor growth and progression through anti-apoptotic effects suggesting that SIRT1 inhibitors are potential compounds for cancer therapy.

As is the case with the role of proteins in altering DNA methylation in cancer, many proteins can alter histone modifications in tumor cells (Chapter 10). The alteration of the expression, activity, or targeting of histone modifying enzymes such as the HDACs is central to both the initiation and progression of cancer. It is through knowledge of the role of histone modifying proteins that has led to many of the extant histone-modifying anticancer drugs that have been developed. Other interventions in reducing cancer risk or progression involve the impact of dietary influences on histone modifications in cancer (Chapter 11). Several dietary ingredients such as butyrate, which is generated in the colon by the fermentation of dietary fiber, have histone deacetylase inhibition properties that inhibit cancer progression. Environmental factors such as occupational exposure to insoluble nickel compounds can inhibit the activity of histone acetyltransferases. This can lead to hypoacetylation of histone H4 and the eventual generation of reactive oxygen species. Thus, as is the case for DNA methylation, histone alterations are also important causes of cancer and the subject of intense interest for a number of potential modes for cancer intervention.

1.4 IMPACT OF OTHER EPIGENETIC PROCESSES ON CANCER

Besides DNA methylation and histone modification alterations in cancer, there are many other epigenetic processes that can contribute to cancer. Noncoding RNAs comprise a significant transcriptional output from eukaryotic genomes and many noncoding RNAs affect transcriptional and posttranscriptional levels of numerous genes that can lead to the development of cancer (Chapter 12). Chromosome position effect is another well-known epigenetic process and chromosome position can affect DNA methylation (Chapter 13). For example, DNA repeats can be subject to chromatin position effects in that hypermethylation of the repeat can occur in cancer while the repeat boundary may either be resistant to hypermethylation or become hypomethylated in cancer. In Chapter 14, the importance of chromatin remodeling in cancer is discussed whereby a combination of chromatin modifying and remodeling activities contribute to the initiation or maintenance of altered gene expression patterns in cancer.

Another epigenetic process is a posttranslational modification of proteins referred to as poly-ADP-ribosylation (PAR). For instance, PAR can lead to ADP-ribose polymer adducts on histone proteins thereby contributing to aberrations of gene expression in cancer (Chapter 15). The polycomb group (PcG) proteins also influence chromatin and can maintain gene silencing through covalent modification of histone tails. Specific PcG proteins are expressed at high levels in various cancers (Chapter 16) and these proteins also appear to be important in cancer stem cells. Both aging and cancer are under epigenetic influence and in Chapter 17 a hypothesis is described whereby age-related epigenetics is influenced by acetate metabolism thereby leading to alterations in histone acetylation and gene activity that are known causes of cancers.

Therefore, there are many different types of epigenetic alterations that influence cancer other than DNA methylation and histone modifications. These epigenetic alterations include but are not limited to noncoding RNAs, chromosome position effects, chromatin remodeling, PAR, PcG proteins, and the aging process itself.

1.5   EPIGENETICS IN THE DIAGNOSIS, PROGNOSIS, AND THERAPY OF CANCER

Early detection of cancer is critical in reducing the impact of this disease. Recent studies detailed in Chapter 18 have shown that changes in DNA methylation can be detected very early in tumor development. Sensitive methylation assays can detect changes in DNA methylation characteristic of cancer in body fluids and in the circulation thereby increasing the probability that DNA methylation analyses may play a key role in the diagnosis of cancer. Detection of DNA methylation profiles is also important in the prognosis of cancer (Chapter 19). It is now possible to detect genome-wide DNA methylation profiles that correlate with the prognosis of cancer in a number of human tissues as well as body fluids and serum. The major challenges facing this area of investigation consist of their inconsistency in allowing a close association between the DNA methylation changes and the clinicopathologic stages of cancer as well as the availability of reliable biomarkers. Histone modifications are also important in cancer and several histone modifications have recently been identified that can serve as diagnostic markers of a number of different types of cancers (Chapter 20). Early detection of tumorigenesis through histone modification analysis could greatly reduce not only the incidence of cancer, but also its morbidity and mortality rates. The prognosis of cancer may also be predictable in some cases through histone modification analyses as prognostic biomarkers of cancer as discussed in Chapter 21. Various histone modifications can be detected in primary tumor tissues using immunohistochemical analysis that appear to be related to clinical outcome in many cases of cancer.

Histone deacetylase inhibitors have rapidly advanced as a mode for effective cancer therapy. These inhibitors are a diverse group of synthetic and natural compounds that have anticancer activities affecting histone-mediated gene expression as well as nonhistone proteins in cancer cells such as signal transduction mediators and transcription factors (Chapter 22). A number of epigenetic drugs have been developed that are either approved for anticancer therapy or are in clinical trials (Chapters 23 and 24). These compounds have been especially promising in cases of hematologic malignancies but are also being evaluated for other cancers as well. The DNA hypomethylating nucleoside analogs, 5-azacytidine (Vidaza) and 5-aza-2'-deoxycytidine (Decitabine, Dacogen), have been approved by the FDA for the treatment of myelodysplastic syndromes and it is hoped that additional DNA methylation-modifying drugs may also advance to FDA approval. HDAC inhibitors such as suberoylanilide hydroxamic acid (Zolinza, vorinostat) have been approved for treatment of cutaneous T-cell lymphoma. Therefore, there is considerable promise that drugs that target not only DNA methylation and histone acetylation but also other epigenetic processes will continue to be developed for anticancer therapy.

1.6 FUTURE DIRECTIONS IN EPIGENETIC CANCER RESEARCH

Since epigenetics is now on the leading edge of cancer research, advances in this field of research are likely to have major implications in the diagnosis, prevention, and treatment of cancer (Chapter 25). Especially promising areas are in the fields of new epigenetic targets for anticancer drugs, epigenetic aspects of cancer stem cells, and advances in genomic approaches in cancer epigenetics. The field of cancer epigenetics is currently in a phase of exponential growth and it is likely that this will continue for quite some time in light of the key roles epigenetic processes play in cancer. With the development of new drugs for cancer, we will be increasingly armed with novel approaches for combating this prevalent and potentially devastating disease. Advances in the epigenetics of cancer stem cells could lead to ways to arrest many cancers well before they place patients at risk. And finally, genomic approaches to cancer epigenetics have considerable potential in many different areas but may hold the most promise in enhancing the diagnosis and prognosis of cancer. Together, these epigenetic approaches as well as many others that are arising provide promise in greatly reducing the incidence and severity of cancer.

REFERENCES

1. Jones, P.A. and Laird, P.W., Cancer epigenetics comes of age. *Nat. Genet.*, 21, 163, 1999.
2. Feinberg, A.P. and Tycho, B., The history of cancer epigenetics. *Nat. Rev. Cancer*, 4, 143, 2004.
3. Liu, L., Wylie, R.C., Andrews, L.G., and Tollefsbol, T.O., Aging, cancer and nutrition: The DNA methylation connection. *Mech. Ageing Dev.*, 124, 989, 2003.
4. Verdone, L., Caserta, M., and Di Mauro, E., Role of histone acetylation in the control of gene expression. *Biochem. Cell Biol.*, 83, 344, 2005.
5. Lehrmann, H., Pritchard, L.L., and Harel-Bellan, A., Histone acetyltransferases and deacetylases in the control of cell proliferation and differentiation. *Adv. Cancer Res.*, 86, 41, 2002.

Part I

DNA Methylation and Cancer

2 DNA Hypomethylation in Cancer

Peter L. Molloy

CONTENTS

2.1 INTRODUCTION

In DNA isolated from normal healthy tissues, about 3.5%–4% of cytosines are methylated [1,2]. Methylation in the genome is not distributed evenly—promoters and CpG islands are preferentially found in the unmethylated compartment, as to a lesser extent are phylogenetically conserved sequences of unidentified function [3]. The methylated domain of the genome comprises satellite and interspersed repeat sequences, intragenic sequences (excluding the first exon), and unannotated single-copy sequence; however, some promoters and CpG islands (biased toward shorter ones) were identified in the methylated DNA fraction of brain DNA [3]. Genome scanning approaches [4,5] have identified that the promoters of a number of genes are methylated in a tissue-specific manner and that lack of methylation correlated with expression of associated genes. Disruption of epigenetic regulation in cancer involves both hypermethylation of specific sequences [considered in Chapters 3 and 6] and hypomethylation of other sequences and regions. The methylated compartment of the genome that may become hypomethylated thus includes all classes of repeated sequence, intragenic and single-copy intergenic sequences, the inactive X chromosome, and imprinted regions, as well as a subset of promoters or CpG islands that show tissue-specific methylation. The specific case of imprinted genes is reviewed in Chapter 4.

Measurements of the total level of 5 methyl cytosine (5meC) initially identified that the genomes of cancer cells and tissues were frequently hypomethylated relative to the DNA of healthy tissues [6,7]. While hypomethylation at specific sites of individual genes was shown in some instances, it soon became apparent that hypomethylation of repeat DNA sequences was the dominant factor in the overall reduction in methylation levels. Studies of DNA methylation over the past 20 years have led to the prevailing view that development of cancer is accompanied by widespread epigenetic changes involving global hypomethylation, particularly of repeat DNA sequences, accompanied by focal hypermethylation of multiple CpG island gene regulatory regions [8–10]. The role of DNA hypermethylation in gene silencing can often be understood in terms of a biological function contributing to cancer development, such as silencing of tumor suppressor genes or DNA repair proteins and in a number of instances its functional importance can be demonstrated. Compared with gene-specific hypermethylation, the role of hypomethylation is less well understood and characterized. In this chapter I will review the current state of our knowledge of the nature and extent of hypomethylation in different cancers and consider our understanding of the underlying molecular mechanisms and the potential impact of demethylation of specific sequences in activating pro-carcinogenic gene expression and of repeat sequence hypomethylation in disrupting chromosome integrity and regulation of gene expression.

2.2 HYPOMETHYLATION OF SINGLE-COPY GENES IN CANCER

2.2.1 PREVALENCE OF SINGLE-COPY GENE HYPOMETHYLATION IN CANCER

Analysis of the 5meC content of the low copy number fraction of tumor DNA (fractionated by reassociation kinetics) provided an early indication that hypomethylation in cancer extended to single-copy genes [6]. Combined with the developing evidence that DNA methylation was associated with gene silencing, this prompted studies of the methylation status of proto-oncogenes. Initially hypomethylation of HhaI and HpaII sites in the H-ras, oncogene was identified in colon and lung cancers [11] and hypomethylation of HpaII sites in the 3′ region H-ras was demonstrated to be common in nonsmall cell lung carcinoma and to be associated with allele loss [12]. While promoter sequences of the c-myc gene were unmethylated in all normal tissues, a HpaII site in the third exon was consistently methylated in normal tissue but found to be unmethylated in tumor cell lines [13] and in hepatocellular carcinoma [14]. Hypomethylation of this specific HpaII site has been observed in a number of cancer types and has been associated with progression in colorectal cancer [15] and with infiltration or metastasis in hepatocellular carcinoma [16]. Despite some evidence that this HpaII site may lie in a regulatory region, there is no clear evidence that methylation in the c-myc third exon downregulates its expression and could have a causative role in cancer progression. Since these original studies there have been many examples described of hypomethylation of individual genes in cancers. In many cases (as for myc and ras genes) hypomethylation involves sequences within the body or 3′ end of genes that are methylated in normal tissue DNA and it is likely that most of these are not related to expression of the genes nor significant in terms of cancer development [10,17]. Indeed, as the concept developed that most promoter regions comprised CpG islands and that these were unmethylated in all tissues, the focus methylation studies shifted heavily toward hypermethylation of CpG islands and silencing of associated genes as key epigenetic events contributing to cancer development.

More recently genome-wide analyses of gene expression and genome scanning methods for detection of DNA methylation differences have expanded considerably the number of examples of hypomethylation accompanied by gene activation in cancer. From gene expression profiling of pancreatic cancers, Goggins and colleagues [18,19] identified promoter demethylation associated with a number of activated genes. Among 32 genes with elevated expression 7 were methylated in normal tissue and hypomethylated in pancreatic cancer. Using an approach normally applied for detection of genes hypermethylated in cancer Nishigaki et al. [20] treated gastric cancer cell lines with 5-azacytidine or trichostatin; among over 1300 genes that were shown to be reactivated in at least one cell line they identified 159 genes that were not expressed in normal gastric mucosa but were expressed in at least 1 of 33 gastric cancers. This list contained a number of potential oncogenes. Restriction enzyme-based approaches using two-dimensional gel display, restriction landmark genomic sequencing [21], and selective amplification methods based on restriction digestion with methylation sensitive enzymes allow sampling of large number of sites across the genome and have been used to identify hypomethylated sites. For example, Frigola et al. [22] identified a greater number of hypomethylation than hypermethylation differences in colorectal cancer DNA when compared with normal DNA, while Yoshida and colleagues [23] identified three genes, CACNA1H, Nogo receptor, and MEL1S as hypomethylated in adult T-cell leukemia. Activation of MEL1S is particularly interesting since it involves hypomethylation of an alternative promoter, while the normal MEL1 gene promoter was hypermethylated. Application of these and emerging approaches to genome-wide methylation analysis [24,25] are likely to significantly expand the list of genes that become hypomethylated and expressed in different cancers. Lists of hypomethylated genes can be found in two recent reviews [17,26]. The genes activated in cancer fall into different classes and informative examples from these are discussed below.

2.2.2 ONCOGENES

For classical oncogenes, there is limited evidence of activation through promoter demethylation. This is not surprising, as their promoters are normally unmethylated and increased activity arises through mutation or gene amplification. Demethylation of HpaII sites at the 5′ end of the Bcl2 gene has been observed in B-cell chronic lymphocytic leukemia (CLL), correlating with Bcl2 expression [27]. Among genes identified in an expression scan by Nishigaki et al. [20], R-ras was shown to be activated in 55% of gastric cancers and expression was related to increased hypomethylation of CpG sites in the first intron; however, methylation in normal tissue was partial and heterogeneous.

These data require further follow-up but could be indicative of a class of oncogene that is expressed in a tissue or developmentally specific manner, methylated in nonexpressing tissues, but subject to hypomethylation-associated reactivation in cancer. Among oncogenes the best characterized example of cancer-associated hypomethylation is of developmentally important paired homeobox gene PAX2 that is also expressed in a number of cancer types [28]. Wu et al. [29] have demonstrated that PAX2 is silent and methylated in normal endometrial epithelial cells but its promoter is demethylated and it becomes active and responsive to estrogen and tamoxifen in a high proportion of endometrial cancers and in isolated cancer endometrial epithelial cells.

Surprisingly some of the genes reported to be hypomethylated in cancers are normally recognized as tumor suppressor genes. For the p53 gene hypomethylation of exons 5–8 is associated with a twofold increased risk of lung cancer, most likely due to increased risk of mutation rather than any effect on gene expression. Chan et al. [30] reported the presence of methylated BRCA2 promoter sequences in normal ovarian tissue but not in tumor tissue; the reduced methylation in tumor tissue correlated with elevated BRCA2 transcription. However, most of the DNA in the normal tissue DNA was also unmethylated. So the significance of the lack of methylation in the tumor DNA is not clear, and may have resulted from the increased transcription. A clearer example is that of the 14-3-3σ gene that has been found to be methylated in a number of cancers including prostate and breast [31]. In contrast 14-3-3σ is not expressed in normal pancreas where its promoter is methylated, but its expression is commonly elevated in pancreatic cancer, suggesting different tissue-specific functional roles.

2.2.3   GENES INVOLVED IN INVASION AND METASTATIC SPREAD

A significant set of genes that become hypomethylated in cancer have an impact in invasion and metastatic spread. This includes proteases such urokinase type plasminogen activator (uPA), the expression of which has been associated with invasiveness in a number of cancer types. Targeted knockdown of uPA by RNA interference has demonstrated its role in cell survival, tumorigenicity, and invasion in prostate cancer xenograft models [32,33]. Its expression is regulated by promoter methylation in prostate and breast cancer models [32,34] and cancer-linked demethylation is associated with progression and metastasis in breast [34] and prostate cancer [33].

Maspin, a serine protease inhibitor of the serpin family, is a multifunctional protein that has a tissue-specific expression profile that is regulated by DNA methylation and chromatin structure [35]. In tissues where maspin is normally expressed, such as breast and prostate epithelium it is implicated in suppression of invasion and metastasis and is recognized as a tumor suppressor. In breast and prostate cancers silencing of maspin expression associated with promoter methylation is a poor prognostic indicator. In contrast, in a number of tissues where maspin is not normally expressed, it is commonly demethylated and expressed in cancers. Its expression is associated with progression in pancreatic and gall bladder [19], thyroid [36], gastric [37], and colorectal [38] cancers where it is found more frequently in cancers with microsatellite instability (MSI). In the case of colorectal cancer maspin is nuclear localized and expression is highest at the invasive front. Given the apparent multifunctional nature of maspin, it is likely that it is acting at a different point in these cases where its activated expression relates to cancer progression.

Gamma synuclein expression is normally restricted to the brain, but it is activated in a wide variety of cancers. Originally identified as being activated in breast and ovarian cancer [39], a CpG island in exon 1 is demethylated and synuclein-γ expressed in a wide variety of other cancers including, liver, esophagus, lung, colon, cervical, prostate [40], pancreatic [41], and gastric cancer [42]. In most of these cancers expression is strongly associated with advanced disease and poor prognosis [40,43]. Synuclein-γ has been implicated in a number of processes that could contribute to oncogenesis including mitotic checkpoint control and stability [44], regulation of mitogen activated protein kinase pathways through activation of ERK1/2 and inhibition of JNK [45], and as a chaperone and activator of estrogen receptor [46].

Placental S100 (S100P) is calcium-binding protein that has a wide but tissue-specific distribution of expression [47]. In a number of cancers, including breast [48] and pancreatic[19], elevated expression is associated with progression and poor prognosis. In pancreatic cancer expression has been shown to correlate with hypomethylation of its promoter.

2.2.4 CANCER/TESTIS ANTIGEN GENES

The cancer/testis (CT) antigens, typified by the MAGE genes (melanoma and germ cell expressed), code for immunogenic proteins that are recognized by cytotoxic T lymphocytes and are commonly activated in a range of cancers. As their name indicates, expression of these genes is normally restricted to testicular germ cells, though some are also expressed in ovary and trophoblast tissue [49,50]. Over 20 different families have been identified by serological analysis and mRNA profiling. Expression of a number of CT antigens has been correlated with promoter hypomethylation in tumors and cell lines and transfection experiments and treatment of cells with 5-azadeoxycytidine have established the role of methylation in regulating their expression [51]. CT antigens shown to be hypomethylated in different cancers include various members of the MAGE family (especially MAGE-A1 and MAGE-A3) in lung [52], pancreatic [53], testicular [51] gastric [54,55], ovarian [56], and colorectal [57] cancers, melanoma [58,59] and hepatocellular carcinoma [60], GAGE1–8 in pancreatic cancer [53], CAGE in gastric [55,61] and cervical cancer [62], SSX in melanoma [63], XAGE in gastric cancer [64], SPAN-X in myeloma [65], BAGE [66] and TRAG-3 in a number of cancer types [67], and HAGE-1 in CML[68]. The expression of different CT antigens varies within and between tumor types; see Refs. [65,69]. Treatment with 5-aza deoxycytidine has been shown to readily induce expression of a number of CT antigens in tumor cells, while not inducing expression of other genes normally not hypomethylated in cancer. Since this induction is specific and appears limited to tumor cells [70] it offers significant promise as an approach for elevating expression of tumor antigens and inducing a more effective T-cell response.

The biological role of CT antigens is poorly understood but there is increasing evidence for roles in transcriptional regulation. The MAGE-A1 protein acts via SKIP to inhibit histone deacetylase 1 and so function as a transcriptional repressor [71]. NY-ESO genes are close homologues of the yeast PCC1 gene that forms part of an evolutionarily conserved transcription complex [72]. Among the most extensively studied examples is that of the SSX genes in synovial carcinoma [73]. SSX proteins contain a conserved strong transcriptional repression domain [74]. Translocation fusion of the SS18 gene with the SSX1, 2, or 4 gene is common in synovial carcinoma; the fusion proteins have been shown to interact with components of the β-catenin/Wnt signaling pathway [75], to regulate a set of genes including IGF2 and CD44 as direct targets [76,77], and to be involved in disruption of epigenetic regulation [76] including loss of imprinting of IGF2 [77].

For CT antigens NY-ESO [78] and MAGE-A1 [79], it has been demonstrated that their expression is activated by the Zn finger protein BORIS (also known as CTCFL), that is a testis-expressed paralogue of the imprinting insulator factor CTCF. BORIS expression was found to be activated in a number of cancer cell lines and primary cancer samples. Data supported a model whereby expression of BORIS competed with the repressive effect of CTCF and promoted demethylation and expression of CT antigen genes (including BORIS itself). While BORIS expression is implicated in activation of a number of CT antigen genes some were not directly affected by its expression. This, and the limited activation of BORIS expression in urogenital cancers [80], indicates that there are additional routes to their activation. It is clear that CT antigen genes in cancer cells are often poised in a state susceptible to activation and that this occurs through a combination of demethylation and interaction with trans-acting factors that probably in part cell type specific.

2.2.5 NONCODING RNAS

Laner et al. [81] investigated the expression of the nonprotein-coding RNA Xist in prostate cancer tissues and cell lines and characterized a partial demethylation accompanied by slight increases in expression; this partial demethylation paralleled that if LINE repeats. The CpG island of the miRNA gene Let-7a-3 was recently identified as being demethylated in some lung adenocarcinomas [82], correlating with increased expression of the miRNA. Ectopic expression of Let-7a-3 was shown to enhance the tumorigenic phenotype of a lung cancer cell line. These data indicate the potential importance of hypomethylation of noncoding RNAs that are developmentally regulated and silenced in normal adult somatic tissues.

2.3 REPEAT SEQUENCE HYPOMETHYLATION IN CANCER

2.3.1 PREVALENCE OF REPEAT SEQUENCE HYPOMETHYLATION

In most types of cancers, significant genome hypomethylation is observed, but its extent is highly variable for individuals within a cancer type as well as between different cancers. Tables in recent reviews [2,17,26] provide lists of examples of global and repeat sequence hypomethylation. In discussing different cancers below, I will refer to "global hypomethylation" where the assay method is not specific for any sequence class; most common methods include direct determination by HPLC or HPCE of the amount of 5meC after digestion of DNA to mononucleotides, determination of the amount of ^{3}H methyl groups that can be added to genomic DNA by the CpG specific methyltransferase M.Sss1, or use of antibodies to 5meC (quantitative immunohistochemistry). Methylation of specific classes of repeat sequences has been determined by methods used commonly for analysis of single-copy genes—using digestion with methylation sensitive restriction enzymes combined with Southern blotting, amplification of bisulfite-treated DNA with primers selective for unmethylated DNA (qualitative USP by gel analysis or quantitative real time USP), or nonselective amplification of bisulfite-treated DNA followed by determination of methylation by clonal sequencing, COBRA, or pyrosequencing.

2.3.1.1 Gastrointestinal Cancer

Genomic hypomethylation has been most extensively studied in colorectal cancer and was first described over 20 years ago [7]. A clear consensus has emerged that hypomethylation occurs at the earliest stages of colorectal carcinogenesis, with benign hyperplastic polyps and small adenomas showing similar levels of hypomethylation as adenocarcinoma [83–85]. Using anti-5meC antibodies and quantitative immunohistochemistry similar levels of hypomethylation were determined for cancers of all Dukes stages A to D [86]. Similar levels of hypomethylation are seen in familial colorectal cancer syndromes (FAP and HNPCC) as for sporadic or common cancer [87]. In direct comparisons of the methylation levels of tumor and matched normal tissue DNAs almost all tumor samples showed decreased methylation (11 of 11 in Feinberg et al. [83] and 21 of 22 in Bariol et al. [84], 13/13 in Hernandez-Blasquez et al. [86]), with the average level of demethylation of tumor relative to normal being about 10% as measured by direct 5meC content, 15% by a methyl acceptance assay, and 23% by immunohistochemistry, respectively. However, the level of methylation of the normal tissue DNA varies considerably (at least partly relating to age) and there is significant overlap between the methylation level of normal colon mucosa and that seen in cancer patients [84]. In examining the colorectal cancer and matched normal colonic mucosa for hypomethylation of the LINE-1 interspersed repeat sequence Suter et al. [88] identified specific hypomethylation of LINE-1 sequences in the normal mucosa of about 1/3 of cancer patients; this hypomethylation was not evident in colonic mucosa of patients without cancer or in other healthy tissue DNAs. The presence of LINE-1 hypomethylation correlated with those DNAs showing the greatest degree of global hypomethylation in normal mucosa. The data suggest that hypomethylation may in some cases be an early precancerous epigenetic change and could contribute to tumorigenesis. A similar "field effect" of hypermethylation of the MGMT promoter in normal tissue adjacent to a cancer has also been seen in a proportion of colorectal cancer patients [89] and higher levels of methylation of a number of genes in the normal mucosa of patients with CIMP+ tumors [90].

Bariol et al. [84] found no association between degree of hypomethylation and stage or grade of tumors or a range of other clinical parameters. An association was observed, however, between the hypomethylation of hyperplastic polyps and small adenomas and their proliferative rate as determined by PCNA staining. Using an assay for unmethylated LINE-1 sequences to monitor global demethylation, Matsuzaki, Deng, and colleagues [91,92] identified a significant correlation between hypomethylation and microsatellite stable (MSS) cancers exhibiting chromosomal instability. Hypomethylation was less evident in microsatellite unstable cancers. This pattern was also evident in a panel of colorectal cancer cell lines. These results were supported and extended to other repeat families in a recent study [93].

The pattern of hypomethylation in gastric cancer is similar to that seen in colorectal cancer. Global hypomethylation of gastric carcinoma DNA relative to DNA from patients without cancer is seen in a high proportion of cancers [94–97]. Cravo et al. [94] reported hypomethylation in premalignant lesions and carcinoma compared with undiseased patients, but did not see a difference between carcinoma and adjacent nonneoplastic tissue. However later reports, [95–97], found consistent hypomethylation relative to paired normal tissue, though there is significant overlap between the levels of methylation in carcinoma and normal DNA when comparing between patients. As with colorectal cancer, the data are consistent with hypomethylation being an early event in cancer development and Suzuki et al. [97] observed an age-dependence in the level of hypomethylation.

2.3.1.2 Prostate Cancer

Genome hypomethylation in prostate cancer has been most extensively studied by following methylation of LINE-1 repetitive sequences. Approximately half of prostate cancers studied show hypomethylation relative to normal tissue [98–100]. Hypomethylation of LINE-1 sequences was found to occur progressively through prostate cancer development and to be significantly associated with stage. This contrasted with hypermethylation of common target genes that was evident in early stage cancers, but continued to accumulate during progression. Using assays for both LINE-1 and Alu elements, Cho et al. [101] found a similar relationship with stage and grade as well as a significant association with preoperative serum PSA levels. Quantitative immunohistochemistry using 5meC-specific antibodies demonstrated lowered levels of methylation in nearly all prostate cancer specimens with the degree of hypomethylation being related to progression [102]. Presumptive precursor lesions, high-grade prostatic intraepithelial neoplasia, were not found to be hypomethylated but rather showed elevated immunoreactivity compared with normal prostate tissue [103].

2.3.1.3 Breast Cancer

Using a DNA methyl acceptance assay Soares et al. [104] demonstrated an approximately two- to threefold greater methyl acceptance capacity in malignant breast tissue compared with its matched normal tissue but noted the wide variation in methylation levels between patients. Direct determination of the level of 5meC in sporadic breast cancer tissue showed substantial decrease (average >50%) compared with normal tissue [87]. Levels of hypomethylation, 42% and 30%, respectively, in BRCA1 and BRCA2 cancers, were slightly lower, but significantly so. Using stricter criteria for hypomethylation Jackson et al. [105] determined that 1/3 of breast cancers showed hypomethylation relative to a panel of normal tissues, with reductions up to about 30%. Genome wide hypomethylation correlated well with hypomethylation of satellite sequences. Hypomethylation of

Sat2 sequences on chromosome 1 in particular was specific for cancer DNA and present to varying degrees in about half of the breast cancers [105,106]. Satα hypomethylation was also observed and correlated with Sat2 hypomethylation, but did not discriminate cancer and nonneoplastic tissues significantly [105]. SATR-1 sequences (located on chromosome 5) have also been found to be significantly hypomethylated in a large proportion (86%) of breast cancer DNAs compared to normal tissue and variable hypomethylation was observed in normal tissue adjacent to breast cancers [107]. While a statistically significant association of global hypomethylation with disease stage, grade, and tumour size was observed by Soares et al. [104], other studies observed satellite repeat sequence and global hypomethylation [105,107] in early breast cancer and it did not correlate with stage, grade, or other clinical parameters.

2.3.1.4 Hepatocellular Carcinoma

While global hypomethylation of hepatocellular carcinoma (HCC) DNA compared with matched nonmalignant tissue is a normal feature of HCC, its extent varies from minimal in about a quarter of cases to strong in another quarter [108] and the degree of hypomethylation is correlated with histological grade and tumor size. This global hypomethylation is reflected in hypomethylation of interspersed LINE-1 elements [109] as well as pericentromeric Sat2 and Sat3 sequences [110,111], and the NBL2 (DE-1) tandem repeat [112]. While LINE-1 hypomethylation was restricted to HCC, demethylation of Sat2 and Sat3 sequences was evident in a significant fraction of adjacent non-tumor tissue. Karyotypic abnormalities are common in hepatocellular carcinoma, particularly chromosome 1 rearrangements that involve pericentromeric breakpoints, and Wong et al. [111] examined their relationship to satellite repeat hypomethylation. They identified a strong correlation between Sat2 hypomethylation at chromosome 1q12 and copy number gain of chromosome 1q as determined by comparative genomic hybridization. The presence of Sat2 hypomethylation in precancerous liver tissue has led to the suggestion that this leads to chromatin decondensation and subsequent chromosomal rearrangement [111,113]. Saito et al. [113] have identified a significant correlation of Sat2 hypomethylation with overexpression of a DNMT3B splice variant, DNMT3B4. Mutation of DNMT3B underlies the rare recessive disorder ICF syndrome (immunodeficiency, centromeric heterochromatin instability, facial abnormalities) and this is associated with specific hypomethylation of pericentromeric satellite sequences and chromosomal rearrangements. It has been postulated that the overexpressed DNMT3B4 may lack enzymatic activity and compete with active isoforms, inhibiting methylation of pericentromeric satellite regions [113]. The possible role of inhibition of DNMT3B activity on hepatitis B virus induced carcinogenesis has been further supported by studies on the role of hepatitis B virus X protein (HBx). Transfection of an HBx expression vector into liver cell lines resulted in hypomethylation of Sat2 sequences and downregulated DNMT3B [114].

2.3.1.5 Hematologic Cancers

Compared with solid tumors there has been considerably less investigation of global and repeat sequence hypomethylation in leukemias. DNA from leukocytes of healthy controls can be used to provide a reference for "normal" methylation, but there is an intrinsic uncertainty about its suitability as the leukemias arise from different progenitor cells that may represent only a small fraction of the normal blood cell population. For the acute leukemias, acute lymphoblastic leukemia, acute myeloid leukemia [115,116], and multiple myeloma [116] no hypomethylation was evident relative to normal leukocytes.

CLL is associated with significant levels of global hypomethylation [116]; methylation levels were observed to be highly variable, not to differ significantly with CLL stage but to be associated with variable heavy chain gene mutation status, a possible prognostic indicator [117].

For chronic myeloid leukemia a modest global reduction in methylation was detected by electrophoresis of HpaII and MspI cut DNAs [116]. Specific analysis of LINE-1 methylation demonstrated hypomethylation that was significantly more evident in both frequency and extent in blast crisis than in chronic phase, in comparison with minimal hypomethylation in normal bone marrow [118]. Hypomethylation of LINE-1 sequences were correlated with poor clinical prognosis and a lack of response to interferon or imatinib therapy. Roman-Gomez et al. [118] examined a number of molecular parameters associated with CML progression, providing possible insights into underlying mechanisms. Hypomethylation of LINE-1 sequences correlated with activation of transcription both of positive sense and antisense transcripts. From their data, a model was proposed where LINE-1 (antisense) transcription promoted over-expression of the c-met oncogene, contributing to CML progression. LINE hypomethylation was also observed to correlate with overexpression of the BCR-ABL fusion protein and with the DNMT3b4 isoform.

2.3.1.6   Ovarian Cancer

Global and satellite repeat sequence hypomethylation has been studied in series of ovarian cancers, ranging from normal tissue to cystadenomas (benign precursor lesions), malignant carcinomas, and carcinomas of low malignant potential (LMP) that have intermediate characteristics [119–121]. Global methylation was found to be equivalent to normal in cystadenomas but significant hypomethylation was evident in both malignant carcinomas and LMP tumors. Strong hypomethylation correlated with tumors of advanced stage and grade. Global hypomethylation was reflected in hypomethylation of the centromeric Satα repeats and juxtacentromeric Sat2 repeats. Hypomethylation of both Satα and Sat2 sequences located on chromosome 1, and particularly Sat2, was identified as a significant marker of poor prognosis, both for relapse and death [121]. Hypomethylation of Alu sequences was also found to be significantly associated with carcinoma compared with LMP tumors [122].

2.3.1.7   Urothelial and Renal Tumors

LINE-1 sequences have been shown to be hypomethylated in a high proportion (~90%) of urothelial cancers [123,124], but within the same study, to not be hypomethylated in renal cancers [123]. Lack of LINE-1 hypomethylation was a significantly positive prognostic indicator [124]. Hypomethylation of the endogenous retrovirus HERV-K sequences was seen to parallel that of LINE-1 sequences [123]. Hypomethylation of Sat2 and Sat3 sequences was identified in about 40% of urothelial cancers and correlated with histological grade and invasion depth [125]. Hypomethylation also distinguished papillary from nodular carcinoma and was significantly correlated with loss of heterozygosity on chromosome 9 [125].

Wilms tumor is a common pediatric tumor that often involves loss of imprinting ([126] and Chapter 4, this volume]) that arises from embryonic kidney structures that are absent in the adult. Ehrlich and colleagues have therefore used a set of normal tissue DNAs to define a lower 5meC content as normal for comparison with Wilms tumor DNA [127]. Using these criteria, about 60% of Wilms tumors show global hypomethylation, with about 30% being "strong" hypomethylation. Tandem repeat sequences including both centromeric Satα and juxtacentromeric Sat2 sequences [128] are commonly hypomethylated in Wilms tumors; centromeric Satα DNA is hypomethylated in >80% of cancers and Sat2 in about 50%. Pericentric chromosome 1 rearrangements were seen in 14% of the studied Wilms tumors and tumors with rearrangements had pericentromeric repeat hypomethylation [129]. A further tandem repeat, NBL2, was shown to undergo methylation changes that included hypomethylation at some sites and hypermethylation at others; clonal analysis indicated that individual DNA molecules carried these divergent methylation marks at specific sites in both Wilms tumors and ovarian cancer [130,131].

2.3.1.8 Glioblastoma

Primary glioblastoma tumors (8 of 10 Grade 4 astrocytomas) and established glioma cell lines show significant reduction in 5meC content compared with normal brain tissue [132]. For the two glioblastoma samples showing greatest global hypomethylation, hypomethylation of specific repeats D4Z4 and Alu paralleled decreased total 5meC content in comparison with two glioblastomas of normal 5meC content, while that of Sat2 repeats was even more severe and evident (to a lesser extent) in tumors with a normal 5meC content.

2.3.1.9 Summation

The frequency of global hypomethylation and hypomethylation of highly repeated sequences ranges from low or absent in some cancer types such as renal carcinomas, acute leukemias, and multiple myeloma, through to nearly all for urothelial cancers. Within any cancer type the absolute level of methylation and the relative level of methylation compared with matched normal tissue vary continuously and there is no evidence for a specific hypomethylated versus non-hypomethylated cancer classification. Reduction in 5meC content varies from insignificant to over 50%, commonly falling within the range of 10%–30%.

In cases where matched tumor and normal DNAs have been compared, there has generally been a significant overlap in the levels of methylation of normal and cancerous tissue. Thus DNA from normal tissue in one patient may be less methylated than tumor DNA from another. For colon and gastric cancers and HCC, hypomethylation can also be observed in the healthy tissue adjacent to these tumors suggesting a role in the initiation of disease. It is possible that this demethylation represents a field effect of premalignant change within which the cancer has arisen. Evidence for premalignant epigenetic change has also come from hypermethylation of peri-tumoral normal tissue DNA in colorectal cancer [89,90]. It is also possible, though perhaps more difficult to envisage in terms of molecular mechanisms, that hypomethylation in adjacent normal tissue is caused by the presence of the tumor, e.g. through a diffusible factor. More extensive studies of normal tissue from subjects without malignancy in comparison with that from patients with malignancy in relation to tumor DNA are needed to understand whether hypomethylation is a feature of premaligant tissue and whether for some cancers it may be a prognostic indicator of risk.

Global hypomethylation and repeat sequence demethylation are observed as an early event in a number of cancer types, e.g. colon, gastric, breast, and CLL, and the extent of demethylation does not correlate with the stage of the cancer. For other cancers, including prostate, hepatocellular, and ovarian, hypomethylation appears to occur progressively through cancer growth and development and increases with stage or histological grade. Thus the relative timing of global demethylation and its role during cancer initiation and progression may differ between cancer types.

Hypomethylation has been observed for almost all classes of repeat sequences—centromeric and pericentromeric simple tandem repeats, Satα, Sat2, and Sat3, other tandem repeats such as NBL2 and D4Z4, dispersed autonomous and nonautonomous retrotransposons (LINE-1 and Alu, respectively) and endogenous retroviral sequences. There have only been a limited number of studies where different repeat types have been studied in the same samples, but there has been a general concordance between hypomethylation of different repeats as well as with global hypo-methylation. Different relative levels of hypomethylation of repeats is, however, observed in individual cancers. For example chromosome 1 Sat2 sequences may be more or less hypomethy-lated than chromosome 1 Satα sequences in individual ovarian tumors [122]. In glioblastomas Sat2 appears more susceptible to demethylation than Alu or D4Z4 repeats [132]. The lack of tight concordance [133] between global DNA methylation levels and methylation of individual repeat classes (Alu, LINE-1, Sat2, and Satα) is also indicative that the process of demethylation is not randomly spread across the genome.

2.3.2 Relationship of Repeat Sequence Hypomethylation to Other Epigenetic Changes in Cancer

2.3.2.1 Relationship to Single-Copy Gene Hypomethylation

It is not unreasonable to expect that factors favoring general genomic hypomethylation and hypomethylation of repeat sequences might also predispose to demethylation of single-copy genes. De Smet et al. [134] first showed in a set of cell lines that demethylation of the MAGE-A1 promoter was strongly correlated with global hypomethylation. In gastric cancer cell lines and primary gastric carcinomas, Kaneda et al. [2004] extended this observation and characterized a strong association between global hypomethylation, LINE hypomethylation, and hypomethylation and expression of six MAGE genes. In contrast, while general hypomethylation and common hypermethylation of a set of 7 CpG island promoters could occur in the same cell lines or primary cancers, the two processes appeared to be independent. Hypomethylation of BAGE loci is observed in many cancers [66] and was found to be associated with hypomethylation of juxtacentromeric Sat2 repeat sequences in ovarian cancer. Since BAGE loci are also located in juxtacentromeric regions, this raises the possibility that their hypomethylation is linked to repeat sequence and global hypomethylation [66].

2.3.2.2 Relationship of Single Gene and Repeat Sequence Hypomethylation with Specific Hypermethylation

The link with specific gene (especially CpG island) hypermethylation has been studied in a number of cancer types. In prostate cancer, correlations between the accumulation of CpG island hyper-methylation and hypomethylation of LINE and Alu sequences in the progression of cancer have been identified [100,101]. Whether this represents parallel processes during progression or reflects a mechanistic link is not yet clear.

In colorectal cancer no relationship was found between the level of global hypomethylation and hypermethylation of a set of CpG islands [84]. Likewise, from studies of arbitrarily selected restriction sites displaying altered methylation in cancer, it was concluded that hyper- and hypo-methylation were independent processes contributing to colorectal carcinogenesis [22,135], with elevated hypomethylation being associated with cancers showing chromosomal instability [135]. In a study focused on cancers displaying MSI or CpG island methylator phenotype (CIMP+), Estecio et al. [93] identified that MSI cancers showed minimal change in the level of LINE methylation, while MSS cancers showed significant LINE hypomethylation, whether or not they were classified as CIMP+. All MSI tumors were CIMP+ and this group contained those with the highest frequency of hypermethylated CpG islands. They concluded that MSI colorectal cancers follow an alternative progression pathway to those showing hypomethylation and that genome-wide hypomethylation and single-copy gene hypermethylation can promote tumorigenesis by independent means. This is consistent with the observation [91] that MSS tumors have a greater level of hypomethylation compared with MSI tumors and that this correlates with increased chromosomal instability.

Studies in a number of other cancers—urothelial [124], ovarian [121,122], and gastric cancers [136], melanoma [137] as well as Wilms tumors [127], and non-Hodgkin's lymphoma [138]—also support general conclusion that gene-specific hypermethylation and global hypomethylation are independent processes. That is, the presence or level of global hypomethylation does not alter the likelihood of CpG island hypermethylation, and they are unlikely to be causally related. The differences in relative timing of hypomethylation of repeat sequences and single-gene hyper-methylation during carcinogenesis in many cancer types also support this conclusion.

In a contrary example of coordinated hyper- and hypomethylation, Park et al. [114] have recently reported the intriguing finding that expression of HbX leads to both hypermethylation of specific genes, e.g. IGFBP3, and hypomethylation of Sat2 repeat sequences.

2.4 WHAT MECHANISMS UNDERLIE HYPOMETHYLATION OF REPEAT SEQUENCES?

Normal DNA methylation patterns, and more generally epigenetic chromatin states, are established and maintained through a complex system that involves DNA methyltransferases, histone modifications, epigenetic regulators (that include polycomb group proteins and histone modifying enzymes), and regulatory RNAs, interactions with the replication machinery and timing of replication within the cell cycle. The general correlation between global hypomethylation and hypomethylation of a variety of different repeat sequence classes as well as single-copy genes such as the CT antigens suggests a disruption that has a pervasive effect, such as on the enzymes involved in common epigenetic marks—DNA methyltransferases, histone (de)acetylases, histone methyltransferases. Within this broader picture significant differences are observed between cancers in the extent of hypomethylation of different repeat classes, for example between Sat2 and Satα in ovarian cancer [121]. These data indicate that hypomethylation is not a stochastic process acting randomly across the genome, but that there is specificity for classes of sequence—this could be the result of active targeting or specific protection of regions, or of a selection process overlaying a random process of demethylation. Below, the possible roles of some of the key candidates are considered in the light of data from clinical materials, cancer cell lines, and model systems.

2.4.1 DNA METHYLTRANSFERASES

Mammalian methylation is controlled by four DNMTs of interchangeable function (reviewed in [139,140]). DNMT1 is the major maintenance DNA methyltransferase and conserves methylation patterns after replication by methylating hemi-methylated CpG sites to produce symmetrically methylated dinucleotides. A number of studies have examined its expression, particularly in relation to CpG island hypermethylation. Though results have varied considerably (see discussion in [17]), levels of DNMT1 mRNA or protein appear to be moderately elevated in many cancers and regulation of activity disrupted [141–145]. While it is clear that reduction of DNMT1 activity, even transiently, can induce stable demethylation [59], the relationship between DNMT1 activity and hypomethylation has been the subject of relatively few studies. No association was seen in hepatocarcinogenesis [110], gastric or colorectal cancer [146], ovarian cancer [122], or Wilms tumors [2]; this combined with knowledge of DNMT1 expression across a wide range of cancers suggests that a direct effect mediated through DNMT1 activity levels is an unlikely causative factor in genome-wide hypomethylation.

The de novo methyltransferases DNMT3A and DNMT3B [139,140] are highly expressed during embryogenesis and at much lower levels in adult somatic tissues. Loss of DNMT3B activity in ICF syndrome leads to demethylation of centromeric and pericentromeric satellite repeats, especially Satα, Sat2, and Sat3 on chromosomes 1, 9, and 16 [147]. Consequently the potential for disruption of its activity to be associated with repeat sequence hypomethylation in cancer has been investigated. However, altered levels of neither DNMT3B nor DNMT3A were seen in liver [110] or ovarian [122] cancers or Wilms tumors [2]. Expression of a splice variant DNMT3B4 that lacks enzymatic activity was found to correlate well with hypomethylation of pericentromeric repeats in hepatocellular carcinoma [113] and with LINE-1 demethylation in CLL [118], but not in ovarian cancer [122]. It was hypothesized that the enzymatically inactive DNMT3B4 isoform may promote hypomethylation through competition with active DNMT3B. An increasing number of splice variants of the DNMTs are being recognized [148] and expression of inactive forms and forms lacking interaction domains with regulatory proteins may significantly effect targeting to particular sequences in cancer.

DNMT3L does not possess intrinsic methyltransferase activity but interacts with DNMT3A and DNMT3B to enhance DNA methylation [149,150]. It is highly expressed in embryogenesis and contributes to the establishment of imprinting [151] and of methylation of repeat sequences. Inactivation of DNMT3L in male germ cells specifically disrupts methylation of LTR and non-LTR retrotransposons [152,153] without affecting methylation of pericentromeric tandem repeats [151]. Ehrlich et al. [122] did not find any association of DNMT3L with hypomethylation in ovarian cancers, while elevated expression was seen in testicular germ cell tumors [154]. As a regulator of methylation of interspersed repeats, the role of DNMT3L in cancer development warrants further investigation.

2.4.2 ACTIVE DEMETHYLATION OF DNA

During normal development active demethylation, that is demethylation in the absence of DNA replication, is most dramatically evident in the early zygote and is probably also significant in the development of primordial germ cells [155]. Postfertilization, genome-wide hypomethylation proceeds through a passive process for the maternal genome, in contrast to the paternal genome that undergoes a rapid and extensive process of active demethylation. The demethylase activity is present in the oocyte, but the molecular mechanism underlying the process is not yet established [156]. It is notable that this epigenetic reprogramming in the mouse embryo is differential between repeat classes [157]. The role of active demethylation in somatic cells is less well understood and characterized but may be more widespread than currently accepted. Myoblast development involves extensive active demethylation of the genome [158] that involves promoter specific demethylation of the myogenin gene in association with its activation [159]. Active demethylation of the IL2 promoter/enhancer has been shown to occur in response to T-cell activation [160]. Accumulating evidence supports the role of DNA glycosylases and DNA repair pathways in the process of active demethylation, whereby base removal to create an abasic site triggers strand nicking and repair through the mismatch or excision repair pathways. In plants, DEMETER and ROS1 are specific 5methyl cytosine glycosylases that are involved in active demethylation [161]. In mammals, thymidine DNA glycosylase (TDG) and the methyl CpG binding domain protein MBD4 both act on T:G mismatched base pairs to initiate repair [162,163]. Nicking adjacent to 5meC is observed during active demethylation triggered by the glucocorticoid receptor [164] and Gadd45 a nuclear protein involved in excision repair has been shown to be important for repair mediated demethylation, probably through recruitment of the XPG excision repair endonuclease [165]. Two cytosine deaminases, AID and APOBEC, that convert 5meC to thymidine are expressed in oocytes, stem cells, and germ cells and it has been proposed that their action to generate T:G mispairs could trigger repair processes and demethylation on the paternal genome in the early embryo [166]. TDG, along with DNMT3a, has been identified in a transcription factor-mediated complex targeting specific demethylation [167]. Recent evidence suggests that transcription is required to promote active demethylation following trichostatin-induced histone acetylation. A similar dependence was seen for both a transfected gene and the endogenous CT antigen gene, GAGE [168]. Two further methyl DNA binding domain proteins MBD2 and MBD3 have also been implicated in active demethylation, MBD2 in valproate-induced demethylation [169], and MBD3 in demethylation of ribosomal RNA promoters [170].

It is not clear yet what part active demethylation might play in cancer-associated hypomethylation and its potential importance is unlikely to be clarified till the molecular mechanisms are more comprehensively understood. There are indications that part of the dynamics of the setting of DNA methylation patterns involves a balance between active demethylation and re-methylation (e.g., of hemimethylated sites) by DNA methyltransferases and imbalance could lead to inappropriate demethylation. It is also of significance that external triggers such as steroid hormones and inhibitors of chromatin modifying enzymes such as trichostatin and valproate can induce active demethylation, indicating a possible impact of environmental agents.

2.4.3 CHROMATIN PROTEINS AND EPIGENETIC REGULATORS

Chromatin structure and re-modelling is intricately linked with the regulation of DNA methylation. States of inactive (heterochromatic) and active (euchromatic) chromatin are characterized by a variety of modification of histones that include acetylation, methylation, phosphorylation, and ubiquitination, each targeted to specific amino acids and regulated by enzymes such histone methyltransferases, acetylases, and de-acetylases. Specific histone modifications such as acetylation of lysine 9 of histone H3 (H3K9) and methylation of lysine 4 (H3K4) are associated with active chromatin and while other modifications, methylations of lysines 9 and 20 (H3K9 and H3K20), are found in inactive chromatin associated with methylated DNA. The pattern of histone modifications has been termed the histone code and its complexity in establishing different states of activity or preparedness is steadily being exposed [171]; related to the processes of histone modification are a series of epigenetic regulators such as polycomb and trithorax group proteins [172]. The relationship of changes in chromatin modifications in cancer development has been recently reviewed [173,174]; Hoffman and Schulz [17] provide a table of such regulatory proteins implicated in influencing DNA methylation in cancers.

De-acetylation and methylation of lysine 9 of histone H3 are key modifications involved in establishment of epigenetically silenced chromatin and have been proposed to act in concert with DNA methylation in a self-re-enforcing loop to maintain the silenced state [175–177]. Specific interactions with DNA methyltransferase 1 [178] and methylated DNA binding proteins [179] have been implicated mechanistically and H3K9 modification has been shown to direct methylation of pericentromeric satellite repeats [180]. Knockout in mice of the H3K9 methyltransferase Suv39h results in genome instability and increases cancer predisposition [181]. The putative oncogene GASCI, a member of the jumonji family of histone demethylases, demethylates H3K9 and is overexpressed in squamous cell carcinomas [182]. Disruption of this heterochromatin regulatory loop thus has a potential causative role widespread genomic hypomethylation.

When assessed at a whole-of-genome level cancer cell lines and primary tumors show significant changes in other histone modification as well as DNA hypomethylation—decreases in histone H4 lysine 6 monoacetylation and histone H4 trimethylation correlate with hypomethylation of the tandem repeats Sat2, NBL2, and D4Z4 [183]. Chromatin immunoprecipitation experiments indicated a specific loss of association of the histone acetyltransferases MOF, MOZ, and MORF with the respective repeats. These data point to the presence of distinct regulatory pathways that regulate the chromatin modification, and perhaps ultimately DNA methylation, of different sequence families and emphasizes that disruption of chromatin modification at many levels could contribute to the range of DNA hypomethylation patterns observed in cancers.

2.4.4   REGULATORY RNA

Emerging evidence from a number of model organisms has demonstrated a role for low-level transcription of repeat sequences as part of a re-enforcing regulatory loop that establishes and maintains heterochromatin [184–186]. The role of noncoding RNA in the epigenetics of imprinting [187,188] and X-inactivation [189] is well established. Components of the RNA interference pathway are implicated in generation of small RNAs, but it is not clear how this process works in mammalian cells in the absence of an RNA-directed RNA polymerase that is a critical component in fission yeast and plant genomes. Bidirectional transcription may be sufficient to produce double-stranded RNAs that could trigger transcriptional epigenetic silencing. The process of X-inactivation in mice is known to depend both on transcription of the noncoding Xist transcript and on the antisense Tsix transcript [189] though it is not clear if this is also the case for the human X-chromosome [190]. The presence of a DNA methylation system in human cells responsive to double-stranded RNA or antisense transcription is supported by the example of an α-thalassemia mutation in which an intergenic deletion resulted in antisense transcription from a neighboring gene, gene silencing, and promoter methylation [191]. Key proteins involved in heterochromatin formation that bind small RNAs are those of the PIWI and HP1 families. Methylation of H3K9 acts to localize HP1 to heterochromatin and in turn HP1 recruits effectors such as histone de-acetylases and H3K9 methylase. In mouse cells, binding of an RNA component to HP1 has been shown to be involved in the organization of pericentromeric repeats into heterochromatin [192]. The do novo methyltransferases DNMT3A and DNMT3B and other methyl CpG binding proteins also bind RNA with high affinity [193]. The involvement of RNA in the silencing process provides a means by which silencing could be coordinated to target related DNA sequences, either tandemly arrayed as for satellite repeats or dispersed sequences such as those of the retrotransposon families. Disruption of such control could then lead to class-specific hypomethylation as is seen in cancers and in relation to DNMT3B in ICF syndrome.

Converse to the role of RNA in promoting gene silencing and heterochromatin formation, Imamura et al. [194] reported the specific demethylation of the CpG island promoter of the Sphk1 gene in response to expression of a noncoding antisense RNA. Early experiments [195] had also demonstrated an RNA dependent gene-specific hypomethylation using cell extracts from different cell types. As evidence for widespread transcription of nonprotein-coding regions of the genome accumulates [196], our understanding of the function of RNA in maintaining epigenetic states and aberrant function in cancer is likely to grow rapidly.

2.4.5 Genetic and Environmental Effects

The methylation of DNA is intrinsically linked to the supply of methyl groups through S-adenosyl methionine (SAM) and a wide range of studies in animal model systems, cell lines, and epidemiological studies in humans have examined links between cancer incidence and methyl group metabolism and supply [197,198]. Critical dietary components in pathways leading to synthesis of SAM are folate, vitamin B12 and B6, methionine, choline, and selenium. In a rat model, a folate/methyl deficient diet leads to increased hepatocarcinogenesis that is associated with genomic hypomethylation [199]. As well as a simple limitation of SAM as a substrate for methylating DNA, James et al. [200] discuss the role of DNA lesions that arise in response to the folate/methyl deficient diet. These include DNA strand breaks, gaps, abasic sites, and incorporated uracils, all of which bind with high affinity to DNMT1 and promote passive demethylation. In discussing the role of methyl group metabolism Hoffmann and Schulz [17] raise the possibility that proliferation rate places an increased stress on the availability of SAM and that this could be reflected in the early genomic hypomethylation of rapidly dividing cancers such as colon and bladder, compared with renal and prostate cancers that are more slowly proliferating.

In addition to dietary supply of nutrients and cofactors, the activity of the enzymes involved in methyl group metabolism contribute significantly to the available pools of SAM, of the methylation reaction product, S-adenosylhomocysteine that is a potent inhibitor of DNMTs, and to levels of dUTP, all of which will impact on DNA hypomethylation. The potential impact of polymorphisms in several genes involved in methyl group metabolism—methionine synthase, thymidylate synthase, cystathione β synthase, methylene tetrahydrofolate dehydrogenase and methionine synthase reductase and especially methylene tetrahydrofolate reductase (MTHFR)—has been examined in relation to cancer risk. The presence of low functioning alleles of MTHFR (C667T) has been associated with hypomethylation status in glioblastomas and other cancers [132], suggesting a contributing role. Friso et al. [201] demonstrated that homozygous (T/T) carriers of the low functioning allele showed much reduced blood DNA methylation levels compared with carriers of wild-type alleles; they further demonstrated a combined relationship of genotype and low folate status. While a number of studies have inferred an altered cancer risk for carriers of the T allele, a note of caution is introduced by a meta-analysis of 22 studies of MTHFR polymorphisms and folate intake in relation to breast cancer risk, that determined that there was no effect of MTHFR alleles on risk nor an interaction with folate intake [202].

A number of additional environmental and exogenous factors have been shown to induce hypomethylation. These include carcinogens such arsenic [203] and benzopyrene [204], procaine and procainamide [205], ultraviolet light [206], and low-dose radiation exposure [207]. Hypomethylation induced by viral infection also has the potential to contribute to carcinogenesis.

Infection by herpes simplex virus inhibits host cell DNA methylation and virus transformed cell lines show global hypomethylation [208]. HBX protein induces both specific hypermethylation and repeat sequence hypomethylation [114].

Thus a range of dietary, environmental, and genetic factors that could contribute to carcinogenesis through disrupting local regions of methylation or lowering distributed methylation across the genome and enhancing the effects of other factors in initiating or spreading hypomethylation changes.

2.5 ROLE OF HYPOMETHYLATION IN CANCER DEVELOPMENT

2.5.1 CHROMOSOMAL INSTABILITY AND GENOME INTEGRITY

Genome-wide hypomethylation has the potential to impact broadly on gene expression, repair, recombination, and chromosome segregation and correlative evidence implicates all of these in cancer development. DNMT mutants in mice have been used to study the impact of a generalized reduction of DNA methylation levels on cancer development. Mice with lowered DNMT1 expression resulting in about 10% of their normal level of 5meC were prone for development of aggressive T-cell lymphomas associated with elevated chromosomal instability, particularly trisomy of chromosome 15 [209]. A similar promotion of T- and B-cell lymphomas was observed when hypomorphic DNMT1 expression was combined with mismatch repair deficiency, MLH−/− [210]. Likewise in a mouse model of soft tissue sarcoma development, hypomorphic expression of DNMT1 promoted earlier onset of sarcoma formation that was associated with LOH of the chromosome 11 region harboring mutations in *Nf1* and *p53* genes [211]. The effect of hypomethylation on intestinal tumor development in the Apc (min)/+ model demonstrated the complexity of unraveling the role of hypomethylation. Reduction in methylation levels reduces the frequency of formation of intestinal polyps [212–215]. However, the much reduced occurrence of macroscopic polyps contrasts with the elevated incidence of the precursor microscopic adenomas that were associated with LOH at the *Apc* locus [216]. This suggests that in this system hypomethylation enhances early oncogenic events involving chromosomal instability, but suppresses cancer progression, presumably through limiting hypermethylation of specific tumor suppressor gene promoters. In the same Apc (min)/+ model, hypomethylation caused multifocal liver tumors [216], reinforcing that the relative importance of hypomethylation and other genetic and epigenetic events varies between cancer types.

The influence of hypomethylation on mutational events has been studied in DNMT1 knockouts in mouse embryonic stem cells. Chen et al. [217] reported an increased rate of mutation, primarily deletion, in both an endogenous and introduced gene. In contrast, a separate study found decreased rates of deletion and missense mutation in an introduced reporter gene [218]. Independent reports demonstrated that DNMT1 deficiency in ES cells resulted in enhanced rates of mismatch repair, as monitored by slippage mutations [219,220]. Analysis of DNA methylation and chromatin surrounding the monitored gene [219] suggested a direct role for the DNMT1 protein rather than hypomethylation per se. Thus, while hypomethylation may affect rates of local mutation the potential impact in cancer development remains unclear.

The role of hypomethylation in larger chromosome rearrangements is clearer. In ICF syndrome, a genetic disorder involving loss of activity of DNMT3B, there is specific loss of methylation of satellite repeat sequences, e.g. Sat2, Sat3, and Satα on chromosomes 1, 9, and 16. In combination with stimulated cell division this leads to chromosomal rearrangements, particularly involving chromosomes 1 and 16.

Experimental disruption of DNA methylation in cellular systems also leads to chromosomal instability. For example, genetic knockout of DNMT1 and DNMT3B in the near-diploid colorectal cancer line HCT116 depletes 5meC by 80% and leads to increased aneuploidy and increased frequency of balanced and unbalanced chromosomal rearrangements [221] distributed across all chromosomes. Specific amplification of two regions at chromosome 11q23 and 18p was also observed; this could indicate an increased propensity local recombination and amplification that could be selected if favorable for growth.

For human cancers, a relationship between hypomethylation of satellite DNA sequences was first identified in lymphoblastoid cell lines [222]. Extension of these observations demonstrated a relationship between hypomethylation of pericentromeric Sat2 sequences on chromosomes 1 and 16 with decondensation and rearrangement; this contrasted with hypomethylated Sat3 (chromosomes 9 and Y) sequences that remained condensed and stable [223]. The relationship between hypomethylation of Sat2 sequences and chromosome instability has been studied in a number of cancer types (Table 2.1). In hepatocellular carcinoma [224], Wilms tumors [147], and glioblastoma [132] hypomethylation is associated with pericentromeric rearrangements involving chromosome 1 and 1q copy number gains. Likewise, satellite repeat hypomethylation is associated with chromosome 9 loss of heterozygosity in urothelial cancer [125]. Hypomethylation is seen to precede chromosomal changes in hepatocellular carcinoma and Wilms tumors and to be evident in normal liver tissue adjacent to carcinoma. In contrast, while copy number changes (aneuploidy) were associated with satellite hypomethylation in breast cancer, there was a negative association with chromosomal rearrangement [225]. The combined observations support a model where demethylation of satellite sequences can lead to chromatin decondensation, but is not sufficient [223], increasing the likelihood of illegitimate recombination or chromosome mis-segregation.

Hypomethylation of interspersed LINE-1 repeat sequences [91,92,99] and dispersed individual sites [135] has also been associated with chromosomal instability in colorectal and prostate cancers. In the case of prostate cancer, Schulz et al. [226] conclude that the apparent relationship is not direct, but a result of convergence of two separate trends each related to advancing stage. Since hypomethylation of interspersed repeat sequences is normally correlated with that of satellite sequences; these associations are probably indirect. It is difficult to assess the extent to which hypomethylation of interspersed sequences such as Alu and LINE-1 are directly involved in somatic rearrangements and deletions through illegitimate recombination. However, there are many examples of germline LINE-1 and Alu mediated deletion mutations in inherited cancers, such as MLH1, MSH2 [227], and APC [228] in colorectal cancer.

The involvement of Alu elements in mediating translocations and rearrangements in leukemias is well documented [229]; these include Philadelphia translocation in some cases of chronic myelogenous leukemia, complex *BCR/ABL1* rearrangements, and partial duplication of the *MLL* gene in acute myeloid leukemia. There are also a number of examples of intragenic deletions involving retrotransposon elements in cancer cell lines and sporadic tumors. An intragenic deletion in the p53 gene in a pancreatic cancer cell line is flanked by Alu elements [230]. Inactivation of the apoptosis-associated caspase DNAse in hepatoma involves specific Alu recombination and is common both in cell lines and primary cancer [231]. Interstitial deletions of 0.1 to >30 Mb at the CDKN2A locus is common in bladder cancer [232]; most deletion breakpoints were located in or close to LINE-1 retrotransposon clusters that flank the locus and may facilitate deletion. While it is to be expected that such recombination events would be facilitated by hypomethylation of the repeat sequences, there is little evidence that hypomethylation precedes the rearrangements and such evidence is difficult to gather [233].

Hypomethylation and consequent transcription of LINE and SINE elements has the potential to lead to retrotransposition and insertional mutagenesis. Such retrotransposition of intracisternal type A particles is commonly observed in leukemias in mice and often affects expression of cytokine genes [234]. In hereditary human cancers de novo Alu insertions in BRCA1 and 2 have been identified in three cases of hereditary breast cancer [235] and in the MSH2 gene in colorectal cancer [236]. Evidence for such insertional mutagenesis in the development of sporadic cancers is very limited. In one example somatic LINE insertion was identified at the translocation breakpoint in a desmoplastic round cell tumor, suggesting that transposition was associated with formation of the breakpoint or its resolution [237]. The small number of LINE elements (80–100) with functional

TABLE 2.1

Hypomethylation and Chromosomal Instability

Cancer Type	Hypomethylation[a]	Chromosomal Instability	Comment	References
Colorectal	LINE-1 (cobra)	Genome-wide CGH	Chromosomal instability associated with LINE hypomethylation and MSS tumors	[91]
		LOH at 5q21, 8p12–22, 17p13, 18q21		[92]
	200 random sites (AIMS)	200 random sites (AP-PCR) Genome wide CGH	Good correlation between chromosomal instability and hypomethylation sites distributed across genome. Correlation not observed with total 5meC level, but small sample numbers	[135]
Breast	Sat2, Chr1; Sat2, Chr16 (Southern blot)	Chr1 and 16 copy number and rearrangements (FISH)	Copy number changes associated with hypomethylation, but rearrangements and breakage show reverse trend	[225]
Wilms	General Satα Chr1 & 10 Satα Chr1 Sat2[a]	LOH G-band karyotype	Pericentromeric rearrangements associated with hypomethylation Satα hypomethylation not correlated with aneuploidy	[128,129]
Hepatocellular	Chr1 Sat2[a]	CGH	Strong correlation of hypomethylation with Chr1q copy number gain. Hypomethylation present in adjacent normal tissue	[111]
Glioblastoma	Sat2 (bisulfite sequencing) Methyl acceptor assay D4Z4, Alu	Array CGH	Chromosome 1 rearrangements associated with most severe hypomethylation	[132]
Urothelial	Sat2 and 3[a]	Chr9 microsatellite markers	Strong correlation of LOH with Sat2 & 3 hypomethylation	[125]
Prostate	LINE-1[a]	CGH	Correlation of common Chr8 loss with hypomethylation. Subsequently demonstrated to be parallel changes and unlikely mechanistically linked	[99,226]

[a] Analysis by Southern blot except where stated.

retrotransposition activity combined with the translation inefficiency of the transposase [238] may limit the extent of transposition [239]. Additionally, LINE-1 retrotransposon activity triggers an apoptotic response and thus expression of transposition-competent LINE-1 elements may be selected against [240].

2.5.2 Effects of Transcription of Repeat Sequences

Hypomethylation is expected to lead to increased transcription of repeat sequences, with the potential to impact in a number of ways on normal cellular processes. Using antibodies targeted to LINE-1 proteins Bratthauer and colleagues demonstrated functional transcription of LINE-1 in adult and pediatric germ cell tumors [241] as well as breast cancers [242], and expression of LINE-1 transcripts associated with LINE-1 hypomethylation has been identified in urothelial carcinomas [123]. However, despite hypomethylation in hepatocellular carcinoma, increased transcription was not seen [108]. Elevated transcription of endogenous retroviral sequences associated with hypomethylation has been reported for a number of cancers [123]. In bladder cancer HERV-k sequences are commonly hypomethylated but this is not associated with increased transcription. Transcription from endogenous repeat sequences would appear to require both hypomethylation and the presence of the appropriate transcription factors that may be cell-type specific.

Though there is little direct evidence for a role in cancer cells, hypomethylation and deregulated transcription of repeats could impact cancer cells in a number of ways:

1. Through expression of proteins produced from transcripts, from viral sequences themselves, or adjacent genes. It is notable that both retroviral and LINE-1 promoters can be active bidirectionally. A possible example is in chronic myeloid leukemia where LINE sequences are commonly hypomethylated and transcribed. Roman-Gomez et al. [118] identified a strong association of LINE hypomethylation with activation of the c-met oncogene and have postulated that its increased transcription is driven from a LINE-1 element in intron 2 (in the antisense direction relative to the LINE-1 sequence)
2. Through direct transcriptional interference with neighboring genes
3. Through production of RNAs producing regulatory "noise," e.g. through binding proteins involved in RNA processing or providing decoy targets for micro RNAs
4. Through production of transcripts antisense to endogenous RNAs, giving rise to dsRNAs and possible downregulated expression
5. Widened accessibility of DNA may lead to altered competition for transcription and other regulatory proteins, leading to dis-regulated expression

2.5.3 Expression of Single-Copy Genes

A growing number of single-copy genes, both with and without CpG island promoters, are being shown to be regulated in a tissue or developmentally specific manner. Reactivation of expression of such genes in cancer can contribute oncogenic function (e.g. R-Ras and PAX2) or promote invasion and metastatic spread (e.g. uPA and maspin). For example, PAX2 expression in endometrial cancer is mediated by hypomethylation of its promoter and its expression has been experimentally demonstrated to promote tumor growth in nude mice [29]. The CT antigens represent a particularly interesting set of genes that are critically involved on germ cell proliferation and differentiation and their ectopic reactivation in somatic cells has the potential to disrupt or override normal cellular regulatory processes.

From the limited cases studied, there appears to be an association between global hypomethylation and demethylation of single-copy genes, but it is not clear whether there is a direct mechanistic link. If a common process underlies both, the observed global and repeat sequence, hypomethylation may reflect selection for activation of specific single-copy genes.

2.6 CAN HYPOMETHYLATION BE USED AS A DIAGNOSTIC OR PROGNOSTIC MARKER IN CANCER?

Because of the relative stability of DNA and methylation marks, the ability to isolate DNA from tissue and bodily fluid samples and the growing number of candidate methylation biomarkers in cancer, there is considerable interest in the development of DNA methylation-based diagnostic assays, see Ref. [243]. The center of most of this focus has been on hypermethylation markers, but there is increasing interest in the potential of hypomethylation markers in cancer diagnosis, classification, and prognosis and following treatment response.

As with any biomarker, hypomethylation of target sequences must demonstrate sufficient sensitivity and specificity for clinical utility and must be able to be assayed with appropriate accuracy and sensitivity. Potential application and targets for diagnostic and prognostic applications are discussed below. For a discussion of detection methods for discriminating methylated DNAs, see Clark et al. [244]. For single-copy genes hypomethylation (like hypermethylation) can be measured by PCR of bisulfite-treated DNA in combination with primers that are specific for unmethylated sequences, USP [245]. For quantification, Real time PCR using specific fluorogenic probes or SybrGreen can be used, in combination with methylation-independent amplification of a reference gene; the melting profile of the amplified product can be used to verify the amplification of unmethylated DNA. The lower melting temperature of hypomethylated amplicons after bisulfite treatment can be readily used to provide sensitive and selective amplification, bisulfite differential denaturation PCR [246] that depends on the average level of methylation within an amplicon, rather than the status of specific CpG sites. The use of the same primer set at selective and nonselective temperatures allows quantification of hypomethylation relative to total DNA level for a single amplicon [Mitchell and Molloy, paper in preparation]. For estimating the level of hypomethylation after nonselective amplification of bisulfite-treated DNA a number of methods including COBRA and pyrosequencing are suitable [244]. Determination of the methylation status of repeated DNA sequences poses specific additional problems because of sequence heterogeneity and the common mutation of CpG sites to TpG; after bisulfite treatment, mutated and unmethylated sites cannot be distinguished. Quantitative analysis of CpG sites after amplification using nonmethylation selective primers can be used to compare methylation levels between samples using COBRA [247] or pyrosequencing [248]; these methods can be used for determinations above about 5% unmethylated sequence. Weisenberger et al. [133] have described a quantitative assay approach using MSP and USP primers and real time PCR (MethyLight) for quantification of Sat2 and LINE-1 methylation.

2.6.1 APPLICATION FOR DIAGNOSIS

Diagnostic assays for hypomethylation must be able to detect cancer DNA-specific methylation changes in a background of normally methylated DNA likely to be present in clinical samples and be applicable to limited quantities of DNA. Sensitive and technically specific methods are available for such selective amplification of hypomethylated sequences from single-copy genes, but current methods are inadequate for detection of low proportions of hypomethylated repeat sequences.

The potential utility of any hypomethylation biomarkers for early cancer detection has yet to be evaluated extensively. One promising example is the use of the Xist gene 5' sequences that are highly methylated in all somatic cell types but hypomethylated in testicular germ cell tumors. Hypomethylated sequences can be detected in the plasma patients [249]. The CT antigens represent a class of genes that has diagnostic potential in a variety of cancers. For example, hypomethylation of BAGE was observed in 96% of samples from a variety of cancers [66]. As is the case with other CT antigens and other cancer markers, their activation and hypomethylation are common to a number of cancer types, so detection of hypomethylated sequences in blood or urine would need to be combined with other markers or clinical information to identify the tissue location of the cancer. However, for a number of common cancers, bowel, prostate, breast, and lung, the use of fecal or bodily fluid samples can be used for diagnosis, providing location specificity. Unlike hypermethylation markers, hypomethylation is generally associated with activation of gene expression and the resultant protein product provides an alternative means of diagnosis. The relative merits of DNA versus protein based assays is likely to be case specific and dependent on factors such as the expression level and stability of the protein, the presence of cross-reacting antigens.

The diagnostic use of repeat sequence hypomethylation for early cancer detection has significant appeal, since the high copy number could provide for very sensitive detection. Hypomethylation of SatR repeats in breast cancer [107] and of LINE-1 repeats in colorectal cancer [88] are examples of early potentially diagnostic epigenetic changes. However, while hypomethylation relative to matched normal tissue is very common in these cancers, the range of methylation levels in normal tissue (either matched normal tissue from patients or tissue from controls without cancer) significantly overlaps with that in cancer tissue. It remains to be seen whether specific subclasses of repeat sequence or sites within repeats are able to provide sufficient sensitivity in identifying cancer with a low background from normal tissue.

It is becoming evident that a number of epigenetic changes occur in premalignant or normal tissues and that these might be predisposing to cancer [250]. For example, hypomethylation of LINE-1 sequences [88] and hypermethylation of the MGMT gene [89] are seen in histologically normal tissues adjacent to colorectal cancer, Sat2 hypomethylation is evident in premalignant liver [113] and loss of IGF2 imprinting may affect large regions of the colon or even be systemic and be associated with increased risk of colorectal cancer [251]. Combined with data from dietary manipulation and DNMT knockout in mice, it is reasonable to consider that systemic genome-wide hypomethylation may be a predisposing risk factor for cancer development. Measurement of systemic DNA methylation levels has been proposed as part of a general assessment of "genome health" [252].

2.6.2 DNA Hypomethylation in Cancer Classification, Disease Prognosis, and Monitoring

Many of the extensive changes in DNA methylation in DNA isolated from tumor tissue have the potential to provide clinically useful information. A range of PCR-based approaches for identification and quantification of such changes are applicable to relatively small amounts of DNA isolated from fresh tissue biopsies and formalin fixed tissues and this allows both prospective and retrospective studies of DNA methylation patterns in relation to cancer classification or disease outcome. Table 2.2 provides examples where hypomethylation of either individual genes or repeat sequences shows clinical promise. Hypomethylation of Sat2 pericentromeric repeats in ovarian cancer is strongly associated with progression and death, relative risks of 4.1 and 9.4, respectively, and was found to be a better indicator than either stage or grade [121]. Expression and hypomethylation of the HAGE gene in CML is strongly associated with lack of response to interferon or imanitib treatment and reduced disease-free survival [68], indicating its potential to be used choosing treatment options. Patterns of DNA methylation have the potential to help classify cancer states and subtypes [253], and the stability of DNA and evenness of signals could provide advantages compared with use of RNA and gene expression profiles. An example of the use of hypomethylation as part of such a gene panel is the hypomethylation of MYOD1 as a discriminator of pilocytic astrocytomas from other glioblastomas [254].

The advent of epigenetic therapies [255,256], such as the demethylating drugs decitabine and zebularine, has led to the use of assays for global methylation levels [257] or PCR assays for repeat sequence demethylation [258] for monitoring the extent of demethylation. These assays were shown to provide a useful predictor of clinical response [258] and could be used to determine the extent and timing of treatment.

To date, studies of both single gene and repeat sequence DNA hypomethylation in cancer have been exploratory in nature, and aimed at characterizing and understanding their role in the disease

TABLE 2.2

Hypomethylation Biomarkers of Potential Prognostic Significance

Cancer	Marker	Prognostic Measure Classification	References
Prostate	Total 5meC	Disease recurrence	[102]
	uPA	Progression and metastasis	[33,32]
Ovarian	Sat2	Relapse and death	[121]
Breast	uPA	Metastatic potential	[34]
CML	LINE-1	Progression-free survival	[118]
	HAGE	Treatment response	[68]
Urothelial	LINE-1	Lack of methylation a good prognostic indicator	[124]
Urothelial	Sat2, Sat3	Grade and invasion depth	[125]
Hepatocellular	LINE-1 (serum)	LINE-1 hypomethylation prognostic of short survival	[259]
Gastric	MAGE-A	Stage, metastasis, and survival	[260]
Glioblastoma	MYOD	Part of classification, especially pilocytic astrocytomas	[254]
Colorectal/hepatocellular	c-myc, 3rd intron		[15,16]
Pancreatic	S100P	Poor prognosis	[19]
Various	Maspin	High grade tumors	[36,38]
Various	Synuclein-γ	High correlation with late stage and metastasis	[40,43]

process. With our current understanding, it is now appropriate to design larger studies specifically aimed at evaluating the use of hypomethylation markers in cancer diagnosis, classification and stratification, and prediction and monitoring of disease response.

REFERENCES

1. Ehrlich, M. et al., Amount and distribution of 5-methylcytosine in human DNA from different types of tissues of cells, *Nucleic Acids Res.*, 10, 2709, 1982.
2. Ehrlich, M. et al., Hypomethylation and hypermethylation of DNA in Wilms tumors, *Oncogene*, 21, 6694, 2002.
3. Rollins, R.A. et al., Large-scale structure of genomic methylation patterns, *Genome Res.*, 16, 157, 2006.
4. Song, F. et al., Association of tissue-specific differentially methylated regions (TDMs) with differential gene expression, *Proc. Natl. Acad. Sci. U S A*, 102, 3336, 2005.
5. Khulan, B. et al., Comparative isoschizomer profiling of cytosine methylation: The HELP assay, *Genome Res.*, 16, 1046, 2006.
6. Gama-Sosa, M.A. et al., The 5-methylcytosine content of DNA from human tumors, *Nucleic Acids Res.*, 11, 6883, 1983.
7. Goelz, S.E. et al., Hypomethylation of DNA from benign and malignant human colon neoplasms, *Science (Wash)*, 228, 187, 1985.
8. Fruhwald, M.C. and Plass, C., Global and gene-specific methylation patterns in cancer: aspects of tumor biology and clinical potential, *Mol. Genet. Metab.*, 75, 1, 2002.
9. Esteller, M. and Herman, J.G., Cancer as an epigenetic disease: DNA methylation and chromatin alterations in human tumours, *J. Pathol.*, 196, 1, 2002.
10. Ehrlich, M., DNA methylation in cancer: Too much, but also too little, *Oncogene*, 21, 5400, 2002.
11. Feinberg, A.P. and Vogelstein, B., Hypomethylation of ras oncogenes in primary human cancers, *Biochem. Biophys. Res. Commun.*, 111, 47, 1983.
12. Vachtenheim, J., Horakova, I., and Novotna, H., Hypomethylation of CCGG sites in the 3' region of H-ras protooncogene is frequent and is associated with H-ras allele loss in non-small cell lung cancer, *Cancer Res.*, 54, 1145, 1994.
13. Cheah, M.S., Wallace, C.D., and Hoffman, R.M., Hypomethylation of DNA in human cancer cells: A site-specific change in the c-myc oncogene, *J. Natl. Cancer Inst.*, 73, 1057, 1984.
14. Nambu, S., Inoue, K., and Saski, H., Site-specific hypomethylation of the c-myc oncogene in human hepatocellular carcinoma, *Jpn. J. Cancer Res.*, 78, 695, 1987.

15. Sharrard, R.M. et al., Patterns of methylation of the c-myc gene in human colorectal cancer progression, *Br. J. Cancer.*, 65, 667, 1992.
16. Shen, I. et al., Correlation between DNA methylation and pathological changes in human hepatocellular carcinoma, *Hepatogastroenterology*, 45, 1753, 1998.
17. Hoffmann, M.J. and Schulz, W.A., Causes and consequences of DNA hypomethylation in human cancer, *Biochem. Cell Biol.*, 83, 296, 2005.
18. Sato, N. et al., Discovery of novel targets for aberrant methylation in pancreatic carcinoma using high-throughput microarrays, *Cancer Res.*, 63, 4158, 2003.
19. Sato, N. et al., Identification of maspin and S100P as novel hypomethylation targets in pancreatic cancer using global gene expression profiling, *Oncogene*, 23, 1531, 2004.
20. Nishigaki, M. et al., Discovery of aberrant expression of R-RAS by cancer-linked DNA hypomethylation in gastric cancer using microarrays, *Cancer Res.*, 65, 2115, 2005.
21. Plass, C. and Smiraglia, D.J., Genome-wide analysis of DNA methylation changes in human malignancies, *Curr. Top. Microbiol. Immunol.*, 310, 179, 2006.
22. Frigola, J. et al., Differential DNA hypermethylation and hypomethylation signatures in colorectal cancer, *Hum. Mol. Genet.*, 14, 319, 2005.
23. Yoshida, M. et al., Aberrant expression of the MEL1S gene identified in association with hypomethylation in adult T-cell leukemia cells, *Blood*, 103, 2753, 2004.
24. Weber, M. et al., Chromosome-wide and promoter-specific analyses identify sites of differential DNA methylation in normal and transformed human cells, *Nat. Genet.*, 37, 853, 2007.
25. Hayashi, H. et al., High-resolution mapping of DNA methylation in human genome using oligonucleotide tiling array, *Hum. Genet.*, 120, 701, 2007.
26. Wilson, A.S., Power, B.E., and Molloy, P.L., DNA hypomethylation and human diseases, *Biochim. Biophys. Acta*, 1775, 138, 2007.
27. Hanada, M. et al., bcl-2 gene hypomethylation and high-level expression in B-cell chronic lymphocytic leukemia, *Blood*, 82, 1820, 1993.
28. Muratovska, A. et al., Paired-box genes are frequently expressed in cancer and often required for cancer cell survival, *Oncogene*, 22, 7989, 2003.
29. Wu, H. et al., Hypomethylation-linked activation of PAX2 mediates tamoxifen-stimulated endometrial carcinogenesis, *Nature*, 438, 981, 2005.
30. Chan, K.Y. et al., Epigenetic factors controlling the BRCA1 and BRCA2 genes in sporadic ovarian cancer, *Cancer Res.*, 62, 4151, 2002.
31. Lodygin, D. and Hermeking, H., Epigenetic silencing of 14-3-3sigma in cancer, *Semin. Cancer Biol.*, 16, 214, 2006.
32. Pakneshan, P., Xing, R.H., and Rabbani, S.A., Methylation status of uPA promoter as a molecular mechanism regulating prostate cancer invasion and growth in vitro and in vivo, *FASEB J.*, 17, 1081, 2003.
33. Pulukuri, S.M. et al., Demethylation-linked activation of urokinase plasminogen activator is involved in progression of prostate cancer, *Cancer Res.*, 67, 930, 2007.
34. Pakneshan, P., Tetu, B., and Rabbani, S.A., Demethylation of urokinase promoter as a prognostic marker in patients with breast carcinoma, *Clin. Cancer Res.*, 10, 3035, 2004.
35. Futscher, B.W. et al., Role for DNA methylation in the control of cell type specific maspin expression, *Nat. Genet.*, 31, 175, 2002.
36. Ogasawara, S. et al., Disruption of cell-type-specific methylation at the maspin gene promoter is frequently involved in undifferentiated thyroid cancers, *Oncogene*, 23, 1117, 2004.
37. Akiyama, Y. et al., Cell-type-specific repression of the maspin gene is disrupted frequently by demethylation at the promoter region in gastric intestinal metaplasia and cancer cells, *Am. J. Pathol.*, 163, 1911, 2003.
38. Bettstetter, M. et al., Elevated nuclear maspin expression is associated with microsatellite instability and high tumour grade in colorectal cancer, *J. Pathol.*, 205, 606, 2005.
39. Gupta, A. et al., Hypomethylation of the synuclein gamma gene CpG island promotes its aberrant expression in breast carcinoma and ovarian carcinoma, *Cancer Res.*, 63, 664, 2003.
40. Liu, H. et al., Loss of epigenetic control of synuclein-gamma gene as a molecular indicator of metastasis in a wide range of human cancers, *Cancer Res.*, 65, 7635, 2005.
41. Li, Z. et al., Overexpression of synuclein-gamma in pancreatic adenocarcinoma, *Cancer*, 101, 58, 2004.

42. Yanagawa, N. et al., Demethylation of the synuclein gamma gene CpG island in primary gastric cancers and gastric cancer cell lines, *Clin. Cancer Res.*, 10, 2447, 2004.

43. Wu, K. et al., Expression of neuronal protein synuclein gamma gene as a novel marker for breast cancer prognosis, *Breast Cancer Res. Treat.*, 101, 259, 2007.

44. Inaba, S. et al., Synuclein gamma inhibits the mitotic checkpoint function and promotes chromosomal instability of breast cancer cells, *Breast Cancer Res. Treat.*, 94, 25, 2005.

45. Pan, Z.Z. et al., Gamma-synuclein promotes cancer cell survival and inhibits stress and chemotherapy drug-induced apoptosis by modulating MAPK pathways, *J. Biol. Chem.*, 277, 35050, 2002.

46. Jiang, Y. et al., Gamma synuclein, a novel heat-shock protein-associated chaperone, stimulates ligand-dependent estrogen receptor alpha signaling and mammary tumorigenesis, *Cancer Res.*, 64, 4539, 2004.

47. Jin, G. et al., Characterization of the tissue-specific expression of the s100P gene which encodes an EF-hand Ca2+-binding protein, *Mol. Biol. Rep.*, 30, 243, 2003.

48. Wang, G. et al., Induction of metastasis by S100P in a rat mammary model and its association with poor survival of breast cancer patients, *Cancer Res.*, 66, 1199, 2006.

49. Scanlan, M.J., Simpson, A.J., and Old, L.J. The cancer/testis genes: Review, standardization and commentary, *Cancer Immun.*, 4, 1, 2004.

50. Zendman, A.J., Ruiter, D.J., and Van Muijen, G.N., Cancer/testis-associated genes: Identification, expression profile and putative function, *J. Cell Physiol.*, 194, 272, 2003.

51. de Smet, C., Loirot, A., and Boon, T., Promoter-dependent mechanism leading to selective hypomethylation within the 5' region of gene MAGE-A1 in tumor cells, *Mol. Cell Biol.*, 24, 4781, 2004.

52. Jang, S.J. et al., Activation of melanoma antigen tumor antigens occurs early in lung carcinogenesis, *Cancer Res.*, 61, 7959, 2001.

53. Bert, T. et al., Expression spectrum and methylation-dependent regulation of melanoma antigen-encoding gene family members in pancreatic cancer cells, *Pancreatology*, 2, 146, 2002.

54. Jung, E.J. et al., Expression of family A melanoma antigen in human gastric carcinoma, *Anticancer Res.*, 25, 2105, 2005.

55. Hong, S.J. et al., Relationship between the extent of chromosomal losses and the pattern of CpG methylation in gastric carcinomas, *J. Korean Med. Sci.*, 20, 790, 2005.

56. Imamura, M. et al., Methylation and expression analysis of 15 genes and three normally-methylated genes in 13 ovarian cancer cell lines, *Cancer Lett.*, 241, 213, 2006.

57. Kim, K.H. et al., Promoter hypomethylation and reactivation of MAGE-A1 and MAGE-A3 genes in colorectal cancer cell lines and cancer tissues, *World J. Gastroenterol.*, 12, 5651, 2006.

58. Sigalotti, L. et al., Promoter methylation controls the expression of MAGE2 3 and 4 genes in human cutaneous melanoma, *J. Immunother.*, 25, 16, 2002.

59. Loirot, A. et al., Transient down-regulation of DNMT1 methyltransferase leads to activation and stable hypomethylation of MAGE-A1 in melanoma cells, *J. Biol. Chem.*, 281, 10118, 2006.

60. Qiu, G., Fang, J., and He, Y., 5' CpG island methylation analysis identifies the MAGE-A1 and MAGE-A3 genes as potential markers of HCC, *Clin. Biochem.*, 39, 259, 2006.

61. Cho, B. et al., Promoter hypomethylation of a novel cancer/testis antigen gene CAGE is correlated with its aberrant expression and is seen in premalignant stage of gastric carcinoma, *Biochem. Biophys. Res. Commun.*, 307, 52, 2003.

62. Lee, T.S. et al., DNA hypomethylation of CAGE promotors in squamous cell carcinoma of uterine cervix, *Ann. N Y Acad. Sci.*, 218, 2006.

63. Sigalotti, L. et al., Intratumor heterogeneity of cancer/testis antigens expression in human cutaneous melanoma is methylation-regulated and functionally reverted by 5-aza-2'-deoxycytidine, *Cancer Res.*, 64, 9167, 2004.

64. Lim, J.H. et al., Activation of human cancer/testis antigen gene, XAGE-1, in tumor cells is correlated with CpG island hypomethylation, *Int. J. Cancer*, 116, 200, 2005.

65. Wang, Z. et al., SPAN-Xb expression in myeloma cells is dependent on promoter hypomethylation and can be upregulated pharmacologically, *Int. J. Cancer*, 118, 1436, 2006.

66. Grunau, C. et al., Frequent DNA hypomethylation of human juxtacentromeric BAGE loci in cancer, *Genes Chromosomes Cancer*, 43, 11, 2005.

67. Yao, X. et al., Epigenetic regulation of the taxol resistance-associated gene TRAG-3 in human tumors, *Cancer Genet. Cytogenet.*, 151, 1, 2004.

68. Roman-Gomez, J. Epigenetic regulation of human cancer/testis antigen gene, HAGE, in chronic myeloid leukemia, *Haematologica*, 92, 153, 2007.

69. Dos Santos, N.R. et al., Heterogeneous expression of the SSX cancer/testis antigens in human melanoma lesions and cell lines, *Cancer Res.*, 60, 1654, 2000.

70. Karpf, A.R. et al., Limited gene activation in tumor and normal epithelial cells treated with the DNA methyltransferase inhibitor 5-aza-2′-deoxycytidine, *Mol. Pharmacol.*, 65, 18, 2004.

71. Laduron, S. et al., MAGE-A1 interacts with adaptor SKIP and the deacetylase HDAC1 to repress transcription, *Nucleic Acids Res.*, 32, 4340, 2004.

72. Kisseleva-Romanova, E., Yeast homolog of a cancer-testis antigen defines a new transcription complex, *EMBO J.*, 25, 3576, 2006.

73. de Bruijn, D.R., Nap, J.P., and van Kessel, A.G., The (Epi)genetics of human synovial sarcoma, *Genes Chromosomes Cancer*, 46, 107, 2007.

74. Lim, F.L. et al., KRAB-related domain and a novel transcription repression domain in proteins encoded by SSX genes that are disrupted in human sarcomas, *Oncogene*, 17, 2013, 1998.

75. Pretto, D. et al., The synovial sarcoma translocation protein SYT-SSX2 recruits beta-catenin to the nucleus and associates with it in an active complex, *Oncogene*, 25, 3661, 2006.

76. de Bruijn, D.R. et al., The synovial sarcoma–associated SS18-SSX2 fusion protein induces epigenetic gene (de)regulation, *Cancer Res.*, 66, 9474, 2006.

77. Sun, Y., et al., IGF2 is critical for tumorigenesis by synovial sarcoma oncoprotein SYT-SSX1, *Oncogene*, 25, 1042, 2006.

78. Hong, J.A. et al., Reciprocal binding of CTCF and BORIS to the NY-ESO-1 promoter coincides with derepression of this cancer-testis gene in lung cancer cells, *Cancer Res.*, 65, 7763, 2005.

79. Vatolin, S. et al., Conditional expression of the CTCF-paralogous transcriptional factor BORIS in normal cells results in demethylation and derepression of MAGE-A1 and reactivation of other cancer-testis genes, *Cancer Res.*, 65, 7751, 2005.

80. Hoffmann, M.F. et al., Epigenetic control of CTCFL/BORIS and OCT4 expression in urogenital malignancies, *Biochem. Pharmacol.*, 72, 1577, 2006.

81. Laner, T. et al., Hypomethylation of the XIST gene promoter in prostate cancer, *Oncol. Res.*, 15, 257, 2005.

82. Brueckner, B., et al., The human let-7a-3 locus contains an epigenetically regulated microRNA gene with oncogenic function, *Cancer Res.*, 67, 1419, 2007.

83. Feinberg, A.P. et al., Reduced genomic 5-methylcytosine content in human colonic neoplasia, *Cancer Res.*, 48, 1159, 1988.

84. Bariol, C., et al., The relationship between hypomethylation and CpG island methylation in colorectal neoplasia, *Am. J. Pathol.*, 162, 1361, 2003.

85. Pufulete, M. et al., Folate status, genomic DNA hypomethylation, and risk of colorectal adenoma and cancer: A case control study, *Gastroenterology*, 124, 1240, 2003.

86. Hernandez-Blazquez, F.J. et al, Evaluation of global DNA hypomethylation in human colon cancer tissues by immunohistochemistry and image analysis, *Gut*, 47, 689, 2000.

87. Esteller, M., et al., DNA methylation patterns in hereditary human cancers mimic sporadic tumorigenesis, *Human Molec. Genet.*, 10, 3001, 2001.

88. Suter, C.M., Martin, D.I., and Ward, R.L., Hypomethylation of L1 retrotransposons in colorectal cancer and adjacent normal tissue, *Int. J. Colorectal Dis.*, 19, 95, 2004.

89. Shen, L. et al., MGMT promoter methylation and field defect in sporadic colorectal cancer, *J. Natl. Cancer Inst.*, 91, 1317, 2005.

90. Kawakami, K. et al., DNA hypermethylation in the normal colonic mucosa of patients with colorectal cancer, *Br. J. Cancer*, 94, 593, 2006.

91. Matsuzaki, K. et al., The relationship between global methylation level, loss of heterozygosity, and microsatellite instability in sporadic colorectal cancer, *Clin. Cancer Res.*, 11, 8564, 2005.

92. Deng, G. et al., Regional hypermethylation and global hypomethylation are associated with altered chromatin conformation and histone acetylation in colorectal cancer, *Int. J. Cancer*, 118, 2999, 2006.

93. Estecio, M.R.H. et al., LINE-1 hypomethylation in cancer is highly variable and inversely correlated with microsatellite instability, *Plos ONE*, 5, e399, 2007.

94. Cravo, M. et al., Global DNA hypomethylation occurs in the early stages of intestinal type gastric carcinoma, *Gut*, 39, 434, 1996.

95. Fang, J.Y., et al., Relationship of plasma folic acid and status of DNA methylation in human gastric cancer, *J. Gastroenterol.*, 32, 171, 1997.

96. Kaneda, A., et al., Frequent hypomethylation in multiple promoter CpG islands is associated with global hypomethylation, but not with frequent promoter hypermethylation, *Cancer Sci.*, 95, 58, 2004.

97. Suzuki, K. et al., Global DNA demethylation in gastrointestinal cancer is age dependent and precedes genomic damage, *Cancer Cell*, 9, 199, 2006.

98. Santourlidis, S. et al., High frequency of alterations in DNA methylation in adenocarcinoma of the prostate, *Prostate*, 39, 166, 1999.

99. Schulz, W.A. et al., Genomewide DNA hypomethylation is associated with alterations on chromosome 8 in prostate carcinoma, *Genes Chromosomes Cancer*, 35, 58, 2002.

100. Florl, A.R. et al., Coordinate hypermethylation at specific genes in prostate carcinoma precedes LINE-1 hypomethylation, *Br. J. Cancer*, 91, 985, 2004.

101. Cho, N.Y. et al., Hypermethylation of CpG island loci and hypomethylation of LINE-1 and Alu repeats in prostate adenocarcinoma and their relationship to clinicopathological features, *J. Pathol.*, 211, 269, 2007.

102. Brothman, A.R., et al., Global hypomethylation is common in prostate cancer cells: A quantitative predictor for clinical outcome? *Cancer Genet. Cytogenet.*, 156, 31, 2005.

103. Mohamed, M.A. et al., Epigenetic events, remodelling enzymes and relationship to chromatin organization in prostatic intraepithelial neoplasia and prostatic adenocarcinoma, *Brit. J. Urol. Int.*, 99, 908, 2007.

104. Soares, J. et al., Global DNA hypomethylation in breast carcinoma, correlation with prognostic factors and tumor progression, *Cancer*, 85, 112, 1998.

105. Jackson, K. et al., DNA Hypomethylation is prevalent even in low-grade breast cancers, *Cancer Biol. Ther.*, 3, 1225, 2004.

106. Narayan, A. et al., Hypomethylation of pericentromeric DNA in breast adenocarcinomas, *Int. J. Cancer*, 77, 833, 1998.

107. Costa, F.F. et al., SATR-1 hypomethylation is a common and early event in breast cancer, *Cancer Genet. Cytogenet.*, 165, 135, 2006.

108. Lin, C.H. et al., Genome-wide hypomethylation in hepatocellular carcinogenesis, *Cancer Res.*, 61, 4238, 2001.

109. Takai, D. et al., Hypomethylation of LINE1 Retrotransposon in human hepatocellular carcinomas, but not in surrounding liver cirrhosis, *Jpn. J. Clin. Oncol.*, 30, 306, 2000.

110. Saito, Y. et al., Expression of mRNA for DNA methyltrandferases and methyl-CpG-binding proteins and DNA methylation status on CpG islands and pericentromeric satellite regions during human hepatocarcinogenesis, *Hepatology*, 33, 561, 2001.

111. Wong, N. et al., Hypomethylation of chromosome 1 heterochromatin DNA correlates with q-arm copy gain in human hepatocellular carcinoma, *Am. J. Pathol.*, 159, 465, 2001.

112. Nagai, H. et al., A novel sperm-specific hypomethylation sequence is a demethylation hotspot in human hepatocellular carcinoma, *Gene*, 237, 15, 1999.

113. Saito, Y. et al., Overexpression of a splice variant of DNA methyltransferase 3b, DNMT3b4, associated with DNA hypomethylation on pericentromeric satellite regions during human hepatocarcinogenesis, *Proc. Natl. Acad. Sci. U S A*, 99, 10060, 2002.

114. Park, I.Y. et al., Aberrant epigenetic modifications in hepatocarcinogenesis induced by hepatitis B virus X protein, *Gastroenterology*, 132, 1476, 2007.

115. Pfeifer, G.P. et al., DNA methylation levels in acute human leukemia, *Cancer Lett.*, 39, 185, 1988.

116. Wahlfors, J. et al., Genomic hypomethylation in human chronic lymphocytic leukemia, *Blood*, 80, 2074, 1992.

117. Lyko, F. et al., Quantitative analysis of DNA methylation in chronic lymphocytic leukemia patients, *Electrophoresis*, 25, 1530, 2004.

118. Roman-Gomez, J. et al, Promoter hypomethylation of the LINE-1 retrotransposable elements activates sense/antisense transcription and marks progression of chronic myeloid leukemia, *Oncogene*, 24, 7213, 2005.

119. Cheng, P. et al., Alterations in DNA methylation are early, but not initial, events in ovarian tumorigenesis, *Br. J. Cancer*, 75, 396, 1997.

120. Qu, G. et al., Satellite DNA hypomethylation vs. overall genomic hypomethylation in ovarian epithelial tumors of different malignant potential, *Mutat. Res.*, 423, 91, 1999.

121. Widschwendter, M. et al., DNA hypomethylation and ovarian cancer biology, *Cancer Res.*, 64, 4472, 2004.

122. Ehrlich, M. et al., Quantitative analysis of associations between DNA hypermethylation, hypomethyla-tion, and DNMT RNA levels in ovarian tumors, *Oncogene*, 25, 2636, 2006.

123. Florl, A.R. et al., DNA methylation and expression of LINE-1 and HERV-K provirus sequences in urothelial and renal cell carcinomas, *Br. J. Cancer*, 80, 1312, 1999.

124. Neuhausen, A. et al., DNA methylation alterations in urothelial carcinomas, *Cancer Biol. Ther.*, 5, 993, 2006.

125. Nakagawa, T. et al., DNA hypomethylation on pericentromeric satellite regions significantly correlates with loss of heterozygosity on chromosome 9 in urothelial carcinomas, *J. Urol.*, 173, 243, 2005.

126. Satoh, Y. et al., Genetic and epigenetic alterations on the short arm of chromosome 11 are involved in a majority of sporadic Wilms' tumours, *Br. J. Cancer*, 95, 541, 2006.

127. Ehrlich, M. et al., Hypomethylation and hypermethylation of DNA in Wilms tumors, *Oncogene*, 21, 6694, 2002.

128. Qu, G.M., et al., Frequent hypomethylation in Wilms tumors of pericentromeric DNA in chromosomes 1 and 16, *Cancer Genet. Cytogenet.*, 109, 34, 1999.

129. Ehrlich, M. et al., Satellite DNA hypomethylation in karyotyped Wilms tumors, *Cancer Genet. Cytogenet.*, 141, 97, 2003.

130. Nishiyama, R. et al., A DNA repeat, NBL2, is hypermethylated in some cancers but hypomethylated in others, *Cancer Biol. Ther.*, 4, 440, 2005.

131. Nishiyama, R. et al., Both hypomethylation and hypermethylation in a 0.2 kb region of a DNA repeat in cancer, *Mol. Cancer Res.*, 3, 617, 2005.

132. Cadieux, B. et al., Genome-wide hypomethylation in human glioblastomas associated with specific copy number alteration, methylenetetrahydrofolate reductase allele status, and increased proliferation, *Clin. Cancer Res.*, 66, 8469, 2006.

133. Weisenberger, D.J. et al., Analysis of repetitive element DNA methylation by MethyLight, *Nucleic Acids Res.*, 33, 6823, 2005.

134. de Smet, C. et al., The activation of human gene MAGE-1 in tumor cells is correlated with genome-wide demethylation, *Proc. Natl. Acad. Sci. U S A*, 93, 7149, 1996.

135. Rodriguez, J. et al., Chromosomal instability correlates with genome-wide DNA demethylation in human primary colorectal cancers, *Cancer Res.*, 66, 8462, 2006.

136. Kaneda, A. et al., Frequent hypomethylation in multiple promoter CpG islands is associated with global hypomethylation, but not with frequent promoter hypermethylation, *Cancer Sci.*, 95, 58, 2004.

137. Furuta, J. et al., Promoter methylation profiling of 30 genes in human malignant melanoma, *Cancer Sci.*, 95, 962, 2004.

138. Pini, J.T. et al., Evidence that general genomic hypomethylation and focal hypermethylation are two independent molecular events of non-Hodgkin's lymphoma, *Oncol. Res.*, 14, 399, 2004.

139. Goll, M.G. and Bestor, T.H., Eukaryotic cytosine methyltransferases, *Annu. Rev. Biochem.*, 74, 481, 2005.

140. Klose, R.J. and Bird, A.P., Genomic DNA methylation: the mark and its mediators, *Trends Biochem. Sci.*, 31, 89, 2006.

141. De Marzo, A.M. et al., Abnormal regulation of DNA methyltransferase expression during colorectal carcinogenesis, *Cancer Res.*, 59, 3855, 1999.

142. Li, S. et al., DNA hypomethylation and imbalanced expression of DNA methyltransferases (DNMT1, 3A, and 3B) in human uterine leiomyoma, *Gynecol. Oncol.*, 90, 123, 2003.

143. Etoh, T. et al., Increased DNA methyltransferase 1 (DNMT1) protein expression correlates significantly with poorer tumor differentiation and frequent DNA hypermethylation of multiple CpG islands in gastric cancers, *Am. J. Pathol.*, 164, 689, 2004.

144. Agoston, A.T. et al., Increased protein stability causes DNA methyltransferase 1 dysregulation in breast cancer, *J. Biol. Chem.*, 280, 18302, 2005.

145. Peng, D.F. et al., Increased DNA methyltransferase 1 (DNMT1) protein expression in precancerous conditions and ductal carcinomas of the pancreas, *Cancer Sci.*, 96, 403, 2005.

146. Kanai, Y. et al., DNA methyltransferase expression and DNA methylation of CPG islands and peri-centromeric satellite regions in human colorectal and stomach cancers, *Int. J. Cancer*, 91 205, 2001.

147. Ehrlich, M., The ICF syndrome, a DNA methyltransferase 3B deficiency and immunodeficiency disease, *Clin. Immunol.*, 109, 17, 2003.

148. Ostler, K.R. et al., Cancer cells express aberrant DNMT3B transcripts encoding truncated proteins, *Oncogene*, 26, 5553, 2007.

149. Chedin, F., Lieber, M.R., and Hsieh, C.L., The DNA methyltransferase-like protein DNMT3L stimulates de novo methylation by Dnmt3a, *Proc. Natl. Acad. Sci. U S A*, 99, 16916, 2002.

150. Chen, Z.X. et al., Physical and functional interactions between the human DNMT3L protein and members of the de novo methyltransferase family, *J. Cell. Biochem.*, 95, 902, 2005.

151. Hata, K. et al., Dnmt3L cooperates with the Dnmt3 family of de novo DNA methyltransferases to establish maternal imprints in mice, *Development*, 129, 1983, 2002.

152. Bourc'his, D. and Bestor, T.H., Meiotic catastrophe and retrotransposon reactivation in male germ cells lacking Dnmt3L, *Nature*, 431, 96, 2004.

153. Hata, K. et al., Meiotic and epigenetic aberrations in Dnmt3L-deficient male germ cells, *Mol. Reprod. Dev.*, 73, 116, 2006.

154. Almstrup, K. et al., Genome-wide gene expression profiling of testicular carcinoma in situ progression into overt tumours, *Br. J. Cancer.*, 92, 1934, 2005.

155. Reik., W., Stability and flexibility of epigenetic gene regulation in mammalian development, *Nature*, 447, 425, 2007.

156. Morgan, H.D. et al., Epigenetic reprogramming in mammals, *Hum. Mol. Genet.*, 14, R47, 2005.

157. Kim, S.H. et al., Differential DNA methylation reprogramming of various repetitive sequences in mouse preimplantation embryos, *Biochem. Biophys. Res. Commun.*, 324, 58, 2004.

158. Jost, J.P. et al., 5-Methylcytosine DNA glycosylase participates in the genome-wide loss of DNA methylation occurring during mouse myoblast differentiation., *Nucleic Acids Res.*, 29, 4452, 2001.

159. Lucarelli, M. et al., The dynamics of myogenin site-specific demethylation is strongly correlated with its expression and with muscle differentiation, *J. Biol. Chem.*, 276, 7500, 2001.

160. Bruniquel, D. and Schwartz, R.H., Selective, stable demethylation of the interleukin-2 gene enhances transcription by an active process, *Nature Immunol.*, 4, 235, 2003.

161. Morales-Ruiz, T. et al., DEMETER and REPRESSOR OF SILENCING 1 encode 5-methylcytosine DNA glycosylases, *Proc. Natl. Acad. Sci. U S A*, 103, 6853, 2006.

162. Hendrich, B. et al., The thymine glycosylase MBD4 can bind to the product of deamination at methylated CpG sites, *Nature*, 401, 301, 1999.

163. Li, Y.Q. et al., Association of Dnmt3a and thymine DNA glycosylase links DNA methylation with base-excision repair, *Nucleic Acids Res.*, 35, 390, 2007.

164. Kress, C., Thomassin, H., and Grange, T., Active cytosine demethylation triggered by a nuclear receptor involves DNA strand breaks, *Proc. Natl. Acad. Sci. U S A*, 103, 11112, 2006.

165. Barreto, G. et al., Gadd45a promotes epigenetic gene activation by repair-mediated DNA demethylation, *Nature*, 445, 671, 2007.

166. Morgan, H. et al., Activation-induced cytosine deaminase deaminates 5-methylcytosine in DNA and is expressed in pluripotent tissues: implications for epigenetic reprogramming, *J. Biol. Chem.*, 279, 52353, 2004.

167. Gallais, R. et al., Deoxyribonucleic acid methyl transferases 3a and 3b associate with the nuclear orphan receptor COUP-TFI during gene activation, *Mol. Endocrinol.*, 21, 2085, 2007.

168. D'Alessio, A.C., Weaver, I.C., and Szyf, M., Acetylation induced transcription is required for active DNA demethylation in methylation silenced genes, *Mol. Cell. Biol.*, 27, 7462, 2007.

169. Deitch, N., Bovenzi, V., and Szyf, M., Valproate Induces replication-independent active DNA demethylation, *J. Biol. Chem.*, 278, 27586, 2003.

170. Brown, S.E. and Szyf, M., Epigenetic programming of the rRNA promoter by MBD3, *Mol. Cell. Biol.*, 27, 4938, 2007.

171. Kouzarides, T., Chromatin modifications and their function, *Cell*, 128, 693, 2007.

172. Schuettengruber, B., Genome regulation by polycomb and trithorax proteins, *Cell*, 128, 735, 2007.

173. Esteller, M., Cancer epigenomics: DNA methylomes and histone-modification maps, *Nat. Rev. Genet.*, 8, 286, 2007.

174. Santos-Rosa, H. and Caldas, C., Chromatin modifier enzymes, the histone code and cancer, *Eur. J. Cancer*, 41, 2381, 2005.

175. Richards, E.J. and Elgin, S.C., Epigenetic codes for heterochromatin formation and silencing: Rounding up the usual suspects, *Cell*, 108, 489, 2002.

176. Fuks, F., DNA methylation and histone modifications: Teaming up to silence genes, *Curr. Opin. Genet. Dev.*, 15, 490, 2005.

177. Bannister, A.J. and Kouzarides, T., Reversing histone methylation, *Nature*, 436, 1103, 2005.

178. Espada, J. et al., Human DNA methyltransferase 1 is required for maintenance of the histone H3 modification pattern, *J. Biol. Chem.*, 279, 37175, 2004.

179. Fuks, F. et al., The methyl-CpG-binding protein MeCP2 links DNA methylation to histone methylation, *J. Biol. Chem.*, 278, 4035, 2003.

180. Lehnertz, B. et al., Suv39h-mediated histone H3 lysine 9 methylation directs DNA methylation to major satellite repeats at pericentric heterochromatin, *Curr. Biol.*, 13, 1192, 2003.

181. Peters, A.H. et al., Loss of the Suv39h histone methyltransferases impairs mammalian heterochromatin and genome stability, *Cell*, 107, 323, 2001.

182. Cloos, P.A., et al., The putative oncogene GASC1 demethylates tri- and dimethylated lysine 9 on histone H3, *Nature*, 442, 307, 2006.

183. Fraga, M.F. et al., Loss of acetylation at Lys16 and trimethylation at Lys20 of histone H4 is a common hallmark of human cancer, *Nat. Genet.*, 37, 391, 2005.

184. Bernstein, E. and Allis, C.D., RNA meets chromatin, *Genes Dev.*, 19, 1635, 2005.

185. Martienssen, R.A., Zaratiegui, M., and Goto, D.B., RNA interference and heterochromatin in the fission yeast *Schizosaccharomyces pombe*, *Trends Genet.*, 21, 450, 2005.

186. Grewal, S.I.V. and Elgin, S.C.R., Transcription and RNA interference in the formation of heterochromatin, *Nature*, 447, 399, 2007.

187. Lewis, A. and Reik, W., How imprinting centres work, *Cytogenet. Genome Res.*, 113, 81, 2006.

188. Royo, H. et al., Small non-coding RNAs and genomic imprinting, *Cytogenet. Genome Res.*, 113, 99, 2006.

189. Migeon, B.R., X chromosome inactivation: Theme and variations, *Cytogenet. Genome Res.*, 99, 8, 2002.

190. Migeon, B.R., Is Tsix repression of Xist specific to mouse? *Nat. Genet.*, 33, 337, 2003.

191. Tufarelli, C. et al., Transcription of antisense RNA leading to gene silencing and methylation as a novel cause of human genetic disease, *Nat. Genet.*, 34, 157, 2003.

192. Maison, C. et al., Higher-order structure in pericentric heterochromatin involves a distinct pattern of histone modification and an RNA component, *Nat. Genet.*, 30, 329, 2002.

193. Jeffery, L. and Nakielny, S., Components of the DNA methylation system of chromatin control are RNA-binding proteins, *J. Biol. Chem.*, 279, 49479, 2004.

194. Imamura, T. et al., Non-coding RNA directed DNA demethylation of Sphk1 CpG island, *Biochem. Biophys. Res. Commun.*, 322, 593, 2004.

195. Fremont, M. et al., Demethylation of DNA by purified chick embryo 5-methylcytosine-DNA glycosylase requires both protein and RNA, *Nucleic Acids Res.*, 25, 2375, 1997.

196. Kapranov, P. et al., RNA maps reveal new RNA classes and a possible function for pervasive transcription, *Science*, 316, 1484, 2007.

197. Davis, C.D. and Ulthus, E.O., DNA methylation, cancer susceptibility, and nutrient interactions, *Exp. Biol. Med. (Maywood)*, 229, 988, 2004.

198. Giovannucci, E., Alcohol, one-carbon metabolism, and colorectal cancer: Recent insights from molecular studies, *J. Nutr.*, 134, 2475S, 2004.

199. Pogribny, I.P. et al., Irreversible global DNA hypomethylation as a key step in hepatocarcinogenesis induced by dietary methyl deficiency, *Mutat. Res.*, 593, 80, 2006.

200. James, S.J. et al., Mechanisms of DNA damage, DNA hypomethylation, and tumor progression in the folate/methyl-deficient rat model of hepatocarcinogenesis, *J. Nutr.*, 133, 3740S, 2003.

201. Friso, S. et al., A common mutation in the 5,10-methylenetetrahydrofolate reductase gene affects genomic DNA methylation through an interaction with folate status, *Proc. Natl. Acad. Sci. U S A*, 99, 5606, 2002.

202. Lewis, S.J. et al., Meta-analyses of observational and genetic association studies of folate intakes or levels and breast cancer risk, *J. Natl. Cancer Inst.*, 98, 1607, 2006.

203. Chen, H. et al., Chronic inorganic arsenic exposure induces hepatic global and individual gene hypomethylation: Implications for arsenic hepatocarcinogenesis, *Carcinogenesis*, 25, 1779, 2004.

204. Sadikovic, B. and Rodenhiser, D.I., Benzopyrene exposure disrupts DNA methylation and growth dynamics in breast cancer cells, *Toxicol. Appl. Pharmacol.*, 216, 458, 2006.

205. Villa-Garea, A. et al., Procaine is a DNA demethylating agent with growth-inhibitory effects in human cancer cells, *Cancer Res.*, 63, 4984, 2003.

206. Mittal, A. et al., Exceptionally high protection of photocarcinogenesis by topical application of (−)-epigallocatechin-3-gallate in hydrophilic cream in SKH-1 hairless mouse model: relationship to inhibition of UVB-induced global hypomethylation, *Neoplasia*, 5, 555, 2003.

207. Pogribny, I. et al., Fractionated low-dose radiation exposure leads to accumulation of DNA damage and profound alterations in DNA and histone methylation in the murine thymus, *Mol. Cancer Res.*, 3, 553, 2005.

208. Macnab, J.C. et al., Hypomethylation of host cell DNA synthesized after infection or transformation of cells by herpes simplex virus, *Mol. Cell. Biol.*, 8, 1443, 1988.

209. Gaudet, F. et al., Induction of tumors in mice by genomic hypomethylation, *Science*, 300, 489, 2003.

210. Trinh, B.N. et al., DNA methyltransferase deficiency modifies cancer susceptibility in mice lacking mismatch repair, *Mol. Cell. Biol.*, 22, 2906, 2002.

211. Eden, A. et al., Chromosomal instability and tumors promoted by DNA hypomethylation, *Science*, 300, 455, 2003.

212. Laird, P. et al., Suppression of intestinal neoplasia by DNA hypomethylation, *Cell*, 81, 197, 1995.

213. Cormier, R.T. and Dove, W.F., Dnmt1N/+ reduces the net growth rate and multiplicity of intestinal adenomas in C57BL/6-multiple intestinal neoplasia (Min)/+ mice independently of p53 but demonstrates strong synergy with the modifier of Min 1(AKR) resistance allele, *Cancer Res.*, 60, 3965, 2000.

214. Eads, C.A., Nickel, A.E., and Laird, P.W., Complete genetic suppression of polyp formation and reduction of CpG-island hypermethylation in Apc(Min/+) Dnmt1-hypomorphic mice, *Cancer Res.*, 62, 1296, 2003.

215. Trasler, J. et al., Impact of Dnmt1 deficiency, with and without low folate diets, on tumor numbers and DNA methylation in Min mice, *Carcinogenesis*, 24, 39 2003.

216. Yamada, Y. et al., Opposing effects of DNA hypomethylation on intestinal and liver carcinogenesis, *Proc. Natl. Acad. Sci. U S A*, 102, 13580, 2005.

217. Chen, R.Z., et al., DNA hypomethylation leads to elevated mutation rates, *Nature*, 395, 89, 1998.

218. Chan, M.F. et al., Reduced rates of gene loss, gene silencing, and gene mutation in Dnmt1-deficient embryonic stem cells, *Mol. Cell. Biol.*, 21, 7587, 2001.

219. Kim, M. et al., Dnmt1 deficiency leads to enhanced microsatellite instability in mouse embryonic stem cells, *Nucleic Acids Res.*, 32, 5742, 2004.

220. Guo, G., Wang, W., and Bradley, A., Mismatch repair genes identified using genetic screens in Blm-deficient embryonic stem cells, *Nature*, 429, 891, 2004.

221. Karpf, A.R. and Matsui, S., Genetic disruption of cytosine DNA methyltransferase enzymes induces chromosomal instability in human cancer cells, *Cancer Res.*, 65, 8635, 2005.

222. Almeida, A. et al., Hypomethylation of classical satellite DNA and chromosome instability in lymphoblastoid cell lines, *Human Genet.*, 91, 538, 1993.

223. Vilain, A. et al., DNA methylation and chromosome instability in lymphoblastoid cell lines, *Cytogenet. Cell Genet.*, 90, 93, 2000.

224. Wong, N. et al., Hypomethylation of chromosome 1 heterochromatin DNA correlates with q-arm copy gain in human hepatocellular carcinoma, *Am. J. Pathol.*, 159, 465, 2001.

225. Tsuda, H. et al., Correlation of DNA hypomethylation at pericentromeric heterochromatin regions of chromosomes 16 and 1 with histological features and chromosomal abnormalities of human breast carcinomas, *Am. J. Pathol.*, 161, 859, 2002.

226. Schulz, W. et al., Factor interaction analysis for chromosome 8 and DNA methylation alterations highlights innate immune response suppression and cytoskeletal changes in prostate cancer, *Mol. Cancer*, 6, 14, 2007.

227. Li, L. et al., Distinct patterns of germ-line deletions in MLH1 and MSH2: the implication of Alu repetitive element in the genetic etiology of Lynch syndrome (HNPCC), *Hum. Mutat.*, 27, 388, 2006.

228. Takahashi, M. et al., Detection of APC gene deletion by double competitive polymerase chain reaction in patients with familial adenomatous polyposis, *Int. J. Oncol.*, 29, 413, 2006.

229. Kolomietz, E. et al., The role of Alu repeat clusters as mediators of recurrent chromosomal aberrations in tumors, *Genes Chromosomes Cancer*, 35, 97, 2002.

230. Slebos, R.J., Resnick, M.A., and Taylor, J.A., Inactivation of the p53 tumor suppressor gene via a novel Alu rearrangement, *Cancer Res.*, 58, 5333, 1998.

231. Hsieh, S.Y. et al., High-frequency Alu-mediated genomic recombination/deletion within the caspase-activated DNase gene in human hepatoma, *Oncogene*, 24, 6584, 2005.

232. Florl, A. and Schulz, W., Peculiar structure and location of 9p21 homozygous deletion breakpoints in human cancer cells, *Genes Chromos. Cancer*, 37, 141, 2003.

233. Schulz, W.A., Steinhoff, C., and Florl, A.R., Methylation of endogenous human retroelements in health and disease, *Curr. Top. Microbiol. Immunol.*, 310, 211, 2006.

234. Wang, X.Y., Steelman, L.S., and McCubrey, J.A., Abnormal activation of cytokine gene expression by intracisternal type A particle transposition: Effects of mutations that result in autocrine growth stimulation and malignant transformation, *Cytokines Cell. Mol. Ther.*, 3, 3, 1997.

235. Teugels, E. et al., De novo Alu element insertions targeted to a sequence common to the BRCA1 and BRCA2 genes, *Hum. Mutat.*, 26, 284, 2005.

236. Kloor, M. et al., A large MSH2 Alu insertion mutation causes HNPCC in a German kindred, *Hum. Genet.*, 115, 432, 2004.

237. Liu, J. et al., LINE-I element insertion at the t(11;22) translocation breakpoint of a desmoplastic small round cell tumor, *Genes Chromosomes Cancer*, 18, 232, 1997.

238. Han, J.S. and Boeke, J.D., A highly active synthetic mammalian retrotransposon, *Nature*, 429, 314, 2004.

239. Brouha, B. et al., Hot L1s account for the bulk of retrotransposition in the human population, *Proc. Natl. Acad. Sci. U S A*, 100, 5280, 2003.

240. Belgnaoui, S.M. et al., Human LINE-1 retrotransposon induces DNA damage and apoptosis in cancer cells, *Cancer Cell Int.*, 6, 13, 2006.

241. Bratthauer, G.L. and Fanning, T.G., Active LINE-1 retrotransposons in human testicular cancer, *Oncogene*, 7, 507, 1992.

242. Bratthauer, G.L., Cardiff, R.D., and Fanning, T.G., Expression of LINE-1 retrotransposons in human breast cancer, *Cancer*, 73, 2333, 1994.

243. Paluszczak, J. and Baer-Dubowska, W. Epigenetic diagnostics of cancer—the application of DNA methylation markers, *J. Appl. Genet.*, 47, 365, 2006.

244. Clark, S.J. et al., Bisulphite modification and analysis, *Nature Protocols*, I, 2353, 2006.

245. Herman, J.G. et al., Methylation-specific PCR: A novel PCR assay for methylation status of CpG islands, *Proc. Natl. Acad. Sci. U S A*, 93, 9821, 1996.

246. Rand, K.N. et al., Bisulphite differential denaturation PCR for analysis of DNA methylation, *Epigenetics*, 1, 94, 2006.

247. Chalitchagorn, K. et al., Distinctive pattern of LINE-1 methylation level in normal tissues and the association with carcinogenesis, *Oncogene*, 23, 8841, 2004.

248. Yang, A.S. et al., A simple method for estimating global DNA methylation using bisulfite PCR of repetitive DNA elements, *Nucleic Acids Res.*, 32, e38, 2004.

249. Kawakami, T. et al., XIST unmethylated DNA fragments in male-derived plasma as a tumour marker for testicular cancer, *Lancet*, 363, 40, 2004.

250. Feinberg, A.P., Ohlsson, R., and Henikoff, S., The epigenetic progenitor origin of human cancer, *Nat. Rev. Genet.*, 7, 21, 2006.

251. Cui, H. et al., Loss of IGF2 imprinting: A potential marker of colorectal cancer risk, *Science*, 299, 1753, 2003.

252. Fenech, M., The genome health clinic and genome health nutrigenomics concepts: Diagnosis and nutritional treatment of genome and epigenome damage on an individual basis, *Mutagenesis*, 20, 255, 2005.

253. Fruhwald, M.F., DNA methylation patterns in cancer: Novel prognostic indicators? *Am. J. Pharmacogenomics*, 3, 245, 2003.

254. Uhlmann, D. et al., Distinct methylation profiles of glioma subtypes, *Int. J. Cancer.*, 106, 52, 2003.

255. Yoo, C.B. and Jones, P.A. Epigenetic therapy of cancer: Past present and future, *Nat. Rev. Drug Discovery*, 5, 37, 2006.

256. Issa, J.P., DNA methylation as a therapeutic target in cancer, *Clin. Cancer Res.*, 13, 1634, 2007.

257. Liu, Z. et al., Characterization of in vitro and in vivo hypomethylating effects of decitabine in acute myeloid leukemia by a rapid LC-MS/MS method, *Nucleic Acids Res.*, 35, e31, 2007.

258. Yang, A.S. et al., DNA methylation changes after 5-aza-2′-deoxycytidine therapy in patients with leukemia, *Cancer Res.*, 66, 5495, 2006.

259. Tangkijvanich, P. et al., Serum LINE-1 hypomethylation as a potential prognostic marker for hepatocellular carcinoma, *Clin. Chim. Acta*, 379, 127, 2007.

260. Jung, E.J. et al., Expression of family A melanoma antigen in human gastric carcinoma, *Anticancer Res.*, 25, 2105, 2005.

3 DNA Hypermethylation and Oncogenesis

J. Tyson DeAngelis, Joel B. Berletch, Lucy G. Andrews, and Trygve O. Tollefsbol

CONTENTS

3.1 INTRODUCTION

DNA methylation comprises an integral part of epigenetics and its gene silencing effects have been implicated in numerous biological processes such as aging, development, memory formation, and oncogenesis. DNA methylation is carried out by the DNA methyltransferase (DNMT) family of enzymes, which transfer methyl groups from *S*-adenosylmethionine to the 5-position of cytosine in CpG dinucleotides [1]. Hypermethylation during oncogenesis usually occurs at CpGs located in CpG islands, which are C–G rich regions of 300–3000 bp in or near 40% of promoters in mammalian genes [1]. About 70% of the promoters of human genes contain CpG islands within close proximity. The product of DNA methylation, 5-methylcytosine, is highly susceptible to mutation due to spontaneous deamination, which results in a cytosine to thymine mutation. CpG conversion to TpG over most of the genome has led to highly conserved regions of DNA in or near promoter regions that contain an unusually high percentage of CpGs. These regions are normally unmethylated and the lack of methylation has prevented accumulation of C→T mutations and has promoted the formation of CpG islands [2]. DNA methylation allows an additional tier of hereditable genetic information to be stored upon the genetic code of DNA without changes in the DNA sequence itself. The accepted paradigm of DNA methylation is that hypermethylation usually leads to silencing of gene activity, while the expression of a few genes may be activated by DNA methylation.

The gene silencing effects of hypermethylation have been linked not only to oncogenesis but also to the ability of tumors to sustain a neoplastic phenotype [2]. Methylation profiling of specific CpG islands in various tumor types has begun to reveal the widespread significance of hypermethylation [3]. Genomic methylation profiling can assist in diagnosis as well as predict prognosis and the elucidation of the importance of DNA hypermethylation in oncogenesis has led to creation of treatments aimed at preventing as well as reversing DNA hypermethylation.

3.2   DNA METHYLTRANSFERASES

In human cells, three major DNA methyltransferases (DNMTs) are responsible for DNA methylation. Methylation patterns are maintained by DNMT1, which preferentially methylates hemimethylated DNA following DNA replication [4]. DNMT3A and DNMT3B are known as the de novo methyltransferases [5]. Although DNMT3A and DNMT3B show no preference for unmethylated DNA over hemi-methylated DNA, the low level of de novo methylation carried out by DNMT1 relative to DNMT3A and DNMT3B have led to the latter two enzymes being designated as the de novo methyltransferases [5,6].

Changes in the expression of the DNMTs are a contributing factor leading to changes in methylation patterns. Correlation of modulations in DNMT expression with changes in methylation patterns was shown by Casillas et al. who observed a decrease in expression of DNMT1 and DNMT3A combined with increased expression of DNMT3B in aging fetal lung fibroblasts [7]. These changes in DNMT expression were shown to correlate with changes in maintenance and de novo methylation activity [8]. DNA methylation has also been implicated in memory formation and the age-related changes in DNMT expression and genomic methylation may also provide insight into age-related memory loss [9].

Fetal lung fibroblasts, neoplastically transformed through the addition of defined genetic elements, show a significant increase in expression of all three DNMTs [7]. A recent report has correlated DNMT upregulation seen in vitro with DNMT overexpression in tumors taken from lung cancer patients [10]. Although all DNMTs were upregulated, the extent to which DNMT1 was overexpressed directly correlated with a poor prognosis, illustrating that sustained DNA methylation may not only contribute to oncogenesis but may also help tumors maintain a malignant phenotype [10].

3.3   METHYL-CpG-BINDING PROTEINS AND RECRUITMENT OF CHROMATIN-REMODELING COMPLEXES

One mechanism by which DNA hypermethylation can regulate gene expression is through the recruitment of methyl-binding proteins containing methyl-CpG-binding domains (MBD). These regions contained in proteins such as MeCP1 and MeCP2 include amino acid residues capable of binding to methylated DNA [11]. In particular, Nan and colleagues found that a short stretch of 85 amino acids in the N-terminal region of the MeCP2 protein retained the ability to bind methylated DNA following deletion analysis [12]. There are currently several proteins that have been identified and are referred to as MBD proteins (MBDPs), which have the capacity to silence transcription by binding to both hemimethylated and fully methylated DNA.

DNA methylation is not the only epigenetic modification known to influence gene expression. Chromatin structure is an important aspect of transcriptional regulation and MBDPs provide the link between DNA methylation and chromatin remodeling-mediated gene silencing. The silencing effect is a result of tight associations between DNA and histone proteins brought on by modifications of histone tails. MeCP2 has been found to be associated with Sin3a, a histone deacetylase (HDAC) corepressor, whereas MeCP1 complexes with HDAC1 and HDAC2 [13,14]. Histone methyltransferases (HMTs) are another family of histone modifying enzymes that repress transcription in part through methylation of lysine 9 on histone H3. HMTs have also been shown to associate with MeCP2, once again linking DNA methylation to chromatin remodeling [15].

By binding to hypermethylated DNA, the involvement of MeCPs in chromatin-remodeling complexes leads to altered chromatin structure and long-term gene silencing; however, without hypermethylation the MeCPs do not normally bind to DNA. The fact that most of methylation association silencing requires DNA methylation to be present leads into perhaps one of the most significant questions in the field of epigenetics: How and why are CpG islands that have been evolutionarily conserved due to a lack of methylation become suddenly hypermethylated? Studies over the last decade have finally begun to unravel a mystery that involves the interactions of a very unique and diverse group of players.

3.4 INITIATION AND PRESERVATION OF ABERRANT METHYLATION PATTERNS

Once hypermethylation of DNA occurs, the MBDPs bind to methylated DNA, thereby preventing the binding of transcriptional activators, and facilitate recruitment of chromatin-remodeling complexes that allow genes to be maintained in a silenced state. Despite genome-wide hypomethylation during oncogenesis, CpG islands located in the 5′ control region of certain tumor suppressor genes are specifically targeted and more frequently methylated than other CpG islands across the genome [3]. The process by which tumor suppressors are targeted is not yet fully understood and elucidation of exactly how this targeting takes place would be a landmark discovery.

De novo methylation by DNMT3A and DNMT3B appears to depend on their association with other proteins, such as transcriptional regulators and chromatin-remodeling proteins. c-Myc is a well-known gene regulator that primarily functions as a transcriptional activator. c-Myc also can repress gene transcription and this ability has been shown to be an effect asserted in part by its association with DNMT3A. Association with other transcriptional regulating proteins allows c-Myc to be recruited to the promoter region of certain tumor suppressors, bringing with it DNMT3A, which in turn will methylate unmethylated CpGs [16].

Another example of how transcription factors can recruit DNMTs to the promoter region of a gene involves the promyelocytic-retinoic acid receptor (PML-RAR) fusion protein. The oncogenetic transcription factor PML-RAR has been shown to interact with DNMT1 and DNMT3A. A gene known to be regulated by PML-RAR is the retinoic acid receptor RARβ2 [17]. The interaction of PML-RAR with DNMTs leads to DNMT localization at the RARβ2 promoter and contributes to methylation-induced repression [17].

Recruitment of DNMTs by oncoproteins is only a piece of the puzzle. Another important factor is the association of DNMTs with chromatin-modifying complexes, a relationship that seems to affect nearly every area of epigenetics. Histone methylation and subsequent chromatin inactivation have been shown to precede de novo DNA methylation, possibly by HMT interaction with DNMT3A [18]. The cyclical association of HMTs with DNMT3a and of MeCP2 with HMTs may also help explain why de novo-methylated CpG islands emerge in close proximity to one another in the genome [19]. Lymphoid-specific helicase (Lsh), a member of the SNF2 subfamily of helicase enzymes, has also been shown to have an effect on DNA methylation in mouse embryonic fibroblasts by interaction with Dnmt3a and 3b but not DNMT1 [20]. No known report has yet to corroborate this study in humans or to illustrate its effect on oncogenesis.

Another molecule has been recently shown to assist in targeting of aberrant methylation patterns is RNA. A study by Castanotto has shown that short hairpin RNAs (shRNAs), containing sequences homologous to the promoter of the *RASSF1A*, mediated a DNA methylation-induced gene silencing effect [21]. The ability of RNA to target genes for methylation opens up an entirely new avenue of epigenetic research. siRNAs might allow researchers, for the first time, to target specific genes for hypermethylation. This would allow a control over gene expression never seen before.

Preservation of methylation patterns is a much simpler process but still involves most of the same players that regulate initiation of aberrant methylation patterns. Recruitment of DNMT1, HDACs, and HMTs by MBDPs allows the chromatin to remain in a condensed state and the methylation patterns to be maintained with a high degree of fidelity [15,22,23].

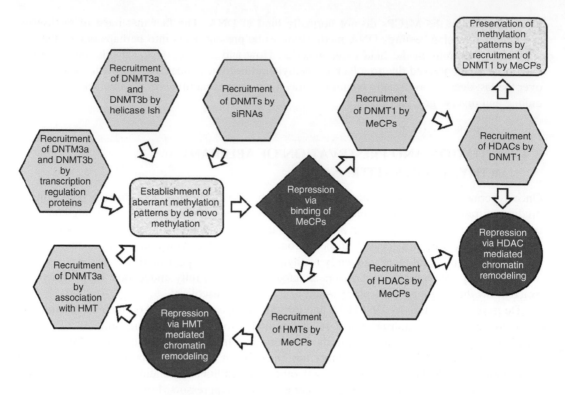

FIGURE 3.1 Possible cyclic mechanisms by which aberrant methylation patterns are initiated and preserved. The interactions of various proteins that are capable of initiating and preserving methylation patterns are depicted. The two rectangles represent instances of DNA methylation where either aberrant methylation patterns are established or preserved. The hexagons represent instances of recruitment that leads to DNA methylation. The dark diamond depicts repression of gene expression as a direct result of methylation and MeCPs. Dark circles represent repression brought on by chromatin-remodeling proteins. This figure illustrates how complex the interactions of epigenetic modulators are and provides insight into why the underlying mechanisms that initiate aberrant methylation patterns have yet to be resolved.

Although multiple proteins and even RNA cooperate to target and effectively silence specific tumor suppressor genes, hypermethylation is the factor that allows genes targeted for silencing to be maintained in a transcriptionally silenced state. Treatment with the demethylating agent, 5-azacytidine (5-aza), can reactivate almost every known methylation-silenced gene, further supporting the vital role of DNA methylation in gene silencing. The mechanism by which 5-aza reverses methylation is discussed later in this chapter.

Figure 3.1 is a schematic representation of the complex interactions between the various proteins that modulate initiation and preservation of hypermethylation. One possible reason why it has been so difficult to elucidate the exact cause of the appearance of aberrant methylation patterns may be that any and all of the players involved may be capable of initiating hypermethylation of a target gene. It also may be possible that one player may initiate DNA methylation in one specific gene but not in another. If this is the case, it would further need to be resolved as to what circumstances are required for a specific methylation modulator to initiate DNA methylation.

3.5 IMPACT OF HYPERMETHYLATION ON ONCOGENESIS

Due to the large number of genes known to be involved in oncogenesis and regulated via methylation, the following sections illustrate a few specific examples of genes across a range of tumor types that are silenced via hypermethylation and contribute to five of the six hallmarks of cancer according to Hanahan et al., which are evading apoptosis, insensitivity to antigrowth signals, sustained angiogenesis, limitless replicative potential, and tissue evasion and metastasis [24]. The sixth hallmark, self-sufficiency in growth signals, has not been shown to be regulated by DNA methylation.

3.5.1 EVADING APOPTOSIS

The internal and external cellular environment is carefully monitored and any abnormalities or changes can trigger a cascade of events leading to apoptosis or programmed cell death. Apoptosis can be triggered by changes in growth rates, contact inhibition, radiation, and even signals from nearby cells and tissues. Multiple proteins are involved in apoptotic pathways, and although their pathways differ in their method of destruction, all lead to cell death.

The *p53* gene is one of the best known apoptosis modulators and it is inactivated by mutations in greater than 50% of cancers [25]. Loss of p53 expression triggers a vital apoptotic pathway, capable of initiating cell death in response to a number of abnormal cellular conditions [25]. The coding region of the *p53* gene is hypermethylated in normal cells and mutations resulting from deamination of 5-methylcytosine have been shown to lead to p53 inactivation [26]. Mouse double minute 2 (MDM2) is a nuclear protein that binds and represses the function of the p53 protein. The cyclin-dependent kinase inhibitor 4a gene (*INK4a*) encodes two proteins: $p16^{INK4a}$ and $p14^{ARF}$, the latter is generated from an alternate reading frame. $p14^{ARF}$ is a nuclear protein that binds and represses the function of MDM2, thus allowing p53 to carry out its apoptotic role. Hypermethylation of the 5′ control region of the *INK4a/ARF* gene results in transcriptional repression of $p14^{ARF}$ and allows MDM2 to bind and repress p53 [27]. Hypermethylation plays an extremely important role in preventing p53 from functioning thereby allowing oncogenesis to occur.

There are two major pathways—mitochondrial- and caspase-mediated pathways—that induce apoptosis-triggered cell death both of which can function independently but also interact with one and other [24]. Activation of the mitochondrial pathway results in various outcomes, one of which is hypoxia-induced cell death. B-cell lymphoma-2 (BCL-2)/adenovirus E1B 19 kDa interacting protein 3 (BNIP3) is a member of the BCL-2 family of apoptotic-linked proteins, and when activated, localizes to the mitochondria. Localization of BNIP3 to the mitochondria results in hypoxia-induced cell death [28]. The promoter of BNIP3 contains a CpG island and is a target of hypermethylation-induced silencing [29]. Silencing of BNIP3 allows cancer cells to avoid cell death induced by hypoxia [28,29].

There are multiple caspase pathways, all of which result in a dismantling of cellular and organelle membranes. Caspase-8, called an initiator caspase, is one of the first caspases activated, and its activation triggers the other caspases to begin the dismantling process [30]. Inactivation of caspase-8 has been shown to be a result of hypermethylation in its 5′ control region and results in inhibition of apoptosis [31].

Apoptosis is a complex process that can be initiated by a multitude of abnormal cellular conditions and induces cell death through numerous pathways. Inactivation of apoptotic pathways is a common trend across a majority of cancers [32]. Epigenetic alterations contribute to the dysregulation of apoptotic pathways. Another consequence of blocking apoptotic pathways is a resistance to anticancer drugs, many of which function by inducing cell death [33].

3.5.2 INSENSITIVITY TO ANTIGROWTH SIGNALS

The propensity of cancer cells to proliferate at rates much greater than normal cells lies in their ability to bypass cell cycle checkpoints. Multiple pathways affect cellular proliferation and one pathway in particular is dysregulated in a majority of cancers. The retinoblastoma (Rb) pathway regulates the G1 to S checkpoint and is one of the effectors of antigrowth signals. Both transition to

S phase and antigrowth signaling pathways involve signal transduction via phosphorylation to induce or inhibit cellular division. Hypermethylation reduces transcription of inhibitors of these pathways allowing the cell to continuously divide despite the lack of sufficient levels of growth signals or the presence of antigrowth signals. A high rate of proliferation is a major contributor to oncogenesis.

Signaling through the TGFβ pathway can lead to a block in cellular division. The Rb pathway is one of the targets of TGFβ signaling and the antigrowth signals function to prevent Rb phosphorylation. $p16^{INK4a}$, the other protein encoded from the *INK4a/ARF* locus, is a well-known tumor suppressor that asserts its effect by binding to and inhibiting cyclin-dependent kinases (CDK) 4 and 6. Binding of $p16^{INK4a}$ to CDK4/6 prevents the CDK complex from binding to cyclin D [34]. When the CDK4/6 complex is bound to cyclin D, the CDK4/6–cyclin D complex will phosphorylate Rb. When Rb is unphosphorylated it is bound to E2F1, a transcription factor that promotes G1 to S transition [35]. When Rb is phosphorylated, E2F1 is free to induce the G1 to S transition. In normal cells, the activity of $p16^{INK4a}$ allows the $p16^{INK4a}$/Rb pathway to act as a cell cycle checkpoint and in combination with TGFβ signaling provides an excellent means to prevent unnecessary proliferation [34]. This inactivation of $p16^{INK4a}$ in cancer cells results from hypermethylation of its promoter and subsequent gene silencing [36]. Inactivation of p16 causes Rb to be constantly phosphorylated despite antigrowth signaling from TGFβ. This is yet another example of how the silencing effects brought on by hypermethylation can allow systems to be bypassed that normally serve to prevent oncogenesis.

A recent study has opened up a new area of oncogenic effects exerted by hypermethylation. The great majority of cases of hypermethylation result in silencing of tumor suppressors; however through silencing microRNA (mirRNA) genes via hypermethylation, oncogenes can become upregulated. mirRNAs are endogenous noncoding RNAs that use RNAi pathways to posttranscriptionally silence expression of a target mRNA [37]. Lujambio et al. have shown that *miR-124a*, a gene that encodes a mirRNA, which contains sequences complementary to CDK6 mRNA, is downregulated in tumors via hypermethylation of its promoter. Downregulation of *miR-124a* causes an upregulation of CDK6 [38]. Upregulation of CDK6 leads to Rb phosphorylation and subsequent increase in cellular proliferation [38]. The functions of mirRNAs are not fully understood, and as their role in cellular biology is revealed, DNA methylation will surely be shown to play a major role in their regulation.

3.5.3 SUSTAINED ANGIOGENESIS

Cancer cells, like normal cells, need a constant supply of oxygen and nutrients to function properly. The process of angiogenesis creates the vessels required to sustain an abnormally high growth rate. For neovascularization to occur, the extracellular matrix (ECM) must be degraded and subsequently restructured. The matrix metalloproteinase (MMP) family of proteases is secreted into the ECM, facilitating its degradation [39]. Another initiator of angiogenesis is the vascular endothelial growth factor (VEGF), which promotes vessel formation through signal transduction-mediated activation of downstream elements [40]. One of the receptors of VEGF, VEGFR2, has been shown to induce angiogenesis through integrin activation [41]. Tissue inhibitor of metalloproteinases-3 (TIMP3) inhibits MMPs preventing them from degrading the ECM [42]. TIMP3 also prevents VEGF-induced angiogenesis by binding to VEGFR2 [43]. Downregulation of TIMP3 is due in part to hypermethylation of its promoter region, and leads to increased angiogenic capabilities [44].

Various signaling systems initiate the formation of the required vessels and here also hypermethylation leads to angiogenesis by silencing genes that serve to counteract the angiogenic process. Another anti-angiogenic protein is thrombospondin-1 (TSP1). Its expression has been shown to inhibit oncogenesis and the downregulation of this protein is a result of methylation of CpG islands in the promoter of the *TSP1* gene [45]. TSP1 downregulation allows the ECM to be restructured during oncogenesis [46].

3.5.4 LIMITLESS REPLICATIVE POTENTIAL

The ability of cancer cells to divide indefinitely is due in part to the multimeric enzyme telomerase. Normal cells display low to undetectable levels of telomerase activity; however, immortal cells such as stem cells and cancer cells display high levels of telomerase activity [47,48]. Telomerase adds repeats of DNA hexamers to the ends of chromosomes and helps circumvent the end-replication problem. The enzyme telomerase is composed of key subunits. One component is hTR, the RNA template that is used to synthesize the hexameric DNA. Another component is human telomerase reverse transcriptase (hTERT). The expression of *hTERT* has been shown to directly correlate with telomerase activity [49]. Due to the role of hTERT in extending the lifespan of cancer cells, *hTERT* gene regulation has been one of the prime targets for anticancer therapies. By downregulating telomerase activity, the chromosomes of cancer cells will eventually become so short that the cells can no longer function properly and will enter into senescence.

The *hTERT* gene has for a long time been one of the few examples of DNA methylation leading to gene activation. Multiple reports have shown that hypermethylation of key CpG dinucleotides in the 5′ control region of the *hTERT* gene leads to its activation in cancer cells. However, recent studies, using more sensitive assays, have shown that an area of the *hTERT* promoter within close proximity of the transcriptional start site must remain unmethylated for transcription to occur [50,51]. These new reports for the first time provide insight into how *hTERT* transcription can occur despite hypermethylation, and illustrate that specific patterns of methylation involving key sites are more likely to control gene expression than total hypermethylation or hypomethylation. Though an area of the *hTERT* regulatory region must remain hypomethylated, it has also been shown that hypermethylation is still necessary for *hTER*T upregulation. CCCTC-binding factor (CTCF), a transcriptional repressor with a zinc finger domain, whose binding is methylation sensitive, has been shown to bind to an unmethylated area within the first exon of *hTERT* [52]. Hypermethylation of *hTERT*, as a result of oncogenesis, prevents the binding of CTCF. Despite this new finding, *hTERT* maintains its distinction as one of the few genes that requires hypermethylation for activation.

3.5.5 TISSUE INVASION AND METASTASIS

Cancer metastasis has been associated with highly invasive tumors and usually results in a poor prognosis. One of the major hallmarks of malignant tumors is the loss of cellular adhesiveness allowing the cells to separate and colonize other tissues. Dysregulation of cytoskeletal signaling proteins may also play a role in promoting cancer metastasis.

Important proteins that work to maintain cell-to-cell junctions are the cadherins and the catenins, which include α-catenin, E-cadherin, and β-catenin. α-Catenin is known to connect E-cadherin and β-catenin to actin filaments, which confer stability to the cell–cell junctions [53]. Through these observations it is suggested that suppression of E-cadherin can cause the release of cancer cells from primary tumors [54,55]. Suppression of E-cadherin expression through DNA hypermethylation was illustrated in a study by Yoshiura et al. where hypermethylation around the promoter was present and treatment with 5-azacytidine induced *E-cadherin* transcription [56]. The *E-cadherin* gene was also examined in vivo and also showed hypermethylation of its promoter region in hepatocellular carcinomas [56].

Focal adhesion proteins localized at ECM contact sites are important in establishing proper cytoskeletal signaling. Recently, integrin linked kinase (ILK) has emerged as a crucial protein in focal adhesion and has been shown to interact with LIM-and-senescent-cell-antigen-like-domains-1 (LIMS1) and parvin to form complexes in ECM sites [57–59]. A protein highly similar to LIMS1, termed LIMS2, has been illustrated to also form complexes with ILK and parvin; however, the LIMS1 and LIMS2 complexes are mutually exclusive suggesting a regulatory role of LIMS2 on the LIMS1–ILK–parvin complex. Recently gastric cancer cell lines have shown increased methylation and associated silencing of LIMS2. siRNA reduction of LIMS2 expression significantly increased cell migration in SNU-484 and SNU-668 gastric cancer cell lines suggesting a regulatory role in cell migration for LIMS2. Thus, silencing of LIMS2 transcription by hypermethylation in gastric cancer may play a key role in invasiveness and metastasis [60].

Table 3.1 summarizes the diverse mechanisms by which hypermethylation promotes oncogenesis. The genes reviewed in this chapter are just a few of the genes regulated by DNA methylation and each year more and more genes are being reported to be modulated by hypermethylation. With the vast range of effects exerted by DNA methylation it comes as no surprise that much focus has been placed on the discovery of cancer therapies aimed at preventing, reversing, and, in the case of oncogenes, inducing DNA methylation. DNA methylation plays such a vital role in oncogenesis, before long it will be considered a hallmark of cancer itself.

3.6   CANCER THERAPEUTICS AND HYPERMETHYLATION

Multiple therapies have been created over the past decade to treat neoplasia by demethylation of DNA. Most of these therapies reduce methylation through inhibition of DNMTs. One class of compounds used as demethylation agents are nucleoside analogues. They have a structure similar to that of cytosine, except they have a modification at their 5 carbon and when incorporated into DNA

TABLE 3.1
Impacts of Hypermethylation on Oncogenesis

Hallmarks of Cancer	Gene	Effect Exerted by Methylation	Contribution to Oncogenesis	References
Evasion of apoptosis	P53	Mutation	Mutation-linked inactivation prevents p53 from initiating apoptotic pathways	[26]
	P14^{ARF}	Silencing	Silencing of p14^{ARF} allows MDM2 to suppress the function of p53	[27]
	BNIP3	Silencing	Silencing of BNIP3 prevents hypoxia-linked apoptosis	[29]
	Caspase-8	Silencing	Silencing of caspase-8 prevents activation of caspase-associated apoptosis	[31]
Insensitivity to antigrowth signals	p16^{INK4a}	Silencing	Silencing of p16^{INK4a} prevents Rb from carrying out its role in cell cycle regulation	[36]
	miR-124a	Silencing	Silencing of miR-124a causes an increase in CDK6 expression, which leads to Rb phosphorylation	[38]
Sustained angiogenesis	TIMP3	Silencing	Silencing of TIMP3 leads to: (1) MMPs to degrade ECM (2) VEGF-mediated growth signals to be efficiently transduced	[44]
	TSP1	Silencing	Silencing of TSP1 promotes vessel formation	[45]
Limitless replicative potential	hTERT	Activation	Activation of hTERT allows telomerase to maintain telomeres by addition of hexameric repeats	[52]
Tissue invasion and metastasis	E-cadherin	Silencing	Silencing of E-cadherin leads to degradation of cell-to-cell junctions	[56]
	LIMS2	Silencing	Silencing of LIMS2 leads to a loss of focal adhesion at ECM contact sites	[60]

they will covalently bind DNMTs to the DNA [61]. The DNMTs will remain bound to the DNA preventing them from carrying out methylation elsewhere. Azacytidine, zebularine, and decitabine have all shown an ability to decrease methylation through DNMT inhibition [62]. Although these compounds are very unstable, superior delivery and handling methods have been designed to increase their efficiency as demethylating agents [63].

The major polyphenol in green tea is (−)-epigallocatechin-3-gallate (EGCG) and it has been shown to reduce hypermethylation through inhibition of DNMTs [64]. EGCG is readily available, making it an ideal compound for cancer prevention and therapy. Other molecules such as genistein, procainamide, antisense oligonucleotides, and siRNAs have all been shown to reduce hypermethylation and are discussed in greater detail later on in this book [65–68].

3.7 CONCLUSION

The rest of this book covers in detail the epigenetics of cancer and in almost every instance DNA methylation will be involved in some capacity. Hypermethylation, in the promoters of tumor suppressor genes, leads to silencing of gene expression and is a primary effect exerted by DNA methylation. Although the exact mechanisms by which these genes are targeted for silencing are yet to be revealed, great progress has been made over the last decade. Contributing to nearly every aspect of carcinogenesis, hypermethylation may provide one of the most promising targets for cancer therapeutics, and may someday be a mainstay of cancer therapy.

REFERENCES

1. Bestor, T.H., The DNA methyltransferases of mammals, *Hum Mol Genet*, 9, 2395, 2000.
2. Egger, G., et al., Epigenetics in human disease and prospects for epigenetic therapy, *Nature*, 429, 457, 2004.
3. Costello, J.F., et al., Aberrant CpG-island methylation has non-random and tumour-type-specific patterns, *Nat Genet*, 24, 132, 2000.
4. Turek-Plewa, J. and Jagodzinski, P.P., The role of mammalian DNA methyltransferases in the regulation of gene expression, *Cell Mol Biol Lett*, 10, 631, 2005.
5. Okano, M., Bell, D.W., Haber, D.A., and Li, E., DNA methyltransferases Dnmt3a and Dnmt3b are essential for de novo methylation and mammalian development, *Cell*, 99, 247, 1999.
6. Hsieh, C.L., The de novo methylation activity of Dnmt3a is distinctly different than that of Dnmt1, *BMC Biochem*, 6, 6, 2005.
7. Casillas, M.A., et al., Transcriptional control of the DNA methyltransferases is altered in aging and neoplastically-transformed human fibroblasts, *Mol Cell Biochem*, 252, 33, 2003.
8. Lopatina, N., et al., Differential maintenance and de novo methylating activity by three DNA methyltransferases in aging and immortalized fibroblasts, *J Cell Biochem*, 84, 324, 2002.
9. Miller, C.A. and Sweatt, J.D., Covalent modification of DNA regulates memory formation, *Neuron*, 53, 857, 2007.
10. Lin, R.K., et al., Alteration of DNA methyltransferases contributes to 5∋XπΓ methylation and poor prognosis in lung cancer, *Lung Cancer*, 55, 205, 2007.
11. Ballestar, E. and Wolffe, A.P., Methyl-CpG-binding proteins. Targeting specific gene repression, *Eur J Biochem*, 268, 1, 2001.
12. Nan, X., Meehan, R.R., and Bird, A., Dissection of the methyl-CpG binding domain from the chromosomal protein MeCP2, *Nucleic Acids Res*, 21, 4886, 1993.
13. Ng, H.H., et al., MBD2 is a transcriptional repressor belonging to the MeCP1 histone deacetylase complex, *Nat Genet*, 23, 58, 1999.
14. Nan, X., et al., Transcriptional repression by the methyl-CpG-binding protein MeCP2 involves a histone deacetylase complex, *Nature*, 393, 386, 1998.
15. Fuks, F., et al., The methyl-CpG-binding protein MeCP2 links DNA methylation to histone methylation, *J Biol Chem*, 278, 4035, 2003.
16. Brenner, C., et al., Myc represses transcription through recruitment of DNA methyltransferase corepressor, *EMBO J*, 24, 336, 2005.

17. Di Croce, L., et al., Methyltransferase recruitment and DNA hypermethylation of target promoters by an oncogenic transcription factor, *Science*, 295, 1079, 2002.

18. Strunnikova, M., et al., Chromatin inactivation precedes de novo DNA methylation during the progressive epigenetic silencing of the RASSF1A promoter, *Mol Cell Biol*, 25, 3923, 2005.

19. Keshet, I., et al., Evidence for an instructive mechanism of de novo methylation in cancer cells, *Nat Genet*, 38, 149, 2006.

20. Zhu, H., et al., Lsh is involved in de novo methylation of DNA, *EMBO J*, 25, 335, 2006.

21. Castanotto, D., et al., Short hairpin RNA-directed cytosine (CpG) methylation of the RASSF1A gene promoter in HeLa cells, *Mol Ther*, 12, 179, 2005.

22. Robert, M.F., et al., DNMT1 is required to maintain CpG methylation and aberrant gene silencing in human cancer cells, *Nat Genet*, 33, 61, 2003.

23. Stirzaker, C., et al., Transcriptional gene silencing promotes DNA hypermethylation through a sequential change in chromatin modifications in cancer cells, *Cancer Res*, 64, 3871, 2004.

24. Hanahan, D. and Weinberg, R.A., The hallmarks of cancer, *Cell*, 100, 57, 2000.

25. Levine, A.J., p53, the cellular gatekeeper for growth and division, *Cell*, 88, 323, 1997.

26. Rideout, W.M., et al., 5-Methylcytosine as an endogenous mutagen in the human LDL receptor and p53 genes, *Science*, 249, 1288, 1990.

27. Yin, D., et al., Methylation, expression, and mutation analysis of the cell cycle control genes in human brain tumors, *Oncogene*, 21, 8372, 2002.

28. Regula, K.M., Ens, K., and Kirshenbaum, L.A., Inducible expression of BNIP3 provokes mitochondrial defects and hypoxia-mediated cell death of ventricular myocytes, *Circ Res*, 91, 226, 2002.

29. Murai, M., et al., Aberrant DNA methylation associated with silencing BNIP3 gene expression in haematopoietic tumours, *Br J Cancer*, 92, 1165, 2005.

30. Thornberry, N.A. and Lazebnik, Y., Caspases: Enemies within, *Science*, 281, 1312, 1998.

31. Martinez, R., et al., CpG island promoter hypermethylation of the pro-apoptotic gene caspase-8 is a common hallmark of relapsed glioblastoma multiforme, *Carcinogenesis*, 28, 1264, 2007.

32. Schmitt, C.A., et al., INK4a/ARF mutations accelerate lymphomagenesis and promote chemoresistance by disabling p53, *Genes Dev*, 13, 2670, 1999.

33. Schmitt, C.A. and Lowe, S.W., Apoptosis and therapy, *J Pathol*, 187, 127, 1999.

34. Rocco, J.W. and Sidransky, D., p16 (MTS-1/CDKN2/INK4a) in cancer progression, *Exp Cell Res*, 264, 42, 2001.

35. Cobrinik, D., et al., The retinoblastoma protein and the regulation of cell cycling, *Trends Biochem Sci*, 17, 312, 1992.

36. Attri, J., et al., Alterations of tumor suppressor gene p16INK4a in pancreatic ductal carcinoma, *BMC Gastroenterol*, 5, 22, 2005.

37. Zeng, Y., Wagner, E.J., and Cullen, B.R., Both natural and designed micro RNAs can inhibit the expression of cognate mRNAs when expressed in human cells, *Mol Cell*, 9, 1327, 2002.

38. Lujambio, A., et al., Genetic unmasking of an epigenetically silenced microRNA in human cancer cells, *Cancer Res*, 67, 1424, 2007.

39. Stetler-Stevenson, W.G., The role of matrix metalloproteinases in tumor invasion, metastasis, and angiogenesis, *Surg Oncol Clin N Am*, 10, 383, 2001.

40. Martiny-Baron, G. and Marme, D., VEGF-mediated tumour angiogenesis: a new target for cancer therapy, *Curr Opin Biotechnol*, 6, 675, 1995.

41. Stupack, D.G. and Cheresh, D.A., Integrins and angiogenesis, *Curr Top Dev Biol*, 64, 207, 2004.

42. Anand-Apte, B., et al., Inhibition of angiogenesis by tissue inhibitor of metalloproteinase-3, *Invest Ophthalmol Vis Sci*, 38, 817, 1997.

43. Qi, J.H., et al., A novel function for tissue inhibitor of metalloproteinases-3 (TIMP3): inhibition of angiogenesis by blockage of VEGF binding to VEGF receptor-2, *Nat Med*, 9, 407, 2003.

44. Feng, H., et al., Down-regulation and promoter methylation of tissue inhibitor of metalloproteinase 3 in choriocarcinoma, *Gynecol Oncol*, 94, 375, 2004.

45. Oue, N., et al., Reduced expression of the TSP1 gene and its association with promoter hypermethylation in gastric carcinoma, *Oncology*, 64, 423, 2003.

46. Lawler, J., Thrombospondin-1 as an endogenous inhibitor of angiogenesis and tumor growth, *J Cell Mol Med*, 6, 1, 2002.

47. Hahn, W.C., Telomere and telomerase dynamics in human cells, *Curr Mol Med*, 5, 227, 2005.

48. Kim, N.W., et al., Specific association of human telomerase activity with immortal cells and cancer, *Science*, 266, 2011, 1994.
49. Nakamura, T.M., et al., Telomerase catalytic subunit homologs from fission yeast and human, *Science*, 277, 955, 1997.
50. Devereux, T.R., et al., DNA methylation analysis of the promoter region of the human telomerase reverse transcriptase (hTERT) gene, *Cancer Res*, 59, 6087, 1999.
51. Zinn, R.L., et al., hTERT is expressed in cancer cell lines despite promoter DNA methylation by preservation of unmethylated DNA and active chromatin around the transcription start site, *Cancer Res*, 67, 194, 2007.
52. Renaud, S., et al., Dual role of DNA methylation inside and outside of CTCF-binding regions in the transcriptional regulation of the telomerase hTERT gene, *Nucleic Acids Res*, 35, 1245, 2007.
53. Hirohashi, S. and Kanai, Y., Cell adhesion system and human cancer morphogenesis, *Cancer Sci*, 94, 575, 2003.
54. Behrens, J., et al., Dissecting tumor cell invasion: epithelial cells acquire invasive properties after the loss of uvomorulin-mediated cell-cell adhesion, *J Cell Biol*, 108, 2435, 1989.
55. Vleminckx, K., et al., Genetic manipulation of E-cadherin expression by epithelial tumor cells reveals an invasion suppressor role, *Cell*, 66, 107, 1991.
56. Yoshiura, K., et al., Silencing of the E-cadherin invasion-suppressor gene by CpG methylation in human carcinomas, *Proc Natl Acad Sci U S A*, 92, 7416, 1995.
57. Wu, C. and Dedhar, S., Integrin-linked kinase (ILK) and its interactors: A new paradigm for the coupling of extracellular matrix to actin cytoskeleton and signaling complexes, *J Cell Biol*, 155 (4), 505–510, 2001.
58. Wu, C., The PINCH-ILK-parvin complexes: Assembly, functions and regulation, *Biochim Biophys Acta*, 1692, 55, 2004.
59. Legate, K.R., et al., ILK, PINCH and parvin: The tIPP of integrin signalling, *Nat Rev Mol Cell Biol*, 7, 20, 2006.
60. Kim, S.K., et al., The epigenetic silencing of LIMS2 in gastric cancer and its inhibitory effect on cell migration, *Biochem Biophys Res Commun*, 349, 1032, 2006.
61. Zhou, L., et al., Zebularine: A novel DNA methylation inhibitor that forms a covalent complex with DNA methyltransferases, *J Mol Biol*, 321, 591, 2002.
62. Kaminskas, E., et al., Approval summary: Azacitidine for treatment of myelodysplastic syndrome subtypes, *Clin Cancer Res*, 11, 3604, 2005.
63. Marcucci, G., et al., Bioavailability of azacitidine subcutaneous versus intravenous in patients with the myelodysplastic syndromes, *J Clin Pharmacol*, 45, 597, 2005.
64. Fang, M.Z., et al., Tea polyphenol (−)-epigallocatechin-3-gallate inhibits DNA methyltransferase and reactivates methylation-silenced genes in cancer cell lines, *Cancer Res*, 63, 7563, 2003.
65. Fang, M.Z., et al., Reversal of hypermethylation and reactivation of p16INK4a, RARbeta, and MGMT genes by genistein and other isoflavones from soy, *Clin Cancer Res*, 11, 7033, 2005.
66. Lee, B.H., Yegnasubramanian, S., Lin, X., and Nelson, W.G., Procainamide is a specific inhibitor of DNA methyltransferase 1, *J Biol Chem*, 280, 40749, 2005.
67. Davis, A.J., et al., Phase I and pharmacologic study of the human DNA methyltransferase antisense oligodeoxynucleotide MG98 given as a 21-day continuous infusion every 4 weeks, *Invest New Drugs*, 21, 85, 2003.
68. Leu, Y.W., et al., Double RNA interference of DNMT3b and DNMT1 enhances DNA demethylation and gene reactivation, *Cancer Res*, 63, 6110, 2003.

4 Imprinting Alterations in Tumorigenesis

Jeremiah Bernier-Latmani and Phillip Shaw

CONTENTS

4.1 INTRODUCTION

Imprinted genes are monoallelically expressed from either the maternally or paternally inherited allele. The crucial nature of this regulation is reflected in phenotypic manifestations in individuals exhibiting improper imprinted gene expression. This is evident in a number of human syndromes, including Beckwith–Wiedemann syndrome (BWS) and Silver–Russell syndrome (SRS), where genetic or epigenetic lesions disrupt appropriate imprinted gene regulation. Clinically, BWS is characterized by large size at birth, neonatal hypoglycemia, minor dysmorphic features, and increased risk of several types of major organ tumors, including Wilms tumor [1]. SRS describes a uniform malformation syndrome characterized by pre- and postnatal growth restriction [2]. Diagnosis can be difficult and at least four of the following criteria should be present: intrauterine growth retardation, poor postnatal growth, relatively normal head circumference, classic facial phenotype, and asymmetry [2]. Further evidence for the necessity of maintaining the delicate balance of imprinted gene regulation is provided by the widespread and frequent occurrence of loss of imprinting (LOI) in human cancers. LOI denotes loss of monoallelic expression, regardless of the mechanism. Therefore, research into the molecular mechanisms that regulate imprinting is not strictly an academic study of expression modulation, but may have clinical benefits to patients with imprinting dysfunction. The precise imprinted genes and tumor types exhibiting LOI are discussed in more detail below.

Before proceeding to a discussion of alterations in imprinting during tumorigenesis, it is appropriate to first briefly define how imprinted gene expression is regulated. Central to the mechanism of imprinted gene regulation are regions marked epigenetically by DNA methylation. Epigenetic modifications include non-sequence changes of both DNA (methylation) and histone proteins (acetylation, methylation, phosphorylation, and ubiquitinylation, for an excellent review of epigenetic modifications, see Ref. [3]). These hereditable and generally reversible modifications are located in precise locations within imprinted gene loci, referred to as imprinting control regions (ICRs) or differentially methylated regions (DMRs). Only one of the two ICR alleles is methylated, constituting the molecular signature by which the two alleles are distinguished. An understanding of the initial marking of imprinted regions during gonocyte development provides insight into possible mechanisms by which LOI may arise in cancer (for review see Ref. [4]).

In the following sections we present the developmental events of imprint marking, "reading" mechanisms of imprint marks, chromatin modifications associated with imprinted loci, biological function of imprinted genes, human syndromes exhibiting altered imprinted gene expression, incidence of LOI in human cancers, and finally lesions leading to LOI.

4.1.1 DEVELOPMENTAL ESTABLISHMENT OF IMPRINTING MARKS

Both maternal and paternal marks are established during the development of their respective gonocytes, oocytes, and sperm. Figure 4.1 provides a temporal scheme of these events in the mouse. At approximately 10.5 days of embryonic development (E10.5), primordial gonocytes have finished migrating to the germinal ridge. Shortly thereafter, all DNA methylation is erased, regardless of parental origin [5]. The primordial gonocytes will then diverge towards either oocyte or sperm development. This decision is largely determined by the presence or absence of the Sry gene on the Y chromosome [6]. Methylation marks within paternally imprinted gene loci are reestablished between E14.5 and E17.5 [5]. The proteins responsible for this methylation are the de novo DNA methyltransferases (DNMTs) 3a, 3b, and L [7]. Once in place, these methylation marks are then stably maintained in all subsequent cellular generations by the maintenance methylation complex comprised primarily of DNA methyltransforase 1 (DNMT1) [8,9]. Maternally imprinted gene loci are methylated within oocytes during postbirth oocyte development (10–25 days postpartum [10]). As with paternal methylation marks, the retention of maternally imprinted gene marks is assured by the maintenance methylation complex during DNA replication. This elaborate system of erasure, ICR methylation, and subsequent monoallelic expression arose very late in evolution and is uniquely found in eutherian (placental) and metatherian (marsupial) mammals [11].

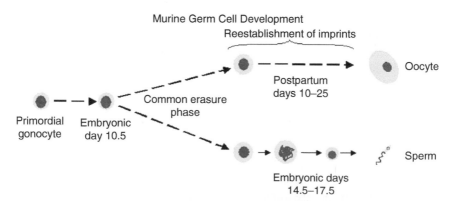

FIGURE 4.1 Developmental cycle of methylation and de-methylation of ICRs, or DMRs in mice. Primordial gonocytes, a common precursor for both the male and female lineages, undergo erasure of all preexisting methylation marks (12.5 days of embryonic development). The development of male and female germ cells diverges thereafter with male germ cell development occurring during embryonic development, while female germ cells mature shortly after birth. The reestablishment of specific paternal methylation in mice occurs during the embryonic interval from 14.5 to 17.5 days. Oocytes reestablish maternal methylation from postnatal day 10 to 25.

4.1.2 Imprinting and Gene Expression: Reading of Imprinted Gene Marks

Typically, imprinted loci are organized in genetic units containing from 2 to 10 genes. Within each locus is an ICR, which, depending on the epigenetic modifications, will stimulate or silence adjacent gene expression. The location of an ICR is indicated by a region of DNA that is differentially methylated between the two parental alleles and is also referred to as a DMR. There are three characterized mechanisms by which the methylation status of an ICR or DMR is "read": (1) chromatin barrier to enhancers, (2) antisense RNA mediated chromatin alterations, and (3) promoter silencing of microimprinted loci. Figure 4.2 presents a scheme of these three regulatory mechanisms.

The *Igf2/H19* gene pair, a paternally imprinted locus, is the best-characterized chromatin barrier locus described to date. The *Igf2* gene encodes a potent fetal growth factor, while the

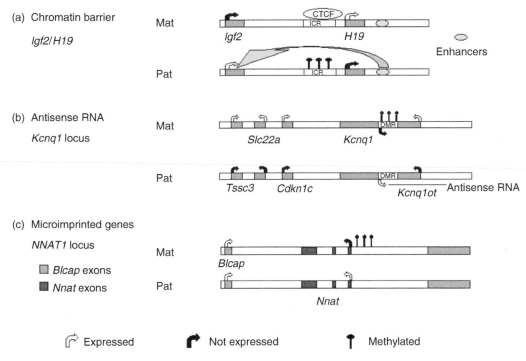

FIGURE 4.2 Imprinted gene regulation. Three documented mechanisms of imprinted gene regulation are presented: (a) chromatin barrier, (b) antisense RNA, and (c) promoter methylation-microimprinting. (a) The methylation status of the *H19* ICR determines expression of *Igf2*. A methylated *H19* ICR prevents the binding of the 11-zinc finger protein CTCF and thereby allows enhancers downstream of the H19 gene to access the paternal *Igf2* promoter. Upon binding the nonmethylated *H19* ICR, CTCF initiates assembly of a chromatin barrier which blocks interaction of the enhancers downstream of the *H19* gene with the *Igf2* promoter. This manner of imprinting is thought to also apply to the *Dlk1/Gtl2* locus. (b) The expression of an antisense RNA is dependent on ICR methylation status. In the featured example, a nonmethylated ICR in the paternal *Kcnq1* allele allows transcription of an antisense RNA, *Kcnq1ot*, which results in silencing of other members of the *Kcnq1* imprinted gene cluster, including *Slc22a*, *Tssc3*, and *Cdkn1c*. The mechanism by which the antisense RNA silences the surrounding genes is not understood. Alternatively, the maternal *Kcnq1* ICR allele is methylated within the *Kcnq1ot* promoter, preventing transcription of the antisense RNA, and the adjacent genes of the imprinted cluster are expressed. (c) Microimprinted genes exhibit the simplest mode of imprinting. The promoters of the maternal allele of other known genes in this category (*Nnat*, *U2af1-rs1*, *Inpp5f_v2*, *Nap1l5*) are methylated resulting in paternal allele expression. Shown is the *Nnat* imprinted gene located in an intron of the *Blcap* gene.

function of the nontranslated transcript of the *H19* gene remains enigmatic [12]. The two genes are separated by approximately 100 kb of DNA, in which lies the *H19* ICR. Downstream of the *H19* gene are a pair of "enhancers" which can, in the absence of a chromatin barrier, stimulate transcription of the *Igf2* gene in *cis* [13]. The maternal *H19* ICR allele is nonmethylated, allowing binding of the ubiquitously expressed factor CTCF [14]. The binding of CTCF is crucial to the assembly of a chromatin barrier, which abrogates the affect of the *H19* enhancers on *Igf2* transcription [15]. Thus, the maternal allele expresses the *H19* gene transcript but there is little to no transcription of *Igf2*. The paternal *H19* ICR is methylated, thereby disrupting CTCF binding, resulting in enhancer stimulation of *Igf2* transcription (Figure 4.2a) [15]. Coincident with *Igf2* expression, paternal *H19* transcription is silenced. This model of ICR function is abundantly supported in the literature, including transgenic mouse models with mutation and deletions of the *H19* ICR [16]. This type of regulation is also thought to be operative at the *Dlk1/Gtl2* locus on mouse chromosome 12 [17], however, extensive experimental confirmation is presently lacking.

The second mechanism of imprint reading is antisense RNA mediated regulation [18]. An example is the maternally imprinted locus centered around the potassium channel gene, *Kcnq1*, on mouse chromosome 7, centromeric to the *Igf2/H19* locus (Figures 4.2b and 4.3). The ICR in this locus is located within an intron of the *Kcnq1* gene and corresponds to the promoter of an antisense RNA transcript. The paternal allele ICR is nonmethylated and the antisense RNA, relative to *Kcnq1* gene transcription, is actively transcribed [19]. This transcript, *Kcnq1ot*, is responsible for allele-specific alteration of the local chromatin by an, as yet, unknown mechanism, rendering the adjacent imprinted genes silent [20]. The maternal allele ICR is methylated, which represses the antisense

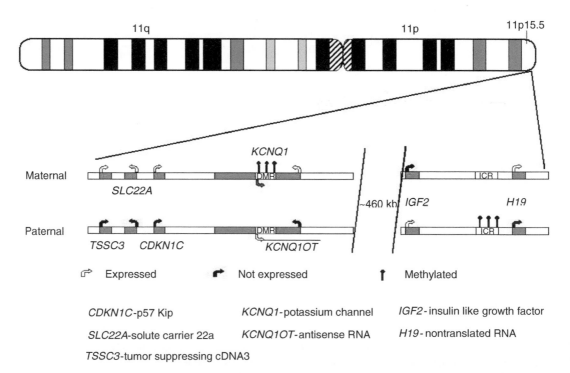

FIGURE 4.3 The two imprinted gene clusters on human chromosome 11p15.5 are shown. The distance between the centromeric *KCNQ1* cluster and the *H19* cluster is approximately 460 Kb. The displayed genes are not drawn to scale, but relative positioning is correct. Black arrows denote silenced alleles, while white arrows indicate active transcription. The abbreviated names of the individual genes are given in full in the lower part of the figure. Orientation relative to the telomere is presented as is the configuration for maternal and paternal chromosomes. Methylation of the *H19* ICR and the *KCNQ1* DMR are indicated by black lollipops.

RNA promoter, abrogating antisense RNA transcription, and the entire locus is found in an active chromatin conformation [20]. On the maternal allele, the imprinted genes, including the human homologues, *KCNQ1* and *CDKN1C*, are actively transcribed (Figures 4.2b and 4.3 [21]).

A third mechanism of imprinted gene regulation is promoter silencing of microimprinted domains (Figure 4.2c). An example is the *Nap1l5* gene, which is expressed uniquely from the paternal allele. Maternal allele expression is silenced by promoter methylation. All microimprinted genes described to date are maternally silenced [22]. Four of the five known microimprinted genes arose in eutherian mammals, suggesting that these genes represent a primordial system of imprinting, which was later elaborated upon during evolution of higher mammals [22,23].

DNA methylation within imprinted gene loci represents one level of allelic distinction and is subsequently accompanied by the establishment of allele-specific chromatin conformations. These chromatin conformations are dependent on additional epigenetic modifications, namely histone methylation and acetylation.

4.1.3   CHROMATIN MODIFICATIONS PRESENT AT ACTIVE AND INACTIVE IMPRINTED GENE ALLELES

Our current understanding is that the mechanisms described above are the first order "reading" of DNA methylation marks present within imprinted loci. To completely understand their impact on imprinted gene expression, it is necessary to look more closely at particular histone modifications found within both active and inactive alleles. It is well established that the combination of particular histone modifications results in a "histone code," which is coincident with particular chromatin conformations and transcriptional activity [24]. Both active and silent chromatin regions have multiple histone modifications which are characteristic of each type and these specifically modified histones provide an interaction platform to which large chromatin modifying complexes can bind [25]. For active chromatin, resulting alterations ensure accessibility of these regions to transcription factors and therefore maintenance of transcriptional activity. In contrast, repressive histone modifications recruit chromatin-modifying complexes responsible for the compaction and resulting inaccessibility of these chromatin regions [26]. Researchers have addressed the question as to whether chromatin regions encompassing imprinted gene loci also follow the histone "code" employed elsewhere in the genome. The sites of DNA methylation within an imprinted gene locus can either be located in a promoter region (antisense RNA regulated loci and microimprinted genes) or in intergenic regions (*Igf2/H19*, *Dlk1/Gtl2*). Thus, the histone modifications characteristic for these types of imprinted gene regulation will be treated separately.

Not surprisingly, trimethylated lysine 27 of histone H3 (H3K27me3) and H3K9me2, histone modifications reported for repressed promoters elsewhere in the genome, are also found at the repressed imprinted gene promoters of the paternal allele of the *Kcnq1* locus [20,27]. The mechanism by which these repressive histone modifications arise within imprinted gene loci is not known, however it has been shown that expression of the entire antisense RNA *Kcnq1ot* is a critical element in the process [19,28].

Specific modified histones are also found on the intergenic ICR of the *Igf2/H19* locus. Typically, acetylated histones are associated with open chromatin regions, and more specifically histone acetylation is significantly enriched in regions of ongoing transcription [29]. H3 and H4 are predominantly acetylated on the maternal ICR allele [30,31]. H4K20me3 and H3K9me3 were found to be significantly enriched on the paternal allele [30]. These latter histone modifications thus appear to correlate with allele specificity and play a role in the overall chromatin conformation of the *H19* ICR.

Recently, our laboratory has reported the presence of a novel histone modification associated with *H19* ICR chromatin, symmetrical dimethyl arginine 3 histone H4 (H4R3me2s) [32]. This modified histone was also shown to be associated with the intergenic ICR found upstream of the imprinted gene *Gtl2* [32]. It remains to be determined if this modification is found associated in other paternal imprinted loci and with the *H19* ICR in tissues other than testis.

A more complete discussion of histone modifications and associated chromatin conformations found in cancers is presented in Part II of this book (Histone Modifications in Cancer).

4.1.4 BIOLOGICAL FUNCTION OF IMPRINTED GENES

Clearly the regulation of imprinted genes to achieve monoallelic expression suggests that their carefully regulated expression is critical for normal development. This is perhaps best demonstrated by observed developmental aberrations in hereditary syndromes that perturb the monoallelic mode of imprinted gene expression. The majority of imprinted genes are thought to be involved in regulation of fetal and placental growth [33]. The most convincing evidence of this is data from studies of mice deficient in *Igf2*, *Igf2r*, and *H19* [34]. Table 4.1 presents the effects of several individual imprinted gene knockout mice on fetal and placental weights. Late gestation stage embryos of *Igf2* knockout mice weigh 50% of wild-type littermates, while embryos of *H19* deficient mice weigh 130% of wild-type littermates. Placental weights follow the same trend (Table 4.1). Collectively, these data highlight the delicate balance achieved by monoallelic expression of these imprinted genes in fetal development. Although precise data of specific gene-deficient mice are not available for many of the presently identified imprinted genes, their involvement in growth deficiencies, or overgrowth syndromes in affected patients is consistent with a role in these processes [35]. Within the *KCNQ1* imprinted gene cluster, for example, the cell cycle regulator, *CDKN1C* ($p57^{KIP2}$), is seen to be mutated in certain cases of BWS. In addition, *KCNQ1* LOI results in a significantly depressed level of expression of this key cell cycle regulator [21].

4.1.5 HEREDITARY SYNDROMES AFFECTING IMPRINTED GENE EXPRESSION

BWS represents a spectrum of disorders including fetal overgrowth, exomphalos, hemihypertrophy, and predisposition to childhood cancers [1]. The majority of BWS arises sporadically, primarily due to uniparental disomy (UPD—two paternal copies of chromosome 11) or epimutations (DNA hyper- or hypomethylation). Additionally, patients with BWS exhibit an elevated risk for Wilms tumor [1]. A Wilms tumor (nephroblastoma) arises from pluripotent embryonic kidney precursor cells and is the most common pediatric kidney cancer in children [1]. Within Wilms tumors, loss of heterozygosity (LOH) in the 11p15.5 region has been documented, with significant bias towards the loss of maternal alleles [36]. Detailed analysis of both LOH and LOI has shown that LOI occurs in early-stage Wilms tumors, while LOH is observed at later stages of tumor progression, indicating that LOI is the first detectable lesion in tumorigenesis [36]. These finding are discussed further below in the context of the role of LOI in the expansion of the stem cell population in the organ in which it occurs.

TABLE 4.1
Effects of Selected Imprinted Gene Knockout Mice on Fetal and Placental Weights

Gene (Expressed Allele)	Function	Knockout Proportion of Normal Weight (%)		Type of KO
		Fetus	Placenta	
Igf2 (paternal)	Fetal growth factor	50	60	*Igf2* gene deletion [72]
Igf2 P0 (paternal)	Fetal growth factor	75	70	Fetal *Igf2* promoter deletion [73]
H19 (maternal)	Noncoding RNA	130	140	*H19* gene deletion [74]
Igf2r (maternal)	*Igf2* clearance receptor	140	140	*Igf2r* gene disruption [75]
Igf2r & *H19*	See above	200	230	*Igf2r* gene disruption and *H19* gene deletion [74,75]

Source: Adapted from Fowden, A.L. et al., *Horm. Res.*, 65, 50, 2006.

Similarly to BWS, the genetic elements responsible for SRS are only partially characterized due to the genetic complexity of the disease. Surprisingly, Gicquel and coworkers recently reported that hypomethylation of the *Igf2/H19* ICR was responsible for at least 30% of SRS [37]. These observations were subsequently confirmed and extended by Bliek et al. [38].

Thus, it is striking that the *Igf2/H19* ICR is the "target" of both biallelic methylation (BWS and Wilms tumors) and hypomethylation (SRS). These two syndromes display the anticipated opposite phenotypes, namely overgrowth in BWS and undergrowth in SRS. The critical role and sensitive dose response to the fetal growth factor, IGF2, in early development is consistent with these observations.

4.1.6 INCIDENCE OF LOI IN CANCER

The minimal definition of LOI is the loss of monoallelic expression, resulting in either biallelic expression or silencing of imprinted genes. Given the crucial role of proteins coded by imprinted genes in stimulating proliferation, IGF2, for example, it is not surprising that researchers began determining the incidence of LOI in diverse human tumor types. As apparent below, virtually all cancers examined to date exhibit LOI.

This section focuses on LOI observed on the short arm of human chromosome 11 which contains 2 imprinted gene loci, paternally imprinted *IGF2/H19* and maternally imprinted *KCNQ1* (Figure 4.3). In the case of *IGF2/H19*, LOI often leads to biallelic expression of *IGF2* and silencing of *H19* expression, while LOI at *KCNQ1* results in biallelic expression of the genes normally silenced on the paternal chromosome, including *CDKN1C*.

4.1.6.1   *IGF2/H19* Loss of Imprinting

LOI at the *IGF2/H19* locus has been documented in 100% of chronic myogenous leukemia [39], 80% of ovarian tumors [40], 70% of Wilms tumors [41], 56% of colorectal cancer [42,43], 56% of Barrett's esophagus [44], 50% of renal cell carcinomas [45], 50% of esophageal cancer [46], 47%–85% of lung adenocarcinoma [47], and 30% of meningiomas [48]. *IGF2* LOI has also been observed in osteosarcoma [49], rhabdomyosarcoma [50], hepatoblastoma [51], Ewing's sarcoma [52], gliomas [53], and laryngeal squamous cell carcinoma [54]. Thus, *IGF2* LOI occurs at a high frequency in a large variety of human tumors and is presently considered to be the most common alteration in cancer [55].

Not only is LOI a prevalent event in human cancers, it has been shown to be a precocious one as well. Thus, a sensitive means of LOI detection could potentially offer a diagnostic tool for early cancer detection. In Wilms tumor, several authors have shown that the earliest lesion observed, genetic or epigenetic, is *IGF2* LOI [36,56–58]. LOI has also been shown to be an early event in ovarian cancer [40] and colorectal cancer [59]. Although the timing of LOI has not been determined in all tumor types exhibiting *IGF2* LOI, it is thought that the epigenetic alterations in normal cells increase tumor risk by expanding the cell population targeted in cancer, tissue stem cells [57,59]. This concept is further discussed in Section 4.1.8.

4.1.6.2   *KCNQ1*-LOH and LOI

LOH within the centromeric imprinted gene cluster *KCNQ1* on human chromosome 11p15.5 (Figure 4.3) has not been studied in as great of detail as *IGF2/H19*, nevertheless it is clear that it occurs at significant frequencies in Wilms [60]. Roughly 50% of BWS patients lose methylation of the maternal DMR of the *KCNQ1* locus [60] and *KCNQ1* LOI has been reported for 40% of colorectal cancer [60,61]. Analysis of liver, breast, cervical, and gastric carcinomas, revealed a 30%–50% loss of maternal *KCNQ1* DMR methylation [62].

Detailed studies of LOI at both the *KCNQ1* and *IGF2* loci in BWS indicate that LOI occurs independently at each locus. While *KCNQ1* LOI is frequent in BWS, *IGF2* LOI is commonly observed in Wilms tumors [61]. Thus, even though they are chromosomal neighbors, all evidence indicates that LOI occurs at individual loci and not globally throughout the genome [61,62].

4.1.7 GENETIC AND EPIGENETIC ALTERATIONS LEADING TO LOI

We have discussed regulated expression of normally imprinted genes and now we can consider the alterations, genetic and epigenetic, operative in the disruption of monoallelic expression of these genes in tumors and hereditary syndromes affecting imprinted gene expression. Table 4.2 outlines the observations made for a number of hereditary syndromes and tumor types that exhibit LOI. On one hand, UPD results in biallelic imprinted gene expression, by providing two transcriptionally active alleles of that gene. For example, paternal chromosome 11 disomy is a common cause (20%) of biallelic expression of *IGF2* in BWS [63]. Another frequently observed epigenetic lesion resulting in LOI is biallelic methylation of the ICR, which presents a similar molecular signature to UPD, and is observed in sporadic Wilms tumors. Another lesion leading to LOI is micro-deletion of the *H19* ICR, which is observed in BWS [64,65]. These deletions result in reduced binding of CTCF, thereby perturbing the assembly of the chromatin barrier, resulting in biallelic expression of *IGF2* [64,65]. Finally, it has been observed that the *IGF2* promoter is hypomethylated in sporadic colon cancer [66]. At present it is not clear whether this hypomethylation is the consequence or cause of upregulated *IGF2* expression, since methylated *IGF2* promoter is observed in the presence of a methylated ICR. Thus, the precise mechanism whereby *IGF2* expression seemingly escapes the mode of regulation described in Figure 4.2a is not clear. Below we entertain two schemes: one to explain the appearance of biallelically methylated ICRs and the other to explain the biallelic expression of *IGF2* in the absence of changes in the methylation status of the *H19* ICR.

As mentioned in Section 4.1 of this chapter, the molecules collaborating in the developmental establishment of paternal imprints have recently been described [32]. A model derived from this work is presented in Figure 4.4. The molecules involved include CTCF-like (CTCFL-BORIS), a protein possessing similar DNA binding specificity to CTCF, a protein arginine methyltransferase, PRMT7, and the de novo DNMT3a, DNMT3b, and DNMT3L. The 11-zinc finger domain of CTCFL binds the ICR and subsequently PRMT7 is recruited to the chromatin region by interaction with the N-terminal region of CTCFL. Thus, PRMT7 is brought into proximity to its substrates, histones H2A and H4, which are subsequently methylated. The methylated arginine residues are then responsible for the recruitment of the de novo DNMTs, although the mechanism of this recruitment is not known. Strikingly, the key molecule in this scenario, CTCFL, is expressed in Wilms tumor (ESTs in Unigene cluster Hs.131543), where biallelic methylation of the *H19* ICR has been documented [67]. In contrast, normal expression of CTCFL is restricted solely to embryonic and adult testis [32]. Thus, the aberrant expression of CTCFL, in collaboration with PRMT7 and the DNMT3s, could promote methylation of the maternal allele ICR, giving rise to a biallelically methylated ICR.

TABLE 4.2
Genetic and Epigenetic Alterations Leading to Loss of Imprinting

Mechanism	Observations	References
UPD	Beckwith–Wiedemann, Prader–Willi, Silver–Russell, and Angelman syndromes	[2,63,76]
Biallelic methylation of ICR	Wilms tumor, gynecological tumors	[49,66,77–79]
Microdeletion of ICR	Wilms tumor, Beckwith–Wiedemann, pseudohypoparathyroidism	[64,65,80,81]
Hypomethylation *H19* ICR	Silver–Russell	[37,38]
Hypomethylation *IGF2* promoter	Colorectal cancer	[82]

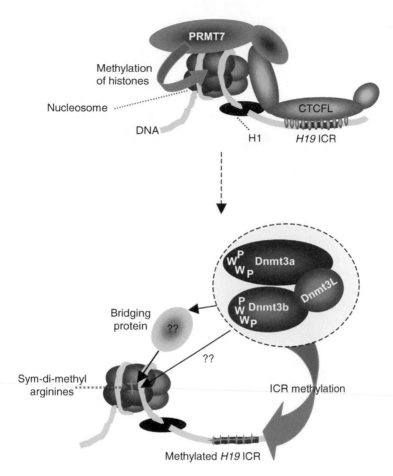

FIGURE 4.4 Model of *H19* ICR methylation. The 11-zinc finger CTCFL, an ortholog of CTCF, binds the ICR. The N-terminal portion of CTCFL interacts with both PRMT7, a protein arginine methyltransferase, and histones H2A and H4. The recruitment of PRMT7 results in symmetrical dimethyl of arginine residues within the adjacent histones H2A and H4. These modified histones recruit the de novo DNMT3a, DNMT3b, and DNMT3L, resulting in ICR-specific DNA methylation. (From Jelinic, P., Stehle, J.-C., and Shaw, P., *PLoS Biol.*, 4, e355, 2006.)

The observation of biallelic expression of *IGF2* in colorectal cancer accompanied by *IGF2* promoter hypomethylation, and a normally imprinted ICR [66], constitutes a more difficult situation to explain mechanistically. As stated above, the hypomethylation of the *IGF2* promoter DMR may not be the causative event of biallelic *IGF2* expression, but simply the consequence. A possible alternative explanation for biallelic *IGF2* expression observed in colorectal cancer comes from the recent description of the zinc finger protein ZAC1 [68]. The protein encoded by the maternally imprinted *Zac1* gene binds in vivo and in vitro to the enhancers downstream of *H19* (Figure 4.2a) and can, when over-expressed in transfected cells, stimulate *Igf2* expression up to 200-fold of normal levels [68]. The precise mechanism whereby *Zac1* could regulate *Igf2* expression, however, remains to be elucidated. *Zac1*-deficient mice evidence a slight down regulation of *Igf2* ($0.73\times$), consistent with a role of *Zac1* in *Igf2* expression [68]. *Zac1*-deficient mice exhibit a 23% weight reduction at birth, reminiscent of other imprinted gene mutant mice (Table 4.2 [68]). It will be very interesting to follow the unraveling of the mechanism responsible for *IGF2* upregulation in colorectal cancer, given the high frequency with which it occurs.

4.1.8 LOI: A STEM CELL LESION?

A recent review by Feinberg, Henikoff and Ohlsson [55] proposed that LOI occurs within the stem cell population of a particular organ in which a tumor appears. LOI would result in upregulation of *IGF2*, thereby giving rise to an expanded stem cell population through stimulated proliferation. Thus, in the absence of an increased mutation rate, a greater number of mutant cells will arise, simply due to the significantly increased number of cells in the population. Subsequent to this initial event, other alterations, including both genetic and epigenetic lesions, will occur, stimulating tumor progression. The best-characterized tumor system exhibiting LOI is Wilms tumor. The evidence is quite convincing that the earliest event to take place is LOI [36]. Additionally, Wilms tumors are often accompanied by nephrogenic rests in healthy adjacent tissue, which are thought to represent an expanded stem cell population [69]. What is of note is that the stem cell theory would explain: (1) the expanded stem population, (2) observations of LOI in healthy portions of a tumor-bearing organ, and (3) an elevated recurrence rate of tumors in patients with a LOI tumor. Mouse models engineered to express "biallelic" levels of *Igf2* in the stem cell population of the colon are also supportive of the concept that LOI can stimulate expansion of the stem cell population. Feinberg and coworkers reported a significant expansion of the stem cell niche in these mice and, when crossed with mice containing the *Min*− mutation (an adeno polyposis coli gene mutant [70]), the resulting mice exhibited a twofold increase in tumor incidence [71]. Thus, evidence is beginning to accumulate in support of the LOI stem cell hypothesis, yet further models and more detailed analyses of tumor systems exhibiting LOI should be undertaken to further substantiate this provocative concept.

ACKNOWLEDGMENTS

This work was funded by the Swiss National Science Foundation, the Muschamp Foundation, and the League Contre le Cancer.

REFERENCES

1. Weksberg, R., Shuman, C., and Smith, A.C., Beckwith-Wiedemann syndrome, *Am. J. Med. Genet. C Semin. Med. Genet.*, 137C(1), 12–23, 2005.
2. Rossignol, S., Silver-Russell syndrome and its genetic origins, *J. Endocrinol. Investig.*, 29(1 Suppl.), 9–10, 2006.
3. Bernstein, B.E., Meissner, A., and Lander, E.S., The mammalian epigenome, *Cell*, 128(4), 669–681, 2007.
4. Jelinic, P. and Shaw, P., Loss of imprinting and cancer, *J. Pathol.*, 211(3), 261–268, 2007.
5. Li, J.Y. et al., Timing of establishment of paternal methylation imprints in the mouse, *Genomics*, 84, 952–960, 2004.
6. Brennan, J. and Capel, B., One tissue, two fates: Molecular genetic events that underlie testis versus ovary development, *Nat. Rev. Genet.*, 5(7), 509–521, 2004.
7. Okano, M. et al., DNA methyltransferases Dnmt3a and Dnmt3b are essential for de novo methylation and mammalian development, *Cell*, 99(3), 247–257, 1999.
8. Okuwaki, M. and Verreault, A., Maintenance DNA methylation of nucleosome core particles, *J. Biol. Chem.*, 279(4), 2904–2912, 2004.
9. Li, E., Beard, C., and Jaenisch, R., Role for DNA methylation in genomic imprinting, *Nature*, 366(6453), 362–365, 1993.
10. Lucifero, D. et al., Gene-specific timing and epigenetic memory in oocyte imprinting, *Hum. Mol. Genet.*, 13(8), 839–849, 2004.
11. Murphy, S.K. and Jirtle, R.L., Imprinting evolution and the price of silence, *BioEssays*, 25(6), 577–588, 2003.
12. Gabory, A. et al., The H19 gene: Regulation and function of a non-coding RNA, *Cytogenet. Genome Res.*, 113(1–4), 188–193, 2006.
13. Leighton, P.A. et al., An enhancer deletion affects both H19 and Igf2 expression, *Genes Dev.*, 9(17), 2079–2089, 1995.

14. Hark, A.T. et al., CTCF mediates methylation-sensitive enhancer-blocking activity at the H19/Igf2 locus, *Nature*, 405(6785), 486–489, 2000.

15. Bell, A.C. and Felsenfeld, G., Methylation of a CTCF-dependent boundary controls imprinted expression of the Igf2 gene, *Nature*, 405(6785), 482–485, 2000.

16. Szabo, P.E. et al., Role of CTCF binding sites in the Igf2/H19 imprinting control region, *Mol. Cell. Biol.*, 24(11), 4791–4800, 2004.

17. Takada, S. et al., Epigenetic analysis of the Dlk1-Gtl2 imprinted domain on mouse chromosome 12: implications for imprinting control from comparison with Igf2-H19, *Hum. Mol. Genet.*, 11(1), 77–86, 2002.

18. Sleutels, F., Zwart, R., and Barlow, D.P., The non-coding Air RNA is required for silencing autosomal imprinted genes, *Nature*, 415(6873), 810–813, 2002.

19. Mancini-DiNardo, D. et al., Elongation of the Kcnq1ot1 transcript is required for genomic imprinting of neighboring genes, *Genes Dev.*, 20(10), 1268–1282, 2006.

20. Umlauf, D. et al., Imprinting along the Kcnq1 domain on mouse chromosome 7 involves repressive histone methylation and recruitment of Polycomb group complexes, *Nat. Genet.*, 36(12), 1296–1300, 2004.

21. Diaz-Meyer, N. et al., Silencing of CDKN1C (p57(KIP2)) is associated with hypomethylation at KvDMR1 in Beckwith-Wiedemann syndrome, *J. Med. Genet.*, 40(11), 797–801, 2003.

22. Wood, A.J. and Oakey, R.J., Genomic imprinting in mammals: Emerging themes and established theories, *PLoS Genet.*, 2(11), 1677–1685, 2006.

23. Evans, H.K. et al., Comparative phylogenetic analysis of Blcap/Nnat reveals eutherian-specific imprinted gene, *Mol. Biol. Evol.*, 22(8), 1740–1748, 2005.

24. Jenuwein, T. and Allis, C.D., Translating the histone code, *Science*, 293(5532), 1074–1080, 2001.

25. Li, B., Carey, M., and Workman, J.L., The role of chromatin during transcription, *Cell*, 128(4), 707–719, 2007.

26. Mellor, J., It takes a PHD to read the histone code, *Cell*, 126(1), 22–24, 2006.

27. Beatty, L., Weksberg, R., and Sadowski, P.D., Detailed analysis of the methylation patterns of the KvDMR1 imprinting control region of human chromosome 11, *Genomics*, 87(1), 46–56, 2006.

28. Kanduri, C., Thakur, N., and Pandey, R.R., The length of the transcript encoded from the Kcnq1ot1 antisense promoter determines the degree of silencing, *EMBO J.*, 25(10), 2096–2106, 2006.

29. Calestagne-Morelli, A. and Ausio, J., Long-range histone acetylation: biological significance, structural implications, and mechanisms, *Biochem. Cell Biol.*, 84(4), 518–527, 2006.

30. Delaval, K. et al., Differential histone modifications mark mouse imprinting control regions during spermatogenesis, *EMBO J.*, 26(3), 720–729, 2007.

31. Fournier, C. et al., Allele-specific histone lysine methylation marks regulatory regions at imprinted mouse genes, *EMBO J.*, 21(23), 6560–6570, 2002.

32. Jelinic, P., Stehle, J.-C., and Shaw, P., The testis-specific factor CTCFL cooperates with the protein methyltransferase PRMT7 in H19 imprinting control region methylation, *PLoS Biol.*, 4(11), e355, 2006.

33. Morison, I.M., Ramsay, J.P., and Spencer, H.G., A census of mammalian imprinting, *Trends Genet.*, 21(8), 457–465, 2005.

34. Fowden, A.L. et al., Imprinted genes, placental development and fetal growth, *Horm. Res.*, 65, 50–58, 2006.

35. Jiang, Y.-H., Bressler, J., and Beaudet, A.L., Epigenetics and human disease, *Annu. Rev. Genomics Hum. Genet.*, 5(1), 479–510, 2004.

36. Yuan, E. et al., Genomic profiling maps loss of heterozygosity and defines the timing and stage dependence of epigenetic and genetic events in Wilms' tumors, *Mol. Cancer Res.*, 3(9), 493–502, 2005.

37. Gicquel, C. et al., Epimutation of the telomeric imprinting center region on chromosome 11p15 in Silver-Russell syndrome, *Nat. Genet.*, 37(9), 1003–1007, 2005.

38. Bliek, J. et al., Hypomethylation of the h19 gene causes not only Silver-Russell syndrome (SRS) but also isolated asymmetry or an SRS-like phenotype, *Am. J. Hum. Genet.*, 78(4), 604–614, 2006.

39. Randhawa, G.S. et al., Loss of imprinting in disease progression in chronic myelogenous leukemia, *Blood*, 91(9), 3144–3147, 1998.

40. Kamikihara, T. et al., Epigenetic silencing of the imprinted gene ZAC by DNA methylation is an early event in the progression of human ovarian cancer, *Int. J. Cancer*, 115(5), 690–700, 2005.

41. Mummert, S.K., Lobanenkov, V.A., and Feinberg, A.R., Association of chromosome arm 16q loss with loss of imprinting of insulin-like growth factor-II in Wilms tumor, *Genes Chromosomes Cancer*, 43(2), 155–161, 2005.

42. Cruz-Correa, M. et al., Loss of imprinting of insulin growth factor II gene: A potential heritable biomarker for colon neoplasia predisposition, *Gastroenterology*, 126(4), 964–970, 2004.

43. Cui, H.M. et al., Loss of imprinting in normal tissue of colorectal cancer patients with microsatellite instability, *Nat. Med.*, 4(11), 1276–1280, 1998.

44. Feagins, L.A. et al., Gain of allelic gene expression for IGF-II occurs frequently in Barrett's esophagus, *Am. J. Physiol. Gastrointest. Liver Physiol.*, 290(5), G871–G875, 2006.

45. Oda, H. et al., Loss of imprinting of igf2 in renal-cell carcinomas, *Int. J. Cancer*, 75(3), 343–346, 1998.

46. Hibi, K. et al., Loss of H19 imprinting in esophageal cancer, *Cancer Res.*, 56(3), 480–482, 1996.

47. Kohda, M. et al., Frequent loss of imprinting of IGF2 and MEST in lung adenocarcinoma, *Mol. Carcinog.*, 31(4), 184–191, 2001.

48. Muller, S. et al., Genomic imprinting of IGF2 and H19 in human meningiomas, *Eur. J. Cancer*, 36(5), 651–655, 2000.

49. Ulaner, G.A. et al., Loss of imprinting of IGF2 and H19 in osteosarcoma is accompanied by reciprocal methylation changes of a CTCF-binding site, *Hum. Mol. Genet.*, 12(5), 535–549, 2003.

50. Zhan, S.L., Shapiro, D.N., and Helman, L.J., Activation of an imprinted allele of the insulin-like growth-factor-II gene implicated in rhabdomyosarcoma, *J. Clin. Investig.*, 94(1), 445–448, 1994.

51. Rainier, S., Dobry, C.J., and Feinberg, A.P., Loss of imprinting in hepatoblastoma, *Cancer Res.*, 55(9), 1836–1838, 1995.

52. Zhan, S.L., Shapiro, D.N., and Helman, L.J., Loss of imprinting of IGF2 in Ewing's sarcoma, *Oncogene*, 11(12), 2503–2507, 1995.

53. Uyeno, S. et al., IGF2 but not H19 shows loss of imprinting in human glioma, *Cancer Res.*, 56(23), 5356–5359, 1996.

54. Grbesa, I. et al., Loss of imprinting and promoter usage of the IGF2 in laryngeal squamous cell carcinoma, *Cancer Lett.*, 238(2), 224–229, 2006.

55. Feinberg, A.P., Ohlsson, R., and Henikoff, S., The epigenetic progenitor origin of human cancer, *Nat. Rev. Genet.*, 7(1), 21–33, 2006.

56. Moulton, T. et al., Epigenetic lesions at the H19 locus in Wilms-tumor patients, *Nat. Genet.*, 7(3), 440–447, 1994.

57. Ravenel, J.D. et al., Loss of imprinting of insulin-like growth factor-II (IGF2) gene in distinguishing specific biologic subtypes of Wilms tumor, *J. Natl. Cancer. Inst.*, 93(22), 1698–1703, 2001.

58. Taniguchi, T. et al., Epigenetic changes encompassing the IGF2/H19 locus associated with relaxation of the IGF2 imprinting and silencing of H19 in Wilms-tumor, *Proc. Natl. Acad. Sci. U S A*, 92(6), 2159–2163, 1995.

59. Kaneda, A. and Feinberg, A.P., Loss of imprinting of IGF2: A common epigenetic modifier of intestinal tumor risk, *Cancer Res.*, 65(24), 11236–11240, 2005.

60. Lee, M.P. et al., Loss of imprinting of a paternally expressed transcript, with antisense orientation to K(V)LQT1, occurs frequently in Beckwith-Wiedemann syndrome and is independent of insulin-like growth factor II imprinting, *Proc. Natl. Acad. Sci. U S A*, 96(9), 5203–5208, 1999.

61. Tanaka, K. et al., Loss of imprinting of long QT intronic transcript 1 in colorectal cancer, *Oncology*, 60(3), 268–273, 2001.

62. Scelfo, R.A.M. et al., Loss of methylation at chromosome 11p15.5 is common in human adult tumors, *Oncogene*, 21(16), 2564–2572, 2002.

63. Cooper, W.N. et al., Molecular subtypes and phenotypic expression of Beckwith-Wiedemann syndrome, *Eur. J. Hum. Genet.*, 13(9), 1025–1032, 2005.

64. Prawitt, D. et al., Microdeletion and IGF2 loss of imprinting in a cascade causing Beckwith-Wiedemann syndrome with Wilms' tumor, *Nat. Genet.*, 37(8), 785–786, 2005.

65. Sparago, A. et al., Microdeletions in the human H19 DMR result in loss of IGF2 imprinting and Beckwith-Wiedemann syndrome, *Nat. Genet.*, 36(9), 958–960, 2004.

66. Cui, H.M. et al., Loss of imprinting of insulin-like growth factor-II in Wilms' tumor commonly involves altered methylation hut not mutations of CTCF or its binding site, *Cancer Res.*, 61(13), 4947–4950, 2001.

67. Feinberg, A.P. and Tycko, B., The history of cancer epigenetics, *Nat. Rev. Cancer*, 4(2), 143–153, 2004.

68. Varrault, A. et al., Zac1 regulates an imprinted gene network critically involved in the control of embryonic growth, *Dev. Cell*, 11(5), 711–722, 2006.

69. Charles, A.K., Brown, K.W., and Berry, P.J., Microdissecting the genetic events in nephrogenic rests and Wilms' tumor development, *Am. J. Pathol.*, 153(3), 991–1000, 1998.

70. Gould, K.A. and Dove, W.F., Localized gene action controlling intestinal neoplasia in mice, *Proc. Natl. Acad. Sci. U S A*, 94(11), 5848–5853, 1997.

71. Sakatani, T. et al., Loss of imprinting of Igf2 alters intestinal maturation and tumorigenesis in mice, *Science*, 307(5717), 1976–1978, 2005.

72. Fowden, A.L., The insulin-like growth factors and feto-placental growth, *Placenta*, 24(8–9), 803–812, 2003.

73. Constancia, M. et al., Placental-specific IGF-II is a major modulator of placental and fetal growth, *Nature*, 417(6892), 945–948, 2002.

74. Leighton, P.A. et al., Disruption of imprinting caused by deletion of the H19 gene region in mice, *Nature*, 375(6526), 34–39, 1995.

75. Ludwig, T. et al., Mouse mutants lacking the type 2 IGF receptor (IGF2R) are rescued from perinatal lethality in Igf2 and Igf1r null backgrounds, *Dev. Biol.*, 177(2), 517–535, 1996.

76. Vogels, A. and Fryns, J.P., The Prader-Villi syndrome and the Angelman syndrome, *Genet. Couns.*, 13(4), 385–396, 2002.

77. Dowdy, S.C. et al., Biallelic methylation and silencing of paternally expressed gene 3 (PEG3) in gynecologic cancer cell lines, *Gynecol. Oncol.*, 99(1), 126–134, 2005.

78. Murphy, S.K. et al., Frequent IGF2/H19 domain epigenetic alterations and elevated IGF2 expression in epithelial ovarian cancer, *Mol. Cancer Res.*, 4(4), 283–292, 2006.

79. Nakagawa, H. et al., Loss of imprinting of the insulin-like growth factor II gene occurs by biallelic methylation in a core region of H19-associated CTCF-binding sites in colorectal cancer, *Proc. Natl. Acad. Sci. U S A*, 98(2), 591–596, 2001.

80. Bastepe, M. et al., Autosomal dominant pseudohypoparathyroidism type Ib is associated with a heterozygous microdeletion that likely disrupts a putative imprinting control element of GNAS, *J. Clin. Investig.*, 112(8), 1255–1263, 2003.

81. Niemitz, E.L. et al., Microdeletion of LIT1 in familial Beckwith-Wiedemann syndrome, *Am. J. Hum. Genet.*, 75(5), 844–849, 2004.

82. Cui, H. et al., Loss of imprinting in colorectal cancer linked to hypomethylation of H19 and IGF2, *Cancer Res.*, 62(22), 6442–6446, 2002.

5 Proteins That Modulate DNA Methylation Aberrations in Cancer

*Sabrina L. Walthall, Sharla M.O. Phipps,
Lucy G. Andrews, and Trygve O. Tollefsbol*

CONTENTS

5.1 INTRODUCTION

In recent years, great strides have been taken toward understanding the establishment of aberrant epigenetic patterns in human cancers. For DNA methylation, it is known that two apparently contrasting phenomena coexist in the cancer cell: hypomethylation and hypermethylation. It has been shown that a profound loss of global 5-methylcytosine genomic content occurs in cancer cells with discrete areas of dense hypermethylation [1–3] occurring in the CpG islands located in the promoters of certain tumor-suppressor genes, such as *p16^INK4a*, *BRCA1*, or *hMLH*, leading to gene silencing [1,3]. In contrast, for cancer cells overall hypomethylation takes place predominantly in DNA repetitive sequences and has been linked to chromosomal instability [2,3].

Epigenetic proteins may contribute to human cancers by silencing tumor suppressor genes via recruitment of transcriptional repressor machinery to the corresponding promoter CpG island, with

65

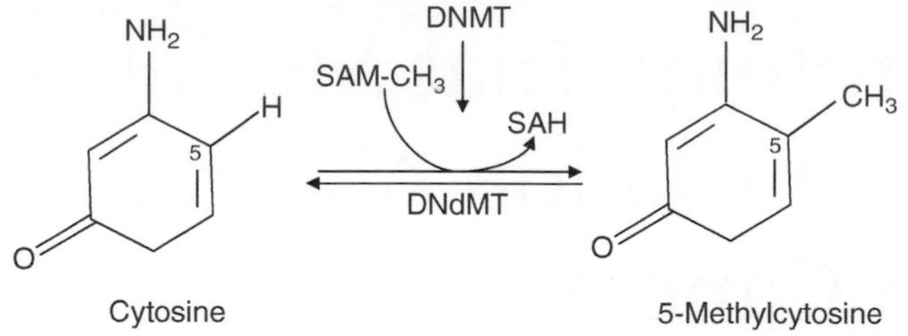

FIGURE 5.1 Schematic of DNA methylation and demethylation. DNA methyltransferases (DNMT 1, 3a, or 3b) catalyze the addition of a methyl group (CH_3) to the 5 carbon of cytosine using S-adenosyl-L-methionine (SAM) as the methyl donor. This reaction yields the product producing 5-methylcytosine. DNA demethyl-transferase (DNdMT) catalyzes the removal of methyl groups from methylcytosine converting it back to cytosine.

members such as the DNA methyltransferases (DNMTs), methyl-CpG-binding proteins (MBDPs), histone deacetylases (HDACs), histone methyltransferases (HMTs) for lysine 9 of histone H3 (HMT K9 H3), and polycomb group (PcG) complexes. This chapter explores these proteins and their roles in regulating DNA methylation aberrations in human cancers.

5.2   DNA METHYLATION

DNA methylation was discovered in the calf thymus by Hotchkiss in 1948 using paper chromatography [4]. In 1964, Gold and Hurwitz identified the first DNA methyltransferase (Mtase) in *Escherichia coli*. Since then, DNA methylation has been found in the evolutionary hierarchy ranging from bacteria to eukaryotes. The first mammalian DNA Mtase was discovered by the Razin group [5] and is known today as DNMT1.

DNA methylation in eukaryotes involves addition of a methyl group to the carbon 5 position of the cytosine ring (Figure 5.1). This reaction is catalyzed by DNA methyltransferase and is found occurring predominantly at short canonical 5'-CG-3' sequences, which are referred to as CpG dinucleotides, and rarely at non-CG sites. DNA methylation is one of the most common eukaryotic DNA modifications and is one of the many epigenetic alterations, which can occur. The genome is not uniformly methylated in eukaryotic organisms but has methylated and unmethylated regions interspersed throughout. As evolution has progressed, it has been shown that the CpG dinucleotides have almost been eliminated from higher eukaryotes and are present at only 20% of their predicted frequency.

DNA methylation in cancer has become the topic of intense investigation. As compared with normal cells, the cancer cells show major disruptions in their DNA methylation patterns. Reports of hypermethylation in cancer far outnumber the reports of hypomethylation in cancer (Table 5.1).

TABLE 5.1
Examples of Genes Hypermethylated in Cancer

Gene	Function	Cancer Type	References
$p16^{INK4a}$	Cyclin-dependent kinase inhibitor	Leukemia	[106]
hMLH1	DNA mismatch repair	Uterine	[107]
$p15^{IKN4a}$	Cyclin-dependent kinase inhibitor	Acute myeloid lymphoma	[108]
MGMT	Repair of alkylated guanine in DNA	Brain, lung, colon	[109]
GSTP1	Prevents oxidative damage in DNA	Breast, prostate	[110]
DAPK	Mediator of apoptosis	Lymphoma	[111]

Hypermethylation involves CpG islands whereas hypomethylation involves repeated DNA sequences, such as long interspersed nuclear elements. In cancerous cells, hypermethylation is often correlated with the repression of tumor suppressor genes that cause familial cancers through germline mutations. Promoter hypermethylation can also lead to loss of function of certain genes, which can result in sporadic tumorigenesis. These instances of hypermethylation and others stress that an imbalance of DNA methylation drives tumorigenesis, leading us to examine closely the enzymes that catalyze methylation of the CpG islands, the DNMTs.

5.3   DNA METHYLTRANSFERASES

DNMTs are enzymes that transfer methyl groups to the cytosine ring and have been characterized in a number of eukaryotes. The target site for DNMTs in DNA is the dinucleotide palindrome CG (commonly referred to as CpG, with p denoting the phosphate group). DNA methylation helps maintain transcriptional silence in nonexpressed or noncoding regions of the genome. Hypermethylation by DNMTs ensures that regions of DNA are either late-replicating, transcriptionally quiescent, or exhibiting suppressed expression. In contrast, regions of DNA that are hypomethylated or unmethylated, mainly in the promoter regions of euchromatin, maintain transcriptional activation. The DNMTs found in mammalian cells include DNMT1, DNMT2, DNMT3a, and DNMT3b.

5.3.1   DNMT1

The first DNA methyltransferase gene was cloned from mouse cells and is now referred to as *Dnmt1* [6]. This gene is highly conserved among eukaryotes with orthologs identified in various species except those lacking DNA methylation, such as *Saccharomyces cerevisiae*, *Caenorhabditis elegans*, and *Drosophila melanogaster*. The translation of the human *Dnmt1* gene results in the DNMT1 protein which consists of 1616 amino acids and is 78% identical to mouse Dnmt1 at the amino acid level [7–9].

Dnmt1 is expressed constitutively in proliferating cells and ubiquitously in somatic tissues throughout mammalian development. Although Dnmt1 can methylate both unmethylated and hemimethylated CpG dinucleotides in vitro, its activity toward hemimethylated substrates is 5- to 50-fold higher, depending on the specific study [10–13]. Dnmt1 is a nuclear protein, which localizes diffusely in the nucleoplasm during G1 and G2 phases but associates with replication foci during S phase [14], suggesting that Dnmt-mediated methylation is coupled to DNA replication. Furthermore, genetic studies have shown that Dnmt1 functions as a maintenance methyltransferase, which is essential for maintaining DNA methylation [15]. Dnmt1 therefore represents the major mammalian enzyme responsible for the maintenance of CpG methylation.

5.3.2   DNMT2

The second member of the DNA methyltransferase family is *Dnmt2*, which appears to be the most conserved methyltransferase gene among eukaryotes. It encodes for a protein of 391 amino acids in the human and 415 amino acids in the mouse [16–18]. Although DNA methyltransferase 2 has all 10 conserved motifs and the ability to bind DNA [19], it was initially thought to have no detectable catalytic activity [16,17,20,22] but later was shown to act as an RNA methyltransferase, which methylates tRNA^{Asp} [21]. However, Dnmt2 messenger RNA (mRNA) is ubiquitously expressed at low levels in most human and mouse adult tissues and in mouse embryonic stem cells [16]. It has also been shown that a higher expression level can be detected in cancer cell lines [16–18]. A weak but reproducible methyltransferase activity was recently demonstrated for the recombinant Dnmt2 protein in vitro [23–25] and in vivo [26]. Although the specificity of the enzyme is not yet known, evidence suggests preferences for CG [23] or CT and CA sites [23]. It remains to be determined whether mammalian Dnmt2 is involved in non-CpG methylation as well. Genetic evidence, however, has demonstrated that Dnmt2 does not play a major role in global de novo or maintenance methylation of CpG sites in mammals.

5.3.3 DNMT3

After Dnmt1 was identified, it was debated as to whether maintenance methylation and de novo methylation in mammals were carried out by Dnmt1 alone or by more distinct enzymes. The first genetic evidence for the existence of independently encoded de novo methyltransferases was found in mouse ES cells homozygous for a null *Dnmt1* mutation that contained residual levels of methylcytosine [27]. These studies eventually led to the identification of *Dnmt3a* and *Dnmt3b* genes in the mouse and human systems [28], which are now known to function primarily as de novo methyltransferases.

The mouse *Dnmt3a* gene encodes for Dnmt3a and Dnmt3a2, which consist of 908 and 689 amino acids, respectively. Their transcripts are initiated from two different promoters and as a result, Dnmt3a2 lacks the N-terminal region of Dnmt3a [28–30]. The two human isoforms of DNMT3A, namely DNMT3A and DNMT3A2, have 912 and 689 amino acids, respectively, and show 98% sequence identity to their mouse counterparts. Both Dnmt3a and Dnmt3a2 show different localization patterns [29] although they are enzymatically active in vitro and in vivo. Dnmt3a is concentrated in heterochromatic foci and shows preference to methylate sites that are flanked by pyrimidines rather than purines [31], whereas Dnmt3a2 localizes diffusely in the nucleus [29,32].

Alternative splicing of the Dnmt3b gene encodes for multiple isoforms. At least four mouse isoforms (Dnmt3b1, Dnmt3b2, Dnmt3b3, and Dnmt3b6) and five human isoforms (DNMT3B1–DNMT3B5) have been identified. The longest forms from mouse (Dnmt3b1) and human (DNMT3B1) systems consist of 859 and 853 amino acids, respectively, and show 94% sequence identity [28–30,33,34]. Dnmt3b1 and Dnmt3b2 are active enzymes whereas the other Dnmt3b isoforms are inactive because they lack one or more motifs in their catalytic domain [28,35–37]. Dnmt3b1 has been shown to localize to heterochromatin [32], where Dnmt3a is also detected.

Although we understand the role of the DNMTs, the mechanisms by which DNA methylation is translated into transcriptionally silent chromatin remain to be fully elucidated. Several hypotheses can be proposed to explain the way the various DNMTs affect methylation in cancer cells. The first possibility is that methylation by the DNMTs leads to the recruitment of proteins that selectively recognize methylated DNA and either impede binding of other proteins or have a direct effect on repressing transcription [38]. Another possibility suggests that the DNMTs bind and methylate CpG islands in regions where transcription factors normally bind, thereby inhibiting their access [39]. This methylation by the DNMTs could also inhibit other proteins with affinity for unmethylated regions and affect transcription. The third possibility is that methylation patterns from the DNMTs may affect nucleosome positioning, leading to the assembly of specialized nucleosomal structures on methylated DNA that silence transcription more effectively than conventional chromatin [40]. Any of these possibilities could lead to silencing of various tumor suppressors and the onset of cancer.

DNA methylation by the methyltransferases can lead to gene silencing but depletion of DNMT1 and DNMT3B but not DNMT3A reactivates the expression of methylation silenced genes and induces apoptosis of human cancer cells [41]. This may suggest that methylation silencing in numerous types of cancer is mainly catalyzed by DNMT1 and DNMT3B [32]. It was reported that DNMT1 and DNMT3B cooperated in order to silence *CDKN2A* and *TIMP3* gene expression in human colon cancer HCT116 cells [43] and increased expression of both methyltransferases correlated with the increase in breast cancer aggressiveness [45]. Overexpression of DNMT1 and DNMT3B were also found in ovarian cancer cell lines HeyA8, HeyC2, SKOV-3, and PA-1 [47], while breast cancer MCF-7 cells compared to normal human mammary epithelial cells (HMECs) revealed increase in DNMT1 protein stability without an increase in

DNMT1 mRNA level [49]. These studies combined suggest that methylation by the DNMTs protein does affect aberrations in cancer.

5.4   METHYL-CpG-BINDING PROTEINS

Although methylation of CpG islands by the DNMTs appears to affect methylation patterns of the DNA, it is also known that the active recruitment of MBDPs is the most widespread mechanism of methylation-dependent repression. The MBDPs are interpreters of the DNA methylation signal [42,44]. These DNA binding proteins target sequence-specific regions that consist of only two base pairs, 5-methylcytosine followed by guanine (5mCpG). MeCP1 and MeCP2 were the first two proteins identified with methyl-CpG-binding activities [46,48]. The protein MeCP1 was originally identified as a large multiprotein complex, whereas MeCP2 is a single polypeptide with an affinity for a single methylated CpG. Later characterization of MeCP2 led to the identification of its methyl-CpG-binding domain (MBD) [50] and its transcriptional repression domain (TRD). At present, there are five known MBDPs in mammals [51–54] MeCP2, MBD1, MBD2, and MBD4, which binds 5mCpG regions through a conserved motif known as MBD [55] and Kaiso which binds methylated DNA through a zinc finger motif [54]. The MBD3 protein has 70% amino acid homology to the MBD2 protein, contains the conserved MBD motif but has lost the ability to specifically bind methylated DNA [52].

5.4.1   MᴇCP1

MeCP1 was originally identified as a nuclear factor that could discriminate between methylated and unmethylated DNA using band-shift assays [50]. Radioactive DNA molecules that contained 12 or more symmetrically methylated CpG pairs specifically complexed with MeCP1, leading to a shift in electrophoretic mobility. MeCP1 appeared to be a large multi-subunit complex that recognizes densely methylated DNA and represses transcription [50,56]. See MBD2 and MBD3 sections for more information.

5.4.2   MᴇCP2

MeCP2 is a 52 kDa, highly abundant single protein chain that colocalizes with methylated DNA in nuclei. It consists of an 80-amino-acid MBD region near the N-terminus which appears sufficient to bind methylated DNA in vitro and in vivo [53,55]. The MBD represents a novel DNA-binding domain and characterization of it in MeCP2 has led to the discovery of other family members of MBD proteins [52]. MeCP2 also contains a second functional domain, TRD, which overlaps a nuclear localization signal and is required for transcriptional repression in vitro and in vivo [57–59].

In vivo, tethering and transient transfection assays have been used to show MeCP2 transcriptional repression functions in mammalian cells. Furthermore, findings that MeCP2 associates with HDACs and the transcriptional corepressor Sin3 via its TRD validated it as a transcriptional repressor. These findings led to the proposal that MeCP2 is recruited to methylated DNA where it forms a protein complex with Sin3 and HDACs, leading to transcriptional repression [60].

5.4.3   MBD1

MBD1 is the largest member of the MBD family. It is a 50–70 kDa protein that has sequence similarity to the binding domain of MeCP2. It has multiple isoforms of the CXXCXXC sequence, which are similar in sequence to the one found in DNMT1. Its MBD motif is located at the amino terminus end and its carboxy terminus is a region involved in transcriptional repression similar to

MeCP2. It has been shown that one of its various CXXC motifs binds to DNA in a methylation-independent manner [61], while the full length protein has an affinity for densely methylated DNA [48,61]. In human cells, MBD1 has been shown to localize to the pericentromeric region and along the chromosome arms during metaphase [48]. Initially, MBD1 was thought to be a member of the MeCP1 complex but later its native form was extracted from HeLa cells and revealed to be a 200–400 kDa protein not consistent with MeCP1 [48]. Transcriptional and transient assays in HeLa nuclear extracts and mouse cells, respectively, have shown that MBD1 represses methylated DNA via its TRD and MBD motifs [48,62]. Further studies have also shown that MBD1 has a third CXXC motif that binds to unmethylated DNA and represses transcription [61]. It has also been shown that MBD repression is dependent on HDAC activity, but the deacetylase involved has not been determined.

5.4.4 MBD2

MBD2, a 44 kDa protein, shares extensive sequence similarity to a large region of MBD3 corresponding roughly to amino acids 140–400 [52]. However, MBD2 mRNA codes for 140 amino acids preceding this conserved region, which contains a repeat consisting of glycine and arginine residues, and has no similarity to any region in MBD3 [52]. MBD2 binds methylated DNA in a manner very similar to the isolated MBD domain of MeCP2 [52,63]. Like MBD1 and MeCP2, it has been shown that MBD2 also has TRD that overlaps with its MBD motif [64]. Another protein MBD2b, an isoform of MBD2, which lacks the amino terminal 140 amino acids has been reported to possess DNA demethylase activity but this result is still questionable [63,65].

It has been reported that in mammalian cells, MBD2 immunoprecipitations show association with the histone deacetylase 1 (HDAC1) and the histone-binding protein RbA p48/p46. These data suggest that MBD2 is the unknown methyl-CpG-binding protein of the MeCP1 complex, although it has not yet been shown that it can co-immunoprecipitate with Sin3 and Mi-2/NURD complexes. The role of MBD2 association with these complexes is still under investigation, however, data do show that the MBD2b associates with Sin3A.

5.4.5 MBD3

MBD3 is biochemically the best-characterized member of the MBD family. It is, however, the smallest protein (30 kDa) with a coding region similar to that of MBD2 [52]. It has been purified by several laboratories and has been reported as a member of the Mi-2 [66], NuRD [67], and HDAC 1 cII complex [68]. Each version of the complex is different but contains a histone deacetylase, a chromatin remodeling ATPase, and other proteins [63,67,68]. MBD3 has a variety of splice variants in several different species. In mammals, the MBD3 spliced isoform results in a deletion of its MBD domain, and the resulting protein is known as MBDΔ [52,67]. This isoform, only under certain conditions, will bind methylated DNA in vitro but the affinity is much less than that of MeCP2 or MBD2 [52,63]. In *Xenopus laevis* the spliced isoform, which inserts 20 amino acids between exon 1 and 2, is known as MBD3 long form. This protein, unlike the mammalian isoform, binds methylated DNA with affinity similar to that of MeCP2 [63]. Studies in *Drosophila* and *C. elegans* have shown that MBD3 is transcriptional corepressor [69]. It is believed that this repression is due to the enzymatic function of its Mi-2 ATPase and the HDAC1/HDAC2 deacetylase activity.

5.4.6 MBD4

MBD4 is the only member of the MBD family that does not appear to be involved in transcriptional repression and does not associate with HDACs. It is a 62 kDa protein with the MBD located close to its N-terminus. MBD4 was first identified through a two-hybrid assay [70] as having a region of similarity to bacterial DNA repair enzymes [52]. It was originally thought to be an endonuclease but was found to be a glycosylase because it had no endonuclease activity [71,72]. The methyl-binding domain of MBD4 can recognize symmetrically methylated CpG base pairs [52], but it has a preference for binding to 5mCpG-TpG mismatches [71], which are the primary product of deamination at 5mCpG. Hence it has been proposed that the role of MBD4 is to minimize mutation at 5-methylcytosine [71]. It has been shown that MBD4 is alternatively spliced but none of the isoforms have an affected MBD region. Mutations in the MBD4 gene have been isolated in carcinomas where DNA mismatch repair activity is defective [73,74].

5.4.7 KAISO

Kaiso was originally identified due to its ability to associate with a subfamily member of armadillo-domain proteins found at cell/cell junctions, p120 catenin [75,76]. It is the most recently identified methyl-CpG-binding protein. Kaiso is different from the other MDB proteins in that it uses a zinc finger motif instead of an MBD to recognize methylated DNA [54]. It also requires at least two symmetrically methylated CpG dinucleotides (5′-5mCpGp5mCpG-3′) in its recognition sequence [54]. Although it has been shown that Kaiso can bind methylated DNA in vitro; no in vivo targets have been found [54]. Kaiso, however, is an interesting MBD because its association with p120 provides a bridge to methyl-binding proteins and cancer-related proteins (p120 is known to be found in metastatic cells) [77].

Though the limited pool of clinical and experimental data does not support a primary role for MBDPs in tumorigenesis, it suggests that decreased levels of MBD2, MeCP2, and MBD4 might be involved in tumor progression. The newest inductee into the methyl-CpG-binding protein family, Kaiso, appears to have the best cancer credentials, but its relationship with p120 needs to be further explored.

Expression studies of MBDPs in cancer cells suggest that the levels of MBD2, MECP2, and MBD4 are decreased in some tumors. In human hepatocellular carcinomas (HCCs), a slight reduction of MBD2 expression as well as a more marked reduction of MECP2 expression was observed while no significant alterations in the levels of MBD1 or MBD3 transcripts were seen [78]. It was also reported that reduced mRNA expression of MBD2 can be found in human colorectal and stomach cancers [80]. Reduction of MBD2 and MECP2 transcript levels might increase the probability that a cell has undergone the first few steps of oncogenic transformation and will become fully cancerous. Though the limited experimental data do not support a primary role for MBDPs in tumorigenesis, they do offer a suggestion that reduced levels of the proteins might be involved in tumor progression. Given that DNA methylation is usually associated with inappropriate silencing of genes in cancer, the contribution that a decrease in the levels of the MBDPs might play in tumorigenesis is still unclear.

5.4.8 HDACs 1 AND 2

Although not directly involved in DNA methylation, histones play an integral role in the regulation of gene expression. HDACs 1 and 2 have been found to work together with DNA methylation machinery to silence tumor suppressors [79,81]. No somatic mutations of HDACs 1 and 2 have been found in tumors but they appear to have altered expression in neoplasias [82]. A recent study that explores the use of DNA methyltransferase inhibitor hydralazine and histone deacetylase inhibitor valproate in cervical cancer cell lines confirms the correlation between DNA hypomethylation in LCR regions and H4 hyperacetylation, indicating that these epigenetic mechanisms are key to the regulation of the E6/E7 oncogenes in HPV-related neoplasias [83]. Moreover, this study illustrated that such inhibitors could be safely administered to patients to reduce tumor load and reactivate the expression of previously silenced tumor suppressors without inadvertently increasing the expression of viral oncoproteins. Targeting of HDACs to tumor suppressor promoter regions, perhaps by aberrant methylation, contributes to the silencing of such genes as the cyclin-dependent kinase inhibitor p21 or pro-apoptotic BCL2-associated X protein (BAX).

5.4.9 OTHER MINOR PROTEINS INVOLVED IN DNA METHYLATION

SIRT1 is a type III histone deacetylase, named because of sequence homology to yeast Sir2. The expression profile of SIRT1 is potentially dysregulated in human cancers as evidenced by the overexpression of SIRT1 in leukemia cell [84]. PcG and Trithorax-group proteins are epigenetic transcriptional repressors and activators and are involved in large complexes with such proteins as the HDACs and HMTs. Perinatal exposure to estrogen-like compounds such as diethylstilbestrol (DES) early in life can perturb the methylation patterns of genes such as epidermal growth factor, *c-jun* and *c-myc*, influencing cancer risk in adulthood [85]. Expression of Lactoferrin, a globular member of the transferrin family proteins, was found to be altered in developing mice treated with DES due to differential methylation patterns at the Lactoferrin promoter and coincided with ~100% of epithelial uterine cancers at 18 months of age [86].

5.5 DEMETHYLATION AND CANCER ABERRATIONS

Thus far, we have discussed the proteins involved in hypermethylation leading to cancer but have yet to discuss those involved in hypomethylation. As stated earlier in the chapter, hypermethylation in cancer cells has been studied more and therefore we have a clearer understanding of its role in carcinogenesis. Although DNA hypomethylation was identified years before hypermethylation in human tumors [87] it is just recently that we have begun to investigate demethylation of DNA outside of CpG islands in many tumor types [88–90].

5.5.1 DNA HYPOMETHYLATION

Multiple observations in numerous tumors have established that tumor tissue is globally hypomethylated relative to its normal counterpart [91,92] (Table 5.2). There have been at least three proposals that explain how global hypomethylation may play a role in cancer. It was first proposed that hypomethylation resulted in activation of demethylation of proto-oncogenes [93,94]. Secondly, hypomethylation has been proposed to predispose cells to chromosomal instability [95,96]. Lastly, hypomethylation has been proposed to increase metastasis of cancer cells [97].

Although it is the source of some controversy, in 1999, the Szyf laboratory discovered a DNA demethyltransferase that catalyzed the removal of methyl groups from methylcytosine, leaving a dCpdG DNA sequence and methanol as a byproduct of the reaction [98]. Soon thereafter they showed that DNA demethylase was a processive enzyme, meaning once demethylation was initiated it proceeded, uninterrupted, at high rate for at least 250 bp [99]. The same group later showed that the DNA demethylase had dinucleotide specificity, could demethylate both fully methylated and hemimethylated DNA and had the ability to demethylate mCpG sites in different sequence contexts [100].

TABLE 5.2
Examples of Genes Hypomethylated in Cancer

Gene	Function	Cancer Type	References
Cytochrome P450 1B1	Responsible for 4-hydroxylation of estrogen	Prostate	[112]
Maspin	Inhibition of tumor invasion and metastasis	Colorectal	[113]
P-cadherin	Transmembrane molecule, cell/cell adhesion	Breast	[114]
Synuclein	Member of neuronal protein family	Breast, ovarian, colon	[115]
TRAG-3	Responsible for resistance to taxol	Colon, liver, lung	[116]
BRCA1 and 2	Regulates cell division	Ovarian	[117]
p53	Transcription factor, regulates cell cycle	Lung	[118]

The DNA demethylase is a member of the family of MBD proteins and is related to the MeCP2 transcriptional repressor [46]. It has been shown that it is a 262 amino acid protein identical to MBD2b and has a 10- to 100-fold selectivity for binding methylated CpG dinucleotides [52]. The enzyme has been fractionated on a sucrose gradient in the 160–190 K range suggesting that it might multimerize or be a part of a demethylase complex [13].

Although DNA demethylase has been discovered and plays a role in hypomethylation, there are also other factors that might aide in decreased methylation seen in cancer cells. The lack of DNA methyltransferase activity seen when an individual DNMT or a combination of DNMTs is knocked out is linked to genomic hypomethylation and an associated chromosomal abnormality and instability [96,101–103]. Recent studies demonstrate the importance of chromatin changes in genome-wide hypomethylation, such as histone modifications and chromatin remodeling. In cancer cells, loss of monoacetylation of H4Lys16 and trimethylation of H4Lys20 correlates with DNA hypomethylation of several tandem repeat sequences in leukemic cells [104]. Mouse knockout experiments have also shown that loss of methylation at H3K9 leads to genomic instability and increases cancer predisposition. Another means of increased hypomethylation implicates RNA. Active demethylation of specific gene sequences in a cell-free system has been shown to be dependent on the presence of RNA with extracts obtained from different cell types [105]. External factors such as diet and environment might also affect DNA demethylation. It is known that tumors develop due to such factors as UV radiation, chemicals, and methyl donor deficiencies. Such effects of environmental factors were noted in children born through assisted reproductive technologies. These children had a greater risk than normal children of developing Wilms tumor of the kidney, retinoblastoma, and Angelman's syndrome, which is associated with hypomethylation.

5.6 CONCLUSION

Cancer is an epigenetic disease. The disruption of epigenetic modifications is the most common feature of all human tumors. There is substantial evidence that DNA methylation plays a critical role in silencing specific genes during development and cell differentiation. The activation of proto-oncogenes through hypomethylation, and transcriptional inactivation of tumor-suppressor genes through hypermethylation, along with failure of de novo methylation may contribute to neoplasia. Therefore selective modulation of DNA methylation may have important clinical implications for the prevention and treatment of cancer.

ACKNOWLEDGMENTS

This work was supported in part by grants from the National Cancer Institute and the Susan G. Komen For the Cure Foundation.

REFERENCES

1. Jones, P.A. and Baylin, S.B., The fundamental role of epigenetic events in cancer, *Nat. Rev. Genet.*, 3, 415, 2002.
2. Feinberg, A.P. and Tycko, B., The history of cancer epigenetics, *Nat. Rev. Cancer*, 4, 143, 2004.
3. Esteller, M., Aberrant DNA methylation as a cancer-inducing mechanism, *Annu. Rev. Pharmacol. Toxicol.*, 45, 629, 2005.
4. Hotchkiss, R.D., The quantitative separation of purines, pyrimidines and nucleosides by paper chromatography, *J. Biol. Chem.*, 175, 315, 1948.
5. Gruenbaum, Y., Cedar, H., and Razin, A., Substrate and sequence specificity of a eukaryotic DNA methylase, *Nature*, 295, 620, 1982.
6. Bestor, T. et al., Cloning and sequencing of a cDNA encoding DNA methyltransferase of mouse cells: The carboxyl-terminal domain of the mammalian enzymes is related to bacterial restriction methyltransferases, *J. Mol. Biol.*, 203, 971, 1988.

7. Tucker, K.L. et al., Complementation of methylation deficiency in embryonic stem cells by DNA methyltransferase minigene, *Proc. Natl. Acad. Sci. U S A,* 93, 12920, 1996.

8. Yen, R.W. et al., Isolation and characterization of the cDNA encoding human DNA methyltransferase, *Nucleic Acids Res.,* 20, 2287, 1992.

9. Yoder, J.A. et al., New 50 regions of the murine and human genes for DNA (cytosine-5)-methyltransferase, *J. Biol. Chem.,* 271, 31092, 1996.

10. Hitt, M.M. et al., De novo and maintenance DNA methylation by a mouse plasmacytoma cell DNA methyltransferase, *J. Biol. Chem.,* 263, 4392, 1988.

11. Pradhan, S. et al., Recombinant human DNA (cytosine-5) methyltransferase. I. Expression, purification, and comparison of de novo and maintenance methylation, *J. Biol. Chem.,* 274, 33002, 1999.

12. Ruchirawat, M., Noshari, J., and Lapeyre, J.N., Kinetic mechanisms and interaction of rat liver DNA methyltransferase with defined DNA substrates, *Mol. Cell. Biochem.,* 76, 45, 1987.

13. Yoder, J.A., et al., DNA (cytosine-5)-methyltransferases in mouse cells and tissues: Studies with a mechanism-based probe, *J. Mol. Biol.,* 270, 385, 1997.

14. Leonhardt, H. et al., A targeting sequence directs DNA methyltransferase to sites of DNA replication in mammalian nuclei, *Cell,* 71, 865, 1992.

15. Howell, C.Y. et al., Genomic imprinting disrupted by a maternal effect mutation in the Dnmt1 gene, *Cell,* 104, 829, 2001.

16. Okano, M., Xie, S., and Li, E., Dnmt2 is not required for de novo and maintenance methylation of viral DNA in embryonic stem cells, *Nucleic Acids Res.,* 26, 2536, 1998.

17. Van den Wyngaert, I. et al., Cloning and analysis of a novel human putative DNA methyltransferase, *FEBS Lett.,* 426, 283, 1998.

18. Yoder, J.A. and Bestor, T.H., A candidate mammalian DNA methyltransferase related to pmt1p of fission yeast, *Hum. Mol. Genet.,* 7, 279, 1998.

19. Dong, A. et al., Structure of human DNMT2, an enigmatic DNA methyltransferase homolog that displays denaturant resistant binding to DNA, *Nucleic Acids Res.,* 29, 439, 2001.

20. Tweedie, S. et al., Vestiges of a DNA methylation system in *Drosophila melanogaster? Nat. Genet.,* 23, 389, 1999.

21. Goll, M.G. et al., Methylation of tRNAAsp by the DNA methyltransferase homolog Dnmt2, *Science,* 311, 395, 2006.

22. Wilkinson, C.R. et al., The fission yeast gene pmt1þ encodes a DNA methyltransferase homologue, *Nucleic Acids Res.,* 23, 203, 1995.

23. Hermann, A., Schmitt, S., and Jeltsch, A., The human Dnmt2 has residual DNA-(cytosine-C5) methyl-transferase activity, *J. Biol. Chem.,* 278, 31717, 2003.

24. Tang, L.Y.et al., The eukaryotic DNMT2 genes encode a new class of cytosine-5 DNA methyltrans-ferases, *J. Biol. Chem.,* 278, 33613, 2003.

25. Kunert, N. et al., A Dnmt2-like protein mediates DNA methylation in *Drosophila, Development,* 130, 5083, 2003.

26. Liu, K. et al., Endogenous assays of DNA methyltransferases: Evidence for differential activities of DNMT1, DNMT2, and DNMT3 in mammalian cells in vivo, *Mol. Cell. Biol.,* 23, 2709, 2003.

27. Lei, H. et al., De novo DNA cytosine methyltransferase activities in mouse embryonic stem cells, *Development,* 122, 3195, 1996.

28. Okano, M., Xie, S., and Li, E., Cloning and characterization of a family of novel mammalian DNA (cytosine-5) methyltransferases, *Nat. Genet.,* 19, 219, 1998.

29. Chen, T. et al., A novel Dnmt3a isoform produced from an alternative promoter localizes to euchromatin and its expression correlates with active de novo methylation, *J. Biol. Chem.,* 277, 38746, 2002.

30. Xie, S. et al., Cloning, expression and chromosome locations of the human DNMT3 gene family, *Gene,* 236, 87, 1999.

31. Lin, I.G. et al., Murine de novo methyltransferase Dnmt3a demonstrates strand asymmetry and site preference in the methylation of DNA in vitro, *Mol. Cell Biol.,* 22, 704, 2002.

32. Bachman, K.E., Rountree, M.R., and Baylin, S.B., Dnmt3a and Dnmt3b are transcriptional repressors that exhibit unique localization properties to heterochromatin, *J. Biol. Chem.,* 276, 32282, 2001.

33. Hansen, R.S. et al., The DNMT3B DNA methyltransferase gene is mutated in the ICF immunodeficiency syndrome, *Proc. Natl. Acad. Sci. U S A,* 96, 14412, 1999.

34. Robertson, K.D. et al., The human DNA methyltransferases (DNMTs) 1, 3a and 3b: Coordinate mRNA expression in normal tissues and overexpression in tumors, *Nucleic Acids Res.*, 27, 2291, 1999.

35. Aoki, A. et al., Enzymatic properties of de novo-type mouse DNA (cytosine-5) methyltransferases, *Nucleic Acids Res.*, 29, 3506, 2001.

36. Chen, T. et al., Establishment and maintenance of genomic methylation patterns in mouse embryonic stem cells by Dnmt3a and Dnmt3b, *Mol. Cell. Biol.*, 23, 5594, 2003.

37. Hsieh, C.L., In vivo activity of murine de novo methyltransferases, Dnmt3a and Dnmt3b, *Mol. Cell. Biol.*, 19, 8211, 1999.

38. Lewis, J.D. et al., Purification, sequence, and cellular localization of a novel chromosomal protein that binds to methylated DNA, *Cell*, 69, 905, 2002.

39. Tate, P.H. and Bird, A.P., Effects of DNA methylation on DNA-binding proteins and gene expression, *Curr. Opin. Genet. Dev.*, 3, 226, 1993.

40. Kass, S.U., Pruss, D., and Wolffe, A.P., How does DNA methylation repress transcription? *Trends Genet.*, 13, 444, 1997.

41. Beaulieu, N. et al., An essential role for DNA methyltransferase DNMT3B in cancer cell survival, *J. Biol. Chem.*, 277, 28176, 2002.

42. Hendrich, B. and Bird, A., Mammalian methyltransferases and methyl-CpG-binding domains: Proteins involved in DNA methylation, *Curr. Top. Microbiol. Immunol.*, 249, 55, 2000.

43. Rhee, I. et al., DNMT1 and DNMT3b cooperate to silence genes in human cancer cells, *Nature,* 416, 552, 2002.

44. Wade, P.A., Methyl CpG binding proteins: Coupling chromatin architecture to gene regulation, *Oncogene,* 20, 3166, 2001.

45. Girault, I. et al., Expression analysis of DNA methyltransferases 1, 3A, and 3B in sporadic breast carcinomas, *Clin. Cancer Res.*, 9, 4415, 2003.

46. Lewis, J.D. et al., Purification, sequence, and cellular localization of a novel chromosomal protein that binds to methylated DNA, *Cell*, 69, 905, 1992.

47. Ahluwalia, A. et al., DNA methylation in ovarian cancer. II. Expression of DNA methyltransferases in ovarian cancer cell lines and normal ovarian epithelial cells, *Gynecol. Oncol.*, 82, 299, 2001.

48. Ng, H.H., Jeppesen, P., and Bird, A., Active repression of methylated genes by the chromosomal protein MBD1, *Mol. Cell. Biol.*, 20, 1394, 2000.

49. Agoston, A.T. et al., Increased protein stability causes DNA methyltransferase 1 dysregulation in breast cancer, *J. Biol. Chem.*, 280, 18302, 2005.

50. Meehan, R.R. et al., Identification of a mammalian protein that binds specifically to DNA containing methylated CpGs, *Cell*, 58, 499, 1989.

51. Cross, S.H. et al., A component of the transcriptional repressor MeCP1 shares a motif with DNA methyltransferase and HRX proteins, *Nat. Genet.*, 13, 256, 1997.

52. Hendrich, B. and Bird, A., Identification and characterization of a family of mammalian methyl-CpG binding proteins, *Mol. Cell. Biol.*, 18, 6538, 1998.

53. Nan, X., Meehan, R.R., and Bird, A., Dissection of the methyl-CpG binding domain from the chromosomal protein MeCP2, *Nucleic Acids Res.*, 21, 4886, 1993.

54. Prokhortchouk, A. et al., The p120 catenin partner Kaiso is a DNA methylation-dependent transcriptional repressor, *Genes Dev.*, 15, 1613, 2001.

55. Nan, X. et al., DNA methylation specifies chromosomal localization of MeCP2, *Mol. Cell. Biol.*, 16, 414, 1996.

56. Boyes, J. and Bird, A., DNA methylation inhibits transcription indirectly via a methyl-CpG binding protein, *Cell*, 64, 1123, 1991.

57. Nan, X., Campoy, F.J., and Bird, A., MeCP2 is a transcriptional repressor with abundant binding sites in genomic chromatin, *Cell*, 88, 471, 1997.

58. Jones, P.L. et al., Methylated DNA and MeCP2 recruit histone deacetylase to repress transcription, *Nat. Genet.*, 19, 187, 1998.

59. Kaludov, N.K. and Wolffe, A.P., MeCP2 driven transcriptional repression in vitro: Selectivity for methylated DNA, action at a distance and contacts with the basal transcription machinery, *Nucleic Acids Res.*, 28, 1921, 2000.

60. Bird, A.P. and Wolffe, A.P., Methylation-induced repression-belts, braces, and chromatin, *Cell*, 99, 451, 1999.

61. Fujita, N. et al., Mechanism of transcriptional regulation by methyl-CpG binding protein MBD1, *Mol. Cell. Biol.*, 20, 5107, 2000.

62. Fujita, N. et al., Methylation-mediated transcriptional silencing in euchromatin by methyl-CpG binding protein MBD1 isoforms, *Mol. Cell. Bio.*, 19, 6415, 1999.

63. Wade, P.A. et al., Mi-2 complex couples DNA methylation to chromatin remodelling and histone deacetylation, *Nat. Genet.*, 23, 62, 1999.

64. Boeke, J. et al., The minimal repression domain of MBD2b overlaps with the methyl-CpG-binding domain and binds directly to sin3A, *J. Biol. Chem.*, 275, 34963, 2000.

65. Ng, H.H. et al., MBD2 is a transcriptional repressor belonging to the MeCP1 histone deacetylase complex, *Nat. Genet.*, 23, 58, 1999.

66. Wade, P.A. et al. A multiple subunit Mi-2 histone deacetylase from *Xenopus laevis* cofractionates with an associated Snf2 superfamily ATPase, *Curr. Biol.*, 8, 843, 1998.

67. Zhang, Y. et al., Analysis of the NuRD subunits reveals a histone deacetylase core complex and a connection with DNA methylation, *Genes Dev.*, 13, 1924, 1999.

68. Humphrey, G.W. et al., Stable histone deacetylase complexes distinguished by the presence of SANT domain protein CoREST/kiaa0071 and Mta-L1, *J. Biol. Chem.*, 276, 6817, 2001.

69. Ahringer, J., NuRD and SIN3 histone deacetylase complexes in development, *Trends Genet.*, 16, 351, 2000.

70. Bellacosa, A., MED1, a novel human methyl-CpG-binding endonuclease, interacts with DNA mismatch repair protein MLH1, *Proc. Natl. Acad. Sci. U S A*, 96, 3969, 1999.

71. Hendrich, B., The thymine glycosylase MBD4 can bind to the product of deamination at methylated CpG sites, *Nature*, 401, 301, 1999.

72. Petronzelli, F. et al., Investigation of the substrate spectrum of the human mismatch-specific DNA N-glycosylase MED1 (MBD4): Fundamental role of the catalytic domain, *J. Cell. Physiol.*, 185, 473, 2000.

73. Bader, S., Somatic frameshift mutations in the MBD4 gene of sporadic colon cancers with mismatch repair deficiency, *Oncogene*, 18, 8044, 1999.

74. Riccio, A. et al., The DNA repair gene MBD4 (MED1) is mutated in human carcinomas with microsatellite instability, *Nat. Genet.*, 23, 266, 1999.

75. Daniel, J.M. and Reynolds, A.B., The catenin p120(ctn) interacts with Kaiso, a novel BTB/POZ domain zinc finger transcription factor, *Mol. Cell. Biol.*, 19, 3614, 1999.

76. Anastasiadis, P.Z. and Reynolds, A.B., The p120 catenin family: Complex roles in adhesion, signaling and cancer, *J. Cell Sci.*, 113, 1319, 2000.

77. van Hengel, J. et al., Nuclear localization of the p120(ctn) Armadillo-like catenin is counteracted by a nuclear export signal and by E-cadherin expression, *Proc. Natl. Acad. Sci. U S A*, 96, 7980, 1999.

78. Saito, Y. et al., Expression of mRNA for DNA methyltransferases and methyl-CpG-binding proteins and DNA methylation status on CpG islands and pericentromeric satellite regions during human hepatocarcinogenesis, *Hepatology*, 33, 561, 2001.

79. Fahrner, J.A. et al., Dependence of histone modifications and gene expression on DNA hypermethylation in cancer, *Cancer Res.*, 62, 7213, 2002.

80. Kanai, Y. et al., Reduced mRNA expression of the DNA demethylase, MBD2, in human colorectal and stomach cancers, *Biochem. Biophys. Res. Commun.*, 264, 962, 1999.

81. Ballestar, E. et al., Methyl-CpG binding proteins identify novel sites of epigenetic inactivation in human cancer, *EMBO J.*, 22, 6335, 2003.

82. Gibbons, R.J., Histone modifying and chromatin remodelling enzymes in cancer and dysplastic syndromes, *Hum. Mol. Genet.*, 14, R85, 2005.

83. de la Cruz-Hernandez, E. et al., The effects of DNA methylation and histone deacetylase inhibitors on human papillomavirus early gene expression in cervical cancer, an in vitro and clinical study, *Virol. J.*, 4, 18, 2007.

84. Bradbury, C.A. et al., Histone deacetylases in acute myeloid leukaemia show a distinctive pattern of expression that changes selectively in response to deacetylase inhibitors, *Leukemia,* 19, 1751, 2005.

85. Li, S. et al., Environmental exposure, DNA methylation, and gene regulation: Lessons from diethylstilbestrol-induced cancers, *Ann. NY Acad. Sci.*, 983, 161, 2003.

86. Li, S. et al., Developmental exposure to diethylstilbestrol elicits demethylation of estrogen-responsive lactoferrin gene in mouse uterus, *Cancer Res.*, 57, 4356, 1997.

87. Feinberg, A.P. and Vogelstein, B., Hypomethylation distinguishes genes of some human cancers from their normal counterparts, *Nature*, 301, 89, 1983.
88. Cravo, M. et al., Global DNA hypomethylation occurs in the early stages of intestinal type gastric carcinoma, *Gut*, 39, 434, 1996.
89. Kim, Y.I. et al., Global DNA hypomethylation increases progressively in cervical dysplasia and carcinoma, *Cancer*, 74, 893, 1994.
90. Soares, J. et al., Global DNA hypomethylation in breast carcinoma: Correlation with prognostic factors and tumor progression, *Cancer,* 85, 112, 1999.
91. Ehrlich, M. et al., Hypomethylation and hypermethylation of DNA in Wilms tumors, *Oncogene*, 21, 6694, 2002.
92. Feinberg, A.P., Reduced genomic 5-methylcytosine content in human colonic neoplasia, *Cancer Res.*, 48, 1159, 1988.
93. Bhave, M.R., c-H-ras and c-K-ras gene hypomethylation in the livers and hepatomas of rats fed methyl-deficient, amino acid-defined diets, *Carcinogenesis*, 9, 343, 1988.
94. Kaneko, Y. et al., Hypomethylation of c-myc and epidermal growth factor receptor genes in human hepatocellular carcinoma and fetal liver, *Jpn. J. Cancer Res.*, 76, 1136, 1997.
95. Ji, W. et al., DNA demethylation and pericentromeric rearrangements of chromosome 1, *Mutat. Res.*, 379, 33, 1997.
96. Eden, A. et al., Chromosomal instability and tumors promoted by DNA hypomethylation, *Science*, 300, 455, 2003.
97. Chen, R.Z. et al., DNA hypomethylation leads to elevated mutation rates, *Nature,* 395, 89, 1998.
98. Bhattacharya, S.K. et al., A mammalian protein with specific demethylase activity for mCpG DNA, *Nature*, 397, 579, 1999.
99. Cervoni, N., Bhattacharya, S., and Szyf, M., DNA demethylase is a processive enzyme, *J. Biol. Chem.*, 274, 8363, 1999.
100. Ramchandani, S., DNA methylation is a reversible biological signal, *Proc. Natl. Acad. Sci. U S A*, 96, 6107, 1999.
101. Dodge, J.E. et al., Inactivation of Dnmt3b in mouse embryonic fibroblasts results in DNA hypomethylation, chromosomal instability, and spontaneous immortalization, *J. Biol. Chem.*, 280, 7986, 2005.
102. Couillard, J., The role of DNA hypomethylation in the control of stromelysin gene expression, *Biochem. Biophys. Res. Commun.*, 342, 1233, 2006.
103. James, S.R., Link, P.A., and Karpf, A.R., Epigenetic regulation of X-linked cancer/germline antigen genes by DNMT1 and DNMT3b, *Oncogene*, 25, 6975, 2006.
104. Fraga, M.F. et al., Loss of acetylation at Lys16 and trimethylation at Lys20 of histone H4 is a common hallmark of human cancer, *Nat. Genet.*, 37, 391, 2005.
105. Fremont, M. et al., Demethylation of DNA by purified chick embryo 5-methylcytosine-DNA glycosylase requires both protein and RNA, *Nucleic Acids Res.*, 25, 2375, 1997.
106. Baylin, S.B. et al., Alterations in DNA methylation: A fundamental aspect of neoplasia, *Adv. Cancer Res.*, 72, 141.
107. Esteller, M. et al., MLH1 promoter hypermethylation is associated with the microsatellite instability phenotype in sporadic endometrial carcinomas, *Oncogene*, 17, 2413, 1998.
108. Herman, J.G. et al., Hypermethylation-associated inactivation indicates a tumor suppressor role for p15INK4B, *Cancer Res.*, 56, 722, 1996.
109. Esteller, M. et al., Inactivation of the DNA repair gene O^6-*methylguanine-DNA methyltransferase* by promoter hypermethylation is a common event in primary human neoplasia, *Cancer Res.*, 59, 793, 1999.
110. Lee, W.H. et al., Cytidine methylation of regulatory sequences near the π-class *glutathione S-transferase* gene accompanies human prostatic carcinogenesis, *Proc. Natl. Acad. Sci. U S A,* 91, 11733, 1994.
111. Katzenellenbogen, R.A., Baylin, S.B., and Herman, J.G., Hypermethylation of the DAP-Kinase CpG island is a common alteration in B-cell malignancies, *Blood*, 93, 4347, 1999.
112. Tokizane, T. et al., Cytochrome *P*450 1B1 is overexpressed and regulated by hypomethylation in prostate cancer, *Clin. Cancer Res.*, 11, 5793, 2005.
113. Bettstetter, M. et al., Elevated maspin expression is associated with microsatellite instability and high tumour grade in colorectal cancer, *J. Pathol.*, 205, 606, 2005.
114. Paredes, J. et al., P-cadherin overexpression is an indicator of clinical outcome in invasive breast carcinomas and is associated with CDH3 promoter hypomethylation, *Clin. Cancer Res.*, 11, 5869, 2005.

115. Gupta, A. et al., Hypomethylation of the synuclein gamma gene CpG island promotes its aberrant expression in breast carcinoma and ovarian carcinoma, *Cancer Res.*, 63, 664, 2003.
116. Yao, X. et al., Epigenetic regulation of the taxol resistance-associated gene TRAG-3 in human tumors, *Cancer Genet. Cytogenet.*, 151, 1, 2004
117. Chan, K.Y. et al., Epigenetic factors controlling the BRCA1 and BRCA2 genes in sporadic ovarian cancer, *Cancer Res.*, 62, 4151, 2002.
118. Woodson, K. et al., Hypomethylation of p53 in peripheral blood DNA is associated with the development of lung cancer, *Cancer Epidemiol. Biomarkers Prev.*, 10, 69, 2001.

6 Role of DNA Methylation in Cancer Progression

Sang-Woon Choi and Simonetta Friso

CONTENTS

6.1 INTRODUCTION

Historically, cancer has been regarded as a genetic disease characterized by mutations or loss of heterozygosity. However, an accumulating body of evidence suggests that cancer is also an epigenetic disease and aberrant epigenetic gene silencing or derepression is thought to play a major role in cancer initiation, progression, and even metastasis. Among many epigenetic phenomena that can alter gene expression through heritable noncoding changes, DNA methylation has been most extensively investigated in cancer.

In 1985 Goelz et al. [1] first reported that DNA from both benign colon polyps and malignant carcinomas was substantially hypomethylated compared to DNA from adjacent normal appearing colonic tissue. Thereafter, in a rodent hepatocellular carcinoma model of chronic dietary methyl deficiency, Pogribny et al. [2] demonstrated a progressive loss of methyl groups at most CpG sites during the first 36 weeks of methyl deficiency. However, after 54 weeks of deficiency, the majority of CpG sites in the DNA of tumor were remethylated. Both *p53* gene-specific and genomic DNA methylation were also increased. In the preneoplastic lesions, the level of *p53* mRNA was increased in association with hypomethylation in the gene. On the other hand in tumor tissues, *p53* mRNA

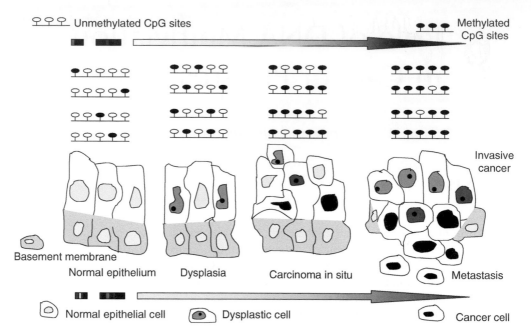

FIGURE 6.1 Progressive changes in promoter methylation at CpG sites during cancer initiation and progression. The number of methylated CpG residues in tumor suppressor genes is in parallel with the stage of carcinogenesis.

was decreased along with relative hypermethylation in the gene. Thereafter, a body of evidence has also accumulated demonstrating that the density of CpG-island methylation gradually increases in parallel with tumor progression [3,4] (Figure 6.1). Collectively, the pattern of DNA methylation is changing during the progression of tumor.

Since reduced genomic DNA methylation, increased DNA promoter methylation, and loss of imprinting are common epigenetic alterations in carcinogenesis and the pattern of DNA methylation is changing during carcinogenesis, investigating the role of DNA methylation in cancer progression may lead to the discovery of important new avenues in cancer progression [5].

6.2   ABERRANT DNA METHYLATION DURING EARLY NEOPLASTIC PROGRESSION

Epigenetic gene silencing by progressive promoter DNA hypermethylation can be considered as a time sequence event in cancer progression (Figure 6.1). Recent evidence has shown that aberrant DNA methylation plays an essential role in the early step of cancer initiation before major genetic changes [6]. Since stem cells are now regarded as responsible cells for cancer initiation and progression, one might speculate that epigenetic changes in stem cells produce a polyclonal population of preneoplastic cells and disturb the balance between undifferentiated progenitor cells and differentiated committed cells in their capacity for aberrant differentiation [7].

6.2.1   Epigenetic Gene Silencing through Promoter Methylation

In the early phase of cancer, epigenetic silencing by promoter DNA hypermethylation is frequently found in a group of genes, so-called gatekeeper and caretaker genes, characterized respectively by their control of net cellular proliferation and maintenance of genomic integrity. The normal epigenetic status of these genes prevents cells from acquiring continuous cell renewal ability as well as allows these genes to be activated when cells need differentiation. On the other hand,

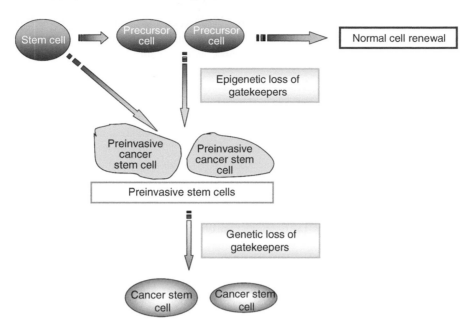

FIGURE 6.2   Epigenetic loss of gatekeeper genes during cancer progression. Epigenetic silencing of gate-keeper genes facilitates abnormal clonal expansion in stem/precursor cells. Normal epigenetic patterns of those genes maintain differentiation of stem/precursor cells for normal adult cell renewal, whereas the abnormal gene silencing by promoter methylation induces abnormal survival of the cells and mutations in those genes exert a stronger force for further tumor progression. (From Jones, P.A. and Baylin, S.B., *Cell*, 128, 683, 2007.)

epigenetic silencing of these genes inhibits their activation, allowing abnormal proliferation and clonal expansion [6] (Figure 6.2).

Early epigenetic changes may predispose cells to the permanent genetic alterations that progress to neoplastic transformation. The silencing of tumor suppressor genes or DNA repair genes in association with promoter DNA hypermethylation is a cellular heterogeneous process that can start at the early neoplastic process and eventually progress to cancer. Silencing of *p16* by promoter DNA methylation could allow mammary epithelial cells to escape senescence, resulting in genetic instability and silencing of O^6-methylguanine-DNA methyltransferase gene (*MGMT*) by promoter DNA methylation also allows cells to acquire specific types of genetic mutation that arise from the inability to repair DNA guanosine adducts. Thus, the loss of critical gene function by this process is probably a fundamental event in the progression of many types of cancer [8] and the role of this type of epigenetic silencing in cancer progression is discussed in the later sections of this chapter.

6.2.2   ACTIVATION OF NORMALLY SILENCED ALLELE BY LOSS OF IMPRINTING

Imprinting is a parent-of-origin gene-silencing mechanism regulated by DNA methylation and chromatin structure. It causes reduced or absent expression of a specific allele of a gene in somatic cells of the offspring [7]. Imprinting is a common feature of eukaryotic cells, affecting genes that regulate cell growth, cell cycle, signaling, and development. During the early carcinogenesis, loss of imprinting activates the normally silenced allele of an imprinted gene such as insulin-like growth factor II (*IGF-2*). Normally, the *IGF-2* gene is paternally expressed and the maternal origin allele is imprinted by DNA methylation. Loss of imprinting of *IGF-2*, which accounts for half of Wilms tumor in children and is associated with an increased risk of colorectal cancer, activates the normally silenced allele and thereby induces cellular proliferation with two active alleles [9]. It appears that loss of imprinting may contribute to the early abnormal expansion of the progenitor cell population in patients with an inherited predisposition to Wilms tumor. This epigenetic change may also accelerate carcinogenesis [9,10] by providing subsequent genetic and epigenetic alterations that further foster tumor progression.

In a chronic myelogenous leukemia study, the imprinting status of *IGF-2* was evaluated in stable phase, accelerated phase, and blast crisis. Interestingly, five of six stable-phase patients showed normal imprinting, while all six cases of advanced disease (three accelerated phase and three blast crisis) demonstrated loss of imprinting ($P < .01$) [11]. In an animal study using mice that are genetically altered in the *IGF-2* and combined with $Apc^{Min/+}$, a rodent model of human familial adenomatous polyposis, mice that express excess *IGF-2* demonstrated a 10-fold increase in the number and the diameter of colon adenoma ($P < .0001$) compared to $Apc^{Min/+}$ littermate controls. On the other hand, mice with reduced *IGF-2* demonstrated a threefold reduction in small intestinal adenoma number ($P < .0001$) and a significant decrease in adenoma diameter ($P < .001$) compared with $Apc^{Min/+}$ littermate controls [12]. These two observations from human and animal studies demonstrate an association between loss of imprinting of *IGF-2* and tumor progression.

6.2.3 ACTIVATION OF ONCOGENES AND CHROMOSOMAL INSTABILITY

Decreased genomic DNA methylation is a common feature in many cancers [1]. However, the role of genomic DNA hypomethylation in tumor progression is not clearly elucidated yet. Evidence indicates that hypomethylation might contribute to genomic instability, structural changes in chromosomes, and enhancement of gene expression. The knockout of *Lsh* (lymphoid-specific helicase), a gene for maintaining normal methylation and heterochromatin organization, leads to genomic DNA hypomethylation as well as genetic instability and chromosomal aberration [13]. The knockout of the DNA methyltransferase 1 gene also induces tumors along with profound genomic DNA hypomethylation in all tissues and genomic instability in thymic tumor DNA [14]. A human colon cancer study using fresh frozen colorectal carcinomas and paired normal tissues demonstrated that genomic DNA hypomethylation is mainly associated with chromosomal instability, which could be explained by genome-wide effects rather than site-specific genetic and epigenetic alterations, supporting a direct connection between genome-wide hypomethylation and chromosomal instability in human colorectal carcinogenesis [15]. This observation is consistent with the above cited studies demonstrating a role of DNA demethylation in inducing chromosomal instability.

Almost 50% of the human genome consists of transposable elements and other repetitive DNA sequences that are highly methylated. Hypomethylation or demethylation in these sequences might induce chromosomal breakage or recombination; hypomethylation of long interspersed nuclear elements 1 (LINE-1) retrotransposons might promote chromosomal rearrangement during carcinogenesis and progressive demethylation of the LINE-1 promoter sequence might lead to abnormal gene expression and accelerate the process of neoplastic progression [16].

In somatic tissues CpG islands of certain genes are normally methylated [17]. These methylated CpG islands can be hypomethylated in cancer and activates those genes, which could be oncogenes such as *HRAS* [18] and the cancer/testis (*CT*) genes that are expressed normally in the testis and aberrantly in tumors [19]. The frequency of hypomethylated genes is also high in other cancers: *cyclin D2* [20] and *maspin* [21] in gastric cancer, *MN/CA9* in renal cell carcinoma [22], *S100A4* (S100 calcium binding protein A4) in colon cancer [23], *14-3-3σ* in pancreas cancer [24], and *HPV16* (human papillomavirus 16) in cervical cancer [25].

6.2.4 ONCOGENE ADDICTION

The term "oncogene addiction" was proposed to describe a phenomenon in which tumor cells become conditioned on a certain oncogenic pathway for their continuous proliferation and survival until the accumulation of multiple genetic alterations [26]. Even though the mechanisms for oncogene addiction are uncertain, it appears that epigenetic abnormalities let preneoplastic cells addicted to a particular oncogenic pathway during the early phase of tumor development. The addiction to these pathways for cell proliferation or survival can also facilitate preneoplastic cells to obtain genetic mutations in the same pathway, providing the cell with selective conditions that promote tumor progression [5].

In the early phase of preinvasive colonic lesions that have a risk of progressing to colon cancer, aberrant epigenetic phenomena might induce abnormal activation of the Wnt pathway, which plays an important role in colonic carcinogenesis until the development of mutations in this pathway. In normal colon epithelial cells, secreted frizzled-related proteins (SFRPs) are antagonists of Wnt signaling by competing with Wnt proteins for binding to their receptor, frizzled (FRZ) (Figure 6.3) [5]. If *SFRPs* are repressed through promoter DNA methylation [27], Wnt signaling inactivates the adenomatosis polyposis coli (APC) complex that phosphorylates β-catenin and leads to its degradation, allowing β-catenin to accumulate in the cytoplasm and finally in the nucleus. In the nucleus, accumulated β-catenin activates transcription of genes such as *MYC, cyclin D*, and other genes that promote cell proliferation. Consequently, this results in the expansion of colonic epithelial stem cells and progenitor cells as well as the formation of aberrant crypt foci (ACF) that contain premalignant hyperplastic cells. Thus, the epigenetic silencing can enhance Wnt signaling pathway, which may addict cells in this pathway toward later mutations in the downstream genes such as *APC* and β-*catenin*, which can fully activate Wnt signaling and accelerate colonic tumorigenesis. Sequential epigenetic and genetic inactivation of this pathway represents an example of oncogene addiction.

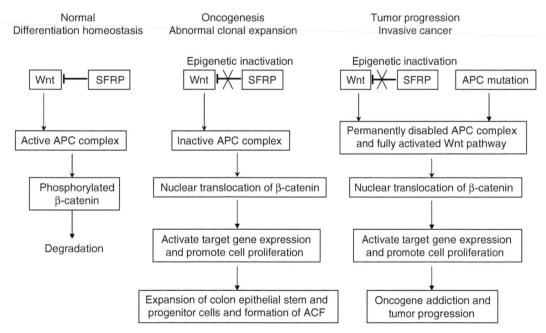

FIGURE 6.3 Inactivation of Wnt pathway for carcinogenesis. SFRPs, antagonists of Wnt proteins for binding to their receptor, activate the APC complex to phosphorylate β-catenin, which leads to its degradation and subsequently inhibits abnormal proliferation of colonic epithelial cells and maintains normal differentiation. In the event of epigenetic silencing of *SFRPs*, activated Wnt signaling inactivates the APC complex, allowing β-catenin to accumulate in the cytoplasm and finally in the nucleus. In the nucleus, β-catenin activates the transcription of genes that promote cell proliferation, resulting in the expansion of colonic epithelial stem and progenitor cells and formation of ACF. Ultimately, persistent epigenetic activation of the Wnt pathway may cause mutations in *APC* or β-*catenin* and these epigenetic and genetic changes fully activate the Wnt pathway to promote cancer progression. (From Baylin, S.B. and Ohm, J.E., *Nat. Rev. Cancer*, 6, 107, 2006.)

It appears that epigenetic silencing of *HIC1* (hypermethylated in cancer 1) gene is also involved in the early phase of tumor progression and leads to the oncogenic addiction pathway. *HIC1* is regarded as a tumor suppressor gene located at chromosome 17p13.3, a region which is frequently deleted in several human cancers and hypermethylated in the preinvasive stage breast [28] and colon tumors [29]. HIC1 actually impairs tumor suppressor activity of *p53*. *HIC1* heterozygote mice can develop late onset tumors in which the wild-type allele is hypermethylated [30]. *HIC1* represses transcription of *SIRT1* (sirtuin 1), a member of the histone deacetylase family of sirtuins, and one of the deacetylase targets of SIRT1 is *p53*; epigenetic silencing of *HIC1* increases SIRT1 levels, resulting in deacetylation of *p53* and reduces its transcriptional activity [31]. In *HIC1* deficient normal and neoplastic cells, this decreased *p53* function results in a reduced apoptotic response to DNA damage, while derepression of *HIC1* restores the apoptotic response of the cell in a *p53*-dependent manner [31].

In summary, epigenetic silencing occurs during the early phase of tumor progression and possibly during tumor progression. Epigenetic processes, which can silence key regulatory genes, participate in genetic aberrations to cause dysregulation of critical genes [6] by permitting pre-neoplastic cells addicted to particular oncogenic pathways.

6.2.5 Interrelationship between DNA Methylation and Histone Modifications

During tumor progression, the rate of epigenetic gene repression increases in parallel with the density of CpG island methylation (Figure 6.1) [3]. This progressive silencing by DNA methylation is closely associated with other epigenetic mechanisms, especially modifications of histone tail amino acid residues, which form transcriptionally repressive chromatin assembles at promoters. In cancer cells, patterns of such modifications are different among genes that are abnormally repressed or transcriptionally active.

Reduced methylation at histone H3 lysine 4 (H3-K4) and increased methylation at histone H3 lysine 9 (H3-K9) play a critical role in the maintenance of promoter DNA methylation-associated gene silencing in colorectal cancer [32]. Trimethylation at histone H4 lysine 20 (H4-K20) is a marker of constitutive heterochromatin, gene silencing, and aging, while loss of this trimethylation is associated with the hypomethylation of repetitive sequences of DNA [33]. Recent studies have shown that the histone methyltransferases that catalyze these histone modifications recruit DNA methyltransferases to gene promoters. Therefore, it is suggested that histone methylation occurs during the initial phases of gene silencing in carcinogenesis and DNA methylation spreads over the promoter [34]. In these repressive chromatin events, DNA methylation seems to play a more dominant role than histone modifications in epigenetic gene silencing. Pharmacological inhibition of DNA methylation can reestablish gene transcription with the appearance of transcriptional activation marks, acetylation at K9 and K14 and methylation K4 of H3, and the loss of the silencing mark, methylation at H3-K9 [35,36].

In a cultured cell line, overexpressing a point mutant (lysine to arginine, H3-K27R) resulted in global reduction of both H3-K27 methylation and DNA methylation as well as increased expression of tumor suppressor genes, indicating an association between H3K27 methylation and DNA methylation on gene expression [37]. However, in the human hepatocellular carcinoma study, H3K27 methylation and DNA methylation were differently associated with expression of critical genes [38].

DNA methylation and histone acetylation are also dynamically linked in the epigenetic control of gene expression in tumorigenesis. In general, acetylated histones are associated with unmethylated DNA and are almost absent in methylated DNA regions [39]. DNA methyltransferase 1 is involved in maintaining histone acetylation status [40] and in vitro experiments have shown that DNA methyltransferase inhibitors and histone deacetylase inhibitors are highly synergistic in the re-induction of silenced genes in cancer [41,42].

6.3 MUTATION INITIATED BY DNA METHYLATION

6.3.1 MUTATION INDUCED BY INACTIVATION OF REPAIR GENES THROUGH PROMOTER METHYLATION

Promoter methylation of *MGMT* is associated with silencing of this gene in cancers of colon, lung, and lymphoid organs [3] as well as the loss of MGMT protein [43], mRNA expression [44], and enzyme activity [45]. MGMT is a DNA repair protein that removes mutagenic and cytotoxic adducts from O^6-guanine in DNA [46], which can result in $G \rightarrow A$ transition mutations. The promoter methylation-associated silencing of *MGMT* that occurs early in human tumorigenesis, such as in small colonic adenomas, appears to be predisposed to mutations; G:C to A:T transitions in critical genes, such as *K-ras* [47] and *p53* [48]. *K-ras* mutation is the most common genetic alteration in human cancer [49] and the distribution of this mutation is similar to the pattern of *MGMT* promoter hypermethylation. Aberrant methylation of *MGMT* occurs in 40% of cases of colorectal carcinoma [44] where *K-ras* mutations are frequent [50], while it is uncommon in breast carcinoma where *K-ras* mutations are extremely rare. The *p53* is one of the most frequently mutated tumor suppressor genes in human cancer and transition mutation is the most common in *p53* mutation. Approximately 52% of the mutational events are missense transitional changes and among them 72% are G:C to A:T transitions [51], which are thought to be associated with the epigenetic silencing of *MGMT*.

The DNA mismatch repair system recognizes and corrects base-pair errors in newly replicated DNA and impairment in this system results in a 100-fold higher mutation rate compared with normal cells [52]. These mutations are particularly evident in microsatellite sequences (short repeated sequences of DNA) and the microsatellite instability is a common finding in mismatch repair gene-deficient cancers. In microsatellite instability-positive cancers, mismatch repair genes are mutated and inactivated, whereas the mismatch repair gene *MLH1* (mutL homologue 1) is commonly hypermethylated in sporadic tumors that have microsatellite instability without mutations in mismatch repair genes. Interestingly, this promoter hypermethylation in *MLH1* can be found in normal-looking colonic epithelium in patients who harbor microsatellite instability-positive colorectal cancers, indicating that this epigenetic silencing of mismatch repair gene precedes the microsatellite instability in cancer [50].

Germline mutations in one allele of the *BRCA1* (breast cancer 1, early onset) gene increase the risk of early onset breast cancer. Since *BRCA1* is crucial for double-strand break repair through the potentially error-free pathway of homologous recombination, survival of *BRCA1* deficient cells seems to be dependent on less accurate checkpoint mechanisms. In such a case tumorigenic traits could be obtained through genomic rearrangements and gene mutations; the association between *BRCA1* deficient cells and *p53* mutations has been reported by Greenblatt et al. [53]. In fact, familial *BRCA1* tumors are associated with *p53* mutations. In contrast to familial *BRCA1* tumors, 9% of sporadic breast cancer has *BRCA1* methylation and substantially elevated frequency of *p53* mutations is found in the subset of *BRCA1* methylated tumors, suggesting that *BRCA1* methylation might lead to alterations in the same molecular pathway, which is commonly found in familial *BRCA1* tumors [54].

6.3.2 MUTATION BY SPONTANEOUS DEAMINATION AT METHYLCYTOSINE IN TUMOR SUPPRESSOR GENES

Five-methylcytosine by itself can be mutagenic through the spontaneous hydrolytic deamination of methylated cytosine, which results in C to T transition mutations at methylated CpG sites [55]. Even though deamination of 5-methylcytosine has been considered one mechanism for the increased transition rate (C to T at CpG sites) at *p53* mutational hotspots in human cancers, the spontaneous deamination model hardly explains the whole process of mutagenesis at CpG sequences in mammalian cells, due to the low deamination rate; only two 5-methylcytosines undergo deamination per day in each cell [56]. This number is much lower than numbers of endogenous or exogenous DNA adducts, which can be between hundreds and several hundred thousands per cell [57]. It suggests the presence of certain unknown endogenous processes that promote the deamination of 5-methylcytosine at CpG sites, for example, methylated CpG-specific base modification by endogenous mutagens [58].

Methylated CpG sequences can also be targets for specific exogenous mutagens. Methyl groups at the CpG residues in the *p53* coding region is highly associated with increased rates of mutation induced by ultraviolet light during the development of skin cancers [59], in which more than 30% of *p53* mutations are C to T and CC to TT transitions and found at mutational hotpots [60]. The methylated CpG residue may serve as a binding site for benzo(a)pyrene diol epoxide and other carcinogens that are found in tobacco smoke, resulting in DNA adducts and subsequent G to T transversion mutations, which are often found in the aerodigestive tumors of smokers [61].

6.3.3 LOSS OF HETEROZYGOSITY AND PROMOTER METHYLATION OF TUMOR SUPPRESSOR GENES

Epigenetically repressed or genetically disrupted tumor suppressor genes have commonly been found in genomic regions that are characterized by chromosomal deletions, causing loss of heterozygosity [3]. To date, direct evidences for the cause and effect relationship between DNA methylation and loss of heterozygosity are few. However, the observation that aberrant DNA methylation precedes loss of heterozygosity on chromosome 16 in noncancerous liver tissues of patients with hepatocellular carcinoma suggests that aberrant DNA methylation might play a certain role in the precancerous stage of hepatocarcinogenesis by preceding, or perhaps causing, loss of heterozygosity [62].

The *RASSF1A* (Ras association domain family protein 1, isoform A) gene is located at chromosome 3p21.3 and found to be inactivated in major human cancers. Hypermethylation of *RASSF1A* is found in a subset of cervical cancers, in which concomitant loss of heterozygosity at 3p21 is common [63]. However, in a human study which compared cervical squamous cell carcinoma with corresponding noncancerous tissues, there was no correlation between promoter methylation status and loss of heterozygosity of the chromosome 3p genes, *RASSF1A*, von Hippel-Lindau disease (*VHL*), retinoic acid receptor β (*RAR-β*), and fragile histidine triad (*FHIT*) [64].

6.4 EPIGENETIC PLASTICITY AND CANCER PROGRESSION

Cancer is a heterogeneous disease with plasticity and alterations of epigenetic status might be responsible for this heterogeneity during the neoplastic process and the ability to evolve its phenotype [7].

As embryonic stem cells differentiate into all cell types during embryogenesis, adult stem cells in different tissues can undergo multipotent differentiation to various cell types of these tissues in specific situations. Epigenetic mechanisms control differentiation of both embryonic and adult stem cells and the plasticity of differentiation in these cells is associated with transcription accessibility to genes expressed in other normal tissues. The plasticity potential of malignant cells has been studied in epithelial tumors [65,66] and it appears that cancer stem cells also have a plasticity of differentiation similar to normal embryonic and adult stem cells.

Cancer is maintained by self-renewing cancer stem cells that can be derived from normal stem cells or more differentiated cells that are reprogrammed to express the properties of stem cells [67] through epigenetic modifications of their DNA or chromatin [68]. Broad plasticity in cancer stem cells might also be associated with acquiring a new phenotype of invasion or migration for cancer progression. Teratocarcinoma develops from defective germ cells in testis and embryonal carcinoma cells from such teratocarcinomas can form normally appearing embryoid bodies in vitro and differentiate to cells of all three germ layers like normal embryonic stem cells. Embryonal cancer cell lines can also differentiate into neuron, skeletal muscle, and cardiac muscle when cultured in the presence of different inducers. Cancer stem cells from leukemia, neuroblastoma, colon carcinoma, and hepatocellular carcinoma can differentiate into other cell types, which are normally found in normal blood, brain, colon, and liver, respectively [67].

When a cell adopts a new gene expression pattern through epigenetic changes, the cell reserves the ability to switch back to previous gene expression patterns and phenotypic plasticity by epigenetic mechanisms can provide cancer cells the ability to alter gene expression and appropriately adapt to its environment during cancer progression [69].

6.5   SPREADING OF ABERRANT METHYLATION IN CANCER PROGRESSION

In 1928, Heitz first distinguished heterochromatin from euchromatin on the basis of different characteristics at interphase [70]. In general euchromatin is less dense, hypomethylated, and easily transcribed, whereas heterochromatin is highly condensed, hypermethylated, and not transcribed. One crucial feature of heterochromatin is its ability to propagate and thereby influence expression of nearby genes. If heterochromatin spreads across the heterochromatin domain boundary, which has been implicated as borders between heterochromatin and nearby euchromatin domains [71], it causes epigenetic repression of contiguous sequences. However, in normal conditions the spreading ability of heterochromatin is limited to prevent inappropriate encroachment into adjacent euchromatin.

Even though exact underlying mechanisms are not yet known, for a certain gene the heterochromatin spread seems to be initiated during the course of aging, the single most important risk factor for cancer. The promoter of the estrogen receptor gene in the colon is unmethylated in young individuals, partially methylated in the elderly, and fully methylated in cancer patients [72]. In an animal study, old mice have significantly increased $p16$ promoter DNA methylation compared with young mice, which have invariable unmethylated $p16$ promoter [73]. This age-dependent progressive promoter hypermethylation as well as widespread de novo methylation of CpG islands in cancer might be explained by the breakdown of DNA boundary elements to allow the spread of methylated heterochromatin to unmethylated euchromatin. When the aberrantly silenced genes were demethylated and derepressed by 5-aza-2'-deoxycytidine treatment in colon cancer cell lines, their promoters remained in a condition of heterochromatic characteristics [74]. This observation supports the hypothesis that the spread of methylated heterochromatin causes the widespread promoter hypermethylation in cancer tissue.

6.6   METASTASIS AND DNA METHYLATION

One of the most complicated problems during cancer progression is the acquisition of a metastatic nature, which is associated with the high mortality rates in cancer by the metastatic spread of tumor cells from the original site. It emerges that this tumor phenotype is due to an acquisition of new traits within the tumor through clonal evolution and selection. This type of tumor heterogeneity and progression might be explained with epigenetic mechanisms [7].

Cavalli et al. [75] analyzed promoter methylation status of *TP16*, *THBS2*, *E-Cadherin* (*ECAD*), *RARβ2*, *MINT1*, *MINT2*, and *MINT31* in six paired primary breast tumors and their matched sentinel lymph nodes. Between the primary tumors and lymph nodes 29% (8/42) of the methylation measurements were different. In a tumor and matched sentinel lymph node set, the primary tumor showed four hypermethylated genes, while the sentinel lymph nodes demonstrated loss of all four hypermethylation. This observation suggested a possible association between early regional metastasis and heterogeneous promoter methylation status of critical genes.

Malignant cells with metastatic potential exist as a subpopulation among the heterogeneous cell population in the primary tumor [3]. Metastasized cells create a new cellular heterogeneity of the metastatic subpopulation at the distant site. Thus, epigenetically mediated gene silencing could be a better explanation for metastatic dynamics than mutation. The heterogeneous promoter status of

CDH1 (E-cadherin gene), a gene that is related to cell invasion, provides good evidence. The loss of *CDH1* function facilitates tumor cells to acquire the invasive properties of metastatic tumor cells and cell heterogeneity by the hypermethylated promoter of this gene can play a major role in different *CDH1* expression in both primary and metastatic breast cancer [76]. *CST6*, a breast tumor suppressor gene that is expressed in normal breast epithelium, can be silenced by promoter DNA hypermethylation in metastatic breast cancer cell lines. In a breast cancer and matched lymph node metastases study, methylation-dependent silencing of *CST6* occurs more frequently in metastatic lesions [77]. In pancreas cancer, expression of *E-cadherin* and *DAP kinase*, which are implicated as invasion suppressor genes, is more frequently reduced in lymph node metastases than in the primary tumor and methylation of the promoter region is associated with this reduction [78].

6.7   TUMOR MICROENVIRONMENT AND CANCER PROGRESSION

Interactions between tumor cells and the microenvironment can play a certain role in cancer progression. The microenvironment consists of stromal cells such as fibroblasts, immune cells, inflammatory cells, adipocytes, glial cells, smooth muscle cells, and vascular cells. It also includes the extracellular matrix, growth factors, and cytokines [79]. During cancer progression, malignant cells obtain an ability to overcome the microenvironmental control by the host, to invade the surrounding tissue, and to reach distant sites from the primary sites [80]. Recent evidence suggests that prostate epithelium and stroma interact in an organ-specific and androgen-dependent manner and this interaction drives reciprocal stromal reactions to prostate tumor epithelium, which further drives tumor epithelium to develop malignant properties [81].

Even though increasing evidence suggests that changes in the cellular environment contribute to tumorigenesis, the molecular basis of these alterations is not yet identified. Even though epigenetic modifications of the neoplastic cells in tumors have been firmly implicated in tumorigenesis, it is not well established whether epigenetic modifications occur in the nonneoplastic stromal cells. However, in a previous breast cancer study, distinct alterations in promoter DNA methylation occur in epithelial and myoepithelial cells and stromal fibroblasts during breast tumorigenesis in a tumor stage-specific and cell type-specific manner, suggesting an epigenetic role in the maintenance of the abnormal cellular microenvironment in breast cancer [82].

When the methylation status of *GSTP1*, *RARβ2*, and *CD44*, three important genes in prostate carcinogenesis, was analyzed from prostatectomy specimens, high levels of gene methylation in the tumor epithelium and stromal cells were found but no methylation was detected in both normal epithelium and stromal cells in normal-appearing tissues located in close proximity to tumors.

TABLE 6.1
Mechanistic Role of DNA Methylation in Cancer Progression

Cancer Progression	Mechanism
Early neoplastic progression	Epigenetic gatekeeper gene silencing [6,7]
	Activation of normally silenced allele by loss of imprinting [7,9]
	Activation of oncogene and chromosomal instability [13–15]
	Oncogene addiction [5,6,31]
	Interactions with histone modifications [32,33]
Mutation	Inactivation of repair gene [47,48]
	Spontaneous deamination at methylcytosine residue [55,59,60]
	Association with loss of heterozygosity [3,62]
Acquiring a new phenotype	Epigenetic plasticity [7,67]
	Spreading of aberrant methylation [7,71]
Tumor invasion and metastasis	Epigenetic plasticity [7]
	Tumor microenvironment [79,82]

This observation indicates that the stromal milieu to create a tumor-prone microenvironment might partially be due to gene methylation changes that occur in tumor-associated stromal cells [83].

Collectively, the interactions between tumors and their microenvironments are dynamic and altered DNA methylation in the tumor and its stroma coevolves during tumor initiation and progression.

6.8 CONCLUSION

The role of DNA methylation in cancer progression is as important as that in cancer initiation, because DNA methylation is involved in early cancer progression and could possibly lead to the occurrence of genetic changes such as mutations and loss of heterozygosity. Even though evidence is limited, it appears that DNA methylation affects the entire process of cancer progression through alterations in critical gene expression, chromosomal instability, and interactions with other epigenetic phenomena (Table 6.1).

Since DNA methylation is not simply a consequence of carcinogenesis but a critical mechanism for the initiation and progression of cancer, this epigenetic phenomenon should be a main target for the prevention and treatment of cancer.

REFERENCES

1. Goelz, S.E., Vogelstein, B., Hamilton, S.R., and Feinberg, A.P., Hypomethylation of DNA from benign and malignant human colon neoplasms, *Science* 228(4696), 187–190, 1985.
2. Pogribny, I.P., Miller, B.J., and James, S.J., Alterations in hepatic p53 gene methylation patterns during tumor progression with folate/methyl deficiency in the rat, *Cancer Lett* 115(1), 31–38, 1997.
3. Jones, P.A. and Baylin, S.B., The fundamental role of epigenetic events in cancer, *Nat Rev Genet* 3(6), 415–428, 2002.
4. Nephew, K.P. and Huang, T.H., Epigenetic gene silencing in cancer initiation and progression, *Cancer Lett* 190(2), 125–133, 2003.
5. Baylin, S.B. and Ohm, J.E., Epigenetic gene silencing in cancer—a mechanism for early oncogenic pathway addiction? *Nat Rev Cancer* 6(2), 107–116, 2006.
6. Jones, P.A. and Baylin, S.B., The epigenomics of cancer, *Cell* 128(4), 683–692, 2007.
7. Feinberg, A.P., Ohlsson, R., and Henikoff, S., The epigenetic progenitor origin of human cancer, *Nat Rev Genet* 7(1), 21–33, 2006.
8. Rocco, J.W. and Sidransky, D., p16(MTS-1/CDKN2/INK4a) in cancer progression, *Exp Cell Res* 264(1), 42–55, 2001.
9. Sakatani, T., Kaneda, A., Iacobuzio-Donahue, C.A., Carter, M.G., de Boom Witzel, S., Okano, H., Ko, M.S., Ohlsson, R., Longo, D.L., and Feinberg, A.P., Loss of imprinting of Igf2 alters intestinal maturation and tumorigenesis in mice, *Science* 307(5717), 1976–1978, 2005.
10. Feinberg, A.P. and Tycko, B., The history of cancer epigenetics, *Nat Rev Cancer* 4(2), 143–153, 2004.
11. Randhawa, G.S., Cui, H., Barletta, J.A., Strichman-Almashanu, L.Z., Talpaz, M., Kantarjian, H., Deisseroth, A.B., Champlin, R.C., and Feinberg, A.P., Loss of imprinting in disease progression in chronic myelogenous leukemia, *Blood* 91(9), 3144–3147, 1998.
12. Hassan, A.B. and Howell, J.A., Insulin-like growth factor II supply modifies growth of intestinal adenoma in Apc(Min/+) mice, *Cancer Res* 60(4), 1070–1076, 2000.
13. Fan, T., Yan, Q., Huang, J., Austin, S., Cho, E., Ferris, D., and Muegge, K., Lsh-deficient murine embryonal fibroblasts show reduced proliferation with signs of abnormal mitosis, *Cancer Res* 63(15), 4677–4683, 2003.
14. Gaudet, F., Hodgson, J.G., Eden, A., Jackson-Grusby, L., Dausman, J., Gray, J.W., Leonhardt, H., and Jaenisch, R., Induction of tumors in mice by genomic hypomethylation, *Science* 300(5618), 489–492, 2003.
15. Rodriguez, J., Frigola, J., Vendrell, E., Risques, R.A., Fraga, M.F., Morales, C., Moreno, V., Esteller, M., Capella, G., Ribas, M., and Peinado, M.A., Chromosomal instability correlates with genome-wide DNA demethylation in human primary colorectal cancers, *Cancer Res* 66(17), 8462–9468, 2006.

16. Suter, C.M., Martin, D.I., and Ward, R.L., Hypomethylation of L1 retrotransposons in colorectal cancer and adjacent normal tissue, *Int J Colorectal Dis* 19(2), 95–101, 2004.

17. Strichman-Almashanu, L.Z., Lee, R.S., Onyango, P.O., Perlman, E., Flam, F., Frieman, M.B., and Feinberg, A.P., A genome-wide screen for normally methylated human CpG islands that can identify novel imprinted genes, *Genome Res* 12(4), 543–554, 2002.

18. Feinberg, A.P. and Vogelstein, B., Hypomethylation of ras oncogenes in primary human cancers, *Biochem Biophys Res Commun* 111(1), 47–54, 1983.

19. Simpson, A.J., Caballero, O.L., Jungbluth, A., Chen, Y.T., and Old, L.J., Cancer/testis antigens, gametogenesis and cancer, *Nat Rev Cancer* 5(8), 615–625, 2005.

20. Oshimo, Y., Nakayama, H., Ito, R., Kitadai, Y., Yoshida, K., Chayama, K., and Yasui, W., Promoter methylation of cyclin D2 gene in gastric carcinoma, *Int J Oncol* 23(6), 1663–1670, 2003.

21. Akiyama, Y., Maesawa, C., Ogasawara, S., Terashima, M., and Masuda, T., Cell-type-specific repression of the maspin gene is disrupted frequently by demethylation at the promoter region in gastric intestinal metaplasia and cancer cells, *Am J Pathol* 163(5), 1911–1919, 2003.

22. Cho, M., Uemura, H., Kim, S.C., Kawada, Y., Yoshida, K., Hirao, Y., Konishi, N., Saga, S., and Yoshikawa, K., Hypomethylation of the MN/CA9 promoter and upregulated MN/CA9 expression in human renal cell carcinoma, *Br J Cancer* 85(4), 563–567, 2001.

23. Nakamura, N. and Takenaga, K., Hypomethylation of the metastasis-associated S100A4 gene correlates with gene activation in human colon adenocarcinoma cell lines, *Clin Exp Metastasis* 16(5), 471–479, 1998.

24. Sato, N., Maitra, A., Fukushima, N., van Heek, N.T., Matsubayashi, H., Iacobuzio-Donahue, C.A., Rosty, C., and Goggins, M., Frequent hypomethylation of multiple genes overexpressed in pancreatic ductal adenocarcinoma, *Cancer Res* 63(14), 4158–4166, 2003.

25. Badal, V., Chuang, L.S., Tan, E.H., Badal, S., Villa, L.L., Wheeler, C.M., Li, B.F., and Bernard, H.U., CpG methylation of human papillomavirus type 16 DNA in cervical cancer cell lines and in clinical specimens: Genomic hypomethylation correlates with carcinogenic progression, *J Virol* 77(11), 6227–6234, 2003.

26. Weinstein, I.B., Cancer. Addiction to oncogenes—the Achilles heal of cancer, *Science* 297(5578), 63–64, 2002.

27. Nojima, M., Suzuki, H., Toyota, M., Watanabe, Y., Maruyama, R., Sasaki, S., Sasaki, Y., Mita, H., Nishikawa, N., Yamaguchi, K., Hirata, K., Itoh, F., Tokino, T., Mori, M., Imai, K., and Shinomura, Y., Frequent epigenetic inactivation of SFRP genes and constitutive activation of Wnt signaling in gastric cancer, *Oncogene* 26(32), 4699–4713, 2007.

28. Fujii, H., Biel, M.A., Zhou, W., Weitzman, S.A., Baylin, S.B., and Gabrielson, E., Methylation of the HIC-1 candidate tumor suppressor gene in human breast cancer, *Oncogene* 16(16), 2159–2164, 1998.

29. Wales, M.M., Biel, M.A., el Deiry, W., Nelkin, B.D., Issa, J.P., Cavenee, W.K., Kuerbitz, S.J., and Baylin, S.B., p53 activates expression of HIC-1, a new candidate tumour suppressor gene on 17p13.3, *Nat Med* 1(6), 570–577, 1995.

30. Chen, W.Y., Zeng, X., Carter, M.G., Morrell, C.N., Chiu Yen, R.W., Esteller, M., Watkins, D.N., Herman, J.G., Mankowski, J.L., and Baylin, S.B., Heterozygous disruption of Hic1 predisposes mice to a gender-dependent spectrum of malignant tumors, *Nat Genet* 33(2), 197–202, 2003.

31. Chen, W.Y., Wang, D.H., Yen, R.C., Luo, J., Gu, W., and Baylin, S.B., Tumor suppressor HIC1 directly regulates SIRT1 to modulate p53-dependent DNA-damage responses, *Cell* 123(3), 437–448, 2005.

32. Kondo, Y., Shen, L., and Issa, J.P., Critical role of histone methylation in tumor suppressor gene silencing in colorectal cancer, *Mol Cell Biol* 23(1), 206–215, 2003.

33. Fraga, M.F., Ballestar, E., Villar-Garea, A., Boix-Chornet, M., Espada, J., Schotta, G., Bonaldi, T., Haydon, C., Ropero, S., Petrie, K., Iyer, N.G., Perez-Rosado, A., Calvo, E., Lopez, J.A., Cano, A., Calasanz, M.J., Colomer, D., Piris, M.A., Ahn, N., Imhof, A., Caldas, C., Jenuwein, T., and Esteller, M., Loss of acetylation at Lys16 and trimethylation at Lys20 of histone H4 is a common hallmark of human cancer, *Nat Genet* 37(4), 391–400, 2005.

34. Bachman, K.E., Park, B.H., Rhee, I., Rajagopalan, H., Herman, J.G., Baylin, S.B., Kinzler, K.W., and Vogelstein, B., Histone modifications and silencing prior to DNA methylation of a tumor suppressor gene, *Cancer Cell* 3(1), 89–95, 2003.

35. Fahrner, J.A., Eguchi, S., Herman, J.G., and Baylin, S.B., Dependence of histone modifications and gene expression on DNA hypermethylation in cancer, *Cancer Res* 62(24), 7213–7218, 2002.

36. Nguyen, C.T., Weisenberger, D.J., Velicescu, M., Gonzales, F.A., Lin, J.C., Liang, G., and Jones, P.A., Histone H3-lysine 9 methylation is associated with aberrant gene silencing in cancer cells and is rapidly reversed by 5-aza-2'-deoxycytidine, *Cancer Res* 62(22), 6456–6461, 2002.

37. Abbosh, P.H., Montgomery, J.S., Starkey, J.A., Novotny, M., Zuhowski, E.G., Egorin, M.J., Moseman, A.P., Golas, A., Brannon, K.M., Balch, C., Huang, T.H., and Nephew, K.P., Dominant-negative histone H3 lysine 27 mutant derepresses silenced tumor suppressor genes and reverses the drug-resistant phenotype in cancer cells, *Cancer Res* 66(11), 5582–5591, 2006.

38. Kondo, Y., Shen, L., Suzuki, S., Kurokawa, T., Masuko, K., Tanaka, Y., Kato, H., Mizuno, Y., Yokoe, M., Sugauchi, F., Hirashima, N., Orito, E., Osada, H., Ueda, R., Guo, Y., Chen, X., Issa, J.P., and Sekido, Y., Alterations of DNA methylation and histone modifications contribute to gene silencing in hepatocellular carcinomas, *Hepatol Res* 37(11), 974–983, 2007.

39. Irvine, R.A., Lin, I.G., and Hsieh, C.L., DNA methylation has a local effect on transcription and histone acetylation, *Mol Cell Biol* 22(19), 6689–6696, 2002.

40. Jones, P.L. and Wolffe, A.P., Relationships between chromatin organization and DNA methylation in determining gene expression, *Semin Cancer Biol* 9(5), 339–347, 1999.

41. Jahangeer, S., Elliott, R.M., and Henneberry, R.C., Beta-adrenergic receptor induction in HeLa cells: Synergistic effect of 5-azacytidine and butyrate, *Biochem Biophys Res Commun* 108(4), 1434–1440, 1982.

42. Cameron, E.E., Bachman, K.E., Myohanen, S., Herman, J.G., and Baylin, S.B., Synergy of demethylation and histone deacetylase inhibition in the re-expression of genes silenced in cancer, *Nat Genet* 21(1), 103–107, 1999.

43. Esteller, M., Sanchez-Cespedes, M., Rosell, R., Sidransky, D., Baylin, S.B., and Herman, J.G., Detection of aberrant promoter hypermethylation of tumor suppressor genes in serum DNA from non-small cell lung cancer patients, *Cancer Res* 59(1), 67–70, 1999.

44. Esteller, M., Hamilton, S.R., Burger, P.C., Baylin, S.B., and Herman, J.G., Inactivation of the DNA repair gene O6-methylguanine-DNA methyltransferase by promoter hypermethylation is a common event in primary human neoplasia, *Cancer Res* 59(4), 793–797, 1999.

45. Herfarth, K.K., Brent, T.P., Danam, R.P., Remack, J.S., Kodner, I.J., Wells, S.A., Jr., and Goodfellow, P.J., A specific CpG methylation pattern of the MGMT promoter region associated with reduced MGMT expression in primary colorectal cancers, *Mol Carcinog* 24(2), 90–98, 1999.

46. Pegg, A.E., Mammalian O6-alkylguanine-DNA alkyltransferase: Regulation and importance in response to alkylating carcinogenic and therapeutic agents, *Cancer Res* 50(19), 6119–6129, 1990.

47. Esteller, M., Toyota, M., Sanchez-Cespedes, M., Capella, G., Peinado, M.A., Watkins, D.N., Issa, J.P., Sidransky, D., Baylin, S.B., and Herman, J.G., Inactivation of the DNA repair gene O6-methylguanine-DNA methyltransferase by promoter hypermethylation is associated with G to A mutations in K-ras in colorectal tumorigenesis, *Cancer Res* 60(9), 2368–2371, 2000.

48. Esteller, M., Risques, R.A., Toyota, M., Capella, G., Moreno, V., Peinado, M.A., Baylin, S.B., and Herman, J.G., Promoter hypermethylation of the DNA repair gene O(6)-methylguanine-DNA methyltransferase is associated with the presence of G:C to A:T transition mutations in p53 in human colorectal tumorigenesis, *Cancer Res* 61(12), 4689–4692, 2001.

49. Barbacid, M., ras genes, *Annu Rev Biochem* 56, 779–827, 1987.

50. Jacinto, F.V. and Esteller, M., Mutator pathways unleashed by epigenetic silencing in human cancer, *Mutagenesis* 22(4), 247–253, 2007.

51. Greenblatt, M.S., Bennett, W.P., Hollstein, M., and Harris, C.C., Mutations in the p53 tumor suppressor gene: Clues to cancer etiology and molecular pathogenesis, *Cancer Res* 54(18), 4855–4878, 1994.

52. Jiricny, J., The multifaceted mismatch-repair system, *Nat Rev Mol Cell Biol* 7(5), 335–346, 2006.

53. Greenblatt, M.S., Chappuis, P.O., Bond, J.P., Hamel, N., and Foulkes, W.D., TP53 mutations in breast cancer associated with BRCA1 or BRCA2 germ-line mutations: Distinctive spectrum and structural distribution, *Cancer Res* 61(10), 4092–4097, 2001.

54. Birgisdottir, V., Stefansson, O.A., Bodvarsdottir, S.K., Hilmarsdottir, H., Jonasson, J.G., and Eyfjord, J.E., Epigenetic silencing and deletion of the BRCA1 gene in sporadic breast cancer, *Breast Cancer Res* 8(4), R38, 2006.

55. Pfeifer, G.P., p53 mutational spectra and the role of methylated CpG sequences, *Mutat Res* 450(1–2), 155–166, 2000.

56. Schmutte, C. and Jones, P.A., Involvement of DNA methylation in human carcinogenesis, *Biol Chem* 379 (4–5), 377–388, 1998.

57. Marnett, L.J. and Burcham, P.C., Endogenous DNA adducts: potential and paradox, *Chem Res Toxicol* 6(6), 771–785, 1993.

58. Pfeifer, G.P., Mutagenesis at methylated CpG sequences, *Curr Top Microbiol Immunol* 301, 259–281, 2006.

59. You, Y.H., Li, C., and Pfeifer, G.P., Involvement of 5-methylcytosine in sunlight-induced mutagenesis, *J Mol Biol* 293(3), 493–503, 1999.

60. Tommasi, S., Denissenko, M.F., and Pfeifer, G.P., Sunlight induces pyrimidine dimers preferentially at 5-methylcytosine bases, *Cancer Res* 57(21), 4727–4730, 1997.

61. Yoon, J.H., Smith, L.E., Feng, Z., Tang, M., Lee, C.S., and Pfeifer, G.P., Methylated CpG dinucleotides are the preferential targets for G-to-T transversion mutations induced by benzo[a]pyrene diol epoxide in mammalian cells: Similarities with the p53 mutation spectrum in smoking-associated lung cancers, *Cancer Res* 61(19), 7110–7117, 2001.

62. Kanai, Y., Ushijima, S., Tsuda, H., Sakamoto, M., and Hirohashi, S., Aberrant DNA methylation precedes loss of heterozygosity on chromosome 16 in chronic hepatitis and liver cirrhosis, *Cancer Lett* 148(1), 73–80, 2000.

63. Yu, M.Y., Tong, J.H., Chan, P.K., Lee, T.L., Chan, M.W., Chan, A.W., Lo, K.W., and To, K.F., Hypermethylation of the tumor suppressor gene RASSFIA and frequent concomitant loss of heterozygosity at 3p21 in cervical cancers, *Int J Cancer* 105(2), 204–209, 2003.

64. Choi, C.H., Lee, K.M., Choi, J.J., Kim, T.J., Kim, W.Y., Lee, J.W., Lee, S.J., Lee, J.H., Bae, D.S., and Kim, B.G., Hypermethylation and loss of heterozygosity of tumor suppressor genes on chromosome 3p in cervical cancer, *Cancer Lett* 255(1), 26–33, 2007.

65. Yaccoby, S., The phenotypic plasticity of myeloma plasma cells as expressed by dedifferentiation into an immature, resilient, and apoptosis-resistant phenotype, *Clin Cancer Res* 11(21), 7599–7606, 2005.

66. Prindull, G. and Zipori, D., Environmental guidance of normal and tumor cell plasticity: Epithelial mesenchymal transitions as a paradigm, *Blood* 103(8), 2892–2899, 2004.

67. Lotem, J. and Sachs, L., Epigenetics and the plasticity of differentiation in normal and cancer stem cells, *Oncogene* 25(59), 7663–7672, 2006.

68. Di Croce, L., Raker, V.A., Corsaro, M., Fazi, F., Fanelli, M., Faretta, M., Fuks, F., Lo Coco, F., Kouzarides, T., Nervi, C., Minucci, S., and Pelicci, P.G., Methyltransferase recruitment and DNA hypermethylation of target promoters by an oncogenic transcription factor, *Science* 295(5557), 1079–1082, 2002.

69. Domann, F.E. and Futscher, B.W., Flipping the epigenetic switch, *Am J Pathol* 164(6), 1883–1886, 2004.

70. Passarge, E., Emil Heitz and the concept of heterochromatin: Longitudinal chromosome differentiation was recognized fifty years ago, *Am J Hum Genet* 31(2), 106–115, 1979.

71. Grewal, S.I. and Jia, S., Heterochromatin revisited, *Nat Rev Genet* 8(1), 35–46, 2007.

72. Issa, J.P., Ottaviano, Y.L., Celano, P., Hamilton, S.R., Davidson, N.E., and Baylin, S.B., Methylation of the oestrogen receptor CpG island links ageing and neoplasia in human colon, *Nat Genet* 7(4), 536–540, 1994.

73. Keyes, M.K., Jang, H., Mason, J.B., Liu, Z., Crott, J.W., Smith, D.E., Friso, S., and Choi, S.W., Old age and dietary folate are determinants of genomic and *p16*-specific DNA methylation in the mouse colon, *J Nutr*, 137(7), 1713–1717, 2007.

74. McGarvey, K.M., Fahrner, J.A., Greene, E., Martens, J., Jenuwein, T., and Baylin, S.B., Silenced tumor suppressor genes reactivated by DNA demethylation do not return to a fully euchromatic chromatin state, *Cancer Res* 66(7), 3541–3549, 2006.

75. Cavalli, L.R., Urban, C.A., Dai, D., de Assis, S., Tavares, D.C., Rone, J.D., Bleggi-Torres, L.F., Lima, R.S., Cavalli, I.J., Issa, J.P., and Haddad, B.R., Genetic and epigenetic alterations in sentinel lymph nodes metastatic lesions compared to their corresponding primary breast tumors, *Cancer Genet Cytogenet* 146(1), 33–40, 2003.

76. Shinozaki, M., Hoon, D.S., Giuliano, A.E., Hansen, N.M., Wang, H.J., Turner, R., and Taback, B., Distinct hypermethylation profile of primary breast cancer is associated with sentinel lymph node metastasis, *Clin Cancer Res* 11(6), 2156–2162, 2005.

77. Rivenbark, A.G., Livasy, C.A., Boyd, C.E., Keppler, D., and Coleman, W.B., Methylation-dependent silencing of CST6 in primary human breast tumors and metastatic lesions, *Exp Mol Pathol* 83(2), 188–197, 2007.

78. Dansranjavin, T., Mobius, C., Tannapfel, A., Bartels, M., Wittekind, C., Hauss, J., and Witzigmann, H., E-cadherin and DAP kinase in pancreatic adenocarcinoma and corresponding lymph node metastases, *Oncol Rep* 15(5), 1125–1131, 2006.
79. Mohla, S., Tumor microenvironment, *J Cell Biochem* 6(9), 1496–1504, 2007.
80. Zigrino, P., Loffek, S., and Mauch, C., Tumor-stroma interactions: Their role in the control of tumor cell invasion, *Biochimie* 87(3–4), 321–328, 2005.
81. Sung, S.Y., Hsieh, C.L., Wu, D., Chung, L.W., and Johnstone, P.A., Tumor microenvironment promotes cancer progression, metastasis, and therapeutic resistance, *Curr Probl Cancer* 31(2), 36–100, 2007.
82. Hu, M., Yao, J., Cai, L., Bachman, K.E., van den Brule, F., Velculescu, V., and Polyak, K., Distinct epigenetic changes in the stromal cells of breast cancers, *Nat Genet* 37(8), 899–905, 2005.
83. Hanson, J.A., Gillespie, J.W., Grover, A., Tangrea, M.A., Chuaqui, R.F., Emmert-Buck, M.R., Tangrea, J.A., Libutti, S.K., Linehan, W.M., and Woodson, K.G., Gene promoter methylation in prostate tumor-associated stromal cells, *J Natl Cancer Inst* 98(4), 255–261, 2006.

Part II

Histone Modifications in Cancer

7 Alterations in Histone Acetylation in Tumorigenesis

Sabita N. Saldanha and Trygve O. Tollefsbol

CONTENTS

7.1   INTRODUCTION

Cellular functions including but not limiting to the production of cell-specific biomolecules, DNA replication and repair, apoptosis, and senescence are intricately and inherently orchestrated by chromosomes [1]. These functions are specified by coding messages in genes. The genetic code, however, is susceptible to alterations, predominantly irreversible mutations, which can result in diseased states at the organismal level. Epigenetic changes involve heritable alterations in gene expression without mutations in the genetic code [2]. Such an outcome is possible as chromatin, the structural component of chromosomes, can undergo changes affecting the condensation states of the chromosome, which allow for or inhibit gene expression [3] (Figure 7.1). These processes are catalyzed by reversible enzymatic modifications. The epigenetics of gene control has gained considerable support and has since been a chosen target for therapeutic intervention for several diseases, in particular cancer.

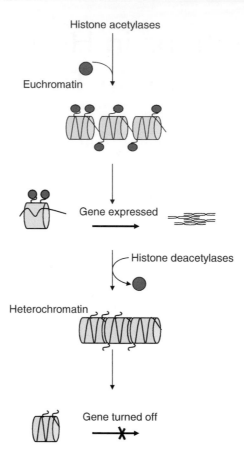

FIGURE 7.1 Histone-mediated gene expression. Histone-mediated chromatin fluxes are depicted in the figure. These changes in histone dynamics are brought about by reversible enzymatic reactions mediated by histone acetyltransferases and histone deacetylases. Addition of an acetyl group changes the charge environment surrounding the DNA, as a result the DNA is freed from the nucleosome core, allowing for the transcriptional complexes to bind and induce transcription. Histone deacetylases on the other hand have apposing effects as the neutral charge induced by the acetyl group is removed, allowing the DNA to complex back with the histones, inhibiting expression.

7.2 NUCLEOSOMES

At the molecular level, the nucleosome constitutes the core of chromatin and consists of basic proteins called histones [2,4]. The 146 bp of DNA complexes with the nucleosome, an octamer, and does so via it is negatively charged phosphate backbone. In the octamer, the dimeric H3 and H4 subunits interact in a tetramer formation, whereas H2A and H2B histones remain as dimers [2,4]. The complex does not include the histone protein H1, but this protein, a linker, appears to facilitate the condensation of the nucleosome units [2,3]. The amino terminal domains of histones protrude out of the complex and are susceptible to enzymatic modification at lysine residues [5–8]. Although predominantly lysine residues are modified, in some cases serine or arginine residues are also altered [8–10].

The altered states largely depend on the enzyme catalyzing the modification and the availability of the residues to the enzymes. The type and position of the modification on the histone dictates a pattern, termed the histone code, which specifies the transcriptional regulation of the complexed DNA [11]. This regulation is based upon the availability and accessibility of transcription factors and other transactivating or repressible elements to the site. Within the same histone subunit, differential alterations can have colossal effects on gene expression under its control. For example, methylation and acetylation of the histone H3 lysine 9 (H3 K9) subunit can have apposing effects on gene expression in conjunction with other histone modifications [8].

7.3 EFFECTS OF NUCLEOSOMAL MODIFICATIONS

There are many modifications known to posttranslationally alter the residues, of which acetylation is the most extensively studied [7,8]. The orientation and modulations of the histones affect the fluidity of the chromatin, resulting in heterochromatin or euchromatin states. In normal cells, the cell cycle is carefully regulated by cell cycle regulatory genes. However, alterations in genes that control tumor suppressors or the cell cycle promote the tumor phenotype. Therefore in abnormal cell growth and division there are three possible outcomes (Figure 7.2): (1) genes that control cell cycle division are turned off, (2) tumor suppressor genes are turned off, or (3) oncogenes are switched on. This can arise by an imbalance in the acetylation/deacetylation states of the histone residues in the

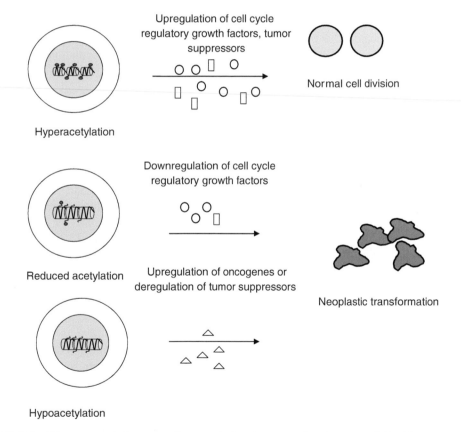

FIGURE 7.2 Histone acetylation toward a neoplastic phenotype. Involvement of the histone acetylation/ deacetylation balance in tumorigenesis. In cells that are normal, genes involved in the cell cycle regulation are expressed and to a certain extent are controlled by histone acetylation of the histones associated with the promoters. [O] represents p21 and the [☐] represents p53; both are tumor suppressors. However, in response to local hypoacetylation and in conjunction with defects in other cellular pathways, the tumor suppressors are turned off and tumor formation may be induced as the cell cycle is deregulated. In some instances hyperacetylation of oncogenes [△], as in the case with HPVE6, occur that can induce a cancer phenotype.

nucleosome, in addition to other factors. This probably explains why most genes crucial to maintaining the cell cycle balance are turned off or deregulated in neoplasms.

7.4   HISTONE ACETYLATION

Histone acetylases, commonly referred to as HATs, recognize the basic substrates and add an acetyl group [8]. The addition of an acetyl group to the lysine residue neutralizes the charge, which relaxes the bound DNA from the histone complex. However, expression of the genes is dependent on the availability of transactivating factors and proteins to the freed up DNA [8]. Moreover it is important which lysine residues are acetylated because at any given time the position of the lysine and the histone subunit involved in the acetylation largely dictates the downstream effects [8].

Histone molecules are maintained in a delicate balance between acetylated and deacetylated states and work in perfect synchronization for optimal gene function. However, deregulation of this posttranslational balance can essentially lead to the upregulation of oncogenes that enable cells to acquire tumorigenic, metastatic, and invasive phenotypes. Re-acetylation probably prevents such an outcome, as genes crucial to cell cycle regulatory functions are upregulated. Attenuation of deacetylation, with an increase in acetylation of histones, has positive consequences in terms of inhibiting neoplastic growth. Increased acetylation tends to slow down cell growth and induce differentiation and apoptosis, as evidenced by a number of studies utilizing compounds that inhibit histone deacetylases in cancer cell lines. The effect is mediated by the upregulation of proteins like p53 by acetylation of their lysine residues [12,13].

7.5   HISTONE ACETYLATION AND CANCER

HATs exhibit substrate preference toward the lysine moieties that they acetylate. For instance, GCN4, a histone acetylase, preferentially acetylates H4 lysine residues at positions 5, 8, 12, and 16 [14,15]. Acetylation of each of these residues has a different effect. Acetylation of H4 K5 and H4 K12 are merely associated with directing the histones to newly formed DNA in the S phase [14,15]. However, H4 K16 enrichment is involved with actively transcribing genes [14,15]. Thus physiological conditions and requirements of the cell most likely dictate the histone marks. Similarly in histone H3, acetylation by GCN5 at positions 9, 14, 16, and 23 is observed. As in the case of H4, in H3 the function of histone deposition to newly synthesized chromatin is mediated by K9 acetylation [14,15]. The remaining acetylated positions are involved with gene expression [14,15]. Therefore mutations in the enzyme that dictate this function can cause mistargeting of acetylation, changes in the acetylation pattern, and deregulation of acetyl enrichment. These events can disrupt the cell cycle balance and create an environment conducive to the development of cellular transformations.

Acetylation events do not act alone in mediating the downstream effects but require the collaborative effort of other pathways, influenced by other histone codes. In some cases the effect of acetylation at a given residue can rescue the function or provide for redundancy for mutated lysine residues. This phenomenon was clearly demonstrated in a study where the substrates of GCN5 were mutated to determine the overall effects on cell growth. Mutations in all the K residues of H4 resulted in accumulation of cells predominantly in the G2 phase [14]. The observed mutated phenotype was rescued when a lysine residue other than the ones mutated was acetylated, providing for the redundancy of function. This clearly indicates that acetylation is required for cell viability. Many neoplastic transformations arise from deregulation in genes that affect particular target genes or several pathways. Some of the genes come under direct epigenetic modifications whereas others are regulated indirectly. We have discussed some of the cancers that have arisen due to changes in histone modifications and the target genes they affect.

7.6   INVOLVEMENT OF HISTONE ACETYLATION IN BREAST CANCER

Aberrant histone acetylation can be one of the mechanisms of many different events that result in cancer of breast tissue. Estradiol, an estrogen steroid, binds to estrogen receptors. This complex associates with the estrogen response element and mediates its downstream effects [16]. Interestingly this association also encourages the binding of protein complexes that posses HAT activity. Breast cancers have been distinguished into two categories, estrogen-dependent and estrogen-independent type [16]. In the estrogen-dependent form, the breast tissue is constantly bathed with the steroid molecule, in part due to the ovulation cycle. However, other mechanisms may also promote estrogen synthesis. Over-expression of the hormone can induce changes in the histone acetylation pattern although there is some debate regarding the type of acetylation patterning that occurs. Some studies indicate that an increase in histone acetylation of H4 by estradiol in breast cancer cells occurs [16] while other studies show a decrease in acetylation [16]. Acetylation of lysine residues can achieve different isoforms—mono, di, tri, and tetra. Therefore, in theory, if the lysine is predominantly mono-acetylated, then histones can be relatively easily deacetylated. However, the higher acetylated isoforms can decrease the rate at which deacetylation occurs. Therefore, the increase in acetylation found in the studies done on breast cancer by estradiol could be mediated by slower deacetylation rates with no change in acetylation rates of higher acetylated isoforms.

All histone subunits are not acetylated/deacetylated at the same rates. Some populations are acetylated faster than the others. Tetra-acetylation of H4 is mediated by rapid acetylation whereas some other histones achieve higher acetylated isoforms much more slowly. A study showed that 60%–70% of histones are acetylated in breast cancer [16]. These histones are classified into three categories (1) those that are always acetylated and remain in that state, (2) those that are acetylated/deacetylated quickly, and (3) those that are acetylated/deacetylated relatively slow [16]. The majority of histones fall into the last category (3). However, 10% of histones belong to category (2) [16]. In addition to acetylated states, if the majority of the isoforms found are of the higher order, then deacetylation of such molecules will be slowed down, culminating in continuous gene expression.

The category (2) histones are the most affected by estradiol. Estradiol increases the rate of histone deacetylation but does not alter the rate of histone acetylation. Hyperacetylated histones H3 and H4 are found to be associated with coding regions of E2-ER (required for HAT activity) and cMyc genes (expressed in proliferating cells) [31]. cMyc is a well-studied oncogene and the upregulation of myc translates into tumor phenotypes. Therefore, histone acetylation does not always correlate with a normal phenotype. Localized hyperacetylation of genes involved in oncogenesis triggered by hormones can precipitate a neoplastic phenotype.

7.7   LUNG AND NASAL CANCERS MEDIATED BY HISTONE ACETYLATION

7.7.1   Nickel Exposure and Histone Acetylation

Occupational exposure to soluble nickel is known to cause lung and nasal cancers [17]. Besides DNA damage and slow repair response, nickel affects chromatin configuration [17]. Nickel has been shown to decrease the acetylation of histone H3 and H4. In fact, inhibition of gene expression has been associated with hypoacetylation of histones H3 and H4 as a result of nickle-mediated inhibition of HAT activity [17]. Nickel induces an overall decrease in acetylation of all histones. In the group of histones, H2b is the most sensitive to deacetylation and H3 is the least sensitive. Lysines at position 5, 12, 15, and 20 are affected in H2b [17]. It has been observed though that the sensitivity to deacetylation of H2b is related to the exposure periods and shorter exposures have no profound effect on the acetylated states of H2b K5 and K15 residues. K12 and K20 however undergo a different fate, and directly correlate to the time and dose of nickel exposure [17].

Therefore genes containing residues that are primarily acetylated at these positions will be affected and would induce a cascade of downstream effects.

In most instances hyperacetylation is associated with gene expression, however, some studies have shown that acetylation of histones can inhibit expression. This inhibition is primarily caused by a blend of hyper/hypoacetylated patterns at specific lysine residues on different histone subunits at promoter and coding locations. These patterns are more gene-specific and do not associate with the overall global hypo- or hyperacetylation states.

7.7.2 LUNG CANCERS ASSOCIATED WITH ACETYLATION OF THE *RARβ* GENE

In most lung cancers, the *RARβ* gene is defective and therefore is not expressed in the presence of retinoic acid (RA) imparting an RA refractory phenotype to the cells [18]. Analysis of lung cancer cell lines has shown that cell lines unresponsive to RA have deacetylation of histone H3 with hyperacetylation of histone H4 [18]. However, in RA-responsive cells, both H3 and H4 are acetylated at the promoter of RARβ receptor, irrespective of the methylation status of the promoter [18]. Therefore, a reduction in H3 acetylation has been shown to correlate with the RA refractoriness in lung cancer cells [18]. In addition to reduced histone acetylation at specific subunits, methylation of a promoter may be a secondary essential mechanism to gene expression. Therefore, some lung cancers are associated with hypoacetylation and hypermethylation of specific moieties that control the expression of RARβ, which is crucial for downstream signaling pathways.

7.8   MODULATION OF HISTONE H4 IN GASTRIC CANCERS

In a majority of gastric and colorectal cancers H4 acetylation is markedly reduced which precipitates invasive and metastatic events [19]. In colon cancers, global hypoacetylation enhances tumor invasiveness and metastasis allowing for a means of possible cancer therapy. Gastric carcinomas are of two types, intestinal and diffuse [19]. In both these types of cancers deacetylation of histone H4 is observed. This reduction is gradual and begins from the early stage where precancerous lesions predominate and progress into the late stages of invasiveness and metastasis [19,20]. However, one should be careful in interpreting the observed results as acetylation alone cannot dictate a cancerous phenotype but relies on a multitude of factors and mechanisms, all of which target genes that control normal cell phenotype.

7.9   CANCERS OF THE THYROID AND HISTONE ACETYLATION

Essentially the process of acetylation or deacetylation of histones is not directly involved in neoplastic transformations. It is a domino effect induced by modulations of transcription factors, which control cell-specific functions that influence oncogenic events. For instance in many thyroid cancers the sodium iodide symporter (NIS) expression is downregulated [21]. The expression of this protein, NIS, is important as it is involved in iodine uptake. In primary solid thyroid tumors the cells are unable to absorb iodine in part due to low or no expression of NIS [22]. Reduced acetylation in conjunction with methylation at the NIS promoter is probably responsible for the decreased NIS expression [21]. Studies have shown that histone reacetylation by inhibition of the deacetylation process stimulates NIS expression [21,22]. Thus, targeting enzymes that inhibit histone deacetylation may have potential as a therapeutic measure against thyroid cancers. This process helps not only in target-specific treatments but also in the induction of differentiation and apoptosis by increasing the expression of cell cycle regulatory genes.

RA treatment is another approach to restore radioactive iodine uptake in a small subset of metastatic thyroid cancers [22]. The promoter region of RARβ, a receptor for RA, appears to be unmethylated and deacetylated at H3 and H4, but one must be cautious to interpretations, as these observations are cell-type specific. In the presence of the HDACs alone or in combined treatments with RA, the acetylation levels can be restored. Therefore alterations in histone patterns can contribute to the refractoriness in differentiated thyroid cancers and may be a suitable tool for therapy. This study highlights that acetylation is not responsible for the direct conversion of a normal cell to neoplastic forms but controls a host of other genes crucial for down-stream effects.

7.10 GENES AFFECTED BY ACETYLATION THAT PLAY A ROLE IN CARCINOGENESIS

A myriad of genes are associated with cancers. In this section, we discuss a few genes that are affected by modulations of histone molecules. These modulations may affect gene expression in several ways: (1) directly affect the promoter region of the genes, (2) affect coactivators or transcription factors that bind to promoter regions, and (3) affect the expression and stability of proteins that bind to other coactivators. We have discussed those genes whose expression is crucial to various stages of the cell cycle (Figure 7.3) and the expression of which is completely deregulated in a majority of cancers.

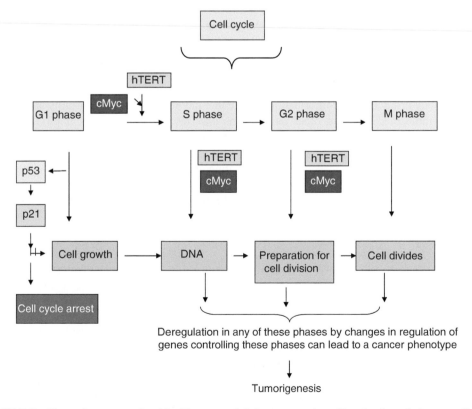

FIGURE 7.3 Genes that are regulated by histone modulations across the cell cycle. Acetylation can affect any of the genes or pathways depicted above. Hypoacetylation of genes that induce apoptosis can induce the cell to enter the S phase for continued growth. Similarly, hyperacetylation of genes required for the S and G phases of the cell cycle can result in uncontrolled growth resulting in transformations and sustained tumor growth.

7.11 p53 AND p21

Cylins and cyclin-dependent kinase genes control the transition of cells in each of the G1, S, and G2 phases of the cell cycle and are essential as they direct cell division in a regulated manner. The cell cycle is constantly monitored and a close surveillance is kept on damage to DNA. When DNA damage is detected, proteins involved in rectifying the damage are induced. These proteins force the cell to exit the cell cycle and undergo apoptosis. p53 and p21 are two very important proteins required for this function. p21 comes under the direct control of p53; however, p21 expression can be induced by p53-independent pathways as well. The induction of p53 stimulates the production of p21 that is a cyclin-dependent kinase. This protein inhibits cdk4/cyclinD and the downstream pathways associated with it resulting in cell cycle arrest.

Transactivators not only have the ability to induce gene expression but some of them, such as PCAF and p300 posses intrinsic HAT-like activity. [23]. In most cases histones associated with promoter regions are susceptible to HAT activity, which alters the residues to modulate expression. In some instances HATs associate with proteins directly and modulate downstream effects. Acetylation of lysine residues in proteins can enhance their binding ability to DNA domains. The affinity to the binding site, which could be a DNA binding domain or a protein domain, is important, because this affinity dictates the exponential downstream effects responsible in maintaining a structurally and functionally stable cell. Such is the case with the protein p53 in that its transcriptional regulation is not directly mediated by acetylation, but the protein itself is acetylated to enhance its stability and binding affinity to its consensus DNA binding regions [23].

p21 can be induced by p53 dependent and independent pathways. Studies have shown that acetylation of p53 specifically at residues K373/382 is important to induce p21 expression [12]. This induction is independent of the phosphorylation of Ser residues at 15, 20, and 392. In the majority of cancers p53 expression is reduced or completely inhibited. p53 regulates its expression via a feed back loop mechanism. Thus, in the absence of p53 or in the presence of deacetylated p53, the tumor suppressor is unable to bind to the p21 promoter thereby attenuating the expression of p21. Acetylation of specific residues is necessary to bring about this effect. Acetylation of K373/382 may bring about a conformation change that increases the binding affinity and specificity of the p53 molecule; however, other modulations may reverse this affinity. If p21 is no longer expressed then the apoptotic pathways controlled by this protein are affected and the cells escape this check and proceed toward a proliferating phenotype. In some malignancies histones conjoining at the promoter of p21 are hypoacetylated, inactivating expression [24]. Therefore, inhibition of p21 expression mediated by hypoacetylation of its promoter or hypoacetylation of the p53 protein can be one of the early mechanisms directing cells toward a more cancer-like phenotype [24–26].

7.12 hTERT

hTERT, the catalytic subunit of telomerase, is upregulated in the majority of cancers and p53 and p21 (a p53-induced protein) are downregulated. Many studies have focused on the promoter region of *hTERT* to determine factors that govern its expression since hTERT is the rate-limiting message for telomerase activity. DNA methylation and histone modulations, primarily acetylations, have been linked to the regulation of *hTERT* both in normal as well as malignant cells [27–31].

In most normal cells, with the exception of the germ cells and adult stem cells as well as a few other cell types such as those of the intestinal lining, hTERT is absent. In normal cells, *hTERT* repression is mediated by negative modulators, which when bound to the *hTERT* promoter can induce deacetylation of the promoter, primarily at histones H3 and H4 [30]. The *hTERT* promoter has binding sites for Mad1, a repressor protein by function. It is believed that Mad1 modulates the expression of *hTERT* via the deacetylation of the histones that complex to the promoter [27,30]. Mad1 binds to the mSin3 corepressor protein that has deacetylase activity. Thus, when these repressor complexes bind to their DNA binding regions at the *hTERT* promoter, the deacetylation switch is turned on and the histones surrounding the promoter are deacetylated [30]. With deacetylation, changes in histone and chromatin dynamics of the *hTERT* promoter ensue, which prevent oncogenic activators of *hTERT* from binding to their DNA-binding domains. This change in histone dynamics may not entirely be brought about by histone acetylation but also by other modulating processes like methylation of CpG islands and specific lysine residues of histone protein with the *hTERT* promoter.

In some cases, normal cells tend to escape the stringent cell cycle checks and instead of following the senescent or apoptotic pathways, proliferate by the induction of the repressed *hTERT* and reactivation of the telomerase molecule. The reactivation of *hTERT* has been linked to many mechanisms [30]. Recently though, epigenetic mechanisms are found to be the hallmark in *hTERT* control [30]. In certain malignancies reactivation of *hETRT* has been linked to the phosphorylation of Ser 10 of the histone H3 subunit via the p38 mitogen-activated kinase pathway [32]. However, this transactivation is dependent on the acetylation of the H3 K14 residue [32]. Histone modulations sometimes function using various histone marks and this is one such example. Ser 10 phosphorylation alone weakly induces *hTERT* expression and telomerase activity. However, only when the Ser 10 moiety of H3 is phosphorylated with the combined acetylation of H3 K14 at the *hTERT* promoter is the *hTERT* specifically transduced. This occurrence may be cell-specific but is proof to the epigenetic mechanisms of *hTERT* control.

Interpretations of the effects of histone acetylation on gene expression should be made carefully as histone modulations may be global but in most tumorigenic cases local hyperacetylation of oncogenes, such as in the case of *HPVE6* is observed. The HPVE6 protein is responsible for a majority of virally transduced cervical cancers. Interestingly this protein is also know to be a modulator of *hTERT* expression and has a DNA-binding domain on the promoter. The E6 protein modulates the expression of *hTERT*, the rate-limiting molecule of telomerase, by the acetylation of histone H3 of the *hTERT* promoter in association with E6-associated protein [33].

7.13 cMYc TRANSCRIPTION ACTIVATOR

cMyc is an oncogene and is notably overexpressed in many tumors including breast, colon, and prostate. About 5% of the total of genes studied so far are affected by cMyc [34]. The mechanism of cMyc-mediated target gene expression is not yet fully understood. Like many other transcription factors, Myc has the ability of affecting target gene expression directly by binding to its promoter residing domains. Indirect pathways can include sequential or combinatorial protein interactions [35]. Myc has the ability to acetylate histones, especially H4 and H3 [35,36] and cMyc interacts with transactivators like TRRAP, a protein that tethers to the GCN5 and Tip60 proteins that have HAT activity [35,36]. Acetylation of cMyc target genes is modulated by this large protein complex; however, actual transactivation of the promoters of its target genes is dependent on other mediators as well [34,35]. Histone acetylation mediated gene expression by cMyc, however, has been the subject of some debate [34]. Some studies have demonstrated that the transcriptional expression of cMyc target genes may be independent of histone acetylation but could be a very essential step in further downstream gene expressions [34,37]. Therefore when cMyc is overexpressed it can essentially cause the hyperacetylation of histones of its target genes which otherwise in normal cells are maintained at optimal acetylation/deacetylation balance. This modulation of histone dynamics can dramatically affect the expression of cMyc-mediated mitogen activated target genes.

7.14 MAD1 REPRESSOR

Mad1 is an antagonist of Myc and is involved in repressing gene expression. Both these transcription factors compete for the same binding site at the E-box as a complex with Max. This binding is seen in the transcriptional control of promoters of genes like *cyclin D2*, *Cad*, and *hTERT* [38]. For these genes, in quiescent cells or differentiated cells, Mad1 attenuates and represses expression by deacetylation of their promoters. This inactivation is brought about by HDACs (histone deacety-lases) recruited to the promoter regions by Mad1 in tandem with corepressor Sin3 [30,38]. Deacetylation recoils the chromatin at these important promoter regions terminating expression of cancer promoting genes in conjunction with reactivation of cell cycle control genes.

7.15   Mnt COTRANSACTIVATOR OR REPRESSOR

Mnt, a transcription repressor, interacts with coactivators such as cMyc and Max and forms a regulatory network that controls downstream genes, important in fine tuning the proliferation, differentiation, and quiescent phases of a cell. Myc and Mnt are coexpressed in proliferating cells and Mnt competes with Myc to dimerize with Max and bind to the E-box elements [39,40]. However Mnt initiates the induction of Myc target genes such as *cyclin D2* important for the S phase of the cell cycle [39]. It has been documented that the functions of Mnt are mediated by the association with corepressors Sin3, which recruits HDAC to deacetylate histones [39]. Mnt func-tions in repressing Myc-mediated tumor formation as well as the activation of Myc-targeted promoters that contain E-boxes. Therefore Mnt repressive or activating function is perhaps regulated by many mechanisms but fluctuations in histone dynamics are a plausible explanation. Thus, when the functions of this transcription factor is deregulated, either by mutations or other such changes, its downstream histone modulations are affected inducing the overexpression of Myc or other such targets resulting in the development of neoplasms.

7.16   Sp1 COACTIVATOR

Sp1 modulates gene activation or repression. This action is solely dependent on the target tran-scription factor to which it binds. For example Sp1 influences the regulation of MCP1 (monocyte chemoattractant protein 1). The MCP1 protein is a chemokine that is involved in recruiting monocytes/macrophages to the areas of inflammation. However, this protein is also linked to prostate neoplasia and prostate adenocarcinomas. Tumor necrosis factor (TNF) modulates the expression of MCP1 and does so by recruiting acetylases CBP/p300 to the promoter regions. This chromatin-modulated induction of MCP1 requires the activity of p65 and Sp1. TNF action acetylates both the proximal and distal regions of MCP1 promoter in conjunction with p65 activity, but Sp1 acetylates only the distal regions of MCP1. p65 and Sp1 perhaps only aid in the regulation of gene activity but may not be involved with direct induction of the gene. Such an observation is interesting as binding sites for Sp1 are present on other transcription factor promoters and genes associated with cancers such as *hTERT*. Any disturbance in the genes involved in regulating gene activity by acetylation may generate imbalances in histone acetylation patterns. This may lead to the induction of continuous oncogenic expression in conjunction with the reduction of cell cycle regulatory proteins.

7.17   FUTURE PROSPECTS

Many genes are important in orchestrating the fine balance between cell proliferation versus tumorigenesis. Even if it is not the primary mechanism of gene control, this balance is mediated by chromatin fold changes via histone modulation. For a given gene expression, several pathways may function in tandem to bring about the expression. Failure in any step along the pathway can have detrimental effects, which is seen in many cancers. This failure occurs when the nucleosome units undergo aberrations, resulting in changes in the histone code and affecting the fluidity of DNA regions blocking access to binding elements. Therapeutics may have potential in utilizing the compounds that can reverse the abnormal histone code bringing the functions to normal.

REFERENCES

1. Bernstein, E. and Hake, S.B., The nucleosome: A little variation goes a long way, *Biochem. Cell Biol.*, 84, 505, 2006.
2. Ducasse, M. and Brown, M.A., Epigenetic aberrations and cancer, *Mol. Cancer*, 5, 60, 2006.
3. Kornberg, R.D. and Lorch, Y., Chromatin-modifying and -remodeling complexes, *Curr. Opin. Genet. Dev.*, 9, 148, 1999.
4. Kornberg, R.D. and Lorch, Y., Twenty-five years of the nucleosome, fundamental particle of the eukaryote chromosome, *Cell*, 98, 285, 1999.
5. Khorasanizadeh, S., The nucleosome: From genomic organization to genomic regulation, *Cell*, 116, 259, 2004.
6. Eberharter, A. and Becker, P.B., Histone acetylation: A switch between repressive and permissive chromatin. Second in review series on chromatin dynamics, *EMBO. Rep.*, 3, 224, 2002.
7. Marushige, K., Activation of chromatin by acetylation of histone side chains, *Proc. Natl. Acad. Sci. U S A*, 73, 3937, 1976.
8. Santos-Rosa, H. and Caldas, C., Chromatin modifier enzymes, the histone code and cancer, *Eur. J. Cancer.*, 41, 2381, 2005.
9. Akbarian, S. et al., Chromatin alterations associated with down-regulated metabolic gene expression in the prefrontal cortex of subjects with schizophrenia, *Arch. Gen. Psychiatry*, 62, 829, 2005.
10. Davie, J.K. and Dent, S.Y., Transcriptional control: An activating role for arginine methylation, *Curr. Biol.*, 12, R59, 2002.
11. Wang, Y. et al., Beyond the double helix: Writing and reading the histone code, *Novartis Found. Symp.*, 259, 3, 2004.
12. Zhao, Y. et al., Acetylation of p53 at lysine 373/382 by the histone deacetylase inhibitor depsipeptide induces expression of p21(Waf1/Cip1), *Mol. Cell. Biol.*, 26, 2782, 2006.
13. Sykes, S.M. et al., Acetylation of the p53 DNA-binding domain regulates apoptosis induction, *Mol. Cell*, 24, 841, 2006.
14. Verdone, L., Caserta, M., and Di Mauro, E., Role of histone acetylation in the control of gene expression, *Biochem. Cell Biol.*, 83, 344, 2005.
15. Verdone, L. et al., Histone acetylation in gene regulation, *Brief Funct. Genomic. Proteomic.*, 5, 209, 2006.
16. Sun, J.M., Chen, H.Y., and Davie, J.R., Effect of estradiol on histone acetylation dynamics in human breast cancer cells, *J. Biol. Chem.*, 276, 49435, 2001.
17. Golebiowski, F. and Kasprzak, K.S., Inhibition of core histones acetylation by carcinogenic nickel(II), *Mol. Cell. Biochem.*, 279, 133, 2005.
18. Suh, Y.A. et al., Loss of retinoic acid receptor beta gene expression is linked to aberrant histone H3 acetylation in lung cancer cell lines, *Cancer Res.*, 62, 3945, 2002.
19. Yasui, W. et al., Histone acetylation and gastrointestinal carcinogenesis, *Ann. N.Y. Acad. Sci.*, 980, 220, 2003.
20. Tahara, E., Histone acetylation and retinoic acid receptor beta DNA methylation as novel targets for gastric cancer therapy, *Drug News Perspect.*, 15, 581, 2002.
21. Puppin, C. et al., Effects of histone acetylation on sodium iodide symporter promoter and expression of thyroid-specific transcription factors, *Endocrinology*, 146, 3967, 2005.
22. Haugen, B.R., Redifferentiation therapy in advanced thyroid cancer, *Curr. Drug Targets Immune Endocr. Metabol. Disord.*, 4, 175, 2004.
23. Liu, L. et al., p53 sites acetylated in vitro by PCAF and p300 are acetylated in vivo in response to DNA damage, *Mol. Cell Biol.*, 19, 1202, 1999.
24. Mitani, Y. et al., Histone H3 acetylation is associated with reduced p21(WAF1/CIP1) expression by gastric carcinoma, *J. Pathol.*, 205, 65, 2005.
25. Archer, S.Y. and Hodin, R.A., Histone acetylation and cancer, *Curr. Opin. Genet. Dev.*, 9, 171, 1999.
26. Fang, J.Y. et al., Epigenetic modification regulates both expression of tumor-associated genes and cell cycle progressing in human colon cancer cell lines: Colo-320 and SW1116, *Cell Res.*, 14, 217, 2004.
27. Cong, Y.S. and Bacchetti, S., Histone deacetylation is involved in the transcriptional repression of hTERT in normal human cells, *J. Biol. Chem.*, 275, 35665, 2000.
28. Lv, J. et al., Molecular cloning of a novel human gene encoding histone acetyltransferase-like protein involved in transcriptional activation of hTERT, *Biochem. Biophys. Res. Commun.*, 311, 506, 2003.

29. Hou, M. et al., The histone deacetylase inhibitor trichostatin A derepresses the telomerase reverse transcriptase (hTERT) gene in human cells, *Exp. Cell Res.*, 274, 25, 2002.

30. Xu, D., Switch from Myc/Max to Mad1/Max binding and decrease in histone acetylation at the telomerase reverse transcriptase promoter during differentiation of HL60 cells, *Proc. Natl. Acad. Sci. U S A*, 98, 3826, 2001.

31. Takakura, M. et al., Telomerase activation by histone deacetylase inhibitor in normal cells, *Nucleic Acids Res.*, 29, 3006, 2001.

32. Ge, Z. et al., Mitogen-activated protein kinase cascade-mediated histone H3 phosphorylation is critical for telomerase reverse transcriptase expression/telomerase activation induced by proliferation, *Mol. Cell Biol.*, 26, 230, 2006.

33. James, M.A., Lee, J.H., and Klingelhutz, A.J., HPV16-E6 associated hTERT promoter acetylation is E6AP dependent, increased in later passage cells and enhanced by loss of p300, *Int. J. Cancer.*, 119, 1878, 2006.

34. Eberhardy, S.R., D'Cunha, C.A., and Farnham, P.J., Direct examination of histone acetylation on Myc target genes using chromatin immunoprecipation, *J. Biol. Chem.*, 275, 33798, 2000.

35. Frank, S.R. et al., Binding of c-Myc to chromatin mediates mitogen-induced acetylation of histone H4 and gene activation, *Genes Dev.*, 15, 2069, 2001.

36. Vervoorts, J. et al., Stimulation of c-MYC transcriptional activity and acetylation by recruitment of the cofactor CBP, *EMBO Rep.*, 4, 484, 2003.

37. Faiola, F. et al., Dual regulation of c-Myc by p300 via acetylation-dependent control of Myc protein turnover and coactivation of Myc-induced transcription, *Mol. Cell Biol.*, 25, 10220, 2005.

38. Bouchard, C. et al., Regulation of cyclin D2 gene expression by the Myc/Max/Mad network: Myc-dependent TRRAP recruitment and histone acetylation at the cyclin D2 promoter, *Genes Dev.*, 15, 2042, 2001.

39. Popov, N. et al., Mnt transcriptional repressor is functionally regulated during cell cycle progression, *Oncogene*, 24, 8326, 2005.

40. Smith, A.G. et al., Expression and DNA-binding activity of MYCN/Max and Mnt/Max during induced differentiation of human neuroblastoma cells, *J. Cell. Biochem.*, 92, 1282–1295, 2004.

8 Histone Methylation and the Initiation of Cancer

Shi Huang

CONTENTS

> A favorite explanation has been that [carcinogens] cause alterations in the genes of cells of the body, somatic mutation as these are termed. But numerous facts, when taken together, decisively exclude this supposition.
>
> **Rous (1966) [1]**

8.1 INTRODUCTION

The classical mutation theory of cancer cannot adequately explain the findings of cancer epidemiology. Several decades of epidemiological studies have consistently shown that most cancers (~70%–90%) are preventable and related to environmental factors [2–5]. Rapid increases in rates of colon cancer among migrants from low- to high-risk areas indicate that large international differences in cancer rate are due to environmental rather than genetic causes. Doll and Peto have suggested that differences in diet may account for 90% of the variation in rates among countries but the specific factors that are responsible have not been established [2]. Mutagens are found to be a relatively minor aspect of the environmental contribution to cancer. Even in cases where a carcinogen has mutagenic effects (smoking, radiation, etc.), it remains debatable as to the relative importance of the cytostatic/toxic effects versus the mutagenic effects of the carcinogen [6,7]. Many mutagens are not carcinogens. Defects in DNA repair systems do not always cause cancer [8,9]. Most incipient tumors and even advanced cancers regress naturally, inconsistent with the irreversibility of mutations [10]. Cancers of the same phenotype often have a different spectrum of gene mutations, whereas cancers with the same mutations have dissimilar clinical features. Normal cells rarely gain growth advantages but instead often undergo cell death or senescence when a mutation occurs or when an oncogene is activated [11,12]. The relatively stable genotypes and phenotypes of end stage cancer seem at odds with an enhanced mutation rate. Number of mutations and timing of onset vary greatly among different cancers. None of the hallmarks of cancer [13] cannot be generated by an epigenetic reprograming of wild-type genome or cannot find a match among the extremely large number of cell types in a human body. Some tumors such as teratomas have few mutations, and implantation of embryonic stem cells into many somatic tissues forms teratomas [14]. Cancer often contains aneuploidy but what causes aneuploidy remains unaccounted for by the mutation theory. Overall, this large body of facts contradicts the mutation theory.

It is widely thought that tumorigenesis is a form of somatic Darwinian evolution. This neo-Darwinian mutation theory of cancer has its roots in the work of Theodor Boveri a century ago [15] but come to dominate mainstream thinking only in the last 40 years. The basic idea is that somatic mutations drive clonal expansion of incipient tumor cells that carry those mutations [13,16–18]. The initiating event in sporadic cancers linked with environmental carcinogens is thought to be a mutation in a single cell that drives the clonal expansion of the single cell. Few question the great success of the mutation theory in explaining familial cancer syndromes, which seem to have a single gene etiology. Many, however, have recognized the limitation of the mutation theory in the development of common cancers.

The neo-Darwinian mutation theory of evolution was synthesized in the 1940s based on the great advances in Mendelian genetics of that era, with, unfortunately, a nearly complete ignorance of epigenetics. The genetics of 1940s studies lower organisms where genetics plays a relatively more dominant role in phenotype determination. But epigenetics is more involved in complex organisms. Therefore, theories that work for lower organisms may not adequately describe complex organisms. Despite the rapidly advancing knowledge that epigenetics and genetics are both equal contributors of heritable phenotypes, the neo-Darwinian evolution theory essentially ignores the role of epigenetic changes in phenotype evolution and transformation. Epigenetic information is not only inherited during mitotic cell division but is also transmitted through the germline to the next generation [19,20]. There is nothing unique about genetics that is not shared by epigenetics that makes genetics uniquely important in phenotype evolution and transformation. The two are the yin and yang opposite sides of the same phenomenon of inheritance and cannot exist independent of each other, just like night and day are the two opposite sides of the same phenomenon of the Sun– Earth orbit. Since the neo-Darwinian theory is the foundation of the mutation theory of cancer that also ignores epigenetics, a complete and more correct formulation of a cancer theory would inevitably lead to a conflict with the neo-Darwinian theory. The study of cancer epigenetics may not only lead to a more complete understanding of cancer but may also more importantly contribute directly to a more complete formulation of an evolution theory in general. Just like the study of genetics of lower organisms in the 1940s led to the synthesis of neo-Darwinism, the study of epigenetics of higher organisms that is going on today may fundamentally change our view of evolution and transformation of phenotypes. This review will discuss a unified evolution theory of cancer that unites both genetics and epigenetics. The new cancer theory may serve as a stepping stone toward a future synthesis of a more complete evolution theory in general that would take into account both genetics and epigenetics.

It has been recognized that the mutation theory cannot exclude alternative hypotheses that do not feature mutations as the initiating cause of cancer. Some have suggested that selection can drive carcinogenesis by allowing growth of incipient tumor cells with preexisting mutations, without a need for genetic instability [12,21,22]. It has been noted that environmental selection can induce mutations, even when the environment contains no mutagens [23]. There are also researchers who believe that epigenetic changes are sufficient to drive tumorigenesis [10]. However, these hypotheses lack molecular pathways and do not adequately address the mechanisms of mitotic inheritance of somatically acquired tumor phenotypes. The strength of the mutation theory is that it does provide a molecular pathway of carcinogenesis in the case of familial cancer syndromes. The initiating cause in this case is an inherited mutation in a cancer gene in all cells of the body. However, neither the mutation theory nor the epigenetic theory at this point in time provides a detailed molecular pathway from the major known causes of cancer (such as the typical Western diet) to the alteration of specific target cancer genes in all cells of a target tissue soon upon exposure to a carcinogen. Without a molecular pathway, the findings of cancer epidemiology cannot by itself establish a causal relationship between environmental factors and cancer. Epidemiology could only show correlation. This weakness of epidemiology coupled with a present lack of molecular pathways has created a confusion as to what specific environmental factor does or does not cause cancer. Almost any dietary nutrients or components could be correlated with cancer by epidemiological studies [5]. In the absence of a molecular pathway, it is hard to make a convincing case for any of the dietary agents. The molecular study of epigenetics holds special promise to distinguish methyl nutrients as the cancer preventive components of a healthy diet, or to distinguish methyl deficiency as the chief carcinogenic element for the typical Western diet.

Epigenetic programs control heritable transmission of phenotypes in a DNA-sequence-independent fashion. DNAs are packaged into chromosomes or chromatin in eukaryotic cells. This packaging of DNA involves complex formation between DNA and core histones and other proteins. Unfolding of chromatin is required for many processes that involve access of proteins to DNA. The state of chromatin is controlled largely by covalent modifications of histone tails. The major modifications include phosphorylation, acetylation, ubiquitination, and methylation. The specific combination of these modifications has been termed the histone code that determines the on or off state of genes or folding/unfolding state of the chromatin [24,25]. The formation of heterochromatin, which represses gene transcription, is controlled by the methylation of histone tails. A different kind of methylation marks transcriptionally active chromatin termed euchromatin. Remarkably, some of the histone modifications and their corresponding chromatin state are highly stable and mitotically inheritable. It is likely that heritable gene expression patterns of specific cell types are largely maintained by histone modifications.

Recently, great advance has been made in understanding a key molecular machinery, the histone methyltransferases or HMTs, that controls epigenetic inheritance. Importantly, alterations in HMTs have been linked with carcinogenesis. Although a large number of protein enzymes involved in chromatin/histone modification have been linked with cancer [26], the finding of HMTs as tumor suppressors is particularly interesting as it may directly explain the molecular mechanism for a major environmental cause of cancer, the typical Western diet [27–29]. Here, I would like to summarize the recent progress in the field of HMTs and cancer. The finding of HMTs in tumor suppression has important implications for a unified pathway of cancer initiation. The epigenetic component of the unified pathway may in fact explain the cause or initiation of most sporadic common cancers.

8.2   EPIGENETIC INHERITANCE CONTROLLED BY HISTONE METHYLATION

Histone methyltransferases are enzymes that transfer the methyl group from the methyl donor S-adenosylmethionine (SAM) to the arginine or lysine residues in histones or other proteins. HMTs are just a subset of the whole methyltransferase superfamily that uses SAM as methyl donor. This superfamily also includes DNA methyltransferases (DNMTs), other protein methyltransferases, small molecule metabolite methyltransferases, RNA methyltransferases, etc. All these SAM-dependent methylation reactions produce a SAM analog, S-adenosylhomocysteine (SAH), which is a competitive product inhibitor of the methylation reaction. So, some methylation enzymes are expected to be sensitive to inhibition by a moderate increase in SAH or decrease in SAM, especially those that have lower affinity for SAM or SAH.

On the basis of the sequence and structure of their catalytic domain, the HMTs can be divided into three classes, protein arginine methyltransferases (PRMTs) [30], SET domain containing lysine methyltransferases [31], and Dot1 class lysine methyltransferases [32,33]. The number of SET domain class genes in mammals is dramatically larger than the other two classes. This family has been strongly linked with mitotic inheritance of gene activity and carcinogenesis. So, the focus of this review is on the SET domain family of HMTs.

Gene expression patterns are controlled by a combination of a variety of chromatin/histone-modifying enzymes. Among the many forms of posttranslational modifications, methylation is the most stable, which makes it better suited for transmitting long-lasting inheritable information. Indeed, SET domain methylation enzymes have been directly linked with cell fate inheritance in Drosophila [34–39]. Only part of the information stored in the network of histone modifications may need to be inherited as long as the inherited information (say, methylation) can serve to direct the retrieval of other information (say, acetylation, phosphorylation, etc.).

It is now well established that acquired phenotypes of a somatic cell can be inherited through cell division without any need of changes in DNA. Drosophila genetic studies have shown that cell fate is mitotically inherited. Such inheritance can be quite stable; when cultured for up to 10 years, hundreds of cell divisions later, wing disc cells can still give rise to adult wing structures [40]. The past decade of genetic and biochemical studies show that SET domain HMTs are the key players in mitotic inheritance of gene expression patterns [34–39,41–45]. In a typical adult human cell, most genes are located in regions of heterochromatin and are not expressed, whereas actively expressed genes are located in euchromatin regions. The heterochromatin and euchromatin status is mitotically inherited. The initiation and maintenance of heterochromatin is carried out by histone H3 lysine 9 methylation, whereas euchromatin is characterized by H3 lysine 4 methylation [46–49]. Methylation of H3 lysine 27 is associated with gene silencing and X-chromosome inactivation [38,50]. Also associated with gene silencing is methylation of H3 lysine 36 [51], H4 lysine 20 [52,53], and H3 lysine 76 [32,33]. A best studied example of histone-mediated inheritance is the control of homeotic genes in *Drosophila* by polycomb and trithorax group HMTs [34–39]. The homeotic genes are activated in the early embryo by the products of segmentation genes. At this stage, transient, localized activators and repressors determine the segmental domains of expression of each homeotic genes. But, after gastrulation, polycomb and trithorax HMTs take over to maintain the segmentation pattern of expression for the rest of development.

8.3 HISTONE METHYLTRANSFERASES IN TUMOR SUPPRESSION

The function of HMTs in mitotic inheritance of cell fate suggests that loss of HMTs may lead to cell fate transformation and carcinogenesis. It is intellectually satisfying that HMT class genes were proposed as tumor suppressors even before the importance of histone methylation was appreciated [54–56]. While a small number of HMTs may be oncogenic, most HMTs seem to function as tumor suppressor genes.

The majority of HMTs contains a conserved catalytic domain termed the SET domain. The SET domain was discovered as a ~120–150 amino acid sequence homology present in several *Drosophila* genes, SU(VAR)3–9, E(Z), and Trithorax [41,42]. A related domain is termed the PR (PRDIBF1 and RIZ1) domain that was independently found as a ~130 amino acid homology present in the human retinoblastoma (Rb) interacting zinc finger (ZF) protein RIZ1 and the human transcriptional repressor PRDI-BF1 (positive regulatory domain I-binding factor 1) [57–59]. Despite the sequence similarity between PR and SET, which is typically ~20%–30% identical in amino acids, they are clearly distinctive because identities among PR domains or among SET domains are usually ~40%. Also, SET domains are primarily found at the carboxyl-termini of proteins, whereas PR domains are mostly located at the amino-termini. PR domain-containing proteins commonly have ZF DNA-binding domains, whereas SET domain proteins lack obvious DNA-binding motifs. The PR-domain containing genes have the Human Gene Mapping Workshop nomenclature committee approved gene symbol *PRDMs* (PR-domain containing, with ZFs). The PR-proteins may have lower affinity for SAM or SAH since one of the conserved regions in the canonical SET domain that contact SAM/SAH is not conserved in the PR domain. The H residue of the highly conserved NH motif in the canonical SET domain forms a hydrogen bond with SAM [60,61], whereas RIZ1 contains NW at these positions. Nonetheless, the PR domain of RIZ1 has a very similar structural fold as the canonical SET domain [62].

By repeated PSI-BLAST analysis of protein sequence database, the PR domain was found to be related to the SET domain [59]. By applying the same PSI-BLAST analysis in a more extensive way, the SET domain was found in several plant protein lysine methyltransferases [63]. This observation in turn led to the identification and characterization of histone H3 methylation activity for the human SET domain protein, SUV39H1 [43]. The SET domain is the catalytic motif of the SUV39H1 HMT. Many other SET domain-containing proteins have since been shown to have HMT activities. RIZ1 is the first PR-protein shown to have HMT activity [64].

Histone methylation machinery is highly conserved during evolution. The canonical SET domain is found in all eukaryotes while the PR domain is found only in multicellular animals. So, the PR domain is more involved in complex metazoan specific biological functions. Consistently, metazoan evolution is associated with a more dramatic increase in the number of PR domain genes than in that of SET genes. There are ~8 SET genes in yeast, more than 30 in *Arabidopsis thaliana*, ~20 in *Drosophila*, ~20 in *Caenorhabditis elegans*, and ~34 in humans. In contrast, there are no PR domain proteins in yeast, none in plants (*A. thaliana*), 2 in *Drosophila*, 2 in *C. elegans*, and 17 in humans.

The first member of the SAM-dependent methyltransferase superfamily to be shown a tumor suppressor is the RIZ1 gene, which is also the best characterized with regard to a tumor suppression role in human cancers. The tumor suppressor function of RIZ1 is mediated primarily by the PR domain itself, thus implicating a similar role for other PR/SET-containing genes. The PR/SET domain was first proposed to have a tumor suppressor role long before its HMT activity is known [54]. While the field of tumor suppressors was initially reluctant to accept the idea of RIZ1 and the PR/SET family as important tumor suppressors, the idea has withstood the test of time and accumulation of data. The following provides a summary of RIZ1 and other members of the PR/SET family (Table 8.1).

8.3.1 PRDM2 (RIZ1, MTBZF, G3в)

The *RIZ1* gene was originally isolated in a functional screening for proteins that bind to the Rb tumor suppressor [58]. The gene produces two mRNA and protein products due to alternative promoter usage [65]. It is only the full-length (1719 amino acids) product RIZ1 that contains the PR domain, whereas the short product RIZ2 lacks the amino terminal 200 residues of RIZ1 but is otherwise identical to RIZ1.

RIZ1 but not RIZ2 has tumor suppressor functions. While the binding to Rb may implicate a tumor suppressor role for RIZ1, much of the data on RIZ1 are independent of Rb. The RIZ gene maps to chromosome 1p36, one of the most commonly deleted regions in human cancers [66]. *RIZ1* gene silencing, but not *RIZ2*, is common in many types of human tumors, including breast cancer, liver cancer, colon cancer, neuroblastoma, melanoma, lung cancer, and osteosarcoma [54,67,68], and is associated with promoter DNA methylation [69,70]. RIZ1 has activities in cell cycle G2/M arrest, apoptosis induction, and suppression of xenograft tumors [54,66,71]. RIZ1 has a Pro704 deletion polymorphism that is preferentially deleted in cancers and is more common in Asians [72]. Two other RIZ1 polymorphisms (+92337G > A and +95701C > A) have been linked with reduced risk of lung cancer [73]. These two polymorphisms locate in an intron of RIZ1 but not RIZ2. Another polymorphism (C99243T) is linked with increased breast cancer risk [74]. Frequent frameshift mutations of the gene are common in microsatellite-unstable tumors [67,75,76]. RIZ1 missense mutations also occur in human cancers and are found within the PR-domain and PR-nearby regions [77]. These mutations decrease the HMT activity of RIZ1 [63]. More importantly, mouse gene knock out models show that *RIZ1* inactivation, while RIZ2 is normal, can indeed cause tumor susceptibility [77]. These studies establish a direct link between a PR-domain and tumorigenesis, and suggest a critical role for the HMT activity of the PR domain in the tumor suppression function of RIZ1.

Heterozygous mutation in RIZ1 is sufficient to cause tumors even in the absence of loss of heterozygosity, suggesting that partial loss of function in RIZ1 is oncogenic [77]. Although it takes a near lifetime for RIZ1-deficient mice to develop tumors, this seemingly un-flattering property of RIZ1 is in fact more relevant to the human reality of sporadic common cancers, which is linked to a near lifetime exposure to environmental carcinogens such as the typical Western diet. It is also more relevant to tumor formation in rodents induced by a methyl-deficient diet, which also takes a near lifetime exposure to the methyl-deficient diet [78–82]. Indeed, if loss of RIZ1 or any other genes would cause tumors early in life, they would be automatically disqualified as the direct early targets for dietary methyl deficiency induced carcinogenesis. Thus, we can safely eliminate most familial tumor suppressor genes such as Rb and p53 as the early targets of inactivation for the typical Western diet. To be a potent, familial gene has often been the gold standard for granting a gene the status of tumor suppressor. But few realized the irony that to be a familial gene also means to be less relevant to the early initiating cause for the majority of sporadic common cancers. It is no surprise that the tremendous focus on studying familial cancer genes in the past 25 years has had little direct impact on understanding the initiating pathway for sporadic common cancers that are caused by environmental factors.

RIZ1 silencing or promoter hypermethylation appears to be very common in human cancers, with incidence rate ranging from 5% to 100% depending on specific tumor types. A survey of literature shows that RIZ1 silencing occurs in 100% thyroid carcinomas, 34%–68% in liver cancer, 39%–69% in gastric cancer, 59% in Barrett's esophagus, 58% in esophageal adenocarcinoma, 60% in nasopharyngeal carcinoma, 42% in breast cancer, 30% in prostate cancer, 23% in ovarian cancer, 30% in colon cancer, 38% in lung cancer (cell lines), 36% in parathyroid tumors, 50% in

TABLE 8.1

Human PR and SET Domain Containing Genes: Properties and Chromosomal Location/Links to Human Cancers

Genes	Chromosomes	Cancer Types	Properties	Genbank ID
PR-domain family				
PRDI-BF1 (PRDM1/BLIMP1)	6q21-q22.1	(−) B-lymphoma, melanoma, stomach cancers	c-Myc repressor, proapoptotic Yin–yang expression	XM_004084
RIZ1 (PRDM2/MTBzf/G3b)	1p36.13-p36.23(−)	Nearly all	H3K9, yin–yang expression G2/M arrest	NM_012231
MDS1-EVI1(PRDM3)	3q26	t Leukemia, (+) ovarian	Yin–yang expression	NM_004991
PRDM4 (PFM1/SC1)	12q23-24.1	(−) Pancreatic, ovarian, stomach cancers	Inhibits DNA synthesis	NM_012406
PRDM5 (PFM2)	4q25-q26	(−) Ovarian, melanoma, liver, lung, and colorectal cancers	Silenced in cancer	NP_061169
PRDM6 (Prism/PFM3)	5q21-q23	(−) Colorectal, lung, ovarian, leukemia, and stomach cancers	H4K20, yin–yang expression Role in smooth muscle	AAF78078
PRDM7 (PFM4)	16q24	(−) Melanoma, prostate, liver	Absent in most tissues	NP_443722
PRDM8 (PFM5)	4q21.1	(−) Colorectal	Highly retina specific expression	NP_064611
PRDM9 (PFM6, *Meisetz*)	5p14	(+) Ovarian, fibrosarcoma	H3K4me3, paralog of PRDM7 Absent in most tissues	NP_064612
PRDM10 (PFM7, Tristanin)	11q25	(−) Melanoma, colorectal, ovarian, oral cancer	?	NP_064613
PRDM11 (PFM8)	11p11.2	(−) Melanoma	?	NP_064614
PRDM12 (PFM9)	9q33-q34.1	(−) Ovarian, bladder, esophagus, lung, leukemia	Role in leukemia	NP_067632
PRDM13 (PFM10)	6q16-q21	(−) Lymphoma, melanoma, stomach	Tumor antigen	NP_067633
PRDM14 (PFM11)	8p12-21	(−) Melanoma, prostate	?	NP_078780
PRDM15 (PFM12/ZNF298)	21q22.3	(−) Leukemia, pancreas	Deleted in pancreatic cancer	AAF78093
PRDM16 (PFM13/MEL1)	1p36.23-p36.33	(−) B-cell lymphoma, liver, colorectal, stomach, breast, etc.	Yin–yang, paralog of MDS1-EVI11	NP_071397 BAB84297
PRDM17 (PFM14)	11p12-p14	(−) Breast, ovarian, prostate, testis Bladder, gastric, lung, Wilms tumor	?	AAK29075

(continued)

TABLE 8.1 (continued)

Human PR and SET Domain Containing Genes: Properties and Chromosomal Location/Links to Human Cancers

Genes	Chromosomes	Cancer Types	Properties	Genbank ID
SET-domain family				
SUV39H1	Xp11	t sarcoma	H3K9, binds Rb	XM_047059
SUV39H2	10p12-p14	(−) Head/neck cancer	Paralog of SUV39H1	XM_083831
EZH1(ENX2)	17q21	(−) Breast cancer, near BRCA1	SET-missing mRNA, yin–yang	XM_008394
EZH2(ENX1)	7q35-q36	Myeloid disorders, uterus	H3K9, H3K27	XM_053967
MLL1 (ALL1/HRX)	11q23	t leukemia; (−) garstric cancer	H3K4, binds Sbf1, yin–yang	NM_005933
MLL2 (ALR)	12p13.1	(−) Breast and lung cancer	H3K4, ALL-1 related or paralog 1	AAC51734
MLL3 (KIAA1506)	7q35-q36	Myeloid disorders, uterus	H3K4, ALL-1 related or paralog 2	NM_021230
MLL4 (KIAA0304)	19q13.1	(+) Pancreatic cancer	H3K4, ALL-1 paralog 3	XM_028760
MLL5 (FLJ10078)	7q23-q31	(−) Breast, colon, etc	H3K4, ALL-1 paralog 4	XP_004843
ESET(SETDB1/KG1)	1q21	(−) Breast cancer	H3K9me3	NP_036564
G9a (EHMT2/Bat8)	6p21.3	(−) Breast cancer; (+) gastric,	H3K9me2	CAA49491
EuHMTase (EHMT1/GLP1)	9q33-34	(−) Bladder cancer	H3K9me2, G9a paralog	BAB56104
ASH1	1q22	(−) Breast cancer	H3K4, H3K9, H4K20, H3K36	NP_060959
NSD1(ARA267)	5q35	(−) Liver cancer; 5q syndrome; t(5;11)(q35p15.5) in leukemia	H3K36, H4K20	AAK92049
MMSET (NSD2 /WHSC1)	4p16	t multiple myeloma, t(4;14)	Yin-yang, NSD1 paralog	NP_579878
REIIBP	4p16	t multiple myeloma	Repressor of IL-5 gene expressed from NSD2 intron	AAK00344
NSD3 (WHISTLE)	8p12	(+) In breast cancer (−) Melanoma, Prostate	H3K4, K27, NSD2 paralog	CAC28351
CLLD8 (CLLL8)	13q14	(−) B-cell, many tumors, near Rb	?	AAK38373
Suv420H1 (CGI-85)	11q13	(−) t leukemia/solid tumor	H4K20me3	NP_057112

Gene	Location	Cancer type	Target/Function	Accession
Suv420H2	19q13	(−) Glioma	H4K20me3	NM_032701
KIAA1757 (SETD5)	3p26.3	(−) Pancreas, bladder, nasopharyngeal, near VHL	?	NP_060657
PR-SET7 (SET8)	13q11-q13	(−) Breast cancer	H4K20	AF287261
SET9 (KIAA1717/SET7)	4q28	(−) Liver and brain tumors / (+) Osteosarcoma	H3K4	AL69901
KIAA1076	12q24	(−) Pancreatic, ovarian, stomach cancers	?	XP_037523
KIAA0339 (SETD1A)	4q21	(−) Colorectal cancer	Paralog of KIAA1076 / Near PRDM8	BAA20797
HYPB (KIAA1732/SETD2)	5q31	(−) Colorectal, lung, ovarian, leukemia and stomach cancers	Binds huntinting	CAC28349
SETMAR	3p25.1-3pter	(−) Pancreas, bladder	SET fused with transposase	NM_006515
SET27H	?	?	Novel EST	AW176331
S-ET domain or SMYD family				
Smyd1 (Bop)	2p11	t lymphoma/leukemia	H3K4	U76374
Smyd2 (HSKM-B)	1q41	(−) Liver, breast cancer	Methylates H3K36 and p53K370	NP_064582
Smyd3 (ZNFN3A1(FLJ21080))	1q44	(−) Liver, breast, prostate cancer	H3K4	NP_073580
Smyd4 (KIAA1936)	17p13	(−) Many cancers	?	BAB67829
NN8-4AG	2p13	(−) Metanephric adenomas	Retinoic acid induced	U50383
Rubisco ISMT homolog (20% identical):				
FLJ21148	16q13	(−) Aipose tissue tumors	?	NM_024860

Note: The table lists, to the best knowledge of the author at this writing, all presently cloned or deposited PR domain genes (17 total or 15 if excluding paralogs), and SET domain genes (33 total or 17 if excluding paralogs). In addition, 5 related human genes contain a divergent SET domain that has an insertion within the domain a MYND ZF motif; these genes are grouped under the name S-ET domain or SMYD family. Finally, one human gene shows ~20% overall identity to the plant lysine methyltransferase Rubisco ISMT and is more related to the plant enzyme than to either PR or SET domain. Alternative gene names are in braces. The cancer types listed are those that carry alterations in the chromosome where a PR or SET gene is located. t, chromosomal translocation; (−) chromosomal deletion or loss of heterozygosity; (+), chromosomal amplification; ?, not known.

pheochromocytomas, 26% in gallbladder carcinomas, less than 5% in multiple myelomas, and less than 5% in medulloblastomas. The papers on RIZ1 promoter methylation are too numerous to be referenced here. The major types of cancers where RIZ1 status have yet to be examined include endometrial cancer, cervical cancer, pancreatic cancer, testis tumors, kidney cancer, glioblastoma, oral cavity cancer, head and neck cancer, and larynx cancer.

The function of RIZ1 may be to repress basal level gene transcription by methylating histone H3 on lysine 9. The target genes of RIZ1 may include IGF-1 and estrogen targets such as pS2 [83,84]. Some RIZ1 targets would presumably be oncogenes but they remain to be identified. RIZ1 has DNA-binding activity and transcription repressor activity. It binds Sp1 element GGGCGG using its N-terminal ZF region ZF1–3 [85]. It also binds the motif CTCATATGAC [86] and the TTGGC motif [87]. RIZ1 expression is not regulated by most cell signaling agents. Although it is phosphorylated on multiple Ser/Thr residues (residues 641T, 642S, 738T, 739S, 740S, 743S, 745S, and 749S), such phosphorylation is not regulated by epidermal growth factor [88].

Induction of gene transcription involves a derepression of RIZ1 mediated repression [84]. One of the RIZ1 target genes is the pS2 gene [84]. RIZ1 normally binds to the pS2 gene promoter and silences it by methylating H3K9. Induction of pS2 gene transcription by estrogen removes RIZ1 from direct binding to the promoter and causes a decrease in H3K9 methylation. RIZ1 may in turn be turned into a coactivator with estrogen receptor [87,89]. The Pro704-plus polymorphism has higher coactivator function and modulates the impact of estrogen on bone mineral density [90]. A role for RIZ1 in maximum estrogen response in vivo is demonstrated in mice lacking RIZ1 [84]. Derepression of RIZ1-mediated repression may also involve demethylation by the histone demethylase LSD1 that is recruited by estrogen treatment [91]. LSD1 antagonizes the function of RIZ1 and vice versa [92]. Knock down of LSD1 represses pS2 transcription while partial knock down of RIZ1 by siRNA is sufficient to confer full activation of pS2 gene transcription by unliganded estrogen alpha receptor [91]. It is unclear, however, whether a small amount of residual RIZ1 protein may act as coactivators in the siRNA experiment. In the complete knock out model of RIZ1 in mice, complete RIZ1 deficiency is not sufficient to confer full activation of most estrogen target genes.

A recent study examined the regulation of several HMTs by methyl nutrients in rats using Western blot analysis of crude tissue extracts [93]. At 2.25 and 4.5 months of treatment with methyl-deficient diet, no significant change was found for Suv39h1 and PR-SET7. A 25% reduction was found for RIZ1 and Suv4-20h2 at 4.5 months. At 9 months of diet treatment, all 4 HMTs examined showed lower level in methyl-deficient diet. Since significant regulation of these HMTs did not occur until 9 months after diet treatment, those changes seem unlikely to represent the early initiating event in dietary carcinogenesis. However, it is likely that the antibody (from Abcam) used in the Pogribny et al. paper was detecting a non-RIZ1 related protein. RIZ1 and most other HMTs are expressed at very low levels and we are not aware of any antibody that can detect endogenous RIZ1 in a Western blot of crude tissue or cell extracts. In our experience working with RIZ1, we could only detect endogenous RIZ1 protein in a two step fashion, first by immunoprecipitation of crude tissue or cell extracts followed by Western blot of the immunoprecipitates using a different RIZ1 antibody (for e.g., see the paper by Steele-Perkins et al.) [77]. In this fashion, we can detect a doublet of both RIZ1 and the alternative shorter product RIZ2. Most commercial antibodies for HMTs remain to be characterized regarding their target specificity and usefulness in Western blot analysis of crude cell extracts. Thus, the regulation of RIZ1 and other HMTs by methyl nutrients remains an important open question.

8.3.2 PRDM3 (MDS1-EVI1)

Like the RIZ1 gene, the MDS1-EVI1 gene at 3q21q26 also produces two mRNA and protein products of different length, MDS1-EVI1 and EVI1, that differ only in the presence or absence of the PR domain located at the amino terminus [94]. This gene is not yet known to have HMT activity.

The 3q21q26 syndrome is a group of diseases with a recurrent translocation, inversion, or insertion between the regions of 3q21 and 3q26 and is associated with MDS or AML. 3q21q26 syndrome has specific clinical features, including normal or elevated platelet counts at the initial diagnosis, hyperplasia with dysplasia of megakaryocytes, poor response to chemotherapy, and poor prognosis. The chromosomal breakpoints at 3q26 are clustered at the 5′ region of the *EVI1* gene (ecotropic virus integration site 1) in t(3;3)(q21;q26) and at the 3′ region in inv(3)(q21q26) [95]. However, the breakpoints at 3q21 in both t(3;3)(q21;q26) and inv(3)(q21q26) are clustered within a 50 kilobase (kb) region near the ribophorin I (*RPN1*) gene, which is a member of membrane proteins of rough endoplasmic reticulum. On the basis of these results, it is suggested that the region of 3q21 with the *RPN1* gene translocated to the q26 region near the *EVI1* gene may activate *EVI1* expression as an enhancer element.

The *Evi1* gene was originally identified as a proto-oncogene activated in murine myeloid leukemias by retroviral insertional mutagenesis [96]. EVI1 plays an important role in mouse development [97,98]. Transcriptional activation of the human *EVI1* gene occurs in myeloid leukemias and myelodysplastic syndromes as a consequence of chromosomal translocations and insertions involving chromosome band 3q26. Abnormal expression of *EVI1* has also been demonstrated in patients with a normal karyotype, suggesting that activation of the *EVI1* gene may occur via other mechanisms as well. The *EVI1* gene encodes a nuclear 145 kDa Cys2His2-type ZF protein that contains two separate domains of DNA-binding ZFs. The N-terminal domain (ZF1–7) recognizes the sequence GA(C/T)AAGA (T/C)AAGATAA. From in vitro studies it is apparent that *EVI1* is a proto-oncogene capable of disturbing normal cellular proliferation and differentiation in certain cell types [99–101].

The mechanism by which *EVI1* exerts these biological activities remains to be elucidated. Although *EVI1* has been described as a possible transcriptional activator, *EVI1*-mediated repressor activity has also been reported. Perturbation of transforming growth factor-signaling has been proposed as a repressive mechanism of action by which EVI1 may contribute to leukemogenesis. EVI1 has also a strong effect on cell proliferation, and it was shown that it accelerates the cell cycle by interaction with BRG1, leading to activation of the E2F promoter [102]. Transgenic mouse models of EVI1 have been developed. One of these models developed normally without signs of leukemia or myelodysplasia, but they showed defects in erythropoiesis [103]. In another line of transgenic model, EVI1 was found to induce a fatal disease of several stages that is characterized by severe pancytopenia [104]. The disease does not progress to acute myeloid leukemia. Thus, this mouse model confirms an association between *EVI1* and myelodysplastic syndrome.

MDS1 (myelodysplastic syndrome 1) was first cloned as one of the partner genes of *AML1* in the t(3;21)(q26;q22), associated with therapy-related acute myeloid leukemia and myelodysplastic syndrome as well as with chronic myeloid leukemia in blast crisis [95]. The MDS1/EVI1 gene, but not the EVI1 gene, contains the PR domain. The PR domain is commonly disrupted by chromosomal translocation and virus integration events in human and murine myeloid leukemia. As a result, the *EVI1* gene becomes overexpressed and is thought to directly contribute to malignant transformation and may act as a dominant negative mutant to inhibit the function of *MDS1-EVI1* [105]. These observations suggest that MDS1-EVI1, unlike EVI1, may play an anti-oncogenic role.

MDS1-EVI1 and *RIZ1* genes thus share remarkable similarities in the differential involvement of their two different products in tumorigenesis. The PR-containing product appears anti-tumorigenic and commonly disabled while the PR-deficient product appears oncogenic and uniformly present or overexpressed in tumor cells. This yin–yang phenomenon is expected to be repeated in other family members [56], and again highlights the critical role for the PR domain in the tumor suppression function of these genes.

8.3.3   PRDM16 (MEL1/PFM13)

MEL1 (1p36.3) (MDS1-EVI1-like 1) encodes a ZF protein that shares 63% sequence similarity to *MDS1-EVI1*, with a similar domain structure [106]. Two different *MEL1* products of 170 and 150 kDa were identified by immunoblotting analysis, designated as full-length MEL1 and short-form MEL1S, respectively [107]. Like *MDS1-EVI1* and *EVI1*, two mRNAs with and without the PR domain would be transcribed from this locus. MEL1 is the PR-containing form, with the PR domain coded from codon ATC91 (exon 2) to codon CCC223 (exon 5) [106], and MEL1S is the PR-lacking form initiated from an internal codon ATG599 (exon 9) [107]. The fusion of MEL1 or MEL1S to GAL4 DNA-binding domain made them GAL4-binding site-dependent transcriptional repressors. Moreover, overexpression of *MEL1S* blocked granulocytic differentiation induced by G-CSF in IL-3-dependent murine myeloid L-G3 cells, while *MEL1* could not block the differentiation. From these results, it was suggested that overexpression of *MEL1*S could be one of the causative factors in the pathogenesis of t(1;3)-positive myeloid leukemia cells [107]. Consistently, overexpression of MEL1S, due to promoter hypomethylation, has been reported in adult T-cell leukemias and confers resistance to TGF-beta signaling [108]. Abnormal activation of MEL1S transcript appears to have immortalization activity in primary bone marrow progenitor cells [109].

There are conflicting reports on the expression of MEL1 in normal tissues. It was first reported that the PR-lacking form *MEL1S* is expressed in leukemia cells with t(1;3) but not in normal bone marrow, suggesting that the ectopic expression of *MEL1S* is specific to the t(1;3)-positive MD͞sAML [104,107]. In contrast, a recent study shows that both MEL1 and MEL1S are widely expressed in normal tissues, including bone marrow [110]. The 1p36 location of PRDM16 suggests a potential role of this gene in tumor suppression. Interestingly, a CpG island within the gene is methylated in 55% of lung cancer [111]. It remains unclear if this gene has HMT enzyme activity.

8.3.4   PRDM1(BLIMP1/PRDI-BF1)

This gene was first cloned as a transcriptional repressor named PRDI-BF1 [112]. It was later independently cloned based on its high-level expression in differentiated cells versus nondifferentiated cells [57,113]. Overexpression of *PRDM1* can drive B-cell maturation into plasma cell [113]. The gene is a repressor of c-myc transcription [114]. The capacity of *PRDM1* in driving cell differentiation, apoptosis, and repressing *c-myc* oncogene expression suggests a potential for *PRDM1* in tumor suppression. Consistently, *PRDM1* maps to chromosome band 6q21-q22.1 that is commonly deleted in several types of human cancers including gastric carcinoma, melanoma, and B-cell non-Hodgkin lymphomas [115].

Interestingly, PRDM1 gene also expresses two different length products that differ only in the presence or absence of an intact PR domain [116]. PRDM1-beta lacks the box A of the PR domain and is produced from an internal promoter. While PRDM1-alpha has transcription repression activities, PRDM1-beta lacks repression function but retains DNA-binding activity. PRDM1-beta is expressed at low levels in normal tissues, but its expression is highly elevated in myeloma cell lines. So, PRDM1 is similar to PRDM2 and PRDM3 in displaying a yin–yang pattern in oncogenesis. PRDM1 has been shown to be mutated or inactivated in defuse large B-cell lymphomas [117,118]. The mutations also often just affect the PR-containing form PRDM1-alpha, consistent with the yin and yang model of PR domain family genes in cancer.

PRDM1 functions as a critical fate-determining factor in mouse germ cells [119]. It has not been shown to have HMT activity but can associate with other SET domain HMTs (G9a) as well as the arginine methyltransferase PRMT5 [120,121]. PRDM1 plays a role in both B and T cells [122,123]. It also functions in epidermal lineage commitment and is important in maintaining the lineage of unipotent progenitor cells [124]. Here, loss of PRDM1 leads to increased cell proliferation and hyperplasia.

8.3.5   PRDM4 (PFM1/SC-1)

The *PRDM4* gene was independently cloned from human tissues based on the sequence homology to the RIZ1's PR domain [125], and from rat tissues based on its ability to bind to the p75 NGF receptor [126]. It maps to a tumor suppressor locus on 12q23-q24.1 that is commonly deleted in ovarian, gastric, and pancreatic cancers [125]. Expression of this gene can be induced by serum starvation and nerve growth factor, suggesting a role in promoting cell growth arrest and differentiation [125,126]. Consistently, overexpression of PRDM4 causes inhibition of DNA synthesis [126]. It remains unclear whether PRDM4 has HMT activity.

8.3.6   PRDM5 (PFM2)

The PRDM5 gene maps to human chromosome 4q26, a region thought to harbor tumor suppressor genes for breast, ovarian, liver, lung, colon, and other cancers [127]. The gene has a CpG island promoter and is silenced in human breast, ovarian, and liver, colon, and gastric cancers through promoter hypermethylation [127,128]. A recombinant adenovirus expressing PRDM5 caused G2/M arrest and apoptosis upon infection of tumor cells [127]. These results suggest that inactivation of PRDM5 may play a role in carcinogenesis. PRDM5 lacks apparent HMT activity but interacts with the G9a HMT [129]. It may play a role in transcriptional control of hematopoiesis [129].

8.3.7   PRDM6 (PRISM)

The PRDM6 or PRISM (PR domain in smooth muscle cell) gene is cloned from genes over-expressed in smooth muscle cells [130]. The gene maps to 5q21-23, a region commonly deleted in colon cancer, lung cancer, ovarian cancer, leukemia, and stomach cancer. PRDM6 acts as a transcriptional repressor by interacting with class I histone deacetylases and the G9a SET domain HMT. Overexpression of PRDM6 in cultured primary smooth muscle cells induces genes associated with the proliferative smooth muscle phenotype while repressing regulators of differentiation, including myocardin and GATA-6. Conversely, small interfering RNA-mediated knockdown of PRDM6 slows cell growth and induces myocardin, GATA-6, and markers of smooth muscle differentiation. These results appear to show that PRDM6 may have oncogenic properties, which is not consistent with the finding on most other PRDM family genes. However, PRDM6 also appear to express a PR domain minus product due to alternative splicing (GenBank accession number AAF78079 and AAF78078). In normal tissues, the PR-minus form is expressed at a higher level than the PR-plus form (unpublished observations). The siRNA experiments of Davis et al. would knock down both forms of PRDM6 [130]. Therefore, it is possible that the growth promoting activity of PRDM6 observed by Davis et al. is mediated by the PR-minus form. In addition, it has been found that PRDM6 inhibits cell proliferation in endothelial cells and has H4K20 methylation activity [131].

8.3.8   PRDM9 (Meisetz)

The PRDM9 or Meisetz (meiosis-induced factor containing a PR/SET domain and ZF motif) is a histone methyltransferase that is important for the progression of early meiotic prophase [132]. The gene is on 5p14 with a nearly identical gene PRDM7 on 16q24. The expressions of both of these genes are extremely low in most tissues. PRDM9 transcripts are detected only in germ cells entering meiotic prophase in female fetal gonads and in postnatal testis. PRDM9/Meisetz has catalytic activity for trimethylation, but not mono- or dimethylation, of lysine 4 of histone H3, and a transactivation activity that depends on its methylation activity. Mice in which the PRDM9/Meisetz gene is disrupted show sterility in both sexes due to severe impairment of the double-stranded break repair pathway, deficient pairing of homologous chromosomes, and impaired sex body formation. In PRDM9/Meisetz-deficient testis, trimethylation of lysine 4 of histone H3 is attenuated and meiotic gene transcription is altered. These findings indicate that meiosis-specific epigenetic events in mammals are crucial for proper meiotic progression.

8.3.9 PRDM14

This gene maps at 8p12-p21 region. While this chromosome region is commonly deleted in prostate cancer and melanoma, it is unclear whether PRDM14 represents the specific target of deletion and many other genes also locate in this region. A CpG island within this gene is methylated in 90% of lung cancer [111]. It is not known if this gene has HMT activity.

8.3.10 MLL1 (ALL-1/TRX1/HRX)

Similar to the PRDM gene family, many genes in the canonical SET domain family are involved in cancer, mostly as candidate tumor suppressors but also some as oncogenes. The MLL1 gene is a human homolog of the *Drosophila trithorax*. The gene is disrupted by chromosomal translocations involving chromosome 11q23 in human acute leukemia [133–136]. MLL1 is a large protein of 3969 amino acids and the SET domain is located at the C-terminal end of the protein (residue 3840–3969). The chromosome translocations result in expression of chimeric proteins composed of the N-terminal ∼1300 residues of *MLL1* linked to a C-terminal polypeptide encoded by any (∼30) of the partner genes. Because the partner genes fused to *MLL1* are diverse, ranging from transcription factors to proteins involved in signal transduction, the underlying mechanism for *MLL1* in tumorigenesis is unknown. It seems implausible that each of the fusion genes in and of itself is oncogenic, which would imply multiple distinct oncogenic mechanisms associated with *MLL1* translocation. On the other hand, none of the overexpressed fusion proteins of *MLL1* contains the SET domain. This observation suggests a possible negative selection for SET domain functions in tumor cells. There are several lines of evidence that the SET-minus chimeric proteins exert a dominant negative effect on the normal ALL-1 protein encoded by the intact allele present in the leukemic cells [137–139]. Also, *MLL1* inactivation may be involved in solid tumors [140]. These observations suggest that *MLL1* may function as a tumor suppressor. However, oncogenic activities for MLL1 fusion transcripts have been described [141]. It is possible that both either loss or gain of MLL1 function may be involved in carcinogenesis. MLL1 is part of a large protein complex and has H3-K4 methylation activity [37,39].

The yeast Dot1 and its human counterpart, hDOT1L, methylate lysine 79 located within the globular domain of histone H3. hDOT1L interacts with AF10, an MLL1 fusion partner involved in acute myeloid leukemia, through the OM-LZ region of AF10 required for MLL1-AF10-mediated leukemogenesis [142]. Direct fusion of hDOT1L to MLL1 results in leukemic transformation in an hDOT1L methyltransferase activity-dependent manner. Transformation by MLL1-hDOT1L and MLL1-AF10 results in upregulation of a number of leukemia-relevant genes, such as Hoxa9, concomitant with hypermethylation of H3-K79. Thus, mistargeting of hDOT1L to Hoxa9 plays an important role in MLL1-AF10-mediated leukemogenesis.

8.3.11 MLL2 (ALR)

This paralog of MLL maps to 19q13.1 and is amplified in solid tumor cell lines [143]. This gene has H3 lysine 4 trimethylation activity and is in a large complex with the tumor suppressor Menin, a product of the MEN1 gene mutated in familial multiple endocrine neoplasia type 1 [144]. Menin is required for MLL2 activity. A subset of tumor-derived Menin mutants lacks the associated histone methyltransferase activity. The data suggest that histone methylation by MLL2 is important in the tumor suppressor pathway of Menin.

8.3.12   EZH2 (ENX1)

This gene is the human homolog of the *Drosophila* polycomb group protein E(Z). The protein exists in a protein complex that has methyltransferase activity on lysine 9 and lysine 27 of histone H3 [35,36,38,145]. However, EZH2 protein alone lacks enzyme activity. EZH2 is required for early mouse development and B-cell development [146,147]. The gene maps to 7q35-q36 and is commonly overexpressed in metastatic prostate cancer [148]. Many genes associated with CpG islands undergo de novo methylation in cancer. Genes methylated in cancer cells are specifically packaged with nucleosomes containing histone H3 trimethylated on Lys27 [149]. This chromatin mark is established on these unmethylated CpG island genes early in development and then maintained in differentiated cell types by the presence of an EZH2-containing Polycomb complex. In cancer cells, as opposed to normal cells, the presence of this complex brings about the recruitment of DNA methyl transferases, leading to de novo methylation. These results suggest that tumor-specific targeting of de novo methylation is pre-programmed by an established epigenetic system that normally has a role in marking embryonic genes for repression.

8.3.13   G9A (BAT8/EHMT2)

The gene G9a and its paralog EuHMTase (GLP) dimethylates H3K9 [150]. Upregulation of G9a and Ezh2 has been found in liver cancers [151]. G9a may play an oncogenic role in silencing tumor suppressor genes [152]. G9a has been found to interact with three PRDM proteins, PRDM1, PRDM5, and PRDM6 [120,129,130].

8.3.14   ESET (SETDB1)

The gene ESET/SETTB1 trimethylates H3K9 [153,154]. It interacts with DNMT3A and DNMT3B and may play a role in gene silencing in cancer [155].

8.3.15   SUV39H1/2

The SUV39H1 and its paralog gene SUV39H2 have H3-K9 methylation activity [43]. The role of the two genes in human cancers remains largely unknown or uncharacterized. However, a tumor suppressor function is suggested by mice deficient in SUV39H1/2 function, which show chromosomal instability and are prone to develop B-cell lymphomas [156]. It is interesting to note that these tumors are similar to those developed in RIZ1 mutant mice [77]. SUV39H1 is also an Rb-binding protein and can be recruited by the Rb-E2F complex for transcriptional repression of E2F-responsive promoters [157]. SUV39H1 methylates histone H3 at lysine 9 and creates a binding site for heterochromatin protein 1 (HP1) [48,49]. The combined action of SUV39H1 and HP1 leads to heterochromatin formation and gene silencing. Loss of SUV39H1 would lead to impaired ability of Rb to repress E2F1-responsive genes, thus contributing to the inactivation of the one of the best-characterized tumor suppressor pathway. In addition, the SET domain of SUV39H1/2 interacts with a phosphatase like oncoprotein Sbf-1 [158]. So, the transforming activity of Sbf-1 may be in part mediated by deregulation of SUV39H1/2 function.

Suv39H1 functions in the tumor suppressor function of cellular senescence. Acute induction of oncogenic Ras provokes cellular senescence involving the Rb pathway. The Rb-mediated silencing of growth-promoting genes by heterochromatin formation associated with methylation of histone H3 lysine 9 was identified as a critical feature of cellular senescence, which may depend on the histone methyltransferase Suv39h1 [159]. It has been shown that H3K9-mediated senescence is a novel Suv39h1-dependent tumor suppressor mechanism whose inactivation permits the formation of aggressive but apoptosis-competent lymphomas in response to oncogenic Ras [11].

8.3.16 Suv4-20H1/2

These two related genes have H4K20 trimethylation activity and play a role in heterochromatin formation [160]. They bind to Rb and mediate the role of Rb in forming heterochromatin [161]. The ability to facilitate Rb function coupled with the fact that H4K20 trimethylation is commonly lost in cancer [162] suggests that Suv4-20H1/2 may function as tumor suppressors. Consistently, reduction in Suv4-20H2 has been found in liver carcinogenesis induced by methyl-deficient diet [93].

8.3.17 SET9 (SET7)

SET9/SET7 methylates H3 lysine 4 and maps to 4q28 [163,164]. Set9 also specifically methylates p53 at one residue 372 within the carboxyl-terminus regulatory region [165]. Methylated p53 is restricted to the nucleus and the modification positively affects its stability. Set9 regulates the expression of p53 target genes in a manner dependent on the p53-methylation site. It is also involved in the stabilization of p53 that is necessary for inducing apoptosis in response to DNA damage [166]. By augmenting p53 activity, SET9 may be expected to function as a tumor suppressor.

8.3.18 PR-Set7 (Set8)

The PR-Set7 gene maps to 13q11 that is often deleted in breast cancer. The gene has H4 lysine 20 monomethylation activity and is involved in mitosis [52,53].

8.3.19 NSD1

NSD1 (nuclear receptor binding SET domain protein 1) was cloned as a coactivator of nuclear receptors [167]. NSD1 protein methylates histone H3 Lys-36 and H4 Lys-20 [167]. NSD1 is fused with NUP98 gene in the t(5;11) (q35;p15.5) in AML [169]. Both fusion transcripts are expressed and it remains unclear whether the fusion transcripts act as an oncogene or as a dominant negative or both. Mutation in NSD1 causes Sotos syndrome, a disorder that is characterized by the overgrowth of neural tissues, heart defects, and increased risk of cancers [170].

8.3.20 NSD2 (MMSET/WHSC1)

This gene shows yin–yang expression in tumor tissues as a result of chromosomal t(4;14) translocation in myeloma [171]. In normal tissues, the MMSET gene expresses two products, MMSET-II and MMSET-I, the latter of which lacks the SET domain due to alternative splicing. The t(4;14) translocation in myeloma results in a greater degree of overexpression of MMSET-1 than MMSET-II or in increase in the ratio of MMSET-I to MMSET-II relative to that in normal tissues, indicating that MMSET-I might function as a dominant-negative regulator of the full-length product. The MMSET gene has also been found fused with IgH in t(4;14) myeloma. The effects of this fusion on the expression of the two transcripts remain unclear.

8.3.21 NSD3 (Whistle)

NSD3/Whistle is a paralog gene of NSD1 and NSD2. It maps to 8p12 and has been found to be fused to the NUP98 gene in AML associated with t(8;11) (p11.2;p15) [171]. NSD3 gene is also amplified in breast cancers [173]. However, the role of NSD3 in these cancers remains unclear. NSD3 methylates H3K4 and K27 and can induce apoptosis and repress transcription by interacting with HDAC1 [174].

8.3.22 Smyd3

The SET and MYND domain containing three gene Smyd3 (also ZNFN3A1 and FLJ21080) maps to 1q44 and is overexpressed in the majority of colorectal carcinomas and hepatocellular carcinomas. Introduction of SMYD3 into NIH3T3 cells enhanced cell growth, whereas genetic knockdown with small-interfering RNAs (siRNAs) in cancer cells resulted in significant growth suppression. The SET domain of SMYD3 showed histone H3-lysine 4-specific trimethyltransferase activity, which was enhanced in the presence of the heat-shock protein HSP90A [175]. Alleles of SMYD3 that confer higher expression levels appear to correlate with higher risk of breast cancer [176].

8.3.23 SMYD2

Like Smyd3, the closely related protein Smyd2 (also HSKM-B) on 1q41 also has oncogene activity [177]. The tumor suppressor p53 is one of only a few nonhistone proteins known to be regulated by lysine methylation. Smyd2 methylates Lys 370 in p53. This methylation site, in contrast to the site Lys 372, is repressing to p53-mediated transcriptional regulation. Smyd2 helps maintain low concentrations of promoter-associated p53. Set9-mediated methylation of Lys 372 inhibits Smyd2-mediated methylation of Lys 370, providing regulatory cross talk between posttranslational modifications. In addition, the inhibitory effect of Lys 372 methylation on Lys 370 methylation is caused, in part, by blocking the interaction between p53 and Smyd2. Thus, similar to histones, p53 is subject to both activating and repressing lysine methylation. In addition to methylating p53, Smyd2 is known to methylate Lys 36 of histone H3 [178].

8.3.24 OTHER *PR/SET* DOMAIN GENES

The human genome is estimated to contain ~17 PRDMs and ~33 SET domain genes (Table 8.1). Location to chromosomal regions commonly deleted in human cancers is a common feature for these genes. These are generally heterozygous deletions but could still be significant due to haplo-insufficiency. While many of these genes have yet to be studied in the context of cancer, those that have been studied so far are remarkably similar in their role as tumor suppressor genes, especially for the PR domain family. The yin–yang pattern of the PR domain was conserved in five PRDMs, PRDM1, 2, 3, 6, and 16. While several canonical SET genes appear to be tumor suppressors, some may be oncogenic such as Ezh2, Smyd2, and Smyd3. Thus, unlike the canonical SET-domain family, the PRDM family is more conserved in having a tumor suppressor function. Overall histone methylation deficiency is common in human cancer [162] but this may not mean a lack of regional or small scale hypermethylation. This may be similar to the situation with DNA, where both overall hypomethylation and regional hypermethylation occur.

It should be noted that the above discussion does not exclude the following possibilities. First, while most HMTs may act in tumor suppression by maintaining the normal epigenetic pattern, it is possible that some PR or SET domain-containing genes function as oncogenes. Some histone methylation activity must be required in the silencing of tumor suppressor genes [179–181] or in the mitotic inheritance of malignant phenotypes. Such activity may come from overexpressed HMTs or from certain house-keeping kind of HMTs. It could also come from the reduced but not absent activity of partially inhibited HMT tumor suppressors. Second, it is possible that some methyltransferase-activity negative SET or PR genes may nevertheless act as tumor suppressors. In such a scenario, other functions of a PR or SET domain may be involved, such as the protein-binding activity [59,182–184]. However, one can never be sure about a negative enzyme activity for a PR/SET protein since one may not have used the correct assay conditions or used the right native protein or protein complexes. Many PR/SET proteins exist natively in huge multiprotein complexes and may lack enzyme activity when assayed alone in vitro. Finally, it is possible that certain PR or SET genes, regardless of enzyme activities, may not be involved in tumorigenesis. Methyltransferase function per se, regardless of substrates, does not necessarily predict a role in cancer, in view of the presence of many such enzymes that clearly do not have any role in cancer. There are 11 PRMTs or protein arginine methyltransferases but few of these have been linked with cancer. The Dot1 enzyme is involved in cancer by way of fusion with the SET-domain protein MLL1. So, the PR/SET family is significantly more involved in carcinogenesis than other types of SAM-dependent methyltransferases.

8.4 HISTONE DEMETHYLASES AS ONCOGENES

If histone lysine methyltransferases mostly function in tumor suppression, it is then expected that demethylases that remove lysine methylation may function as oncogenes. This appears to be the case for at least some members of the large family of demethylases [185,186], most of which have yet to be studied in relationship to cancer.

8.4.1 LSD1 (KIAA0601, P110B, BHC110, NPAO)

The lysine specific demethylase 1 is the first known histone lysine demethylase [185]. It appears to antagonize the function of RIZ1 in gene repression [91], and may therefore be expected to function as an oncogene. Indeed, high levels of LSD1 are commonly found in high-risk prostate cancers [187]. Knock down of LSD1 gene expression in MCF7 cells is associated with G2/M arrest and growth inhibition [188]. MCF7 cells express RIZ1. The G2/M arrest induced by knocking down LSD1 is consistent with an elevated activity of RIZ1, which is known to cause G2/M arrest. Inhibitors of LSD1 may be useful as cancer drugs. But if LSD1 works by antagonizing RIZ1, inhibitors of LSD1 would be expected to be the most effective in RIZ1 positive tumors.

8.4.2 GASC1 (JMJD2C)

The gene amplified in squamous cell carcinoma 1 (GASC1) was amplified and overexpressed in several esophageal squamous cell carcinomas cell lines. It is within a region that is commonly amplified in tumors. The deduced amino acid sequence of GASC1 contains two PHD-finger motifs and a JMJD2 demethylase domain or jumonji domain containing (JmjC) 2 subfamily of the jumonji family [189]. GASC1 demethylates H3K9me3/me2. Ectopic expression of GASC1 or other JMJD2 members markedly decreases H3K9me3/me2 levels, increases H3K9me1 levels, delocalizes HP1, and reduces heterochromatin in vivo. Inhibition of GASC1 expression decreases cell proliferation [165].

8.4.3 PLU-1 (JARID1B)

The Plu-1/JARID1B gene was first identified as an overexpressed gene in breast cancer [190]. The gene was later found to be an H3K4 demethylase of the JmjC domain family [191]. PLU-1-mediated H3K4 demethylase activity plays an important role in the proliferative capacity of breast cancer cells through repression of tumor suppressor genes, including BRCA1.

8.4.4 RBP2

The RB-binding protein RBP2 is a member of the JMJD2 demethylase family and removes trimethylated H3K4 [192,193]. Promotion of differentiation by RB involves inhibition of RBP2 function [194].

8.5 EPIGENETIC ORIGIN OF SPORADIC CANCERS

Alteration in epigenetic enzymes provides a mechanism for mitotic inheritance of carcinogenic effects independent of changes in DNA. Mutational inactivation is essentially similar to a persistent inhibition of protein function at the gene product level. Since mutations in HMTs cause cancer [77,156,170], persistent inhibition of HMTs at the protein level or down regulation of HMT expression will likely do the same. Regardless of mutations or not, reduced HMT function may lead to mitotically inheritable changes in chromatin methylation and gene expression patterns. Such abnormal chromatin patterns may result in the generation of neoplastic phenotypes, which may be inherited from one cell to its progeny.

Mechanisms that would persistently inactivate HMTs can be envisioned for most carcinogenic events. Some may in fact best account for carcinogenesis by the typical Western diet. The typical Western diet is rich in meat and low in vegetables/fruits, which is linked with one-third of human cancer death in the United States [2]. Numerous dietary factors have been hypothesized to influence cancer risks in Western countries, but for the moment the evidence for any one of these remains unclear, except for the methyl donor folic acid and meat (and in turn protein/methionine) [4,195,196]. High consumption of meat and low intake of folic acid in the form of fruits/vegetables seems to be the chief carcinogenic element of diet [4]. Methyl donor nutrients that need to be supplied by diet include methionine, folic acid, vitamin B6, and B12. Among these nutrients, methionine is special because too much or too little of it are both harmful to methylation. Methionine is needed to make SAM and too little intake of protein/methionine can reduce SAM levels. But too high intake of methionine coupled with low folate can reduce SAM levels and raise SAH levels [197]. The typical Western diet or any diet high in animal protein/methionine but low in plant foods/folate can be expected to raise SAH levels [27,29]. High methionine intake (threefold more than control) in adult rats has been shown to raise SAM by 10% but SAH by threefold [198]. This effect of high methionine intake may be specific for adult animals since young weanling rats did not show similar response to high methionine intake [199,200]. Inhibition of DNA methylation seems to be more significantly linked with an increase in SAH than a decrease in SAM or SAM/SAH ratio [201] but this may not be universally true [202]. Aging is associated with an increase in SAH, which correlates well with the fact that cancer is a disease of aging [203]. Human aging is also known to be associated with increased levels of homocysteine and decreased levels of plasma folate and B12 [204,205]. High methionine intake can increase homocysteine levels [205]. Several amino acids, including cysteine, glycine, serine, and threonine, can be metabolized to form homocysteine. High plasma homocysteine level correlates with elevated SAH levels and methyltransferase inhibition [206]. Mild increases in plasma total homocysteine have been correlated with a fivefold increase in lymphocyte SAH levels [207].

The authors of two prospective studies have interpreted their data to mean that red meat (beef, pork, and lamb) but not white meat (poultry) is carcinogenic [195,196]. This message has been widely publicized. But this may represent a misinterpretation of the actual data by the authors. A careful look at the actual data shows that people who ate red meat actually ate approximately twofold more meat in quantity than people who ate poultry [195,196]. The category of men who ate the most red meat ate 129.5 g/day whereas the category of men who ate the most poultry ate only 63.1 g/day [195]. Similarly for women, the category who ate the most red meat ate 134 g/day whereas the category who ate the most chicken and fish ate only 65 g/day [196]. Thus it may not be red meat per se that is oncogenic but may be the overconsumption of meat in general, and in turn of protein. Indeed, comparing the category of men who ate 64.1 g/day red meat versus those who ate 18.5 g/day showed no difference in cancer risk, just like the situation for poultry where men who ate 63.1 g/day had similar cancer risk as men who ate 15.8 g/day [195].

Animal meat proteins have approximately twofold higher methionine contents than plant proteins (data from the publication by the Food and Agricultural Organization of the United States titled "Amino-acid content of foods and biological data on proteins"). So overconsumption of animal meat is much more likely than overconsumption of plant food to result in excess-intake of methionine. A strict vegetarian diet has not been found to be cancer preventive [208], likely because such a diet may not provide enough protein-derived methionine to be able to maintain normal SAM and SAH levels. Most prospective studies in the Western countries have not found a significant association between protein and cancer but such studies may not be informative because of the limited range of variations in protein intake in Western countries [209]. Much wider range of variation in protein intake is found in international comparisons (such as the United States vs. China), which do show a strong link between excess intake of protein and cancer [210].

It has been a well-established phenomenon known for 40 years that excess dietary protein is carcinogenic in rodent liver [211]. But the mechanism has remained unclear to this day. Over-consumption of animal milk protein (20% casein diet) was carcinogenic in rodent livers while the same high amount of plant protein (20% gluten or soy protein diet) was not [212]. Neither was a 5% casein diet. The methionine content of casein is ~5 times higher than gluten or soy protein. So, the carcinogenic effect of overconsumption of milk protein is likely mediated by excess methionine intake, leading to lower SAM or higher SAH levels.

These past studies, together with the recent understanding of HMTs in tumor suppression, suggest that a cancer preventive diet should consist of a properly balanced amount of both animal and plant foods in order to maintain normal levels of SAM and SAH. While the exact amount of each food may need to be carefully determined by future studies, it is safe to recommend for most people in the United States today to cut their meat and milk intake by half, while increase their plant food intake by as much as necessary to meet the calorie needs. This is based on the fact that most people in the United States consume approximately threefold more proteins than necessary for normal homeostasis. Relative to total calorie intake, only 5%–6% dietary protein is required to replace the protein regularly excreted by the body (as amino acids). About 9%–10% protein, however, is the amount that has been recommended for the past 50 years to be assured that most people at least get their 5%–6% "requirement" [210]. Almost all Americans exceed this 9%–10% recommendation with an average of about 15%–16%. The average American dietary intake of protein (91 g/day) is much higher than the Chinese (64 g/day, standardized for a body weight of 65 kg/143 pounds) [210]. The protein source for Americans is mostly animals, while for Chinese mostly plants. The cancer incidence in China is 2.2 million in year 2002 with a total population of 1300 million whereas the cancer incidence in the United States is 1.1 million in 2002 with a total population of 281 million. So, the cancer incidence rate in the United States is 2.4 times higher than in China, which correlates well with the much higher intake of animal foods in the United States. So simply reducing animal food intake by half may handily reduce the United States cancer incidence rate by at least half. The identification of the molecular target (most likely an HMT enzyme) for methyl nutrients in cancer prevention would soon provide the theoretical rational for doing that. The present lack of a sound rational together with a confusing interpretation of various confusing epidemiological studies may be the main reason for why few people have recognized that too much animal protein as well as too little are both carcinogenic or harmful to methylation.

Among all vitamins that have been examined, only methyl donor vitamins are found to have some impact on cancer prevention (see http://www.hsph.harvard.edu/nutritionsource/vitamins.html). Notably, vitamins thought to protect DNA from oxidation-induced mutation fail to provide cancer protection. Since the message of methyl nutrients is about not too much and not too little (e.g., methionine), taking these nutrients in pill forms are unlikely to be as cancer preventive as eating balanced whole foods.

An increase in SAH has been shown to inhibit RIZ1 and to reduce histone H3-K9 methylation much more so than H3K36 and H3K4 methylation [64]. Such inhibition of RIZ1 may be critical to carcinogenesis because it is also the biochemical outcome of naturally occurring mutations in RIZ1 [64]. These observations suggest a molecular pathway of dietary carcinogenesis as follows:

(1) Typical Western diet—chronic increase in SAH or decrease in SAM in most or all cells of the target organ tissue
(2) Persistent partial inactivation of some HMT enzymes (especially H3K9 methylase) in most or all cells of the target organ tissue—heritable changes in chromatin methylation and gene expression pattern in most or all cells of the target organ tissue

(3) Survival and clonal expansion of one of the cells in the target tissue that has acquired extra carcinogenic hits that are either genetic or epigenetic—acquisition and inheritance of additional carcinogenic hits in one of the partial malignant cells

(4) Clonal expansion of the single cell carrying multiple hits that are either genetic or epigenetic—end stage cancer (Figure 8.1)

In this pathway, the initiating event of carcinogenesis is the inactivation of some HMT tumor suppressors in most or all cells of a target tissue, just like the initiating event in familial cancers is the heterozygous mutation of a cancer gene in all cells of the body (Figure 8.1). Such inactivation reduces the normalness of most or all cells of a target tissue so that when a second hit randomly occurs in one of those semi-normal cells, the single cell that carries the second hit can escape the fate of cell death or senescence to go on to become malignant. Heterozygous mutations or deletions in most familial cancer genes such as p53 and Apc are known to be carcinogenic due to haplo-insufficiency [213]. All cells in a person with a germline heterozygous p53 mutation are partially malignant or less than fully normal. Such a semi-normal cell may proliferate in response to a

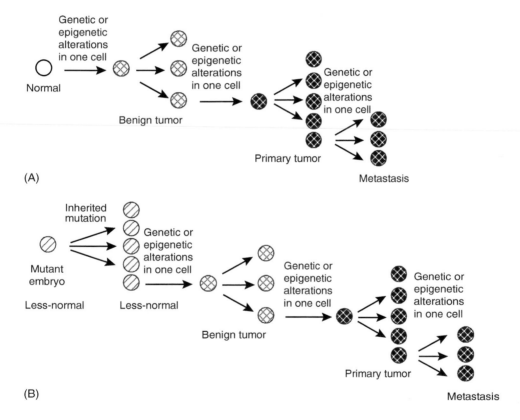

FIGURE 8.1 The multicell origin model of cancer initiation versus the classical single-cell origin model. (A) The classical model of sporadic cancers. (Modified from Feinberg, A.P., Ohlsson, R., and Henikoff, S., *Nat Rev Genet* 7, 21, 2006.) Here, the initiating event is some genetic or epigenetic alterations in a single normal cell that can directly trigger clonal proliferation. (B) The multicell origin model for familial cancers. Here, the initiating event is a germline mutation (either heterozygous or homozygous as in p53 or RIZ1 knockout mice) that results in all cells of the organism being less than fully normal but does not cause clonal proliferation. One of these less normal cells later acquires other genetic or epigenetic alterations and consequently proliferates clonally.

(continued)

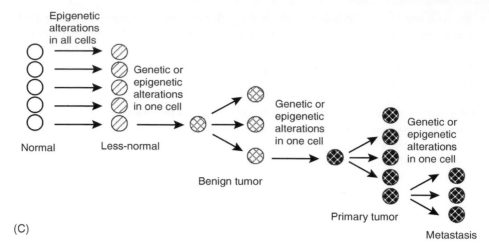

(C)

FIGURE 8.1 (continued) (C) The multicell origin model for sporadic cancers linked with environmental factors. In this model, cancer is initiated by environmental effects on a specific epigenetic enzyme(s) at the enzyme activity level or gene expression level in most or all cells of a target tissue resulting in most or all cells becoming less than fully normal, being in a state similar to all the cells in an organism with inherited cancer mutation. Just like the rate-limiting step in familial cancers is an inherited mutation resulting in all cells being less normal, the rate-limiting step for sporadic cancers may be the partial reduction in epigenetic control in most or all cells of a target tissue. Once that happens over prolonged period of time, cancer is almost guaranteed sooner or later due to other stochastic genetic or epigenetic events. Such random oncogenic events normally would only cause apoptosis or senescence or nothing in a fully normal cell but would cause clonal proliferation in a less normal cell. Hence the key to prevent sporadic cancers may be at the initiation or rate-limiting step to maintain high epigenetic complexity that controls the apoptosis or senescence response to oncogenic events.

complete loss in p53 whereas a fully normal cell may undergo cell death or senescence in response to complete p53 loss.

In addition to inactivation by methyl donor deficiency, HMT activity can be chronically suppressed by other mechanisms, including down regulation of mRNA expression, gene silencing by promoter hypermethylation, mutations, posttranslational modifications, and protein–protein interactions. A carcinogenic condition may not inactivate all HMTs equally, so that most HMTs can remain active and maintain mitotic inheritance of a neoplastic phenotype. Also, the reduced enzyme activity of a tumor suppressor HMT may still be responsible in part for the epigenetic pattern of the tumor cells. Also, other epigenetic enzymes may be similarly affected. A general pathway of HMTs or other epigenetic proteins in oncogenesis may be as follows: (1) exposure of all cells of a target tissue to an environmental carcinogen, (2) inactivation of a subset of HMTs or alterations in other epigenetic proteins in most or all cells of the target tissue due to enzyme inhibition or down regulation of gene expression, (3) genetic or epigenetic alterations in other genes including HMTs in one of the cells of the target tissue, (4) survival and clonal expansion of the single cell, (5) additional cycles of hits and clonal expansion, (6) clonal expansion of a single cell with multiple hits that are either genetic or epigenetic, (7) establishment and maintenance of a cancer-specific genetic and epigenetic pattern (Figure 8.1C). The initial events in carcinogenesis involving HMTs are more likely to be inactivation of HMTs rather than activation, since most HMTs act as tumor suppressors and also since it is methylation deficiency rather than excess that causes cancers and accompanies aging, the number one risk factor of sporadic common cancers.

The Darwinian mutation theory of cancer commonly considers the initiating cause of cancer as the event that occurs in only one single cell of a tissue organ (Figure 8.1A). This view is also held by researchers who study DNA methylation, considering DNA hypermethylation of a tumor suppressor gene to be equivalent to mutations. In this view, an environmental carcinogen initiates cancer by causing a mutation (or DNA methylation) in a single cell that later expands clonally (Figure 8.1A). But the phenotypes of homozygous knockouts of p53 and other tumor suppressor genes show that the initiating cause such as loss of p53 is present in all cells despite not causing all cells to become cancer and despite taking a long time before one of the cells growing into a cancer by clonal expansion. For both familial and sporadic cancers, the first or initial hit that is either genetic or epigenetic must occur in most or all cells of a target tissue (Figure 8.1B and C). Without such a large pool of less than fully normal cells that have survived a first hit, there is little statistical chance for a second oncogenic hit to randomly occur to the same single cell that has previously acquired the first oncogenic hit. A large pool of less than fully normal cells guarantees within the lifetime of the organism that one of them will acquire and survive an oncogenic mutation by purely normal errors in DNA-related processes without any need of exposure to a mutagen or any need of an increase in mutation rate or genetic instability. An environmental carcinogen that is not a mutagen such as the typical Western diet may not induce mutations per se but may merely act to allow more spontaneous normal errors in DNA-sequence to be compatible with cell survival and clonal expansion. A mutation in a potent cancer gene in a fully normal cell does not usually cause proliferation but rather causes cell death or senescence. The reason for this may be that some HMTs like Suv39H1/2 are key players in the senescence response of cells to oncogene activation [11], and that some HMTs like SET9 augments p53 in apoptosis induced by DNA damage [165,166] and some like RIZ1 can directly cause apoptosis and G2/M arrest [211].

Unlike the neo-Darwinian mutation theory, the HMT epigenetic pathway suggests that an environmental carcinogen initiates cancer by reducing the epigenetic normalness of most or all cells in a target tissue or the whole organism, which allows subsequently one of these cells that has randomly acquired additional hits (either genetic or epigenetic) to escape the fate of cell death or senescence to be able to expand clonally (Figure 8.1C). The epigenetic pattern of the clonal pool of partially malignant cells is maintained in part by the persistent exposure to the environmental carcinogen. If this epigenetic pattern is reversed back to near normal state by withdrawing the carcinogen, the pool of partial malignant cells may not be able to continue to survive or proliferate. This explains why lifetime exposure to a carcinogen is necessary for carcinogenesis and why stopping carcinogen exposure after having been exposed early in life does have cancer preventive benefits. In contrast, if the initiating event is a carcinogen-caused mutation in a single cell that subsequently expands clonally, then stopping carcinogen exposure after having been exposed early in life should be expected to have little cancer preventive benefits. Indeed, familial cancer syndromes that are initiated by a cancer gene mutation are not known to be preventable by manipulating exposure to environmental carcinogens such as diet and smoking. If smoking or diet initiates cancer by causing a mutation in a familial cancer gene, then quitting smoking or changing from a carcinogenic diet early in life to a healthy diet later in life should not be expected to prevent sporadic cancers. But they do.

The multicell origin model of cancer (Figure 8.1C) was implied by the proposal that change in SAM or SAH level may directly inactivate HMT tumor suppressors [27], since SAM/SAH is present in every cell. This model, like the epigenetic progenitor model of Feinberg et al. [214], suggests that the initiating event for sporadic cancers linked with environment is epigenetic change in multiple cells. But unlike the Feinberg model, the multicell model suggests that most or all cells (such as all hepatocytes of liver) of a target organ tissue are epigenetically changed rather than just a small population of progenitor or stem cells. The multicell model does not require clonal proliferation of epigenetically changed progenitor cells. It is therefore a more general model that covers tumorigenesis originating from both de-differentiation of differentiated cells and blockage of differentiation of progenitor stem cells. Hepatocytes are not stem cells but are clearly the cells that give rise to hepatomas. Any progenitor or stem cells of any cell types are partially differentiated cells and may not give rise to tumors without some loss of its differentiated characters or reduction in its epigenetic complexity. The multicell model can be tested by comparing global gene expression pattern of an organ tissue soon after exposure to a carcinogen such as a methyl-deficient diet.

The epigenetic target gene(s) of the carcinogen should be among the genes whose expression pattern or enzyme activity is changed by the carcinogen, and should be responsible for the change in a subset of those genes that changed pattern. Indeed, if we know the target gene pattern of an HMT enzyme such as RIZ1 by way of a microarray analysis of RIZ1 knockout cells, we could examine if there is any specific enrichment of RIZ1 targets among changed genes in tissues exposed to a carcinogen. If there is, we would conclude that RIZ1 is altered by the carcinogen at least at the enzyme activity level (if we do not see changes in RIZ1 expression) in most cells of the target tissue soon after exposure to the carcinogen. Thus, the multicell model offers a practical approach to identify early epigenetic changes that initiate carcinogenesis. It is simple and practical because it does not require the difficult, if not impossible in many cases, isolation of a few progenitor cells from a target organ tissue as would be required by the epigenetic progenitor model.

Since methyl nutrient deficiency is well known to cause cancer based on both human and animal studies [27,29], it is clear that methyl nutrients can prevent cancer. About 70%–90% of human cancers are preventable and dietary methyl nutrients may account for a large portion of those preventable cancers. The establishment of a convincing case for methyl nutrients in cancer prevention could have a huge impact on the cancer statistics of the Western countries. Thus, a major challenge today in the big picture of cancer is to identify the direct molecular targets for methyl nutrients in cancer prevention. Such a target should, in the best-case scenario, satisfy several minimal criteria. It should be among the genes that are activated or upregulated by methyl nutrients. It should be inactivated or down regulated by a methyl-deficient diet relatively early in the process of carcinogenesis by methyl deficiency. Its early inactivation in most or all cells of a target organ tissue should be compatible with organism survival through most of its lifetime. It should require methyl nutrients to function and may have little methyl independent functions. It should have a relatively lower affinity for SAM and SAH and thus be more sensitive to inhibition by a moderate decrease in SAM or increase in SAH compared to most other methyltransferases. Its absence should attenuate nearly all of the tumor preventive benefits of methyl nutrients. Its absence alone should be tumorigenic but should not be more tumorigenic than a methyl-deficient diet. Its heterozygous mutation or partial loss of function should be in itself tumorigenic. Finally, a methyl-deficient diet should have little other carcinogenic effects other than inactivating the prime target. The study of gene knockout mice fed either a methyl-complete diet or a methyl-deficient diet provides a realistic approach to unequivocally identify the prime target for methyl nutrients in cancer prevention. Using this approach, we have studied the RIZ1 knockout mice over the past 3 years. The results suggest that RIZ1 satisfy nearly all of the essential criteria for the prime target for methyl nutrients in cancer prevention [215].

8.6   ROLE OF DNA METHYLATION

DNA methylation is inherited during mitotic division, and therefore should be able to underlie a mutation-independent cancer mechanism. Hypomethylation of DNA is common in cancer and may act via DNA mutation-based mechanisms [216,217], but it seems to also prevent cancer formation [218]. Hypermethylation of tumor suppressor genes has been well documented in cancer and is presently a major focus in the field of cancer epigenetics [219,220]. However, most major research and review papers on DNA hypermethylation fail to mention an obvious contradiction to the importance of DNA hypermethylation: it is methyl deficiency rather than excess that causes cancer. These papers rarely mention dietary carcinogenesis despite the primary importance of diet in the big picture of cancer or cancer epigenetics. It is likely that DNA hypermethylation, while important in some steps of the cancer progression pathway, may not provide the answer to one of the most important problems in cancer or cancer epigenetics, the problem of how one-third of cancer death in the United States is linked to the typical Western diet that is methyl deficient. In contrast, it is relatively easier to use the phenomenon of DNA hypomethylation to explain, albeit only in a speculative way, carcinogenesis by methyl-deficient diet. But the problem here is that few cancer genes are known to be affected in a significant and meaningful way by DNA hypomethylation. If there is a molecular pathway from moderate DNA hypomethylation to cancer, no one has found it yet. Unfavorable to the hypothesis of DNA hypomethylation in dietary carcinogenesis, methyl-deficient diet has been linked with increased DNMT activity or increased DNMT expression, in contrast to expectations [221,222]. In a summary of research on DNA hypomethylation and methyl-deficient diet, Kim states that "collectively, currently available evidence indicates that genomic DNA hypomethylation in the colorectum is not a probable mechanism by which folate deficiency enhances colorectal carcinogenesis." [223].

While DNA hypomethylation induced by DNMT1 knockouts in mice can indeed lead to cancer [224,225], it remains unclear whether moderate hypomethylation in the presence of normal DNMT expression does play any role in the reality of human cancer. This is similar to the situation with chemical carcinogens. While extremely high doses of a chemical carcinogen such as nitrosamine can indeed cause cancer in rodents, there is no evidence that the amount of nitrosamine typically encountered by humans does play any role in the cancer statistics of humans. This is also similar to the situation with some familial cancer genes such as BRCA1. While mutations in BRCA1 cause familial breast cancer, they are rarely involved in most sporadic breast cancers. The past 40 years of extensive study has clearly shown that cancer can be caused by many diverse mechanisms or agents but most of these may not be actually responsible for the cancer statistics of humans. The focus in cancer research may need to be shifted from the easy task of discovering what can cause cancer under unusual conditions to the more difficult task of discovering what does cause the common cancers under the human conditions of today. Only by doing so, can we hope to achieve a dramatic change in cancer statistics.

Silencing of tumor suppressor genes appears to be initiated by HMTs and can occur without DNA methylation [181]. Although eliminating DNA methylation affects histone methylation, it does not suggest a causal role for DNA methylation in silencing. This is similar to X-inactivation where silencing occurs after histone methylation but prior to DNA methylation [226]. It has also been shown that hypermethylation of promoters arises secondarily to transcriptional inactivation [227]. Thus, silencing of tumor suppressor genes such as RIZ1 may encompass the following sequential steps: transcriptional down regulation of RIZ1 in most or all cells of a target tissue due to alterations in HMTs or other epigenetic proteins—recruitment of DNMTs and hypermethylation of RIZ1 in one single cell—clonal expansion of the single cell. Without some initiating event that reduces the normalness of all cells in a target tissue, the hypermethylation of a tumor suppressor gene in a single cell may, like a mutation, not be compatible with cell survival or proliferation. Overall, existing data consistently suggest that DNA methylation mainly acts to reinforce a pattern that has been established and maintained by HMTs [228].

One way to determine whether DNA hypermethylation plays an initiating role in cancer is to see whether any normal adult human or animal tissue organ shows DNA hypermethylation in nearly all the cells of the tissue organ early in its exposure to a carcinogen. Since a precancerous normal tissue can clearly tolerate loss of a tumor suppressor gene in all the cells long before the tissue develops tumors (e.g., mice with homozygous p53 or RIZ1 knockout), the cancer-initiating event could easily be among those that occur in most or all cells of a tissue organ soon after exposure to a carcinogen. But there is no data showing that DNA hypermethylation of any gene can occur in most cells of an organ tissue soon after exposure to a carcinogen [229]. Even in normal tissues of normal adults that presumably have been exposed to environmental factors over very long time, only a small portion of the cells of the normal tissue sometimes shows hypermethylation [230]. So, while DNA hyper-methylation may occur relatively early in the cancer forming process such as already present in preinvasive tumors and may be responsible for the clonal expansion of some premalignant cell [214,231], it is still much too late to qualify as the initiating event by the standard demanded by the HMT pathway or the familial cancer pathway where the first event must occur to most or all normal cells of an apparently normal target organ tissue (Figure 8.1B and C). The HMT pathway predicts that inactivation or down regulation of a target HMT by methyl deficiency should occur in most or all cells of a target organ soon after the organism is exposed to a methyl-deficient diet and long before a tumor is formed or before clonal proliferation of any cells in the target tissue. In contrast, there is not even a perceivable molecular mechanism for any environmental carcinogen to directly cause DNA hypermethylation or hypomethylation of a specific target gene in most or all cells of a target organ tissue in the absence of clonal proliferation. The finding of DNA hypermethyation of a specific gene in a small pool of cells in a premalignant or preinvasive tissue is almost certainly a result of clonal expansion of a single premalignant cell that has acquired DNA hypermethylation rather than a result of all these cells independently and simultaneously acquired hypermethylation of the same gene. Just like the situation where a mutation or oncogene activation in a fully normal cell often causes apoptosis or senescence rather than clonal proliferation, DNA hypermethylation of a tumor suppressor gene in a fully normal cell may trigger the same response.

8.7 ROLE OF MUTATION AND ANEUPLOIDY

If epigenetic changes as a result of HMT alteration are equally effective for inheriting newly acquired somatic phenotypes, why most end stage cancers contain mutations and aneuploidy? Part of the reason may be that HMT inactivation will inevitably drive mutagenesis by causing chromatin/genome instability [156]. Indeed, HMT inactivation may be directly responsible for aneuploidy. Extensive efforts in search for a gene mutation that causes chromosomal instability have left the mechanisms unclear. Mutations in HMTs are generally rare in human cancers, in part perhaps because of the high number of paralog genes or of genes with similar functions. The elusiveness of a mutational mechanism as a cause for aneuploidy suggests that the likely solution lies in the persistent inactivation of HMTs at the protein level. This in turn suggests a mechanism for environmental selection to drive aneuploidy and genetic instability.

How to account for point mutations in cancer? HMTs are likely intimately involved in DNA replication, transcription, and repair, given that such DNA-based processes inevitably involve histones [232]. Because spontaneous mutation is tightly coupled to DNA-linked cellular processes, abnormal HMT functions may affect somatic mutagenesis. However, most point mutations detected in full-blown cancer may arise during tumor progression (not initiation), from spontaneous mutations due to replication errors and defects in DNA repair. These mutations would not be able to cause clonal expansion of premalignant cells where these cells are not already abnormal in HMTs or other epigenetic enzymes.

The epigenetic pathway by no means excludes a mutation-based pathway of cancer. It allows a role for mutations in solidifying phenotype and in further endowing an incipient tumor cell with additional growth advantages. The acquisitions of mutations could be a rate-limiting step under some circumstances, such as in the case of a germ line mutation affecting a tumor susceptibility gene. However, the new hypothesis does suggest that the driving force in the formation of common cancers of today is not mutations. Indeed, without epigenetic mechanisms in reducing the normalness of a precancerous cell, a cell may not survive a drastic, qualitative change like a cancer gene mutation [11,12]. Recent success stories of cancer therapy based on mutated or overexpressed cancer genes do not contradict the view that those genes may not play a causal role in the early stage of tumor formation. Such successes merely suggest that those genes may play a rate-limiting role in maintaining late stage cancer phenotypes.

8.8 INVERSE RELATIONSHIP BETWEEN GENETIC DIVERSITY/MUTATION AND EPIGENETIC COMPLEXITY

Cancer is clearly a disease of both genetics and epigenetics, just like normal phenotypes are determined by a unity of both genetics and epigenetics. How is genetics related to epigenetics? Does mutation in DNA affect epigenetic programming and vice versa? Here, I propose a hypothesis for an inverse relationship between genetic diversity/mutation and epigenetic programs/complexity.

The hypothesis suggests that too many mutations or to much genetic diversity would interfere with epigenetic programs or complexity while a relaxation in epigenetic programs or complexity would also allow more mutations to accumulate. The degree of epigenetic programming complexity of an organism is directly linked to the number of cell types of the organism. Each cell type represents a distinct epigenetic program and the more the cell types the more epigenetically complex the organism. The phenotype of a particular cell type in an epigenetically complex organism is largely determined by complex epigenetic programming. Cancer cell represents a partial loss of epigenetic complexity or programming since the highly organized epigenetic programming of a normal cell is usually lost rather than enhanced in cancer cells. HMTs are usually lost or downregulated rather than activated in cancer cells. It is gross methylation deficiency rather than excess that characterizes cancer cells. Complex epigenetic programming or epigenetic complexity requires adequate methylation and deficiency in methylation could only be expected to lead to reduced epigenetic complexity. Cancer is a direct result of epigenetic transformation where the normal complex epigenetic programs are reduced to a cancer specific program that is less complex. Too many mutations or too much genetic diversity would lead to lower epigenetic complexity leading to cancer. Reduction in epigenetic complexity as a result of harmful environmental impact on the epigenetic enzyme machinery would also allow more mutations to accumulate as well as directly lead to phenotype transformation or cancer. The hypothesis therefore provides a unified view of cancer as a disease of both genetics and epigenetics. In short, the cancer phenotype is maintained by a less complex epigenetic program that is either derived directly from a harmful environmental effect on the epigenetic enzymes or indirectly from a mutation-induced reduction in epigenetic complexity.

From a theoretical perspective, the more complex the system, the more restriction would be placed on the choice of building blocks. The number of choices of different materials for constructing a toy bicycle is much greater than that for a space shuttle. Likewise, the more simple the organism in epigenetic programs, the more variation it can tolerate in its building blocks or genotypes and the more dependent it is on mutational variants for adapting to changing environments. The genetic diversity of bacteria is greater than any other types of organisms but bacteria are the least complex in epigenetic programs. Bacteria adapt to environments primarily via gene mutations. Simple organisms are built more by the primary function of a gene rather than by a specific expression pattern of the gene. A gene may only have one expression pattern in simple organisms and many variants of the gene may be able to fit within that one expression pattern. In contrast, when an organism is built by multiple distinct gene expression patterns or cell types, the variation in gene sequence would be necessarily restricted. The reason is easy to understand. If cell type A is determined by expression pattern X and cell type B by pattern Y of the same gene, a mutational variant of the gene must be compatible with both expression pattern X and pattern Y. Such multilevel compatibility reduces the number of variants of the gene that can meet the multiple requirements. If ten mutational variants can fit with expression pattern X, then may be only three of the ten would fit with both patterns X and Y. The more expression patterns or cell types or functional pathways/networks a gene is involved with, the more restriction would be placed on the number of variants of the gene. Genetic diversity is restricted by epigenetic complexity and vice versa. It is impossible to build complex epigenetic programs if the DNAs are constantly changing. To compensate for the loss in the range of genetic diversity, complex organisms use different epigenetic programming of the same gene set, in addition to mutation, to adapt to environments and to evolve new phenotypes. Fish and human share nearly identical gene sets and the evolution from fish to human is in a large part a process of epigenetic programming. Therefore, organisms that are constructed in part by epigenetic programming are necessarily more restricted in genetic diversity than organisms that are less dependent on epigenetic programs.

Research on epigenetic programs is still at its infancy. On the basis of the limited knowledge of today, we can still envision several ways by which epigenetic programs may restrict genetic diversity or mutation. First, most genes are needed for the proper functioning of multiple adult tissues. A mutant variant of these genes needs to be compatible with multiple tissue types. Thus, the number of viable mutant variants is limited by the number of tissue types with which the gene is involved. Second, some genes are only expressed in one adult tissue type, such as hemoglobin in red blood cells. These genes however are still expressed in several different cell types or exhibit different expression patterns at different time points during development from fertilized egg to an adult organism. The gene expression pattern of fetal red blood cells is different from adult red blood cells. So these genes still need to be compatible with several different developmental gene expression patterns. Furthermore, these genes need to be repressed in most cell types during development and during normal adult life. They need to be packaged in a chromatin state that silences gene expression. Some mutant variants may interfere with such chromatin mediated repression and would be negatively selected. Third, some genes are expressed in only one adult cell type but the function of the gene is needed for most cell types of an organism. The function of hemoglobin is needed for the oxygen supply of every cell type. Many house-keeping genes such as actin are needed for most cell types. Such general function of a protein like hemoglobin and actin may be fine-tuned for the need of multiple tissues. Most mutant variants that may alter the oxygen-binding capacity of hemoglobin may not be compatible with the task of supplying oxygen to multiple tissues while they may not matter if the task is to do so for only a small number of tissue types. Most mutant variants of actin may not be compatible with the task of organizing cytoskeleton of multiple tissues while they may not matter if the task is to do so for only a small number of tissue types. A house-keeping gene may also exhibit new functions or connections with new networks in complex organisms that are absent in simple organisms, such as the apoptosis function of cytochrome c. Also, for a complex organism to evolve a new cell type, it is necessary to keep the house keeping genes unchanged so that new cell types can evolve with the least amount of unnecessary disruption to existing cell types. It may not matter much as to which specific version of a house keeping gene is used but it is important to stick with one once it is selected by an organism. Fourth, the coding region of every gene in complex organisms encodes not only amino acids but also epigenetic information such as the nucleosome code [233]. A nucleosome code allows the nucleosome to locate in the right position in the genome. A silent mutation may nevertheless affect the nucleosome code and alters the chromatin packaging state of the gene, which may affect either gene repression or activation. Fifth, the noncoding and the nonexpressed regions of the genome are nevertheless packaged into chromatin and encode the nucleosome code and other information necessary for gene expression and organization, and are therefore not free from epigenetic restrictions.

Since epigenetic complexity has a way of limiting the incidence of mutations as discussed above, a relaxation in epigenetic control may be expected to allow more mutations to occur and to accumulate, which in turn would cause more harm to epigenetic programs, finally resulting in cancer. Indeed, epigenetic programs are often deregulated in cancer and methylation deficiency is a hallmark of cancer [27,220]. Loss of epigenetic control as indicated by loss of DNA methylation occurs during aging and precedes mutations in cancer [234]. Conversely, the inverse relationship between DNA diversity/mutation and epigenetic programs predicts that high genetic diversity or too many mutations would interfere with epigenetic programming. Indeed, too many mutations, either germ line or somatic, are well known to cause cancer, which is essentially a disease where the normal epigenetic programs have been replaced by a cancer specific program that is less complex. Mutations in Rb would definitely affect HMTs since it directly binds to several of them.

Changes in epigenetic complexity may involve changes that affect large regions of the genome, such as amplification or deletion of long stretches of DNA. Thus, such copy number changes may be expected to be a common behavior of the genome just like point mutations are. Indeed, copy number variations are observed to be common in the human genome [235]. Within a specific level of epigenetic complexity, a certain range of neutral and random copy number changes are allowed that may affect slightly epigenetic programs, just like a certain range of random point mutations are allowed. Relaxation of epigenetic programs is expected to allow more abnormal copy number changes to occur. Indeed, cancer often exhibits aneuploidy and amplifications or deletions of long stretch of DNA.

The fact of rare familiar cancer mutations in humans is sufficient to prove the thesis that there is an upper limit to the amount of genetic diversity in an organism that is imposed by epigenetic programs. The fact that those rare germline cancer mutations cause mostly tissue specific lethal diseases is consistent with the notion that the upper limit is set up by the complexity of epigenetic programs. If humans lack the retina cell type or the retina specific epigenetic program, most of the mutations in the Rb gene would have been tolerated as normal variations and the genetic diversity of humans would have been in turn expanded. Also, numerous disease alleles in humans correspond to normal alleles in rhesus macaques [236]. Thus, many alleles or mutant variants that can be tolerated in a less complex organism in fact cause lethal diseases in humans.

To appreciate the power of high epigenetic complexity in restricting mutations, it is informative to compare the genetic diversity of different kinds of species that have vastly different epigenetic complexity but have evolved for about the same amount of time. Although flowering plants and mammals appeared about the same time in evolution ($\sim$125 million years ago) and so have had the same amount of time to accumulate mutations or genetic diversity, the genetic diversity of mammals is much less than that of flowering plants [237]. Of course, the epigenetic complexity or the number of cell types of mammals is much greater than that of flowering plants. In addition, two interbreeding populations of the medaka fish that separated 4 million years ago have reached a genetic distance that is threefold higher than that of two distinct primate species, humans and chimpanzees, that separated 5–7 million years ago [238]. If genetic diversity has an upper limit as imposed by epigenetic complexity, then the genetic distance between two species must also have an upper limit and simply cannot increase linearly and indefinitely with time. The neo-Darwinian gradual mutation theory predicts that the genetic distance between two distinct species is smaller in the past than in the present. This has been contradicted by the analysis of DNA sequences from Neanderthal fossils and of collagen sequences from fossils of dinosaurs and mastodons [239]. These observations suggest that the extremely low genetic diversity of humans is not because humans have not had enough time to accumulate mutations, or that the extremely high genetic diversity of bacteria is not because bacteria have had the longest time to accumulate mutations. The hypothesis of inverse relationship between genetic diversity and epigenetic complexity strongly challenges the popular neo-Darwinian molecular clock theory where genetic diversity is thought to be strictly a linear function of time and is not related to organism phenotypes or epigenetic complexity.

High epigenetic complexity of a species eliminates harmful mutations from degrading the species by causing tissue specific lethal diseases in those individuals that harbor those mutations, thus preventing the clonal expansion of those mutant individuals. Likewise, high epigenetic complexity of a cell type eliminates harmful mutations from degrading the cell type by causing lethal consequences in those cells that harbor those mutations, thus preventing the clonal expansion of those mutant cells to form tumors. Mutations in key cancer genes in a normal cell often cause cell death or senescence [11,12], but would be compatible with cell survival or proliferation if the cells have become less complex in epigenetic programs. High epigenetic complexity allows cells to undergo cell death or senescence whenever a mutation is sensed to interfere with epigenetic complexity. A reduction in epigenetic complexity would allow more mutations to be found as not interfering with epigenetic programming, which would in turn allow those cells with those mutations to escape the fate of death or senescence. Indeed, disrupting the Suv39H1 histone methyltransferase allows cells with activated Ras oncogene to escape senescence to become cancer [11]. The HMT-controlled senescence or apoptosis response to oncogenic events is an important tumor suppression pathway. Alberts et al. comments in their popular textbook: "In a lifetime, a typical gene is likely to have undergone mutation on about 10^{10} separate occasions in any individual. From this point of view, the problem of cancer seems to be not why it occurs, but why it occurs so infrequently" [240]. Epigenetic restriction of mutations is a common mechanism involved in preventing clonal expansions of both a mutant individual organism and a mutant cell. To prevent those cells with mutations to survive and proliferate, it is therefore essential to maintain epigenetic programming at a high complexity level. Having a healthy diet balanced in methyl nutrients is the easiest thing to do to maintain high epigenetic complexity.

8.9 CONCLUSION

Loss of HMT function, rather than gain, is more commonly linked with carcinogenesis and may be directly linked to dietary carcinogenesis in the Western countries. The hypothesis of inverse relationship between epigenetic complexity and genetic mutations/diversity offers a unified evolution theory of cancer as a disease of both genetics and epigenetics and strongly challenges the prevailing neo-Darwinian mutation theory of cancer and the neo-Darwinian evolution theory in general. Cancer is a disease of reduced epigenetic complexity. The inverse relationship allows DNA mutations to result from reduced epigenetic complexity and vice versa. The HMT-based epigenetic pathway differs from the mutation theory of cancer because it does not require DNA mutations to be the initiating cause of cancer. It is similar to other cancer hypotheses that feature a primary role for environmental selection. But, unlike those previously proposed selection hypotheses, the pathway presented here provides for the first time proven molecular targets for environmental factors and solves the issue of DNA independent inheritance during clonal expansion of tumor cells. These molecular targets, the histone methylation enzymes, can function to transmit environmental effects into heritable changes in cell phenotypes. Chronic changes in their enzyme activity may lead to heritable changes in cell phenotypes and may be the primary cause of sporadic common cancers. While the initiating event in familial cancers is the mutation of a cancer gene in all cells of the body, the equivalent initiating event in sporadic cancers caused by the typical Western diet may be the partial inactivation of an HMT in most or all cells of a target tissue (Figure 8.1). The HMT-based pathway of dietary carcinogenesis provides specific candidate molecular targets for methyl nutrients in cancer prevention, which are presently being verified using gene knockout animals fed with methyl complete or deficient diet. The success of such efforts may bring about great impact on the cancer statistics in the Western countries.

ACKNOWLEDGMENT

Work from the author's laboratory is supported by the NIH.

REFERENCES

1. Rous, P., *The Challenge to Man of the Neoplastic Cell. In Les Prix Nobel 1966, page 162–171* Almqvist & Wiksell International, Stockholm, 1966.
2. Doll, R. and Peto, R., The causes of cancer: Quantitative estimates of avoidable risks of cancer in the United States today, *J Nation Cancer Inst* 66(6), 1191–1308, 1981.
3. Trichopoulos, D., Li, F.P., and Hunter, D.J., What causes cancer? *Sci Am* 275, 80–87, 1996.
4. Willett, W.C., Diet and cancer: One view at the start of the millennium, *Cancer Epidemiol Biomarkers Prev* 10(1), 3–8, 2001.
5. Greenwald, P., Diet and cancer, perspectives of prevention, *Adv Nutrit Cancer 2* edited by Zappia et al., 1–19, 1999.
6. Blagosklonny, M.V., Oncogenic resistance to growth-limiting conditions, *Nat Rev Cancer* 2(3), 221–225, 2002.
7. Schuller, H.M., Mechanisms of smoking-related lung and pancreatic adenocarcinoma development, *Nat Rev Cancer* 2, 455–463, 2002.
8. Friedberg, E.C., How nucleotide excision repair protects against cancer, *Nat Rev Cancer* 1, 22–33, 2001.
9. Cairns, J., The interface between molecular biology and cancer research, *Mutat Res* 462(2–3), 423–428, 2000.
10. Prehn, R.T., Cancers beget mutations versus mutations beget cancers, *Cancer Res* 54(20), 5296–5300, 1994.

11. Braig, M., Lee, S., Loddenkemper, C., Rudolph, C., Peters, A.H., Schlegelberger, B., Stein, H., Dorken, B., Jenuwein, T., and Schmitt, C.A., Oncogene-induced senescence as an initial barrier in lymphoma development, *Nature* 436(7051), 660–665, 2005.

12. Tomlinson, I. and Bodmer, W., Selection, the mutation rate and cancer: ensuring that the tail does not wag the dog, *Nat Med* 5(1), 11–12, 1999.

13. Hanahan, D. and Weinberg, R.A., The hall marks of cancer, *Cell* 100, 57–70, 2000.

14. Stevens, L.C., The biology of teratomas, *Advances in Morphogenesis* 6, 1–28, 1967.

15. Boveri, T., *The Origin of Malignant Tumors* The Williams and Wilkins Co., Baltimore, 1929.

16. Loeb, L.A., Springgate, C.F., and Battula, N., Errors in DNA replication as a basis of malignant changes, *Cancer Res* 34(9), 2311–2321, 1974.

17. Nowell, P.C., The clonal evolution of tumor cell populations, *Science* 194, 23–28, 1976.

18. Cahill, D.P., Kinzler, K.W., Vogelstein, B., and Lengauer, C., Genetic instability and darwinian selection in tumours, *Trends Cell Biol* 9(12), M57–60, 1999.

19. Hitchins, M.P., Wong, J.J., Suthers, G., Suter, C.M., Martin, D.I., Hawkins, N.J., and Ward, R.L., Inheritance of a cancer-associated MLH1 germ-line epimutation, *N Engl J Med* 356(7), 697–705, 2007.

20. Cropley, J.E., Suter, C.M., Beckman, K.B., and Martin, D.I., Germ-line epigenetic modification of the murine A vy allele by nutritional supplementation, *Proc Natl Acad Sci U S A* 103(46), 17308–17312, 2006.

21. Rubin, H., The role of selection in progressive neoplastic transformation, *Adv Cancer Res* 83, 159–207, 2001.

22. Tomlinson, I.P., Novelli, M.R., and Bodmer, W.F., The mutation rate and cancer, *Proc Natl Acad Sci U S A* 93(25), 14800–14803, 1996.

23. Cairns, J., Mutation and cancer: The antecedents to our studies of adaptive mutation, *Genetics* 148(4), 1433–1440, 1998.

24. Strahl, B.D. and Allis, C.D., The language of covalent histone modifications, *Nature* 403(6765), 41–45, 2000.

25. Jenuwein, T. and Allis, C.D., Translating the histone code, *Science* 293, 1074–1080, 2001.

26. Lund, A.H. and van Lohuizen, M., Epigenetics and cancer, *Genes Dev* 18, 2315–2335, 2004.

27. Huang, S., Histone methyltransferases, diet nutrients, and tumor suppressors, *Nat Rev Cancer* 2, 469–476, 2002.

28. Kim, K.-C. and Huang, S., Histone methyltransferases in tumor suppression, *Cancer Biol Ther* 2, 491–499, 2003.

29. Davis, C.D. and Ross, S.A., Dietary components impact histone modifications and cancer risk, *Nutr Rev* 65(2), 88–94, 2007.

30. Zhang, Y. and Reinberg, D., Transcription regulation by histone methylation: interplay between different covalent modifications of the core histone tails, *Genes Dev* 15, 2343–2360, 2001.

31. Jenuwein, T., Re-SET-ting heterochromatin by histone methyltransferases, *Trends Cell Biol* 11(6), 266–273, 2001.

32. van Leeuwen, F., Gafken, P.R., and Gottschling, D.E., Dot1p modulates silencing in yeast by methylation of the nucleosome core, *Cell* 109, 745–756, 2002.

33. Feng, Q., Wang, H., Ng, H.H., Erdjument-Bromage, H., Tempst, P., Struhl, K., and Zhang, Y., Methylation of H3-lysine 79 is mediated by a new family of HMTases without a SET domain, *Curr Biol* 12(12), 1052–1058, 2002.

34. Cavalli, G. and Paro, R., Epigenetic inheritance of active chromatin after removal of the main transactivator, *Science* 286(5441), 955–958, 1999.

35. Czermin, B., Melfi, R., McCabe, D., Seitz, V., Imhof, A., and Pirrotta, V., Drosophila enhancer of Zeste/ESC complexes have a histone H3 methyltransferase activity that marks chromosomal Polycomb sites, *Cell* 111(2), 185–196, 2002.

36. Muller, J., Hart, C.M., Francis, N.J., Vargas, M.L., Sengupta, A., Wild, B., Miller, E.L., O'Connor, M.B., Kingston, R.E., and Simon, J.A., Histone methyltransferase activity of a *Drosophila* Polycomb group repressor complex, *Cell* 111(2), 197–208, 2002.

37. Milne, T.A., Briggs, S.D., Brock, H.W., Martin, M.E., Gibbs, D., Allis, C.D., and Hess, J.L., MLL targets SET domain methyltransferase activity to Hox gene promoters, *Mol Cell* 10(5), 1107–1117, 2002.

38. Cao, R., Wang, L.J., Wang, H.B., Xia, L., Erdjument-Bromage, H., Tempst, P., Jones, R.S., and Zhang, Y., Role of histone H3 lysine 27 methylation in polycomb-group silencing, *Science* 298, 1039–1043, 2002.

39. Nakamura, T., Mori, T., Tada, S., Krajewski, W., Rozovskaia, T., Wassell, R., Dubois, G., Mazo, A., Croce, C.M., and Canaani, E., ALL-1 is a histone methyltransferase that assembles a supercomplex of proteins involved in transcriptional regulation, *Mol Cell* 10(5), 1119–1128, 2002.

40. Hadorn, E., Dynamics of determination, *Symp Dev Biol* 25, 83, 1967.

41. Tschiersch, B., Hofmann, A., Krauss, V., Dorn, R., Korge, G., and Reuter, G., The protein encoded by the Drosophila position-effect variegation suppressor gene Su(var)3–9 combines domains of antagonistic regulators of homeotic gene complexes, *EMBO J* 13(16), 3822–3831, 1994.

42. Jones, R.S. and Gelbart, W.M., The Drosophila Polycomb-group gene Enhancer of zeste contains a region with sequence similarity to trithorax, *Mol Cell Biol* 13(10), 6357–6366, 1993.

43. Rea, S., Elsenhaber, F., O'Carroll, D., Strahl, B., Zu-Wen, S., Manfred, S., Opravil, S., Mechtler, K., Ponting, C., Allis, C., and Jenuwein, T., Regulation of chromatin structure by site-specific histone H3 methyltransferases, *Nature* 406, 593–599, 2000.

44. Unhavaithaya, Y., Shin, T.H., Miliaras, N., Lee, J., Oyama, T., and Mello, C.C., MEP-1 and a homolog of the NURD complex component Mi-2 act together to maintain germline-soma distinctions in *C. elegans*, *Cell* 111(7), 991–1002, 2002.

45. Beisel, C., Imhof, A., Greene, J., Kremmer, E., and Sauer, F., Histone methylation by the *Drosophila* epigenetic transcriptional regulator Ash1, *Nature* 419(6909), 857–862, 2002.

46. Noma, K.I., Allis, C.D., and Grewal, S.I.S., Transitions in distinct histone H3 methylation patterns at the heterochromatin domain boundaries, *Science* 293, 1150–1155, 2001.

47. Litt, M., Simpson, M., Gaszner, M., Allis, C.D., and Felsenfeld, G., Correlation between histone lysine methylation and developmental changes at the chicken beta-globan locus, *Science* 293, 2453–2455, 2001.

48. Bannister, A.J., Zegerman, P., Partridge, J.F., Miska, E.A., Thomas, J.O., Allshire, R.C., and Kouzarides, T., Selective recognition of methylated lysine 9 on histone H3 by the HP1 chromo domain, *Nature* 410 (March), 120–124, 2001.

49. Lachner, M., O'Carroll, D., Rea, S., Mechtler, K., and Jenuwein, T., Methylation of histone H3 lysine 9 creates a binding site for HP1 proteins, *Nature* 410 (March), 116–120, 2001.

50. Plath, K., Fang, J., Mlynarczyk-Evans, S.K., Cao, R., Worringer, K.A., Wang, H., de la Cruz, C.C., Otte, A.P., Panning, B., and Zhang, Y., Role of histone H3 lysine 27 methylation in X inactivation, *Science* 300, 131–135, 2003.

51. Strahl, B.D., Grant, P.A., Briggs, S.D., Sun, Z.W., Bone, J.R., Caldwell, J.A., Mollah, S., Cook, R.G., Shabanowitz, J., Hunt, D.F., and Allis, C.D., Set2 is a nucleosomal histone H3-selective methyltransferase that mediates transcriptional repression, *Mol Cell Biol* 22(5), 1298–1306, 2002.

52. Nishioka, K., Rice, J.C., Sarma, K., Erdjument-Bromage, H., Werner, J., Wang, Y., Chuikov, S., Valenzuela, P., Tempst, P., Steward, R., Lis, J.T., Allis, C.D., and Reinberg, D., PR-Set7 is a nucleosome-specific methyltransferase that modifies lysine 20 of histone H4 and is associated with silent chromatin, *Mol Cell* 9(6), 1201–1213, 2002.

53. Fang, J., Feng, Q., Ketel, C.S., Wang, H., Cao, R., Xia, L., Erdjument-Bromage, H., Tempst, P., Simon, J.A., and Zhang, Y., Purification and functional characterization of SET8, a nucleosomal histone H4-lysine 20-specific methyltransferase, *Curr Biol* 12(13), 1086–1099, 2002.

54. He, L., Yu, J.X., Liu, L., Buyse, I.M., Wang, M.-S., Yang, Q.-C., Nakagawara, A., Brodeur, G.M., Shi, Y.E., and Huang, S., RIZ1, but not the alternative RIZ2 product of the same gene, is underexpressed in breast cancer, and forced RIZ1 expression causes G_2-M cell cycle arrest and/or apoptosis, *Cancer Res* 58, 4238–4244, 1998.

55. Huang, S., The retinoblastoma protein-interacting zinc finger gene RIZ in 1p36-linked cancers, *Front Biosci* 4, D528–532, 1999.

56. Jiang, G.-L. and Huang, S., The *yin-yang* of PR-domain family genes in tumorigenesis, *Histol Histopathol* 15(1), 109–117, 2000.

57. Huang, S., Blimp-1 is the murine homolog of the human transcriptional repressor PRDI-BF1, *Cell* 78, 9, 1994.

58. Buyse, I.M., Shao, G., and Huang, S., The retinoblastoma protein binds to RIZ, a zinc finger protein that shares an epitope with the adenovirus E1A protein, *Proc Natl Acad Sci U S A* 92, 4467–4471, 1995.

59. Huang, S., Shao, G., and Liu, L., The PR domain of the Rb-binding zinc finger protein RIZ1 is a protein binding interface and is related to the SET domain functioning in chromatin-mediated gene expression, *J Biol Chem* 273, 15933–15940, 1998.

60. Wilson, J.R., Jing, C., Walker, P.A., Martin, S.R., Howell, S.A., Blackburn, G.M., Gamblin, S.J., and Xiao, B., Crystal structure and functional analysis of the histone methyltransferase SET7/9, *Cell* 111(1), 105–115, 2002.

61. Jacobs, S.A., Harp, J.M., Devarakonda, S., Kim, Y., Rastinejad, F., and Khorasanizadeh, S., The active site of the SET domain is constructed on a knot, *Nat Struct Biol* 9(11), 833–838, 2002.

62. Briknarova, K., Zhou, X., Satterthwait, A., Hoyt, D.W., Ely, K.R., and Huang, S., Structure studies of the SET domain from RIZ1 tumor suppressor. *Biochem Biophys Res Commun*, 366, 807–813, 2008.

63. Schultz, J., Copley, R.R., Doerks, T., Ponting, C.P., and Bork, P., SMART: A web-based tool for the study of genetically mobile domains, *Nucleic Acids Res* 28(1), 231–234, 2000.

64. Kim, K.-C., Geng, L., and Huang, S., Inactivation of a histone methyltransferase by mutations in human cancers, *Cancer Res* 63, 7619–7623, 2003.

65. Liu, L., Shao, G., Steele-Perkins, G., and Huang, S., The retinoblastoma interacting zinc finger gene RIZ produces a PR domain lacking product through an internal promoter, *J Biol Chem* 272, 2984–2991, 1997.

66. Buyse, I.M., Takahashi, E., and Huang, S., Physical mapping of the retinoblastoma-interacting zinc finger gene RIZ to D1S228 on chromosome 1p36, *Genomics* 34, 119–121, 1996.

67. Jiang, G.-L., Liu, L., Buyse, I.M., Simon, D., and Huang, S., Decreased RIZ1 expression but not RIZ2 in hepatoma and suppression of hepatoma tumorigenicity by RIZ1, *Int J Cancer* 83(4), 541–547, 1999.

68. Chadwick, R.B., Jiang, G.-L., Bennington, G.A., Yuan, B., Johnson, C.K., Stevens, M.W., Niemann, T.H., Peltomaki, P., Huang, S., and de la Chapelle, A., Candidate tumor suppressor RIZ is frequently involved in colorectal carcinogenesis, *Proc Natl Acad Sci U S A* 97, 2662–2667, 2000.

69. Du, Y., Carling, T., Fang, W., Piao, Z., Sheu, J.C., and Huang, S., Hypermethylation in human cancers of the RIZ1 tumor suppressor gene, a member of a histone/protein methyltransferase superfamily, *Cancer Res* 61, 8094–8099, 2001.

70. Chang, H.W., Chan, A., Kwong, D.L., Wei, W.I., Sham, J.S., and Yuen, A.P., Detection of hypermethylated RIZ1 Gene in primary tumor, mouth, and throat rinsing fluid, nasopharyngeal swab, and peripheral blood of nasopharyngeal carcinoma patient, *Clin Cancer Res* 9, 1033–1038, 2003.

71. Jiang, G.L. and Huang, S., Adenovirus expressing RIZ1 in tumor suppressor gene therapy of micro-satellite-unstable colorectal cancers, *Cancer Res* 61, 1796–1798, 2001.

72. Fang, W., Piao, Z., Simon, D., Sheu, J.-C., Perucho, M., and Huang, S., Preferential loss in human cancer of polymorphic RIZ allele that is more common in Asians than Caucasians, *Br J Cancer* 84, 743–747, 2001.

73. Yoon, K.A., Park, S., Hwangbo, B., Shin, H.D., Cheong, H.S., Shin, H.R., and Lee, J.S., Genetic polymorphisms in the Rb-binding zinc finger gene RIZ and the risk of lung cancer, *Carcinogenesis*, 28, 1971–1977, 2007.

74. Cebrian, A., Pharoah, P.D., Ahmed, S., Ropero, S., Fraga, M.F., Smith, P.L., Conroy, D., Luben, R., Perkins, B., Easton, D.F., Dunning, A.M., Esteller, M., and Ponder, B.A., Genetic variants in epigenetic genes and breast cancer risk, *Carcinogenesis* 27(8), 1661–1669, 2006.

75. Piao, Z., Fang, W., Malkhosyan, S., Kim, H., Horii, A., Perucho, M., and Huang, S., Frequent frameshift mutations of RIZ in human gastrointestinal and endometrial carcinomas with microsatellite instability, *Cancer Res* 60, 4701–4704, 2000.

76. Sakurada, K., Furukawa, T., Kato, Y., Kayama, T., Huang, S., and Horii, A., RIZ, the retinoblastoma protein interacting zinc finger gene, is mutated in genetically unstable cancers of the pancreas, stomach, and colorectum, *Genes, Chromosomes Cancer* 30, 207–211, 2001.

77. Steele-Perkins, G., Fang, W., Yang, X.H., Van Gele, M., Carling, T., Gu, J., Buyse, I.M., Fletcher, J., Liu, J., Bronson, R., Chadwick, R., de la Chapelle, A., Zhang, X.K., Speleman, F., and Huang, S., Tumor formation and inactivation of RIZ1, an Rb-binding member of a nuclear protein-methyltransferase superfamily, *Genes Dev* 15, 2250–2262, 2001.

78. Mikol, Y.B., Hoover, K.L., Creasia, D., and Poirier, L.A., Hepatocarcinogenesis in rats fed methyl-deficient, amino acid-defined diets, *Carcinogenesis* 4(12), 1619–1629, 1983.

79. de Camargo, J.L., Punyarit, P., and Newberne, P.M., Early stages of nodular transformation of the B6C3F1 mouse liver induced by choline deficiency, *Toxicol Pathol* 13(1), 10–17, 1985.

80. Ghoshal, A.K. and Farber, E., The induction of liver cancer by dietary deficiency of choline and methionine without added carcinogens, *Carcinogenesis* 5(10), 1367–1370, 1984.

81. Shinozuka, H., Katyal, S.L., and Perera, M.I., Choline deficiency and chemical carcinogenesis, *Adv Exp Med Biol* 206, 253–267, 1986.

82. Yokoyama, S., Sells, M.A., Reddy, T.V., and Lombardi, B., Hepatocarcinogenic and promoting action of a choline-devoid diet in the rat, *Cancer Res* 45(6), 2834–2842, 1985.

83. Pastural, E., Takahashi, N., Dong, W.F., Bainbridge, M., Hull, A., Pearson, D., Huang, S., Lowsky, R., DeCoteau, J.F., and Geyer, C.R., RIZ1 repression is associated with insulin-like growth factor-1 signaling activation in chronic myeloid leukemia cell lines, *Oncogene* 26(11), 1586–1594, 2007.

84. Carling, T., Kim, K.C., Yang, X.H., Gu, J., Zhang, X.K., and Huang, S., A histone methyltransferase is required for maximal response to female sex hormones, *Mol Cell Biol* 24(16), 7032–7042, 2004.

85. Xie, M., Shao, G., Buyse, I.M., and Huang, S., Transcriptional repression mediated by the PR domain zinc finger gene RIZ, *J Biol Chem* 272, 26360–26366, 1997.

86. Muraosa, Y., Takahashi, K., Yoshizawa, M., and Shibahara, S., cDNA cloning of a novel protein containing two zinc-finger domains that may function as a transcription factor for the human heme-oxygenase-1 gene, *Eur J Biochem* 235, 471–479, 1996.

87. Medici, N., Abbondanza, C., Nigro, V., Rossi, V., Piluso, G., Belsito, A., Gallo, L., Roscigno, A., Bontempo, P., Puca, A.A., Molinari, A.M., Moncharmont, B., and Puca, G.A., Identification of a DNA binding protein cooperating with estrogen receptor as RIZ (retinoblastoma interacting zinc finger protein), *Biochem Biophys Res Commun* 264, 983–989, 1999.

88. Olsen, J.V., Blagoev, B., Gnad, F., Macek, B., Kumar, C., Mortensen, P., and Mann, M., Global, in vivo, and site-specific phosphorylation dynamics in signaling networks, *Cell* 127(3), 635–648, 2006.

89. Abbondanza, C., Medici, N., Nigro, V., Rossi, V., Gallo, L., Piluso, G., Belsito, A., Roscigno, A., Bontempo, P., Puca, A.A., Molinari, A.M., Moncharmont, B., Puca, G.A.B., and Puca, G.A., The retinoblastoma-interacting zinc-finger protein RIZ is a downstream effector of estrogen action, *Proc Natl Acad Sci U S A* 97(7), 3130–3135, 2000.

90. Grundberg, E., Akesson, K., Kindmark, A., Gerdhem, P., Holmberg, A., Mellstrom, D., Ljunggren, O., Orwoll, E., Mallmin, H., Ohlsson, C., and Brandstrom, H., The impact of estradiol on bone mineral density is modulated by the specific estrogen receptor-alpha cofactor retinoblastoma-interacting zinc finger protein-1 insertion/deletion polymorphism, *J Clin Endocrinol Metab* 92(6), 2300–2306, 2007.

91. Garcia-Bassets, I., Kwon, Y.S., Telese, F., Prefontaine, G.G., Hutt, K.R., Cheng, C.S., Ju, B.G., Ohgi, K.A., Wang, J., Escoubet-Lozach, L., Rose, D.W., Glass, C.K., Fu, X.D., and Rosenfeld, M.G., Histone methylation-dependent mechanisms impose ligand dependency for gene activation by nuclear receptors, *Cell* 128(3), 505–518, 2007.

92. Rosenfeld, M.G., Lunyak, V.V., and Glass, C.K., Sensors and signals: a coactivator/corepressor/epigenetic code for integrating signal-dependent programs of transcriptional response, *Genes Dev* 20 (11), 1405–1428, 2006.

93. Pogribny, I.P., Tryndyak, V.P., Muskhelishvili, L., Rusyn, I., and Ross, S.A., Methyl deficiency, alterations in global histone modifications, and carcinogenesis, *J Nutr* 137(1 Suppl), 216S–222S, 2007.

94. Fears, S., Mathieu, C., Zeleznik-Le, N., Huang, S., Rowley, J.D., and Nucifora, G., Intergenic splicing of *MDS1* and *EVI1* occurs in normal tissues as well as in myeloid leukemia and produces a new member of the PR domain family, *Proc Natl Acad Sci U S A* 93, 1642–1647, 1996.

95. Nucifora, G., Begy, C.R., Kobayashi, H., Roulston, D., Claxton, D., Pedersen-Bjergaard, J., Parganas, E., Ihle, J.N., and Rowley, J.D., Consistent intergenic splicing and production of multiple transcripts between AML1 at 21q22 and unrelated genes at 3q26 in (3;21)(q26;q22) translocations, *Proc Natl Acad Sci U S A* 91, 4004–4008, 1994.

96. Morishita, K., Parker, D.S., Mucenski, M.L., Jenkins, N.A., Copeland, N.G., and Ihle, J.N., Retroviral activation of a novel gene encoding a zinc finger protein in IL-3-dependent myeloid leukemia cell lines, *Cell* 54, 831–840, 1988.

97. Perkins, A.S., Mercer, J.A., Jenkins, N.A., and Copeland, N.G., Patterns of Evi-1 expression in embryonic and adult tissues suggest that Evi-1 plays an important regulatory role in mouse development, *Development* 111, 479–487, 1991.

98. Hoyt, P.R., Bartholomew, C., Davis, A.J., Yutzey, K., Gamer, L.W., Potter, S.S., Ihle, J.N., and Mucenski, M.L., The Evi1 proto-oncogene is required at midgestation for neural, heart, and paraxial mesenchyme development, *Mech Dev* 65(1–2), 55–70, 1997.

99. Kreider, B.L., Orkin, S.H., and Ihle, J.N., Loss of erythropoietin responsiveness in erythroid progenitors due to expression of the Evi-1 myeloid-transforming gene, *Proc Natl Acad Sci U S A* 90, 6454–6458, 1993.

100. Morishita, K., Parganas, E., Matsugi, T., and Ihle, J.N., Expression of the Evi-1 zinc finger gene in 32Dc13 myeloid cells blocks granulocytic differentiation in response to granulocyte colony-stimulating factor, *Mol Cell Biol* 12, 183–189, 1992.

101. Sitailo, S., Sood, R., Barton, K., and Nucifora, G., Forced expression of the leukemia-associated gene EVI1 in ES cells: Model for myeloid leukemia with 3q26 rearrangements, *Leukemia* 13, 1639–1645, 1999.

102. Chi, Y., Senyuk, V., Chakraborty, S., and Nucifora, G., EVI1 promotes cell proliferation by interacting with BRG1 and blocking the repression of BRG1 on E2F1 activity, *J Biol Chem* 278, 49806–49811, 2003.

103. Louz, D., van den Broek, M., Verbakel, S., Vankan, Y., van Lom, K., Joosten, M., Meijer, D., Lowenberg, B., and Delwel, R., Erythroid defects and increased retrovirally-induced tumor formation in Evi1 transgenic mice, *Leukemia* 14, 1876–1884, 2000.

104. Buonamici, S., Li, D., Chi, Y., Zhao, R., Wang, X., Brace, L., Ni, H., Saunthararajah, Y., and Nucifora, G., EVI1 induces myelodysplastic syndrome in mice, *J Clin Invest* 114, 713–719, 2004.

105. Soderholm, J., Kobayashi, H., Mathieu, C., Rowley, J.D., and Nucifora, G., The leukemia-associated gene MDS1/EVI1 is a new type of GATA-binding transactivator, *Leukemia* 11(3), 352–358, 1997.

106. Mochizuki, N., Shimizu, S., Nagasawa, T., Tanaka, H., Taniwaki, M., Yokota, J., and Morishita, K., A novel gene, MEL1, mapped to 1p36.3 is highly homologous to the MDS1/EVI1 gene and is transcriptionally activated in t(1;3)(p36;q21)-positive leukemia cells, *Blood* 96(9), 3209–3214, 2000.

107. Nishikata, I., Sasaki, H., Iga, M., Tateno, Y., Imayoshi, S., Asou, N., Nakamura, T., and Morishita, K., A novel EVI1 gene family, MEL1, lacking a PR domain (MEL1S) is expressed mainly in t(1;3) (p36;q21)-positive AML and blocks G-CSF-induced myeloid differentiation, *Blood* 102, 3323–3332, 2003.

108. Yoshida, M., Nosaka, K., Yasunaga, J., Nishikata, I., Morishita, K., and Matsuoka, M., Aberrant expression of the MEL1S gene identified in association with hypomethylation in adult T-cell leukemia cells, *Blood* 103, 2753–2760, 2004.

109. Du, Y., Jenkins, N.A., and Copeland, N.G., Insertional mutagenesis identifies genes that promote the immortalization of primary bone marrow progenitor cells, *Blood* 106(12), 3932–3939, 2005.

110. Lahortiga, I., Agirre, X., Belloni, E., Vazquez, I., Larrayoz, M.J., Gasparini, P., Lo Coco, F., Pelicci, P.G., Calasanz, M.J., and Odero, M.D., Molecular characterization of a t(1;3)(p36;q21) in a patient with MDS. MEL1 is widely expressed in normal tissues, including bone marrow, and it is not overexpressed in the t(1;3) cells, *Oncogene* 23, 311–316, 2004.

111. Shiraishi, M., Sekiguchi, A., Terry, M.J., Oates, A.J., Miyamoto, Y., Chuu, Y.H., Munakata, M., and Sekiya, T., A comprehensive catalog of CpG islands methylated in human lung adenocarcinomas for the identification of tumor suppressor genes, *Oncogene* 21(23), 3804–3813, 2002.

112. Keller, A.D. and Maniatis, T., Identification and characterization of a novel repressor of beta-interferon gene expression, *Genes Dev* 5, 868–879, 1991.

113. Turner, C.A., Jr., Mack, D.H., and Davis, M.M., Blimp-1, a novel zinc finger-containing protein that can drive the maturation of B lymphocytes into immunoglobulin-secreting cells, *Cell* 77, 297–306, 1994.

114. Lin, Y., Wong, K.-K., and Calame, K., Repression of c-*myc* transcription by Blimp-1, an inducer of terminal B cell differentiation, *Science* 276, 596–598, 1997.

115. Mock, B.A., Liu, L., Le Paslier, D., and Huang, S., The B-lymphocyte maturation promoting transcription factor BLIMP1/PRDI-BF1 maps to D6S447 on human chromosome 6q21-q22.1 and the syntenic region of mouse chromosome 10, *Genomics* 37, 24–28, 1996.

116. Gyory, I., Fejer, G., Ghosh, N., Seto, E., and Wright, K.L., Identification of a functionally impaired positive regulatory domain I binding factor 1 transcription repressor in myeloma cell lines, *J Immun* 170, 3125–3133, 2003.

117. Pasqualucci, L., Compagno, M., Houldsworth, J., Monti, S., Grunn, A., Nandula, S.V., Aster, J.C., Murty, V.V., Shipp, M.A., and Dalla-Favera, R., Inactivation of the PRDM1/BLIMP1 gene in diffuse large B cell lymphoma, *J Exp Med* 203(2), 311–317, 2006.

118. Tam, W., Gomez, M., Chadburn, A., Lee, J.W., Chan, W.C., and Knowles, D.M., Mutational analysis of PRDM1 indicates a tumor-suppressor role in diffuse large B-cell lymphomas, *Blood* 107(10), 4090–4100, 2006.

119. Ohinata, Y., Payer, B., O'Carroll, D., Ancelin, K., Ono, Y., Sano, M., Barton, S.C., Obukhanych, T., Nussenzweig, M., Tarakhovsky, A., Saitou, M., and Surani, M.A., Blimp1 is a critical determinant of the germ cell lineage in mice, *Nature* 436(7048), 207–213, 2005.

120. Gyory, I., Wu, J., Fejer, G., Seto, E., and Wright, K.L., PRDI-BF1 recruits the histone H3 methyltransferase G9a in transcriptional silencing, *Nat Immunol* 5(3), 299–308, 2004.

121. Ancelin, K., Lange, U.C., Hajkova, P., Schneider, R., Bannister, A.J., Kouzarides, T., and Surani, M.A., Blimp1 associates with Prmt5 and directs histone arginine methylation in mouse germ cells, *Nat Cell Biol* 8(6), 623–630, 2006.

122. Martins, G.A., Cimmino, L., Shapiro-Shelef, M., Szabolcs, M., Herron, A., Magnusdottir, E., and Calame, K., Transcriptional repressor Blimp-1 regulates T cell homeostasis and function, *Nat Immunol* 7(5), 457–465, 2006.

123. Kallies, A., Hawkins, E.D., Belz, G.T., Metcalf, D., Hommel, M., Corcoran, L.M., Hodgkin, P.D., and Nutt, S.L., Transcriptional repressor Blimp-1 is essential for T cell homeostasis and self-tolerance, *Nat Immunol* 7(5), 466–474, 2006.

124. Horsley, V., O'Carroll, D., Tooze, R., Ohinata, Y., Saitou, M., Obukhanych, T., Nussenzweig, M., Tarakhovsky, A., and Fuchs, E., Blimp1 defines a progenitor population that governs cellular input to the sebaceous gland, *Cell* 126(3), 597–609, 2006.

125. Yang, X.-H. and Huang, S., *PFM1 (PRDM4)*, a new member of the PR-domain family, maps to a tumor suppressor locus on human chromosome 12q23-q24.1, *Genomics* 61, 319–325, 1999.

126. Chittka, A. and Chao, M.V., Identification of a zinc finger protein whose subcellular distribution is regulated by serum and nerve growth factor, *Proc Natl Acad Sci U S A* 96, 10705–10710, 1999.

127. Deng, Q. and Huang, S., PRDM5 is silenced in human cancers and has growth suppressive activities, *Oncogene* 23(28), 4903–4910, 2004.

128. Watanabe, Y., Toyota, M., Kondo, Y., Suzuki, H., Imai, T., Ohe-Toyota, M., Maruyama, R., Nojima, M., Sasaki, Y., Sekido, Y., Hiratsuka, H., Shinomura, Y., Imai, K., Itoh, F., and Tokino, T., PRDM5 identified as a target of epigenetic silencing in colorectal and gastric cancer, *Clin Cancer Res* 13(16), 4786–4794, 2007.

129. Duan, Z., Person, R.E., Lee, H.H., Huang, S., Donadieu, J., Badolato, R., Grimes, H.L., Papayannopoulou, T., and Horwitz, M.S., Epigenetic regulation of protein-coding and microRNA genes by the Gfi1-interacting, tumor suppressor PRDM5, *Mol Cell Biol*, 19, 6889–6902, 2007.

130. Davis, C.A., Haberland, M., Arnold, M.A., Sutherland, L.B., McDonald, O.G., Richardson, J.A., Childs, G., Harris, S., Owens, G.K., and Olson, E.N., PRISM/PRDM6, a transcriptional repressor that promotes the proliferative gene program in smooth muscle cells, *Mol Cell Biol* 26(7), 2626–2636, 2006.

131. Wu, Y., Ferguson, J.E., 3rd, Wang, H., Kelley, R., Ren, R., McDonough, H., Meeker, J., Charles, P.C., Wang, H., and Patterson, C., PRDM6 is enriched in vascular precursors during development and inhibits endothelial cell proliferation, survival, and differentiation, *J Mol Cell Cardiol*, 44, 47–58, 2008.

132. Hayashi, K. and Matsui, Y., Meisetz, a novel histone tri-methyltransferase, regulates meiosis-specific epigenesis, *Cell Cycle* 5(6), 615–620, 2006.

133. Ziemin-van der Poel, S., McCabe, N.R., Gill, H.J., Espinosa, R.I., Patel, Y., Harden, A., Rubinelli, P., Smith, S.D., LeBeau, M.M., Rowley, J.D., and Diaz, M., Identification of a gene, MLL, that spans the breakpoint in 11q23 translocations associated with human leukemias, *Proc Natl Acad Sci U S A* 88, 10735–10739, 1991.

134. Gu, Y., Nakamura, T., Alder, H., Prasad, R., Canaani, O., Cimino, G., Croce, C.M., and Canaani, E., The t(4;11) chromosome translocation of human acute leukemias fuses the ALL-1 gene, related to *Drosophila trithorax*, to the AF-4 gene, *Cell* 71(4), 701–708, 1992.

135. Tkachuk, D.C., Kohler, S., and Cleary, M.L., Involvement of a homolog of Drosophila trithorax by 11q23 chromosomal translocations in acute leukemias, *Cell* 71(4), 691–700, 1992.

136. Djabali, M., Selleri, L., Parry, P., Bower, M., Young, B.D., and Evans, G.A., A trithorax-like gene is interrupted by chromosome 11q23 translocations in acute leukaemias, *Nat Genet* 2(2), 113–118, 1992.

137. Arakawa, H., Nakamura, T., Zhadanov, A.B., Fidanza, V., Yano, T., Bullrich, F., Shimizu, M., Blechman, J., Mazo, A., Canaani, E., and Croce, C.M., Identification and characterization of the ARP1 gene, a target for the human acute leukemia ALL1 gene, *Proc Natl Acad Sci U S A* 95(8), 4573–4578, 1998.

138. Prasad, R., Leshkowitz, D., Gu, Y., Alder, H., Nakamura, T., Saito, H., Huebner, K., Berger, R., Croce, C.M., and Canaani, E., Leucine-zipper dimerization motif encoded by the AF17 gene fused to ALL-1 (MLL) in acute leukemia, *Proc Natl Acad Sci U S A* 91(17), 8107–8111, 1994.

139. Schichman, S.A., Caligiuri, M.A., Gu, Y., Strout, M.P., Canaani, E., Bloomfield, C.D., and Croce, C.M., ALL-1 partial duplication in acute leukemia, *Proc Natl Acad Sci U S A* 91(13), 6236–6239, 1994.

140. Baffa, R., Negrini, M., Schichman, S.A., Huebner, K., and Croce, C.M., Involvement of the ALL-1 gene in a solid tumor, *Proc Natl Acad Sci U S A* 92(11), 4922–4926, 1995.

141. Ayton, P.M. and Cleary, M.L., Molecular mechanisms of leukemogenesis mediated by MLL fusion proteins, *Oncogene* 20, 5695–5707, 2002.

142. Okada, Y., Feng, Q., Lin, Y., Jiang, Q., Li, Y., Coffield, V.M., Su, L., Xu, G., and Zhang, Y., hDOT1L links histone methylation to leukemogenesis, *Cell* 121(2), 167–178, 2005.

143. Huntsman, D.G., Chin, S.F., Muleris, M., Batley, S.J., Collins, V.P., Wiedemann, L.M., Aparicio, S., and Caldas, C., MLL2, the second human homolog of the Drosophila trithorax gene, maps to 19q13.1 and is amplified in solid tumor cell lines, *Oncogene* 18, 7975–7984, 2001.

144. Hughes, C.M., Rozenblatt-Rosen, O., Milne, T.A., Copeland, T.D., Levine, S.S., Lee, J.C., Hayes, D.N., Shanmugam, K.S., Bhattacharjee, A., Biondi, C.A., Kay, G.F., Hayward, N.K., Hess, J.L., and Meyerson, M., Menin associates with a trithorax family histone methyltransferase complex and with the hoxc8 locus, *Mol Cell* 13(4), 587–597, 2004.

145. Kuzmichev, A., Nishioka, K., Erdjument-Bromage, H., Tempst, P., and Reinberg, D., Histone methyltransferase activity associated with a human multiprotein complex containing the Enhancer of Zeste protein, *Genes Dev* 16(22), 2893–2905, 2002.

146. O'Carroll, D., Erhardt, S., Pagani, M., Barton, S.C., Surani, M.A., and Jenuwein, T., The polycomb-group gene Ezh2 is required for early mouse development, *Mol Cell Biol* 21, 4330–4336, 2001.

147. Su, I.H., Basavara, J.A., Krutchinsky, A.N., Hobert, O., Ullrich, A., Chait, B.T., and Tarakhovsky, A., Ezh2 controls B cell development through histone H3 methylation and Igh rearrangement, *Nat Immun* 4, 124–131, 2003.

148. Varambally, S., Dhanasekaran, S.M., Zhou, M., Barrette, T.R., Kumar-Sinha, C., Sanda, M.G., Ghosh, D., Pienta, K.J., Sewalt, R.G., Otte, A.P., Rubin, M.A., and Chinnaiyan, A.M., The polycomb group protein EZH2 is involved in progression of prostate cancer, *Nature* 419, 624–629, 2002.

149. Schlesinger, Y., Straussman, R., Keshet, I., Farkash, S., Hecht, M., Zimmerman, J., Eden, E., Yakhini, Z., Ben-Shushan, E., Reubinoff, B.E., Bergman, Y., Simon, I., and Cedar, H., Polycomb-mediated methylation on Lys27 of histone H3 pre-marks genes for de novo methylation in cancer, *Nat Genet* 39(2), 232–236, 2007.

150. Tachibana, M., Sugimoto, K., Fukushima, T., and Shinkai, Y., Set domain-containing protein, G9a, is a novel lysine-preferring mammalian histone methyltransferase with hyperactivity and specific selectivity to lysines 9 and 27 of histone H3, *J Biol Chem* 276(27), 25309–25317, 2001.

151. Kondo, Y., Shen, L., Suzuki, S., Kurokawa, T., Masuko, K., Tanaka, Y., Kato, H., Mizuno, Y., Yokoe, M., Sugauchi, F., Hirashima, N., Orito, E., Osada, H., Ueda, R., Guo, Y., Chen, X., Issa, J.P., and Sekido, Y., Alterations of DNA methylation and histone modifications contribute to gene silencing in hepatocellular carcinomas, *Hepatol Res* 37, 974–983, 2007.

152. Wozniak, R.J., Klimecki, W.T., Lau, S.S., Feinstein, Y., and Futscher, B.W., 5-Aza-2′-deoxycytidine-mediated reductions in G9A histone methyltransferase and histone H3 K9 di-methylation levels are linked to tumor suppressor gene reactivation, *Oncogene* 26(1), 77–90, 2007.

153. Yang, L., Xia, L., Wu, D.Y., Wang, H., Chansky, H.A., Schubach, W.H., Hickstein, D.D., and Zhang, Y., Molecular cloning of ESET, a novel histone H3-specific methyltransferase that interacts with ERG transcription factor, *Oncogene* 21(1), 148–152, 2002.

154. Schultz, D.C., Ayyanathan, K., Negorev, D., Maul, G.G., and Rauscher, F.J., 3rd, SETDB1: a novel KAP-1-associated histone H3, lysine 9-specific methyltransferase that contributes to HP1-mediated silencing of euchromatic genes by KRAB zinc-finger proteins, *Genes Dev* 16(8), 919–932, 2002.

155. Li, H., Rauch, T., Chen, Z.X., Szabo, P.E., Riggs, A.D., and Pfeifer, G.P., The histone methyltransferase SETDB1 and the DNA methyltransferase DNMT3A interact directly and localize to promoters silenced in cancer cells, *J Biol Chem* 281(28), 19489–19500, 2006.

156. Peters, A., O'Carroll, D., Scherthan, H., Mechtler, K., Sauer, S., Schofer, C., Weipoltshammer, K., Pagani, M., Lachner, M., Kohlmaier, A., Opravil, S., Doyle, M., Sibilia, M., and Jenuwein, T., Loss of the Suv39h histone methyltransferases impairs mammalian heterochromatin and genome stability, *Cell* 107, 323–337, 2001.

157. Nielsen, S.J., Schneider, R., Bauer, U.M., Bannister, A.J., Morrison, A., O'Carroll, D., Firestein, R., Cleary, M., Jenuwein, T., Herrera, R.E., and Kouzarides, T., Rb targets histone H3 methylation and HP1 to promoters. [see comments], *Nature* 412(6846), 561–565, 2001.

158. Firestein, R., Cui, X., Huie, P., and Cleary, M.L., Set domain-dependent regulation of transcriptional silencing and growth control by SUV39H1, a mammalian ortholog of Drosophila Su(var)3–9, *Mol Cell Biol* 20, 4900–4909, 2000.

159. Narita, M., Nunez, S., Heard, E., Narita, M., Lin, A.W., Hearn, S.A., Spector, D.L., Hannon, G.J., and Lowe, S.W., Rb-mediated heterochromatin formation and silencing of E2F target genes during cellular senescence, *Cell* 113(6), 703–716, 2003.

160. Schotta, G., Lachner, M., Sarma, K., Ebert, A., Sengupta, R., Reuter, G., Reinberg, D., and Jenuwein, T., A silencing pathway to induce H3-K9 and H4-K20 trimethylation at constitutive heterochromatin, *Genes Dev* 18(11), 1251–1262, 2004.

161. Gonzalo, S., Garcia-Cao, M., Fraga, M.F., Schotta, G., Peters, A.H., Cotter, S.E., Eguia, R., Dean, D.C., Esteller, M., Jenuwein, T., and Blasco, M.A., Role of the RB1 family in stabilizing histone methylation at constitutive heterochromatin, *Nat Cell Biol* 7(4), 420–428, 2005.

162. Fraga, M.F., Ballestar, E., Villar-Garea, A., Boix-Chornet, M., Espada, J., Schotta, G., Bonaldi, T., Haydon, C., Ropero, S., Petrie, K., Iyer, N.G., Perez-Rosado, A., Calvo, E., Lopez, J.A., Cano, A., Calasanz, M.J., Colomer, D., Piris, M.A., Ahn, N., Imhof, A., Caldas, C., Jenuwein, T., and Esteller, M., Loss of acetylation at Lys16 and trimethylation at Lys20 of histone H4 is a common hallmark of human cancer, *Nat Genet* 37(4), 391–400, 2005.

163. Wang, H., Cao, R., Xia, L., Erdjument-Bromage, H., Borchers, C., Tempst, P., and Zhang, Y., Purification and functional characterization of a histone H3-lysine 4-specific methyltransferase, *Mol Cell* 8(6), 1207–1217, 2001.

164. Nishioka, K., Chuikov, S., Sarma, K., Erdjument-Bromage, H., Allis, C.D., Tempst, P., and Reinberg, D., Set9, a novel histone H3 methyltransferase that facilitates transcription by precluding histone tail modifications required for heterochromatin formation, *Genes Dev* 16(4), 479–489, 2002.

165. Chuikov, S., Kurash, J.K., Wilson, J.R., Xiao, B., Justin, N., Ivanov, G.S., McKinney, K., Tempst, P., Prives, C., Gamblin, S.J., Barlev, N.A., and Reinberg, D., Regulation of p53 activity through lysine methylation, *Nature* 432(7015), 353–360, 2004.

166. Ivanov, G.S., Ivanova, T., Kurash, J., Ivanov, A., Chuikov, S., Gizatullin, F., Herrera-Medina, E.M., Rauscher, F., III, Reinberg, D., and Barlev, N.A., Methylation-acetylation interplay activates p53 in response to DNA damage, *Mol Cell Biol* 27, 6756–6769, 2007.

167. Huang, N., vom Baur, E., Garnier, J.M., Lerouge, T., Vonesch, J.L., Lutz, Y., Chambon, P., and Losson, R., Two distinct nuclear receptor interaction domains in NSD1, a novel SET protein that exhibits characteristics of both corepressors and coactivators, *EMBO J* 17(12), 3398–3412, 1998.

168. Rayasam, G.V., Wendling, O., Angrand, P.O., Mark, M., Niederreither, K., Song, L., Lerouge, T., Hager, G.L., Chambon, P., and Losson, R., NSD1 is essential for early post-implantation development and has a catalytically active SET domain, *EMBO J* 22, 3153–3163, 2003.

169. Jaju, R.J., Fidler, C., Haas, O.A., Strickson, A.J., Watkins, F., Clark, K., Cross, N.C., Cheng, J.F., Aplan, P.D., Kearney, L., Boultwood, J., and Wainscoat, J.S., A novel gene, NSD1, is fused to NUP98 in the t(5;11)(q35;p15.5) in de novo childhood acute myeloid leukemia, *Blood* 98(4), 1264–1267, 2001.

170. Kurotaki, N., Imaizumi, K., Harada, N., Masuno, M., Kondoh, T., Nagai, T., Ohashi, H., Naritomi, K., Tsukahara, M., Makita, Y., Sugimoto, T., Sonoda, T., Hasegawa, T., Chinen, Y., Tomita, H., Kinoshita, A., Mizuguchi, T., Yoshiura Ki, K., Ohta, T., Kishino, T., Fukushima, Y., Niikawa, N., and Matsumoto, N., Haploinsufficiency of NSD1 causes Sotos syndrome, *Nat Genet* 30, 365–366, 2002.

171. Chesi, M., Nardini, E., Lim, R.S., Smith, K.D., Kuehl, W.M., and Bergsagel, P.L., The t(4;14) translocation in myeloma dysregulates both FGFR3 and a novel gene, MMSET, resulting in IgH/MMSET hybrid transcripts, *Blood* 92(9), 3025–3034, 1998.

172. Rosati, R., La Starza, R., Veronese, A., Aventin, A., Schwienbacher, C., Vallespi, T., Negrini, M., Martelli, M.F., and Mecucci, C., NUP98 is fused to the NSD3 gene in acute myeloid leukemia associated with t(8;11)(p11.2;p15), *Blood* 99, 3857–3860, 2002.

173. Angrand, P.O., Apiou, F., Stewart, A.F., Dutrillaux, B., Losson, R., and Chambon, P., NSD3, a new SET domain-containing gene, maps to 8p12 and is amplified in human breast cancer cell lines, *Genomics* 74(1), 79–88, 2001.

174. Kim, S.M., Kee, H.J., Choe, N., Kim, J.Y., Kook, H., Kook, H., and Seo, S.B., The histone methyltransferase activity of WHISTLE is important for the induction of apoptosis and HDAC1-mediated transcriptional repression, *Exp Cell Res* 313(5), 975–983, 2007.

175. Hamamoto, R., Furukawa, Y., Morita, M., Iimura, Y., Silva, F.P., Li, M., Yagyu, R., and Nakamura, Y., SMYD3 encodes a histone methyltransferase involved in the proliferation of cancer cells, *Nat Cell Biol* 6(8), 731–740, 2004.

176. Tsuge, M., Hamamoto, R., Silva, F.P., Ohnishi, Y., Chayama, K., Kamatani, N., Furukawa, Y., and Nakamura, Y., A variable number of tandem repeats polymorphism in an E2F-1 binding element in the 5′ flanking region of SMYD3 is a risk factor for human cancers, *Nat Genet* 37(10), 1104–1107, 2005.

177. Huang, J., Perez-Burgos, L., Placek, B.J., Sengupta, R., Richter, M., Dorsey, J.A., Kubicek, S., Opravil, S., Jenuwein, T., and Berger, S.L., Repression of p53 activity by Smyd2-mediated methylation, *Nature* 444(7119), 629–632, 2006.

178. Brown, M.A., Sims, R.J., III, Gottlieb, P.D., and Tucker, P.W., Identification and characterization of Smyd2: A split SET/MYND domain-containing histone H3 lysine 36-specific methyltransferase that interacts with the Sin3 histone deacetylase complex, *Mol Cancer* 5, 26, 2006.

179. Fahrner, J.A., Eguchi, S., Herman, J.G., and Baylin, S.B., Dependence of histone modifications and gene expression on DNA hypermethylation in cancer, *Cancer Res* 62, 7213–7218, 2002.

180. Kondo, Y., Shen, L.L., and Issa, J.P.J., Critical role of histone methylation in tumor suppressor gene silencing in colorectal cancer, *Mol Cell Biol* 23, 206–215, 2003.

181. Bachman, K.E., Park, B.H., Rhee, I., Rajagopalan, H., Herman, J.G., Baylin, S.B., Kinzler, K.W., and Vogelstein, B., Histone modifications and silencing prior to DNA methylation of a tumor suppressor gene, *Cancer Cell* 3(1), 89–95, 2003.

182. Cui, X., De Vivo, I., Slany, R., Miyamoto, A., Firestein, R., and Cleary, M.L., Association of SET domain and myotubularin-related proteins modulates growth control, *Nat Genet* 18(4), 331–337, 1998.

183. Cardoso, C., Timsit, S., Villard, L., Khrestchatisky, M., Fontes, M., and Colleaux, L., Specific interaction between the XNP/ATR-X gene product and the SET domain of the human EZH2 protein, *Hum Mol Genet* 7(4), 679–684, 1998.

184. Rozenblatt-Rosen, O., Rozovskaia, T., Burakov, D., Sedkov, Y., Tillib, S., Blechman, J., Nakamura, T., Croce, C.M., Mazo, A., and Canaani, E., The C-terminal SET domains of ALL-1 and TRITHORAX interact with the INI1 and SNR1 proteins, components of the SWI/SNF complex, *Proc Natl Acad Sci U S A* 95(8), 4152–4157, 1998.

185. Shi, Y. and Whetstine, J.R., Dynamic regulation of histone lysine methylation by demethylases, *Mol Cell* 25(1), 1–14, 2007.

186. Klose, R.J. and Zhang, Y., Regulation of histone methylation by demethylimination and demethylation, *Nat Rev Mol Cell Biol* 8(4), 307–318, 2007.

187. Kahl, P., Gullotti, L., Heukamp, L.C., Wolf, S., Friedrichs, N., Vorreuther, R., Solleder, G., Bastian, P.J., Ellinger, J., Metzger, E., Schule, R., and Buettner, R., Androgen receptor coactivators lysine-specific histone demethylase 1 and four and a half LIM domain protein 2 predict risk of prostate cancer recurrence, *Cancer Res* 66(23), 11341–11347, 2006.

188. Scoumanne, A. and Chen, X., The lysine-specific demethylase 1 is required for cell proliferation in both p53-dependent and -independent manners, *J Biol Chem* 282(21), 15471–15475, 2007.

189. Yang, Z.Q., Imoto, I., Fukuda, Y., Pimkhaokham, A., Shimada, Y., Imamura, M., Sugano, S., Nakamura, Y., and Inazawa, J., Identification of a novel gene, GASC1, within an amplicon at 9p23-24 frequently detected in esophageal cancer cell lines, *Cancer Res* 60(17), 4735–4739, 2000.

190. Lu, P.J., Sundquist, K., Baeckstrom, D., Poulsom, R., Hanby, A., Meier-Ewert, S., Jones, T., Mitchell, M., Pitha-Rowe, P., Freemont, P., and Taylor-Papadimitriou, J., A novel gene (PLU-1) containing highly conserved putative DNA/chromatin binding motifs is specifically up-regulated in breast cancer, *J Biol Chem* 274(22), 15633–15645, 1999.

191. Yamane, K., Tateishi, K., Klose, R.J., Fang, J., Fabrizio, L.A., Erdjument-Bromage, H., Taylor-Papadimitriou, J., Tempst, P., and Zhang, Y., PLU-1 is an H3K4 demethylase involved in transcriptional repression and breast cancer cell proliferation, *Mol Cell* 25(6), 801–812, 2007.

192. Christensen, J., Agger, K., Cloos, P.A., Pasini, D., Rose, S., Sennels, L., Rappsilber, J., Hansen, K.H., Salcini, A.E., and Helin, K., RBP2 belongs to a family of Demethylases, specific for tri- and dimethylated lysine 4 on histone 3, *Cell* 128, 1063–1076, 2007.

193. Klose, R.J., Yan, Q., Tothova, Z., Yamane, K., Erdjument-Bromage, H., Tempst, P., Gilliland, D.G., Zhang, Y., and Kaelin, W.G., Jr., The retinoblastoma binding protein RBP2 Is an H3K4 demethylase, *Cell* 128, 889–900, 2007.

194. Benevolenskaya, E.V., Murray, H.L., Branton, P., Young, R.A., and Kaelin, W.G., Jr., Binding of pRB to the PHD protein RBP2 promotes cellular differentiation, *Mol Cell* 18(6), 623–635, 2005.

195. Giovannucci, E., Rimm, E.B., Stampfer, M.J., Colditz, G.A., Ascherio, A., and Willett, W.C., Intake of fat, meat, and fiber in relation to risk of colon cancer in men, *Cancer Res* 54(9), 2390–2397, 1994.

196. Willett, W.C., Stampfer, M.J., Colditz, G.A., Rosner, B.A., and Speizer, F.E., Relation of meat, fat, and fiber intake to the risk of colon cancer in a prospective study among women, *N Engl J Med* 323(24), 1664–1672, 1990.

197. Farrar, C.E. and Clarke, S., Diet-dependent survival of protein repair-deficient mice, *J Nutr Biochem* 16(9), 554–561, 2005.

198. Finkelstein, J.D. and Martin, J.J., Methionine metabolism in mammals. Adaptation to methionine excess, *J Biol Chem* 261(4), 1582–1587, 1986.

199. Regina, M., Korhonen, V.P., Smith, T.K., Alakuijala, L., and Eloranta, T.O., Methionine toxicity in the rat in relation to hepatic accumulation of *S*-adenosylmethionine: Prevention by dietary stimulation of the hepatic transsulfuration pathway, *Arch Biochem Biophys* 300(2), 598–607, 1993.

200. Rowling, M.J., McMullen, M.H., Chipman, D.C., and Schalinske, K.L., Hepatic glycine N-methyltransferase is up-regulated by excess dietary methionine in rats, *J Nutr* 132(9), 2545–2550, 2002.

201. Caudill, M.A., Wang, J.C., Melnyk, S., Pogribny, I.P., Jernigan, S., Collins, M.D., Santos-Guzman, J., Swendseid, M.E., Cogger, E.A., and James, S.J., Intracellular *S*-adenosylhomocysteine concentrations predict global DNA hypomethylation in tissues of methyl-deficient cystathionine beta-synthase heterozygous mice, *J Nutrit* 131(11), 2811–2818, 2001.

202. Baric, I., Fumic, K., Glenn, B., Cuk, M., Schulze, A., Finkelstein, J.D., James, S.J., Mejaski-Bosnjak, V., Pazanin, L., Pogribny, I.P., Rados, M., Sarnavka, V., Scukanec-Spoljar, M., Allen, R.H., Stabler, S., Uzelac, L., Vugrek, O., Wagner, C., Zeisel, S., and Mudd, S.H., *S*-adenosylhomocysteine hydrolase deficiency in a human: a genetic disorder of methionine metabolism, *Proc Natl Acad Sci U S A* 101(12), 4234–4239, 2004.

203. Stramentinoli, G., Gualano, M., Catto, E., and Algeri, S., Tissue levels of *S*-adenosylmethionine in aging rats, *J Gerontol* 32(4), 392–394, 1977.

204. Brattstrom, L., Lindgren, A., Israelsson, B., Andersson, A., and Hultberg, B., Homocysteine and cysteine: Determinants of plasma levels in middle-aged and elderly subjects, *J Intern Med* 236(6), 633–641, 1994.

205. Andersson, A., Brattstrom, L., Israelsson, B., Isaksson, A., Hamfelt, A., and Hultberg, B., Plasma homocysteine before and after methionine loading with regard to age, gender, and menopausal status, *Eur J Clin Invest* 22(2), 79–87, 1992.

206. Clarke, S. and Banfield, K., Can elevated plasma homocysteine levels result in the inhibition of intracellular methyltransferases? In *Chemistry and Biology of Pteridines and Folates*, Milstein, S., Kapatos, G., Levine, R.A., and Shane, B. Kluwer Academic Press, Boston, 2002, pp. 557–562.

207. Melnyk, S., Pogribna, M., Pogribny, I.P., Yi, P., and James, S.J., Measurement of plasma and intracellular *S*-adenosylmethionine and *S*-adenosylhomocysteine utilizing coulometric electrochemical detection: Alterations with plasma homocysteine and pyridoxal 5′-phosphate concentrations, *Clin Chem* 46(2), 265–272, 2000.

208. Key, T.J., Fraser, G.E., Thorogood, M., Appleby, P.N., Beral, V., Reeves, G., Burr, M.L., Chang-Claude, J., Frentzel-Beyme, R., Kuzma, J.W., Mann, J., and McPherson, K., Mortality in vegetarians and nonvegetarians: detailed findings from a collaborative analysis of 5 prospective studies, *Am J Clin Nutr* 70 (3 Suppl), 516S–524S, 1999.

209. Campbell, T.C., Animal protein and ischemic heart disease, *Am J Clin Nutr* 71(3), 849–851, 2000.

210. Campbell, T.C. and Campbell, T.M., *The China Study* BenBella Books, Dallas, TX 75206, 2005.

211. Madhavan, T.V. and Gopalan, C., The effect of dietary protein on carcinogenesis of aflatoxin, *Arch Pathol* 85(2), 133–137, 1968.

212. Schulsinger, D.A., Root, M.M., and Campbell, T.C., Effect of dietary protein quality on development of aflatoxin B1-induced hepatic preneoplastic lesions, *J Natl Cancer Inst* 81(16), 1241–1245, 1989.

213. Venkatachalam, S., Shi, Y.P., Jones, S.N., Vogel, H., Bradley, A., Pinkel, D., and Donehower, L.A., Retention of wild-type p53 in tumors from p53 heterozygous mice: Reduction of p53 dosage can promote cancer formation, *EMBO J* 17(16), 4657–4667, 1998.

214. Feinberg, A.P., Ohlsson, R., and Henikoff, S., The epigenetic progenitor origin of human cancer, *Nat Rev Genet* 7(1), 21–33, 2006.

215. Zhou, W., Alonso, S., Takai, D., Lu, S.C., Yamamoto, F., Perucho, M., and Huang, S. Requirement of RIZ1 for cancer prevention by methyl-balanced diet. Submitted, 2008. Preprint available at *Nature Precedings* http://hdl.handle.net/10101/npre.2008.1732.1.
216. Gaudet, F., Hodgson, J.G., Eden, A., Jackson-Grusby, L., Dausman, J., Gray, J.W., Leonhardt, H., and Jaenisch, R., Induction of tumors in mice by genomic hypomethylation, *Science* 300, 489–492, 2003.
217. Wilson, A.S., Power, B.E., and Molloy, P.L., DNA hypomethylation and human diseases, *Biochim Biophys Acta* 1775(1), 138–162, 2007.
218. Laird, P.W., Jackson-Grusby, L., Fazeli, A., Dickinson, S.L., Jung, W.E., Li, E., Weinberg, R.A., and Jaenisch, R., Suppression of intestinal neoplasia by DNA hypomethylation, *Cell* 81(2), 197–205, 1995.
219. Jones, P.A. and Baylin, S.B., The epigenomics of cancer, *Cell* 128(4), 683–692, 2007.
220. Feinberg, A.P. and Tycko, B., The history of cancer epigenetics, *Nat Rev Cancer* 4(2), 143–153, 2004.
221. Lopatina, N.G., Vanyushin, B.F., Cronin, G.M., and Poirier, L.A., Elevated expression and altered pattern of activity of DNA methyltransferase in liver tumors of rats fed methyl-deficient diets, *Carcinogenesis* 19(10), 1777–1781, 1998.
222. Ghoshal, K., Li, X., Datta, J., Bai, S., Pogribny, I., Pogribny, M., Huang, Y., Young, D., and Jacob, S.T., A folate- and methyl-deficient diet alters the expression of DNA methyltransferases and methyl CpG binding proteins involved in epigenetic gene silencing in livers of F344 rats, *J Nutr* 136(6), 1522–1527, 2006.
223. Kim, Y.I., Folate and DNA methylation: A mechanistic link between folate deficiency and colorectal cancer? *Cancer Epidemiol Biomarkers Prev* 13(4), 511–519, 2004.
224. Gaudet, F., Hodgson, J.G., Eden, A., Jackson-Grusby, L., Dausman, J., Gray, J.W., Leonhardt, H., and Jaenisch, R., Induction of tumors in mice by genomic hypomethylation, *Science* 300(5618), 489–492, 2003.
225. Eden, A., Gaudet, F., Waghmare, A., and Jaenisch, R., Chromosomal instability and tumors promoted by DNA hypomethylation, *Science* 300(5618), 455, 2003.
226. Heard, E., Rougeulle, C., Arnaud, D., Avner, P., Allis, C., and Spector, D.L., Methylation of histone H3 at Lys-9 Is an early mark on the X chromosome during X inactivation, *Cell* 107, 727–738, 2001.
227. Mutskov, V. and Felsenfeld, G., Silencing of transgene transcription precedes methylation of promoter DNA and histone H3 lysine 9, *EMBO J* 23(1), 138–149, 2004.
228. Tamaru, H. and Selker, E.U., A histone H3 methyltransferase controls DNA methylation in *Neurospora crassa*, *Nature* 414(1), 277–283, 2002.
229. Pogribny, I.P. and James, S.J., De novo methylation of the p16INK4A gene in early preneoplastic liver and tumors induced by folate/methyl deficiency in rats, *Cancer Lett* 187(1–2), 69–75, 2002.
230. Holst, C.R., Nuovo, G.J., Esteller, M., Chew, K., Baylin, S.B., Herman, J.G., and Tlsty, T.D., Methylation of p16(INK4a) promoters occurs in vivo in histologically normal human mammary epithelia, *Cancer Res* 63(7), 1596–1601, 2003.
231. Baylin, S.B. and Ohm, J.E., Epigenetic gene silencing in cancer—a mechanism for early oncogenic pathway addiction? *Nat Rev Cancer* 6(2), 107–116, 2006.
232. Corda, Y., Schramke, V., Longhese, M.P., Smokvina, T., Paciotti, V., Brevet, V., Gilson, E., and Geli, V., Interaction between Set1p and checkpoint protein Mec3p in DNA repair and telomere functions, *Nat Genet* 21(2), 204–208, 1999.
233. Segal, E., Fondufe-Mittendorf, Y., Chen, L., Thastrom, A., Field, Y., Moore, I.K., Wang, J.P., and Widom, J., A genomic code for nucleosome positioning, *Nature* 442(7104), 772–778, 2006.
234. Suzuki, K., Suzuki, I., Leodolter, A., Alonso, S., Horiuchi, S., Yamashita, K., and Perucho, M., Global DNA demethylation in gastrointestinal cancer is age dependent and precedes genomic damage, *Cancer Cell* 9(3), 199–207, 2006.
235. Redon, R., Ishikawa, S., Fitch, K.R., Feuk, L., Perry, G.H., Andrews, T.D., Fiegler, H., Shapero, M.H., Carson, A.R., Chen, W., Cho, E.K., Dallaire, S., Freeman, J.L., Gonzalez, J.R., Gratacos, M., Huang, J., Kalaitzopoulos, D., Komura, D., Macdonald, J.R., Marshall, C.R., Mei, R., Montgomery, L., Nishimura, K., Okamura, K., Shen, F., Somerville, M.J., Tchinda, J., Valsesia, A., Woodwark, C., Yang, F., Zhang, J., Zerjal, T., Zhang, J., Armengol, L., Conrad, D.F., Estivill, X., Tyler-Smith, C., Carter, N.P., Aburatani, H., Lee, C., Jones, K.W., Scherer, S.W., and Hurles, M.E., Global variation in copy number in the human genome, *Nature* 444(7118), 444–454, 2006.

236. Gibbs, R.A., Rogers, J., Katze, M.G., Bumgarner, R., Weinstock, G.M., Mardis, E.R., Remington, K.A., Strausberg, R.L., Venter, J.C., Wilson, R.K., Batzer, M.A., Bustamante, C.D., Eichler, E.E., Hahn, M.W., Hardison, R.C., Makova, K.D., Miller, W., Milosavljevic, A., Palermo, R.E., Siepel, A., Sikela, J.M., Attaway, T., Bell, S., Bernard, K.E., Buhay, C.J., Chandrabose, M.N., Dao, M., Davis, C., Delehaunty, K.D., Ding, Y., Dinh, H.H., Dugan-Rocha, S., Fulton, L.A., Gabisi, R.A., Garner, T.T., Godfrey, J., Hawes, A.C., Hernandez, J., Hines, S., Holder, M., Hume, J., Jhangiani, S.N., Joshi, V., Khan, Z.M., Kirkness, E.F., Cree, A., Fowler, R.G., Lee, S., Lewis, L.R., Li, Z., Liu, Y.S., Moore, S.M., Muzny, D., Nazareth, L.V., Ngo, D.N., Okwuonu, G.O., Pai, G., Parker, D., Paul, H.A., Pfannkoch, C., Pohl, C.S., Rogers, Y.H., Ruiz, S.J., Sabo, A., Santibanez, J., Schneider, B.W., Smith, S.M., Sodergren, E., Svatek, A.F., Utterback, T.R., Vattathil, S., Warren, W., White, C.S., Chinwalla, A.T., Feng, Y., Halpern, A.L., Hillier, L.W., Huang, X., Minx, P., Nelson, J.O., Pepin, K.H., Qin, X., Sutton, G.G., Venter, E., Walenz, B.P., Wallis, J.W., Worley, K.C., Yang, S.P., Jones, S.M., Marra, M.A., Rocchi, M., Schein, J.E., Baertsch, R., Clarke, L., Csuros, M., Glasscock, J., Harris, R.A., Havlak, P., Jackson, A.R., and Jiang, H., Evolutionary and biomedical insights from the rhesus macaque genome, *Science* 316(5822), 222–234, 2007.

237. Huang, S., The genetic equidistance result of molecular evolution is independent of mutation rates. Submitted, 2008. Preprint available at *Nature Precedings* http://hdl.handle.net/10101/npre.2008.1733.1.

238. Kasahara, M., Naruse, K., Sasaki, S., Nakatani, Y., Qu, W., Ahsan, B., Yamada, T., Nagayasu, Y., Doi, K., Kasai, Y., Jindo, T., Kobayashi, D., Shimada, A., Toyoda, A., Kuroki, Y., Fujiyama, A., Sasaki, T., Shimizu, A., Asakawa, S., Shimizu, N., Hashimoto, S., Yang, J., Lee, Y., Matsushima, K., Sugano, S., Sakaizumi, M., Narita, T., Ohishi, K., Haga, S., Ohta, F., Nomoto, H., Nogata, K., Morishita, T., Endo, T., Shin, I.T., Takeda, H., Morishita, S., and Kohara, Y., The medaka draft genome and insights into vertebrate genome evolution, *Nature* 447(7145), 714–719, 2007.

239. Huang, S., Ancient fossil specimens are genetically more distant to an outgroup than extant sister species are. *Riv. Biol.* In press, 2008. Preprint available at *Nature Precedings* http://hdl.handle.net/10101/npre.2008.1676.1.

240. Alberts, B., Bray, D., Lewis, J., Raff, M., Roberts, K., and Watson, J.D., *Molecular Biology of the Cell* Garland Publishing, Inc., New York, 1983.

9 Role of Sirtuins in Aging and Tumorigenesis

Xiaohong Zhang, Tung I. Hsieh, Santo V. Nicosia, and Wenlong Bai

CONTENTS

9.1 INTRODUCTION: SIR2 AND SIRTUINS AS CLASS III HISTONE DEACETYLASES

9.1.1 HISTONE DEACETYLASE FAMILY

Eukaryotic gene expression is tightly controlled by chromatin remodeling through the modification of core histones. Acetylation is one of the most intensively studied modifications, which is highly regulated by two classes of enzymes, histone acetyltransferases (HATs) and histone deacetylases (HDACs). HATs deliver acetyl groups to the lysine residues in the amino terminal tail of core histones, neutralize the positive charge, and result in the unwinding of chromatin. Acetylation provides docking sites for regulatory proteins to activate transcription. HDACs, on the other hand, induce transcriptional repression and gene silencing by catalyzing the removal of the acetyl groups from core histones. In humans, 18 HDACs have been identified, which are divided into 4 subclasses based on their homology to yeast proteins [1]. The class I HDACs, including 1, 2, 3, and 8, are ubiquitously expressed nuclear proteins homologous to the yeast RPD3. Class II HDACs (4, 5, 6, 7, 9, and 10) are homologous to yeast Hda1 and can be found in both the nucleus and the cytoplasm. HDAC11 is remotely similar to class I and II HDACs but the sequence identity is too weak to be placed in either class [2]. It is thus designated as the sole member of class IV. All members of class I, II, and IV HDACs are sensitive to the inhibition by trichostatin A (TSA).

9.1.2 MEMBERS OF SIR2 FAMILY: THE CLASS III HDACS

The class III HDACs, namely sirtuins (sir-two-ns) or SIRT (silent information regulator two) [3,4], are homologues of the yeast silent information regulator 2 (SIR2). Different from the other three classes of HDACs, the activity of Sir2 family is NAD^+-dependent. They convert acetyl lysine and NAD^+ to O-acetyl-ADP-ribose and nicotinamide [5] (Figure 9.1). Due to its dependency on NAD^+, the activity of sirtuins is regulated by NAD^+/NADH ratio and thus sensitive to the status of cellular metabolism and redox. Instead of being inhibited by TSA, the activity of SIR2 family is subjected to feedback inhibition by nicotinamide, a product of sirtuin-catalyzed deacetylation and other reactions that consume NAD^+ in cells.

Sirtuins are highly conserved across a variety of organisms ranging from yeast to human [6]. Interestingly, sirtuins are also present in prokaryotes. For example, CobB, a Salmonella Sir2, regulates the synthesis of a cobalamin biosynthetic intermediate [7] and deacetylates the acetyl-CoA synthetase in an NAD^+-dependent manner to activate the enzyme [8]. The archaeal *Sulfolobus solfataricus* P2 encodes the SIR2 homologue, namely ssSIR2 [9], which functions as an active NAD^+-dependent deacetylase as well as a mono-ADP-ribosyltransferase (ART). It interacts with and deacetylates Alba, resulting in an increase in its affinity to DNA and an enhancement of its activity to repress transcription [10]. On the basis of the molecular phylogenetic analysis of the conserved core deacetylase domain sequence of 60 sirtuins, five different classes [6] were categorized with class V identified only in gram positive bacteria and *Thermotoga maritima*.

In mammals, there are seven homologues of yeast SIR2, namely SIRT1 to SIRT7 (Table 9.1 and Figure 9.2). Although it is debatable whether any one of the seven mammalian sirtuins is the true homologue of yeast SIR2, SIRT1 appears to be the closest one [11]. It is a deacetylase and localized

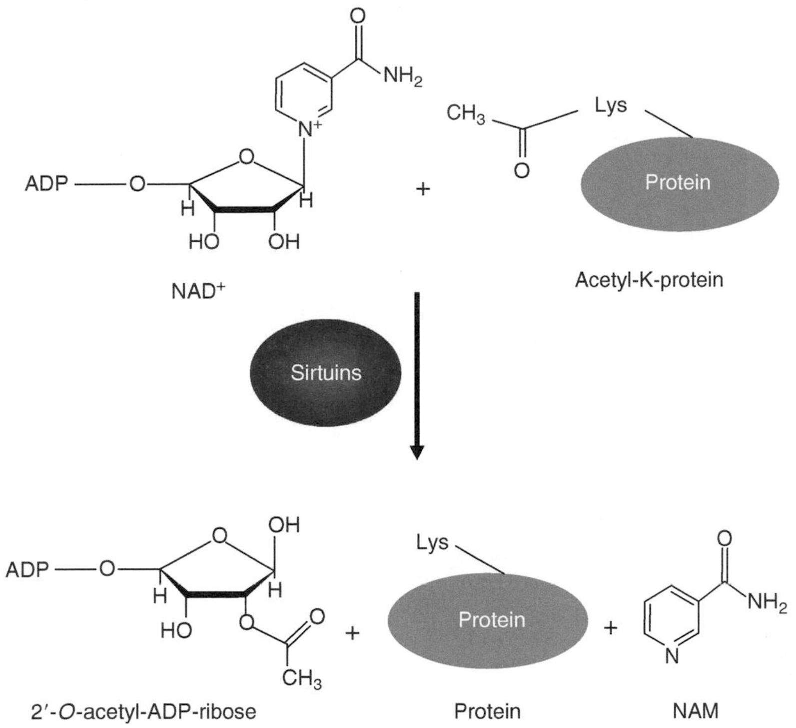

FIGURE 9.1 Biochemical reactions of sirtuin-mediated deacetylation and its dependency on NAD^+. The protein deacetylation is accomplished by the transfer of the acetyl group (red) from substrate to ADP-ribose coupled to NAD^+ cleavage that gives rise to nicotinamide (NAM). It is important to point out that not all sirtuins are active deacetylase. Some members only function as an ART (e.g., SIRT4).

mainly in nucleus whereas other members of the family are found to be either localized to nonnuclear compartment or lack a deacetylase activity. Mammalian SIRT1 (Figure 9.2) has a long N-terminal sequence that shows little conservation with yeast SIR2 but contains functional nuclear localization sequences that are conserved among SIRT1 from different mammals (Figure 9.3a and b). Together with nuclear exporting sequences located in the catalytic domain, the nuclear localization sequences control the nucleus–cytoplasm shuttling of SIRT1 [12]. The C-terminus of mammalian SIRT1 is largely missing in yeast SIR2 and other mammalian sirtuins (Figure 9.2). It contains multiple high probability sites for posttranslational modifications, which are conserved among mammalian SIRT1 molecules from different species (Figure 9.3c). Thus, the C-terminus of SIRT1 appears to represent a mammalian-specific regulatory domain. SIRT1 is the best characterized sirtuin in mammalian cells and its functions in cancer cells are the main subject of the present discussion.

SIRT2 and SIRT3 are closely related to SIRT1 and belong to the same branch as yeast SIR2 on phylogenic tree [6]. SIRT2 is a cytoplasmic deacetylase that co-localizes with and deacetylates tubulin [13,14]. It is as a regulator of mitotic progression that modulates mitotic exit and subsequent cytokinesis. Overexpression of SIRT2 delays cell cycle progression through mitosis. In the presence of mitotic cyclins, SIRT2 becomes ubiquitinated and is degraded via the 26S proteasome pathway [15]. SIRT3 is considered to be the closest paralog of SIRT2 and localized in mitochondria. It only functions as a deacetylase after cleavage of the signal peptide [16,17]. SIRT3 deacetylates and activates acetyl-CoA synthetase II [18]. A VNTR polymorphism (72 bp repeat core) in intron 5 functions as an enhancer for SIRT3 expression [19]. The allele completely lacking enhancer activity is virtually absent in males older than 90 years [19], suggesting that the insufficient expression of SIRT3 may be detrimental for longevity. Recently, it has been shown that SIRT3 is

TABLE 9.1

Mammalian Sirtuins and Their Enzymatic Activity, Chromosomal and Subcellular Localizations, Substrates, and Tissue Distribution

Homo sapiens:

Sirtuin Gene	Sirtuin Subclass	HDAC Activity	Chromosome Location	Predominant Subcellular Location	Cellular Substrates	Major Tissue of Expression	Relative in Yeast	Relative in Bacteria
SIRT1	I	Active	10	Nucleus	H1, H3, H4, p53, FOXO, Ku70, p300, E2F1, etc.	Ubiquitous	Yes	No
SIRT2	I	Active	19q	Cytoplasm	Tubulin	Bone, Skeleton, testes, heart	Yes	No
SIRT3	I	After cleavage	11p15.5	Nucleus, mitochondria	Acetyl-CoA synthetase	Liver, kidney, muscle, heart	Yes	No
SIRT4	II	Inactive	12q	Mitochondria	Glutamate dehydrogenase	Kidney, brain, liver, heart	No	Yes
SIRT5	III	Active	6p23	Mitochondria	ND	Brain, testes, heart	Yes	Yes
SIRT6	IV	Active	19p13.3	Nucleus	DNA polβ, histones	Bone, ovaries	No	No
SIRT7	IV	?	17q	Nucleoli	ND	Liver	No	No

Note: ND: Not determined. Please note that multiple sirtuins, including SIRT2, SIRT3, SIRT4, and SIRT6, are likely to act as an ART.

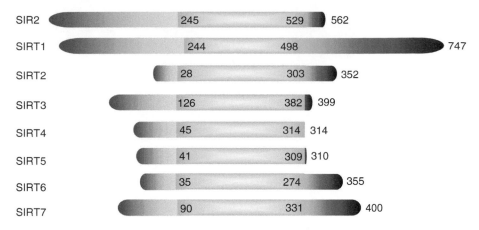

FIGURE 9.2 Structure of yeast SIR2 and human sirtuins. The protein structure of yeast SIR2 and the seven human sirtuins, SIRT1 to SIRT7, is shown. The catalytic domain in the middle is shown in purple. The N- and C-terminal sequences are shown in green and red, respectively. The amino acids flanking the catalytic domain and the residue at the C-terminal end were numbered. Please note that yeast SIR2 and SIRT1 contain a long N-terminal region with a similar length and that SIRT1 has a unique long C-terminal region.

a nuclear deacetylase of which the localization to mitochondria is induced by cellular stresses or overexpression [20].

SIRT4 and SIRT5 are distant from SIR2 and SIRT1 in phylogeny tree and their homologous proteins found in prokaryotes. SIRT4 protein is a mitochondrial ART that apparently lacks

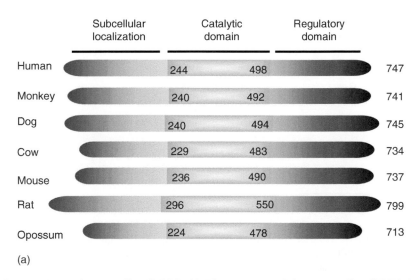

(a)

FIGURE 9.3 Members of mammalian SIRT1. (a) The structure of the mammalian SIRT1 from different species is shown. The N-terminal region controls subcellular localization. The middle region is the catalytic domain. The C-terminal region is predicted to function as a mammalian specific regulatory domain. It contains highly probable sites for phosphorylation and sumoylation. *Homo sapiens* (Human); *Macaca mulatto* (Monkey); *Bos Taurus* (Cow); *Rattus norvegicus* (Rat); *Canis familiaris* (Dog); *Mus musculus* (Mouse); *Monodelphis domestica* (Opossum).

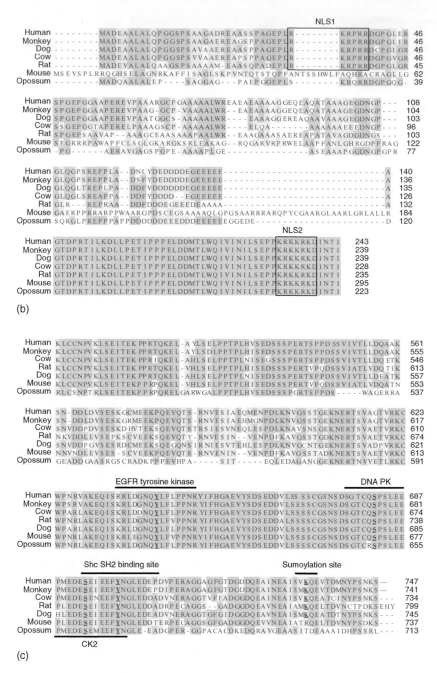

FIGURE 9.3 (continued) (b) (**See color insert following page 272.**) Sequence alignment of the N-terminal region of mammalian SIRT1 proteins. The conserved nuclear localization sequences are placed in a square. Perfectly conserved residues are shown in blue and those that are functionally conserved shown in yellow. (c) (**See color insert following page 272.**) Sequence alignment of the C-terminal region of mammalian SIRT1 proteins. Perfectly conserved residues are shown in blue and those that are functionally conserved shown in yellow. Conserved residues predicted with Scansite software (http://scansite.mit.edu/motifscan_seq.phtml) at high stringency and Sumoplot (www.abgent.com/sumoplot.html) as probable phosphorylation and sumoylation sites are shown in red. EGFR: epidermal growth factor receptor; DNA-PK: DNA dependent protein kinase; SH2; Src homology 2; CK2: casein kinase 2.

(continued)

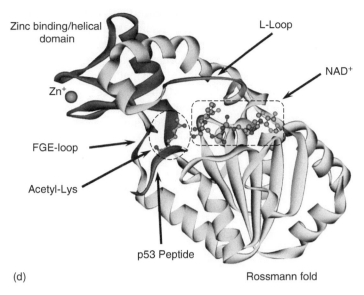

(d)   Rossmann fold

FIGURE 9.3 (continued)   (d) (**See color insert following page 272.**) 3D structure of the catalytic domain of sirtuins. The ribbon structure of SIR2-TM in complex with acetylated p53 peptide and NAD^+ is shown. The X-ray data of SIR2-TM [28] was downloaded from Protein Data Bank (PDB) (http://www.rcsb.org) and visualized with Accelrys ViewerLite Version 5.0. The colors for different domains are cyan for the Rossmann fold, green for the helical module, red for the zinc-binding module, blue for the FGE loop that makes contact with acetyl-Lys of the substrate, pink for the K382 acetyl-p53 peptide and brown for the L-1B loop that serves as a ceiling for NAD^+ binding.

deacetylase activity. It interacts with and inhibits glutamate dehydrogenase (GDH) [21]. SIRT5 is the closest homologue of the bacterial *CobB* and the most distant one from SIR2 and SIRT1 [6]. It is a mitochondrial deacetylase with unknown functions. Because of the mitochondrial localization, it is speculated that SIRT5 may also regulate metabolic and aging processes in humans [22].

SIRT6 is a nuclear, chromatin-associated protein that promotes resistance to DNA damage. It suppresses genomic instability in mouse cells, in association with a role in base excision repair [23]. It remains to be established whether SIRT6 is an NAD^+-dependent ART or a deacetylase. Previous studies have shown that SIRT6 acts as an ART [24]. However, Mostoslavsky et al. [23] have recently demonstrated that SIRT6 deacetylates core histones and DNA polymerase β of which the activity is inhibited by acetylation in vivo [25]. SIRT6-deficient mice are small and develop abnormalities including profound lymphopenia, loss of subcutaneous fat, lordokyphosis, and severe metabolic defects at 2–3 weeks of age, and eventually the mice die around 4 weeks [23]. It is thus likely that one function of SIRT6 is to promote normal DNA repair. The loss of SIRT6 leads to abnormalities in mice, which overlaps with aging-associated degenerative processes.

Ford et al. [26] have reported that SIRT7 is a nucleolar protein associated with active rDNA that interacts with RNA polymerase I (Pol 1) and histones. Overexpression of SIRT7 increases Pol I mediated transcription, whereas knockdown of SIRT7 or inhibition of its catalytic activity results in decreased association of Pol I with rDNA and a reduction of Pol I transcription. Depletion of SIRT7 stops cell proliferation and triggers apoptosis. These findings suggest that SIRT7 is a positive regulator of Pol I transcription, which is required for cell viability in mammals.

9.1.3 STRUCTURE OF SIRTUINS AND ITS IMPLICATION ON SUBSTRATES SPECIFICITY

Among the members of SIR2 family, the three-dimensional structure has been published for those that basically contain the core deacetylase domain with very short N- and C-terminal sequences [27–36]

On the basis of these published structures, the deacetylase region of sirtuins is composed of a larger Rossmann fold domain and a smaller Zn^+ binding/helical domain (Figure 9.3d). The two domains form a hydrophobic cleft that binds the substrates and the coenzyme NAD^+. The large domain is composed of six parallel strands that form a central β sheet sandwiched between four α helices on one side and two on the other side. The small domain consists of a three-stranded antiparallel β sheet, two α helices, and a large portion of a long loop (L Loop) connecting the two domains. The loop segment forms a flap above the NAD binding pocket to serve as a "ceiling" and is likely to be dynamic in solution. When the ceiling is raised, sirtuins adopt an open state with a more accessible binding pocket and vice verse they change to a close state with reduced accessibility of the binding pocket.

The zinc-binding motif in the smaller domain contains a conserved Cys-X-X-Cys-$(X)_{15-20}$-Cys-X-X-Cys sequence and protrudes from the main body of sirtuin structure, forming an "overhang" over the hydrophilic cleft between the two domains that binds the substrates. The zinc motif may be responsible for creating and maintaining a substrate binding cleft, in which the nonconserved residues may determine the substrate specificity of the different sirtuins.

The structural studies conclude that sirtuins bind the acetylated peptides primarily through β sheet interactions with the main chain atoms of the substrate, forming an enzyme–substrate β sheet. The substrate peptide comprises the middle strand of a staggered antiparallel three-stranded β sheet. The bottom strand is contributed by the Rossmann fold domain, and the top strand is contributed by part of a conserved loop structure, named FGE loop because it contains a conserved Phe–Gly–Glu motif that makes contact with acetyl-lysine of the substrate. The acetyl-lysine side chain of the substrate inserts into a hydrophobic tunnel located at the interface between the Rossmann fold and the FGE loop. NAD^+ binding to the flexible loop region (the ceiling for the NAD^+ binding pocket) favors a closing of the cofactor binding cleft, which in turn causes the FGE loop to shift to its substrate binding conformation, suggesting that NAD^+ and protein substrates bind sirtuins in a cooperative fashion. The burial of the acetyl-lysine side chain within the enzyme active site positions the acetyl group for reaction with the ribose ring of the nicotinamide moiety of NAD^+ and puts it in contact with the catalytic site of sirtuins. The hydrophobic character of the tunnel and the residues that line its entrance are likely to favor binding of acetyl-lysine over that of an unacetylated, charged lysine side chain.

Because the substrate peptide is inserted between β strands belonging to two different domains of sirtuins and joins the two domains through the formation of a staggered enzyme–substrate β sheet, the substrates of sirtuins are predicted to be either unstructured peptides or segments of proteins that can unfold upon binding to sirtuins. Consistent with this prediction, the C-terminus of p53 deacetylated by human SIRT1 in vivo has been shown by NMR studies to be unstructured in solution [37]. Acetylated histone tails, which are substrates for sirtuins, are similarly unstructured in the context of the nucleosome [38,39]. In addition, the structural analyses suggest that the peptide backbone hydrogen bonds play a dominant role in substrate binding and limited extent of substrate side chain is buried. Thus, sirtuins are predicted to show weak selectivity for particular peptide side chains flanking the acetyl-lysine. This is in agreement with the observations that a variety of sirtuins are capable of deacetylating a broad array of substrates [27] and that SIRT1 lacks substrate specificity in vitro [40]. This raises the possibility that most sirtuins may depend upon domains outside the conserved catalytic core for precise substrate targeting in the cell, either directly or through their involvement in the formation of multiprotein complexes that mediate substrate recognition. However, a recent structural analysis of SIR2 homologue from *Thermatoga maritima* (Sir2Tm) with several acetyl peptide substrates showed that the residues in -1 and $+2$ positions of the acetyl–lysine played an important role in determining substrate specificity [28], suggesting that the catalytic core of SIR2 proteins can discriminate substrates. This is supported by a recent demonstration that human SIRT1 discriminates among peptide substrates in an acetyl peptide library by as many as 20-fold [41].

9.2 SIRTUINS: CALORIC RESTRICTION AND AGING

Physiological functions of proteins are the foundations for their roles in pathological processes. Before the discussion about the potential role of sirtuins in tumorigenesis, it is important to review the current knowledge about their physiological functions. Multiple cellular functions have been attributed to members of SIR2 family, including transcriptional silencing [42], rDNA recombination [43], glucose homeostasis and metabolism [3], DNA repair and cell survival [44–46], etc. Consistent with these cellular functions, mammalian sirtuins are shown to deacetylate and regulate the activity of histones [47], p300 [48], TAF_I68, HES-1 and HEY-2 [49], MyoD [50], CTIP2 [51], PPARγ [52], etc. In terms of their physiological functions, the most notable one is the involvement of sirtuins in regulating aging and lifespan.

9.2.1 Positive Effect of SIR2 on Yeast Replicative Lifespan

The first evidence for the involvement of sirtuins in aging came from studies in yeast. Kennedy et al. [53] found that specific mutations in a component of the SIR complex (SIR2, SIR3, and SIR4) increased the replicative lifespan of yeast. At that time, the transcriptional repression at telomeres by the SIR complex was thought to be responsible for the antiaging effect since the SIR complex represses transcription of genes placed near telomeres [54]. However, the predominant view at the present time is that SIR2 regulates yeast lifespan through the suppression of rDNA recombination. The localization of the SIR complex to the nucleolus correlates with its effect on lifespan [55] and SIR2 inhibits rDNA recombination, a process that leads to the formation of extrachromosomal rDNA circles (ERCs) [54]. The ERCs contain replication initiation sites but lack centromeres. They undergo replication in S phase but remain in the mother cells during mitosis, resulting in their accumulation in and the aging of the mother cells [56]. Yeast cells without SIR2 are short-lived because of their increased rate of ERC formation. SIR2 overexpression decreases the levels of ERCs, resulting in the extension of yeast lifespan [57].

9.2.2 SIR2: Caloric Restriction and the Extension of Yeast Replicative Lifespan

One theory proposes that the exposure of an organism to chronic small doses of stress will condition the organism to stress resistance, contributing to extended lifespan [58]. Calorie restriction (CR) might be one such stress that enables an organism to produce fewer free radicals, to accumulate lower levels of oxidatively damaged macromolecules, and to maintain elevated antioxidants and detoxification capacities [52].

Indeed, CR has been known for a long time to enhance longevity of organisms from yeast to mammals. The idea that a simple reduction in energy intake might make organisms including humans live longer is fascinating. However, the molecular mechanisms remain elusive until the realization that yeast SIR2 is required for the replicative lifespan extension upon CR [59]. By either reducing the glucose in the growth medium or creating mutant yeast strains with altered glucose metabolism, Lin et al. [59] have shown that SIR2-deleted yeast strains are short-lived and do not respond to CR for lifespan extension. Similarly, the SIR2 homologue in *Drosophila* and *C. elegans* regulates the lifespan [60,61] and small molecules that enhance SIR2 enzymatic activity prolong lifespan of yeast and worm [62], suggesting a conservation of this function during evolution.

Although the sirtuin field appears to be in agreement that the increased SIR2 activity extends lifespan, questions have been raised about the SIR2 dependency of CR-induced lifespan extension. Two separate studies have shown that CR causes robust lifespan extension in the absence of SIR2 if FOB1, a protein that has replication fork blocking activity and promotes rDNA recombination [63], is also deleted [64,65]. However, these studies were performed mainly with 0.05% (40× reduction) glucose, which is 10 times lower than the dose used by studies showing SIR2-dependency (0.5% glucose, 4× reduction). The latter dose is considered more ideal because it has minimal effect on yeast growth rate and is more physiologically similar to levels of CR employed in other model systems. It is thus argued that CR induced by 0.5% glucose and 0.05% glucose may use SIR2 dependent and independent mechanisms, respectively. Existing evidences also suggest that other yeast sirtuins, such as Hst1 and Hst2 [65], may act redundantly to mediate SIR2-independent lifespan extension by CR.

9.2.3 Negative Effect of SIR2 on Yeast Chronological Lifespan

Different from replicative lifespan that is measured by successive removal of the smaller and easily distinguishable daughter cells to determine the capacity of a mother yeast cell to divide, the chronological lifespan measures the survival of yeast population in a nondividing state. The chronological lifespan appears to be regulated by the level of superoxide instead of ERCs. Overexpression of the superoxide dismutases *SOD1* or *SOD2* extends yeast chronological lifespan [66] and the frequency of spontaneous DNA mutation increases with chronological age [67].

Severe CR induced by switching yeast from glucose/ethanol containing medium to water causes decrease in metabolic rates and lifespan extension by two- to threefold. Opposite to the positive role of SIR2 in replicative lifespan, deletion of SIR2 shows no effect on the chronological lifespan of wild-type yeast in glucose medium but further extends the lifespan induced by severe CR in water or the mutation of nutrient responsive kinases such as PKA and SCH9 [68]. Different from the involvement of SIR2-mediated suppression of ERC formation in replicative lifespan, the positive effect of SIR2 deletion on the extension of chronological lifespan by severe CR was found to be associated with increased expression of stress resistance genes and alcohol dehydrogenase Ach2 [68]. It is reasoned that the loss of SIR2 may promote the entry into a state that protects cells against aging during starvation. Overall, it appears that SIR2 is important for the regulation of aging processes but it may play either an anti- or a pro-aging role, depending on the availability of nutrients and the activity of glucose signaling pathways.

9.2.4 Roles of Mammalian Sirtuins in Aging and Food Restriction

Similar to yeast, increased SIR2 in worms [60] and flies [61] was found to extend lifespan. Consistent with the positive effect of SIR2 in lower organisms, mammalian SIRT1 has also been shown to exhibit antiaging activity. The transgenic expression of deltaNp63alpha, a p53 related protein, in the skin produces mice with an accelerated aging phenotype. The increased deltaNp63alpha expression induces cellular senescence that is rescued by SIRT1 [69], suggesting that deltaNp63alpha regulates aging in mammals, at least in part, through SIRT1.

Similar to yeast and *Drosophila* SIR2, mammalian SIRT1 has also been implicated in mediating the lifespan extension of mammals in response to CR [46,70]. CR promotes a large increase in physical activity in wild type but not in SIRT1 knockout mice [71]. SIRT1 knockout mice exhibit a high degree of embryonic and postnatal lethality. However, survivors progress to adulthood and exhibit phenotypes such as decreased body weight, blood glucose, and insulin, all of which occur in CR mice. Although the interpretation of the data has been questioned [72] and a direct demonstration of SIRT1 deficiency on mouse lifespan is currently lacking, the study can be viewed as the first direct evidence for conserved roles between SIR2 in lower organisms and mammalian SIRT1 in mediating physiological effects of food restriction. Two subsequent studies show that resveratrol, an activator of SIRT1 rich in grape extracts, prolongs lifespan of mice by activating SIRT1 [73,74].

SIRT1 has been shown to control hepatic glucose metabolism through the deacetylation and activation of PPARγ coactivator 1 (PGC-1α), a transcriptional coactivator that controls metabolism at the level of gene transcription [75]. PGC-1α is projected as the key molecular target for SIRT1 to regulate glucose metabolism and aging processes in mice. However, it is important to point out that SIRT1 in mammalian cells appears to act in a cell-specific fashion, which does not always lead to PGC-1α activation. Although increased SIRT1 expression promotes fat mobilization in the white adipose tissues [52] and gluconeogenesis in the liver [76], it is the downregulation of SIRT1 in pancreatic β cells that accounts for the decreased insulin level in response to fasting [77]. The expression of SIRT1 in β cells has been found to increase ATP production and insulin secretion [78], a result opposite to what would have been expected if SIRT1 in beta cells acts through PGC-1α activation. In PC-12 cells, instead of activation, SIRT1 suppresses PGC-1α activity [79].

Besides the potential involvement of SIRT1 in regulating mammalian lifespan, evidences are also accumulating for the involvement of other mammalian sirtuins in regulating aging. It is possible that multiple mammalian sirtuins act cooperatively to accomplish the tasks of yeast SIR2. The knockout of SIRT6 in mice results in genomic instability due to deficiency in DNA base excision repair pathway, leading to aging-associated degenerative phenotypes [23]. The data suggest a positive role of SIRT6 on mammalian aging. Although SIRT6-null mice have not been tested under food restriction conditions, the facts that SIRT6-null mice exhibit a severe reduction in circulating IGF-1 levels and that SIRT6 requires its enzymatic activity to function argue that it has the potential to serve as an NAD^+ sensor to link nutrient status to genome stability. In contrast to SIRT6, SIRT4 [21] functions in the mitochondria of β cell to repress the activity of GDH by ADP-ribosylation and to downregulate insulin secretion. The activity opposes the effects of CR mediated through SIRT1 in β cells that positively regulates glucose-stimulated insulin secretion [78] by repressing the mitochondrial uncoupling protein UCP-2 [77]. SIRT3, another mitochondrial sirtuin, has been reported to be upregulated by CR and cold exposure in brown fat, where it may alter the metabolic activity [80]. Perturbations in IGF-1 and insulin signaling have been linked to alterations in the rate of aging in multiple organisms [81]. All sirtuin-deficient mouse models published thus far exhibit alterations in insulin/IGF-1 signaling, suggesting that they all contribute to certain aspects of mammalian aging. Since the function of SIRT4 in β cells is through its ART activity, the functions of sirtuins in aging are apparently not exclusively mediated through their deacetylase activity.

9.3   CANCER AS AN AGING DISEASE

Cancer is an aging-associated disease and develops primarily in older adults. As a matter of fact, aging is considered the most potent of all carcinogens [82]. In humans, the incidence of cancer rises exponentially in the final decades of life, beginning at about the mid-point of the maximum lifespan [82,83]. This dramatic age-dependent escalation in cancer risk is fueled largely by a marked increase in epithelial cancers, as opposed to cancers of mesenchymal or hematopoietic origin. It is generally believed that the cancer-prone phenotype of old humans might reflect the combined effects of cumulative mutational load, increased epigenetic gene silencing, telomere dysfunction, and altered stromal milieu [82].

Cancer and aging are also linked because the molecules that control tumorigenesis, e.g., tumor suppressors, also regulate aging and lifespan. In general, tumor suppressors are divided into caretakers and gatekeepers [84,85]. Caretakers protect the genome from damage or mutation to ensure genomic stability but usually do not directly stimulate cell proliferation. Candidates for caretaker tumor suppressors are BRCA1 and BRCA2. In contrast to caretakers, p53 and Rb act as gatekeepers to eliminate potential cancer cells or to suppress their proliferation through apoptosis or cellular senescence. Regardless of whether a tumor suppressor is a caretaker or a gatekeeper, inactivating mutations increase the risk of developing cancer. Therefore, all tumor suppressors in theory should directly promote the longevity of the organisms by preventing the development of malignant tumors. In reality, however, the relationship is more complex. Tumor suppression mechanisms, particularly apoptosis and cellular senescence, suppress the development of cancer but promote the development of specific aging phenotypes. First, both apoptosis and cellular senescence contribute to the depletion of the renewal capacity of tissues by exhausting the supply of progenitors or stem cells. Second, senescent cells, resembling carcinoma-associated (activated) fibroblasts, secrete degradative enzymes, cytokines, and growth factors [86] and contribute to aging by actively disrupting the integrity, function, and homeostasis of tissues as they accumulate.

Tumor suppressor genes might show antagonistic pleiotropy [87], an evolutionary hypothesis proposing that traits benefiting young organisms can have unselected deleterious effects later in life. Although they protect organisms from cancer early in life, they may promote aging phenotypes, including late life cancer, in older organisms. Moreover, when caretaker mechanisms fail, the aging phenotypes that develop might derive not only from the loss of genomic integrity, but also from the apoptosis or cellular senescence that can occur in response to the accumulated damage. So, the caretaker and gatekeeper tumor suppression mechanisms can interact. In addition to tumor suppressors, the overexpression of oncogenes is also linked to the cellular senescence through oncogenic stresses that activate tumor suppressors, further complicating the already sophisticated relationship between aging and tumorigenesis. Overall, aging and tumorigenesis appear not separable at molecular level and molecules that regulate one are likely also to regulate the other.

9.4 SIRTUINS AND TUMORIGENESIS

The antagonistic pleiotropy theory explains why molecules suppressing tumorigenesis can act as a double-edged sword to promote aging. If tumor suppressors promote aging, is it also true then that antiaging molecules like sirtuins promote tumorigenesis? Currently, the information for mammalian sirtuins in tumorigenesis is limited and SIRT1 is the one most studied in cancer cells. Thus, the following discussion about sirtuins in cancer will mainly focus on SIRT1.

Analogous to the role of SIRT1 in aging, its role in tumorigenesis appear to be complex. So far, there is no solid evidence to suggest SIRT1 as an oncogene or a tumor suppressor. However, a significant number of studies show that SIRT1 functionally interacts with tumor suppressors and oncogenes. These molecules are either substrates/interacting proteins of which the functions are regulated by the enzymatic activity of SIRT1 or proteins that modulate SIRT1 expression. Apparently by acting through these proteins, SIRT1 can have an effect on tumorigenesis. Tumorigenesis is a multiple-step process that involves initiation, growth, progression, and the development of resistance to therapies. The function of SIRT1 could vary at different stages and has to be analyzed in a context-specific manner.

9.4.1 ROLE OF SIRTUINS IN CELLULAR SENESCENCE AND TUMOR INITIATION

As discussed above, aging is the most potent carcinogen. With the projected antiaging effect of sirtuins, increased activity of sirtuins could delay aging and tumorigenesis together. In addition, SIRT1 appears regulating hormonal actions and thus may influence tumor initiation from hormone-sensitive tissues. On the other hand, mitotic activity is a positive parameter associated with tumorigenesis and gatekeeper tumor suppressors typically use cellular senescence as the major mechanism to suppress tumorigenesis. Several studies have shown that SIRT1 negatively regulates cellular senescence, raising the possibility that SIRT1 may promote tumor initiation through its negative effect on cellular senescence.

9.4.1.1 Effects of SIRT1 on Tumor Suppressors and Senescence

Sasaki et al. [88] analyzed the level of SIRT1 in human lung fibroblasts and MEFs from mice with normal, accelerated, and delayed aging. They found that SIRT1 protein, but not mRNA, decreased significantly with serial cell passage in both human and murine cells. SIRT1 decreased rapidly in prematurely senescent MEFs, remained high in MEFs with delayed senescence, and was positively correlated with the S phase marker proliferating cell nuclear antigen but inversely correlated with senescence-activated β-galactosidase activity. Reacquisition of mitotic capability after spontaneous immortalization of late passage MEFs restored SIRT1 level to that of early passage, proliferative ones. In the animal, SIRT1 was found to be decreased with age in tissues in which mitotic activity also declines with aging but not in tissues such as the brain in which there is little change in mitotic activity throughout life. Loss of SIRT1 with age was found to be accelerated in mice with accelerated aging phenotype but was not observed in growth hormone receptor knockout mice that are long lived. These analyses positively correlate SIRT1 to mitotic activity and cell's ability to proliferate.

Consistent with the above correlation analyses, Langley et al. [89] showed that human SIRT1 was recruited to the promyelocytic leukemia protein (PML) nuclear bodies and became colocalized with p53 in nuclear bodies upon overexpression of either PML or active Ras. When overexpressed in primary MEFs, SIRT1 antagonized PML-induced acetylation of p53 and rescued PML-mediated premature cellular senescence. The data show that SIRT1 deacetylase is a negative regulator of the cellular senescence induced by oncogenic stress through the activation of p53.

Different from the view that SIRT1 may negatively regulate tumor suppressors to promote tumor initiation, Chua et al. [90] showed that murine SIRT1 was involved in the induction of $p19^{ARF}$ tumor suppressor and cellular senescence induced by chronic oxidative stress. MEFs from SIRT1 null mice showed increased resistance to replicative senescence. Their extended replicative lifespan was correlated with enhanced proliferative capacity and diminished upregulation of either the $p19^{ARF}$ senescence regulator or its downstream target p53 tumor suppressor by chronic stress. The studies raise the possibility that the mammalian SIRT1 may suppress tumor initiation by inducing the expression of tumor suppressors and cellular senescence.

However, in the same study that showed the positive role of SIRT1 in senescence induced by chronic oxidative stress, SIRT1-deficient MEFs were found to show normal $p19^{ARF}$ induction and cell cycle arrest upon acute DNA damage or oncogene expression [90]. In contrast to wild-type MEFs, the immortalized SIRT1-deficient MEFs showed resistance to transformation by active K-Ras. Wild-type MEF cultures of passage 50 had lost p53-dependent induction of the cell cycle regulator p21 or $p19^{ARF}$ in response to active Ras whereas SIRT1 deficient MEF cultures at the same passage showed normal $p19^{ARF}$ and p21 induction. The data support the earlier view that SIRT1 may promote tumor initiation. The data could also be interpreted to suggest that the SIRT1 may function as a "co-oncogene" that cooperates with active oncogenes in cellular transformation.

9.4.1.2   Effect of SIRT1 on Hormonal Signaling

Li et al. [91] recently generated a conditional knockout mouse model and found that the loss of SIRT1 increased the basal level of IGFBP-1, which suppresses the action of IGF-1. Exogenous estrogen decreased the expression of IGFBP-1 to the level of wild-type mice. Because estrogens are a risk factor for breast cancer, the studies suggest that SIRT1 positively modulates the efficacy of the estrogen–IGF-1 signal and its activity could ultimately influence the risk of breast cancer.

Opposite to the potential positive effect of SIRT1 on estrogens and breast cancer, Fu et al. [92] recently showed that human SIRT1 bound to and inhibited androgen action in prostate cancer cells through deacetylation of the androgen receptor. The studies further showed that the expression of SIRT1 suppressed the growth of LNCaP cells induced by dihydrotestosterone. Interestingly, the suppressive effect of SIRT1 on prostate cancer growth was found to be limited to AR-positive cells, implying a prostate cancer specific beneficial effect of sirtuin activators. If the information can be translated into clinics, sirtuin activators might not only prolong the lifespan in men but also suppress prostate cancer growth, owing to the negative effect of SIRT1 on androgen signaling. However, the situation in prostate cancer appears not to be that straightforward. As described subsequently, SIRT1 may play a positive role in prostate cancer progression and reoccurrence.

9.4.2   ROLE OF SIRTUINS IN TUMOR PROGRESSION

Cancer mortalities are largely due to the invasion of the tumors to new places. Heterochromatin formation and epigenetic silencing of genes associated with DNA and protein methylation are shown to contribute to both aging and tumor progression. Growing evidence suggest that epigenetic mechanisms and chromatin remodeling regulate genes in an age-dependent manner and that aging is associated with a gradual failure of the proper control of chromatin structure. For instance, de novo hypermethylation of CpG islands nested in gene promoters has been shown to be a common means of silencing tumor suppressor genes in aging cells and cancer [93,94], which contribute to the progression towards full malignant transformation [95]. Besides DNA methylation, the silencing of tumor suppressor expression could also be due to the aging associated epigenetic changes in histone modifications. For example, aging is known to be associated with an increase in heterochromatin formation and histone trimethylation of K20H4 [96]. SIRT1 is involved in transcriptional silencing and appears also promoting tumor progression through epigenetic silencing and heterochromatin formation. In addition, SIRT1 may also be involved in transmitting the extracellular cues triggered by tumor invasion into nucleus to suppress the functions of proapoptotic transcription factors.

9.4.2.1   SIRT1 and Histone H1 Deacetylation in Heterochromatin Formation and Tumor Progression

It is known that aging is associated with changes in the distribution of H1 subtypes [97–99], deamination of H1 molecules [100], and a loss of α-amino acetylation of serine 1 [101]. Vaquero et al. [47] showed that SIRT1 mediates heterochromatin formation, which include deacetylation of histone tail, recruitment and deacetylation of histone H1, and the spreading of hypermethylated K79H3 with resultant silencing. In the same studies, SIRT1 was shown to deacetylate K9H3 and K16H4, which is accompanied by increased methylation at K20H4.

The K9H3 methylation recruits HP1, a structural protein that plays an important role in heterochromatin formation and gene silencing. Cells exhibiting altered levels of HP1 are predicted to show a loss of silencing at genes regulating cancer progression. Although it remains to be demonstrated, SIRT1-mediated deacetylation of K9H3 should facilitate its methylation, which should enhance HP1 binding and DNA methylation. Thus, SIRT1 might promote tumor progression through its positive effect on the formation of heterochromatin and gene silencing.

9.4.2.2   SIRT1 and Polycomb Group Complex in Heterochromatin Formation and Cancer Progression

Polycomb group (PcG) proteins are important gene regulatory molecules that determine cell fate during normal and pathogenic development. K27H3 methylation recruits a Polycomb complex involved in regulating stem cell pluripotency, silencing of developmentally regulated genes, and controlling cancer progression. Kotake et al. [102] have recently shown that the pRB family proteins are required for K27H3 trimethylation and Polycomb repression complexes (PRCs) binding to and silencing p16INK4alpha tumor suppressor gene.

Multiple studies suggest that SIR2 family and PcG proteins are coupled for transcriptional repression and epigenetic silencing. Furuyama et al. [103] showed that SIR2 is required for polycomb silencing and associated with an E(Z) histone methyltransferase complex. Using protein purification approaches, Kuzmichev et al. [104] identified a protein complex, Polycomb repressive complex 4 (PRC4), which was shown to contain SIRT1 and the isoform 2 of the PRC component Eed [104]. PRC4 appears to be present only in undifferentiated pluripotent cells such as ES cells as well as in cells that have lost their "normal" regulation. SIRT1, together with other subunits of the PRC complexes, was found to be overexpressed in breast, colon, and prostate cancers as compared to normal tissue controls.

More importantly, the formation of PRC4 was promoted by the overexpression of Ezh2, a histone–lysine methyltransferase of which the overexpression was found to be associated with prostate cancer progression [105,106]. Ezh2 was found to be expressed at a much higher level in metastatic prostate and breast cancer as compared to primary tumors. Interestingly, decreased acetylation of histone H3 and H4 at specific lysine residues (K9H3, K18H3, and K16H4) was shown to be associated with reoccurrence of prostate cancers [107]. Fraga et al. [108] reported that acetylation of K16H4 and trimethylation of K20H4 are reduced at repetitive DNA sequences in multiple cancer types. These studies suggest that loss of acetylation at K16H4, a known SIRT1 site, is the hallmark of the development of multiple human cancers.

9.4.2.3 Role of SIRT1 in the Suppression of FOXO1 by FHL2 in Response to Lysophosphatic Acid, a Serum Factor Involved in Cancer Growth and Progression

Rho family GTPases are well known for their ability to regulate actin cytoskeletal remodeling in response to extracellular signals, thereby promoting changes in cell morphology, adhesion, and motility [109]. In addition, by affecting multiple signaling pathways, Rho family members regulate cellular transformation and metastasis, and have also been implicated in transcriptional regulation [110]. Müller et al. [111] showed that the stimulation of the Rho signaling pathway induces translocation of the transcriptional coactivator FHL2 to the nucleus and subsequent activation of FHL2- and androgen receptor-dependent genes. Prostate tumors overexpress Rho GTPases and display altered cellular localization of FHL2 concomitant with tumor dedifferentiation. Sphingosine-1-phosphate-induced FHL2 activation is mediated by Rho GTPases and depends on Rho-kinase. The data suggest that FHL2 acts as a novel molecular transmitter of the Rho signaling pathway to integrate extracellular cues triggered by tumor invasion into altered gene expression.

Yang et al. [112] demonstrated that FHL2 is a FOXO suppressor and that the suppression of FOXO activity involves SIRT1. In response to lysophosphatic acid (LPA) stimulation, FHL2 moves into nucleus, binds to FOXO1, and inhibits its transcriptional activity. It is important to note that LPA stimulates cancer cell proliferation, migration, and survival, and the aberrant LPA production, receptor expression, and signaling are believed to contribute to cancer progression and metastasis [113]. The studies suggest that SIRT1-mediated FOXO1 repression may be involved in prostate tumor invasion and metastasis. Moreover, FOXO1 was recently identified as a tumor suppressor [114–116] and microarray analyses showed that FOXO1 mRNA is decreased in metastatic prostate cancers as compared to primary tumors [105]. Overall, the analyses show that SIRT1, besides the involvement in histone methylation as a component of PRC complexes, may also regulate prostate cancer progression through the suppression of FOXO factors.

9.4.3 Anti-Apoptotic Function of SIRT1 in Cancer Cells and the Response of Cancer Cells to Genotoxic Stress

Different from the role of SIRT1 in tumor initiation and progression for which the evidences are largely circumstantial, many studies consistently point to a role for SIRT1 in suppressing cancer cell apoptosis and reducing their sensitivity to agents that cause DNA damage. The linkage of sirtuins to cellular response to DNA damage can be traced back to studies with yeast SIR2. In proliferating yeast, SIR2 protects against DNA damage by inducing nonhomologous end joining and homologous recombination pathways [68,117] or by mediating the asymmetric inheritance of oxidatively damaged proteins to mother cells, ensuring the generation of relatively damage-free daughters, and continued colony proliferation [118]. However, with a yet-to-be defined mechanism, SIR2 appears also promoting age-dependent genomic instability. The accumulation of spontaneous DNA mutations with chronological age is dramatically reduced in yeast strains lacking both SCH9 and SIR2 [68].

Similar to SIR2 in yeast, mammalian sirtuins also regulate genomic stability and cellular response to cell death induced by DNA damage. As discussed earlier, SIRT6 null mice exhibit genomic instability due to deficiency in DNA base excision repair, leading to the appearance of an aging-associated degenerative phenotype. For SIRT1, most of the substrates identified in cancer cells turn out to be proteins with established role in apoptosis.

9.4.3.1   Suppression of p53 Mediated Apoptosis by SIRT1

p53 is the best known gatekeeper tumor suppressor that induces cancer cell apoptosis and inhibits cell cycle progression. Three separate studies showed that p53 activity was negatively regulated by SIRT1-mediated deacetylation [89,119,120]. These studies found that SIRT1 localized in the nucleus, where it physically interacted with, and deacetylated the K382 of p53. Each of the reports demonstrated that SIRT1-mediated deacetylation decreased p53-mediated transcriptional activation as measured by artificial reporter gene readouts or reduced endogenous p21 protein levels. Besides the studies by Lanley et al. [89] that linked the inhibition of p53 by SIRT1 to PML-induced senescence, the other studies suggest that the inhibition is associated with the resistance of cells to apoptotic induction. Luo et al. [120] showed that the overexpression of SIRT1 in H1299 human cancer cells suppressed p53 dependent apoptosis and increased cell survival of MEFs and IMR-90 human fibroblasts after exposure to etoposide or hydrogen peroxide. Vaziri et al. [119] showed that the overexpression of a catalytically impaired variant, SIRT1^{H363Y}, decreased cell survival following DNA damage. Importantly, the effects of SIRT1^{H363Y} could be reversed by overexpression of dominant-negative p53, demonstrating that the effect of SIRT1^{H363Y} on cell viability requires active p53. SIRT1 overexpression therefore seems to mediate the survival of cells during periods of severe stress through the inhibition of apoptosis. In addition, Dai et al. [121] recently showed that p73 was deacetylated by SIRT1 and that apoptosis mediated by p73 was partially relieved by SIRT1. In a study related to cellular senescence, the overexpression of another p53 related protein, DeltaNp63alpha, was found to downregulate SIRT1 and to cause an accelerated aging phenotype in the mouse [69]. These studies show that functional cross talk with SIRT1 may be a common feature for members of p53 family.

Although the published studies are in agreement that human SIRT1 inhibits apoptosis through deacetylation of p53, it is controversial whether this is also true for SIRT1 in mouse cells. Different from the data by Luo et al. [120] showing that the overexpression of mouse SIRT1 in MEFs causes resistance to apoptosis induced by etoposide, later studies found that fibroblasts and embryonic stem cells obtained from SIRT1 knockout mice were not sensitized to genotoxic stress in spite of increased p53 acetylation [122,123]. Studies with SIRT1-null mice suggest that SIRT1 has little effect on p53-mediated biological outcomes in mice [124,125]. It was also reported that the p53 activity was not altered in response to DNA damage in MEFs engineered so that endogenous p53 could not be acetylated because six lysine residues were mutated, including K379 (the mouse equivalent of human K382) [126].

9.4.3.2   Suppression of FOXO-Mediated Apoptosis by SIRT1

FOXO factors are negative aging regulators in worms and flies. Mammalian FOXO factors are known to cause cancer cell apoptosis, cell cycle arrest, and resistance to oxidative stresses. Recent studies [114,115] identified FOXO factors, particularly FOXO1 [114], as tumor suppressors. FOXO factors are also identified as the key mediators underlying stem cell's ability to handle physiological level of oxidative stresses [127].

Genetic analyses in worms have shown that the *Sir.2-1* and *daf-16*, the respective *C. elegans* homologues of yeast SIR2 and mammalian FOXO factors, are functionally connected in regulating lifespan [60,128–130]. Several laboratories have shown that SIRT1 deacetylates mammalian FOXO1 and FOXO3a [112,131,132] and the deacetylation was consistently found to suppress their proapoptotic activity. However, acetylation of FOXO4 by CBP was found to inhibit FOXO4 activity and deacetylation by SIRT1 relieved the inhibition and enhanced the expression of CDK inhibitor, p27^{Kip1} [133]. Similarly, deacetylation of FOXO1 by SIRT1 was found to potentiate the FOXO1-dependent expression of p27^{Kip1} and manganese superoxide dismutase [134]. Brunet et al.[131] showed that SIRT1 suppressed the induction of proapoptotic genes by FOXO3a, but enhanced the induction of genes involved in growth inhibition (e.g., p27^{Kip1}) and stress responses (e.g., GADD45) [131]. These studies lead to the hypothesis that the deacetylation may suppress FOXO-induced apoptosis but induce the activation of cell cycle arresting genes and stress-resistant factors, shifting the balance from apoptosis to stress resistance and survival [131,135]. It appears very likely that SIRT1 can either lead to activation or repression of FOXO-dependent transcription depending on the promoters, cellular context, the external signals, or the proteins that regulate the SIRT1-FOXO interaction [112,136]. For example, the role of SIRT1-mediated deacetylation in oxidative stress-induced FOXO1 action in HepG2 [134] and 293 [131] cells may differ from FOXO1 action in prostate cancer cells after stimulation with LPA, which was shown to be mediated through FHL2 as an adaptor [112].

9.4.3.3 Ku70 Deacetylation and the Suppression of Bax Release into Mitochondria by SIRT1

Bax is a proapoptotic member of Bcl-2 family whose localization to mitochondria triggers cytochrome c release and apoptosis. Under normal conditions, Bax is sequestered in the cytoplasm by its tight association with Ku70, a DNA repair factor [137,138]. In response to lethal dose of DNA damage or stress, two lysines in Ku70, K539 and K542, become acetylated by CBP or PCAF, which disrupts the Ku70–Bax interaction, allowing Bax to localize to mitochondria and initiate apoptosis [138]. In 293T cells, Cohen et al. [46] showed that SIRT1 deacetylates the Ku70 at K539 and K542 sites. Such an action strengthens the ability of Ku70 to sequester Bax away from mitochondria and inhibited cell death mediated by Bax. CR induced SIRT1 expression and protected 293T cells from Bax-induced apoptosis. It remains to be determined whether the effect of SIRT1 on Bax localization can be generalized to other cells because many studies have pointed prosurvival function of SIRT1 to its nuclear action. However, a recent study [139] showed that SIRT1 is cleaved by caspase during the apoptosis of N2a neuroblastoma cells, which is associated with increased cytoplasmic localization of SIRT1. The data suggest the possible involvement of cytoplasmic action of SIRT1 in cell death. The studies however did not make it clear whether the cleavage and increased cytoplasmic localization is a consequence of cell apoptosis or a cause. It is also unclear whether the cleavage product still functions as an NAD^+-dependent deacetylase.

9.4.3.4 E2F1 Deacetylation and the Suppression of E2F1-Mediated Apoptosis by SIRT1

Members of the E2F family are transcription factors whose activity is controlled by Rb tumor suppressor during the cell cycle. Besides cell cycle regulation, some E2F members, particularly E2F1, also have important roles in regulating apoptosis. For instance, E2F1 stimulates the transcription of several genes in the apoptotic pathways [140] and its overexpression induces premature S-phase entry and apoptosis [141,142]. In addition, DNA damage stabilizes E2F1 and the resultant activation may be involved in apoptosis of cells such as thymocytes induced by etoposide [143].

Wang et al. [144] recently showed that DNA damage, such as treatment with etoposide, induced SITR1 expression through E2F1 in H1299 cells. Furthermore, SIRT1 deacetylated E2F1, suppressed its transcriptional activity, and proapoptotic function [144]. SIRT1 also appeared to reduce the capability of E2F1 to promote cell cycle progression. Etoposide is a chemotherapeutic drug being used clinically. The data suggest that overexpression of SIRT1 may cause resistance to chemotherapy.

9.4.3.5 Deacetylation of Bcl6 and the Survival of B-Cell Lymphoma

Bcl6 is an oncogenic transcriptional repressor belonging to the BTB/POZ zinc finger family of transcription factors. It binds to the same DNA sequence recognized by STAT6 to repress genes involved in the formation of the germinal center. The transcriptional repressor function of Bcl6 requires interaction of two repressor domains with corepressor complexes containing HDACs. Bcl6 is normally downregulated at transcriptional level by a number of signals. In leukemia, chromosomal translocations place heterologous promoters in front of the coding exons of Bcl6. This promoter substitution disrupts the downregulation and causes its deregulated expression.

Bereshchenko et al. [145] showed that Bcl6 acetylation disrupts its ability to recruit HDACs, to repress transcription and to induce cellular transformation. Using different HDAC inhibitors, the authors found that Bcl6 acetylation was controlled at least partly by sirtuins. Mutant Bcl6 in which the acetylation sites were changed to glutamines lost the ability to transform Rat-1 fibroblasts, implying that the deacetylation is required for its oncogenic activity. Mentioned as data not shown, the authors stated that pharmacological inhibition of sirtuins in B-cell lymphoma leads to the accumulation of the inactive acetylated Bcl6, which caused cell cycle arrest and apoptosis.

Because the repression function of Bcl6 is required for the B-cell lymphoma, the studies suggest that the interference of Bcl6 repressor activity by the inhibition of sirtuins and other HDACs may be a promising therapeutic treatment for B-cell lymphoma.

9.4.3.6 NFκB Deacetylation and the Promotion of Cancer Cell Apoptosis by SIRT1

Nuclear factor-kappa B (NFκB) is a transcription factor composed of a heterodimeric protein complex that controls the expression of genes of which the protein products function in important cellular processes including adhesion, angiogenesis, cell cycle, and apoptosis. The most common and also the best-studied form of NFκB is a heterodimer between p50 and RelA/p65. IκB proteins normally sequester NFκB in the cytoplasm. Following stimulation, IκB proteins become phosphorylated by the IκB kinase, which subsequently targets IκB for ubiquitination and degradation through the 26S proteasome and triggers the release of NFκB into the nucleus to regulate transcription.

Studies by Yeung et al. [146] showed that SIRT1 directly interacted with RelA/p65 and deacetylated K310. Consistent with the SIRT1-mediated deacetylation, nicotinamide inhibited the transcriptional activity of NFκB whereas SIRT1 activator resveratrol enhanced the activity as well as TNFα-induced expression of cIAP-2 gene expression, which is known to be mediated by NFκB. Overexpression of SIRT1 in 293 cells and small cell lung carcinoma cells sensitized them to apoptosis induction by TNFα. The knockdown of SIRT1 in H1299 cells reversed the suppressive effect of resveratrol on TNFα-induced cIAP-2 gene expression and inhibited TNFα-induced apoptosis following treatment with resveratrol. The study identifies the RelA/p65 subunit of NFκB as an SIRT1 deacetylase target and has profound implications in cell survival in response to TNFα signaling or the signaling of other death receptors.

In support of the functional connection between SIRT1 and NFκB, Gao et al. [147] showed that the breast cancer associated protein 3 (BCA-3) suppressed the activity of NFκB through the recruitment of SIRT1 to the transcriptional complex in a neddylation-dependent manner. Yang et al. [148] showed that cigarette smoke extracts decreases SIRT1 activity, decreased its interaction with RelA/p65, and increased NFκB-dependent release of proinflammatory mediators. Because inflammation is coupled to tumorigenesis, the studies imply a repressive role for SIRT1 in inflammation and tumorigenesis of the lungs.

The above studies are obviously opposite to the previously described findings that suggest SIRT1 as a pro-survival protein. The opposite cellular responses are likely due to differences in apoptotic signaling instead of cell types because the same H1299 cell was used for the studies with both p53 and NFκB. TNFα-induced apoptosis occurs through extrinsic death receptor-mediated pathways that activate the Fas-associated death domain protein and caspase-8. The death response to DNA damage mediated by p53 and FOXO factors go through the mitochondria-dependent intrinsic apoptotic pathway. Therefore, the ability of SIRT1 to induce apoptosis or cell survival may depend on the types of apoptotic stimuli. In support of this hypothesis, a recent work demonstrates that treatment of neuroblastoma cell lines with resveratrol sensitizes these cells to TNF-related apoptosis-inducing ligand-induced apoptosis in the absence of a functional p53 pathway [149].

Furthermore, it is also reported that NFκB activity is modulated by p53 and ARF tumor suppressors to inhibit tumor growth [150]. In the presence of p53 and ARF, NFκB represses, rather than activate, the expression of tumor-promoting genes, suggesting a dual function of NFκB during tumor progression. In the early stages, NFκB inhibits tumor growth but, as further mutations lead to a loss of tumor suppressor expression, the oncogenic functions of NFκB become unleashed, allowing it to actively contribute to tumorigenesis [150]. Therefore, in cells expressing wild-type p53, SIRT1-mediated deacetylation of NFκB may contribute to cancer cell survival.

9.4.4   TUMOR-SPECIFIC ACTIVITY OF SIRT1 AND THE TARGETS OF SIRT1 FOR CANCER THERAPY

Given the role of SIRT1 in cancer cell survival and the effect of SIRT1-mediated deacetylation on the activity of tumor suppressors and oncogenes, it is expected that SIRT1 will have a cancer specific role in cell survival.

9.4.4.1   Increased SIRT1 Expression in Tumor Cells

The level of SIRT1 expression is known to decrease in fibroblasts that undergo senescence. To the opposite, the SIRT1 expression is increased during tumorigenesis [104]. As mentioned earlier, Kuzmichev et al. [104] found that SIRT1, together with other components of PRC complexes, is overexpressed in breast, colon, and prostate cancers as compared to normal tissue controls. Using *Nkx3.1* $^{+/-}$ *Pten* $^{+/-}$ compound mutant mice that develop prostate intraepithelial neoplasm (PIN) lesions by 6 months of age and adenocarcinoma by 12 months, their analysis reveal low levels of SIRT1 in normal prostate, which was restricted to epithelial cells with some expression to the stroma. PIN displays a moderate elevation in cells expressing Ezh2 and SIRT1, and further elevated levels of expression were observed in cancer-derived tissues.

9.4.4.2   Regulation of SIRT1 Promoter Activity by Tumor Suppressors

Consistent with the increased level of SIRT1 protein in tumors, the promoter activity of SIRT1 expression was found to be negatively regulated by tumor suppressors. P53 has been shown to bind to the SIRT1 promoter to repress its expression [70]. Under nutritional restriction conditions, FOXO3a is induced which binds and relieves p53-mediated repression, allowing SIRT1 expression to increase. Because the function of p53 is often compromised in tumors, its decreased activity may contribute to the increased expression of SIRT1 in tumor cells. Similar to p53, tumor suppressor HIC1 was found to form a transcriptional repression complex with SIRT1, which directly binds to the SIRT1 promoter and represses its transcription. The loss of HIC1 in tumors results in increased SIRT1 expression and the deacetylation/inactivation of p53, allowing cells to bypass apoptosis and to survive DNA damage [69]. As discussed earlier, E2F1 increases the expression of SIRT1 by binding to its promoter and directly stimulating SIRT1 transcription, which is thought to be a negative feedback mechanism to suppress E2F1-induced apoptosis [144].

9.4.4.3   Regulation of SIRT1 mRNA Stability by Chk2

Besides transcriptional control, SIRT1 expression is also regulated at the posttranscriptional level. The RNA-binding protein HuR regulates the stability of many target mRNAs. It was recently reported that HuR associated with the 3′ untranslated region of SIRT1 mRNA stabilized the mRNA and increased SIRT1 expression [151]. The studies found that oxidative stress triggered the dissociation of the HuR from SIRT1 mRNA, in turn promoting SIRT1 mRNA decay, reducing SIRT1 abundance and lowering cell survival. This occurs by the activation of the cell cycle checkpoint kinase, Chk2, which interacts with HuR and possibly phosphorylates HuR, an event that appears to be important for HuR dissociation after hydrogen peroxide treatment. Because Chk2 is an effector of p53 tumor suppressor, the findings imply that altered function of Chk2 during tumorigenesis may lead the loss of a negative regulation mechanism to control SIRT1 expression. ChK2 function is known to be altered in both hereditary and sporadic human cancers.

9.4.4.4   Cancer-Specific Activity of SIRT1 and the Potential of SIRT1 Inhibitors as Anticancer Agents

Consistent with the increased SIRT1 expression in cancer cells, cancer-specific functions of SIRT1 has been reported. Using synthetic small interfering RNA to silence SIRT1 gene expression, Ford et al. [152] showed that the SIRT1 RNAi induced growth arrest and apoptosis in human epithelial cancer cells in the absence of applied stress. In contrast, normal human epithelial cells and normal human diploid fibroblasts appear to be refractory to SIRT1 silencing. Interestingly, the studies suggest that the SIRT1-regulated pathway is independent of Bcl2, p53, Bax, and caspase-2. Instead, SIRT1 appears to suppress apoptosis downstream from these apoptotic factors in a manner dependent on FOXO4, but not FOXO3, as the proapoptotic mediator [152].

The cancer-specific cell survival function points to SIRT1 as a novel molecular target for selective killing of cancer versus noncancerous epithelial cells. Consistently, SIRT1 inhibitors have been shown to exhibit antitumor activity [153]. Treatment of Bcl6-expressing lymphoma cells with cambinol, an inhibitor of the enzymatic activity of SIRT1 and SIRT2, induces apoptosis [153], whereas sirtinol induces senescence-like growth arrest in MCF7 human breast and H1299 lung cancer cells [154]. More importantly, Prutt et al. [155] show that SIRT1 localizes to the promoters of aberrantly silenced tumor suppressor genes, but not to the same promoters in cell lines where these genes are expressed. Inhibition of SIRT1 reactivates the silenced tumor suppressor genes without loss of promoter DNA hypermethylation [155].

Consistent with the cancer-specific cell survival function, SIRT1 expression also increases the resistance of cancer cells to therapeutic treatments with ionizing radiation [119], etoposide [144], or cisplatin treatment [44]. In addition, the expression of the SIRT1 is increased both at the RNA and protein levels in five drug-resistant cell lines when compared to their drug-sensitive counterparts [156]. Biopsies from cancer patients treated with chemotherapeutic agents also express high levels of SIRT1 but not other sirtuins. More importantly, SIRT1 siRNA significantly reverses the resistance phenotype and reduces expression of the multidrug resistance molecule P-glycoprotein. Ectopic SIRT1 expression induces expression of P-glycoprotein, rendered cells resistant to doxorubicin. Collectively, these studies suggest that SIRT1 is a molecular target to reverse the resistance of cancers to chemotherapy.

9.5   CONCLUSION REMARKS: THE POTENTIAL OF SIRTUINS AS MOLECULAR TARGET FOR THE INTERVENTION OF AGING AND TUMORIGENESIS

As discussed in the text, sirtuins may increase replicative lifespan but decrease chronological lifespan, suggesting a complex role of sirtuins in aging. Similarly, the role of sirtuins in tumorigenesis is also complex and appears to be context-dependent. In healthy cells, such as stem cells, sirtuins promote their survival and maintain their youthfulness locally by reducing the accumulation of free radicals and damaged macromolecules in the cells and systematically by suppressing the action of aging-promoting hormones and metabolism. By shielding normal cells away from excessive oxidative stresses that damage the genome, sirtuins might act in a way similar to the "caretaker" tumor suppressors to prevent tumorigenesis. For example, FOXO factors are identified as critical mediators for the resistance of hematopoietic stem cells to physiologic oxidative stress

[127] and, as discussed earlier in the text, multiple studies have shown that SIRT1 selectively enhances FOXO-mediated expression of genes involved in stress resistance. Through the action of FOXO factors and their downstream targets, SIRT1 may shield the genome of stem cells from ROS-induced damages (Figure 9.4). On the other hand, in long-lived cells in which DNA damage has accumulated to a degree beyond repair or in cells that have suffered acute oncogenic or excessive genotoxic stress, they normally should die or go through senescence due to the increased proapoptotic and senescence-inducing activity of "gatekeeper" tumor suppressors. Because SIRT1 suppresses apoptosis and senescence induced by tumor suppressors, these cells may survive and form tumors. Under this situation, SIRT1 may promote tumorigenesis. Therefore, the effect of sirtuins on tumorigenesis may depend on the cellular context, particularly the status of genomic integrity and tumor suppressor/oncogene expression.

Furthermore, in cancer cells, SIRT1 expression is increased due to the loss of functions of tumor suppressors such as p53, HIC1, Chk2, etc., which normally suppress SIRT1 expression. The increased SIRT1, by acting through its positive effects on epigenetic silencing and heterochromatin formation, may promote tumor invasion and progression. More importantly, SIRT1 predominantly exerts antiapoptotic effects by deacetylating and inactivating key molecules involved in cancer cell apoptosis, including gatekeeper tumor suppressors (Figure 9.5). Thus, SIRT1 in cancer cells may promote tumor growth and progression and cause resistance of cancer cells to therapeutic treatments. Except the studies with NFκB, most studies suggest that SIRT1 confers resistance of cancer cells to DNA damage-induced apoptosis, arguing that SIRT1 inhibitors are potential compounds for cancer therapy.

For the projected role of SIRT1 in delaying aging and prolong lifespan, SIRT1 activators are being investigated as potential aging defying medicine. On the other hand, for its positive role in cancer progression and the resistance of cancer cells to therapies, SIRT1 inhibitors, similar to the inhibitors of class I and II HDACs, are being considered as compounds with potential for cancer

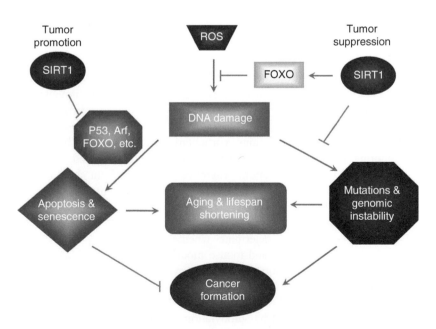

FIGURE 9.4 The complex and context-specific functions of SIRT1 in aging and tumorigenesis. SIRT1 may suppress the reactive oxygen species (ROS)-induced induction of genomic instability, senescence, and apoptosis to delay aging and prolong lifespan. For cancer formation, however, the protective effect on genomic instability may be preventive while the inhibition of apoptosis and cellular senescence may be promotional.

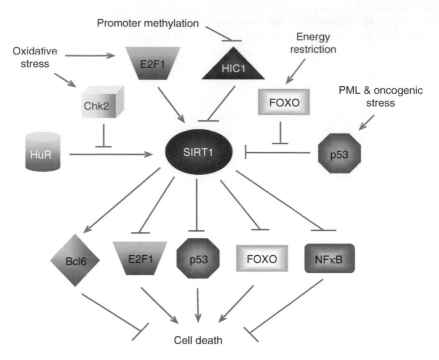

FIGURE 9.5 Regulation of SIRT1 expression and the effect of SIRT1-mediated deacetylation of apoptotic proteins on the survival of cancer cells as well as their resistance to DNA damage-induced apoptosis. The expression of SIRT1 is controlled at transcriptional and posttranscriptional levels by tumor suppressors. Through these tumor suppressors, oxidative and oncogenic stress and energy restriction regulate SIRT1 levels in the cells. The loss of function of these tumor suppressors during transformation results in increased expression of SIRT1 in cancer cells, which in turn suppresses cancer cell apoptosis and causes resistance to DNA damage-based therapeutic treatment.

treatment. Is it possible, then, to activate mammalian sirtuins to prolong lifespan without increasing cancer mortality or to inhibit their activity for cancer prevention/treatment without inducing premature aging? An accurate answer to these questions requires a much better understanding about the role of mammalian sirtuins in tumorigenesis in animal models. First, tissue-specific transgenic animals that overexpress sirtuins, particularly SIRT1, need to be generated to assess their tumor-promoting (or tumor suppression) potential in vivo. Second, the potential context specific roles of sirtuins in tumor initiation, progression, and treatment should be further defined in sirtuin-deficient mice that have been crossed to mouse tumor models engineered by transgenic expression of oncogenes (e.g., NFκB) or knockout of tumor suppressors (such as PML/p53 double null mice). Third, sirtuin inhibitors with beneficial activity detected in cellular assays have to be tested in animal models to determine whether they cause premature aging. Finally, the information obtained in cell lines and mouse models need to be validated in cancer patients to determine whether sirtuin inhibitors/activators provide a clear benefit for their well being. The last point is particularly important because results obtained from mouse models are not always directly translatable into clinics. Yang et al. [157] have recently shown that human SIRT1 activity is positively regulated by sumoylation in the C-terminal region. SENP1-mediated desumoylation in response to genotoxic stress reduces the ability of SIRT1 to deacetylate members of p53 family and sensitizes cancer cells to apoptosis. Interestingly, the sumoylation motif is conserved among SIRT1 proteins from rat to human but not in mouse SIRT1, suggesting the existence of differences between the functions of human and mouse sirtuins and their mechanism of regulation.

REFERENCES

1. Thiagalingam, S., Cheng, K.H., Lee, H.J., Mineva, N., Thiagalingam, A., and Ponte, J.F., Histone deacetylases: Unique players in shaping the epigenetic histone code, *Ann N Y Acad Sci* 983, 84–100, 2003.

2. Gao, L., Cueto, M.A., Asselbergs, F., and Atadja, P., Cloning and functional characterization of HDAC11, a novel member of the human histone deacetylase family, *J Biol Chem* 277(28), 25748–25755, 2002.

3. Haigis, M.C. and Guarente, L.P., Mammalian sirtuins—emerging roles in physiology, aging, and calorie restriction, *Genes Dev* 20(21), 2913–2921, 2006.

4. Blander, G. and Guarente, L., The Sir2 family of protein deacetylases, *Annu Rev Biochem* 73, 417–435, 2004.

5. Sauve, A.A., Wolberger, C., Schramm, V.L., and Boeke, J.D., The biochemistry of sirtuins, *Annu Rev Biochem* 75, 435–465, 2006.

6. Frye, R.A., Phylogenetic classification of prokaryotic and eukaryotic Sir2-like proteins, *Biochem Biophys Res Commun* 273(2), 793–798, 2000.

7. Tsang, A.W. and Escalante-Semerena, J.C., CobB, a new member of the SIR2 family of eucaryotic regulatory proteins, is required to compensate for the lack of nicotinate mononucleotide:5,6-dimethyl-benzimidazole phosphoribosyltransferase activity in cobT mutants during cobalamin biosynthesis in *Salmonella typhimurium* LT2, *J Biol Chem* 273(48), 31788–31794, 1998.

8. Starai, V.J., Celic, I., Cole, R.N., Boeke, J.D., and Escalante-Semerena, J.C., Sir2-dependent activation of acetyl-CoA synthetase by deacetylation of active lysine, *Science* 298(5602), 2390–2392, 2002.

9. She, Q., Singh, R.K., Confalonieri, F., Zivanovic, Y., Allard, G., Awayez, M.J., Chan-Weiher, C.C., Clausen, I.G., Curtis, B.A., De Moors, A., Erauso, G., Fletcher, C., Gordon, P.M., Heikamp-de Jong, I., Jeffries, A.C., Kozera, C.J., Medina, N., Peng, X., Thi-Ngoc, H.P., Redder, P., Schenk, M.E., Theriault, C., Tolstrup, N., Charlebois, R.L., Doolittle, W.F., Duguet, M., Gaasterland, T., Garrett, R.A., Ragan, M.A., Sensen, C.W., and Van der Oost, J., The complete genome of the crenarchaeon *Sulfolobus solfataricus* P2, *Proc Natl Acad Sci U S A* 98(14), 7835–7840, 2001.

10. Bell, S.D., Botting, C.H., Wardleworth, B.N., Jackson, S.P., and White, M.F., The interaction of Alba, a conserved archaeal chromatin protein, with Sir2 and its regulation by acetylation, *Science* 296(5565), 148–151, 2002.

11. North, B.J. and Verdin, E., Sirtuins: Sir2-related NAD-dependent protein deacetylases, *Genome Biol* 5(5), 224, 2004.

12. Tanno, M., Sakamoto, J., Miura, T., Shimamoto, K., and Horio, Y., Nucleocytoplasmic shuttling of the NAD^+-dependent histone deacetylase SIRT1, *J Biol Chem* 282(9), 6823–6832, 2007.

13. North, B.J., Marshall, B.L., Borra, M.T., Denu, J.M., and Verdin, E., The human Sir2 ortholog, SIRT2, is an NAD^+-dependent tubulin deacetylase, *Mol Cell* 11(2), 437–444, 2003.

14. Inoue, T., Hiratsuka, M., Osaki, M., Yamada, H., Kishimoto, I., Yamaguchi, S., Nakano, S., Katoh, M., Ito, H., and Oshimura, M., SIRT2, a tubulin deacetylase, acts to block the entry to chromosome condensation in response to mitotic stress, *Oncogene* 26(7), 945–957, 2007.

15. Dryden, S.C., Nahhas, F.A., Nowak, J.E., Goustin, A.S., and Tainsky, M.A., Role for human SIRT2 NAD-dependent deacetylase activity in control of mitotic exit in the cell cycle, *Mol Cell Biol* 23(9), 3173–3185, 2003.

16. Schwer, B., North, B.J., Frye, R.A., Ott, M., and Verdin, E., The human silent information regulator (Sir)2 homologue hSIRT3 is a mitochondrial nicotinamide adenine dinucleotide-dependent deacetylase, *J Cell Biol* 158(4), 647–657, 2002.

17. Onyango, P., Celic, I., McCaffery, J.M., Boeke, J.D., and Feinberg, A.P., SIRT3, a human SIR2 homologue, is an NAD-dependent deacetylase localized to mitochondria, *Proc Natl Acad Sci U S A* 99 (21), 13653–13658, 2002.

18. Hallows, W.C., Lee, S., and Denu, J.M., Sirtuins deacetylate and activate mammalian acetyl-CoA synthetases, *Proc Natl Acad Sci U S A* 103(27), 10230–10235, 2006.

19. Bellizzi, D., Rose, G., Cavalcante, P., Covello, G., Dato, S., De Rango, F., Greco, V., Maggiolini, M., Feraco, E., Mari, V., Franceschi, C., Passarino, G., and De Benedictis, G., A novel VNTR enhancer within the SIRT3 gene, a human homologue of SIR2, is associated with survival at oldest ages, *Genomics* 85(2), 258–263, 2005.

20. Scher, M.B., Vaquero, A., and Reinberg, D., SirT3 is a nuclear NAD$^+$-dependent histone deacetylase that translocates to the mitochondria upon cellular stress, *Genes Dev* 21(8), 920–928, 2007.

21. Haigis, M.C., Mostoslavsky, R., Haigis, K.M., Fahie, K., Christodoulou, D.C., Murphy, A.J., Valenzuela, D.M., Yancopoulos, G.D., Karow, M., Blander, G., Wolberger, C., Prolla, T.A., Weindruch, R., Alt, F.W., and Guarente, L., SIRT4 inhibits glutamate dehydrogenase and opposes the effects of calorie restriction in pancreatic beta cells, *Cell* 126(5), 941–954, 2006.

22. Michishita, E., Park, J.Y., Burneskis, J.M., Barrett, J.C., and Horikawa, I., Evolutionarily conserved and nonconserved cellular localizations and functions of human SIRT proteins, *Mol Biol Cell* 16(10), 4623–4635, 2005.

23. Mostoslavsky, R., Chua, K.F., Lombard, D.B., Pang, W.W., Fischer, M.R., Gellon, L., Liu, P., Mostoslavsky, G., Franco, S., Murphy, M.M., Mills, K.D., Patel, P., Hsu, J.T., Hong, A.L., Ford, E., Cheng, H.L., Kennedy, C., Nunez, N., Bronson, R., Frendewey, D., Auerbach, W., Valenzuela, D., Karow, M., Hottiger, M.O., Hursting, S., Barrett, J.C., Guarente, L., Mulligan, R., Demple, B., Yanco-poulos, G.D., and Alt, F.W., Genomic instability and aging-like phenotype in the absence of mammalian SIRT6, *Cell* 124(2), 315–329, 2006.

24. Liszt, G., Ford, E., Kurtev, M., and Guarente, L., Mouse Sir2 homolog SIRT6 is a nuclear ADP-ribosyltransferase, *J Biol Chem* 280(22), 21313–21320, 2005.

25. Hasan, S., El-Andaloussi, N., Hardeland, U., Hassa, P.O., Burki, C., Imhof, R., Schar, P., and Hottiger, M.O., Acetylation regulates the DNA end-trimming activity of DNA polymerase beta, *Mol Cell* 10(5), 1213–1222, 2002.

26. Ford, E., Voit, R., Liszt, G., Magin, C., Grummt, I., and Guarente, L., Mammalian Sir2 homolog SIRT7 is an activator of RNA polymerase I transcription, *Genes Dev* 20(9), 1075–1080, 2006.

27. Avalos, J.L., Celic, I., Muhammad, S., Cosgrove, M.S., Boeke, J.D., and Wolberger, C., Structure of a Sir2 enzyme bound to an acetylated p53 peptide, *Mol Cell* 10(3), 523–535, 2002.

28. Cosgrove, M.S., Bever, K., Avalos, J.L., Muhammad, S., Zhang, X., and Wolberger, C., The structural basis of sirtuin substrate affinity, *Biochemistry* 45(24), 7511–7521, 2006.

29. Finnin, M.S., Donigian, J.R., and Pavletich, N.P., Structure of the histone deacetylase SIRT2, *Nat Struct Biol* 8(7), 621–625, 2001.

30. Marmorstein, R., Structure and chemistry of the Sir2 family of NAD$^+$-dependent histone/protein deactylases, *Biochem Soc Trans* 32(Pt 6), 904–909, 2004.

31. Min, J., Landry, J., Sternglanz, R., and Xu, R.M., Crystal structure of a SIR2 homolog-NAD complex, *Cell* 105(2), 269–279, 2001.

32. Tanny, J.C. and Moazed, D., Recognition of acetylated proteins: Lessons from an ancient family of enzymes, *Structure* 10(10), 1290–1292, 2002.

33. Zhao, K., Chai, X., Clements, A., and Marmorstein, R., Structure and autoregulation of the yeast Hst2 homolog of Sir2, *Nat Struct Biol* 10(10), 864–871, 2003.

34. Zhao, K., Chai, X., and Marmorstein, R., Structure of the yeast Hst2 protein deacetylase in ternary complex with 2′-*O*-acetyl ADP ribose and histone peptide, *Structure* 11(11), 1403–1411, 2003.

35. Zhao, K., Chai, X., and Marmorstein, R., Structure of a Sir2 substrate, Alba, reveals a mechanism for deacetylation-induced enhancement of DNA binding, *J Biol Chem* 278(28), 26071–26077, 2003.

36. Zhao, K., Chai, X., and Marmorstein, R., Structure and substrate binding properties of cobB, a Sir2 homolog protein deacetylase from *Escherichia coli*, *J Mol Biol* 337(3), 731–741, 2004.

37. Rustandi, R.R., Baldisseri, D.M., and Weber, D.J., Structure of the negative regulatory domain of p53 bound to S100B(betabeta), *Nat Struct Biol* 7(7), 570–574, 2000.

38. Luger, K., Mader, A.W., Richmond, R.K., Sargent, D.F., and Richmond, T.J., Crystal structure of the nucleosome core particle at 2.8 A resolution, *Nature* 389(6648), 251–260, 1997.

39. Luger, K. and Richmond, T.J., The histone tails of the nucleosome, *Curr Opin Genet Dev* 8(2), 140–146, 1998.

40. Blander, G., Olejnik, J., Krzymanska-Olejnik, E., McDonagh, T., Haigis, M., Yaffe, M.B., and Guarente, L., SIRT1 shows no substrate specificity in vitro, *J Biol Chem* 280(11), 9780–9785, 2005.

41. Garske, A.L. and Denu, J.M., SIRT1 top 40 hits: Use of one-bead, one-compound acetyl-peptide libraries and quantum dots to probe deacetylase specificity, *Biochemistry* 45(1), 94–101, 2006.

42. Braunstein, M., Rose, A.B., Holmes, S.G., Allis, C.D., and Broach, J.R., Transcriptional silencing in yeast is associated with reduced nucleosome acetylation, *Genes Dev* 7(4), 592–604, 1993.

43. Gottlieb, S. and Esposito, R.E., A new role for a yeast transcriptional silencer gene, SIR2, in regulation of recombination in ribosomal DNA, *Cell* 56(5), 771–776, 1989.

44. Matsushita, N., Takami, Y., Kimura, M., Tachiiri, S., Ishiai, M., Nakayama, T., and Takata, M., Role of NAD-dependent deacetylases SIRT1 and SIRT2 in radiation and cisplatin-induced cell death in vertebrate cells, *Genes Cells* 10(4), 321–332, 2005.

45. Alcendor, R.R., Kirshenbaum, L.A., Imai, S., Vatner, S.F., and Sadoshima, J., Silent information regulator 2alpha, a longevity factor and class III histone deacetylase, is an essential endogenous apoptosis inhibitor in cardiac myocytes, *Circ Res* 95(10), 971–980, 2004.

46. Cohen, H.Y., Miller, C., Bitterman, K.J., Wall, N.R., Hekking, B., Kessler, B., Howitz, K.T., Gorospe, M., de Cabo, R., and Sinclair, D.A., Calorie restriction promotes mammalian cell survival by inducing the SIRT1 deacetylase, *Science* 305(5682), 390–392, 2004.

47. Vaquero, A., Scher, M., Lee, D., Erdjument-Bromage, H., Tempst, P., and Reinberg, D., Human SirT1 interacts with histone H1 and promotes formation of facultative heterochromatin, *Mol Cell* 16(1), 93–105, 2004.

48. Bouras, T., Fu, M., Sauve, A.A., Wang, F., Quong, A.A., Perkins, N.D., Hay, R.T., Gu, W., and Pestell, R.G., SIRT1 deacetylation and repression of p300 involves lysine residues 1020/1024 within the cell cycle regulatory domain 1, *J Biol Chem* 280(11), 10264–10276, 2005.

49. Takata, T. and Ishikawa, F., Human Sir2-related protein SIRT1 associates with the bHLH repressors HES1 and HEY2 and is involved in HES1- and HEY2-mediated transcriptional repression, *Biochem Biophys Res Commun* 301(1), 250–257, 2003.

50. Hisahara, S., Chiba, S., Matsumoto, H., and Horio, Y., Transcriptional regulation of neuronal genes and its effect on neural functions: NAD-dependent histone deacetylase SIRT1 (Sir2alpha), *J Pharmacol Sci* 98(3), 200–204, 2005.

51. Senawong, T., Peterson, V.J., Avram, D., Shepherd, D.M., Frye, R.A., Minucci, S., and Leid, M., Involvement of the histone deacetylase SIRT1 in chicken ovalbumin upstream promoter transcription factor (COUP-TF)-interacting protein 2-mediated transcriptional repression, *J Biol Chem* 278(44), 43041–43050, 2003.

52. Picard, F., Kurtev, M., Chung, N., Topark-Ngarm, A., Senawong, T., Machado De Oliveira, R., Leid, M., McBurney, M.W., and Guarente, L., Sirt1 promotes fat mobilization in white adipocytes by repressing PPAR-gamma, *Nature* 429(6993), 771–776, 2004.

53. Kennedy, B.K., Austriaco, N.R., Jr., Zhang, J., and Guarente, L., Mutation in the silencing gene SIR4 can delay aging in *S. cerevisiae*, *Cell* 80(3), 485–496, 1995.

54. Rusche, L.N., Kirchmaier, A.L., and Rine, J., The establishment, inheritance, and function of silenced chromatin in *Saccharomyces cerevisiae*, *Annu Rev Biochem* 72, 481–516, 2003.

55. Kennedy, B.K., Gotta, M., Sinclair, D.A., Mills, K., McNabb, D.S., Murthy, M., Pak, S.M., Laroche, T., Gasser, S.M., and Guarente, L., Redistribution of silencing proteins from telomeres to the nucleolus is associated with extension of life span in *S. cerevisiae*, *Cell* 89(3), 381–391, 1997.

56. Sinclair, D.A. and Guarente, L., Extrachromosomal rDNA circles—a cause of aging in yeast, *Cell* 91(7), 1033–1042, 1997.

57. Kaeberlein, M., McVey, M., and Guarente, L., The SIR2/3/4 complex and SIR2 alone promote longevity in *Saccharomyces cerevisiae* by two different mechanisms, *Genes Dev* 13(19), 2570–2580, 1999.

58. Arumugam, T.V., Gleichmann, M., Tang, S.C., and Mattson, M.P., Hormesis/preconditioning mechanisms, the nervous system and aging, *Ageing Res Rev* 5(2), 165–178, 2006.

59. Lin, S.J., Defossez, P.A., and Guarente, L., Requirement of NAD and SIR2 for life-span extension by calorie restriction in *Saccharomyces cerevisiae*, *Science* 289(5487), 2126–2128, 2000.

60. Tissenbaum, H.A. and Guarente, L., Increased dosage of a sir-2 gene extends lifespan in *Caenorhabditis elegans*, *Nature* 410(6825), 227–230, 2001.

61. Rogina, B. and Helfand, S.L., Sir2 mediates longevity in the fly through a pathway related to calorie restriction, *Proc Natl Acad Sci U S A* 101(45), 15998–6003, 2004.

62. Howitz, K.T., Bitterman, K.J., Cohen, H.Y., Lamming, D.W., Lavu, S., Wood, J.G., Zipkin, R.E., Chung, P., Kisielewski, A., Zhang, L.L., Scherer, B., and Sinclair, D.A., Small molecule activators of sirtuins extend *Saccharomyces cerevisiae* lifespan, *Nature* 425(6954), 191–196, 2003.

63. Kobayashi, T. and Horiuchi, T., A yeast gene product, Fob1 protein, required for both replication fork blocking and recombinational hotspot activities, *Genes Cells* 1(5), 465–474, 1996.

64. Kaeberlein, M., Andalis, A.A., Liszt, G.B., Fink, G.R., and Guarente, L., *Saccharomyces cerevisiae* SSD1-V confers longevity by a Sir2p-independent mechanism, *Genetics* 166(4), 1661–1672, 2004.

65. Lamming, D.W., Latorre-Esteves, M., Medvedik, O., Wong, S.N., Tsang, F.A., Wang, C., Lin, S.J., and Sinclair, D.A., HST2 mediates SIR2-independent life-span extension by calorie restriction, *Science* 309 (5742), 1861–1864, 2005.

66. Fabrizio, P., Liou, L.L., Moy, V.N., Diaspro, A., Valentine, J.S., Gralla, E.B., and Longo, V.D., SOD2 functions downstream of Sch9 to extend longevity in yeast, *Genetics* 163(1), 35–46, 2003.

67. Fabrizio, P., Battistella, L., Vardavas, R., Gattazzo, C., Liou, L.L., Diaspro, A., Dossen, J.W., Gralla, E.B., and Longo, V.D., Superoxide is a mediator of an altruistic aging program in *Saccharomyces cerevisiae*, *J Cell Biol* 166(7), 1055–1067, 2004.

68. Fabrizio, P., Gattazzo, C., Battistella, L., Wei, M., Cheng, C., McGrew, K., and Longo, V.D., Sir2 blocks extreme life-span extension, *Cell* 123(4), 655–667, 2005.

69. Sommer, M., Poliak, N., Upadhyay, S., Ratovitski, E., Nelkin, B.D., Donehower, L.A., and Sidransky, D., DeltaNp63alpha overexpression induces downregulation of Sirt1 and an accelerated aging phenotype in the mouse, *Cell Cycle* 5(17), 2005–2011, 2006.

70. Nemoto, S., Fergusson, M.M., and Finkel, T., Nutrient availability regulates SIRT1 through a forkhead-dependent pathway, *Science* 306(5704), 2105–2108, 2004.

71. Chen, D., Steele, A.D., Lindquist, S., and Guarente, L., Increase in activity during calorie restriction requires Sirt1, *Science* 310(5754), 1641, 2005.

72. Pani, G., Fusco, S., and Galeotti, T., Smaller, hungrier mice, *Science* 311 (5767), 1553–1554; author reply 1553–1554, 2006.

73. Baur, J.A., Pearson, K.J., Price, N.L., Jamieson, H.A., Lerin, C., Kalra, A., Prabhu, V.V., Allard, J.S., Lopez-Lluch, G., Lewis, K., Pistell, P.J., Poosala, S., Becker, K.G., Boss, O., Gwinn, D., Wang, M., Ramaswamy, S., Fishbein, K.W., Spencer, R.G., Lakatta, E.G., Le Couteur, D., Shaw, R.J., Navas, P., Puigserver, P., Ingram, D.K., de Cabo, R., and Sinclair, D.A., Resveratrol improves health and survival of mice on a high-calorie diet, *Nature* 444(7117), 337–342, 2006.

74. Lagouge, M., Argmann, C., Gerhart-Hines, Z., Meziane, H., Lerin, C., Daussin, F., Messadeq, N., Milne, J., Lambert, P., Elliott, P., Geny, B., Laakso, M., Puigserver, P., and Auwerx, J., Resveratrol improves mitochondrial function and protects against metabolic disease by activating SIRT1 and PGC-1alpha, *Cell* 127(6), 1109–1122, 2006.

75. Lin, J., Handschin, C., and Spiegelman, B.M., Metabolic control through the PGC-1 family of transcription coactivators, *Cell Metab* 1(6), 361–370, 2005.

76. Rodgers, J.T., Lerin, C., Haas, W., Gygi, S.P., Spiegelman, B.M., and Puigserver, P., Nutrient control of glucose homeostasis through a complex of PGC-1alpha and SIRT1, *Nature* 434(7029), 113–118, 2005.

77. Bordone, L., Motta, M.C., Picard, F., Robinson, A., Jhala, U.S., Apfeld, J., McDonagh, T., Lemieux, M., McBurney, M., Szilvasi, A., Easlon, E.J., Lin, S.J., and Guarente, L., Sirt1 regulates insulin secretion by repressing UCP2 in pancreatic beta cells, *PLoS Biol* 4(2), e31, 2006.

78. Moynihan, K.A., Grimm, A.A., Plueger, M.M., Bernal-Mizrachi, E., Ford, E., Cras-Meneur, C., Permutt, M.A., and Imai, S., Increased dosage of mammalian Sir2 in pancreatic beta cells enhances glucose-stimulated insulin secretion in mice, *Cell Metab* 2(2), 105–117, 2005.

79. Nemoto, S., Fergusson, M.M., and Finkel, T., SIRT1 functionally interacts with the metabolic regulator and transcriptional coactivator PGC-1{alpha}, *J Biol Chem* 280(16), 16456–16460, 2005.

80. Shi, T., Wang, F., Stieren, E., and Tong, Q., SIRT3, a mitochondrial sirtuin deacetylase, regulates mitochondrial function and thermogenesis in brown adipocytes, *J Biol Chem* 280(14), 13560–13567, 2005.

81. Kenyon, C., The plasticity of aging: Insights from long-lived mutants, *Cell* 120(4), 449–460, 2005.

82. DePinho, R.A., The age of cancer, *Nature* 408(6809), 248–254, 2000.

83. Campisi, J., Cancer and ageing: Rival demons? *Nat Rev Cancer* 3(5), 339–349, 2003.

84. Kinzler, K.W. and Vogelstein, B., Lessons from hereditary colorectal cancer, *Cell* 87(2), 159–170, 1996.

85. Campisi, J., Senescent cells, tumor suppression, and organismal aging: Good citizens, bad neighbors, *Cell* 120(4), 513–522, 2005.

86. Krtolica, A. and Campisi, J., Cancer and aging: A model for the cancer promoting effects of the aging stroma, *Int J Biochem Cell Biol* 34(11), 1401–1414, 2002.

87. Kirkwood, T.B. and Austad, S.N., Why do we age? *Nature* 408(6809), 233–238, 2000.

88. Sasaki, T., Maier, B., Bartke, A., and Scrable, H., Progressive loss of SIRT1 with cell cycle withdrawal, *Aging Cell* 5(5), 413–422, 2006.

89. Langley, E., Pearson, M., Faretta, M., Bauer, U.M., Frye, R.A., Minucci, S., Pelicci, P.G., and Kouzarides, T., Human SIR2 deacetylates p53 and antagonizes PML/p53-induced cellular senescence, *EMBO J* 21(10), 2383–2396, 2002.

90. Chua, K.F., Mostoslavsky, R., Lombard, D.B., Pang, W.W., Saito, S., Franco, S., Kaushal, D., Cheng, H.L., Fischer, M.R., Stokes, N., Murphy, M.M., Appella, E., and Alt, F.W., Mammalian SIRT1 limits replicative life span in response to chronic genotoxic stress, *Cell Metab* 2(1), 67–76, 2005.

91. Li, H., Rajendran, G.K., Liu, N., Ware, C., Rubin, B.P., and Gu, Y., SirT1 modulates the estrogen-insulin-like growth factor-1 signaling for postnatal development of mammary gland in mice, *Breast Cancer Res* 9(1), R1, 2007.

92. Fu, M., Liu, M., Sauve, A.A., Jiao, X., Zhang, X., Wu, X., Powell, M.J., Yang, T., Gu, W., Avantaggiati, M.L., Pattabiraman, N., Pestell, T.G., Wang, F., Quong, A.A., Wang, C., and Pestell, R.G., Hormonal control of androgen receptor function through SIRT1, *Mol Cell Biol* 26(21), 8122–8135, 2006.

93. Baylin, S.B. and Herman, J.G., DNA hypermethylation in tumorigenesis: Epigenetics joins genetics, *Trends Genet* 16(4), 168–174, 2000.

94. Jones, P.A. and Laird, P.W., Cancer epigenetics comes of age, *Nat Genet* 21(2), 163–167, 1999.

95. Issa, J.P., Aging, DNA methylation and cancer, *Crit Rev Oncol Hematol* 32(1), 31–43, 1999.

96. Vaquero, A., Loyola, A., and Reinberg, D., The constantly changing face of chromatin, *Sci Aging Knowledge Environ* 2003(14), RE4, 2003.

97. Lennox, R.W. and Cohen, L.H., The histone H1 complements of dividing and nondividing cells of the mouse, *J Biol Chem* 258(1), 262–268, 1983.

98. Parseghian, M.H., Newcomb, R.L., Winokur, S.T., and Hamkalo, B.A., The distribution of somatic H1 subtypes is non-random on active vs. inactive chromatin: Distribution in human fetal fibroblasts, *Chromosome Res* 8(5), 405–424, 2000.

99. Parseghian, M.H., Newcomb, R.L., and Hamkalo, B.A., Distribution of somatic H1 subtypes is non-random on active vs. inactive chromatin II: Distribution in human adult fibroblasts, *J Cell Biochem* 83(4), 643–659, 2001.

100. Lindner, H., Sarg, B., Grunicke, H., and Helliger, W., Age-dependent deamidation of H1(0) histones in chromatin of mammalian tissues, *J Cancer Res Clin Oncol* 125(3–4), 182–186, 1999.

101. Sarg, B., Helliger, W., Hoertnagl, B., Puschendorf, B., and Lindner, H., The N-terminally acetylated form of mammalian histone H1(o), but not that of avian histone H5, increases with age, *Arch Biochem Biophys* 372(2), 333–339, 1999.

102. Kotake, Y., Cao, R., Viatour, P., Sage, J., Zhang, Y., and Xiong, Y., pRB family proteins are required for H3K27 trimethylation and Polycomb repression complexes binding to and silencing p16INK4alpha tumor suppressor gene, *Genes Dev* 21(1), 49–54, 2007.

103. Furuyama, T., Banerjee, R., Breen, T.R., and Harte, P.J., SIR2 is required for polycomb silencing and is associated with an E(Z) histone methyltransferase complex, *Curr Biol* 14(20), 1812–1821, 2004.

104. Kuzmichev, A., Margueron, R., Vaquero, A., Preissner, T.S., Scher, M., Kirmizis, A., Ouyang, X., Brockdorff, N., Abate-Shen, C., Farnham, P., and Reinberg, D., Composition and histone substrates of polycomb repressive group complexes change during cellular differentiation, *Proc Natl Acad Sci U S A* 102(6), 1859–1864, 2005.

105. Varambally, S., Dhanasekaran, S.M., Zhou, M., Barrette, T.R., Kumar-Sinha, C., Sanda, M.G., Ghosh, D., Pienta, K.J., Sewalt, R.G., Otte, A.P., Rubin, M.A., and Chinnaiyan, A.M., The polycomb group protein EZH2 is involved in progression of prostate cancer, *Nature* 419(6907), 624–629, 2002.

106. Kleer, C.G., Cao, Q., Varambally, S., Shen, R., Ota, I., Tomlins, S.A., Ghosh, D., Sewalt, R.G., Otte, A. P., Hayes, D.F., Sabel, M.S., Livant, D., Weiss, S.J., Rubin, M.A., and Chinnaiyan, A.M., EZH2 is a marker of aggressive breast cancer and promotes neoplastic transformation of breast epithelial cells, *Proc Natl Acad Sci U S A* 100(20), 11606–11611, 2003.

107. Seligson, D.B., Horvath, S., Shi, T., Yu, H., Tze, S., Grunstein, M., and Kurdistani, S.K., Global histone modification patterns predict risk of prostate cancer recurrence, *Nature* 435(7046), 1262–1266, 2005.

108. Fraga, M.F., Ballestar, E., Villar-Garea, A., Boix-Chornet, M., Espada, J., Schotta, G., Bonaldi, T., Haydon, C., Ropero, S., Petrie, K., Iyer, N.G., Perez-Rosado, A., Calvo, E., Lopez, J.A., Cano, A., Calasanz, M.J., Colomer, D., Piris, M.A., Ahn, N., Imhof, A., Caldas, C., Jenuwein, T., and Esteller, M.,

Loss of acetylation at Lys16 and trimethylation at Lys20 of histone H4 is a common hallmark of human cancer, *Nat Genet* 37(4), 391–400, 2005.

109. Hall, A., Rho GTPases and the actin cytoskeleton, *Science* 279(5350), 509–514, 1998.

110. Bar-Sagi, D. and Hall, A., Ras and Rho GTPases: A family reunion, *Cell* 103(2), 227–238, 2000.

111. Muller, J.M., Metzger, E., Greschik, H., Bosserhoff, A.K., Mercep, L., Buettner, R., and Schule, R., The transcriptional coactivator FHL2 transmits Rho signals from the cell membrane into the nucleus, *EMBO J* 21(4), 736–748, 2002.

112. Yang, Y., Hou, H., Haller, E.M., Nicosia, S.V., and Bai, W., Suppression of FOXO1 activity by FHL2 through SIRT1-mediated deacetylation, *EMBO J* 24(5), 1021–1032, 2005.

113. Mills, G.B. and Moolenaar, W.H., The emerging role of lysophosphatidic acid in cancer, *Nat Rev Cancer* 3(8), 582–591, 2003.

114. Dong, X.Y., Chen, C., Sun, X., Guo, P., Vessella, R.L., Wang, R.X., Chung, L.W., Zhou, W., and Dong, J.T., FOXO1A is a candidate for the 13q14 tumor suppressor gene inhibiting androgen receptor signaling in prostate cancer, *Cancer Res* 66(14), 6998–7006, 2006.

115. Paik, J.H., Kollipara, R., Chu, G., Ji, H., Xiao, Y., Ding, Z., Miao, L., Tothova, Z., Horner, J.W., Carrasco, D.R., Jiang, S., Gilliland, D.G., Chin, L., Wong, W.H., Castrillon, D.H., and DePinho, R.A., FoxOs are lineage-restricted redundant tumor suppressors and regulate endothelial cell homeostasis, *Cell* 128(2), 309–323, 2007.

116. Arden, K.C., FoxOs in tumor suppression and stem cell maintenance, *Cell* 128(2), 235–237, 2007.

117. Lee, S.E., Paques, F., Sylvan, J., and Haber, J.E., Role of yeast SIR genes and mating type in directing DNA double-strand breaks to homologous and non-homologous repair paths, *Curr Biol* 9(14), 767–770, 1999.

118. Aguilaniu, H., Gustafsson, L., Rigoulet, M., and Nystrom, T., Asymmetric inheritance of oxidatively damaged proteins during cytokinesis, *Science* 299(5613), 1751–1753, 2003.

119. Vaziri, H., Dessain, S.K., Ng Eaton, E., Imai, S.I., Frye, R.A., Pandita, T.K., Guarente, L., and Weinberg, R.A., hSIR2(SIRT1) functions as an NAD-dependent p53 deacetylase, *Cell* 107(2), 149–159, 2001.

120. Luo, J., Nikolaev, A.Y., Imai, S., Chen, D., Su, F., Shiloh, A., Guarente, L., and Gu, W., Negative control of p53 by Sir2alpha promotes cell survival under stress, *Cell* 107(2), 137–148, 2001.

121. Dai, J.M., Wang, Z.Y., Sun, D.C., Lin, R.X., and Wang, S.Q., SIRT1 interacts with p73 and suppresses p73-dependent transcriptional activity, *J Cell Physiol* 210(1), 161–166, 2007.

122. Cheng, H.L., Mostoslavsky, R., Saito, S., Manis, J.P., Gu, Y., Patel, P., Bronson, R., Appella, E., Alt, F.W., and Chua, K.F., Developmental defects and p53 hyperacetylation in Sir2 homolog (SIRT1)-deficient mice, *Proc Natl Acad Sci U S A* 100(19), 10794–10799, 2003.

123. McBurney, M.W., Yang, X., Jardine, K., Hixon, M., Boekelheide, K., Webb, J.R., Lansdorp, P.M., and Lemieux, M., The mammalian SIR2alpha protein has a role in embryogenesis and gametogenesis, *Mol Cell Biol* 23(1), 38–54, 2003.

124. Kamel, C., Abrol, M., Jardine, K., He, X., and McBurney, M.W., SirT1 fails to affect p53-mediated biological functions, *Aging Cell* 5(1), 81–88, 2006.

125. Solomon, J.M., Pasupuleti, R., Xu, L., McDonagh, T., Curtis, R., DiStefano, P.S., and Huber, L.J., Inhibition of SIRT1 catalytic activity increases p53 acetylation but does not alter cell survival following DNA damage, *Mol Cell Biol* 26(1), 28–38, 2006.

126. Feng, L., Lin, T., Uranishi, H., Gu, W., and Xu, Y., Functional analysis of the roles of posttranslational modifications at the p53 C terminus in regulating p53 stability and activity, *Mol Cell Biol* 25(13), 5389–5395, 2005.

127. Tothova, Z., Kollipara, R., Huntly, B.J., Lee, B.H., Castrillon, D.H., Cullen, D.E., McDowell, E.P., Lazo-Kallanian, S., Williams, I.R., Sears, C., Armstrong, S.A., Passegue, E., DePinho, R.A., and Gilliland, D.G., FoxOs are critical mediators of hematopoietic stem cell resistance to physiologic oxidative stress, *Cell* 128(2), 325–339, 2007.

128. Ahmed, S., Uncoupling of pathways that promote postmitotic life span and apoptosis from replicative immortality of *Caenorhabditis elegans* germ cells, *Aging Cell* 5(6), 559–563, 2006.

129. Berdichevsky, A., Viswanathan, M., Horvitz, H.R., and Guarente, L., *C. elegans* SIR-2.1 interacts with 14-3-3 proteins to activate DAF-16 and extend life span, *Cell* 125(6), 1165–1177, 2006.

130. Wang, Y. and Tissenbaum, H.A., Overlapping and distinct functions for a *Caenorhabditis elegans* SIR2 and DAF-16/FOXO, *Mech Ageing Dev* 127(1), 48–56, 2006.

131. Brunet, A., Sweeney, L.B., Sturgill, J.F., Chua, K.F., Greer, P.L., Lin, Y., Tran, H., Ross, S.E., Mostoslavsky, R., Cohen, H.Y., Hu, L.S., Cheng, H.L., Jedrychowski, M.P., Gygi, S.P., Sinclair, D.A., Alt, F.W., and Greenberg, M.E., Stress-dependent regulation of FOXO transcription factors by the SIRT1 deacetylase, *Science* 303(5666), 2011–2015, 2004.

132. Motta, M.C., Divecha, N., Lemieux, M., Kamel, C., Chen, D., Gu, W., Bultsma, Y., McBurney, M., and Guarente, L., Mammalian SIRT1 represses forkhead transcription factors, *Cell* 116(4), 551–563, 2004.

133. van der Horst, A., Tertoolen, L.G., de Vries-Smits, L.M., Frye, R.A., Medema, R.H., and Burgering, B.M., FOXO4 is acetylated upon peroxide stress and deacetylated by the longevity protein hSir2(SIRT1), *J Biol Chem* 279(28), 28873–28879, 2004.

134. Daitoku, H., Hatta, M., Matsuzaki, H., Aratani, S., Ohshima, T., Miyagishi, M., Nakajima, T., and Fukamizu, A., Silent information regulator 2 potentiates Foxo1-mediated transcription through its deacetylase activity, *Proc Natl Acad Sci U S A* 101(27), 10042–10047, 2004.

135. Giannakou, M.E. and Partridge, L., The interaction between FOXO and SIRT1: Tipping the balance towards survival, *Trends Cell Biol* 14(8), 408–412, 2004.

136. Greer, E.L. and Brunet, A., FOXO transcription factors at the interface between longevity and tumor suppression, *Oncogene* 24(50), 7410–7425, 2005.

137. Sawada, M., Sun, W., Hayes, P., Leskov, K., Boothman, D.A., and Matsuyama, S., Ku70 suppresses the apoptotic translocation of Bax to mitochondria, *Nat Cell Biol* 5(4), 320–329, 2003.

138. Cohen, H.Y., Lavu, S., Bitterman, K.J., Hekking, B., Imahiyerobo, T.A., Miller, C., Frye, R., Ploegh, H., Kessler, B.M., and Sinclair, D.A., Acetylation of the C terminus of Ku70 by CBP and PCAF controls Bax-mediated apoptosis, *Mol Cell* 13(5), 627–638, 2004.

139. Ohsawa, S. and Miura, M., Caspase-mediated changes in Sir2alpha during apoptosis, *FEBS Lett* 580(25), 5875–5879, 2006.

140. Nahle, Z., Polakoff, J., Davuluri, R.V., McCurrach, M.E., Jacobson, M.D., Narita, M., Zhang, M.Q., Lazebnik, Y., Bar-Sagi, D., and Lowe, S.W., Direct coupling of the cell cycle and cell death machinery by E2F, *Nat Cell Biol* 4(11), 859–864, 2002.

141. Johnson, D.G., Schwarz, J.K., Cress, W.D., and Nevins, J.R., Expression of transcription factor E2F1 induces quiescent cells to enter S phase, *Nature* 365(6444), 349–352, 1993.

142. Wu, X. and Levine, A.J., p53 and E2F-1 cooperate to mediate apoptosis, *Proc Natl Acad Sci U S A* 91(9), 3602–3606, 1994.

143. Lin, W.C., Lin, F.T., and Nevins, J.R., Selective induction of E2F1 in response to DNA damage, mediated by ATM-dependent phosphorylation, *Genes Dev* 15(14), 1833–1844, 2001.

144. Wang, C., Chen, L., Hou, X., Li, Z., Kabra, N., Ma, Y., Nemoto, S., Finkel, T., Gu, W., Cress, W.D., and Chen, J., Interactions between E2F1 and SirT1 regulate apoptotic response to DNA damage, *Nat Cell Biol* 8(9), 1025–1031, 2006.

145. Bereshchenko, O.R., Gu, W., and Dalla-Favera, R., Acetylation inactivates the transcriptional repressor BCL6, *Nat Genet* 32(4), 606–613, 2002.

146. Yeung, F., Hoberg, J.E., Ramsey, C.S., Keller, M.D., Jones, D.R., Frye, R.A., and Mayo, M.W., Modulation of NF-kappaB-dependent transcription and cell survival by the SIRT1 deacetylase, *EMBO J* 23(12), 2369–2380, 2004.

147. Gao, F., Cheng, J., Shi, T., and Yeh, E.T., Neddylation of a breast cancer-associated protein recruits a class III histone deacetylase that represses NFkappaB-dependent transcription, *Nat Cell Biol* 8(10), 1171–1177, 2006.

148. Yang, S.R., Wright, J., Bauter, M., Seweryniak, K., Kode, A., and Rahman, I., Sirtuin regulates cigarette smoke-induced proinflammatory mediator release via RelA/p65 NF-kappaB in macrophages in vitro and in rat lungs in vivo: Implications for chronic inflammation and aging, *Am J Physiol Lung Cell Mol Physiol* 292(2), L567–576, 2007.

149. Fulda, S. and Debatin, K.M., Sensitization for anticancer drug-induced apoptosis by the chemopreventive agent resveratrol, *Oncogene* 23(40), 6702–6711, 2004.

150. Perkins, N.D., NF-kappaB: Tumor promoter or suppressor? *Trends Cell Biol* 14(2), 64–69, 2004.

151. Abdelmohsen, K., Pullmann, R., Jr., Lal, A., Kim, H.H., Galban, S., Yang, X., Blethrow, J.D., Walker, M., Shubert, J., Gillespie, D.A., Furneaux, H., and Gorospe, M., Phosphorylation of HuR by Chk2 regulates SIRT1 expression, *Mol Cell* 25(4), 543–557, 2007.

152. Ford, J., Jiang, M., and Milner, J., Cancer-specific functions of SIRT1 enable human epithelial cancer cell growth and survival, *Cancer Res* 65(22), 10457–10463, 2005.

153. Heltweg, B., Gatbonton, T., Schuler, A.D., Posakony, J., Li, H., Goehle, S., Kollipara, R., Depinho, R.A., Gu, Y., Simon, J.A., and Bedalov, A., Antitumor activity of a small-molecule inhibitor of human silent information regulator 2 enzymes, *Cancer Res* 66(8), 4368–4377, 2006.

154. Ota, H., Tokunaga, E., Chang, K., Hikasa, M., Iijima, K., Eto, M., Kozaki, K., Akishita, M., Ouchi, Y., and Kaneki, M., Sirt1 inhibitor, sirtinol, induces senescence-like growth arrest with attenuated Ras-MAPK signaling in human cancer cells, *Oncogene* 25(2), 176–185, 2006.

155. Pruitt, K., Zinn, R.L., Ohm, J.E., McGarvey, K.M., Kang, S.H., Watkins, D.N., Herman, J.G., and Baylin, S.B., Inhibition of SIRT1 reactivates silenced cancer genes without loss of promoter DNA hypermethylation, *PLoS Genet* 2(3), e40, 2006.

156. Chu, F., Chou, P.M., Zheng, X., Mirkin, B.L., and Rebbaa, A., Control of multidrug resistance gene mdr1 and cancer resistance to chemotherapy by the longevity gene sirt1, *Cancer Res* 65(22), 10183–10187, 2005.

157. Yang, Y., Fu, W., Chen, J., Olashaw, N., Zhang, X., Nicosia, S.V., Bhalla, K., and Bai, W., SIRT1 sumoylation regulates its deacetylase activity and cell's response to genotoxic stress, *Nat Cell Biol* 9, 1253–1262, 2007.

10 Proteins That Alter Histone Modifications in Cancer

Ada Ao and Jianrong Lu

CONTENTS

10.1 BRIEF INTRODUCTION TO POSTTRANSLATIONAL COVALENT MODIFICATIONS OF HISTONE TAILS

Epigenetic regulation centers on the chromatin structure and the effects of remodeling and modifying enzymes upon it. Chromatin can be altered at the DNA and protein levels (e.g., via DNA methylation and histone modifications). Changes in DNA methylation can have profound effects on gene expression. The histone octamer, forming the core of the nucleosome, is subject to transformation by two types of enzymes: remodelers and modifiers. Remodelers are classified as ATP-dependent proteins that alter DNA–histone interactions by repositioning histones, making regulatory *cis* elements on a DNA sequence more accessible to transcription factors; whereas modifiers are defined as ATP-independent proteins that perform posttranslational covalent modifications on amino-terminus tails of histones. This can either facilitate or hinder the binding of regulatory proteins on DNA and thus affect transcriptional activity.

A number of histone-modifying enzymes have been identified and characterized that are capable of attaching different moieties to specific residues on histone tails. These modifications include: acetylation, methylation, phosphorylation, SUMOlation, ubiquitination, poly-ADP-ribosylation, biotination, citrullination, and proline isomerization. Only a few of these are understood in detail, whereas others have only been observed and their biological function is unclear. While all histone subunits can be modified, most reports have concentrated on the canonical core histones octamer: H2A, H2B, H3, and H4. Each modification can be achieved by a plethora of enzymes that favor particular residues on various histone subunits. These proteins are arranged into families and classes based on their structural motif and mechanism. There are also classes of enzymes that reverse most modification, rendering histone modification dynamic and adaptable.

The "histone code" theory was formulated to correlate distinct covalent modifications of histone tails with specific gene expression states [1–3]. This hypothesis supposes unique patterns of post-translational modifications on one or more histone tails as a molecular code to guide recruitment of corresponding chromatin-regulatory effectors to the chromatin fiber and bring about a specific chromatin state. Distinct histone modifications can act sequentially or in concert. Understanding how the histone marks are regulated and interpreted is at the heart of chromatin biology.

10.2 HISTONE ACETYLTRANSFERASE/DEACETYLASE IN CANCER

10.2.1 HISTONE ACETYLTRANSFERASES

Acetylation possesses many levels of function in the nucleus, including transcription activation, DNA repair, and cell cycle regulation. It is generally associated with transcription activation because acetyl addition to specific residues loosens the negatively charged DNA from the positively charged amino group of the histone tail. The acetyl moiety can also be a target for transcription factor recruitment. This modification is carried out by histone acetyltransferases (HATs), which are divided into three major families: GNAT, MYST, and CBP/P300 [4,5]. These enzymes can generally acetylate more than one lysine but display some specificity. Aside from histone acetylation, some of them can also bind nonhistone cancer-related proteins like p53, as well as nuclear hormone receptors like androgen receptor (AR) [5]. PCAF is a member of the GNAT family that is correlated with epithelial cancer when mutated (Table 10.1). P300 and CBP are highly related and serve mostly as coactivators in complex with transcription factors, nuclear hormone receptors, and tumor suppressors. Mutation or rearrangement of the two proteins is linked to an assortment of cancers and Rubinstein–Taybi syndrome (RTS) whose patients have increased susceptibly to cancer due to abnormal gene expression (Table 10.1) [6]. MOZ and MORF both belong in the MYST family of HATs. Their disruption by translocation has been implicated in certain hematological malignancies as well as uterine leiomyomata, resulting in deregulation of proto-oncogene c-fos (Table 10.1) [5,7].

10.2.2 HISTONE DEACETYLASES

Acetylation is reversible via histone deacetylases (HDACs), which are generally divided into three classes. They are associated with global transcriptional silencing by promoting a condensed chromatin structure. When mutated, they are linked to cancer by aberrantly silencing proliferation inhibitors such as p21 [8]. They also act as corepressors for oncogenic translocation-produced chimera transcription factors like PML-RAR, which results in hematopoietic malignancies [8]. Class I HDACs consisting of HDAC1, HDAC2, HDAC3, and HDAC8 are the most abundant. They are homologous to yeast HDAC yRPD3 and are expressed in the nuclei of most cell lines and tissue types [9]. HDAC1 and HDAC2 are part of the core subunit of Sin3 and NuRD chromatin

TABLE 10.1

HAT/HDAC in Cancer

HAT/HDAC	Substrate	Mutation	Cancer Type	References
PCAF	H3/H4, tyrosine aminotransferase, p53, AR	Missense	Epithelial	[26]
P300	H2A/H2B/H3/H4, p53, AR	Bi-allelic mutation + LOH	Gastric carcinoma	[27]
		Stop codon mutation	Colon, breast	[28]
		Missense	Colorectal, gastric, epithelial	[29,30]
		MOZ/p300 gene fusion	Acute monocytic leukemia	[31,32]
		MLL/p300 gene fusion	Acute myeloid leukemia (AML)	[33]
		Homozygous deletion	Cervical	[34]
		Point mutation	RTS	[35]
CBP	H2A/H2B/H3/H4, transcription factors	Stop codon mutation	Colon, epithelial	[26,36]
		In-frame deletion	Epithelial, lung	[26,37]
		Homozygous deletion	Lung	[37]
		Missense	Lung	[37]
		MYST4/CBP gene fusion	AML	[38]
		MOZ/CBP gene fusion	AML	[39–41]
		MORF/CBP gene fusion	AML	[42]
		Internal tandem duplication + LOH	Esophageal carcinoma	[43]
		MLL/CBP gene fusion	Therapy-related leukemia	[44–46]
		Deletions	RTS	[35,47]
		Intragenic duplications	RTS	[35]
		Point mutations	RTS	[48]
MOZ	H3/H4	MOZ/TIF2 gene fusion	AML	[49]
		MOZt(2;8)(p23;pl 1)	Myelodysplastic syndrome	[50]
MORF	H3/H4	MORFt(10;17)(q22;q21)	Uterine leiomyomata	[51]
NCOA1	H3/H4	PAX3/NCOA1 gene fusion	Rhabdomyosarcoma	[52]
HDAC1	Histone, p53, E2F1	Overexpression	Prostate, gastric	[53,54]
		Translocation	Leukemia, lymphoma	

Source: Modified from Santos-Rosa, H. and Caldas, C., *Eur. J. Cancer*, 41, 2381, 2005.

remodeling complex, and are transcriptional silencers in other repressive complexes as well. Class II HDACs include HDAC4–7, HDAC9, and HDAC10. They are derived from yeast yHFA1 and are expressed in only a few cell types [9]. Unlike class I HDACs, class II members are present in large complexes found in both the nucleus and the cytoplasm [9]. Both class I and class II HDACs are sensitive to trichostatin A (TSA), which has been used as an antitumor drug albeit with side effects [10]. Class III HDACs, also called Sirtuins, are derived from yeast silent information repressor 2 (Sir2). They are unique due to their NAD-dependent catalytic mechanism, which shields them from TSA inhibition. Functionally, they are associated with transcriptional silencing in yeast, and regulation of mitotic checkpoint in mammals [11].

10.2.3 Acetylation Status of Histone and Cancer

There is no linear correlation between global histone acetylation and cancer. Besides histones, HATs/HDACs have a wide range of substrates, such as transcription factors NF-κB and the p53 tumor suppressor, and chaperon proteins like HSP90 [12]. However, acetylation is connected to tumor formation through a multitude of complex interactions. A theme emerges from the ever-growing collection of studies concerning histone acetylation and cancer (Table 10.1). It has been suggested that there exists an acetylation equilibrium maintained by HATs/HDACs in normal cells. In tumor cells, acetylation patterns are disrupted by altered HAT/HDAC expression. This results in aberrant gene expression, including tumor suppressors and oncogenes [6,12–15]. The oncogenic function of acetylation enzymes are determined by the protein complexes they associate with. These complexes can direct acetylation activity to specific promoters. For example, abnormal fusion proteins can target HAT/HDAC to oncogenic genes (Table 10.1). In acute promyelocytic leukemia, fusion proteins like PML-RAR are unable to dissociate from HDAC complexes upon ligand binding and result in aberrant transcriptional silencing of RAR-targeted genes [12]. Other links between acetylation and cancer are related to crosstalk within epigenetic pathways. MTA, a component of the chromatin remodeling and deacetylation NuRD complex, is tied to breast tumor metastasis and its overexpression has been correlated with hypomethylation and hypoacetylation of H4 in esophageal squamous cells (reviewed in Refs. [16,17]).

10.2.4 THE SIRTUIN FAMILY AND CANCER

The sirtuin HDAC family (Sirt1-7) is a well-known longevity factor in yeast. It mediates DNA repair through nonhomologous end joining and homologous recombination pathways. In mammals, Sirt1 is associated with preserving genomic stability. There is substantial evidence that Sirt1 downregulates p53 through deacetylation (p53 is stabilized by acetylation upon cellular stress) to allow cell-cycle restart after recovery from stress [18,19]. Sirt1 also modifies other DNA damage repair factors like Ku70 to facilitate repairs [20]. There is some evidence that suggests overexpression of Sirt1 is oncogenic. In a preliminary study, treatment of lung and breast cancer cells with a Sirt1 inhibitor induces senescent-like growth arrest (reviewed in Ref. [21]). Such cell cycle arrest is associated with oncogene-induced senescence where DNA-damage response is triggered by accumulation of DNA breaks, resulted from oncogene-induced hyper-replication [22]. Sirt2 may be another member of the Sirtrins that maintains genome stability. It normally resides in the cytoplasm but translocates into the nucleus during mitosis, and deacetylases lysine 16 of histone H4 (H4K16). Its overexpression delays cells from exiting mitosis, while its loss has been characterized in gliomas. Thus, Sirt2 is proposed to contribute as a mitotic checkpoint protein through global deacetylation of H4K16 and delaying cell division [23]. Sirt3–5 are specific for the mitochondria. Their known functions are related to energy metabolism and neutralization of oxidative stress in aging cells. Sirt6 is a necessary component in the DNA repair process called base-excision repair [24], indicating it as an important contributor to genomic stability and tumor suppression. Sirt7 is necessary for cell survival as its deletion leads to apoptosis, although the mechanism is unclear. Its known function is to activate ribosomal RNA synthesis [25], which is essential for downstream protein synthesis and provides a clue to its role in cell proliferation.

10.3 HISTONE METHYLTRANSFERASE/DEMETHYLASE IN CANCER

10.3.1 INTRODUCTION TO HISTONE METHYLATION

Methylation is more complicated than acetylation due to the types and levels of modification that can occur. It is specific for lysine (K) and arginine (R) residues of histone H3 and H4. Arginine can accept two methyl groups that can be symmetrical or asymmetrical, while lysine can be mono- (Me1), di-(Me2), or trimethylated (Me3) (Tables 10.2 and 10.3). Histone methylation causes no change in overall protein charge. Therefore, its functional effects on transcriptional regulation and chromatin structure are elicited through recruitment of effector proteins. Specific histone lysine methylation marks are commonly associated with either the active or the repressed state.

TABLE 10.2
Arginine HMTs

Arginine HMTs	Type	Substrate
PRMT1	PRMT Type 1 (asymmetrical)	H4R3
PRMT4 (CARM1)	PRMT Type 1 (asymmetrical)	H3R2, R8, R17, R26
PRMT5	PRMT Type 2 (symmetrical)	H2A, H4
PRMT7	PRMT Type 2 (symmetrical)	H4

Source: Adapted from Santos-Rosa, H. and Caldas, C., *Eur. J. Cancer*, 41, 2381, 2005; Bedford, M.T. and Richard, S., *Mol. Cell*, 18, 263, 2005.

Methylation of H3K9, H3K27, and H4K20 is generally considered a repressive feature as it is common in heterochromatin formation, X-inactivation, genomic imprinting, transcriptional silencing, as well as DNA methylation. Methylation at H3K4, H3K36, and H3K79 are related to transcription activation in conjunction with other coactivator proteins. Also, certain methylation patterns on H3K4, H3K36, and H3K79 appear to physically interact with and facilitate the elongating RNA Polymerase II (which transcribes RNA from DNA template).

10.3.2 Histone Lysine Methylation

Histone methyltransferases (HMTs) are classified into families and groups based on substrates and structural motifs, and each individual protein can methylate multiple substrates. Histone lysine methyltransferases (HKMTs) families (Table 10.3) include: SET1/2, SUV39, EZH, SMYD, PRDM, alternative SETs, and non-SET HKMTs [55]. The SET family, SUV39 family, SMYD family, EZH family, and PRDM family are the more noteworthy since they have been correlated with cancer when mutated (Table 10.4). The SET family generally methylates H3K4 and has coactivator properties [55]. It includes four MLL isoforms, which is involved in many types of leukemia when oncogenic chromosomal translocation occurs. The resulting MLL is fused to transcription factors and other cellular proteins [56]. There are more than 30 MLL fusion partners (Table 10.4), amongst which is a non-SET HKMT. It is called Dot1L and is specific for H3K79 [56].

SUV39 members are responsible for tri- and dimethylation of H3K9 (Table 10.3), which is required for the binding of the repressive protein HP1 and is associated with heterochromatin formation and transcriptional repression. Centromeric heterochromatin formation and maintenance is critical for genomic integrity. Indeed, SUV39h1 and h2 mutations in mice lead to increased

TABLE 10.3
Lysine HMTs

Lysine HMTs	Family	Substrates
MLLs, SET1, NSD1–3, ASH1L, SET8, SET 7/9 SUV420H1/H2	SET	H3K4, K9, K36, H4K20
SUV39H1/H2, EHMT, SETDB1, SETDB2, SETMAR	SUV39	H3K4, H3K9, K27
EZH1, EZH2	EZH	H3K9, K27
SMYD1-5	SMYD	H3K4
PRDM 1-16, RIZ1	PRDM	H3K4, H3K9
DOT1L	Non-SET	H3K79

Source: Adapted from Volkel, P. and Angrand, P.O., *Biochimie*, 89, 1, 2007.

TABLE 10.4

HMTs in Cancer

HMT	Substrate	Mutation/Alteration	Cancer Type	Reference
MLL1/ALL1	H3K4	Partial tandem duplication	AML	[53,68,69]
		Partial nontandem duplication	ALL	[70]
		Amplification	AML, ALL, MDS, RAEB	[71–74]
		Trisomy 11+	ALL, MDS	[45,69,75]
		Rearrangements	MDS, Acute leukemia	[76–78]
		MLL fusion (30 + forms)	Acute leukemia, AML, ALL	N/A
MLL2	H3K4	19q13.1 amplification	Solid tumors	[79]
MLL3	H3K4	7q36 deletion	AML	[80]
SMYD3	H3K4	Overexpression	Breast	[81]
SUV39H1/H2	H3K9	Overexpression	Lymphoma	[59]
SUV4-20	H4K20	Overexpression	NA	[63]
RIZ1	H3K9	Repression	MLL, thyroid, breast, prostate	[82–84]
EZH2	H3K4, H3K27	Amplification	Primary breast tumor, prostate	[85,86]
		Aberrant expression		
		Overexpression		
PRMT1	H4R3	Overexpression	Prostate, breast	[87,88]
PRMT4 (CARM1)	H3R2, R17, R26	Overexpression	Prostate carcinomas	[87–89]
PRMT5	H2A, H4	Overexpression	Prostate, breast	[90,91]

Source: Adapted from Santos-Rosa, H. and Caldas, C., *Eur. J. Cancer*, 41, 2381, 2005; Bedford, M.T. and Richard, S., *Mol. Cell*, 18, 263, 2005; Volkel, P. and Angrand, P.O., *Biochimie*, 89, 1, 2007.

genomic instability relating to the loss of H3K9 methylation and collaborate with oncogenes in inducing lymphomas [7,57]. Furthermore, Human SUV39h1/2 and HP1 associates with the Rb tumor suppressors, which may result in recruitment of heterochromatinizing activities to the E2F-responsive promoters and hence silencing of E2F-regulated growth-promoting genes during cellular senescence [58]. Senescence is a stable form of cell cycle arrest and is a key barrier guarding cells against unlimited proliferation and cancer [59–61].

Attention is being directed to Suv4-20h1/2, which di- and trimethylates H4K20 (Table 10.3) [55,59,62,63]. Suv4–20 may be recruited by the Rb tumor suppressor through direct protein–protein interactions. Mouse cells deficient for all Rb family members exhibit a marked genomic instability and decreased H4K20 trimethylation [64]. Interestingly, studies on H4 modifications in cancer have suggested that global loss of trimethylation at H4K20, in addition to the loss of monoacetylation at H4K16, is a common hallmark of human tumor cells [63]. This histone H4 "cancer signature" is associated with the hypomethylation of DNA repetitive sequences, a well-known characteristic of cancer cells [63]. Together, it is likely that Suv4-20-mediated H4K20 trimethylation contributes to Rb-dependent heterochromatin formation and tumor suppression.

SMYDs include five related proteins that contain the catalytic SET domain and a zinc-finger motif that shows DNA sequence specificity (Table 10.3) [55]. SMYD3 is known to methylate H3K4. Overexpression of SMYD3 is noted in colorectal carcinomas and breast cancer, possibly resulted from hypermethylation and silencing of tumor suppressing genes [65]. The EZH family members are structurally similar to SUV39 and methylates H3K9 and K27 [55]. EZH2 (Table 10.3) together with SUZ12 forms the Polycomb Group (PcG) repressor complex, and overexpression is known to promote metastasis in breast cancer, prostate cancer, and lymphomas (Table 10.4) [66–68]. The Rb/E2F pathway has been shown to regulate the EZH2 promoter and EZH2 expression is tied to cell proliferation, prompting some to term EZH2 an oncogene [66,67].

10.3.3 Histone Arginine Methylation

Compared to HKMTs, there is less known concerning arginine methyltransferases (PRMTs) (Tables 10.2 and 10.4). As a whole, they can modify nonhistone substrates that have a GAR-motif [62]. These substrates can take part in DNA repair, RNA processing, and signal transduction and may be indirectly involved in carcinogenesis. PRMTs are capable of mono- or dimethylation. The latter is divided into two classes: type 1 (asymmetrical) and type 2 (symmetrical). There are nine PRMT isoforms identified to date, and would most likely be more [92]. PRMTs are not directly connected to cancer, as there are no mutants that suggest a causal link. However, because it serves as coactivators for many nuclear hormone receptors, they are often overexpressed in prostate and breast cancer. For example, CARM1 (Table 10.2), a Type 1 PRMT specific for H3R2, R17, and R26, is noted in prostate cancer as a coactivator required for AR function [93]; while PRMT5 is also found to be overexpressed in various types of lymphomas [94]. It is believed to inhibit tumor suppressors and allow uncontrolled cell growth.

10.3.4 Histone Demethylation

Histone methylation was thought to be static until recent efforts found many enzymes that remove the methyl group by demethylimination or demethylation. The difference between the two is strictly biochemical. The demethylimination reaction is catalyzed by PAD1, which converts methylarginine into citrullin. It is not considered a true demethylase as its product is not an unconverted peptide [95–97]. Nor is PAD1 specific since it converts multiple arginines on histone H3 and H4. LSD1 is the first true lysine demethylase found. It is a FAD-dependent amino oxidase specific for H3K4me1/me2. LSD1 is found to affect androgen responsive genes but the mechanism is unclear [96,97]. JmjC-domain family is another class of demethylases that can demethylate mono-/di-/trimethylated lysines on histone tails. It requires α-ketoglutarate and Fe (II) as cofactors [96,97].

One of its members, GASC1/JMJD2C (gene amplified in squamous cell carcinoma 1), is capable of demethylating H3K9me3/me2 [98], an epigenetic mark generated by Suv39h and recognized by the effector HP1. Ectopic expression of GASC1 in cells by transient transfection not only causes a marked decrease of global H3K9me3/me2 levels but also displaces HP1 from chromatin and reduces heterochromatin formation/maintenance as measured by immunofluorescence [98]. Consistent with its potential role in antagonizing the Rb-Suv39h tumor suppressive activities, GASC1 was originally identified as a gene amplified in several cell lines derived from esophageal squamous carcinomas and is frequently overexpressed in various cancer types [98]. In agreement with a contribution of GASC1 to tumor development, inhibition of GASC1 expression caused a significant reduction of cell proliferation [98]. However, tumor cells with amplification of GASC1 do not show an overall reduction of H3K9me3 levels, and inhibition of GASC1 by short hairpin RNA (shRNA) does not lead to changes in the global levels of H3K9me3 either [98]. Although it remains possible that GASC1 might participate in control of heterochromatin formation and maintenance, these observations indicate that the GASC1 demethylase expressed at physiological levels may contribute to cell proliferation capacity by regulating the H3K9me3 levels at specific subsets of genes rather than globally. It is interesting to identify which genes are regulated and how they are targeted by GASC1, and determine whether they may be silenced during senescence and whether GASC1-overexpressing cells may have impaired their capability to initiate and maintain senescence.

PLU-1/JARID1B is a histone demethylase enzyme that has the ability to reverse the trimethyl H3K4 mark [99], and antagonizes its effect as a transcriptional activator. PLU-1 is overexpressed in human breast cancer cells, and promotes cancer cell proliferation in soft-agar colony formation and in xenograft mouse mammary tumors [99]. It directly represses potential negative regulators of cell growth, such as CAV1, HOXA5, and BRCA1, and facilitates the G1 to S transition [99]. This PLU-1-mediated transcriptional repression relies on its histone demethylase activity. Reduction of PLU-1 expression, which seems not to affect global H3K4 methylation, results in increased H3K4me3 levels at these target genes. A PLU-1-related H3K4 demethylase, RBP2/JARID1A [95,100,101], is known to physically interact with Rb (which inhibits cell growth). It remains to be determined whether PLU-1 (and RBP2) may elicit its effect on G1/S transition through Rb, although the identified PLU-1 targets are not the typical S-phase genes controlled by Rb/E2F.

10.4   HISTONE KINASE IN CANCER

10.4.1   INTRODUCTION TO HISTONE PHOSPHORYLATION

Histone phosphorylation has important functions in mitosis and gene regulation as defects in histone kinases can have severe effects on genomic stability and cell proliferation. Abnormal histone phosphorylation has been observed in many cancers such as breast cancer, prostate cancer, and colorectal cancer. However, it is unclear whether histone phosphorylation is the cause or the effect.

10.4.2   HISTONE KINASES AND CANCER

Histone phosphorylation has been observed on H1, H2B, H3, and histone variant H2AX. These modifications are originally identified as mitosis-specific, associated with chromatin condensation during cell cycle G2-M phase transition [7]. Now further study has linked them with cell proliferation and transformation in cancer cells. H1 phosphorylation is carried out by many kinases and has been a standard marker for mitotic cells. The major kinase involved is CDK2, which peaks in metaphase; and correlates with H1 phosphorylation when chromatin is most condensed. H1 phosphorylation is observed to be elevated in tumors and cancer cell culture, but it is more likely to be an effect of tumor formation than the cause [102]. During mitosis, phosphorylation of H3 on serine (S) 10, 28, and threonine (T) 3 are also elevated. Serine phosphorylation is catalyzed by Aurora family kinases (A, B, and C). The different kinases exhibit different functions within the cell, primarily due to their subcellular localization [103]. Aurora A regulates centrosome separation, mitotic spindle assembly, and G2-M transition. Aurora B mediates histone H3 modification, chromatin separation, and cytokinesis. Both Aurora A and B are tumorigenic when overexpressed, but the mechanism is unclear.

10.4.3   CONNECTIONS BETWEEN HISTONE PHOSPHORYLATION AND METHYLATION

Histone phosphorylation is connected to gene induction in conjunction with methylation. It is stipulated that phosphorylation removes repressive marks generated by histone methylation via the "meth-phos switch" [104]. This model suggests phosphorylation of specific residues on the histone tail eject effector proteins like HP-1, thus preventing downstream repressive effects during mitosis. One study has shown depletion of Aurora B kinase during M-phase is sufficient to retain HP-1 on H3K9me3. This study concludes H3S10 phosphorylation by Aurora B is necessary for transient dissociation of HP-1 from H3K9 during mitosis [105]. Another study extended this idea to include HDAC3 in complex with Aurora B kinase during mitosis to enhance the meth-phos switch [106]. In this case, histone phosphorylation is an intermediate in global histone acetylation changes and promotes heterochromatin formation during mitosis.

10.4.4   HISTONE PHOSPHORYLATION DURING DNA REPAIR

Histone phosphorylation also maintains genomic integrity by mediating DNA double-strand break (DSB) repair via homologous-recombination (HR). H2AX, a histone variant defined by a C-terminal SQE motif in yeast, is enriched soon after DSB recognition. Its chief function during HR repair is to recruit proteins that facilitate the radical chromatin structure changes that accompany

HR. It is phosphorylated by ATM/ATR kinases as they activate a number of repair complexes. γ-H2AX, the S129 phosphorylated form of H2AX, is created in the vicinity of DSB site but not within 1–2 kb of the break itself. It is thought the phosphorylated H2AX spread recruits scaffolding-molecule Cohesion to the broken ends, which is crucial for chromatin adhesion during HR. γ-H2AX then recruits other repair complexes like histone acetyltransferase NuA4, and chromatin remodelers INO80. These complexes relax the chromosome to allow binding of DSB repair factors. Upon successful recombination, γ-H2AX is removed by an unknown mechanism. It is suspected that TIP60 HAT acetylation catalyzes removal of γ-H2AX via the chromatin remodeling complex SWR1 [107,108].

10.5   HISTONE UBIQUITINATION AND TRANSCRIPTION REGULATION

H2A and H2B are targets for ubiquitination which brings opposite effects on transcription. In humans H2A K119 ubiquitination is mediated by Bmi-1 and RING1/2, core subunits of the Polycomb repression complex PRC1. The RING finger protein RING1/2 acts as an E3 ubiquitin ligase that participates in the mono-ubiquitination of H2AK119, and the binding of Bmi-1 stimulates the E3 ligase activity. This ubiquitination is critical for PRC1-mediated gene silencing activity [109,110]. Bmi-1 is a known oncogene. It is required for the proliferation and self-renewal of normal and cancer stem cells partly by repressing the expression of tumor suppressors Ink4a and Arf [111]. The Bmi-1-associated PcG pathway plays an essential role in prostate cancer metastasis. Increased Bmi-1 expression in metastatic cancer cells is associated with elevated levels of H2AK119 ubiquitination [112].

Ubiquitination of H2B on H2BK123 is required transcription in both yeast and human and shown to act in concert with H3K4 and K79 methylation, which are also associated with transcription activation [113]. In fact, ubiquitination is required during transcription elongation by facilitating histone chaperon protein FACT [114]. In short, transcription activation in yeast is maintained by a combination of ubiquitination and methylation acting on the basal transcription machinery; as mono-ubiquitination of H2BK123, H3K4 methylation, and elongation factors such as PAF are found in complex with elongating RNA Pol II [114,115].

Both ubiquitination and deubiquitination has been shown to activate transcription. There are two histone de-ubiquitinating enzymes identified in yeast, Ubp8 and Ubp10. Both enzymes antagonize H2BK123 ubiquitination. However, Ubp8 has been shown to activate transcription when acting in complex with acetylating enzymes. Ubp10 functions primarily as a transcription silencer in the heterochromatic region of budding yeast [116].

10.6   OTHER HISTONE MODIFIERS

Histones can undergo many other posttranslational modifications; however, the functional significance of these modifications is less clear. These modifications appear primarily to regulate methylation and acetylation, or to modulate their effects. Therefore, these modifiers are crucial epigenetic regulators as well, and it is interesting to decipher their relation to cancer.

10.6.1   HISTONE SUMOLATION

SUMOlation shares a similar mechanism with ubiquitination and the two may compete for the same lysines as substrate [117]. Their differences reside mainly in their protein-binding partners and may be marking specific chromatin regions for the recruitment of activating or silencing factors. The two modifications have been found on H2A, H2B, H3, and H4. SUMOlation of lysine residues is generally associated with transcriptional repression by blocking acetylation and other positive-acting histone modifications [118].

10.6.2 HISTONE POLY-ADP-RIBOSYLATION

Poly-ADP-ribosylation, mediated by polymer of ADP-ribose polymerases (PARPs), has a multitude of functions within the cell including DNA repair, modulating chromatin structure, and cell division [119]. It is reversible by poly (ADP-ribose) glycohydrolase (PARG). PARPs appear to promote global activation by relaxing chromatin structure, which indicates coactivator functions. However, they may also have a role in conferring insulator properties to insulator protein CTCF, which indicates PARPs can act as a transcriptional silencer [119]. ADP-ribosylation can also occur in the mono-form, mediated by mono (ADP-ribose) transferases (MARTs). There are also studies that show mono-ADP-ribosylation of histones H1 increase upon DNA damage [120]. The multiplicity of its activity makes the function of ribosylated histones ambiguous.

10.6.3 HISTONE BIOTINATION

Relatively little is known about histone biotination and its cellular function. Biotination occurs on multiple lysine residues on H2A, H3, and H4 catalyzed by biotinidase and holocarboxylase synthetase. Biotination of H4 is concentrated on pericentromeric regions and it is associated with gene silencing, mitotic condensation, and DNA damage response. However, they are general observations and the mechanism is unknown. There is no de-biotinidase identified and reversal of biotination is dependent on exogenous biotin [121,122].

10.6.4 HISTONE CITRULLINATION

Citrullination acts mostly as a corepressor in cooperation with nuclear steroid receptors and other transcriptional coregulators. It is catalyzed by deaminating enzymes, like PAD4, which hydrolyze guanidinium side chains and produce citrulline. It irreversibly demethylates arginine residues on histones to produce citrulline. Therefore, it is considered to be an inhibitor of histone methylation rather than a true demethylase [123].

10.6.5 HISTONE PROLINE ISOMERIZATION

Proline isomerization is a novel non-covalent modification that regulates transcription [124]. This modification regulates lysine methylation H3K36 via proline P38. Fpr4 is a proline isomerase belonging to the FK506 binding-protein family (FKBP) in yeast. Fpr4 controls the conformational state of P38. Fpr4 is found to bind the N-terminal tail of histone H3 and H4, and catalyses the isomerization of H3P30 and H3P38 in vitro. The isomerization of P38 inhibits binding of the methyltransferase SET2 and prevents methylation of H3K36. This study illustrates the crosstalk between lysine methylation and proline isomerization [124].

10.7 CONCLUDING REMARKS

The contribution of epigenetic regulation to cancer development is receiving increased attention, as it constitutes another set of "hits" in the generation of aberrant gene expression. Altered expression, activity, or targeting of histone modifying enzymes is critical in tumor initiation [125] and progression. As epigenetic modifications can potentially be reversed, identification of epigenetic events involved in malignant transformation has far-reaching implications for the development of cancer biomarkers and new therapeutics. Drugs that inhibit DNA methylation have been applied in cancer clinical treatment. Recently, suberoylanilide hydroxamic acid (SAHA; vorinostat (Zolinza)) has become the first FDA-approved HDAC inhibitor for the treatment of cutaneous T-cell lymphoma [126]. SAHA reacts with and blocks the catalytic site of HDACs, and has shown significant anticancer activity against both hematologic and solid tumors in clinical trials [126]. As we continue to advance our understanding of epigenetic alterations and their contributions to cancer, more inhibitors of HDACs and other histone modifiers, in particular the newly discovered histone demethylases (given the prevalence of altered histone methylation profiles in cancer), are expected to be developed as important cancer therapeutic drugs.

REFERENCES

1. Strahl, B.D. and C.D. Allis, The language of covalent histone modifications. *Nature*, 2000. 403(6765): 41–45.
2. Turner, B.M., Histone acetylation and an epigenetic code. *Bioessays*, 2000. 22(9): 836–845.
3. Jenuwein, T. and C.D. Allis, Translating the histone code. *Science*, 2001. 293(5532): 1074–1080.
4. Biel, M., V. Wascholowski, and A. Giannis, Epigenetics—an epicenter of gene regulation: Histones and histone-modifying enzymes. *Angew Chem Int Ed Engl*, 2005. 44(21): 3186–3216.
5. Santos-Rosa, H. and C. Caldas, Chromatin modifier enzymes, the histone code and cancer. *Eur J Cancer*, 2005. 41(16): 2381–2402.
6. Iyer, N.G., H. Ozdag, and C. Caldas, p300/CBP and cancer. *Oncogene*, 2004. 23(24): 4225–4231.
7. Zhang, K. and S.Y. Dent, Histone modifying enzymes and cancer: Going beyond histones. *J Cell Biochem*, 2005. 96(6): 1137–1148.
8. Senese, S., et al., Role for histone deacetylase 1 in human tumor cell proliferation. *Mol Cell Biol*, 2007. 27 (13): 4784–4795.
9. Gregoretti, I.V., Y.M. Lee, and H.V. Goodson, Molecular evolution of the histone deacetylase family: Functional implications of phylogenetic analysis. *J Mol Biol*, 2004. 338(1): 17–31.
10. Suzuki, T. and N. Miyata, Non-hydroxamate histone deacetylase inhibitors. *Curr Med Chem*, 2005. 12(24): 2867–2880.
11. Michan, S. and D. Sinclair, Sirtuins in mammals: Insights into their biological function. *Biochem J*, 2007. 404(1): 1–13.
12. Minucci, S. and P.G. Pelicci, Histone deacetylase inhibitors and the promise of epigenetic (and more) treatments for cancer. *Nat Rev Cancer*, 2006. 6(1): 38–51.
13. Seligson, D.B., et al., Global histone modification patterns predict risk of prostate cancer recurrence. *Nature*, 2005. 435(7046): 1262–1266.
14. Bolden, J.E., M.J. Peart, and R.W. Johnstone, Anticancer activities of histone deacetylase inhibitors. *Nat Rev Drug Discov*, 2006. 5(9): 769–784.
15. Conley, B.A., J.J. Wright, and S. Kummar, Targeting epigenetic abnormalities with histone deacetylase inhibitors. *Cancer*, 2006. 107(4): 832–840.
16. Kumar, R., Another tie that binds the MTA family to breast cancer. *Cell*, 2003. 113(2): 142–143.
17. Toh, Y., et al., Expression of the metastasis-associated MTA1 protein and its relationship to deacetylation of the histone H4 in esophageal squamous cell carcinomas. *Int J Cancer*, 2004. 110(3): 362–367.
18. Ford, J., M. Jiang, and J. Milner, Cancer-specific functions of SIRT1 enable human epithelial cancer cell growth and survival. *Cancer Res*, 2005. 65(22): 10457–10463.
19. Chen, W.Y., et al., Tumor suppressor HIC1 directly regulates SIRT1 to modulate p53-dependent DNA-damage responses. *Cell*, 2005. 123(3): 437–448.
20. Cohen, H.Y., et al., Calorie restriction promotes mammalian cell survival by inducing the SIRT1 deacetylase. *Science*, 2004. 305(5682): 390–392.
21. Longo, V.D. and B.K. Kennedy, Sirtuins in aging and age-related disease. *Cell*, 2006. 126(2): 257–268.
22. Di Micco, R., et al., Oncogene-induced senescence is a DNA damage response triggered by DNA hyper-replication. *Nature*, 2006. 444(7119): 638–642.
23. Saunders, L.R. and E. Verdin, Sirtuins: Critical regulators at the crossroads between cancer and aging. *Oncogene*, 2007. 26(37): 5489–5504.
24. Mostoslavsky, R., et al., Genomic instability and aging-like phenotype in the absence of mammalian SIRT6. *Cell*, 2006. 124(2): 315–329.
25. Ford, E., et al., Mammalian Sir2 homolog SIRT7 is an activator of RNA polymerase I transcription. *Genes Dev*, 2006. 20(9): 1075–1080.
26. Ozdag, H., et al., Mutation analysis of CBP and PCAF reveals rare inactivating mutations in cancer cell lines but not in primary tumours. *Br J Cancer*, 2002. 87(10): 1162–1165.
27. Koshiishi, N., et al., p300 gene alterations in intestinal and diffuse types of gastric carcinoma. *Gastric Cancer*, 2004. 7(2): 85–90.

28. Gayther, S.A., et al., Mutations truncating the EP300 acetylase in human cancers. *Nat Genet*, 2000. 24(3): 300–303.

29. Ionov, Y., S. Matsui, and J.K. Cowell, A role for p300/CREB binding protein genes in promoting cancer progression in colon cancer cell lines with microsatellite instability. *Proc Natl Acad Sci U S A*, 2004. 101(5): 1273–1278.

30. Muraoka, M., et al., p300 gene alterations in colorectal and gastric carcinomas. *Oncogene*, 1996. 12(7): 1565–1569.

31. Kitabayashi, I., et al., Fusion of MOZ and p300 histone acetyltransferases in acute monocytic leukemia with a t(8;22)(p11;q13) chromosome translocation. *Leukemia*, 2001. 15(1): 89–94.

32. Chaffanet, M., et al., MOZ is fused to p300 in an acute monocytic leukemia with t(8;22). *Genes Chromosomes Cancer*, 2000. 28(2): 138–144.

33. Ida, K., et al., Detection of chimeric mRNAs by reverse transcriptase-polymerase chain reaction for diagnosis and monitoring of acute leukemias with 11q23 abnormalities. *Med Pediatr Oncol*, 1997. 28(5): 325–332.

34. Ohshima, T., T. Suganuma, and M. Ikeda, A novel mutation lacking the bromodomain of the transcriptional coactivator p300 in the SiHa cervical carcinoma cell line. *Biochem Biophys Res Commun*, 2001. 281(2): 569–575.

35. Roelfsema, J.H., et al., Genetic heterogeneity in Rubinstein-Taybi syndrome: Mutations in both the CBP and EP300 genes cause disease. *Am J Hum Genet*, 2005. 76(4): 572–580.

36. Sanders, S.L., et al., Methylation of histone H4 lysine 20 controls recruitment of Crb2 to sites of DNA damage. *Cell*, 2004. 119(5): 603–614.

37. Kishimoto, M., et al., Mutations and deletions of the CBP gene in human lung cancer. *Clin Cancer Res*, 2005. 11(2 Pt 1): 512–519.

38. Murati, A., et al., Variant MYST4-CBP gene fusion in a t(10;16) acute myeloid leukaemia. *Br J Haematol*, 2004. 125(5): 601–604.

39. Panagopoulos, I., et al., Genomic characterization of MOZ/CBP and CBP/MOZ chimeras in acute myeloid leukemia suggests the involvement of a damage-repair mechanism in the origin of the t(8;16) (p11;p13). *Genes Chromosomes Cancer*, 2003. 36(1): 90–98.

40. Schmidt, H.H., et al., RT-PCR and FISH analysis of acute myeloid leukemia with t(8;16)(p11;p13) and chimeric MOZ and CBP transcripts: Breakpoint cluster region and clinical implications. *Leukemia*, 2004. 18(6): 1115–1121.

41. Rozman, M., et al., Type I MOZ/CBP (MYST3/CREBBP) is the most common chimeric transcript in acute myeloid leukemia with t(8;16)(p11;p13) translocation. *Genes Chromosomes Cancer*, 2004. 40(2): 140–145.

42. Vizmanos, J.L., et al., t(10;16)(q22;p13) and MORF-CREBBP fusion is a recurrent event in acute myeloid leukemia. *Genes Chromosomes Cancer*, 2003. 36(4): 402–405.

43. So, C.K., et al., Loss of heterozygosity and internal tandem duplication mutations of the CBP gene are frequent events in human esophageal squamous cell carcinoma. *Clin Cancer Res*, 2004. 10(1 Pt 1): 19–27.

44. Sugita, K., et al., MLL-CBP fusion transcript in a therapy-related acute myeloid leukemia with the t(11;16)(q23;p13) which developed in an acute lymphoblastic leukemia patient with Fanconi anemia. *Genes Chromosomes Cancer*, 2000. 27(3): 264–269.

45. Satake, N., et al., Novel MLL-CBP fusion transcript in therapy-related chronic myelomonocytic leukemia with a t(11;16)(q23;p13) chromosome translocation. *Genes Chromosomes Cancer*, 1997. 20(1): 60–63.

46. Sobulo, O.M., et al., MLL is fused to CBP, a histone acetyltransferase, in therapy-related acute myeloid leukemia with a t(11;16)(q23;p13.3). *Proc Natl Acad Sci U S A*, 1997. 94(16): 8732–8737.

47. Coupry, I., et al., Analysis of CBP (CREBBP) gene deletions in Rubinstein-Taybi syndrome patients using real-time quantitative PCR. *Hum Mutat*, 2004. 23(3): 278–284.

48. Kalkhoven, E., et al., Loss of CBP acetyltransferase activity by PHD finger mutations in Rubinstein-Taybi syndrome. *Hum Mol Genet*, 2003. 12(4): 441–450.

49. Deguchi, K., et al., MOZ-TIF2-induced acute myeloid leukemia requires the MOZ nucleosome binding motif and TIF2-mediated recruitment of CBP. *Cancer Cell*, 2003. 3(3): 259–271.

50. Imamura, T., et al., Rearrangement of the MOZ gene in pediatric therapy-related myelodysplastic syndrome with a novel chromosomal translocation t(2;8)(p23;p11). *Genes Chromosomes Cancer*, 2003. 36(4): 413–419.

51. Moore, S.D., et al., Uterine leiomyomata with t(10;17) disrupt the histone acetyltransferase MORF. *Cancer Res*, 2004. 64(16): 5570–5577.

52. Wachtel, M., et al., Gene expression signatures identify rhabdomyosarcoma subtypes and detect a novel t(2;2)(q35;p23) translocation fusing PAX3 to NCOA1. *Cancer Res*, 2004. 64(16): 5539–5545.

53. Marcucci, G., K. Mrozek, and C.D. Bloomfield, Molecular heterogeneity and prognostic biomarkers in adults with acute myeloid leukemia and normal cytogenetics. *Curr Opin Hematol*, 2005. 12(1): 68–75.

54. Choi, J.H., et al., Expression profile of histone deacetylase 1 in gastric cancer tissues. *Jpn J Cancer Res*, 2001. 92(12): 1300–1304.

55. Volkel, P. and P.O. Angrand, The control of histone lysine methylation in epigenetic regulation. *Biochimie*, 2007. 89(1): 1–20.

56. Eguchi, M., M. Eguchi-Ishimae, and M. Greaves, Molecular pathogenesis of MLL-associated leukemias. *Int J Hematol*, 2005. 82(1): 9–20.

57. Peters, A.H., et al., Loss of the Suv39h histone methyltransferases impairs mammalian heterochromatin and genome stability. *Cell*, 2001. 107(3): 323–337.

58. Narita, M., et al., Rb-mediated heterochromatin formation and silencing of E2F target genes during cellular senescence. *Cell*, 2003. 113(6): 703–716.

59. Braig, M., et al., Oncogene-induced senescence as an initial barrier in lymphoma development. *Nature*, 2005. 436(7051): 660–665.

60. Chen, Z., et al., Crucial role of p53-dependent cellular senescence in suppression of Pten-deficient tumorigenesis. *Nature*, 2005. 436(7051): 725–730.

61. Michaloglou, C., et al., BRAFE600-associated senescence-like cell cycle arrest of human naevi. *Nature*, 2005. 436(7051): 720–724.

62. Bedford, M.T. and S. Richard, Arginine methylation an emerging regulator of protein function. *Mol Cell*, 2005. 18(3): 263–272.

63. Fraga, M.F., et al., Loss of acetylation at Lys16 and trimethylation at Lys20 of histone H4 is a common hallmark of human cancer. *Nat Genet*, 2005. 37(4): 391–400.

64. Gonzalo, S., et al., Role of the RB1 family in stabilizing histone methylation at constitutive heterochromatin. *Nat Cell Biol*, 2005. 7(4): 420–428.

65. Tsuge, M., et al., A variable number of tandem repeats polymorphism in an E2F-1 binding element in the 5′ flanking region of SMYD3 is a risk factor for human cancers. *Nat Genet*, 2005. 37(10): 1104–1107.

66. Schlesinger, Y., et al., Polycomb-mediated methylation on Lys27 of histone H3 pre-marks genes for de novo methylation in cancer. *Nat Genet*, 2007. 39(2): 232–236.

67. Kotake, Y., et al., pRB family proteins are required for H3K27 trimethylation and Polycomb repression complexes binding to and silencing p16INK4alpha tumor suppressor gene. *Genes Dev*, 2007. 21(1): 49–54.

68. Sambani, C., et al., Partial duplication of the MLL oncogene in patients with aggressive acute myeloid leukemia. *Haematologica*, 2004. 89(4): 403–407.

69. Vey, N., et al., Identification of new classes among acute myelogenous leukaemias with normal karyotype using gene expression profiling. *Oncogene*, 2004. 23(58): 9381–9391.

70. Whitman, S.P., et al., The partial nontandem duplication of the MLL (ALL1) gene is a novel rearrangement that generates three distinct fusion transcripts in B-cell acute lymphoblastic leukemia. *Cancer Res*, 2001. 61(1): 59–63.

71. Espinet, B., et al., MLL intrachromosomal amplification in a pre-B acute lymphoblastic leukemia. *Haematologica*, 2003. 88(2): EIM03.

72. Calabrese, G., et al., Chromosome 11 rearrangements and specific MLL amplification revealed by spectral karyotyping in a patient with refractory anaemia with excess of blasts (RAEB). *Br J Haematol*, 2003. 122(5): 760–763.

73. Poppe, B., et al., Expression analyses identify MLL as a prominent target of 11q23 amplification and support an etiologic role for MLL gain of function in myeloid malignancies. *Blood*, 2004. 103(1): 229–235.

74. Zatkova, A., et al., Distinct sequences on 11q13.5 and 11q23-24 are frequently coamplified with MLL in complexly organized 11q amplicons in AML/MDS patients. *Genes Chromosomes Cancer*, 2004. 39(4): 263–276.

75. Bernasconi, P., et al., Translocation (8;16) in a patient with acute myelomonocytic leukemia, occurring after treatment with fludarabine for a low-grade non-Hodgkin's lymphoma. *Haematologica*, 2000. 85(10): 1087–1091.

76. Barouk-Simonet, E., et al., Role of multiplex FISH in identifying chromosome involvement in myelo-dysplastic syndromes and acute myeloid leukemias with complex karyotypes: A report on 28 cases. *Cancer Genet Cytogenet*, 2005. 157(2): 118–126.

77. Martinez-Ramirez, A., et al., Cytogenetic profile of myelodysplastic syndromes with complex karyotypes: An analysis using spectral karyotyping. *Cancer Genet Cytogenet*, 2004. 153(1): 39–47.

78. Morerio, C., et al., MLL-MLLT10 fusion in acute monoblastic leukemia: Variant complex rearrangements and 11q proximal breakpoint heterogeneity. *Cancer Genet Cytogenet*, 2004. 152(2): 108–112.

79. Huntsman, D.G., et al., MLL2, the second human homolog of the *Drosophila trithorax* gene, maps to 19q13.1 and is amplified in solid tumor cell lines. *Oncogene*, 1999. 18(56): 7975–7984.

80. Dohner, K., et al., Molecular cytogenetic characterization of a critical region in bands 7q35-q36 commonly deleted in malignant myeloid disorders. *Blood*, 1998. 92(11): 4031–4035.

81. Hamamoto, R., et al., Enhanced SMYD3 expression is essential for the growth of breast cancer cells. *Cancer Sci*, 2006. 97(2): 113–118.

82. Tam, K.F., et al., Methylation profile in benign, borderline and malignant ovarian tumors. *J Cancer Res Clin Oncol*, 2007. 133(5): 331–341.

83. Pastural, E., et al., RIZ1 repression is associated with insulin-like growth factor-1 signaling activation in chronic myeloid leukemia cell lines. *Oncogene*, 2007. 26(11): 1586–1594.

84. Hasegawa, Y., et al., DNA methylation of the RIZ1 gene is associated with nuclear accumulation of p53 in prostate cancer. *Cancer Sci*, 2007. 98(1): 32–36.

85. Varambally, S., et al., The polycomb group protein EZH2 is involved in progression of prostate cancer. *Nature*, 2002. 419(6907): 624–629.

86. Bracken, A.P., et al., EZH2 is downstream of the pRB-E2F pathway, essential for proliferation and amplified in cancer. *EMBO J*, 2003. 22(20): 5323–5335.

87. Cheng, D., et al., Small molecule regulators of protein arginine methyltransferases. *J Biol Chem*, 2004. 279(23): 23892–23899.

88. Klinge, C.M., et al., Estrogen response element-dependent regulation of transcriptional activation of estrogen receptors alpha and beta by coactivators and corepressors. *J Mol Endocrinol*, 2004. 33(2): 387–410.

89. Hong, H., et al., Aberrant expression of CARM1, a transcriptional coactivator of androgen receptor, in the development of prostate carcinoma and androgen-independent status. *Cancer*, 2004. 101(1): 83–89.

90. Pal, S., et al., Human SWI/SNF-associated PRMT5 methylates histone H3 arginine 8 and negatively regulates expression of ST7 and NM23 tumor suppressor genes. *Mol Cell Biol*, 2004. 24(21): 9630–9645.

91. Liang, J.J., et al., The expression and function of androgen receptor coactivator p44 and protein arginine methyltransferase 5 in the developing testis and testicular tumors. *J Urol*, 2007. 177(5): 1918–1922.

92. Krause, C.D., et al., Protein arginine methyltransferases: Evolution and assessment of their pharmacological and therapeutic potential. *Pharmacol Ther*, 2007. 113(1): 50–87.

93. Majumder, S., et al., Involvement of arginine methyltransferase CARM1 in androgen receptor function and prostate cancer cell viability. *Prostate*, 2006. 66(12): 1292–1301.

94. Pal, S., et al., Low levels of miR-92b/96 induce PRMT5 translation and H3R8/H4R3 methylation in mantle cell lymphoma. *EMBO J*, 2007. 26(15): 3558–3569.

95. Klose, R.J., et al., The retinoblastoma binding protein RBP2 is an H3K4 demethylase. *Cell*, 2007. 128(5): 889–900.

96. Shi, Y. and J.R. Whetstine, Dynamic regulation of histone lysine methylation by demethylases. *Mol Cell*, 2007. 25(1): 1–14.

97. Klose, R.J. and Y. Zhang, Regulation of histone methylation by demethylimination and demethylation. *Nat Rev Mol Cell Biol*, 2007. 8(4): 307–318.

98. Cloos, P.A., et al., The putative oncogene GASC1 demethylates tri- and dimethylated lysine 9 on histone H3. *Nature*, 2006. 442(7100): 307–311.

99. Yamane, K., et al., PLU-1 is an H3K4 demethylase involved in transcriptional repression and breast cancer cell proliferation. *Mol Cell*, 2007. 25(6): 801–812.

100. Christensen, J., et al., RBP2 belongs to a family of demethylases, specific for tri-and dimethylated lysine 4 on histone 3. *Cell*, 2007. 128(6): 1063–1076.

101. Benevolenskaya, E.V., et al., Binding of pRB to the PHD protein RBP2 promotes cellular differentiation. *Mol Cell*, 2005. 18(6): 623–635.

102. Deshpande, A., P. Sicinski, and P.W. Hinds, Cyclins and cdks in development and cancer: A perspective. *Oncogene*, 2005. 24(17): 2909–2915.

103. Fu, J., et al., Roles of Aurora kinases in mitosis and tumorigenesis. *Mol Cancer Res*, 2007. 5(1): 1–10.

104. Hirota, T., et al., Histone H3 serine 10 phosphorylation by Aurora B causes HP1 dissociation from heterochromatin. *Nature*, 2005. 438(7071): 1176–1180.

105. Fischle, W., et al., Regulation of HP1-chromatin binding by histone H3 methylation and phosphorylation. *Nature*, 2005. 438(7071): 1116–1122.

106. Li, Y., et al., A novel histone deacetylase pathway regulates mitosis by modulating Aurora B kinase activity. *Genes Dev*, 2006. 20(18): 2566–2579.

107. Thiriet, C. and J.J. Hayes, Chromatin in need of a fix: Phosphorylation of H2AX connects chromatin to DNA repair. *Mol Cell*, 2005. 18(6): 617–622.

108. Fillingham, J., M.C. Keogh, and N.J. Krogan, GammaH2AX and its role in DNA double-strand break repair. *Biochem Cell Biol*, 2006. 84(4): 568–577.

109. Wang, H., et al., Role of histone H2A ubiquitination in Polycomb silencing. *Nature*, 2004. 431(7010): 873–878.

110. Cao, R., Y. Tsukada, and Y. Zhang, Role of Bmi-1 and Ring1A in H2A ubiquitylation and Hox gene silencing. *Mol Cell*, 2005. 20(6): 845–854.

111. Pardal, R., et al., Stem cell self-renewal and cancer cell proliferation are regulated by common networks that balance the activation of proto-oncogenes and tumor suppressors. *Cold Spring Harb Symp Quant Biol*, 2005. 70: 177–185.

112. Berezovska, O.P., et al., Essential role for activation of the Polycomb group (PcG) protein chromatin silencing pathway in metastatic prostate cancer. *Cell Cycle*, 2006. 5(16): 1886–1901.

113. Ng, H.H., et al., Ubiquitination of histone H2B by Rad6 is required for efficient Dot1-mediated methylation of histone H3 lysine 79. *J Biol Chem*, 2002. 277(38): 34655–34657.

114. Pavri, R., et al., Histone H2B monoubiquitination functions cooperatively with FACT to regulate elongation by RNA polymerase II. *Cell*, 2006. 125(4): 703–717.

115. Shilatifard, A., Chromatin modifications by methylation and ubiquitination: Implications in the regulation of gene expression. *Annu Rev Biochem*, 2006. 75: 243–269.

116. Kouzarides, T., Chromatin modifications and their function. *Cell*, 2007. 128(4): 693–705.

117. Gill, G., SUMO and ubiquitin in the nucleus: Different functions, similar mechanisms? *Genes Dev*, 2004. 18(17): 2046–2059.

118. Nathan, D., et al., Histone sumoylation is a negative regulator *in Saccharomyces cerevisiae* and shows dynamic interplay with positive-acting histone modifications. *Genes Dev*, 2006. 20(8): 966–976.

119. Schreiber, V., et al., Poly(ADP-ribose): Novel functions for an old molecule. *Nat Rev Mol Cell Biol*, 2006. 7(7): 517–528.

120. Hassa, P.O., et al., Nuclear ADP-ribosylation reactions in mammalian cells: Where are we today and where are we going? *Microbiol Mol Biol Rev*, 2006. 70(3): 789–829.

121. Kothapalli, N., et al., Biological functions of biotinylated histones. *J Nutr Biochem*, 2005. 16(7): 446–448.

122. Hassan, Y.I. and J. Zempleni, Epigenetic regulation of chromatin structure and gene function by biotin. *J Nutr*, 2006. 136(7): 1763–1765.

123. Thompson, P.R. and W. Fast, Histone citrullination by protein arginine deiminase: Is arginine methylation a green light or a roadblock? *ACS Chem Biol*, 2006. 1(7): 433–441.

124. Nelson, C.J., H. Santos-Rosa, and T. Kouzarides, Proline isomerization of histone H3 regulates lysine methylation and gene expression. *Cell*, 2006. 126(5): 905–916.

125. Feinberg, A.P., R. Ohlsson, S. Henikoff, The epigenetic progenitor origin of human cancer. *Nat Rev Genet*, 2006. 7(1): 21–33.

126. Marks, P.A., R. Breslow, Dimethyl sulfoxide to vorinostat: Development of this histone deacetylase inhibitor as an anticancer drug. *Nat Biotechnol*, 2007. 25(1): 84–90.

11 Dietary and Environmental Influences on Histone Modifications in Cancer

*Sabita N. Saldanha, Vijayalakshmi Nandakumar,
Ada Elgavish, and Trygve O. Tollefsbol*

CONTENTS

11.1 INTRODUCTION

Many factors intricately regulate the expression of genes responsible for development and growth. Aberrations in their expression can result in growth and developmental abnormalities, including cancer. Initially, the understanding was that mutations in the genome, i.e., variations in the DNA sequence, were the primary reason [1–3]. However, recent advances have proven, otherwise. Epigenetic mechanisms modifying gene expression without affecting the DNA sequence are heritable and can perpetuate altered phenotypes as well.

11.1.1 CHROMATIN STRUCTURE: THE BASIS OF EPIGENETIC MECHANISMS

Chromatin is the basic organizational form of DNA in the eukaryotic nucleus. The repeat unit of chromatin is the core nucleosome in which 145 base pairs of DNA are wrapped around a histone octamer consisting of two molecules each of the core histones H2A, H2B, H3, and H4. Core histones are susceptible to enzymatic modifications of their tails. The tails that protrude out of the complex are primarily lysine, arginine, and serine residues which can be acetylated, methylated, ubiquitinated, biotinylated, phosphorylated, and sumoylated to influence many different functions pertaining to gene expression and transcriptome stability [4,5].

Nucleosomal arrays along the DNA are believed to fold into a 30 nm fiber upon incorporation of the linker histone H1 [5]. Given its folded structure, chromatin does not generally allow extraneous access. Ionic interactions between the charged DNA and histones have to be overcome to unfold the chromatin structure and allow access of the transcription machinery to the DNA and, finally, gene expression. Moreover, the original chromatin structure has to be reinstated after the tasks have been completed. These chromatin-expression states can be maintained through multiple rounds of cell division that occur during development [6–8]. Given their nature, these chromatin states are prone to intrinsic, intracellular, as well as external cues such as diet, environmental carcinogens and pollutants, resulting in cells with identical DNA sequences but phenotypic differences between them.

11.1.2 TYPES OF EPIGENETIC MODIFICATIONS

The most widely investigated and understood epigenetic changes that contribute to altered gene expression are genomic methylation of CpG islands, histone modifications, and RNA-associated silencing [5–9]. These modifications seem to be interpreted by proteins that recognize specific modifications and facilitate the appropriate downstream effects. Such modifications which influence normal development, growth, and aging can, under certain conditions, initiate diseased states.

Epigenetic imprinting, established during gametogenesis, is faithfully passed on from one generation to the next [10]. A wide variety of dietary components including grains, legumes, fruits, vegetables, tea, and wine which consist mostly of isoflavones possess anticarcinogenic activities [11]. Although the antioxidant and antiproliferative activities of these compounds have been attributed to the action of specific phytochemicals, evidence suggests that the optimal anticarcinogenic effect is obtained by consuming the complex food containing them rather than the isolated putative active ingredients. Section 11.3 discusses epigenetic mechanisms and the influence of diet on each of them.

11.2 GENOMIC METHYLATION OF CPG ISLANDS AND THE DIET

The activity of DNA methyltransferase 1 (DNMT1), an enzyme which methylates CpG residues, is affected by the level of S-adenosylmethionine (SAM). SAM function requires the availability of the methyl moiety, the absence or presence of which can affect DNMT1 activity. The source of the methyl moiety is in methyl-rich diets. Thus, the methyl-content of the diet may affect the activity of DNMT1, methylation of CpG islands and, indirectly, the epigenomic code.

The reversible histone-tail modifications discussed in Section 2.1 can result in higher or lower affinity interactions between DNA and core histones, inducing silenced or activated states of gene expression, respectively. Moreover, they can also affect the modification of other histones to co-mediate effects on gene expression (Table 11.1) [5]. Bioactive molecules and nutrients derived from dietary sources such as dairy products, soyabeans, vegetables, and green tea [10–13] are known to mediate such effects (Figure 11.1).

TABLE 11.1

Histone Tail Modifications and the Corresponding Effects on Chromatin Complex

Histone Modified	Residue Modified	Position Modified	Effect Mediated	Effects on Histones/DNA–Histone Complex	References
H3	Lysine	4	Mono-, di-, and trimethylation	Trimethylation mediates transcriptional expression	[5,99]
H3	Lysine	4	Methylation	Euchromatin; Inhibits methylation of H3 K-9	[5,158]
H3	Lysine	79	Methylation	Euchromatin	[5,158]
H3	Lysine	9	Mono-, di-, and trimethylation	Trimethylation mediates transcriptional silencing	[5,99,158]
H3	Lysine	9	Methylation	Heterochromatin; Inhibits methylation of H3 K4	[5,99]
H3	Lysines	27, 36	Mono-, di-, and trimethylation	K27 methylation involved in X chromosome inactivation. K36 methylation is also target for acetylation and functions as a chromatin switch in the regulation of gene expression	[5,159,160]
H4	Lysine	20	Mono-, di-, and trimethylation	Associated with heterochromatin domains	[5]
H3	Arginine	8, 17	Methylation	Possibly involved in gene activation. Further studies required to clarify the role. H3 R8 methylation may be blocked by H3K9 acetylation	[5,126,161,162]
H4	Arginine	3	Methylation	Role not clear	[5]
H3	Serine	10	Phosphorylation	Inhibits methylation of H3-K9	[5]
H3	Lysines	9, 14	Acetylation	Associated with euchromatin; Activates phosphorylation of H3-S10	[5,158]
H3	Lysines	5, 10, 20	Biotinylation	May be involved in DNA repair, cell cycle proliferation	[99]
H2A	Lysines	10, 15, 125, 127, 129	Biotinylation	Role not clear	[99]
H4	Lysines	8, 12	Biotinylation	Involved in heterochromatin structures and gene silencing	[99]

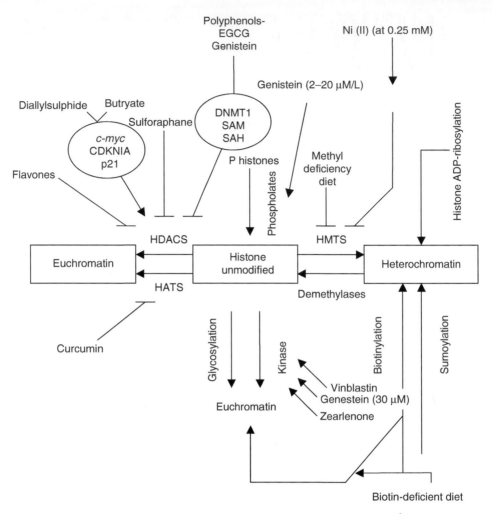

FIGURE 11.1 Dietary components and their influence on histone modification. Acetylation, phosphoryl-ation, and glycosylation of histones are mostly associated with euchromatin formation whereas methylation, sumoylation, ADP-ribosylation, and biotinylation of histones are mostly associated with heterochromatin formation. Diallyl sulfide and butyrate act as HDAC inhibitors in part by decreasing *c-myc* levels and increasing p21 levels. Flavones and sulforanes are also HDAC inhibitors, while curcumin inhibits HMTs. Polyphenols such as EGCG and genistein inhibit HDACs by downregulating *DNMT1* associated with increased levels of *S*-adenosyl methionine. Methyl-deficient diets result in hypomethylation by inhibiting HMTs. Nickel at very low concentrations causes mono-ubiquitination of histones which in turn inhibits HMT. Vinblastine, zearalenone, and genistein (at >30 μM/L) increase kinase activity and promote histone phosphorylation while genistein (at 2–20 μM/L) causes hypophosphorylation of phosphorylated histones (p histones) associated with angiogenesis.

11.3 HISTONE MODIFICATIONS

11.3.1 HISTONE ACETYLATION/DEACETYLATION AND THE EFFECT OF DIET

Acetylation of the ε amino group of the lysine residues of histones neutralizes their positive charge and reduces the strength of the interaction between DNA and the acetylated histones thereby leading to chromatin unfolding and euchromatin formation. As a result, histone acetylation usually causes transcriptional activation [5]. Histone acetylation is regulated by the action of two antagonistic enzymes: histone acetyltransferases (HATs) and histone deacetylases (HDACs). Interestingly, inhibitors of these enzymatic activities interfere with the cell cycle, at least in part by triggering changes in the acetylation/deacetylation balance of histones. This induces growth arrest, differentiation, and apoptosis [14].

Several dietary ingredients known to possess anticarcinogenic activities [12,15] have the ability to inhibit HDAC and HAT enzymes. For example, butyrate (Figure 11.1), generated in the colon by the fermentation of dietary fiber, is a histone deacetylase inhibitor. Butyrate promotes acetylation of histones, leading to expression of genes involved in cellular differentiation and apoptosis of cancer cells [16]. Further supporting the link between histone acetylation and the activation of gene expression, butyrate increases the expression of the *CDKNIA* gene in human colon adenocarcinoma (Colo-320) and human colon cancer (SW116) cell lines, inducing expression of the p21 protein. Interestingly, the latter is due to the hyperacetylation of H3 and H4 histones associated with the promoter of the p21 gene [14].

Flavones (Figure 11.1) are potent cell-specific triggers of apoptosis in a variety of cells [17–45]. Flavopiridol, a flavonoid derived from an indigenous plant from India, induces apoptosis in human leukemia cells by the disruption of cell cycle progression. The result is more pronounced when flavopiridol is used in combination with sodium butyrate as demonstrated in U937 leukemia cells [46]. In combination, these compounds inhibit leukemic cell differentiation and promote mitochondrial damage and cell death by the induction of multiple perturbations in cell cycle and apoptosis regulatory proteins, further supporting the possibility that butyrate effects on histone acetylation may be necessary for an optimal anticarcinogenic effect [47].

The Coxsackie and adenovirus receptor (CAR) is downregulated in several types of cancer [48–55]. Phytoestrogens in combination with FK228, a depsipeptide HDAC, increase the expression of CAR. Pong et al. [54] have reported the potency of phytoestrogen in combination with FK228 to be: genistein > biochanin A > ipriflavone > daidzein. Interestingly, this order correlates with the degree of acetylated H4 levels associated with the CAR promoter, supporting the possibility that the combined effect of phytoestrogens and FK228 on the expression of CAR is regulated by histone acetylation [47].

Diallyl disulfide (DADS) (Figure 11.1), present in garlic and other allium vegetables, has been shown to induce G_2/M cell cycle arrest in HT-29 and Caco-2 human colon cancer cell lines and this effect also seems to be mediated by the state of histone acetylation. Thus, treating these cells with 200 μM DADS increases *CDKNIA* mRNA expression and p21 protein levels, accompanied by an increase in H4 or H3 acetylation within the *CDKNIA* gene promoter region. Increase in H4 or H3 acetylation patterns are also observed in vitro with other organosulfur compounds found in garlic such as *S*-allyl-mercaptocysteine [56–60].

Sulforaphane (Figure 11.1), a compound found in cruciferous vegetables, acts in vitro as an HDAC inhibitor in the range of 3–15 μM, producing an increase in acetylation of histones in human embryonic kidney 293 cells, HCT116 human colorectal cancer cells, and prostate epithelial cells lines (e.g., BPH-1, LnCaP, and PC-3). Cells treated with histone deacetylase inhibitors have been shown to have increased acetylation of histones that specifically regulate genes involved in differentiation and apoptotic pathways [61–65]. HDAC inhibitors work by inactivating the deacetylases thereby changing the acetylation/deacetylation balance and increasing the expression of genes that they control.

Although the molecular pathways that are affected by HDACs are poorly understood, in most cancer phenotypes the use of HDAC inhibitors are associated with the induction of cell differentiation, apoptosis, and death of the tumor cells [66]. For example, induction of G2/M cell cycle arrest and apoptosis by HDAC inhibitors, e.g., sulforaphane, correlates with elevated expression of p21, a protein known to be involved in cell cycle arrest and apoptosis [16,61,63,64]. Probably the acetylation levels of histones, in this case histone H4 associated with the p21 promoter, increases [63,64]. Furthermore, increased acetylated levels of histones H3 or H4 were found in intestinal polyps of Apc^{min} mice treated with sulforaphane, which also displayed increased p21 expression and decreased multiplicity of the polyps when compared to the Apc[min] mice fed with a control diet without sulforaphane [58,63].

Curcumin (diferuloylmethane) (Figure 11.1), an active ingredient derived from the rhizome of the plant *Curcuma longa*, has also been shown to have anticancer activity both in vitro and in vivo. At 20 μM, curcumin induces histone hypoacetylation by inhibiting HAT activity in brain cancer cells in vitro [67]. The inhibitory effect of curcumin on cell proliferation was associated in a time- and concentration-dependent manner with histone acetylation [67]. Interestingly, curcumin upregulates or downregulates histone acetylation, depending on the cell type. For example, in brain cancer cells treated with curcumin, H4 histones are hypoacetylated [67]. In contrast, H3 and H4 histone subunits are hyperacetylated in androgen-dependent and -independent prostrate cancer cells treated with curcumin [68]. This cell-specific effect may be attributed to different genes affected downstream of the histone modification in each of these cell types. Alternatively, the opposite findings may be due to the different concentrations of curcumin used in the two studies.

Recently, a small polyisoprenylated benzophenone molecule, garcinol from the *Garcinia indica* (coccum) fruit rind, has been found to be a naturally occurring inhibitor of HAT [68]. Therefore, it is possible that treatment of cells with garcinol may induce hypoacetylation by inactivating HAT. Further studies will be necessary to determine the efficacy of using compounds that target HAT activity.

11.3.2 HISTONE METHYLATION/DEMETHYLATION AND THE EFFECT OF DIET

11.3.2.1 Histone Methylation

Evidence exists that histone methylation regulates fundamental processes such as heterochromatin formation, X chromosome inactivation, genomic imprinting, transcriptional regulation, and DNA repair [69]. The most heavily methylated histone is H3, followed by H4. Histone methylation is a covalent modification that occurs on the side chain nitrogen atoms of key histone lysines and arginines. Arginine can be either mono- or dimethylated and lysine can be mono-, di-, or trimethylated. The multiple possible states of methylation of a particular histone residue, determined by different histone methyltransferases (HMTs) [5,70,71], lead to different biological outcomes; thus, histone methylation has greater combinatorial potential as compared to other histone modifications.

Trimethylation of the histone H4 lysine 20 residue (H4-Lys20), the only lysine that undergoes methylation in H4, has been associated with constitutive heterochromatin, gene silencing, and aging, and serves as a marker for them [69,72]. Fraga et al. observed that cancer cells have a loss of trimethylated forms of histone H4 [72]. This loss occurs early and accumulates during the carcinogenic process in the skin cancer model with which Fraga worked. The resulting overall hypomethylation may be responsible for gene silencing observed at various stages of carcinogenesis [72].

Further evidence for the involvement of histone methylation in cancer was provided by studies in mice, in which knockout of the enzymes that govern H3-Lys9 methylation resulted in genomic instability and formation of B-cell lymphomas in 28% of the mice between 9 and 15 months of age [73].

11.3.2.2 Histone Demethylation

The removal of a methyl group from histones is achieved by arginine and lysine demethylases, e.g., by deimination and amino-oxidase reactions [69,74]. In contrast to acetylation or phosphorylation, which has fast turnover rates and fit the expected features of a regulatory modification, histone methylation has been demonstrated to have a slow turnover rate. Modifications such as methylation are more stable and are responsible for long-term expression status of certain regions of the genome. Active histone demethylation is involved in either transcriptional expression or repression depending upon the histone residues that are demethylated.

11.3.2.3   Dietary Influence on Histone Methylation and Demethylation

11.3.2.3.1   Effect of a Methyl-Deficient Diet

HMTs require methyl-rich sources to catalyze the transfer of methyl groups to histone lysines or arginines. In the normal cell, SAM, a product of methionine metabolism, serves as a methyl-donor, resulting in methylated CpG islands. Diet is a major source of methyl groups [75–77]. A diet low in methyl-donors may lead to changes in trimethylation and acetylation patterns of H4-Lys 20 and H3-Lys 9, respectively, resulting in the formation of relaxed DNA–histone complexes. Such changes have been observed during hepatocarcinogenesis [77].

Folate is an important dietary methyl-donor, which has dual effects in cancer inhibition and progression [78,79]. Folate is required for 1-C carbon metabolism as well as nucleotide synthesis, all of which are essential for cell growth. Folate is also metabolized to 5 methyltetrahydrofolate, which is a source of methyl groups for methionine and SAM synthesis.

Carcinogenesis in several organs such as colon, prostate, and lung [78,80] has been associated with folate-deficient diets. In contrast, the incidence of breast cancer has been associated with diets high in folate. The key factors that tip the balance towards tumor progression or regression are the timing and dosage of folate administered. This may be a reason that a folate-deficient diet or a diet high in folate may increase the risk of tumor incidence. DNA methylation has been associated with carcinogenesis in these organs. Determining the genes silenced by this mechanism could provide much needed markers for the diagnosis of these cancers [79]. Understanding the molecular mechanism of folate action in different tissues is, therefore, crucial to the understanding of its chemopreventive potential.

11.3.2.3.2   Dietary Polyphenols

The catechol polyphenols inhibit DNMT1 activity in part by increasing intracellular SAH (S-adenosyl-L-homocysteine), a potent inhibitor of DNMTs [81] (Figure 11.1). The effective in vitro apoptotic and cell proliferation inhibitory concentrations of epigallocatechin gallate (EGCG) (10–50 μM), a key component of green tea, are $\sim$50 times higher than the plasma and tissue levels of EGCG generally observed after ingestion of tea [82,83]. The effective high in vitro concentrations may be attributed to the inactivation of bioactive metabolites of EGCG by the culture media used in studies in vitro. Therefore more of the compound is required to bring about an effective response. However, in vivo, a host of factors could be involved in stabilizing the EGCG molecule, allowing low plasma doses to be more effective. Discrepancies have also been found between the in vitro and in vivo effective concentrations for genistein and may have similar causes. For example, in vitro, genistein concentrations (5–20 μM) that trigger DNA demethylation are orders of magnitude higher than plasma levels of genistein ($\sim$270 nM) [84,85].

11.3.3   Histone Phosphorylation and the Effect of Diet

The core histones and histone H1 undergo phosphorylation on specific serine (Ser) and threonine (Thr) residues by H1 and H3 kinases [86,87]. H1 is phosphorylated on the N- and C-terminal domains of Ser/Thr residues, whereas H3 is phosphorylated only on the N-terminal domains of Ser/Thr residues. Both phosphorylation processes are cell cycle dependent [88]. H1 phosphorylation weakens H1 binding to DNA, promoting free access of transcriptional replication factors to the DNA, which facilitates gene expression. H3 phosphorylation at Ser-10, a more frequent process, has been associated with the transcriptional activation of the early response genes c-fos and c-jun. Thus, chromatin decondensation is possibly coordinated by H3 phosphorylation [89]. These findings are consistent with the possibility that H3 phosphorylation may play a role in cell cycle progression [89].

The activation of checkpoints in response to DNA damage leads to cell cycle arrest but when the damage is very severe, it results in apoptotic cell death [90,91]. Sulforaphane, found in mustard seeds, although not an antioxidant itself, is an effective inducer of enzymes that enhance the activity of the crucial intracellular antioxidant, glutathione. In vitro, 20 μM sulforaphane increases phosphorylation at Ser-139 of H2A.X., an isoprotein of histone 2A (H2A) and is a sensitive marker for the presence of DNA double-strand breaks. The ensuing accumulation of cell cycle damage may be the mechanism that triggers apoptosis [89].

Histone phosphorylation has been associated with the anticarcinogenic activity of several flavones [92]. For example, H2A.X phosphorylation known to be induced at double-strand break sites has been observed in MCF-7 cells treated with 50 nM aminoflavones. Moreover, the aminoflavone has a dose-dependent effect on histone phosphorylation [92–94]. The ability of aminoflavone to generate phosphorylated H2A.X selectively in replicating cells makes it a suitable candidate for the treatment of cancers.

The chemopreventive effect of genistein has been associated with increased H1 histone phosphorylation in cancer cells [95]. Genistein-treated breast cancer cell lines such as MDA-MB-231 and BT20 display an increase in kinase activity. This study primarily focused on the expression of cyclin and cyclin-dependent kinases. The increase in kinase activity in genistein-treated cancer cell lines is accompanied by histone phosphorylation, transcriptional activation, and finally G_2M arrest [95]. In contrast to this observation genistein also blocks the recruitment of histone H3 kinase MSK1, thereby inhibiting H3 phosphorylation. This observation is gene specific, however. Interleukin-6, an inflammatory response gene, and a growth factor for many tumors [96], is transcriptionally regulated by nuclear factor-κB (NF-κB) and requires the activation of the mitogen-activated protein kinase (MAPK)/MSK kinase pathway, which phosphorylates NF-κB, p65 and histone H3. In the presence of genistein, as seen in the breast cancer cell line MDA-MB-231, histone H3 kinase MSK1 recruitment to the IL-6 promoter is blocked leading to the loss of H3 phosphorylation and acetylation and reduced IL-6 expression. However, the activity of MSK1 is reduced but not completely inhibited. Although genistein plays a significant role in H3 phosphorylation, the contradictory results obtained by the two studies were primarily due to the different pathways analyzed and the concentrations of genistein used were different (1–30 and 200 μM, respectively).

Zearalenone, a fungal metabolite found in dairy sources, meat, and grains, is a nonsteroidal estrogen molecule [97]. Ingestion of zearalenone increases the endogenous estrogen load thereby stimulating estrogen receptor (ER)-positive cells to undergo mitosis by activating cyclin-dependent kinases [97]. It is, therefore, not surprising that phosphorylation of histone H3, a marker of mitotic activity, is also increased [97].

Another molecule that influences the phosphorylation status of histones H3 is vinblastine, an alkaloid derived from the Madagascar Periwinkle plant. This compound has been used in the treatment of cancers [98]. In vitro treatment of cancer cell lines with this alkaloid shows that the proportion of phosphorylated versus nonphosphorylated H3 molecules progressively increases during the cell cycle arrest in the metaphase; but in the interphase chromatin, the fraction of phosphorylated H3 histones is several-fold lower compared to that in mitotic cells [98].

11.3.4 Histone Biotinylation and Diet

Recent studies have shown that histone modulations involving the covalent attachment of biotin to specific lysine residues, catalyzed by the enzymes biotinidase and holocarboxylase synthetase, occur on histones H2A, H3, and H4 [99]. Biotinylation of histone H4 at lysines 8 and 12 has been associated with heterochromatin structures, gene silencing, mitotic condensation of chromatin, and DNA repair. Histone biotinylation is a reversible process, although debiotinylases have not been characterized [99]. Biotinylation appears to be important in DNA repair and chromatin structure and is more predominant in proliferating than in quiescent cells [99,100]. Biotinylation of histones is known to occur in the presence of DNA double-strand breaks. In some cases, this occurs at specific lysine residues of H4, probably associated with genes involved in the DNA repair machinery [99]. Dietary supplementation of biotin is required for biotinylation and biotin deficiency may have profound effects on chromatin structures [99,100]. Further studies will be necessary to test the possibility that dietary biotin may have epigenetic effects via histone biotinylation.

11.3.5 HISTONE UBIQUITINATION

Recent evidence has implicated histone ubiquitination in gene transcription control [101]. Ubiquitination occurs at lysine residues of histones H2A and H2B at positions 119 and 120, respectively; H1 and H3 are not yet mapped. Mono-ubiquitination of histone H2B is required for methylation of histone H3 at K4 and K79. Histone ubiquitination seems to promote methylation by recruiting proteosomal ATPases by ubiquitin-modified H2B [102].

Nickel is an essential nutrient. However, a nickel overdose can be deleterious, as insoluble nickel compounds, and soluble nickel compounds to a lesser extent, are carcinogenic [103]. The exact mechanism by which nickel induces carcinogenesis is not clear. An interesting find is that, once in the cell, nickel compounds can exert epigenetic effects and deregulate gene expression. This is a possible mechanism underlying the etiology of nickel-induced cancers. Nickel compounds affect three types of histone modifications (1) deacetylation of histones H2A, H2B, H3, and H4; (2) dimethylation of histone H3 at lysine 9; and (3) ubiquitination of histone H2A and H2B. At lower concentration, Ni(II) stimulates mono-ubiquitination of histones H2B and H2A while at higher concentrations ubiquitination is suppressed, apparently because of the presence of a truncated H2N which lacks the K120 ubiquitination site. This modification affects gene expression and DNA repair leading to cell transformation [104].

11.3.6 OTHER DIET-MEDIATED HISTONE MODIFICATIONS

Histone ADP-ribosylation occurs mostly on the glutamate residues of histone H2A during DNA double-strand breaks. In the thyroid, ADP-ribosylation is mainly associated with transcriptional inactivation of chromatin. Reduction of histone ADP-ribosylation is associated with various forms of cancer such as lymphoma [105]. Phosphorylation of S14 of histone H2B and poly ADP-ribosylation of glutamate residues on histone H2A are known markers of double-strand DNA breaks, which are mainly associated with transcriptional inactivation [106].

The ε-amino group of histone lysine residues is also subject to modification by ubiquitin-like proteins such as small ubiquitin-related modifier (SUMO). In contrast to histone ubiquitination, histone modifications by SUMO are generally associated with decreased gene expression, and are reversible. Sumoylation occurs on histones H2A, H2B, H3, and H4 and blocks histone acetylation and ubiquitination [107].

Histones have been reported to be glycosylated. O-linked N-acetylglucosamine (O-GlcNAc) is thought to act in a manner analogous to protein phosphorylation. O-GlcNAc additions on nuclear and cytosolic proteins possess intrinsic HAT activity in vitro [108–110]. Further studies are necessary to elucidate the mechanisms and the relevance of diet-related glycosylation of histones in cancers.

In vivo histone carbonylation occurs to variable extents on all histones except histone H4. In contrast, in vitro, carbonylation has been observed on all histones, including histone H4. Histones H1 and H2A/H2B are predominantly more carbonylated than histone H3. Carbonylation is irreversible and addition of this moiety to histones rich in basic amino acids, like lysine and arginine residues, could mask the positive charge of the nucleosome–DNA complex. This may allow the chromatin to relax, a phenomenon that normally allows gene expression. Interestingly, carbonylation of histones decreases with age, but caloric restriction has been shown to increase the levels of carbonylation [111].

11.4 EFFECT OF ENVIRONMENTAL FACTORS ON EPIGENETIC MODIFICATIONS

As shown above, diet, which may be considered an "environmental factor," is a major trigger of epigenetic modifications. Plants and animals are the major source of nourishment to humans. Therefore, the conditions in which plants and animals are reared affect the human diet. For example, if the soil in which crops are grown has a high metal content or is sprayed with pesticides that are not easily degradable, the latter make their way into the foodchain, resulting in detrimental changes to humans, including effects on the epigenome. Below, we analyze a few environment-mediated mechanisms that are well studied and are proven to contribute to epigenetic changes in cancer.

11.4.1 Pollutants: Smoking and Particulate Matter

11.4.1.1 Mechanisms Underlying the Effect of Cigarette Smoke and Other Particulate Matter

Lung and throat cancers, as well as other respiratory problems such as chronic obstructive pulmonary disease (COPD), ensue due to chronic exposure to cigarette smoke [112,113]. The molecular pathways underlying inflammation of these tissues by cigarette smoke are well studied. [113,114]. One of the most influential factors mediating this response is histone modification of proinflammatory genes [113,114].

Particulate matter in the air of <10 μM (PM10) diameter has also been associated with chronic lung and cardiovascular disease, including severe asthma attacks [115–121]. Just like cigarette smoke, the mechanism underlying their deleterious effect involves acetylation of H4 associated with the promoter region of inflammatory genes, e.g., IL-8 [112].

Cigarette smoke induces oxidative stress, recruits proinflammatory cells [113], and causes the overexpression of proinflammatory genes such as NF-$\kappa\beta$ and AP-1 by affecting the balance of histone-tail acetylation/deacetylation [113]. Inflammatory molecules such as interleukins and cytokines produced by the inflammatory cells induce and maintain the chronic inflamed condition [112,113]. Prolonged inflammation of the lungs increases the risk of developing COPD, which can give rise to lung cancers [122]. The mechanistic action of cigarette smoke in the induction of COPD has been investigated and chromatin changes have been implicated in its etiology. Acrolein, a reactive aldehyde, is a component of cigarette smoke which has also been shown to have epigenetic effects by interfering with the function of enzymes such as HDACs [114].

Of the HDAC class of HDACs, HDAC2 appears to be an important deacetylase involved in the pathogenesis of COPD. Studies have shown that levels of HDAC1 remain unaffected but HDAC2 levels are restored to normal after prolonged exposure to smoke [112]. Although HDAC2 levels are restored, hyperphospho-acetylation of H4 and acetylation of H3 are observed. This may be possible as the HDAC2 levels, though at optimal levels, may be inactive due to the effect of the reactive aldehydes present in the smoke component. These aldehydes have affinities for histidine and lysines, and react with the histidine groups at the HDAC2 active site. Therefore, drugs that are anti-inflammatory may reverse the acetylation status of the genes by increasing HDAC levels and activity [113,114].

Another example of histone acetylation/deacetylation imbalance induced by cigarette smoke, leading to altered gene expression, has been demonstrated for the inflammatory genes TNF-α and IL-8 mediates [112,113]. Cigarette smoke leads to hyperacetylation of the histone H4 associated with the promoters of TNF-α and IL-8 [121,123]. Thus, under normal conditions, NF-$\kappa\beta$ is complexed with HDACs. A component of NF-$\kappa\beta$, p65, has HAT-like activity. When complexed with HDAC, the p65 component of NF-$\kappa\beta$ is rendered inactive and hypoacetylation of histones ensues [113,121]. In contrast, under conditions of oxidative stress, NF-$\kappa\beta$ is released from the HDAC complex. NF-$\kappa\beta$ becomes phosphorylated at serine residues, which transactivates p65 and leads to the hyperacetylation of histone H4. These events trigger the expression of the proinflammatory genes IL-8 and IL-6 [113,114,121].

11.4.1.2 Factors Modulating the Magnitude of Epigenetic Effects of Cigarette Smoke and Other Particulate Matter

It is presumed that individuals who smoke are at a greater risk of developing various pulmonary diseases as discussed above. However, not all smokers develop lung cancers or other respiratory problems. The reason for the susceptibility of some but not all smokers is not well understood. This question was addressed in a Korean population of smokers versus nonsmokers [124]. This study provided strong evidence for a role of epigenetics in lung cancer development in susceptible populations.

Single nucleotide polymorphisms (SNPs) could be an alternative mechanism that modulates the epigenetic effects of cigarette smoke. The HMTs, SMYD3, methylates histone H3 at the K4 residue [124]. Single nucleotide polymorphisms of the gene SMYD3 are associated with greater risk for developing colorectal cancers, hepatocarcinomas, and breast cancers [124]. It is possible that polymorphisms may cause structure and activity modification of the catalytic domain of this enzyme, affecting the methylation status of histone residues. This would be likely to affect heterochromatin states and gene repression.

Yoon et al. [124] showed that SUV39H2, a mammalian HMT, exhibits eight different SNPs. The polymorphism G to C at 1624 in the 3′ UTR is critical to the development of lung cancers. This enzyme methylates histone H3 at K9 and is involved in transcriptional repression. The addition of the methyl moiety to the K9 residue creates a binding site for heterochromatin protein 1 (HP1) [124]. SUV39H2 and HP1 both mediate the repressive functions of Rb protein that is important in controlling the cell cycle through cyclin E [124]. Therefore, histone modification and the enzymes that modulate the covalent modification may act in tandem to influence gene expression and the susceptibility to cancer.

11.4.2 OCCUPATIONAL EXPOSURE TO POLLUTANTS, HISTONE MODIFICATION, AND CANCER

In mines and refineries, workers become exposed to harmful particulate or soluble metal ions [125]. This occupational hazard has been associated with increased susceptibility to cancer. For example, occupational exposure to insoluble nickel compounds plays a significant role in the development of lung and nasal cancers [125]. When insoluble nickel enters cells it is converted to Ni^{2+} ion, which induces the production of reactive oxygen species (ROS), and may repress tumor suppressor gene expression [125]. One pathway underlying Ni^{2+} effects on gene expression is epigenetic. Ni^{2+} inhibits the activity of HATs, causing histone hypoacetylation primarily of the H4 subunit. Ni^{2+} binds to the histidine residue at position 18 of the amino terminal domain of H4, which is in close proximity to the acetylated lysine residue [125]. This triggers the generation of ROS which associates with HATs and may interfere with their catalytic activity, inducing hypoacetylation of associated genes. In addition to inactivating HATs, Ni^{2+} bound to the histidine residue may prevent HAT binding to the H4 subunit and may consequently induce H4 hypoacetylation, which silences gene expression due to the induction of heterochromatin states [125]. ROS have a strong affinity for proteins rich in cysteine and histine residues [125]. The catalytic domains of certain proteins, which possess HAT activity, are rich in these residues. When these HATs undergo oxidative modifications, mediated by ROS, the proteins are rendered inactive and thereby reduce histone acetylation leading to transcriptional suppression of genes associated with tumor suppression.

11.4.3 EXPOSURE TO DRUGS AND CHEMICALS, HISTONE MODIFICATION, AND CANCER

In the late 1940s to the early 1970s expectant mothers, especially in their first trimester, were given diethylstilbestrol (DES). This synthetic estrogen was given to prevent miscarriages. Moreover, animal livestock and cattle were fed with DES to increase growth rate, providing the human population with another source of this estrogenic compound. Since then, DES has been shown to induce developmental defects of the cervix, uterus, and vagina, as well as rare cancers. The observed effect is trans-generational in nature, i.e., it skips the mothers and affects the daughters and granddaughters. This outcome is speculated to have a basis in epigenetic processes [126].

Some DES-mediated effects are mediated by DNA methylation and in certain instances methylation of CpG residues affects the modification of histones surrounding the genes. DES is a synthetic estrogen molecule. In the absence of the DES ligand or estrogen, the ER is complexed to heat-shock protein 90 (Hsp90), and is inactive because Hsp90 inhibits ER–chromatin complex formation [126]. However, in the presence of DES, ER dimerizes with DES and enters the cell nucleus to mediate its downstream effects [126].

Steroid hormones effect the expression of *WNT* genes, which are involved in the development and regulation of the female reproductive system. Therefore, it is not surprising that administration of synthetic estrogen molecules like DES to pregnant women alters the expression of Wnt [126]. The association of Hsp90 with estrogen has opposite effects on *WNT*-associated gene expressions. In the presence of the chaperon Hsp90, optimal activity of histone 3 lysine 4 methyltransferases, SYMD3 is observed which is required for the activation of *WNT* genes [126]. However, DES may deregulate this activity by quenching Hsp90, thereby disrupting the association of Hsp90 with *WNT* genes [126].

The proposed model by Ruden et al. [126] attempts to explain the trans-generational effects of DES-induced uterine abnormalities. Under normal conditions, it is proposed that during development, specific methylation of CpG residues in confined regions of the chromatin in germline stem cells may occur and possibly enhancers of genes associated with uterine cancers such as c-fos are turned off [126]. However, exposure to DES may trigger hypomethylation of the enhancers of these genes allowing regulatory elements to bind to the transcriptome and induce activation of uterine cancer promoting genes [126].

Wnt signaling is also affected by DES and the expression of Wnt is crucial to maintain the methylation pattern and to replenish the uterine stem cells [126]. Since DES has opposing effects on ER-regulated genes and Wnt-mediated gene expression, the presence of this molecule can induce epigenetic effects on the genes that promote cancer phenotypes by modulating methylation patterns and histone modifications.

11.5   CONCLUSIONS

Evidence exists that dietary components affect epigenetic processes such as histone modifications [10,14–17,22,26,45,47,61,66,67,75,99,127–133]. Evidence also exists that histone modifications play a role in the development of cancer [14,74,77,134–157]. Therefore, epigenetic modulation of histone modifications by dietary components may be a useful preventive and therapeutic approach against cancer. To optimize these cancer preventive effects, it will be necessary to explore agents derived from natural sources, to determine the conditions for their optimal effects, e.g., optimal concentration, timing, administration of the natural complex or its putative active ingredients. Mechanisms underlying the effect of dietary ingredients will also need to be elucidated. Studies discussed above suggest that epigenetic mechanisms are likely to be a major mechanism underlying the chemopreventive action of some naturally occurring agents.

Studies discussed in this review also reveal that the negative effects of some environmental pollutants may be mediated by epigenetic mechanisms. Moreover, some pollutants and the diet act synergistically via mechanisms affecting the epigenetic code [10]. Under normal conditions, epigenetic marks maintain proper gene function. However, diet and pollutants-mediated modifications of epigenetic marks affect gene expression, resulting in aberrant cellular division and triggering the neoplastic phenotype.

Elucidating the epigenetic code and mechanisms that underlie its modification by the diet and the environment may serve as a tool to predict the susceptibility of an individual to cancer and the necessary precautions or treatment required to reverse epigenetic changes. Given the diversity of histone modifications, this may not be an easy task.

REFERENCES

1. Vastag, B., Genome analysis yields mutations linked to hereditary prostate cancer, *JAMA*, 287, 827, 2002.
2. Wang, C.Y. et al., Somatic mutations of mitochondrial genome in early stage breast cancer, *Int. J. Cancer*, 121, 1253, 2007.

3. Zhao, H. et al., Genome-wide characterization of gene expression variations and DNA copy number changes in prostate cancer cell lines, *Prostate*, 63, 187, 2005.
4. Peterson, C.L. and Laniel, M.A., Histones and histone modifications, *Curr. Biol.*, 14, R546, 2004.
5. Lusser, A., Acetylated, methylated, remodeled: Chromatin states for gene regulation, *Curr. Opin. Plant Biol.*, 5, 437, 2002.
6. Wolffe, A.P., Inheritance of chromatin states, *Dev. Genet.*, 15, 463, 1994.
7. Vermaak, D., Ahmad, K., and Henikoff, S., Maintenance of chromatin states: An open-and-shut case, *Curr. Opin. Cell Biol.*, 15, 266, 2003.
8. Santoro, R. and De Lucia, F., Many players, one goal: How chromatin states are inherited during cell division, *Biochem. Cell Biol.*, 83, 332, 2005.
9. Henikoff, S., McKittrick, E., and Ahmad, K., Epigenetics, histone H3 variants, and the inheritance of chromatin states, *Cold Spring Harb. Symp. Quant. Biol.*, 69, 235, 2004.
10. Feil, R., Environmental and nutritional effects on the epigenetic regulation of genes, *Mutat. Res.*, 600, 46, 2006.
11. Milner, J.A., Molecular targets for bioactive food components, *J. Nutr.*, 134, 2492S, 2004.
12. Milner, J.A. et al., Molecular targets for nutrients involved with cancer prevention, *Nutr. Cancer*, 41, 1, 2001.
13. Junien, C., Impact of diets and nutrients/drugs on early epigenetic programming, *J. Inherit. Metab. Dis.*, 29, 359, 2006.
14. Davis, C.D. and Ross, S.A., Dietary components impact histone modifications and cancer risk, *Nutr. Rev.*, 65, 88, 2007.
15. Dashwood, R.H. and Ho, E., Dietary histone deacetylase inhibitors: From cells to mice to man, *Semin. Cancer Biol.*, 17, 363, 2007.
16. Myzak, M.C., Ho, E., and Dashwood, R.H., Dietary agents as histone deacetylase inhibitors, *Mol. Carcinog.*, 45, 443, 2006.
17. Hirano, T. et al., Citrus flavone tangeretin inhibits leukaemic HL-60 cell growth partially through induction of apoptosis with less cytotoxicity on normal lymphocytes, *Br. J. Cancer*, 72, 1380, 1995.
18. Habtemariam, S., Flavonoids as inhibitors or enhancers of the cytotoxicity of tumor necrosis factor-alpha in L-929 tumor cells, *J. Nat. Prod.*, 60, 775, 1997.
19. Kuntz, S., Wenzel, U., and Daniel, H., Comparative analysis of the effects of flavonoids on proliferation, cytotoxicity, and apoptosis in human colon cancer cell lines, *Eur. J. Nutr.*, 38, 133, 1999.
20. Wang, I.K., Lin-Shiau, S.Y., and Lin, J.K., Induction of apoptosis by apigenin and related flavonoids through cytochrome c release and activation of caspase-9 and caspase-3 in leukaemia HL-60 cells, *Eur. J. Cancer*, 35, 1517, 1999.
21. Iwashita, K. et al., Flavonoids inhibit cell growth and induce apoptosis in B16 melanoma 4A5 cells, *Biosci. Biotechnol. Biochem.*, 64, 1813, 2000.
22. Wenzel, U. et al., Dietary flavone is a potent apoptosis inducer in human colon carcinoma cells, *Cancer Res.*, 60, 3823, 2000.
23. Sakagami, H. et al., Induction of apoptosis by flavones, flavonols (3-hydroxyflavones) and isoprenoid-substituted flavonoids in human oral tumor cell lines, *Anticancer Res.*, 20, 271, 2000.
24. Mittra, B. et al., Luteolin, an abundant dietary component is a potent anti-leishmanial agent that acts by inducing topoisomerase II-mediated kinetoplast DNA cleavage leading to apoptosis, *Mol. Med.*, 6, 527, 2000.
25. Gupta, S., Afaq, F., and Mukhtar, H., Selective growth-inhibitory, cell-cycle deregulatory and apoptotic response of apigenin in normal versus human prostate carcinoma cells, *Biochem. Biophys. Res. Commun.*, 287, 914, 2001.
26. Birt, D.F., Hendrich, S., and Wang, W., Dietary agents in cancer prevention: Flavonoids and isoflavonoids, *Pharmacol. Ther.*, 90, 157, 2001.
27. Choi, J. et al., Flavones from Scutellaria baicalensis Georgi attenuate apoptosis and protein oxidation in neuronal cell lines, *Biochem. Biophys. Acta*, 1571, 201, 2002.
28. Ko, W.G. et al., Effects of luteolin on the inhibition of proliferation and induction of apoptosis in human myeloid leukaemia cells, *Phytother. Res.*, 16, 295, 2002.
29. Chen, Y.C. et al., Flavone inhibition of tumor growth via apoptosis in vitro and in vivo, *Int. J. Oncol.*, 25, 661, 2004.
30. Monasterio, A. et al., Flavonoids induce apoptosis in human leukemia U937 cells through caspase- and caspase-calpain-dependent pathways, *Nutr. Cancer*, 50, 90, 2004.

31. Quiney, C. et al., Flavones and polyphenols inhibit the NO pathway during apoptosis of leukemia B-cells, *Leuk. Res.*, 28, 851, 2004.

32. Way, T.D., Kao, M.C., and Lin, J.K., Apigenin induces apoptosis through proteasomal degradation of HER2/neu in HER2/neu-overexpressing breast cancer cells via the phosphatidylinositol 3-kinase/Akt-dependent pathway, *J. Biol. Chem.*, 279, 4479, 2004.

33. Brachmann, S.M. et al., Phosphoinositide 3-kinase catalytic subunit deletion and regulatory subunit deletion have opposite effects on insulin sensitivity in mice, *Mol. Cell. Biol.*, 25, 1596, 2005.

34. Brusselmans, K. et al., Induction of cancer cell apoptosis by flavonoids is associated with their ability to inhibit fatty acid synthase activity, *J. Biol. Chem.*, 280, 5636, 2005.

35. Chen, D. et al., Dietary flavonoids as proteasome inhibitors and apoptosis inducers in human leukemia cells, *Biochem. Pharmacol.*, 69, 1421, 2005.

36. Cheng, A.C. et al., Induction of apoptosis by luteolin through cleavage of Bcl-2 family in human leukemia HL-60 cells, *Eur. J. Pharmacol.*, 509, 1, 2005.

37. Daskiewicz, J.B. et al., Effects of flavonoids on cell proliferation and caspase activation in a human colonic cell line HT29: An SAR study, *J. Med. Chem.*, 48, 2790, 2005.

38. Duraj, J. et al., Flavonoid quercetin, but not apigenin or luteolin, induced apoptosis in human myeloid leukemia cells and their resistant variants, *Neoplasma*, 52, 273, 2005.

39. Elsisi, N.S. et al., Ibuprofen and apigenin induce apoptosis and cell cycle arrest in activated microglia, *Neurosci. Lett.*, 375, 91, 2005.

40. Zheng, P.W., Chaing, L.C., and Lin, C.C., Apigenin induced apoptosis through p53-dependent pathway in human cervical carcinoma cells, *Life Sci.*, 76, 1367, 2005.

41. Wu, K., Yuan, L.H., and Xia, W., Inhibitory effects of apigenin on the growth of gastric carcinoma SGC-7901 cells, *World J. Gastroenterol.*, 11, 4461, 2005.

42. Chiang, L.C. et al., Anti-proliferative effect of apigenin and its apoptotic induction in human Hep G2 cells, *Cancer Lett.*, 237, 207, 2006.

43. Horinaka, M. et al., The dietary flavonoid apigenin sensitizes malignant tumor cells to tumor necrosis factor-related apoptosis-inducing ligand, *Mol. Cancer Ther.*, 5, 945, 2006.

44. Khan, T.H. and Sultana, S., Apigenin induces apoptosis in Hep G2 cells: Possible role of TNF-alpha and IFN-gamma, *Toxicology*, 217, 206, 2006.

45. Lim do, Y. et al., Induction of cell cycle arrest and apoptosis in HT-29 human colon cancer cells by the dietary compound luteolin, *Am. J. Physiol. Gastrointest. Liver Physiol.*, 292, G66, 2007.

46. Dasmahapatra, G., Almenara, J.A., and Grant, S., Flavopiridol and histone deacetylase inhibitors promote mitochondrial injury and cell death in human leukemia cells that overexpress Bcl-2, *Mol. Pharmacol.*, 69, 288, 2006.

47. Rosato, R.R. et al., The cyclin-dependent kinase inhibitor flavopiridol disrupts sodium butyrate-induced p21WAF1/CIP1 expression and maturation while reciprocally potentiating apoptosis in human leukemia cells, *Mol. Cancer Ther.*, 1, 253, 2002.

48. Hemmi, S. et al., The presence of human coxsackievirus and adenovirus receptor is associated with efficient adenovirus-mediated transgene expression in human melanoma cell cultures, *Hum. Gene Ther.*, 9, 2363, 1998.

49. Li, Y. et al., Loss of adenoviral receptor expression in human bladder cancer cells: A potential impact on the efficacy of gene therapy, *Cancer Res.*, 59, 325, 1999.

50. Li, D. et al., Variability of adenovirus receptor density influences gene transfer efficiency and therapeutic response in head and neck cancer, *Clin. Cancer Res.*, 5, 4175, 1999.

51. Matsumoto, K. et al., Loss of coxsackie and adenovirus receptor expression is associated with features of aggressive bladder cancer, *Urology*, 66, 441, 2005.

52. Okegawa, T. et al., The dual impact of coxsackie and adenovirus receptor expression on human prostate cancer gene therapy, *Cancer Res.*, 60, 5031, 2000.

53. Okegawa, T. et al., The mechanism of the growth-inhibitory effect of coxsackie and adenovirus receptor (CAR) on human bladder cancer: A functional analysis of car protein structure, *Cancer Res.*, 61, 6592, 2001.

54. Pong, R.C. et al., Mechanism of increased coxsackie and adenovirus receptor gene expression and adenovirus uptake by phytoestrogen and histone deacetylase inhibitor in human bladder cancer cells and the potential clinical application, *Cancer Res.*, 66, 8822, 2006.

55. Sachs, M.D. et al., Integrin alpha(v) and coxsackie adenovirus receptor expression in clinical bladder cancer, *Urology*, 60, 531, 2002.

56. Lea, M.A. et al., Induction of histone acetylation and inhibition of growth of mouse erythroleukemia cells by *S*-allylmercaptocysteine, *Nutr. Cancer*, 43, 90, 2002.

57. Lea, M.A., Randolph, V.M., and Hodge, S.K., Induction of histone acetylation and growth regulation in erythroleukemia cells by 4-phenylbutyrate and structural analogs, *Anticancer Res.*, 19, 1971, 1999.

58. Lea, M.A. and Randolph, V.M., Induction of histone acetylation in rat liver and hepatoma by organo-sulfur compounds including diallyl disulfide, *Anticancer Res.*, 21, 2841, 2001.

59. Lea, M.A. et al., Induction of histone acetylation in mouse erythroleukemia cells by some organosulfur compounds including allyl isothiocyanate, *Int. J. Cancer*, 92, 784, 2001.

60. Lea, M.A., Randolph, V.M., and Patel, M., Increased acetylation of histones induced by diallyl disulfide and structurally related molecules, *Int. J. Oncol.*, 15, 347, 1999.

61. Myzak, M.C. and Dashwood, R.H., Histone deacetylases as targets for dietary cancer preventive agents: Lessons learned with butyrate, diallyl disulfide, and sulforaphane, *Curr. Drug Targets*, 7, 443, 2006.

62. Myzak, M.C. and Dashwood, R.H., Chemoprotection by sulforaphane: Keep one eye beyond Keap1, *Cancer Lett.*, 233, 208, 2006.

63. Myzak, M.C. et al., Sulforaphane inhibits histone deacetylase in vivo and suppresses tumorigenesis in Apc-minus mice, *FASEB. J.*, 20, 506, 2006.

64. Myzak, M.C. et al., Sulforaphane inhibits histone deacetylase activity in BPH-1, LnCaP and PC-3 prostate epithelial cells, *Carcinogenesis*, 27, 811, 2006.

65. Myzak, M.C. et al., Sulforaphane retards the growth of human PC-3 xenografts and inhibits HDAC activity in human subjects, *Exp. Biol. Med. (Maywood)*, 232, 227, 2007.

66. Dashwood, R.H., Myzak, M.C., and Ho, E., Dietary HDAC inhibitors: Time to rethink weak ligands in cancer chemoprevention? *Carcinogenesis*, 27, 344, 2006.

67. Kang, S.K., Cha, S.H., and Jeon, H.G., Curcumin-induced histone hypoacetylation enhances caspase-3-dependent glioma cell death and neurogenesis of neural progenitor cells, *Stem Cells Dev.*, 15, 165, 2006.

68. Shankar, S. and Srivastava, R.K., Involvement of Bcl-2 family members, phosphatidylinositol 3′-kinase/AKT and mitochondrial p53 in curcumin (diferulolylmethane)-induced apoptosis in prostate cancer, *Int. J. Oncol.*, 30, 905, 2007.

69. Tian, X. and Fang, J., Current perspectives on histone demethylases, *Acta. Biochem. Biophys. Sin (Shanghai)*, 39, 81, 2007.

70. van Dijk, K. et al., Monomethyl histone H3 lysine 4 as an epigenetic mark for silenced euchromatin in Chlamydomonas, *Plant Cell*, 17, 2439, 2005.

71. Rice, J.C. et al., Histone methyltransferases direct different degrees of methylation to define distinct chromatin domains, *Mol. Cell*, 12, 1591, 2003.

72. Fraga, M.F. et al., Loss of acetylation at Lys16 and trimethylation at Lys20 of histone H4 is a common hallmark of human cancer, *Nat. Genet.*, 37, 391, 2005.

73. Peters, A.H. et al., Loss of the Suv39h histone methyltransferases impairs mammalian heterochromatin and genome stability, *Cell*, 107, 323, 2001.

74. Santos-Rosa, H. and Caldas, C., Chromatin modifier enzymes, the histone code and cancer, *Eur. J. Cancer*, 41, 2381, 2005.

75. Brunaud, L. et al., Effects of vitamin B12 and folate deficiencies on DNA methylation and carcinogenesis in rat liver, *Clin. Chem. Lab. Med.*, 41, 1012, 2003.

76. Niculescu, M.D. and Zeisel, S.H., Diet, methyl donors and DNA methylation: Interactions between dietary folate, methionine and choline, *J. Nutr.*, 132(Suppl. 8), 2333S, 2002.

77. Pogribny, I.P. et al., Methyl deficiency, alterations in global histone modifications, and carcinogenesis, *J. Nutr.*, 137(Suppl. 1), 216S, 2007.

78. Kim, Y.I., Folate, colorectal carcinogenesis, and DNA methylation: Lessons from animal studies, *Environ. Mol. Mutagen.*, 44, 10, 2004.

79. Kim, Y.I., Folate and DNA methylation: A mechanistic link between folate deficiency and colorectal cancer? *Cancer Epidemiol. Biomarkers Prev.*, 13, 511, 2004.

80. Jang, H., Mason, J.B., and Choi, S.W., Genetic and epigenetic interactions between folate and aging in carcinogenesis, *J. Nutr.*, 135(Suppl. 12), 2967S, 2005.

81. Lee, W.J. and Zhu, B.T., Inhibition of DNA methylation by caffeic acid and chlorogenic acid, two common catechol-containing coffee polyphenols, *Carcinogenesis*, 27, 269, 2006.

82. Lee, L.T. et al., Blockade of the epidermal growth factor receptor tyrosine kinase activity by quercetin and luteolin leads to growth inhibition and apoptosis of pancreatic tumor cells, *Anticancer Res.*, 22, 1615, 2002.

83. Lambert, J.D. et al., Dose-dependent levels of epigallocatechin-3-gallate in human colon cancer cells and mouse plasma and tissues, *Drug Metab. Dispos.*, 34, 8, 2006.

84. Adlercreutz, C.H. et al., Soybean phytoestrogen intake and cancer risk, *J. Nutr.*, 125(Suppl. 3), 757S, 1995.

85. Xu, X. et al., Bioavailability of soybean isoflavones depends upon gut microflora in women, *J. Nutr.*, 125, 2307, 1995.

86. Deshpande, A., Sicinski, P., and Hinds, P.W., Cyclins and Cdks in development and cancer: A perspective, *Oncogene*, 24, 2909, 2005.

87. Nowak, S.J. and Corces, V.G., Phosphorylation of histone H3: A balancing act between chromosome condensation and transcriptional activation, *Trends Genet.*, 20, 214, 2004.

88. Huang, X. et al., Sequential phosphorylation of Ser-10 on histone H3 and ser-139 on histone H2AX and ATM activation during premature chromosome condensation: Relationship to cell-cycle phase and apoptosis, *Cytometry A*, 69, 222, 2006.

89. Davie, J.R. and Chadee, D.N., Regulation and regulatory parameters of histone modifications, *J. Cell Biochem. Suppl.*, 30, 203, 1998.

90. Tang, D. et al., ERK activation mediates cell cycle arrest and apoptosis after DNA damage independently of p53, *J. Biol. Chem.*, 277, 12710, 2002.

91. Zhou, B.B. and Elledge, S.J., The DNA damage response: Putting checkpoints in perspective, *Nature*, 408, 433, 2000.

92. Meng, L.H. et al., DNA-protein cross-links and replication-dependent histone H2AX phosphorylation induced by aminoflavone (NSC 686288), a novel anticancer agent active against human breast cancer cells, *Cancer Res.*, 65, 5337, 2005.

93. Yang, C.S. et al., Molecular targets for the cancer preventive activity of tea polyphenols, *Mol. Carcinog.*, 45, 431, 2006.

94. Yang, C.S. et al., Tea and cancer prevention: Molecular mechanisms and human relevance, *Toxicol. Appl. Pharmacol.*, 224, 265, 2007.

95. Cappelletti, V. et al., Genistein blocks breast cancer cells in the G(2)M phase of the cell cycle, *J. Cell Biochem.*, 79, 594, 2000.

96. Vanden Berghe, W. et al., Attenuation of mitogen- and stress-activated protein kinase-1-driven nuclear factor-kappaB gene expression by soy isoflavones does not require estrogenic activity, *Cancer Res.*, 66, 4852, 2006.

97. Ahamed, S. et al., Signal transduction through the Ras/Erk pathway is essential for the mycoestrogen zearalenone-induced cell-cycle progression in MCF-7 cells, *Mol. Carcinog.*, 30, 88, 2001.

98. Juan, G. et al., Histone H3 phosphorylation and expression of cyclins A and B1 measured in individual cells during their progression through G2 and mitosis, *Cytometry*, 32, 71, 1998.

99. Hassan, Y.I. and Zempleni, J., Epigenetic regulation of chromatin structure and gene function by biotin, *J. Nutr.*, 136, 1763, 2006.

100. Crisp, S.E. et al., Biotin supply affects rates of cell proliferation, biotinylation of carboxylases and histones, and expression of the gene encoding the sodium-dependent multivitamin transporter in JAr choriocarcinoma cells, *Eur. J. Nutr.*, 43, 23, 2004.

101. Conaway, R.C., Brower, C.S., and Conaway, J.W., Emerging roles of ubiquitin in transcription regulation, *Science*, 296, 1254, 2002.

102. Ezhkova, E. and Tansey, W.P., Proteasomal ATPases link ubiquitylation of histone H2B to methylation of histone H3, *Mol. Cell*, 13, 435, 2004.

103. Ke, Q. et al., Alterations of histone modifications and transgene silencing by nickel chloride, *Carcinogenesis*, 27, 1481, 2006.

104. Kacian, D.L. and Fultz, T.J., Nucleic acid sequence amplification methods, U.S. Patent, 5, 399, 1995.

105. Jaylata Devi, B. and Sharan, R.N., Progressive reduction in poly-ADP-ribosylation of histone proteins during Dalton's lymphoma induced ascites tumorigenesis in mice, *Cancer Lett.*, 238, 135, 2006.

106. Althaus, F.R., Poly ADP-ribosylation: A histone shuttle mechanism in DNA excision repair, *J. Cell Sci.*, 102, 663, 1992.

107. Gill, G., SUMO and ubiquitin in the nucleus: Different functions, similar mechanisms? *Genes Dev.*, 18, 2046, 2004.

108. Toleman, C. et al., Characterization of the histone acetyltransferase (HAT) domain of a bifunctional protein with activable *O*-GlcNAcase and HAT activities, *J. Biol. Chem.*, 279, 53665, 2004.

109. Yang, X., Zhang, F., and Kudlow, J.E., Recruitment of *O*-GlcNAc transferase to promoters by corepressor mSin3A: Coupling protein *O*-GlcNAcylation to transcriptional repression, *Cell*, 110, 69, 2002.

110. Yao, D. et al., Methylglyoxal modification of mSin3A links glycolysis to angiopoietin-2 transcription, *Cell*, 124, 275, 2006.

111. Goto, S. et al., Beneficial biochemical outcomes of late-onset dietary restriction in rodents, *Ann. N.Y. Acad. Sci.*, 1100, 431, 2007.

112. Marwick, J.A. et al., Cigarette smoke-induced oxidative stress and TGF-beta1 increase p21waf1/cip1 expression in alveolar epithelial cells, *Ann. N.Y. Acad. Sci.*, 973, 278, 2002.

113. Marwick, J.A. et al., Cigarette smoke alters chromatin remodeling and induces proinflammatory genes in rat lungs, *Am. J. Respir. Cell Mol. Biol.*, 31, 633, 2004.

114. Moodie, F.M. et al., Oxidative stress and cigarette smoke alter chromatin remodeling but differentially regulate NF-kappaB activation and proinflammatory cytokine release in alveolar epithelial cells, *FASEB. J.*, 18, 1897, 2004.

115. Donaldson, K., Gilmour, M.I., and MacNee, W., Asthma and PM10, *Respir. Res.*, 1, 12, 2000.

116. Donaldson, K. et al., Oxidative stress and calcium signaling in the adverse effects of environmental particles (PM10), *Free Radic. Biol. Med.*, 34, 1369, 2003.

117. Li, X.Y. et al., Free radical activity and pro-inflammatory effects of particulate air pollution (PM10) in vivo and in vitro, *Thorax*, 51, 1216, 1996.

118. Gilmour, P.S. et al., Adverse health effects of PM10 particles: Involvement of iron in generation of hydroxyl radical, *Occup. Environ. Med.*, 53, 817, 1996.

119. Peden, D.B., Air pollution in asthma: Effect of pollutants on airway inflammation, *Ann. Allergy Asthma Immunol.*, 87, 12, 2001.

120. Rahman, I. et al., Oxidative stress and TNF-alpha induce histone acetylation and NF-kappaB/AP-1 activation in alveolar epithelial cells: Potential mechanism in gene transcription in lung inflammation, *Mol. Cell Biochem.*, 234, 239, 2002.

121. Gilmour, P.S. et al., Histone acetylation regulates epithelial IL-8 release mediated by oxidative stress from environmental particles, *Am. J. Physiol. Lung Cell Mol. Physiol.*, 284, L533, 2003.

122. Brody, J.S. and Spira, A., State of the art chronic obstructive pulmonary disease, inflammation, and lung cancer, *Proc. Am. Thorac. Soc.*, 3, 535, 2006.

123. Barnes, P.J., Adcock, I.M., and Ito, K., Histone acetylation and deacetylation: Importance in inflammatory lung diseases, *Eur. Respir. J.*, 25, 552, 2005.

124. Yoon, K.A. et al., Novel polymorphisms in the SUV39H2 histone methyltransferase and the risk of lung cancer, *Carcinogenesis*, 27, 217, 2006.

125. Kang, J. et al., Nickel-induced histone hypoacetylation: The role of reactive oxygen species, *Toxicol. Sci.*, 74, 279, 2003.

126. Ruden, D.M. et al., Hsp90 and environmental impacts on epigenetic states: A model for the transgenerational effects of diethylstibesterol on uterine development and cancer, *Hum. Mol. Genet.*, 14, R149, 2005.

127. Galvez, A.F. et al., Chemopreventive property of a soybean peptide (lunasin) that binds to deacetylated histones and inhibits acetylation, *Cancer Res.*, 61, 7473, 2001.

128. Gozzini, A. and Santini, V., Butyrates and decitabine cooperate to induce histone acetylation and granulocytic maturation of t(8;21) acute myeloid leukemia blasts, *Ann. Hematol.*, 84, 54, 2005.

129. Hinnebusch, B.F. et al., The effects of short-chain fatty acids on human colon cancer cell phenotype are associated with histone hyperacetylation, *J. Nutr.*, 132, 1012, 2002.

130. Jung, M., Inhibitors of histone deacetylase as new anticancer agents, *Curr. Med. Chem.*, 8, 1505, 2001.

131. Moyers, S.B. and Kumar, N.B., Green tea polyphenols and cancer chemoprevention: Multiple mechanisms and endpoints for phase II trials, *Nutr. Rev.*, 62, 204, 2004.

132. Chen, J. et al., Valproic acid and butyrate induce apoptosis in human cancer cells through inhibition of gene expression of Akt/protein kinase B, *Mol. Cancer*, 5, 71, 2006.

133. Ohnishi, H. et al., Inhibition of cell proliferation by nobiletin, a dietary phytochemical, associated with apoptosis and characteristic gene expression, but lack of effect on early rat hepatocarcinogenesis in vivo, *Cancer Sci.*, 95, 936, 2004.

134. Altucci, L. et al., Acute myeloid leukemia: Therapeutic impact of epigenetic drugs, *Int. J. Biochem. Cell Biol.*, 37, 1752, 2005.

135. Kobayashi, H., Tan, E.M., and Fleming, S.E., Acetylation of histones associated with the p21WAF1/CIP1 gene by butyrate is not sufficient for p21WAF1/CIP1 gene transcription in human colorectal adenocarcinoma cells, *Int. J. Cancer*, 109, 207, 2004.

136. Toyota, M. and Issa, J.P., Epigenetic changes in solid and hematopoietic tumors, *Semin. Oncol.*, 32, 521, 2005.

137. Kothapalli, N., Sarath, G., and Zempleni, J., Biotinylation of K12 in histone H4 decreases in response to DNA double-strand breaks in human JAr choriocarcinoma cells, *J. Nutr.*, 135, 2337, 2005.

138. Perez-Plasencia, C. and Duenas-Gonzalez, A., Can the state of cancer chemotherapy resistance be reverted by epigenetic therapy? *Mol. Cancer*, 5, 27, 2006.

139. Oligny, L.L., Cancer and epigenesis: A developmental perspective, *Adv. Pediatr.*, 50, 59, 2003.

140. Momparler, R.L., Cancer epigenetics, *Oncogene*, 22, 6479, 2003.

141. Plass, C., Cancer epigenomics, *Hum. Mol. Genet.*, 11, 2479, 2002.

142. Ting, A.H., McGarvey, K.M., and Baylin, S.B., The cancer epigenome—components and functional correlates, *Genes Dev.*, 20, 3215, 2006.

143. Esteller, M., Cancer epigenomics: DNA methylomes and histone-modification maps, *Nat. Rev. Genet.*, 8, 286, 2007.

144. Esteller, M., CpG island methylation and histone modifications: Biology and clinical significance, *Ernst Schering Res. Found. Workshop*, 57, 115, 2006.

145. Schafer, S. and Jung, M., Chromatin modifications as targets for new anticancer drugs, *Arch. Pharm. (Weinheim)*, 338, 347, 2005.

146. Zhang, K. and Dent, S.Y., Histone modifying enzymes and cancer: Going beyond histones, *J. Cell Biochem.*, 96, 1137, 2005.

147. Herranz, M. and Esteller, M., DNA methylation and histone modifications in patients with cancer: Potential prognostic and therapeutic targets, *Methods Mol. Biol.*, 361, 25, 2007.

148. Ducasse, M. and Brown, M.A., Epigenetic aberrations and cancer, *Mol. Cancer*, 5, 60, 2006.

149. Kondo, Y. and Issa, J.P., Epigenetic changes in colorectal cancer, *Cancer Metastasis Rev.*, 23, 29, 2004.

150. Li, L.C., Carroll, P.R., and Dahiya, R., Epigenetic changes in prostate cancer: Implication for diagnosis and treatment, *J. Natl. Cancer. Inst.*, 97, 103, 2005.

151. Sugimura, T. and Ushijima, T., Genetic and epigenetic alterations in carcinogenesis, *Mutat. Res.*, 462, 235, 2000.

152. Verma, M. and Srivastava, S., Epigenetics in cancer: Implications for early detection and prevention, *Lancet Oncol.*, 3, 755, 2002.

153. Mielnicki, L.M., Asch, H.L., and Asch, B.B., Genes, chromatin, and breast cancer: An epigenetic tale, *J. Mammary Gland Biol. Neoplasia*, 6, 169, 2001.

154. Gallinari, P. et al., HDACs, histone deacetylation and gene transcription: From molecular biology to cancer therapeutics, *Cell Res.*, 17, 195, 2007.

155. Gray, S.G. and Teh, B.T., Histone acetylation/deacetylation and cancer: An "open" and "shut" case? *Curr. Mol. Med.*, 1, 401, 2001.

156. Gui, C.Y. et al., Histone deacetylase (HDAC) inhibitor activation of p21WAF1 involves changes in promoter-associated proteins, including HDAC1, *Proc. Natl. Acad. Sci. U S A*, 101, 1241, 2004.

157. Orr, J.A. and Hamilton, P.W., Histone acetylation and chromatin pattern in cancer, a review, *Anal. Quant. Cytol. Histol.*, 29, 17, 2007.

158. Espino, P.S. et al., Histone modifications as a platform for cancer therapy, *J. Cell Biochem.*, 94, 1088, 2005.

159. Morris, S.A. et al., Identification of histone H3 lysine 36 acetylation as a highly conserved histone modification, *J. Biol. Chem.*, 282, 7632, 2007.

160. Plath, K. et al., Role of histone H3 lysine 27 methylation in X inactivation, *Science*, 300, 131, 2003.

161. Bauer, U.M. et al., Methylation at arginine 17 of histone H3 is linked to gene activation, *EMBO. Rep.*, 3, 39, 2002.

162. Pal, S. et al., Human SWI/SNF-associated PRMT5 methylates histone H3 arginine 8 and negatively regulates expression of ST7 and NM23 tumor suppressor genes, *Mol. Cell Biol.*, 24, 9630, 2004.

Part III

Other Epigenetic Aspects of Cancer

12 Noncoding RNAs in Cancer

Maciej Szymański and Jan Barciszewski

CONTENTS

12.1 NONCODING RNAs

Over the last two decades, the concepts concerning the role of RNA molecules in the cell changed dramatically. It is now evident that RNAs are not only passive players providing templates and decoding machinery for translation of the genetic information contained in the DNA into proteins, but also perform a number of functions crucial for many other cellular processes. In the beginning of 1980s, the catalytic properties of self-splicing introns and RNase P RNAs were discovered [1]. These findings prompted formulation of the hypothesis of a primordial "RNA world" which preceded today's "DNA–protein world" [2]. According to this concept, in the prebiotic RNA world, RNA molecules were responsible not only for transmission of genetic information but also carried out all structural and enzymatic functions. Subsequently, a more stable DNA replaced RNA as the primary carrier of genetic information, and the majority of structural and catalytic functions were taken over by proteins.

In contemporary organisms, messenger RNAs (mRNA) act as intermediates providing a link between genetic information stored in genomic DNA and protein. For many years it was believed that protein-coding RNAs represent the major class of transcripts.

Other, nonprotein-coding RNAs play an accessory role in decoding the message into protein sequence. The processing and decoding information contained in mRNA requires participation of several RNA species. Ribosomal RNAs (rRNA) and transfer RNAs (tRNA) are key elements of the machinery responsible for protein biosynthesis on ribosomes. Small nuclear RNAs (snRNA), small nucleolar RNAs (snoRNA), and RNase P RNA participate in maturation and modification of primary transcripts. Several other RNAs have been implicated in other cellular processes including protein transport, synthesis of telomeres, and quality control of protein biosynthesis.

Thus, based on the criterion of their utilization as templates in protein biosynthesis, all transcripts produced in a cell can be divided into two groups, coding and noncoding. The collective name "noncoding RNAs" (ncRNAs) may, therefore, be used to describe all transcripts or their fragments that do not constitute parts of open reading frames. According to this definition ncRNAs would include all housekeeping RNAs (tRNAs, rRNAs, snRNAs, snoRNAs, etc.) as well as nonprotein-coding regions of protein-coding genes (introns, 3'- and 5'-UTRs). In this text, we use ncRNA in a narrow sense referring to RNAs with confirmed or suspected regulatory activity.

12.1.1 REGULATORY RNAS

For a long time it was believed that nonmessenger RNAs play only an accessory role in protein biosynthesis providing means for deciphering the genetic code and structural scaffolds of ribonucleoprotein particles including the ribosome. As such they could be viewed as molecular fossils preserved from the ancient RNA-world. The discovery of catalytic RNAs implicated in self-splicing of group I introns and tRNA maturation shed a new light on the role of RNAs in the cell and demonstrated that enzymatic functions are not exclusively limited to proteins [3,4]. The scope of RNA catalysis has been expanded when it was realized that role of rRNAs and snRNAs goes beyond mere structural function, but they are crucial for the catalytic activity of both ribosome and spliceosome. In fact, the catalytic centers of both the peptidyl transferase and spliceosome are composed of RNA residues with proteins playing accessory roles [5,6].

Another breakthrough in our understanding of the functions of RNA in the cell came with the identification of ncRNAs that are directly or indirectly involved in the regulation of gene expression, a function that also seemed to be an exclusive domain of proteins. It turned out that RNA regulation is widespread in both prokaryotic as well as eukaryotic organisms, and contrary to initial assessments it has a profound impact on the patterns of gene expression. In the last decade, there were a significant number of findings, which demonstrated the ubiquity of RNA-dependent regulatory mechanisms and their involvement in the precise modulation of expression of many genes. The growing interest in the field can be illustrated by the increase in the number of known ncRNAs. In 1999, when the first compilation of ncRNAs was published, there were only a handful of mammalian ncRNA sequences known [7]. Currently, the number of sequenced noncoding transcripts exceeds 30,000 [8], but recent data indicate that the variety of ncRNAs may be much larger [9].

The fundamental difference between regulatory ncRNAs and the housekeeping RNAs is their pattern of expression. Whereas, the infrastructural RNAs are expressed constitutively at constant levels in all cells, the vast majority of ncRNAs represent cell- or tissue-specific transcripts, and their expression is tightly regulated during growth and differentiation. There are also several examples of bacterial and eukaryotic ncRNA genes expression of which is associated with cell's response to stress conditions. The adaptative response to temperature stress and changes in the availability of nutrients involving small untranslated RNAs regulating proteins' functions or translation of certain mRNAs is particularly widespread in bacteria [10], but upregulation of certain noncoding transcripts upon heat shock, oxidative stress, or viral infection is also observed in eukaryotes [11–13].

Eukaryotic ncRNAs constitute a very heterogeneous group of transcripts in terms of their size and biogenesis. The lengths of mature eukaryotic ncRNAs vary from ~20 nt long microRNAs to over 100 kb long transcripts involved in regulation of genetic imprinting in mammals. In contrast to housekeeping RNAs transcribed by RNA polymerases I and III (Pol I and Pol III), the majority of eukaryotic regulatory ncRNAs known to date are products of RNA polymerase II (Pol II) and possess mRNA-like features including hypermodified 5'-cap structures and poly(A) tails. The primary transcripts from ncRNA genes can be spliced and alternative splicing is often used to generate multiple isoforms of mature RNA. However, among the eukaryotic regulatory RNAs, there are several examples of Pol III transcripts that, unlike housekeeping Pol III-dependent genes (e.g., 5S rRNA or tRNA), controlled by internal, constitutive promoters, depend on upstream promoter elements and transcription factors involved in expression Pol II genes [14]. The number of such Pol III transcripts with regulatory functions is difficult to estimate. Due to lack of poly(A) tails, they are missed by strategies based on large scale sequencing of cDNAs. However, there are several classes of Pol III transcripts expressed as abundant tissue-specific small RNAs [15] or transcripts mobilized in response to stress conditions [13,16].

Noncoding RNAs can be transcribed from independent transcription units, but they can also originate from introns excised during splicing [17]. Such, intron-derived RNAs may serve as elements of regulatory networks providing information about a gene expression status [18]. Moreover, snoRNAs and microRNAs are usually embedded within introns of host genes and their biogenesis includes splicing and subsequent processing of intronic sequences to yield mature active RNAs [19–21].

12.1.2 Noncoding RNAs in Mammalian Genomes

The significance of ncRNAs is particularly apparent when one looks at the contents of sequenced genomes. In mammals, the protein-coding part represents approximately 2% of total nuclear DNA [22,23]. Similar proportions are also observed in sequences that are transcribed. In human and mouse transcriptomes, the contribution of coding sequences is 2.1% and 2.3%, respectively. The remaining portion consists of ncRNAs, introns, and untranslated regions of protein-coding genes. This contrasts with the situation observed in the invertebrates in which 40%–70% of transcribed sequences are translated [24].

Although the initial estimates of the number of the RNA-coding genes were based primarily on the analysis of genomic sequences, there are experimental data suggesting that the ncRNAs constitute a significant part of transcriptional output from mammalian genomes. A systematic microarray screening of gene expression from human chromosomes 21 and 22 and the analysis of transcription factor binding sites revealed that nonprotein-coding RNAs account for a significant fraction of transcripts produced from these chromosomes [25]. Similar conclusion was drawn from the analysis of the male specific region of the human Y chromosome, where half of 156 transcription units encodes ncRNAs [26].

The significance of noncoding transcripts has also been confirmed in the large scale sequencing projects of full-length cDNAs from mouse and human [27,28]. In the FANTOM3 database, over one-third of annotated mouse cDNAs have been described as noncoding. In fact, it is estimated that ncRNAs constitute at least half of the mammalian transcriptome [29]. However, the cDNA sequencing projects are limited to the mRNA-like transcripts with poly(A) tails that make up only a fraction of the transcriptional output in mammalian cells, and there is a large part, accounting for approximately 40% of human transcripts that is not polyadenylated [30].

12.2 RNA-DEPENDENT REGULATORY MECHANISMS

For most of ncRNAs identified in eukaryotes their biological significance remains a mystery. Only a small number of these transcripts have been at least partially characterized in terms of activity or expression patterns. In fact, the question remains open if all of these transcripts are really functional or are just representing a transcriptional noise [31]. There are, however, several well-documented examples of mechanisms implicated in the regulation of gene expression in which ncRNAs play an essential role and as in the case of many transcripts implicated in genetic imprinting, their aberrant expression may lead to severe developmental disorders, neurobehavioral diseases, and cancer [32].

RNA-dependent regulation is observed on practically every level of transmission of genetic information from DNA to proteins. ncRNAs have been implicated in the processes associated with chromatin remodeling and setting up epigenetic marks in X-chromosome inactivation and silencing of imprinted genes [33]. RNAs involved in transcriptional regulation have been shown to affect directly the activity of transcription factors [34,35] and RNA polymerase II [36]. On a posttranscriptional level, specific ncRNAs were shown to affect RNA modification and editing [37] as well as mRNA stability and translation [38].

12.2.1 REGULATION OF DNA METHYLATION

The most straightforward way in which RNAs can participate in epigenetic regulation is by direct involvement in DNA methylation. It has been demonstrated that RNA can participate in DNA demethylation in vitro [39]. The double-stranded (ds) RNA-directed DNA methylation is widely used in plants in setting up regions of transcriptionally inactive heterochromatin. In this process methylation is guided by viral, transposon, or transgene sequences, as well as dsRNAs produced from single-stranded RNA templates by RNA-dependent RNA polymerases methylation guides [40]. The short interfering RNA (siRNA) directed DNA methylation was also demonstrated in human cells [41].

The first example suggesting that endogenous ncRNAs may be involved in the regulation of DNA methylation in mammalian cells is tissue-specific expression of sphingosine kinase 1 (Sphk1), an important enzyme responsible for biosynthesis of sphingosine 1-phosphate, an extracellular mediator influencing many biological processes, including cell proliferation and apoptosis. In rat, *Sphk1* gene is expressed in a tissue-specific manner and produces several mRNA isoforms which result from the utilization of alternative transcription initiation sites and alternative splicing of the untranslated first exons located within a 3.7 kb CpG island containing a tissue-specific, differentially methylated region (T-DMR) approximately 0.8 kb upstream of the first exon of *Sphk1a*. The methylation status of the T-DMR varies in different tissues and changes during development [42,43].

In addition to the protein-coding transcripts, a series of noncoding antisense (Khps1) RNAs of varying lengths (600–20,000 nt) is produced. One of the antisense transcripts initiates within the CpG island and overlaps the T-DMR. It has been demonstrated that the antisense RNA is involved in demethylation of CpG and methylation of non-CpG sites within the T-DMR [43]. This observation suggested that the expression of tissue-specific variants of Sphk1 mRNA might depend on different methylation patterns of the T-DMR regulated by expression of antisense RNAs. The preservation of gene structure and tissue-specific methylation patterns between rat, mouse, and human suggests that the regulation of Sphk1 expression is an evolutionarily conserved feature [44].

Deregulation of Sphk1 expression has been implicated in several cancers. In contrast to Sphk2, that plays a role in promoting apoptosis, Sphk1 has prosurvival activity [45]. In mouse and human cancers, overexpression of Sphk1 resulting in altered metabolism of sphingolipids is associated with increased proliferation and resistance to apoptosis [46–49]. It is not known what changes in the methylation of the T-DMR occur during oncogenic transformation, but the altered patterns of expression of antisense RNAs may be responsible for increased expression of Sphk1.

Regulatory mechanisms of tissue-specific gene expression involving antisense RNA-dependent DNA methylation may be more widespread in mammals. Alternative, noncoding first exons overlapping CpG islands are present in many other genes with tissue-specific mRNA isoforms. Such a gene structure is often phylogenetically conserved [50,51]. There are also numerous CpG islands featuring T-DMRs [52–54] and differential DNA methylation that results in tissue-specific expression was confirmed for maspin [55], a tumor suppressor in human breast, involved in cell adhesion and motility, apoptosis, and angiogenesis [56]. Although, for the majority of predicted T-DMRs, there is no direct evidence of similar mechanisms it seems possible that antisense RNAs may be involved in tissue-specific expression of other genes. It is estimated that antisense transcripts are associated with approximately 15% of human protein-coding genes, but the functions of most of these RNAs are unknown [57].

12.2.2 Chromatin Remodeling

Noncoding RNAs have been shown to play a key role in gene silencing through chromatin remodeling. The most spectacular example of such an activity is transcriptional inactivation of one of the mammalian X-chromosomes in XX cells, required for dosage equalization of X-linked genes between males and females [58]. The X-inactivation follows the "n − 1 rule" and results in shutting off transcription of all but one X-chromosome in the cell. The key genetic element responsible for X-chromosome inactivation is *Xic* (X-inactivation center) locus encoding long, spliced, and polyadenylated Xist RNA (X-inactive specific transcript) transcribed from the inactive X-chromosome [59]. The expression of *Xist* is essential for the initiation of silencing and its targeted deletion results in an inability of deficient X-chromosome to undergo inactivation [60].

Association of Xist RNA with the chromatin is followed by an exclusion of Pol II and associated transcription factors from the Xist RNA-coated chromatin resulting in creation of a transcriptionally silent nuclear compartment [61]. Although the mechanism underlying Xist RNA coating is not clear, it is assumed that there exist high affinity binding sites for Xist RNA [53]. A good candidate for such a task is LINE-1 (L1) repetitive element [62] enriched on the X-chromosomes. The existence of X-linked genes that escape inactivation at sites with lower density of L1 seems to support this hypothesis [63]. X-chromosome coating by Xist RNA triggers chromatin modification involving both CpG methylation of the DNA [64] and loss of histone modifications characteristic for euchromatic, transcriptionally active chromatin, followed by global hypoacetylation of histone H4 [65,66]. Xist RNA-coated X-chromosomes are enriched in several histone modifications including dimethylation of H3 Lys-9 [67], trimethylation of H3 Lys-27 [68,69], and dimethylation of H4 Lys-20 [70]. Inactive X-chromosome is also enriched in ubiquitinated histone H2A [71] and a histone H2A variant, macroH2A1.2 [72,73] a transcriptional repressor interfering with histone acetylation [74,75]. The changes in histone composition and modifications depend on expression of Xist RNA that recruits macroH2A1.2 [76] and may play a role in recruitment of chromodomain-containing members of Polycomb group (PcG) repressive complexes 1 (PRC1) and 2 (PRC2) to the X-chromosome [77–80].

Noncoding RNAs were also shown to play a pivotal role in chromatin modifications and gene silencing of imprinted genes. In mammals, there are approximately 70 imprinted genes identified so far [81], but computational analysis of the mouse genome predicted about 600 genes that can potentially show imprinted expression [82]. The imprinted genes clusters frequently include, in addition to protein-coding genes, imprinted ncRNA genes [83]. Within such clusters, the protein-coding and RNA-coding genes usually show reciprocal imprinted expression patterns suggesting a possible role of ncRNAs in establishing or maintaining the imprinted status. The exact functions of the majority of imprinted ncRNAs are unknown. However, in two cases the essential role of ncRNAs in imprinted expression of the genes has been established.

In human, imprinting status of one of the domains within the Beckwith–Wiedeman and Angelman syndromes region at 11p15.5 is controlled by a differentially methylated CpG island (*KvDMR1*) within intron 10 of the *KCNQ1* (*KvLQT1*) gene. Expression of the maternal genes is associated with methylation of this region on the maternal allele. The unmethylated paternal allele produces an ncRNA (LIT1/Kcnq1ot1) antisense to the KCNQ1 that is expressed in most human tissues. Deletion of the CpG island abolishes transcription of the antisense RNA and results in activation of paternal alleles of several maternally expressed genes [84]. Similar dependence of imprinted expression of maternal genes on paternal ncRNA was observed in imprinted domain on mouse chromosome 17 harboring a gene for Igf2r (insulin-like growth factor type 2 receptor). The second intron of *Igf2r* gene, a 3.7 kb imprinting control element (region 2), contains maternally methylated CpG island [85]. From the unmethylated paternal allele a 108 kb long, unspliced noncoding Air RNA (antisense *Igf2r* RNA), overlapping 30 kb of the *Igf2r*, is transcribed. The expression of Air RNA is essential for the silencing not only of the *Igf2r*, but also of the two other maternally expressed genes *Slc22a2* and *Slc22a3* located 110 and 155 kb downstream of the *Igf2r* [86,87].

The precise mechanisms of gene silencing at these imprinted domains are not known. There are, however, certain parallels with X-chromosome inactivation. The expression of ncRNAs serves as a signal that initiates chromatin silencing. Xist-, LIT1-, and Air-induced silencing is bidirectional and affects genes located both downstream and upstream of the RNA gene and is associated with changes in histone modifications [88,89]. Although there is no direct evidence, the structural features of imprinted ncRNAs may serve as beacons for chromatin modifying factors, but it has also been suggested that the process is initiated by a mere act of their transcription [90,91]. Given the number of ncRNAs and the extent of antisense transcription in mammals, it seems plausible that the mechanisms of RNA-dependent chromatin modifications are more widespread and can also affect other than imprinted loci.

12.2.3 REGULATION OF TRANSCRIPTION

Mammalian ncRNAs have been implicated in the regulation of activity of transcription factors and RNA polymerase. The regulatory mechanisms involve either direct interactions with the components of transcriptional machinery or a modulation of transcription factors availability. Protein binding resulting in the change of transcriptional activity has been demonstrated in the case of steroid receptor activator RNA (SRA RNA) that plays a role of a coactivator of several nuclear receptors of steroid hormones [34,92]. Transcriptional regulation of steroid hormones-responsive genes is mediated by the interactions of SRA RNA with SRC-1 (steroid receptor coactivator 1) [34] and hormone induced transcriptional repressors: SHARP, (SMRT/HDAC1 associated repressor protein) [93], and SLIRP (SRA stem-loop interacting RNA-binding protein) [94]. Modulation of the activity of both transcriptional activators and repressors suggests a role of SRA in a fine-tuning of gene expression through a competition of proteins for the RNA component. Interestingly, the usage of alternative transcription start sites of the SRA-encoding gene yields either an ncRNA or a protein-coding mRNA [95].

Transcriptional coactivation by ncRNA was also demonstrated for neuronal genes containing promoters with 20–23 bp long neuron-restrictive silencer element/repressor element 1 (NRSE/RE1). Transcriptional silencing of these genes and restriction of their expression to neurons depend on NRSF/REST (neuron-restrictive silencing factor/repressor element 1 silencing transcription factor). Upon binding to NRSE/RE1, a complex of NRSF/REST with cofactor proteins transcription of respective genes is repressed [96]. In adult neurons, the inhibitory effect of NRSE/RE1 is suppressed by an NRSE small modulatory RNA (NRSE smRNA). This short (~20 bp long) double-stranded RNA has the nucleotide sequence identical to NRSE/RE1. NRSE smRNA binds NRSF/REST and prevents its association with co-repressor proteins, thus converting it into transcriptional activator [97].

Direct involvement of ncRNA in regulation of transcription factor's activity has been demonstrated for a Pol III-transcribed 7SK RNA that binds to and inhibits positive transcription elongation factor b (P-TEFb), a cofactor of RNA polymerase II required for a transition from abortive to productive elongation [98,99]. 7SK RNA and HEXIM1/MAQ1 protein form a complex with P-TEFb and suppress its cyclin-dependent kinase 9 (Cdk9) activity responsible for phosphorylation of the Pol II largest subunit's C-terminal domain. [100]. Under stress conditions, P-TEFb is released from the complex and made available for transcription of stress-induced genes [101].

Regulation of nuclear trafficking of transcription factor was demonstrated in the case on NRON RNA that affects expression of genes activated by the nuclear factor of activated T cells (NFAT). NRON RNA is responsible for inhibition of NFAT's nuclear import by competition for binding to importin-β family members essential for this process [102].

12.2.4 Regulation of Translation

Translational regulation by ncRNAs that affect the availability of ribosome binding site in mRNA is common in bacteria [10]. In mammals, there are two classes of RNAs that are implicated in the regulation of translation. Brain specific cytoplasmic RNAs BC1 (brain cytoplasmic RNA 1) in rodents and BC200 (brain cytoplasmic RNA 200 nt) in primates are Pol III transcripts that show specific neuronal expression [103] and their levels depend on neuronal activity [104]. Translational regulation exerted by these RNAs follows two separate pathways. Nonspecific inhibition of translation is associated with the BC1/BC200 RNA interaction with a poly(A)-binding protein (PABP). The binding of BC1/BC200 RNAs to PABP interferes with the formation of the pre-initiation complex and global suppression of translation [105]. The second pathway affects translation of mRNAs regulated by the fragile X mental retardation protein (FMRP). BC1/BC200 RNAs recognize complementary sequences within target mRNAs and bind FMRP that in turn suppresses translation by as yet unknown mechanism [106–108].

The best studied and most commonly used mechanism of translational regulation in eukaryotes involves a class of small, 17–25 nt long microRNAs. MicroRNAs are the smallest known functional RNAs processed from longer transcripts transcribed by polymerase II [109]. A two-step maturation of microRNAs involves cleavage by double-stranded RNA specific ribonucleases Drosha and Dicer. In a first step, Drosha releases 60–110 nt long hairpin precursors (pre-miRNA) from primary transcripts (pri-miRNA). The pre-miRNAs are exported to the cytoplasm, where mature micro-RNAs are released by Dicer [110].

MicroRNAs are involved in posttranscriptional regulation of gene expression in all complex organisms (REF). However, the mechanisms of action of plant and animal microRNAs are different. The majority of mammalian microRNAs are involved in inhibition of translation while plant microRNAs influence mRNAs stability guiding specific hydrolysis of target mRNAs. Target recognition depends on complementary base pairing between microRNA and mRNA and the mechanism of microRNA's action depends on the extent of complementarity between microRNA and a corresponding binding site. The vast majority of animal microRNAs recognize partially complementary cis-regulatory elements within 3′-UTRs while plant microRNAs recognize fully complementary sites usually located within open reading frames [111]. However, both in plants and in animals, there are exceptions to this general rule. Mammalian miR-196 and Epstein–Barr virus miR-BART2 bind fully complementary sites and induce cleavage of target mRNAs encoding a homeobox protein HOX8B and viral DNA polymerase, respectively [112,113].

Translational repression by microRNAs affects the initiation step and target mRNAs must possess a 7-methyl guanosine cap [114]. The cap structure is bound by Argonaute family protein AGO2 interfering with eIF4A (eukaryotic translation initiation factor 4A) dependent initiation [115]. Another protein contributing to miRNA-induced silencing is a ribosome dissociation factor eIF6 that inhibits joining of 40S and 60S ribosomal subunits and the formation of 80S ribosome [116]. A multiprotein complex containing eIF6 was shown to interact with RISC and is recruited by mRNA-bound microRNAs [117].

12.3 NONCODING RNAs IN CANCER

Altered expression of ncRNAs is observed in many forms of cancer. With the exception of microRNAs, the functions of the majority of these transcripts are unknown and in most cases it is difficult to judge if their overexpression or lack of expression is the cause or a consequence of the changes in cells' gene expression program which ultimately bring about cancer.

12.3.1 MicroRNAs

MicroRNAs are the most extensively studied class of ncRNAs that have been shown to regulate expression of many genes crucial for development and differentiation. There is accumulating evidence that microRNAs, together with protein transcription factors, are key elements of regulatory pathways determining the patterns of gene expression in eukaryotic cells [118]. The significance of microRNA-based regulation is also supported by its conservation throughout evolution [119,120].

Exceptional regulatory potential of microRNAs results from the fact that a single mRNA can be targeted by multiple microRNAs and any single microRNA may participate in the regulation of many different targets and influence many pathways affecting cell's behavior. Although the majority of microRNAs seem to participate in the regulation of a small subset of genes, there are microRNA species for which potential binding sites have been found in hundreds of mRNAs. A systematic analysis of human mRNAs revealed that over 6000 genes could be targeted by more than one microRNAs. Among these, 255 genes show target sites for 15 or more individual microRNAs. This set is enriched in genes encoding nuclear proteins involved in regulation of transcription and proteins associated with development and their targeting by multiple microRNAs suggests that expression of these genes is particularly tightly controlled. This is also consistent with the enrichment of cancer-related genes within the subset of genes with over 30 putative microRNA target sites [121]. It is, therefore, not surprising that the malignant growth is often accompanied by alterations of microRNAs' expression.

The significance of microRNAs for cancer development was also inferred from the analysis of genomic distribution of microRNA-coding regions. MicroRNA-coding genes are often present within genomic regions, aberrations at which correlate with various forms of cancer. The association of microRNAs with both minimal regions of loss of heterozygosity and minimal regions of amplification as well as breakpoint regions suggested that microRNAs could act as oncogenes or tumor suppressors [122]. Similar correlation was also found in the mouse genome, where microRNAs often co-localize with known mouse cancer susceptibility genes [123].

The expression profiles of microRNAs differ between various cell lines. It has been, however, noted that oncogenic transformation is generally associated with decreased expression of microRNAs when compared with healthy tissues. This observation is consistent with their role in the maintenance of differentiated state [124]. In some cases, reduced microRNA expression may be due to downregulation of proteins involved in microRNA biogenesis, including Dicer nuclease responsible for the final step of microRNA precursors' maturation [125]. In fact, silencing of the genes encoding key components of microRNA processing machinery followed by decreased mature microRNA levels promotes tumorigenesis [126]. A deficiency in posttranscriptional processing has been implicated in significant downregulation of mature miR-143 and miR-145 in colorectal cancers. Among potential targets of these RNAs, there are genes that have been implicated in oncogenesis including RAF1 kinase and G-protein γ7. Therefore, reduced levels of these microRNAs could contribute to increased expression of oncogenes and cancer development [127].

Another factor that can influence expression levels of microRNAs is the stability of either precursors or mature RNA. MiR-29a and miR-29b show differential expression during the cell cycle. Although, both microRNAs are expressed as a single polycistronic precursor RNA only miR-29a is detectable at all stages of the cell cycle. Significant enrichment of mature miR-29b is observed only during mitosis and is due to its increased stability in mitotic cells [128].

Systematic analysis of microRNA expression profiles demonstrated that there are significant differences between normal and tumor tissues. In colorectal cancer, it has been shown that oncogenic transformation is associated with changes in expression of 13 microRNAs and that levels of miR-31 correlate with the stage of tumor [129]. A comprehensive analysis of microRNA expression in normal breast tissues and breast tumors revealed significant differences not only between normal and malignant cells but also between different subtypes of cancer [130]. In recent years, microRNA profiling of cancer tissues has become one of the most widely used methods of tumors' analysis [131–133]. The results of these studies demonstrated that microRNAs can serve as excellent molecular markers for classification of human cancers and their expression profiles provide very specific signatures for virtually every type of cancer that in many cases are superior to commonly used mRNA profiles [124].

Although global analysis of microRNAs' expression can be employed for diagnostic purposes or classification of tumors, the understanding of their role in oncogenesis calls for a functional analysis. In recent years, we observed a remarkable progress in characterization of functions of particular microRNAs in cancer development. Despite overall downregulation of microRNAs expression in cancer cells, individual microRNA species can show either elevated or reduced levels when compared with normal tissues. On the basis of expression levels and specific functions of cancer-related microRNAs they can be classified as tumor suppressors or oncogenes [134].

Expression of several microRNAs has been linked to cell proliferation. One of the most extensively studied tumor suppressor microRNAs is let-7. The genes encoding that members of let-7 family are often associated with chromosomal regions associated with cancer [122] and its expression is significantly downregulated in lung cancer cells [135]. Its potential tumor suppressor activity depends on the inhibition of the human RAS genes [136] as well as other proto-oncogenes associated with cell cycle regulation [137]. Thus, high let-7 expression in normal tissues is responsible for control of cell proliferation and its aberrations may lead to uncontrolled growth. In neuroblastoma, overexpression of three microRNAs, miR-9, miR-125a, and miR-125b correlates with decreased growth. One of target genes for these microRNAs is a truncated isoform of the signal-transducing neurotrophin receptor tropomyosin-related kinase C (trkC), which when overexpressed stimulates cell growth [138]. Let-7 microRNA is also involved in the regulation of high mobility group A2 protein (HMGA2) involved in chromatin remodeling and aberrantly expressed in a number of human tumors. Disruption of let-7-dependent regulation by chromosomal rearrangement or mutation of the let-7 binding sites within the 3′-UTR of HMGA2-encoding mRNA results in oncogenic transformation [139,140]. Tumor suppressor activity has also been suggested for miR-16 family members that act as negative regulators of cell growth and cell cycle progression. Members of this microRNA family include miR-16, miR-15a, miR-15b, miR-195, miR-103, and miR-107 and target a number of genes responsible for cell cycle progression from G0/G1 to S phase. Among potential targets of these microRNAs, there are regulators of G1 phase CDK6, CDC27, an activator of NF-κB signaling CARD10 [141], and an antiapoptotic factor BCL2 [142]. Interestingly, although there is a functional redundancy among the family members a phenotypic effect can be brought about even by a small change in miR-16 copy number. This may explain the effects of deletions or point mutations at the miR-15a-miR-16-1 locus at 13q14 observed in chronic lymphocytic leukemias [143] that rarely affect both alleles. Thus, at least in some cases, the dosage of particular microRNAs in the cell may play an important role in the regulation of complex gene networks [141].

In certain breast cancer cell lines, there is marked overexpression of miR-206 involved in the regulation of a ligand-activated oncogene, estrogen receptor-α (ER-α). The inhibitory effect of miR-206 was observed on both mRNA and protein levels and could be reversed by an antago-miR. Interestingly, the expression of miR-206 itself is specifically downregulated by an ER-α agonist E_2 (17β-estradiol). Suppression of ER-α expression results in increased apoptosis and decreased cell growth [144]. These results also suggested a role of miR-206 in normal breast tissue. Hormonal regulation of its expression and in consequence regulation of ER-α may be responsible for induction of ductal growth in response to high estrogen production during pregnancy.

Oncogenic properties have been shown for miR-21 and its overexpression was observed in many forms of cancer [145–148]. The aberrant expression of miR-21 glioblastoma multiforme and breast cancer is responsible for suppression of apoptosis. The inhibition of miR-21 function using synthetic complementary antago-miRs triggered apoptosis by activation of caspases and resulted in decreased proliferation [147,149]. Recently it has been shown that one of the targets regulated by miR-21 is tumor suppressor protein programmed cell death 4 (PDCD4) loss of which is frequently observed in cancer cells [150]. Overexpression of miR-21 was also found in human hepatocellular cancer where it is responsible for repression of PTEN tumor suppressor (phosphatase and tensin homolog) involved in pathways regulating proliferation, cell migration, and invasion [148]. Another tumor suppressor gene targeted by miR-21 is tropomyosin 1 (TPM1), downregulation of which promotes proliferation and tumor progression [151].

The distinction between oncogenic and tumor suppressor properties of microRNAs is not always unambiguous. The human 13q31-32 region contains a *c13orf25* gene that is amplified and overexpressed in certain human lymphomas and lung cancer cell lines [152–154]. The *c13orf25* is a host gene encoding a cluster (miR-17-92) of seven microRNAs. There is a correlation between overexpression of miR-17-92 cluster and cancer development. Thus, the miR-17-92 cluster acts as an oncogene. On the other hand in lymphoma cells overexpressing c-Myc oncogene, miR-17-5p and miR-20a encoded by miR-17-92 seem to suppress tumor development, downregulating expression of transcription factor E2F1, which is involved in the regulation of cell cycle and which is induced by c-Myc. Interestingly, the c-Myc also stimulates expression of *c13orf25* and itself is upregulated by E2F1. Thus miR-17-5p and miR-20a are responsible for balanced expression of both c-Myc and E2F1 [155]. MiR-17-5p was also shown to stimulate proliferation by targeting a tumor suppressor, Rbl2 [124].

The altered profiles of microRNAs' expression in tumors can at least in some cases be attributed to altered DNA methylation profiles, that may reflect global changes in methylation observed in cancer cells. MiR-127 is embedded within a CpG island. Methylation of the CpG island in prostate and bladder tumors results in downregulation of microRNA's expression that is highly expressed in normal tissues. Treatment with demethylating agents significantly increased miR-127 expression [156]. Another microRNA associated with a normally methylated CpG island is let-7a-3. Hypomethylation of this CpG island in some lung adenocarcinomas results in overexpression of let-7a-3, which in turn leads to aberrant expression of genes involved in cell proliferation, adhesion, and differentiation [157]. Hypermethylation of a promoter region is also responsible for the epigenetic silencing of miR-124a and the activation of Cdk6 (cyclin D kinase 6) oncogene [158].

Altered expression of microRNAs in cancer cells may also depend on other factors. It has been proposed that the efficiency of precursor RNA processing may play a role in the availability of mature microRNAs and that the maturation may somehow be regulated developmentally. It has recently been shown that the imprinted H19 RNA serves as a host for miR-675 [159]. However, when expressed, full-length H19 RNA can be detected in the cytoplasm [160]. This suggests that the microRNA processing is rather inefficient. Since H19 RNA seems to play a role during embryogenesis, it has been proposed that the efficiency of maturation may vary at different developmental stages [159]. Developmental regulation at the level of Dicer cleavage was observed in the expression of miR-138. Although, precursor pre-miR-138-2 is expressed and exported from the nucleus in various tissues, the mature miR-138 is present only in brain. It suggests that the final step of pre-microRNA shows tissue specificity and may depend on cell-specific expression of inhibitors or activators of Dicer activity [161].

12.3.2 mRNA-Like Noncoding Transcripts

The first human noncoding transcript that has been linked to cancer is a product of *H19* gene on chromosome 11p15.5 adjacent to the *IGF2* (insulin-like growth factor 2) gene. *IGF2* and *H19* genes are reciprocally imprinted showing paternal and maternal expression, respectively. The expression levels of H19 change during development showing maximum in most fetal tissues. Changes in the methylation status of differentially methylated regions (DMR) upstream of the *H19* gene resulting in the loss of imprinting of *H19* and *IGF2* and biallelic expression of either gene can cause malignant cell growth [162], but the role of H19 RNA in the origin of different forms of cancer is not clear and may depend on the tissue type. Some data suggested that H19 is a tumor suppressor that reduces tumorigenicity and growth [163]. However, its elevated expression in other tumors implies that it has oncogenic properties [164,165]. Recent finding that H19 RNA is also a precursor of miR-675 suggests that to some extent the effects of its expression may be linked to microRNA-dependent pathway and translational regulation [159].

The imprinting abnormalities within *CDKN1C/LIT1* domain have also been linked to cancer. The analysis of the imprinting status of the LIT1 intronic RNA gene demonstrated that the epigenetic changes affecting the histone H3 modifications and CpG methylation within the *KvDMR1* and loss of imprinted expression of the ncRNA may play a role in the origin of colorectal and esophageal cancers [166,167].

Noncoding RNAs are associated with prostate cancers. Elevated expression of the *DD3/PCA3* (prostate cancer antigen 3) gene was observed in over 90% of prostate tumor cases [168]. Another upregulated prostate-specific gene is an androgens-responsive *PCGEM1* [169] and its overexpression has been correlated with increased proliferation [170]. The *OCC-1* (overexpressed in colon carcinoma 1) gene transcripts, which are absent or expressed at low levels in normal mucosa, show elevated expression in colon carcinomas [171]. Altered expression of other ncRNAs have also been linked to ovarian cancer [172], lung adenocarcinoma [173], and B-cell chronic lymphocytic leukemia [174] but the functions of these RNAs are not known.

Overexpression of ncRNA CUDR (cancer upregulated drug resistant) was reported in human squamous carcinoma cell lines resistant to treatment with doxorubicin. Cells expressing the transfected *CUDR* gene are resistant to doxorubicin and etoposide and show downregulation of caspase 3 resulting in suppression of apoptosis. Among normal human tissues, *CUDR* is expressed only in placenta. However, its expression could be detected in a number of cancer cell lines including hepatocarcinoma, breast, colon, and lung tumors. The mechanisms by which CUDR RNA decreases drug sensitivity and affects expression of caspase 3 are unknown [175]. CUDR is one of the few strong RNA markers of cancerogenesis and its highly specific expression makes it potentially a good candidate for development of anticancer drugs.

The number of potential cancer-related ncRNA genes is probably much larger. A comprehensive analysis of ESTs from head, neck, and thyroid cancers revealed over 2000 new antisense ncRNAs mapped to introns of protein-coding genes [176]. It has been noted that the expression levels of intronic antisense transcripts is correlated with degree of tumor differentiation in prostate tumors but the exact role of this phenomenon is unknown [177]. The number of potential cancer-related noncoding transcripts is constantly growing. Recently, a large number of ncRNAs originating from ultraconserved genomic regions [178] and previously unidentified transcriptionally active sites [179] were identified and their expression profiles suggest at least some of them may contribute to oncogenesis.

12.4 CONCLUDING REMARKS

With the deciphering of many cellular pathways that can be affected by aberrant expression of ncRNAs to in cancer development it has become obvious that the classical models of tumorigenesis have to be revised. The terms oncogene and tumor suppressor gene, earlier used only in relation to protein-coding genes, now include also ncRNAs. As demonstrated in case of many microRNAs, their correct expression is equally important for the maintenance of differentiated state and cell growth control.

In recent years, there was a tremendous growth in the number of new ncRNAs identified in mammalian cells. For the majority of these transcripts the mechanisms of their action or their significance remain a mystery. Therefore, the most important challenge for the future research will be functional characterization and identification of molecular pathways in which ncRNAs take part. Since deregulation of some of ncRNA-coding genes seems to contribute to the origin of many human diseases it is also necessary to reveal the regulatory mechanisms that control their expression.

Another important issue that has to be addressed is a problem of RNA modifications. The information on modified nucleotides in mRNAs and ncRNAs is usually lost in the process of isolation, cloning, and sequencing of cDNAs. In fact, the extent of RNA modifications affecting other than certain housekeeping RNAs is unknown. It seems reasonable to assume that the functions of some protein-coding and ncRNAs may depend on posttranscriptional modification. $2'$-O-methylation was identified in Piwi-interacting RNAs (piRNA) [180]. Pseudouridylation of SRA

RNA was shown to affect its function and thus contribute to the regulation of steroid hormones-regulated genes [181]. Adenosine to inosine (A-to-I) editing inhibits processing of the precursor miR-151 by Dicer [182] and has been shown in several other human microRNAs [183]. It has also been proposed that A-to-I editing at microRNA target sites may also contribute to the regulation of translation of a small subset of mRNAs [184]. On the other hand, A-to-I editing of serotonin 5-HT2C receptor mRNA is regulated by MBII-52 snoRNA-mediated 2-O-methylation [185]. Thus, the editing and modifications of RNA, some of which may depend on expression of snoRNAs, represent an additional level of regulation that we only begin to understand. Indeed, computational analysis of human genome revealed that there are likely hundreds of new snoRNAs with as yet unidentified targets [186].

From the medical point of view, ncRNAs present an attractive research subject for two main reasons. First, the expression profiles of noncoding transcripts can be used as very sensitive molecular markers. This applies to both microRNAs as well as longer mRNA-like transcripts that show differential expression in healthy and malignant tissues. Moreover, in both cases it has been shown that for certain tumors, ncRNAs expression profiles can differentiate cancer subtypes [124,173,175]. Second, direct involvement of certain ncRNAs in cancer development makes them potential targets for anticancer drug development. The application of antago-miRs in cell cultures demonstrated the feasibility of such an approach for inactivation of overexpressed microRNA oncogenes [144,147,187] and RNA interference strategy could be applied to target longer ncRNAs.

ACKNOWLEDGMENT

This work was supported by a grant N301 128 31/3864 from the Polish Ministry of Science and Higher Education.

REFERENCES

1. Cropp, T.A. and Chin, J.W., Expanding nucleic acid function in vitro and in vivo, *Curr. Opin. Chem. Biol.*, 10, 601, 2006.
2. Michalak, P., RNA world—the dark matter of evolutionary genomics, *J. Evol. Biol.*, 19, 1768, 2006.
3. Kruger, K. et al., Self-splicing RNA: Autoexcision and autocyclization of the ribosomal RNA intervening sequence of *Tetrahymena*, *Cell*, 31, 147, 1982.
4. Kole, R. et al., *E. coli* RNAase P has a required RNA component, *Cell*, 19, 881, 1980.
5. Rodnina, M.V., Beringer, M., and Wintermeyer, W., How ribosomes make peptide bonds, *Trends Biochem. Sci.*, 32, 20, 2007.
6. Valadkhan, S., snRNAs as the catalysts of pre-mRNA splicing, *Curr. Opin. Chem. Biol.*, 9, 603, 2005.
7. Erdmann, V.A. et al., Collection of mRNA-like non-coding RNAs, *Nucleic Acids Res.*, 27, 192, 1999.
8. Szymanski, M., Erdmann, V.A. and Barciszewski, J., Noncoding RNAs database (ncRNAdb), *Nucleic Acids Res.*, 35(Database issue), D162, 2007.
9. Kapranov, P. et al., RNA maps reveal new RNA classes and a possible function for pervasive transcription, *Science*, 316, 1484, 2007.
10. Majdalani, N., Vanderpool, C.K., and Gottesman, S., Bacterial small RNA regulators, *Crit. Rev. Biochem. Mol. Biol.*, 40, 93, 2005.
11. Wang, Y., et al., Characterization of *adapt33*, a stress-inducible riboregulator, *Gene Expr.*, 11, 85, 2003.
12. Hollander, M.C., Alamo, I., and Fornace, A.J. Jr., A novel DNA damage-inducible transcript, *gadd7*, inhibits cell growth, but lacks a protein product, *Nucleic Acids Res.*, 24, 1589, 1996.
13. Espinoza, C.A., Goodrich, J.A., and Kugel, J.F., Characterization of the structure, function, and mechanism of B2 RNA, an ncRNA repressor of RNA polymerase II transcription, *RNA*, 13, 583, 2007.
14. Szymanski, M. and Barciszewski, J., RNA regulation in mammals, *Ann. N.Y. Acad. Sci.*, 1067, 461, 2006.
15. Skryabin, B.V. et al., The BC200 RNA gene and its neural expression are conserved in Anthropoidea (Primates), *J. Mol. Evol.*, 47, 677, 1998.
16. Valgardsdottir, R. et al., Structural and functional characterization of noncoding repetitive RNAs transcribed in stressed human cells, *Mol. Biol. Cell.*, 16, 2597, 2005.

17. Mattick, J.S., Non-coding RNAs: The architects of eukaryotic complexity, *EMBO Rep.*, 2, 986, 2001.

18. Mattick, J.S., Challenging the dogma: The hidden layer of non-protein-coding RNAs in complex organisms, *Bioessays*, 25, 930, 2003.

19. Runte, M. et al., The IC-SNURF-SNRPN transcript serves as a host for multiple small nucleolar RNA species and as an antisense RNA for UBE3A, *Hum. Mol. Genet.*, 10, 2687, 2001.

20. Rodriguez, A., Griffiths-Jones, S., Ashurst, J.L. et al., Identification of mammalian microRNA host genes and transcription units, *Genome Res.*, 14, 1902–1910, 2004.

21. Ying, S.Y. and Lin, S.L., Intron-derived microRNAs-fine tuning of gene functions, *Gene*, 342, 25, 2004.

22. Venter, J.C. et al., The sequence of the human genome, *Science*, 291, 1304, 2001

23. Gregory, S.G. et al., Initial sequencing and comparative analysis of the mouse genome, *Nature*, 420, 520, 2002.

24. Frith, M.C., Pheasant, M., and Mattick, J.S., Genomics: The amazing complexity of the human transcriptome, *Eur. J. Hum. Genet.*, 13, 894, 2005.

25. Cawley, S. et al., Unbiased mapping of transcription factor binding sites along human chromosomes 21 and 22 points to widespread regulation of noncoding RNAs, *Cell*, 116, 499, 2004.

26. Skaletsky, H. et al., The male-specific region of the human Y chromosome is a mosaic of discrete sequence classes, *Nature*, 423, 825, 2003.

27. Maeda, N. et al., Transcript annotation in FANTOM3: Mouse gene catalog based on physical cDNAs, *PLoS Genet.*, 2, e62, 2006.

28. Ota, T. et al., Complete sequencing and characterization of 21,243 full-length human cDNAs, *Nat. Genet.*, 36, 40, 2004.

29. Carninci, P. and Hayashizaki, Y., Noncoding RNA transcription beyond annotated genes, *Curr. Opin. Genet. Dev.*, 17, 139, 2007.

30. Cheng, J. et al., Transcriptional maps of 10 human chromosomes at 5-nucleotide resolution, *Science*, 308, 1149, 2005.

31. Ponjavic, J., Ponting, C.P., and Lunter, G., Functionality or transcriptional noise? Evidence for selection within long noncoding RNAs, *Genome Res.*, 17, 556, 2007

32. Szymanski, M. et al., A new frontier for molecular medicine: Noncoding RNAs, *Biochim. Biophys. Acta.*, 1756, 65, 2005.

33. Yang, P.K. and Kuroda, M.I., Noncoding RNAs and intranuclear positioning in monoallelic gene expression, *Cell*, 128, 777, 2007.

34. Lanz, R. et al., A steroid receptor coactivator, SRA, functions as an RNA and is present in an SRC-1 complex, *Cell*, 97, 7, 1999.

35. Yik, J.H., et al., Inhibition of P-TEFb (CDK9/Cyclin T) kinase and RNA polymerase II transcription by the coordinated actions of HEXIM1 and 7SK snRNA, *Mol. Cell*, 12, 971, 2003.

36. Allen, T.A. et al., The SINE-encoded mouse B2 RNA represses mRNA transcription in response to heat shock, *Nat. Struct. Mol. Biol.*, 11, 816, 2004.

37. Vitali, P. et al., ADAR2-mediated editing of RNA substrates in the nucleolus is inhibited by C/D small nucleolar RNAs, *J. Cell. Biol.*, 169, 745, 2005.

38. Pillai, R.S., Bhattacharyya, S.N., and Filipowicz, W., Repression of protein synthesis by miRNAs: How many mechanisms? *Trends Cell Biol.*, 17, 118, 2007.

39. Weiss, A. et al., DNA demethylation in vitro: Involvement of RNA, *Cell*, 86, 709, 1996.

40. Mathieu, O. and Bender, J., RNA-directed DNA methylation, *J. Cell Sci.*, 117, 4881, 2004.

41. Morris, K.V. et al., Small interfering RNA-induced transcriptional gene silencing in human cells, *Science*, 305, 1289, 2004.

42. Imamura, T. et al., CpG island of rat sphingosine kinase-1 gene: Tissue-dependent DNA methylation status and multiple alternative first exons, *Genomics*, 76, 117, 2001.

43. Imamura, T. et al., Non-coding RNA directed DNA demethylation of *Sphk1* CpG island, *Biochem. Biophys. Res. Commun.*, 322, 593, 2004.

44. Imamura, T. et al., Identification of genetic and epigenetic similarities of *SPHK1/Sphk1* in mammals, *J. Vet. Med. Sci.*, 66, 1387, 2004.

45. Maceyka, M. et al., SphK1 and SphK2, sphingosine kinase isoenzymes with opposing functions in sphingolipid metabolism, *J. Biol. Chem.*, 280, 37118, 2005.

46. Le Scolan, E. et al., Overexpression of sphingosine kinase 1 is an oncogenic event in erythroleukemic progression, *Blood*, 106, 1808, 2005.

47. Van Brocklyn, J.R. et al., Sphingosine kinase-1 expression correlates with poor survival of patients with glioblastoma multiforme: Roles of sphingosine kinase isoforms in growth of glioblastoma cell lines, *J. Neuropathol. Exp. Neurol.*, 64, 695, 2005.

48. Kohno, M. et al., Intracellular role for sphingosine kinase 1 in intestinal adenoma cell proliferation, *Mol. Cell. Biol.*, 26, 7211, 2006.

49. Akao, Y. et al., High expression of sphingosine kinase 1 and S1P receptors in chemotherapy-resistant prostate cancer PC3 cells and their camptothecin-induced up-regulation, *Biochem. Biophys. Res. Commun.*, 342, 1284, 2006.

50. Barradeau, S. et al., Alternative 5'-exons of the mouse cAMP-dependent protein kinase subunit RIα gene are conserved and expressed in both a ubiquitous and tissue-restricted fashion, *FEBS Lett.*, 476, 272, 2000.

51. Turner, J.D. et al., Tissue specific glucocorticoid receptor expression, a role for alternative first exon usage? *Biochem. Pharmacol.*, 72, 1529, 2006.

52. Shiota, K. et al., Epigenetic marks by DNA methylation specific to stem, germ and somatic cells in mice, *Genes Cells*, 7, 961, 2002.

53. Shiota, K., DNA methylation profiles of CpG islands for cellular differentiation and development in mammals, *Cytogenet. Genome. Res.*, 105, 325–334, 2004.

54. Kremenskoy, M. et al., Genome-wide analysis of DNA methylation status of CpG islands in embryoid bodies, teratomas, and fetuses, *Biochem. Biophys. Res. Commun.*, 311, 884, 2003.

55. Futscher, B.W. et al., Aberrant methylation of the maspin promoter is an early event in human breast cancer, *Neoplasia*, 6, 380, 2004.

56. Khalkhali-Ellis, Z., Maspin: The new frontier, *Clin. Cancer Res.*, 12, 7279, 2006.

57. Lapidot, M. and Pilpel, Y., Genome-wide natural antisense transcription: Coupling its regulation to its different regulatory mechanisms, *EMBO Rep.*, 7, 1216, 2006.

58. Lyon, M.F., Gene action in the X-chromosome of the mouse (*Mus musculus* L.), *Nature*, 190, 372, 1961.

59. Brockdorff, N. et al., The product of the mouse *Xist* gene is a 15 kb inactive X-specific transcript containing no conserved ORF and located in the nucleus, *Cell*, 71, 515, 1992

60. Newall, A.E. et al., Primary non-random X inactivation associated with disruption of *Xist* promoter regulation, *Hum. Mol. Genet.*, 10, 581, 2001.

61. Chaumeil, J. et al., A novel role for Xist RNA in the formation of a repressive nuclear compartment into which genes are recruited when silenced, *Genes Dev.*, 20, 2223, 2006.

62. Lyon, M.F., X-chromosome inactivation: A repeat hypothesis, *Cytogenet. Cell. Genet.*, 80, 133, 1998.

63. Bailey, J.A. et al., Molecular evidence for a relationship between LINE-1 elements and X chromosome inactivation: The Lyon repeat hypothesis, *Proc. Natl. Acad. Sci. U S A*, 97, 6634, 2000.

64. Norris, D.P., Brockdorff, N. and Rastan, S., Methylation status of CpG-rich islands on active and inactive mouse X chromosomes, *Mamm. Genome*, 1, 78, 1991.

65. Chaumeil, J. et al., Integrated kinetics of X chromosome inactivation in differentiating embryonic stem cells, *Cytogenet. Genome Res.*, 99, 75, 2002.

66. Keohane, A.M. et al., X inactivation and histone H4 acetylation in ES cells, *Dev. Biol.*, 180, 618, 1996.

67. Boggs, B.A. et al., Differentially methylated forms of histone H3 show unique association patterns with inactive human X chromosomes, *Nat. Genet.*, 30, 73, 2002.

68. Silva, J. et al., Establishment of histone h3 methylation on the inactive X chromosome requires transient recruitment of Eed-Enx1 polycomb group complexes, *Dev. Cell*, 4, 481, 2003.

69. Plath, K. et al., Role of histone H3 lysine 27 methylation in X inactivation, *Science*, 300, 131, 2003.

70. Nishioka, K. et al., PR-Set7 is a nucleosome specific methyltransferase that modifies lysine 20 of histone H4 and is associated with silent chromatin, *Mol. Cell*, 9, 1201, 2002.

71. Smith, K.P. et al., Ubiquitinated proteins including uH2A on the human and mouse inactive X chromosome: Enrichment in gene rich bands, *Chromosoma*, 113, 324, 2004.

72. Mermoud, J.E. et al., Histone macroH2A1.2 relocates to the inactive X chromosome after initiation and propagation of X-inactivation, *J. Cell. Biol.*, 147, 1399, 1999.

73. Changolkar, L.N. and Pehrson, J.R., macroH2A1 histone variants are depleted on active genes but concentrated on the inactive X chromosome, *Mol. Cell. Biol.*, 26, 4410, 2006.

74. Doyen, C.M. et al., Mechanism of polymerase II transcription repression by the histone variant macroH2A, *Mol. Cell. Biol.*, 26, 1156, 2006.

75. Changolkar, L.N. et al., Developmental changes in histone macroH2A1-mediated gene regulation, *Mol. Cell. Biol.*, 27, 2758, 2007.

76. Choo, J.H. et al., Allele-specific deposition of macroH2A1 in imprinting control regions, *Hum. Mol. Genet.*, 15, 717, 2006.

77. Plath, K. et al., Developmentally regulated alterations in Polycomb repressive complex 1 proteins on the inactive X chromosome, *J. Cell. Biol.*, 167, 1025, 2004.

78. de Napoles, M. et al., Polycomb group proteins Ring1A/B link ubiquitylation of histone H2A to heritable gene silencing and X inactivation, *Dev. Cell*, 7, 663, 2004.

79. Fang, J. et al., Ring1b-mediated H2A ubiquitination associates with inactive X chromosomes and is involved in initiation of X inactivation, *J. Biol. Chem.*, 279, 52812, 2004.

80. Bernstein, E. et al., Mouse polycomb proteins bind differentially to methylated histone H3 and RNA and are enriched in facultative heterochromatin, *Mol. Cell. Biol.*, 26, 2560, 2006.

81. Morison, I.M., Ramsay, J.P., and Spencer, H.G., A census of mammalian imprinting, *Trends Genet.*, 21, 457, 2005.

82. Luedi, P.P., Hartemink, A.J., and Jirtle, R.L., Genome-wide prediction of imprinted murine genes, *Genome Res.*, 15, 875, 2005.

83. Peters, J. and Beechey, C., Identification and characterisation of imprinted genes in the mouse, *Brief. Funct. Genomic Proteomic*, 2, 320, 2004.

84. Horike, S. et al., Targeted disruption of the human LIT1 locus defines a putative imprinting control element playing an essential role in Beckwith–Wiedemann syndrome, *Hum. Mol. Genet.*, 9, 2075, 2000.

85. Wutz, A. et al., Imprinted expression of the Igf2r gene depends on an intronic CpG island, *Nature*, 389, 745, 1997.

86. Sleutels, F., Zwart, R., and Barlow, D.P., The non-coding Air RNA is required for silencing autosomal imprinted genes, *Nature*, 415, 810, 2002.

87. Sleutels, F. et al., Imprinted silencing of Slc22a2 and Slc22a3 does not need transcriptional overlap between Igf2r and Air, *EMBO J.*, 22, 3696, 2003.

88. Umlauf, D. et al., Imprinting along the Kcnq1 domain on mouse chromosome 7 involves repressive histone methylation and recruitment of Polycomb group complexes, *Nat. Genet.*, 36, 1296, 2004.

89. Yamasaki, Y. et al., Neuron-specific relaxation of Igf2r imprinting is associated with neuron-specific histone modifications and lack of its antisense transcript Air, *Hum. Mol. Genet.*, 14, 2511, 2005.

90. Mancini-Dinardo, D. et al., Elongation of the Kcnq1ot1 transcript is required for genomic imprinting of neighboring genes, *Genes Dev.*, 20, 1268, 2006.

91. Kanduri, C., Thakur, N., and Pandey, R.R., The length of the transcript encoded from the Kcnq1ot1 antisense promoter determines the degree of silencing, *EMBO J.*, 25, 2096, 2006.

92. Lanz, R.B. et al., Distinct RNA motifs are important for coactivation of steroid hormone receptors by steroid receptor RNA activator (SRA), *Proc. Natl. Acad. Sci. U S A*, 99, 16081, 2002.

93. Shi, Y. et al., Sharp, an inducible cofactor that integrates nuclear receptor repression and activation, *Genes Dev.*, 15, 1140, 2001.

94. Hatchell, E.C. et al., SLIRP, a small SRA binding protein, is a nuclear receptor corepressor, *Mol. Cell*, 22, 657, 2006.

95. Emberley, E. et al., Identification of new human coding steroid receptor RNA activator isoforms, *Biochem. Biophys. Res. Commun.*, 301, 509, 2003.

96. Chen, Z.F., Paquette, A.J., and Anderson, D.J., NRSF/REST is required in vivo for repression of multiple neuronal target genes during embryogenesis, *Nat. Genet.*, 20, 136, 1998.

97. Kuwabara, T. et al., A small modulatory dsRNA specifies the fate of adult neural stem cells, *Cell*, 116, 779, 2004.

98. Nguyen, V.T. et al., 7SK small nuclear RNA binds to and inhibits the activity of CDK9/cyclin T complexes, *Nature*, 414, 322, 2001.

99. Yang, Z. et al, The 7SK small nuclear RNA inhibits the CDK9/cyclin T1 kinase to control transcription, *Nature*, 414, 317, 2001.

100. Li, Q. et al., Analysis of the large inactive P-TEFb complex indicates that it contains one 7SK molecule, a dimer of HEXIM1 or HEXIM2, and two P-TEFb molecules containing Cdk9 phosphorylated at threonine 186, *J. Biol. Chem.*, 280, 28819, 2005.

101. Yik, J.H. et al., Inhibition of P-TEFb (CDK9/Cyclin T) kinase and RNA polymerase II transcription by the coordinated actions of HEXIM1 and 7SK snRNA, *Mol. Cell*, 12, 971, 2003.

102. Willingham, A.T. et al., A strategy for probing the function of noncoding RNAs finds a repressor of NFAT, *Science*, 309, 1570, 2005.

103. Tiedge, H., Chen, W., and Brosius, J., Primary structure, neural-specific expression, and dendritic location of human BC200 RNA, *J. Neurosci.*, 13, 2382, 1993.

104. Muslimov, I.A. et al., Activity-dependent regulation of dendritic BC1 RNA in hippocampal neurons in culture, *J. Cell Biol.*, 141, 1601, 1998.

105. Wang, H. et al., Dendritic BC1 RNA in translational control mechanisms, *J. Cell Biol.*, 171, 811, 2005.

106. Zalfa, F. et al., The fragile X syndrome protein FMRP associates with BC1 RNA and regulates the translation of specific mRNAs at synapses, *Cell*, 112, 317, 2003.

107. Zalfa, F. et al., Fragile X mental retardation protein (FMRP) binds specifically to the brain cytoplasmic RNAs BC1/BC200 via a novel RNA-binding motif, *J. Biol. Chem.*, 280, 33403, 2005

108. Wells, D.G., RNA-binding proteins: A lesson in repression, *J. Neurosci.*, 26, 7135, 2006.

109. Cai, X., Hagedorn, C.H., and Cullen, B.R., Human microRNAs are processed from capped, polyadenylated transcripts that can also function as mRNAs, *RNA*, 10, 1957, 2004.

110. Zeng, Y., Principles of micro-RNA production and maturation, *Oncogene*, 25, 6156, 2006.

111. Bartel, D.P., MicroRNAs: Genomics, biogenesis, mechanism, and function, *Cell*, 116, 281, 2004.

112. Yekta, S., Shih, I., and Bartel, D.P., MicroRNA-directed cleavage of HOXB8 mRNA, *Science*, 304, 594, 2004.

113. Pfeffer, S. et al., Identification of virus-encoded microRNAs, *Science*, 304, 734, 2004.

114. Pillai, R.S. et al., Inhibition of translational initiation by Let-7 microRNA in human cells, *Science*, 309, 1573, 2005.

115. Kiriakidou, M. et al., An mRNA m(7)G cap binding-like motif within human Ago2 represses translation, *Cell*, 129, 1141, 2007.

116. Ceci, M. et al., Release of eIF6 (p27BBP) from the 60S subunit allows 80S ribosome assembly, *Nature*, 426, 579, 2003.

117. Chendrimada, T.P. et al., MicroRNA silencing through RISC recruitment of eIF6, *Nature*, 447, 823, 2007.

118. Chen, K. and Rajewsky, N., The evolution of gene regulation by transcription factors and microRNAs, *Nat. Rev. Genet.*, 8, 93, 2007.

119. Mahony, S. et al., Regulatory conservation of protein coding and microRNA genes in vertebrates: Lessons from the opossum genome, *Genome Biol.*, 8, R84, 2007.

120. Lee, C.T., Risom, T., and Strauss, W.M., Evolutionary conservation of microRNA regulatory circuits: An examination of microRNA gene complexity and conserved microRNA-target interactions through metazoan phylogeny, *DNA Cell. Biol.*, 26, 209, 2007.

121. Hon, L.S. and Zhang, Z., The roles of binding site arrangement and combinatorial targeting in microRNA repression of gene expression, *Genome Biol.*, 8, R166, 2007.

122. Calin, G.A. et al., Human microRNA genes are frequently located at fragile sites and genomic regions involved in cancers, *Proc. Natl. Acad. Sci. U S A*, 101, 2999, 2004.

123. Sevignani, C. et al., MicroRNA genes are frequently located near mouse cancer susceptibility loci, *Proc. Natl. Acad. Sci. U S A*, 104, 8017, 2007.

124. Lu, J. et al., MicroRNA expression profiles classify human cancers, *Nature*, 435, 834, 2005.

125. Karube, Y. et al., Reduced expression of Dicer associated with poor prognosis in lung cancer patients, *Cancer Sci.*, 96, 111, 2005.

126. Kumar, M.S. et al., Impaired microRNA processing enhances cellular transformation and tumorigenesis, *Nat. Genet.*, 39, 673, 2007.

127. Michael, Z.M. et al., Reduced accumulation of specific microRNAs in colorectal neoplasia, *Mol. Cancer Res.*, 1, 882, 2003.

128. Hwang, H.W., Wentzel, E.A., and Mendell, J.T., A hexanucleotide element directs microRNA nuclear import, *Science*, 315, 97, 2007.

129. Bandrés, E. et al., Identification by Real-time PCR of 13 mature microRNAs differentially expressed in colorectal cancer and non-tumoral tissues, *Mol. Cancer*, 5, 29, 2006.

130. Blenkiron, C. et al., MicroRNA expression profiling of human breast cancer identifies new markers of tumour subtype, *Genome Biol.*, 8, R214, 2007.

131. Roldo, C. et al., MicroRNA expression abnormalities in pancreatic endocrine and acinar tumors are associated with distinctive pathological features and clinical behavior, *J. Clin. Oncol.*, 24, 4677, 2006.

132. Ciafre, S.A. et al., Extensive modulation of a set of microRNAs in primary glioblastoma, *Biochem. Biophys. Res. Commun.*, 334, 1351, 2005.

133. Murakami, Y., Comprehensive analysis of microRNA expression patterns in hepatocellular carcinoma and non-tumorous tissues, *Oncogene*, 25, 2537, 2006.

134. Kent, O.A. and Mendell, J.T., A small piece in the cancer puzzle: MicroRNAs as tumor suppressors and oncogenes, *Oncogene*, 25, 6188, 2006.

135. Takamizawa, J. et al., Reduced expression of the let-7 microRNAs in human lung cancers in association with shortened postoperative survival, *Cancer Res.*, 64, 3753, 2004.

136. Johnson, S.M. et al., RAS is regulated by the let-7 microRNA family, *Cell*, 120, 635, 2005.

137. Johnson, S.M. et al., The let-7 MicroRNA represses cell proliferation pathways in human cells, *Cancer Res.*, 67, 7713, 2007.

138. Laneve, P. et al., The interplay between microRNAs and the neurotrophin receptor tropomyosin-related kinase C controls proliferation of human neuroblastoma cells, *Proc. Natl. Acad. Sci. U S A*, 104, 7957, 2007.

139. Mayr, C., Hemann, M.T., and Bartel, D.P., Disrupting the pairing between let-7 and Hmga2 enhances oncogenic transformation, *Science*, 315, 1576, 2007.

140. Lee, Y.S. and Dutta, A., The tumor suppressor microRNA let-7 represses the *HMGA2* oncogene, *Genes Dev.*, 21, 1025, 2007.

141. Linsley, P.S. et al., Transcripts targeted by the microRNA-16 family cooperatively regulate cell cycle progression, *Mol. Cell. Biol.*, 27, 2240, 2007.

142. Cimmino, A. et al., miR-15 and miR-16 induce apoptosis by targeting BCL2, *Proc. Natl. Acad. Sci. U S A*, 102, 13944, 2005.

143. Calin, G.A. et al., Frequent deletions and down-regulation of micro- RNA genes miR15 and miR16 at 13q14 in chronic lymphocytic leukemia, *Proc. Natl. Acad. Sci. U S A*, 99, 15524, 2002.

144. Adams, B.D., Furneaux, H., and White, B.A., The micro-ribonucleic acid (miRNA) miR-206 targets the human estrogen receptor-a (ERa) and represses ERa messenger RNA and protein expression in breast cancer cell lines, *Mol. Endocrinol.*, 21, 1132, 2007.

145. Wang, T. et al., A micro-RNA signature associated with race, tumor size, and target gene activity in human uterine leiomyomas, *Genes Chromosomes Cancer*, 46, 336, 2007.

146. Volinia, S. et al., A microRNA expression signature of human solid tumors defines cancer gene targets, *Proc. Natl. Acad. Sci. U S A*, 103, 2257, 2006.

147. Chan, J.A., Krichevsky, A.M., and Kosik, K.S., MicroRNA-21 is an antiapoptotic factor in human glioblastoma cells, *Cancer Res.*, 65, 6029, 2005.

148. Meng, F. et al., MicroRNA-21 regulates expression of the PTEN tumor suppressor gene in human hepatocellular cancer, *Gastroenterology*, 133, 647, 2007.

149. Si, M.L. et al., miR-21-mediated tumor growth, *Oncogene*, 26, 2799, 2006.

150. Frankel, L.B. et al., Programmed cell death 4 (PDCD4) is an important functional target of the microRNA miR-21 in breast cancer cells, *J. Biol. Chem.*, 283, 1026, 2008.

151. Zhu, S. et al., MicroRNA-21 targets the tumor suppressor gene tropomyosin1 (TPM1), *J. Biol. Chem.*, 282, 14328, 2007.

152. Hayashita, Y. et al., A polycistronic microRNA cluster, miR-17-92, is overexpressed in human lung cancers and enhances cell proliferation, *Cancer Res.*, 65, 9628, 2005.

153. Ota, A. et al., Identification and characterization of a novel gene, C13orf25, as a target for 13q31-q32 amplification in malignant lymphoma, *Cancer Res.*, 64, 3087, 2004.

154. He, L. et al., A microRNA polycistron as a potential human oncogene, *Nature*, 435, 828, 2005.

155. O'Donnell, K.A. et al., c-Myc-regulated microRNAs modulate E2F1 expression, *Nature*, 435, 839, 2005.

156. Saito, Y. et al., Specific activation of microRNA-127 with downregulation of the proto-oncogene BCL6 by chromatin-modifying drugs in human cancer cells, *Cancer Cell*, 9, 435, 2006.

157. Brueckner, B. et al., The human let-7a-3 locus contains an epigenetically regulated microRNA gene with oncogenic function, *Cancer Res.*, 67, 1419, 2007.

158. Lujambio, A. et al., Genetic unmasking of an epigenetically silenced microRNA in human cancer cells, *Cancer Res.*, 67, 1424–1429, 2007.

159. Cai, X. and Cullen, B.R., The imprinted H19 noncoding RNA is a primary microRNA precursor, *RNA*, 13, 313, 2007.

160. Bartolomei, M.S., Zemel, S., and Tilghman, S.M., Parental imprinting of the mouse H19 gene, *Nature*, 351, 153, 1991.

161. Obernosterer, G. et al., Post-transcriptional regulation of microRNA expression, *RNA*, 12, 1161, 2006.

162. Manoharan, H., Babcock, K., and Pitot, H.C., Changes in the DNA methylation profile of the rat H19 gene upstream region during development and transgenic hepatocarcinogenesis and its role in the imprinted transcriptional regulation of the H19 gene, *Mol. Carcinog.*, 41, 1, 2004.

163. Hao, Y. et al., Tumour-suppressor activity of H19 RNA, *Nature*, 365, 764, 1993.

164. Fellig, Y. et al., H19 expression in hepatic metastases from a range of human carcinomas, *J. Clin. Pathol.*, 58, 1064, 2005.

165. Berteaux, N. et al., H19 mRNA-like noncoding RNA promotes breast cancer cell proliferation through positive control by E2F1, *J. Biol. Chem.*, 280, 29625, 2005.

166. Nakano, S. et al., Expression profile of *LIT1/KCNQ1OT1* and epigenetic status at the KvDMR1 in colorectal cancers, *Cancer Sci.*, 97, 1147, 2006.

167. Soejima, H. et al., Silencing of imprinted *CDKN1C* gene expression is associated with loss of CpG and histone H3 lysine 9 methylation at DMR-LIT1 in esophageal cancer, *Oncogene*, 23, 4380, 2004.

168. Bussemakers, M.J. et al., DD3: A new prostate-specific gene, highly overexpressed in prostate cancer, *Cancer. Res.*, 59, 5975, 2000.

169. Srikantan, V. et al., PCGEM1, a prostate-specific gene, is overexpressed in prostate cancer, *Proc. Natl. Acad. Sci. U S A*, 97, 12216, 2000.

170. Petrovics, G. et al., Elevated expression of PCGEM1, a prostate-specific gene with cell growth-promoting function, is associated with high-risk prostate cancer patients, *Oncogene*, 23, 605, 2004.

171. Pibouin, L. et al., Cloning of the mRNA of overexpression in colon carcinoma-1: A sequence over-expressed in a subset of colon carcinomas, *Cancer Genet. Cytogenet.*, 133, 55, 2002.

172. Rangel, L.B. et al., Characterization of novel human ovarian cancer-specific transcripts (HOSTs) identi-fied by serial analysis of gene expression, *Oncogene*, 22, 7225, 2003.

173. Ji, P. et al., MALAT-1, a novel noncoding RNA, and thymosin beta4 predict metastasis and survival in early-stage non-small cell lung cancer, *Oncogene*, 22, 8031, 2003.

174. Wolf, S. et al., B-cell neoplasia associated gene with multiple splicing (BCMS): The candidate B-CLL gene on 13q14 comprises more than 560 kb covering all critical regions, *Hum. Mol. Genet.*, 10, 1275, 2001.

175. Tsang, W.P. et al., Induction of drug resistance and transformation in human cancer cells by the noncoding RNA CUDR, *RNA*, 13, 890, 2007.

176. Reis, E.M. et al., Antisense intronic non-coding RNA levels correlate to the degree of tumor differenti-ation in prostate cancer, *Oncogene*, 23, 6684, 2004.

177. Reis, E.M. et al., Large-scale transcriptome analyses reveal new genetic marker candidates of head, neck, and thyroid cancer, *Cancer Res.*, 65, 1693, 2005.

178. Calin, G.A. et al., Ultraconserved regions encoding ncRNAs are altered in human leukemias and carcinomas, *Cancer Cell*, 12, 215, 2007.

179. Perez, D.S. et al., Long, abundantly-expressed non-coding transcripts are altered in cancer, *Hum. Mol. Genet.*, 17, 642, 2008.

180. Kirino, Y. and Mourelatos, Z., 2′-O-methyl modification in mouse piRNAs and its methylase, *Nucleic Acids Symp. Ser. (Oxf).*, 51, 417, 2007.

181. Zhao, X. et al., Pus3p- and Pus1p-dependent pseudouridylation of steroid receptor RNA activator controls a functional switch that regulates nuclear receptor signaling, *Mol. Endocrinol.*, 21, 686, 2007.

182. Kawahara, Y. et al., RNA editing of the microRNA-151 precursor blocks cleavage by the Dicer-TRBP complex, *EMBO Rep.*, 8, 763, 2007.

183. Blow, M.J. et al., RNA editing of human microRNAs, *Genome Biol.*, 7, R27, 2006.

184. Liang, H. and Landweber, L.F., Hypothesis: RNA editing of microRNA target sites in humans? *RNA*, 13, 463, 2007.

185. Vitali, P. et al., ADAR2-mediated editing of RNA substrates in the nucleolus is inhibited by C/D small nucleolar RNAs, *J. Cell Biol.*, 169, 745, 2005.

186. Hertel, J., Hofacker, I.L., and Stadler, P.F., SnoReport: Computational identification of snoRNAs with unknown targets, *Bioinformatics*, 24, 158, 2008.

187. Krützfeldt, J. et al., Silencing of microRNAs in vivo with "antagomirs", *Nature*, 438, 685, 2005.

13 Cancer-Linked DNA Hypermethylation and Hypomethylation: Chromosome Position Effects

Melanie Ehrlich

CONTENTS

13.1   INTERACTIONS OF DNA AND CHROMATIN EPIGENETICS IN NORMAL TISSUES

There are intimate relationships between the two faces of vertebrate epigenetics, methylation of DNA cytosine residues at the 5-position and changes in the structure or non-DNA composition of chromatin that are inherited from cell to cell and, sometimes, parent to progeny. Chromatin epigenetics for vertebrates and higher plants includes histone modification (acetylation, methylation, phosphorylation, ubiquitation, and sumoylation) [1–4]; the composition of tightly bound nonhistone proteins, special noncoding RNAs, and histone variants [5–9]; chromatin higher-order structure [10–13]; and the distribution of chromatin in the nucleus [14–17]. There are interrelationships between all of these aspects of chromatin structure.

DNA methylation and chromatin structure interact at various levels. Histone modification and DNA methylation show strong positive and negative associations [18–20]. DNA methylation is correlated locally with more histone H3 lysine-9, -20, and -27 mono-, di-, or trimethylation, more H4 lysine-20 trimethylation, less histone acetylation, and less H3 lysine-4 methylation [21–23]. This is observed at the level of individual genes in relation to their expression, comparisons of constitutive heterochromatin and euchromatin, and developmental changes in normal or cloned mammalian embryos [24]. Part of the link between DNA methylation and modification of histones is that histone deacetylases or H3 lysine-9 (H3 K9) methyltransferases can be recruited by proteins which bind preferentially to methylated DNA, proteins that bind to DNA methyltransferases (DNMTs), and DNA methyltransferases themselves [25–30].

DNMTs not only catalyze DNA methylation, which can impact chromatin structure, but also can have a more direct influence on chromatin epigenetics as shown by many of their specific protein–protein interactions. In addition to binding to various histone deacetylases and the H3 K9 methyl-transferases SUV39H1 and G9a, DNMTs bind specifically to heterochromatin-binding proteins and heterochromatin itself [29,31–33]. Importantly, DNMTs can also recruit chromatin proteins such as chromatin remodeling proteins (hSNF2H, Lsh), transcription factors (Rb, EF1, PU.1, Sp1), repressors (MeCP1, DMAP1, Brg1 complex, Mbd3, and EZH2), replication proteins (PCNA), condensins, and chromatin-remodeling proteins [28,30,32,34–39]. In addition, an N-terminal domain of Dnmt3a and Dnmt3b directs these enzymes to preferentially associate with pericentromeric heterochromatin [40].

Not only does DNA methylation influence histone modification, but also repressive histone modifications can predispose to DNA methylation. Evidence for this is found in a murine double-knockout of the repression-associated histone H3 K9 methyltransferase Suv39h [41]. This knockout mouse lost much of the DNA methylation from pericentromeric (juxtacentromeric) but not from centromeric or endogenous C-type retroviral regions. In addition, double-knockout of the euchromatic H3 K9 and K27 methyltransferase G9a in embryonic stem cells caused partial demethylation of a minor part of the genome with demethylation extending several kb [42]. However, *Dnmt1* single- or *Dnmt3a/Dnmt3b* double-deficient embryonic stem cells did not lose H3 K9 trimethylation from pericentromeric heterochromatin. Depending on the DNA/chromatin target and the cell population, DNA methylation and histone modifications may only partially overlap in function and association [22].

Sometimes, DNA methylation influences histone modification and other aspects of chromatin epigenetics by its effect on gene expression feeding back indirectly into changes in chromatin epigenetics [19]. Other times, DNA methylation seems to directly influence chromatin structure by recruitment of chromatin proteins whose binding to DNA is favored by CpG methylation either in a sequence-specific manner or irrespective of neighboring sequence context [26,43,44]. In addition, DNA methyltransferase binding to transcription repressors can suffice to downregulate expression of certain genes even when the catalytic activity of the DNA methylating enzyme is mutationally lost [38].

A very efficient means of silencing genes and perpetuating the silenced status involves DNMT-repressor interactions on chromatin, DNMT-catalyzed methylation of promoters or enhancers, recruitment of methylated DNA-dependent proteins, and downregulatory histone modifications, all of which lead to the local establishment of transcription-repressive structures. Vire and coworkers [37] refer to this as an epigenetic memory module and give the example of the complex of DNMT1, DNMT3A, or DNMT3B with the polycomb protein EZH2. Downregulation of EZH2 by RNAi decreases methylation of EZH2-repressed promoters, which indicates the importance of EZH2 in recruiting methyltransferases to methylate this set of promoters. Moreover, *WNT1* and *MYT1* have their EZH2-dependent silencing reversed by treatment of the cells with the DNA methylation inhibitor 5-azadeoxycytidine. Therefore, a repressor protein may recruit a DNA methyltransferase to help establish and maintain epigenetic silencing.

The concept of memory is critical to understand the interface of DNA epigenetics and chromatin epigenetics and a considerable portion of the role of DNA methylation in vertebrates and higher plants. This was illustrated in a tour-de-force study by Feng and coworkers [45]. They constructed a reporter transgene in a CRE-inversion cassette in a murine cell line in its normal form or with mutational elimination of all the CpG's from the reporter gene and its transcription control element. Integration of the reporter gene was targeted to chromosomal sites. At one of these sites, they found that silencing of the gene occurred in one orientation but not the other. This silencing occurred irrespective of whether CpG's were present on the transgene. As expected, silencing was accompanied by the loss of acetylation of H3 and H4, loss of H4 K4 methylation, and gain of H3 K9 trimethylation. After establishment of silencing, they flipped the orientation of the cassette by transfection with a Cre expression plasmid and tested whether the transgene in the expression-nonpermissive orientation became reactivated upon being moved to the expression-permissive orientation. The CpG-free transgene had its expression activated with the usual changes in histone modification associated with induction of expression but the normal CpG-containing plasmid did not. The authors conclude that, in their experimental system, DNA methylation was not necessary for repression but that it gave a long-term memory which the repression-associated chromatin epigenetic markers could not.

13.2 DNA SEQUENCES PRONE TO HYPERMETHYLATION OR HYPOMETHYLATION IN CANCER

Cancers usually display both increases and decreases in methylation of different genomic sequences [46–48]. Tumor progression is often accompanied by an increased frequency of DNA hyper- and hypomethylation, which can include the spreading in *cis* of DNA methylation changes [49–51]. Moreover, cancers can exhibit a DNA hypermethylator phenotype [52] analogous to the mutator phenotype. The high frequency of inherited changes found in cancers seems to result from an acquired tumor-related inability to maintain the stable genotype and epigenotype [53]. This instability results in continuous changes in gene expression patterns that can favor tumor formation, increasingly aggressive tumor behavior, and metastasis. Moreover, there are interactions between cancer-linked epigenetic and genetic instability [54,55]

Cancer-associated hypermethylation of CpG islands overlapping promoters contributes to carcinogenesis usually by downregulating expression of anti-tumorigenesis genes [56]. Some of these genes prone to cancer-linked hypermethylation are involved in maintaining gross chromosomal stability or in minimizing point mutagenesis. Hypermethylation of DNA in cancer is not confined to promoters of genes with a cancer-related function nor to promoter regions but the DNA hypermethylation that is most implicated in carcinogenesis is that in promoters of tumor suppressor genes [57–59]. Occasionally cancer-linked hypermethylation of gene regulatory elements abnormally upregulates, rather than downregulates, gene expression [60] thereby contributing to tumor formation or progression.

Decreases in DNA methylation in cancer are often much more numerous than increases in this methylation, leading to a net deficiency in genomic 5-methylcytosine (m^5C) [61]. DNA repeats, both tandem and interspersed, appear to be particularly susceptible to demethylation in cancer [47,62,63]. To test whether cancer-associated satellite DNA demethylation might be an inducer of de novo methylation of transcription control regions of tumor suppressor genes or, alternatively, a response to prior de novo methylation during tumorigenesis, we looked in Wilms tumors and ovarian epithelial cancers for a positive association between this satellite DNA hypomethylation and hypermethylation of CpG islands at the 5' ends of many genes that are prone to cancer-linked hypermethylation [48,51,64]. There was no general positive association between CpG island hypermethylation and hypomethylation of satellite DNA. Therefore, it is unlikely that cancer-linked satellite DNA hypomethylation acts only as an inducer of or responder to cancer-linked hypermethylation in multiple gene regions, and its prevalence in cancer and the large size of the hypomethylated satellite DNA regions suggest that this hypomethylation facilitates carcinogenesis.

Descriptions of cancer-linked methylation changes generally involve either decreases in methylation of normally heavily methylated sequences or increases in methylation of sequences which normally have little or no methylation [47]. However, recently we demonstrated that, relative to normal somatic tissues, cancers can display both hypo- and hypermethylation. These opposite DNA methylation changes were seen in NBL2, a tandem 1.4 kb G + C-rich repeat with a complex sequence. NBL2 is a primate-specific sequence [65] present in about 200–400 copies per haploid human genome, mostly in the vicinity of the centromeres of four of the five acrocentric chromosomes [66]. By Southern blot analysis, we found that NBL2 exhibits either predominant hyper- or hypomethylation at *Hha*I sites in ovarian carcinomas and Wilms tumors relative to normal somatic tissues [66]. Itano et al. [50] demonstrated that hypomethylation at *Not*I sites was an independent prognostic indicator in hepatocellular carcinoma patients.

Studies of NBL2 methylation in cancer in the past were essentially confined to *Not*I digests of NBL2, which give only high-molecular weight fragments from normal tissues, thus not allowing hypermethylation to be visualized [50,65,67]. That we found only a much smaller percentage of tumors showing NBL2 hypomethylation, even at *Not*I sites, in ovarian carcinomas and Wilms tumors [66] than were seen for hepatocellular carcinomas and neuroblastomas [50,65,67] probably reflects tumor-type specificity in epigenetic changes [68,69].

Unexpectedly by hairpin-bisulfite PCR, a genomic sequencing method that detects m^{5}C on covalently linked complementary strands of a DNA fragment [70], 56% of DNA clones from the cancers had decreased methylation at some normally methylated CpG sites in a 0.2 kb region as well as increased methylation at normally unmethylated sites. This study included controls demonstrating that the bisulfite reaction went to completion and that there was no detectable bias in amplification of methylated versus unmethylated DNA that might have produced artifacts. Previous bisulfite-based genomic sequencing studies of cancer DNA usually involved unmethylated CpG-rich promoter regions that become hypermethylated mostly homogeneously [71–73]. There may be several reasons for NBL2 displaying surprisingly complex, nonrandom patterns of altered methylation during carcinogenesis. It is apparently not a gene, and its methylation status probably confers no selective advantage to a developing tumor. This is unlike the situation with promoters of tumor suppressor genes whose almost complete methylation can benefit the growing tumor by repressing transcription or stabilizing this repression and thus be heavily influenced by selection during tumorigenesis. In addition, unlike most DNA regions from cancers analyzed by genomic sequencing, NBL2 normally has very low levels of methylation at some CpG's and complete methylation at many others so that both cancer-linked increases and decreases of DNA methylation can be observed. Furthermore, it seems to be an unusually frequent target for multiple methylation changes during carcinogenesis.

NBL2 is not unique in being subject to either predominant hyper- or hypomethylation in individual specimens of cancers of a given type. We recently observed this for another virtually untranscribed tandem repeat, D4Z4, which is present in the subtelomeric regions of the long arms of chromosomes 4 and 10 [74]. Moreover, we recently detected a CpG site in satellite 2 DNA showing cancer-linked hypermethylation almost adjacent to a CpG site with hypomethylation (C. Shao et al., unpublished data).

13.3 EFFECTS OF CHANGES IN DNA METHYLATION ON CHROMOSOME STRUCTURE: PERICENTROMERIC REGIONS

DNA hypomethylation in the juxtacentromeric region of chromosomes 1 and 16 (1qh and 16qh) has been implicated in pericentromeric rearrangements from studies of the immunodeficiency, centromeric region instability, and facial anomalies syndrome (ICF), a rare disease with targeted chromosome breakage [75]. ICF is usually caused by mutations in *DNMT3B* [76,77] and involves instability almost exclusively of chromosomes 1 and 16, and sometimes 9 in the juxtacentromeric heterochromatin (qh region). Chromosomes 1, 16, 2, and 10 have satellite 2 DNA (Sat2) in this centromere-adjacent constitutive heterochromatin [78]. This qh region is usually larger for Chr1 than for Chr16, which is in turn much larger than that of Chr2 or Chr10. Chr9 has a large juxtacentromeric heterochromatin region predominantly containing satellite 3 (Sat3), which is distantly related to Sat2. Both Sat2 and Sat3 are hypomethylated in all studied ICF tissues and cell cultures [79,80].

ICF is a recessive B-cell deficiency disease that has been described in fewer than 60 patients world-wide in the last several decades. The cytologically detectable rearrangements targeted to the qh region of chromosomes 1 and 16 and sometimes 9 in mitogen-stimulated lymphocytes are an invariable characteristic [75,81]. The hypomethylation of satellite DNA sequences in ICF cells is usually, but not always, limited to the juxtacentromeric DNA without extending to centromeric DNA [80,82]. By HPLC analysis of ICF and control DNA digests, we demonstrated that the hypomethylation of the genome in ICF involved only a rather small percentage of the m^5C residues, namely 7% hypomethylation in brain DNA [80]. We confirmed that the methylation abnormality of ICF is confined to a small percentage of the genome by two-dimensional electrophoresis of restriction digests of DNA from four ICF versus four control lymphoblastoid cell lines (LCLs) [83]. Only 13 of the approximately 1000 spots displayed consistent ICF-specific differences and all but one of these was derived from tandem copies of two unrelated repeats present in several chromosomal locations. Consistent DNA hypomethylation among ICF patients is restricted to a rather small portion of the genome, especially certain tandem DNA repeats. These results indicate that the ICF syndrome is a good model implicating hypomethylation of certain DNA sequences favoring chromosome instability in *cis*.

The ICF-linked 1qh and 16qh rearrangements are predominantly chromosome breaks, whole-arm deletions, multibranched chromosomes, translocations, and isochromosomes (usually containing two 1q arms fused in the pericentromeric region) [75,81]. In addition, there is frequent decondensation in the pericentromeric region of Chr1 in these cells [80,84]. Although ICF patients display no consistently increased cancer incidence, fewer than 60 patients (mostly children) have been identified, and their usually very short average lifespan would preclude detection of a cancer predisposition that was not very high and did not result in tumors rather quickly.

ICF is usually linked to mutations in both alleles of *DNMT3B* [75]. Most biallelic *DNMT3B* mutations in ICF patients are found in the part of the gene encoding the catalytically active C-terminal portion of the protein, namely, one of 10 motifs conserved among all cytosine-C5 methyltransferases. These mutations generally give low-level residual enzymatic activity [76]. The mutant proteins are still able to engage in normal protein–protein interactions [35]. Although DNMT3B has many specific protein-interaction domains, the described protein-interaction domains do not overlap the methyltransferase domain [28,85]. From these findings, it is inferred that the loss of DNA methyltransferase activity and not some other function of the DNMT3B is responsible for the syndrome and its attendant Chr1 and Chr16 instability in lymphoid cells.

ICF-specific chromosome rearrangements may sometimes be a contributing factor for carcinogenesis. ICF-like Chr1/Chr16 multiradial chromosomes, which are expected to be very short-lived structures [80], and 1qh decondensation have been observed in multiple myeloma and hepatocellular carcinomas [86,87]. In urothelial carcinomas, hypomethylation of Sat2 (mostly in Chr1 and Chr16) and Sat3 (mostly in Chr9) was found to be significantly associated with loss of heterozyosity (LOH) on Chr9 [88]. In addition, a recent study of overall genomic hypomethylation in glioblastomas provided evidence for an association of strong hypomethylation with chromosomal rearrangements in the vicinity of Chr1 Sat2 [89]. Indeed, unbalanced Chr1 and Chr16 pericentromeric rearrangements are overrepresented in a wide variety of cancers [90]. This finding and the very frequent observation of hypomethylation of the pericentromeric regions of chromosomes 1 and 16 in diverse cancers suggest that Sat2 hypomethylation predisposes to local rearrangements [47,91].

We examined 52 Wilms tumors by quantitative loss-of-heterozygosity analysis and found no significant relationship of 1q gain to Chr1 Sat2 hypomethylation [55]. However, we did observe a significant association of 1q gain with Chr1 Satα hypomethylation and of Chr16 Sat2 hypomethylation with 16q loss. Nonetheless, the distribution of satellite hypomethylation among cancers does not parallel chromosome rearrangements in *cis* because there was a very much higher frequency of Chr1 Sat2 and Chr1 Satα hypomethylation than of 1q or 1p imbalances. This was also seen in our subsequent analysis of a different set of 35 karyotyped Wilms tumors [92]. In both studies, about half of the tumors displayed Chr1 Sat2 hypomethylation and about 90% exhibited Chr1 Satα hypomethylation. In the second study, we determined the methylation status of Satα throughout the centromeres, and not just in Chr1 Satα. There was a very high degree of concordance between

Chr1 Satα hypomethylation and hypomethylation of Satα throughout the centromeres just as we had found for Chr1 Sat2 and Chr16 Sat2 in cancers [55,62]. Out of thirty-five Wilms tumors, seven had cytogenetically identified, clonal pericentromeric rearrangements, with five affecting Chr1. These five had extra copies of 1q. Four of the five tumors with pericentromeric Chr1 rearrangements displayed hypomethylation of satellite DNA in 1qh and one displayed hypomethylation in only the centromeric satellite DNA of Chr1. The explanation for the much higher frequencies of satellite DNA hypomethylation than of pericentromeric rearrangements is probably that DNA hypomethylation in the pericentromeric regions predisposes to, but does not suffice for, rearrangements in this region via an indirect mechanism.

In a comparative genomic hybridization analysis of hepatocellular carcinomas, Wong et al. [87] found a significant association between 1qh gain and Chr1 Sat2 hypomethylation in contrast to the lack of a statistically significant association of Sat2 hypomethylation and rearrangements in Wilms tumors [92] and breast carcinomas [93]. This difference might reflect specificity for the type of cancer or the cell type. Even in the ICF syndrome, which involves greatly decreased DNMT3B activity in all cells from the patients due to germline mutations in *DNMT3B*, cell-type specific affects on chromosome stability are observed. ICF-type pericentromeric Chr1 or Chr16 rearrangements have been seen in bone marrow cells from only one of four studied ICF patients, but were absent from most examined ICF fibroblast cultures despite the constitutive hypomethylation of Sat2 DNA in 1qh and 16qh in ICF tissues [75]. The rearrangements observed in mitogen-stimulated ICF lymphocytes and in untreated ICF LCLs may occur in vivo, albeit at a very low rate, as deduced from studies of micronucleus formation in unstimulated bone marrow and lymphocytes from ICF patients [81,94,95]. Furthermore, there appears to be a special relationship between the pericentromeric rearrangements and in vitro mitogen stimulation of lymphocytes that is independent of induction of cell cycling per se. A much higher frequency of pericentromeric rearrangements of Chr1 and Chr16 per metaphase is seen 72 or 96 h after mitogen stimulation of ICF lymphocytes than at 48 h, although the frequent abnormal decondensation of 1qh and 16qh can be observed in metaphases at 48 h [81,96,97]. These observations suggest that lymphocytes have a propensity to undergo the ICF-specific rearrangements and that the artificial conditions of mitogen stimulation or short-term culture in vitro enhance the formation of these rearrangements. The viability of ICF patients and cell-type specificity of the disease, mostly an immunodeficiency disease, indicates that a generalized breakdown in 1qh and 16qh chromatin stability is not manifest throughout the tissues of ICF patients.

Other studies also indicate that DNA hypomethylation at 1qh and 16qh does not suffice to produce an ICF-like cellular phenotype of 1qh and 16qh abnormalities. At early passages, untreated cultures from normal chorionic villus (CV) showed few chromosomal abnormalities despite their hypomethylation in Sat2 attributable to their extraembryonic mesodermal (extraembryonic) origin [98]. However, by passage 8 or 9, an average of 82% of the CV metaphases from all eight studied samples exhibited 1qh or 16qh decondensation and 25% had rearrangements in these regions with no other consistent chromosomal abnormality at any passage. This suggests that other epigenetic changes occurred with increasing cell passage that was necessary to obtain the high frequencies of 1qh and 16qh abnormalities. This is reminiscent of the influence of the time of incubation of lymphocyte cultures on chromosome abnormalities in ICF blood cultures, as described above. Untreated amniotic fluid-derived (AF) samples also displayed dramatic cell passage-dependent increases in ICF-like chromosomal aberrations [98]. There was a high degree of methylation of Sat2 in AF cells at all studied passages, which is due to their derivation from embryonic fibroblasts. Nonetheless, all six analyzed late-passage AF cultures displayed 1qh decondensation and recombination in an average of 54% and 3% of the metaphases, respectively. Late-passage skin fibroblasts did not show these aberrations. The frequent 1qh decondensation observed in AF cells at high passage, despite their retention of high levels of Chr1 Sat2 methylation indicates that DNA hypomethylation at 1qh is not necessary for 1qh decondensation.

Nonetheless, the results of our study of AF and CV cultures [98] suggested that Sat2 hypomethylation favors 1qh and 16qh anomalies. In contrast to AF and fibroblast cultures, CV cultures, with their Sat2 hypomethylation, displayed 1qh and 16qh decondensation and rearrangements at significantly lower passage numbers than did AF cultures. Also, in chromosomes 1 and 16, CV cultures exhibited many more ICF-like rearrangements. Similarly, a study of four human LCLs which spontaneously underwent Chr1 Sat2, Chr16 Sat2, Chr9 Sat3, and ChrY Sat3 hypomethylation upon very prolonged culture also showed correlations between this hypomethylation and decondensation and rearrangements in Chr1 Sat2 or Chr16 Sat2 [99]. However, we have also observed high levels of Chr1 Sat2 decondensation in several normal LCLs with little or no hypomethylation in this satellite DNA (M. Ehrlich and C. Tuck-Muller, unpublished data). We propose that this can be explained by only partial overlap of DNA hypomethylation and chromatin modification pathways leading to decondensation at the tandem Sat2 higher-order-repeat units at 1qh and 16qh.

13.4 EFFECTS OF CHANGES IN DNA METHYLATION ON CHROMOSOME STRUCTURE: INTERSPERSED REPEATS AND NONREPEATED REGIONS

Cancer-associated DNA methylation changes may also affect DNA recombination at interspersed repeats, lymphogenesis-related recombination signals, and at various unique DNA sequences. Interspersed as well as tandem repeats are prone to cancer-linked hypomethylation [63]. With few exceptions, the correlation between abnormal DNA methylation changes and chromosome rearrangements is less methylation, more rearrangements. A notable exception is a study on an artificially reconstructed transposon (SB) in mouse cells, which provided evidence for increased SB transposition in the germ line and in transfected embryonal stem cells when it was highly methylated [100].

Interspersed DNA repeats are good candidates for somatic-cell recombination hotspots by homologous recombination or for insertional mutagenesis by retrotransposition. In human DNA, the most numerous of these interspersed repeats are the LINE1 (or L1) repeats and Alu repeats, which constitute ~17% and 10% of the genome, respectively. Retrotransposons or retroviral-derived elements can have their transcription upregulated in vivo by DNA demethylation, as seen in studies of *Dnmt1* knockout mouse embryos, interspecies mammalian hybrids, and mice with an inherited epigenetically controlled phenotype whose expression is regulated by a genetically linked retrotransposon (IAP) [101–103]. In addition, there is evidence for frequent activation of expression of full-length transcripts from retrotransposons in certain types of murine cancer [104]. Suppression of retrotransposition during murine spermatogenesis involves the Dnmt3a- and Dnmt3b-interactive protein Dnmt3L [105]. Dnmt3L stimulates de novo methylation by Dnmt3a [106]. A caveat in consideration of the role of retrotransposition of endogenous elements in human disease is that retrotransposition in humans seems to be much less frequent that in mice [107].

The human genome's LINE1 repeats are up to 6 kb in length, although usually much shorter. They are retrotransposon-derived sequences, but of the $\sim 4 \times 10^5$ copies of LINE1 elements in the human genome, only about 30–60 are estimated to be competent for transposition [108]. There have been occasional reports of cancer-associated retrotransposition-like insertions involving LINE1 sequences [109,110], and they may mobilize cellular RNAs at low frequencies [111]. LINE1 hypomethylation was observed in chronic lymphocytic leukemia versus normal mononuclear blood cells [112], urinary bladder carcinomas compared with normal bladder [113], hepatocellular carcinomas versus nontumorous "normal" or cirrhotic tissue [114], and prostate carcinomas versus normal prostate and other normal tissues [115]. Frequent hypomethylation of LINE repeats throughout rodent and human genomes in cancer probably contributes to recombination at these sequences in solid tumors [116]. However, it should be noted that LINE1 hypomethylation might just be associated with hypomethylation in satellite DNAs or other genomic regions. Of even higher copy number than the LINE1 repeats in the human genome are the Alu repeats (size, ~0.3 kb; copy number, $\sim 1.1 \times 10^6$), which also can be mobilized, thus leading occasionally to cancer-associated gene insertions [117,118] or other types of Alu–Alu recombination [119]. Alu repeats, like LINE1 repeats, frequently exhibit cancer-associated hypomethylation [63].

Retrotransposition might be favored by cancer-associated hypomethylation of human endogenous retroviruses, especially the HERV-K family. However, there are only about 30–50 full-length HERV-K sequences in the human genome, as compared to an estimated 10,000 solitary long terminal repeats from HERV-K [120]. These repeats are usually highly methylated. In a study of urinary bladder cancers [69], cancer-associated hypomethylation of the HERV-K sequences was seen by Southern blotting with an HERV-K *gag* probe. It is possible that such hypomethylation might favor retrotransposition and alter the accessibility of chromatin to RAG1/RAG2 immunoglobulin region recombinase [121].

Dnmt1 knockout or hypomorphic mutant mice or cell cultures derived from them have provided evidence for the involvement of abnormal DNA hypomethylation (or DNA methyltransferase deficiency) in aberrant recombination. Jaenisch et al. showed that homozygous knockout of *Dnmt1*, which caused global DNA hypomethylation, increased abnormal DNA recombination and, thereby, deletion mutagenesis at the *Hprt* locus and a transgenic viral *tk* locus in murine embryonal stem cells [122]. The increase in mutation rate (predominantly deletions) associated with the loss of Dnmt1 activity was about 10-fold. However, using different stably transfected murine embryonal stem cells, Chan et al. unexpectedly found that homozygous knockout of *Dnmt1* decreased gene loss and point mutagenesis from a chimeric *tk-neo* transgene [123]. The differences between these studies might be due to chromosome position effects on hypomethylated DNA sequences.

In another model system, transgenic mice carrying a hypomorphic *Dnmt1* allele and a null *Dnmt1* allele, most of the *Dnmt1* expression was lost [124]. This resulted in a large extent of global and centromeric DNA hypomethylation in the runted transgenic mice. All of these mice developed T-cell lymphomas. Four of ten analyzed tumors had a predominant DJ rearrangement, which suggests a monoclonal origin. This indicates that although oncogenic transformation in these mice was frequent at the level of the individual, it was rare at the cellular level. The lack of RNA for a tested endogenous IAP retrovirus and for c-*myc* and the absence of insertional inactivation of the c-*myc* locus suggest that the loss of Dnmt1 activity did not promote oncogenesis by inducing retrotransposition of proto-oncogenes. By comparative genomic hybridization, lymphoma DNA from the Dnmt1-deficient mice and MMLV transgenic mice were compared. There was significantly more gain of Chr14 and Chr15 in the former mice, which could have resulted from a whole-chromosome gain or an unbalanced translocation in the pericentromeric regions of these acrocentric chromosomes.

In a different transgenic mouse model, mice doubly heterozygous for the mutant *Nf1* and *Tp53* genes were studied [125]. These genes are closely linked on Chr11 and both are often involved together in LOH in murine soft tissue sarcomas. When these mice were also made transgenic for a null allele and a hypomorphic allele of *Dnmt1*, they tended to develop sarcomas at an earlier age. Furthermore, 77% displayed LOH at the *Nf1* and *Tp53* loci in sarcomas of the Dnmt1-deficient mice compared with 45% in isogenic mice that did not have *Dnmt1* mutations. By fluctuation analysis of fibroblasts from the Dnmt1-deficient mice versus the isogenic mice without *Dnmt1* mutations, there was a significant increase, but only twofold, in the development of LOH at the *Nf1* and *Tp53* loci. An LOH analysis of five markers along Chr11 suggested either whole-chromosome loss or unequal translocation at the acrocentric centromere. Therefore, these murine models indicate that DNA hypomethylation plays a significant, but modest, role in chromosome instability during carcinogenesis.

13.5   SOME MECHANISTIC CONSIDERATIONS FOR EFFECTS OF DNA METHYLATION ON CHROMOSOME STABILITY

Interrelationships of DNA hypomethylation and chromatin epigenetics can explain how DNA demethylation could predispose to the generation of chromatin abnormalities in *cis*, including, but not limited to chromosomal rearrangements. A downstream consequence of DNA hypomethylation can be an increase in histone acetylation, a decrease in histone H3 K9 di- or trimethylation, and a decrease in H3 serine 10 phosphorylation [5,26,28,29,31,126,127]. These alterations might predispose to pericentromeric chromatin decondensation and then, a percentage of these decondensed regions could undergo rearrangements. This would be consistent with the partial overlap of DNA demethylation and histone acetylation or H3 K9 demethylation pathways for inducing localized decompaction of euchromatin in promoter regions in the human genome [41,126,128,129]. During the formation of some tumors, changes in histone modification [130] in concert with hypomethylation of DNA [47] might favor chromatin instability.

Another intersection of DNA methylation and chromatin epigenetics in the vicinity of the centromeres involves Rb protein. Rb interacts directly with DNMT1 in vivo and in vitro in a complex containing the tumor suppressor E2F1 and histone deacetylase HDAC1 [32]. Mouse embryo fibroblasts triply deficient for three Rb1 family genes displayed large decreases in H4 K20 trimethylation in pericentromeric (juxtacentromeric) and telomeric chromatin, a small decrease in methylation of pericentromeric DNA, and an increase in aneuploidy and centromeric fusions [131,132]. These Rb1 family members interact with H4 K20 trimethylating enzymes Suv4 20-h1 and Suv4 20-h2. Rb1 seems to have a preference stimulating H4 K20 trimethylation at juxtacentromeric heterochromatin [132].

One of the juxtacentromeric heterochromatin interactions with chromatin proteins is counterintuitive. Normally, juxtacentromeric heterochromatin is enriched in heterochromatin proteins, the associated H3 K9 trimethylation, and DNA methylation. However, ICF lymphoblastoid cell lines exhibit abnormally high levels of the heterochromatin proteins HP1α, HP1β, and HP1γ, which colocalize with the Sat2-rich 1qh and 16qh despite the ICF-linked Sat2 hypomethylation. This may reflect the decondensation of these long regions of constitutive heterochromatin [133]. The chromatin instability specifically in 1qh and 16qh may also be related to the finding that the DNMT3B normally binds to components of the condensing complex and to the chromatin-remodeling enzyme hSNF2H [35] and DNMT3B is usually mutated in the ICF syndrome. Moreover, the colocalization in ICF cells of Sat2 DNA and the proteins HP1, BRCA1, ATRX, and DAXX within the PML-type nuclear bodies suggests that these might be involved in condensation of 1qh and 16qh [134].

13.6 EFFECTS OF DNA SEQUENCE AND CHROMOSOME POSITION ON CANCER-LINKED DNA METHYLATION CHANGES

An unmethylated CpG dyad within the vicinity of a methylation-rich sequence seems to be more prone to methylation than if it is surrounded by mostly unmethylated CpG's. This is probably partly a consequence of the tendency of DNA methylation to spread linearly along DNA [57,135–137]. For example, cancer-associated DNA methylation of CpG islands overlapping a promoter can apparently spread from methylation of CpG islands in the body of a gene [57]. There is evidence for DNA methylation spreading through a CpG island-promoter especially in conjunction with transcriptional repression [138]. Recently, it was shown that cancer-associated methylation can spread even over a 1 MB region although the hypermethylated regions could be dotted with CpG's that escaped methylation [139].

In our detailed studies of methylation of both strands of a small region (0.2 kb) of the tandem repeat NBL2 [140] in cancer by hairpin-bisulfite PCR, we saw evidence of spreading of altered DNA methylation patterns in only some, but not most, of the copies of NBL2. Nonetheless, in most of the clones, the results suggested multiple hits of demethylation and de novo methylation within a 0.2 kb region during carcinogenesis. As an example of the discontinuous nature of cancer-linked hypermethylation in these untranscribed repeat sequences, two CpG sites in NBL2, which are separated by only 6 bp, often displayed opposite methylation changes. In general, hemimethylated CpG dyads in cancer and control DNA clones usually did not occur as runs but rather had the closest CpG on either side as an M/M or U/U dyad. Furthermore, of the 27 cancer clones containing more than one hemimethylated CpG site, 15 (56%) had hemimethylated dyads of opposite polarity with respect to which strand was unmethylated. These U/M and M/U dyads were usually less than 77 bp apart.

On the other hand, some DNA clones from the studied Wilms tumors and ovarian carcinomas had methylation patterns suggesting spreading of methylation or demethylation, e.g., those with all 14 CpG dyads unmethylated or all methylated. Others had the first five or six CpG sites unmethylated on at least one strand. The frequency of clones with no methylation in the first five CpG sites was significantly higher than expected if the methylation at each site was independent, as was the combined frequency of fully methylated or fully unmethylated clones (both *p*-values < .0001). In summary, there seems to be spreading of altered DNA methylation patterns in some, but not most, of the copies of NBL2 in the examined cancers.

The heterogeneous and nonrandom nature of many of the DNA methylation changes that we observed by hairpin genomic sequencing and Southern blot analyses indicate that losses and gains of methylation in tandem DNA repeats (NBL2, D4Z4, and Sat2) during carcinogenesis are often targeted to CpG positions in specific patterns within the repeat and that there are strong preferences for different levels of methylation at different CpG's within a tandem repeat unit. Some of these differences tracked with restriction sites, e.g., *Hha*I sites (GCGC) were more frequently de novo methylated than *Hpy*CH4IV (ACGT) at NBL2 in ovarian carcinomas and Wilms tumors [140]. Other differences were seen at a given type of restriction site at different positions within a repeat unit, e.g., the first versus the second *Bsa*AI site in the D4Z4 repeat unit of somatic control tissues [74]. Genomic sequencing analysis of NBL2 and Sat2 revealed CpG sites that tended to have much lower levels of methylation at both CpG's of the dyad than for other sites within the repeat unit [140] and C. Shao et al. (unpublished data).

Our recent study of changes in methylation of the D4Z4 array of subtelomeric tandem repeats at 4q and 10q revealed a strong chromatin position effect on DNA methylation changes [74]. We found inhomogeneous DNA methylation at the beginning of the D4Z4 array and within each array unit, especially in cancers with overall D4Z4 hypermethylation. The results suggest atypical chromatin structures in vivo that may be critical for organizing D4Z4 chromatin to give the special relationship between D4Z4 array size (less than eleven 3.3 kb units on one allelic 4q35) and facioscapulohumeral muscular dystrophy (FSHD). Most cancers that had overall D4Z4 hypermethylation at the many examined sites within the array displayed no detectable methylation at a *Hpy*CH4IV site (ACGT) 0.3 kb inside the array at the proximal end. The difference in methylation at this first *Hpy*CH4IV site and the analogous sites in the other D4Z4 repeat units further into the array cannot be due to differences in surrounding sequences because the *Hpy*CH4IV site and adjacent sequences are essentially same in all the 3.3 kb repeat units of D4Z4 in the repeat arrays in the subtelomeric regions of 4q and 10q. Therefore, we infer differences in chromatin structure that must be preventing or reversing DNA hypermethylation specifically at the start of the D4Z4 array.

This postulated special chromatin structure at the proximal end of D4Z4 may be most pronounced near the D4Z4–nonD4Z4 border. Therefore, spreading of methylation from the array during carcinogenesis might usually stop at the very beginning of the array. This would explain why the first *Hpy*CH4IV site 0.3 kb into the array was more resistant to cancer-linked hypermethylation than the first *Bsa*AI site (YACGTR) 2.1 kb into the array. The postulated unusual chromatin structure at the junction of the D4Z4 array and its immediately proximal sequence is consistent with the dramatic differences in their G + C contents, 73% for D4Z4 and 43.5% for the most proximal 200 bp outside the array. The latter composition is more like that of the human genome's overall G + C content of 42% [141].

A special chromatin structure at the proximal D4Z4–nonD4Z4 border could also explain the hypomethylation that we observed at *Hpa*II, *Bst*BI, and *Hpy*CH4IV sites from 0.2 to 2.3 kb proximal to the array in most of the cancers, even those tumors displaying overall D4Z4 hypermethylation. Despite the proclivity of the most proximal *Hpa*II site outside the array to cancer-linked hypomethylation, a *Bst*UI site that is only 0.2 kb away was tenaciously methylated in cancers, control somatic tissues, and even in sperm. The constitutive methylation of this

*Bst*UI site is likely to be due to the unusual primary structures nearby, namely $(T_{2-5})(A_{4-5})$ and $(TG)_8$, which could give hairpin structures that may help recruit DNA methyltransferases [142].

The last unexpected relationship of DNA sequence context to epigenetic behavior in this study of D4Z4 methylation in cancer was observed in a subregion of the 3.3 kb D4Z4 repeat unit that was less prone to cancer-linked hypermethylation than surrounding sequences. It contains two potential intramolecular guanine quadruplex structures that are only 48 bp apart. G-quadruplexes are non-B DNA structures that rely on Hoogstein base-pairing between four G residues arranged in a plane. They can be involved in hydrogen-bonded interactions within one local region on a DNA strand or between duplexes [143]. They apparently can exist in vivo in DNA as well as in RNA and mediate transcription control, genomic stability, and special chromatin packing, e.g., at telomeres [144]. Analysis by circular dichroism and nondenaturing gel electrophoresis of single-stranded oligonucleotides with these potential quadruplex structures confirmed their ability to form such structures.

In summary, in our study of D4Z4 methylation in cancer, we found three examples of effects of unusual DNA structures or special inferred chromatin structures on DNA methylation. These were the resistance to cancer-linked hypermethylation of CpG's near G-quadruplex structures, resistance to DNA hypermethylation at a boundary-type element in the beginning of a very highly conserved tandem repeat array, and the invariant retention of methylation near an inverted repeat despite cancer-linked hypomethylation nearby. From this study and others [41,139,142,145], it is clear that the position of a given CpG within chromatin or its proximity to atypical DNA secondary structures can have a major effect on its methylation status.

REFERENCES

1. Boulikas, T. 1992 Poly(ADP-ribose) synthesis in blocked and damaged cells and its relation to carcinogenesis. *Anticancer Res.*, 12, 885–898.
2. Li, B., Carey, M., and Workman, J.L. 2007 The role of chromatin during transcription. *Cell*, 128, 707–719.
3. Iniguez-Lluhi, J.A. 2006 For a healthy histone code, a little SUMO in the tail keeps the acetyl away. *ACS Chem. Biol.*, 1, 204–206.
4. Villar-Garea, A. and Imhof, A. 2006 The analysis of histone modifications. *Biochim. Biophys. Acta*, 1764, 1932–1939.
5. Fuks, F., Hurd, P.J., Wolf, D., Nan, X., Bird, A.P., and Kouzarides, T. 2003 The methyl-CpG-binding protein MeCP2 links DNA methylation to histone methylation. *J. Biol. Chem.*, 278, 4035–4040.
6. Raff, J.W., Kellum, R., and Alberts, B. 1994 The *Drosophila* GAGA transcription factor is associated with specific regions of heterochromatin throughout the cell cycle. *EMBO J.*, 13, 5977–5983.
7. Ryan, R.F., Schultz, D.C., Ayyanathan, K., Singh, P.B., Friedman, J.R., Fredericks, W.J., and Rauscher, F.J., III 1999 KAP-1 corepressor protein interacts and colocalizes with heterochromatic and euchromatic HP1 proteins: A potential role for Kruppel-associated box-zinc finger proteins in heterochromatin-mediated gene silencing. *Mol. Cell Biol.*, 19, 4366–4378.
8. Shareef, M.M., Badugu, R., and Kellum, R. 2003 HP1/ORC complex and heterochromatin assembly. *Genetica*, 117, 127–134.
9. Goldberg, A.D., Allis, C.D., and Bernstein, E. 2007 Epigenetics: A landscape takes shape. *Cell*, 128, 635–638.
10. Bochar, D.A., Savard, J., Wang, W., Lafleur, D.W., Moore, P., Cote, J., and Shiekhattar, R. 2000 A family of chromatin remodeling factors related to Williams syndrome transcription factor. *Proc. Natl. Acad. Sci. U S A*, 97, 1038–1043.
11. Dennis, K., Fan, T., Geiman, T., Yan, Q., and Muegge, K. 2001 Lsh, a member of the SNF2 family, is required for genome-wide methylation. *Genes Dev.*, 15, 2940–2944.
12. Gibbons, R.J., McDowell, T.L., Raman, S., O'Rourke, D.M., Garrick, D., Ayyub, H., and Higgs, D.R. 2000 Mutations in ATRX, encoding a SWI/SNF-like protein, cause diverse changes in the pattern of DNA methylation. *Nat. Genet.*, 24, 368–371.
13. Gregory, R.I., Khosla, S., and Feil, R. 2001 Probing chromatin structure with nuclease sensitivity assays. *Methods Mol. Biol.*, 181, 269–284.

14. Alcobia, I., Quina, A.S., Neves, H., Clode, N., and Parreira, L. 2003 The spatial organization of centromeric heterochromatin during normal human lymphopoiesis: Evidence for ontogenically determined spatial patterns. *Exp. Cell Res.*, 290, 358–369.

15. Cai, S. and Kohwi-Shigematsu, T. 1999 Intranuclear relocalization of matrix binding sites during T cell activation detected by amplified fluorescence in situ hybridization. *Methods*, 19, 394–402.

16. Brown, K.E., Baxter, J., Graf, D., Merkenschlager, M., and Fisher, A.G. 1999 Dynamic repositioning of genes in the nucleus of lymphocytes preparing for cell division. *Mol. Cell*, 3, 207–217.

17. Gasser, S.M. 2001 Positions of potential: Nuclear organization and gene expression. *Cell*, 104, 639–642.

18. Espada, J., Ballestar, E., Fraga, M.F., Villar-Garea, A., Juarranz, A., Stockert, J.C., Robertson, K.D., Fuks, F., and Esteller, M. 2004 Human DNA methyltransferase 1 is required for maintenance of the histone H3 modification pattern. *J. Biol. Chem.*, 279, 37175–37184.

19. Johnson, L., Cao, X., and Jacobsen, S. 2002 Interplay between two epigenetic marks. DNA methylation and histone H3 lysine 9 methylation. *Curr. Biol.*, 12, 1360–1367.

20. Gendrel, A.V., Lippman, Z., Yordan, C., Colot, V., and Martienssen, R.A. 2002 Dependence of heterochromatic histone H3 methylation patterns on the Arabidopsis gene DDM1. *Science*, 297, 1871–1873.

21. Nakao, M. 2001 Epigenetics: Interaction of DNA methylation and chromatin. *Gene*, 278, 25–31.

22. McGarvey, K.M., Fahrner, J.A., Greene, E., Martens, J., Jenuwein, T., and Baylin, S.B. 2006 Silenced tumor suppressor genes reactivated by DNA demethylation do not return to a fully euchromatic chromatin state. *Cancer Res.*, 66, 3541–3549.

23. James, S.R., Link, P.A., and Karpf, A.R. 2006 Epigenetic regulation of X-linked cancer/germline antigen genes by DNMT1 and DNMT3b. *Oncogene*, 25, 6975–6985.

24. Santos, F., Zakhartchenko, V., Stojkovic, M., Peters, A., Jenuwein, T., Wolf, E., Reik, W., and Dean, W. 2003 Epigenetic marking correlates with developmental potential in cloned bovine preimplantation embryos. *Curr. Biol.*, 13, 1116–1121.

25. Deplus, R., Brenner, C., Burgers, W.A., Putmans, P., Kouzarides, T., de Launoit, Y., and Fuks, F. 2002 Dnmt3L is a transcriptional repressor that recruits histone deacetylase. *Nucleic Acids Res.*, 30, 3831–3838.

26. Jones, P.L., Veenstra, G.J., Wade, P.A., Vermaak, D., Kass, S.U., Landsberger, N., Strouboulis, J., and Wolffe, A.P. 1998 Methylated DNA and MeCP2 recruit histone deacetylase to repress transcription. *Nat. Genet.*, 19, 187–191.

27. Ng, H.H., Zhang, Y., Hendrich, B., Johnson, C.A., Turner, B.M., Erdjument-Bromage, H., Tempst, P., Reinberg, D., and Bird, A. 1999 MBD2 is a transcriptional repressor belonging to the MeCP1 histone deacetylase complex. *Nat. Genet.*, 23, 58–61.

28. Rountree, M.R., Bachman, K.E., and Baylin, S.B. 2000 DNMT1 binds HDAC2 and a new co-repressor, DMAP1, to form a complex at replication foci. *Nat. Genet.*, 25, 269–277.

29. Fuks, F., Burgers, W.A., Godin, N., Kasai, M., and Kouzarides, T. 2001 Dnmt3a binds deacetylases and is recruited by a sequence-specific repressor to silence transcription. *EMBO J.*, 20, 2536–2544.

30. Esteve, P.O., Chin, H.G., Smallwood, A., Feehery, G.R., Gangisetty, O., Karpf, A.R., Carey, M.F., and Pradhan, S. 2006 Direct interaction between DNMT1 and G9a coordinates DNA and histone methylation during replication. *Genes Dev.*, 20, 3089–3103.

31. Fuks, F., Hurd, P.J., Deplus, R., and Kouzarides, T. 2003 The DNA methyltransferases associate with HP1 and the SUV39H1 histone methyltransferase. *Nucleic Acids Res.*, 31, 2305–2312.

32. Robertson, K.D., Ait-Si-Ali, S., Yokochi, T., Wade, P.A., Jones, P.L., and Wolffe, A.P. 2000 DNMT1 forms a complex with Rb, E2F1 and HDAC1 and represses transcription from E2F-responsive promoters. *Nat. Genet.*, 25, 338–342.

33. Bachman, K.E., Rountree, M.R., and Baylin, S.B. 2001 Dnmt3a and Dnmt3b are transcriptional repressors that exhibit unique localization properties to heterochromatin. *J. Biol. Chem.*, 276, 32282–32287.

34. Wade, P.A. 2001 Methyl CpG binding proteins: Coupling chromatin architecture to gene regulation. *Oncogene*, 20, 3166–3173.

35. Geiman, T.M., Sankpal, U.T., Robertson, A.K., Chen, Y., Mazumdar, M., Heale, J.T., Schmiesing, J.A., Kim, W., Yokomori, K., Zhao, Y., and Robertson, K.D. 2004 Isolation and characterization of a novel DNA methyltransferase complex linking DNMT3B with components of the mitotic chromosome condensation machinery. *Nucleic Acids Res.*, 32, 2716–2729.

36. Suzuki, M., Yamada, T., Kihara-Negishi, F., Sakurai, T., Hara, E., Tenen, D.G., Hozumi, N., and Oikawa, T. 2006 Site-specific DNA methylation by a complex of PU.1 and Dnmt3a/b. *Oncogene*, 25, 2477–2488.

37. Vire, E., Brenner, C., Deplus, R., Blanchon, L., Fraga, M., Didelot, C., Morey, L., Van Eynde, A., Bernard, D., Vanderwinden, J.M., Bollen, M., Esteller, M., Di Croce, L., de Launoit, Y., and Fuks, F. 2006 The Polycomb group protein EZH2 directly controls DNA methylation. *Nature*, 439, 871–874.

38. Datta, J., Majumder, S., Bai, S., Ghoshal, K., Kutay, H., Smith, D.S., Crabb, J.W., and Jacob, S.T. 2005 Physical and functional interaction of DNA methyltransferase 3A with Mbd3 and Brg1 in mouse lymphosarcoma cells. *Cancer Res.*, 65, 10891–10900.

39. Zhu, H., Geiman, T.M., Xi, S., Jiang, Q., Schmidtmann, A., Chen, T., Li, E., and Muegge, K. 2006 Lsh is involved in de novo methylation of DNA. *EMBO J.*, 25, 335–345.

40. Chen, T., Tsujimoto, N., and Li, E. 2004 The PWWP domain of Dnmt3a and Dnmt3b is required for directing DNA methylation to the major satellite repeats at pericentric heterochromatin. *Mol. Cell Biol.*, 24, 9048–9058.

41. Lehnertz, B., Ueda, Y., Derijck, A.A., Braunschweig, U., Perez-Burgos, L., Kubicek, S., Chen, T., Li, E., Jenuwein, T., and Peters, A.H. 2003 Suv39h-mediated histone H3 lysine 9 methylation directs DNA methylation to major satellite repeats at pericentric heterochromatin. *Curr. Biol.*, 13, 1192–1200.

42. Ikegami, K., Iwatani, M., Suzuki, M., Tachibana, M., Shinkai, Y., Tanaka, S., Greally, J.M., Yagi, S., Hattori, N., and Shiota, K. 2007 Genome-wide and locus-specific DNA hypomethylation in G9a deficient mouse embryonic stem cells. *Genes Cells*, 12, 1–11.

43. Boyes, J. and Bird, A. 1992 Repression of genes by DNA methylation depends on CpG density and promoter strength: Evidence for involvement of a methyl-CpG binding protein. *EMBO J.*, 11, 327–333.

44. Zhang, X.Y., Jabrane-Ferrat, N., Asiedu, C.K., Samac, S., Peterlin, B.M., and Ehrlich, M. 1993 The major histocompatibility complex class II promoter-binding protein RFX (NF-X) is a methylated DNA-binding protein. *Mol. Cell. Biol.*, 13, 6810–6818.

45. Feng, Y.Q., Desprat, R., Fu, H., Olivier, E., Lin, C.M., Lobell, A., Gowda, S.N., Aladjem, M.I., and Bouhassira, E.E. 2006 DNA methylation supports intrinsic epigenetic memory in mammalian cells. *PLoS Genet.*, 2, e65.

46. Baylin, S.B., Hoppener, J.W., de Bustros, A., Steenbergh, P.H., Lips, C.J., and Nelkin, B.D. 1986 DNA methylation patterns of the calcitonin gene in human lung cancers and lymphomas. *Cancer Res.*, 46, 2917–2922.

47. Ehrlich, M. 2002 DNA methylation in cancer: Too much, but also too little. *Oncogene*, 21, 5400–5413.

48. Ehrlich, M. 2006 Cancer-linked DNA hypomethylation and its relationship to hypermethylation. *Curr. Top Microbiol. Immunol.*, 310, 251–274.

49. Laird, P.W. 2003 The power and the promise of DNA methylation markers. *Nat. Rev. Cancer*, 3, 253–266.

50. Itano, O., Ueda, M., Kikuchi, K., Hashimoto, O., Hayatsu, S., Kawaguchi, M., Seki, H., Aiura, K., and Kitajima, M. 2002 Correlation of postoperative recurrence in hepatocellular carcinoma with demethylation of repetitive sequences. *Oncogene*, 21, 789–797.

51. Widschwendter, M., Jiang, G., Woods, C., Muller, H.M., Fiegl, H., Goebel, G., Marth, C., Holzner, E.M., Zeimet, A.G., Laird, P.W., and Ehrlich, M. 2004 DNA hypomethylation and ovarian cancer biology. *Cancer Res.*, 64, 4472–4480.

52. Issa, J.P. 2000 Hypermethylator phenotypes in aging and cancer. In Ehrlich, M. (Ed.), *DNA Alterations in Cancer: Genetic and Epigenetic Alterations*. Eaton Publishing, Natick, pp. 311–322.

53. Ushijima, T., Watanabe, N., Shimizu, K., Miyamoto, K., Sugimura, T., and Kaneda, A. 2005 Decreased fidelity in replicating CpG methylation patterns in cancer cells. *Cancer Res.*, 65, 11–17.

54. Kokalj-Vokac, N., Almeida, A., Viegas-Pequignot, E., Jeanpierre, M., Malfoy, B., and Dutrillaux, B. 1993 Specific induction of uncoiling and recombination by azacytidine in classical satellite-containing constitutive heterochromatin. *Cytogenet. Cell Genet.*, 63, 11–15.

55. Qu, G., Grundy, P.E., Narayan, A., and Ehrlich, M. 1999 Frequent hypomethylation in Wilms tumors of pericentromeric DNA in chromosomes 1 and 16. *Cancer Genet. Cytogenet.*, 109, 34–39.

56. Baylin, S.B. and Herman, J.G. 2000 Epigenetics and loss of gene function in cancer. In Ehrlich, M. (Ed.), *DNA Alterations in Cancer: Genetic and Epigenetic Alterations*. Eaton Publishing, Natick, pp. 293–309.

57. Nguyen, C., Liang, G., Nguyen, T.T., Tsao-Wei, D., Groshen, S., Lubbert, M., Zhou, J.H., Benedict, W.F., and Jones, P.A. 2001 Susceptibility of nonpromoter CpG islands to de novo methylation in normal and neoplastic cells. *J. Natl. Cancer Inst.*, 93, 1465–1472.

58. Jones, P.A. and Baylin, S.B. 2002 The fundamental role of epigenetic events in cancer. *Nat. Rev. Genet.*, 3, 415–428.

59. Powell, M.A., Mutch, D.G., Rader, J.S., Herzog, T.J., Huang, T.H., and Goodfellow, P.J. 2002 Ribosomal DNA methylation in patients with endometrial carcinoma: An independent prognostic marker. *Cancer*, 94, 2941–2952.

60. Kelavkar, U.P., Harya, N.S., Hutzley, J., Bacich, D.J., Monzon, F.A., Chandran, U., Dhir, R., and O'Keefe, D.S. 2007 DNA methylation paradigm shift: 15-lipoxygenase-1 upregulation in prostatic intraepithelial neoplasia and prostate cancer by atypical promoter hypermethylation. *Prostaglandins Other Lipid Mediat.*, 82, 185–197.

61. Gama-Sosa, M.A., Slagel, V.A., Trewyn, R.W., Oxenhandler, R., Kuo, K.C., Gehrke, C.W., and Ehrlich, M. 1983 The 5-methylcytosine content of DNA from human tumors. *Nucleic Acids Res.*, 11, 6883–6894.

62. Narayan, A., Ji, W., Zhang, X.-Y., Marrogi, A., Graff, J.R., Baylin, S.B., and Ehrlich, M. 1998 Hypomethylation of pericentromeric DNA in breast adenocarcinomas. *Int. J. Cancer*, 77, 833–838.

63. Weisenberger, D.J., Campan, M., Long, T.I., Kim, M., Woods, C., Fiala, E., Ehrlich, M., and Laird, P.W. 2005 Analysis of repetitive element DNA methylation by MethyLight. *Nucleic Acids Res.*, 33, 6823–6836.

64. Ehrlich, M., Jiang, G., Fiala, E.S., Dome, J.S., Yu, M.S., Long, T.I., Youn, B., Sohn, O.-S., Widschwendter, M., Tomlinson, G.E., Chintagumpala, M., Champagne, M., Parham, D.M., Liang, G., Malik, K., and Laird, P.W. 2002 Hypomethylation and hypermethylation of DNA in Wilms tumors. *Oncogene*, 21, 6694–6702.

65. Thoraval, D., Asakawa, J., Wimmer, K., Kuick, R., Lamb, B., Richardson, B., Ambros, P., Glover, T., and Hanash, S. 1996 Demethylation of repetitive DNA sequences in neuroblastoma. *Genes Chromosomes Cancer*, 17, 234–244.

66. Nishiyama, R., Qi, L., Tsumagari, K., Dubeau, L., Weissbecker, K., Champagne, M., Sikka, S., Nagai, H., and Ehrlich, M. 2005 A DNA repeat, NBL2, is hypermethylated in some cancers but hypomethylated in others. *Cancer Biol. Ther.*, 4, 440–448.

67. Nagai, H., Kim, Y.S., Yasuda, T., Ohmachi, Y., Yokouchi, H., Monden, M., Emi, M., Konishi, N., Nogami, M., Okumura, K., and Matsubara, K. 1999 A novel sperm-specific hypomethylation sequence is a demethylation hotspot in human hepatocellular carcinomas. *Gene*, 237, 15–20.

68. Costello, J.F., Fruhwald, M.C., Smiraglia, D.J., Rush, L.J., Robertson, G.P., Gao, X., Wright, F.A., Feramisco, J.D., Peltomaki, P., Lang, J.C., Schuller, D.E., Yu, L., Bloomfield, C.D., Caligiuri, M.A., Yates, A., Nishikawa, R., Su Huang, H., Petrelli, N.J., Zhang, X., O'Dorisio, M.S., Held, W.A., Cavenee, W.K., and Plass, C. 2000 Aberrant CpG-island methylation has non-random and tumour-type-specific patterns. *Nat. Genet.*, 24, 132–138.

69. Florl, A.R., Lower, R., Schmitz-Drager, B.J., and Schulz, W.A. 1999 DNA methylation and expression of LINE-1 and HERV-K provirus sequences in urothelial and renal cell carcinomas. *Br. J. Cancer*, 80, 1312–1321.

70. Laird, C.D., Pleasant, N.D., Clark, A.D., Sneeden, J.L., Hassan, K.M., Manley, N.C., Vary, J.C., Jr., Morgan, T., Hansen, R.S., and Stoger, R. 2004 Hairpin-bisulfite PCR: Assessing epigenetic methylation patterns on complementary strands of individual DNA molecules. *Proc. Natl. Acad. Sci. U S A*, 101, 204–209.

71. Melki, J.R., Vincent, P.C., and Clark, S.J. 1999 Concurrent DNA hypermethylation of multiple genes in acute myeloid leukemia. *Cancer Res.*, 59, 3730–3740.

72. Rush, L.J., Raval, A., Funchain, P., Johnson, A.J., Smith, L., Lucas, D.M., Bembea, M., Liu, T.H., Heerema, N.A., Rassenti, L., Liyanarachchi, S., Davuluri, R., Byrd, J.C., and Plass, C. 2004 Epigenetic profiling in chronic lymphocytic leukemia reveals novel methylation targets. *Cancer Res.*, 64, 2424–2433.

73. Amoreira, C., Hindermann, W., and Grunau, C. 2003 An improved version of the DNA Methylation database (MethDB). *Nucleic Acids Res.*, 31, 75–77.

74. Tsumagari, K., Jackson, K., Shao, C., Lacey, M., Sowden, J., Tawil, R., Vedanarayan, V., and Ehrlich, M. 2008 Epigenetics of a tandem DNA repeat: Chromatin DNaseI sensitivity and opposite methylation changes in cancers. *Nucleic Acids Res.*, 36, 2196–2207.

75. Ehrlich, M. 2003 The ICF syndrome, a DNA methyltransferase 3B deficiency and immunodeficiency disease. *Clin. Immunol.*, 109, 17–28.

76. Gowher, H. and Jeltsch, A. 2002 Molecular enzymology of the catalytic domains of the Dnmt3a and Dnmt3b DNA methyltransferases. *J. Biol. Chem.*, 277, 20409–20414.

77. Hansen, R.S., Wijmenga, C., Luo, P., Stanek, A.M., Canfield, T.K., Weemaes, C.M., and Gartler, S.M. 1999 The DNMT3B DNA methyltransferase gene is mutated in the ICF immunodeficiency syndrome. *Proc. Natl. Acad. Sci. U S A*, 96, 14412–14417.

78. Tagarro, I., Fernandez-Peralta, A.M., and Gonzalez-Aguilera, J.J. 1994 Chromosomal localization of human satellites 2 and 3 by a FISH method using oligonucleotides as probes. *Hum. Genet.*, 93, 383–388.

79. Jeanpierre, M., Turleau, C., Aurias, A., Prieur, M., Ledeist, F., Fischer, A., and Viegas-Pequignot, E. 1993 An embryonic-like methylation pattern of classical satellite DNA is observed in ICF syndrome. *Hum. Mol. Genet.*, 2, 731–735.

80. Tuck-Muller, C.M., Narayan, A., Tsien, F., Smeets, D., Sawyer, J., Fiala, E.S., Sohn, O., and Ehrlich, M. 2000 DNA hypomethylation and unusual chromosome instability in cell lines from ICF syndrome patients. *Cytogenet. Cell Genet.*, 89, 121–128.

81. Smeets, D.F.C.M., Moog, U., Weemaes, C.M.R., Vaes-Peeters, G., Merkx, G.F.M., Niehof, J.P., and Hamers, G. 1994 ICF syndrome: A new case and review of the literature. *Hum. Genet.*, 94, 240–246.

82. Jiang, Y.L., Rigolet, M., Bourc'his, D., Nigon, F., Bokesoy, I., Fryns, J.P., Hulten, M., Jonveaux, P., Maraschio, P., Megarbane, A., Moncla, A., and Viegas-Pequignot, E. 2005 DNMT3B mutations and DNA methylation defect define two types of ICF syndrome. *Hum. Mutat.*, 25, 56–63.

83. Kondo, T., Comenge, Y., Bobek, M.P., Kuick, R., Lamb, B., Zhu, X., Narayan, A., Bourc'his, D., Viegas-Pequinot, E., Ehrlich, M., and Hanash, S. 2000 Whole-genome methylation scan in ICF syndrome: Hypomethylation of non-satellite DNA repeats D4Z4 and NBL2. *Hum. Mol. Gen.*, 9, 597–604.

84. Sumner, A.T. 1991 Scanning electron microscopy of mammalian chromosomes from prophase to telophase. *Chromosoma*, 100, 410–418.

85. Kim, G.D., Ni, J., Kelesoglu, N., Roberts, R.J., and Pradhan, S. 2002 Co-operation and communication between the human maintenance and de novo DNA (cytosine-5) methyltransferases. *EMBO J.*, 21, 4183–4195.

86. Sawyer, J., Tricot, G., Mattox, S., Jagannath, S., and Barlogie, B. 1998 Jumping translocations of chromosome 1 q in multiple myeloma: Evidence for a mechanism involving decondensation of pericentromeric heterochromatin. *Blood*, 91, 1732–1741.

87. Wong, N., Lam, W.C., Lai, P.B., Pang, E., Lau, W.Y., and Johnson, P.J. 2001 Hypomethylation of chromosome 1 heterochromatin DNA correlates with q-arm copy gain in human hepatocellular carcinoma. *Am. J. Pathol.*, 159, 465–471.

88. Nakagawa, T., Kanai, Y., Ushijima, S., Kitamura, T., Kakizoe, T., and Hirohashi, S. 2005 DNA hypomethylation on pericentromeric satellite regions significantly correlates with loss of heterozygosity on chromosome 9 in urothelial carcinomas. *J. Urol.*, 173, 243–246.

89. Cadieux, B., Ching, T.T., Vandenberg, S.R., and Costello, J.F. 2006 Genome-wide hypomethylation in human glioblastomas associated with specific copy number alteration, methylenetetrahydrofolate reductase allele status, and increased proliferation. *Cancer Res.*, 66, 8469–8476.

90. Mitelman, F., Mertens, F., and Johansson, B. 1997 A breakpoint map of recurrent chromosomal rearrangements in human neoplasia. *Nat. Genet.*, 15, Spec No, 417–474.

91. Ehrlich, M., Woods, C., Yu, M., Dubeau, L., Yang, F., Campan, M., Weisenberger, D., Long, T.I., Youn, B., Fiala, E., and Laird, P. 2006 Quantitative analysis of association between DNA hypermethylation, hypomethylation, and DNMT RNA levels in ovarian tumors. *Oncogene*, 25, 2636–2645.

92. Ehrlich, M., Hopkins, N., Jiang, G., Dome, J.S., Yu, M.S., Woods, C.B., Tomlinson, G.E., Chintagumpala, M., Champagne, M., Diller, L., Parham, D.M., and Sawyer, J. 2003 Satellite hypomethylation in karyotyped Wilms tumors. *Cancer Genet. Cytogenet.*, 141, 97–105.

93. Tsuda, H., Takarabe, T., Kanai, Y., Fukutomi, T., and Hirohashi, S. 2002 Correlation of DNA hypomethylation at pericentromeric heterochromatin regions of chromosomes 16 and 1 with histological features and chromosomal abnormalities of human breast carcinomas. *Am. J. Pathol.*, 161, 859–866.

94. Fasth, A., Forestier, E., Holmberg, E., Holmgren, G., Nordenson, I., Soderstrom, T., and Wahlstrom, J. 1990 Fragility of the centromeric region of chromosome 1 associated with combined immunodeficiency in siblings: A recessively inherited entity? *Acta. Paediatr. Scand.*, 79, 605–612.

95. Sawyer, J.R., Swanson, C.M., Wheeler, G., and Cunniff, C. 1995 Chromosome instability in ICF syndrome: Formation of micronuclei from multibranched chromosome 1 demonstrated by fluorescence in situ hybridization. *Am. J. Med. Genet.*, 56, 203–209.

96. Tiepolo, L., Maraschio, P., Gimelli, G., Cuoco, C., Gargani, G.F., and Romano, C. 1979 Multibranched chromosomes 1, 9, and 16 in a patient with combined IgA and IgE deficiency. *Hum. Genet.*, 51, 127–137.

97. Brown, D.C., Grace, E., Summer, A.T., Edmunds, A.T., and Ellis, P.M. 1995 ICF syndrome (immunodeficiency, centromeric instability and facial anomalies): Investigation of heterochromatin abnormalities and review oaf clinical outcome. *Hum. Genet.*, 96, 411–416.

98. Tsien, F., Youn, B., Fiala, E.S., Laird, P., Long, T.I., Weissbecker, K., and Ehrlich, M. 2002 Prolonged culture of chorionic villus cells yields ICF syndrome-like chromatin decondensation and rearrangements. *Cytogen. Genome Res.*, 98, 13–21.

99. Vilain, A., Bernardino, J., Gerbault-Seureau, M., Vogt, N., Niveleau, A., Lefrancois, D., Malfoy, B., and Dutrillaux, B. 2000 DNA methylation and chromosome instability in lymphoblastoid cell lines. *Cytogenet. Cell Genet.*, 90, 93–101.

100. Yusa, K., Takeda, J., and Horie, K. 2004 Enhancement of sleeping beauty transposition by CpG methylation: Possible role of heterochromatin formation. *Mol. Cell Biol.*, 24, 4004–4018.

101. Walsh, C.P., Chaillet, J.R., and Bestor, T.H. 1998 Transcription of IAP endogenous retroviruses is constrained by cytosine methylation. *Nat. Genet.*, 20, 116–117.

102. O'Neill, R., O'Neill, M., and Graves, J.A. 1998 Undermethylation associated with retroelement activation and chromosome remodelling in an interspecific mammalian hybrid. *Nature*, 393, 68–72.

103. Morgan, H.D., Sutherland, H.G., Martin, D.I., and Whitelaw, E. 1999 Epigenetic inheritance at the agouti locus in the mouse. *Nat. Genet.*, 23, 314–318.

104. Dupressoir, A. and Heidmann, T. 1997 Expression of intracisternal A-particle retrotransposons in primary tumors of oncogene-expressing transgenic mice. *Oncogene*, 14, 2951–2958.

105. Webster, K.E., O'Bryan, M.K., Fletcher, S., Crewther, P.E., Aapola, U., Craig, J., Harrison, D.K., Aung, H., Phutikanit, N., Lyle, R., Meachem, S.J., Antonarakis, S.E., de Kretser, D.M., Hedger, M.P., Peterson, P., Carroll, B.J., and Scott, H.S. 2005 Meiotic and epigenetic defects in Dnmt3L-knockout mouse spermatogenesis. *Proc. Natl. Acad. Sci. U S A*, 102, 4068–4073.

106. Chedin, F., Lieber, M.R., and Hsieh, C.L. 2002 The DNA methyltransferase-like protein DNMT3L stimulates de novo methylation by Dnmt3a. *Proc. Natl. Acad. Sci. U S A*, 99, 16916–16921.

107. Kazazian, H.H., Jr. and Moran, J.V. 1998 The impact of L1 retrotransposons on the human genome. *Nat. Genet.*, 19, 19–24.

108. Sassaman, D.M., Dombroski, B.A., Moran, J.V., Kimberland, M.L., Naas, T.P., DeBerardinis, R.J., Gabriel, A., Swergold, G.D., and Kazazian, H.H., Jr. 1997 Many human L1 elements are capable of retrotransposition. *Nat. Genet.*, 16, 37–43.

109. Miki, Y., Nishisho, I., Horii, A., Miyoshi, Y., Utsunomiya, J., Kinzler, K.W., Vogelstein, B., and Nakamura, Y. 1992 Disruption of the APC gene by a retrotransposal insertion of L1 sequence in a colon cancer. *Cancer Res.*, 52, 643–645.

110. Morse, B., Rotherg, P.G., South, V.J., Spandorfer, J.M., and Astrin, S.M. 1988 Insertional mutagenesis of the myc locus by a LINE-1 sequence in a human breast carcinoma. *Nature*, 333, 87–90.

111. Wei, W., Gilbert, N., Ooi, S.L., Lawler, J.F., Ostertag, E.M., Kazazian, H.H., Boeke, J.D., and Moran, J.V. 2001 Human L1 retrotransposition: Cis preference versus trans complementation. *Mol. Cell. Biol.*, 21, 1429–1439.

112. Dante, R., Dante-Paire, J., Rigal, D., and Roizes, G. 1992 Methylation patterns of long interspersed repeated DNA and alphoid repetitive DNA from human cell lines and tumors. *Anticancer Res.*, 12, 559–563.

113. Jurgens, B., Schmitz-Drager, B.J., and Schulz, W.A. 1996 Hypomethylation of L1 LINE sequences prevailing in human urothelial carcinoma. *Cancer Res.*, 56, 5698–5703.

114. Takai, D., Yagi, Y., Habib, N., Sugimura, T., and Ushijima, T. 2000 Hypomethylation of LINE1 retrotransposon in human hepatocellular carcinomas, but not in surrounding liver cirrhosis. *Jpn. J. Clin. Oncol.*, 30, 306–309.

115. Santourlidis, S., Florl, A., Ackermann, R., Wirtz, H.C., and Schulz, W.A. 1999 High frequency of alterations in DNA methylation in adenocarcinoma of the prostate. *Prostate*, 39, 166–174.

116. Schulz, W.A. 1998 DNA methylation in urological malignancies. *Int. J. Oncol.*, 13, 151–167.

117. Schichman, S.A., Caligiuri, M.A., Strout, M.P., Carter, S.L., Gu, Y., Canaani, E., Bloomfield, C.D., and Croce, C.M. 1994 ALL-1 tandem duplication in acute myeloid leukemia with a normal karyotype involves homologous recombination between Alu elements. *Cancer Res.*, 54, 4277–4280.

118. Rothberg, P.G., Ponnuru, S., Baker, D., Bradley, J.F., Freeman, A.I., Cibis, G.W., Harris, D.J., and Heruth, D.P. 1997 A deletion polymorphism due to Alu-Alu recombination in intron 2 of the retinoblastoma gene: Association with human gliomas. *Mol. Carcinog.*, 19, 69–73.

119. Schmid, C.W. 1996 Alu: Structure, origin, evolution, significance and function of one-tenth of human DNA. *Prog. Nucleic Acid Res. Mol. Biol.*, 53, 283–319.

120. Leib-Mosch, C., Haltmeier, M., Werner, T., Geigl, E.M., Brack-Werner, R., Francke, U., Erfle, V., and Hehlmann, R. 1993 Genomic distribution and transcription of solitary HERV-K LTRs. *Genomics*, 18, 261–269.
121. Nakase, H., Takahama, Y., and Akamatsu, Y. 2003 Effect of CpG methylation on RAG1/RAG2 reactivity: Implications of direct and indirect mechanisms for controlling V(D)J cleavage. *EMBO Rep.*, 4, 774–780.
122. Chen, B., Dias, P., Jenkins, J.J., III, Savell, V.H., and Parham, D.M. 1998 Methylation alterations of the MyoD1 upstream region are predictive of subclassification of human rhabdomyosarcomas. *Am. J. Pathol.*, 152, 1071–1079.
123. Chan, M.F., van Amerongen, R., Nijjar, T., Cuppen, E., Jones, P.A., and Laird, P.W. 2001 Reduced rates of gene loss, gene silencing, and gene mutation in Dnmt1-deficient embryonic stem cells. *Mol. Cell. Biol.*, 21, 7587–7600.
124. Gaudet, F., Hodgson, J.G., Eden, A., Jackson-Grusby, L., Dausman, J., Gray, J.W., Leonhardt, H., and Jaenisch, R. 2003 Induction of tumors in mice by genomic hypomethylation. *Science*, 300, 489–492.
125. Eden, A., Gaudet, F., Waghmare, A., and Jaenisch, R. 2003 Chromosomal instability and tumors promoted by DNA hypomethylation. *Science*, 300, 455.
126. Coombes, M.M., Briggs, K.L., Bone, J.R., Clayman, G.L., El-Naggar, A.K., and Dent, S.Y. 2003 Resetting the histone code at CDKN2A in HNSCC by inhibition of DNA methylation. *Oncogene*, 22, 8902–8911.
127. Monier, K., Mouradian, S., and Sullivan, K.F. 2007 DNA methylation promotes Aurora-B-driven phosphorylation of histone H3 in chromosomal subdomains. *J. Cell Sci.*, 120, 101–114.
128. Yasui, W., Oue, N., Ono, S., Mitani, Y., Ito, R., and Nakayama, H. 2003 Histone acetylation and gastrointestinal carcinogenesis. *Ann. N.Y. Acad. Sci.*, 983, 220–231.
129. Tryndyak, V.P., Kovalchuk, O., and Pogribny, I.P. 2006 Loss of DNA methylation and histone H4 lysine 20 trimethylation in human breast cancer cells is associated with aberrant expression of DNA methyltransferase 1, Suv4–20h2 histone methyltransferase and methyl-binding proteins. *Cancer Biol. Ther.*, 5, 65–70.
130. Cervoni, N., Detich, N., Seo, S.B., Chakravarti, D., and Szyf, M. 2002 The oncoprotein Set/TAF-1beta, an inhibitor of histone acetyltransferase, inhibits active demethylation of DNA, integrating DNA methylation and transcriptional silencing. *J. Biol. Chem.*, 277, 25026–25031.
131. Gonzalo, S., Garcia-Cao, M., Fraga, M.F., Schotta, G., Peters, A.H., Cotter, S.E., Eguia, R., Dean, D.C., Esteller, M., Jenuwein, T., and Blasco, M.A. 2005 Role of the RB1 family in stabilizing histone methylation at constitutive heterochromatin. *Nat. Cell Biol.*, 7, 420–428.
132. Isaac, C.E., Francis, S.M., Martens, A.L., Julian, L.M., Seifried, L.A., Erdmann, N., Binne, U.K., Harrington, L., Sicinski, P., Berube, N.G., Dyson, N.J., and Dick, F.A. 2006 The retinoblastoma protein regulates pericentric heterochromatin. *Mol. Cell Biol.*, 26, 3659–3671.
133. Luciani, J.J., Depetris, D., Missirian, C., Mignon-Ravix, C., Metzler-Guillemain, C., Megarbane, A., Moncla, A., and Mattei, M.G. 2005 Subcellular distribution of HP1 proteins is altered in ICF syndrome. *Eur. J. Hum. Genet.*, 13, 41–51.
134. Luciani, J.J., Depetris, D., Usson, Y., Metzler-Guillemain, C., Mignon-Ravix, C., Mitchell, M.J., Megarbane, A., Sarda, P., Sirma, H., Moncla, A., Feunteun, J., and Mattei, M.G. 2006 PML nuclear bodies are highly organised DNA-protein structures with a function in heterochromatin remodelling at the G2 phase. *J. Cell Sci.*, 119, 2518–2531.
135. Toth, M., Lichtenberg, U., and Doerfler, W. 1989 Genomic sequencing reveals a 5-methylcytosine-free domain in active promoters and the spreading of preimposed methylation patterns. *Proc. Natl. Acad. Sci. U S A*, 86, 3728–3732.
136. Turker, M.S. 2002 Gene silencing in mammalian cells and the spread of DNA methylation. *Oncogene*, 21, 5388–5393.
137. Yan, P.S., Shi, H., Rahmatpanah, F., Hsiau, T.H., Hsiau, A.H., Leu, Y.W., Liu, J.C., and Huang, T.H. 2003 Differential distribution of DNA methylation within the RASSF1A CpG island in breast cancer. *Cancer Res.*, 63, 6178–6186.
138. Stirzaker, C., Song, J.Z., Davidson, B., and Clark, S.J. 2004 Transcriptional gene silencing promotes DNA hypermethylation through a sequential change in chromatin modifications in cancer cells. *Cancer Res.*, 64, 3871–3877.

139. Frigola, J., Song, J., Stirzaker, C., Hinshelwood, R.A., Peinado, M.A., and Clark, S.J. 2006 Epigenetic remodeling in colorectal cancer results in coordinate gene suppression across an entire chromosome band. *Nat. Genet.*, 38, 540–549.

140. Nishiyama, R., Qi, L., Lacey, M., and Ehrlich, M. 2005 Both hypomethylation and hypermethylation in a 0.2 kb region of a DNA repeat in cancer. *Molec. Cancer Res.*, 3, 617–626.

141. Ehrlich, M., Gama-Sosa, M., Huang, L.-H., Midgett, R.M., Kuo, K.C., McCune, R.A., and Gehrke, C. 1982 Amount and distribution of 5-methylcytosine in human DNA from different types of tissues or cells. *Nucleic Acids Res.*, 10, 2709–2721.

142. Kho, M.R., Baker, D.J., Laayoun, A., and Smith, S.S. 1998 Stalling of human DNA (cytosine-5) methyltransferase at single-strand conformers from a site of dynamic mutation. *J. Mol. Biol.*, 275, 67–79.

143. Huppert, J.L. and Balasubramanian, S. 2007 G-quadruplexes in promoters throughout the human genome. *Nucleic Acids Res.*, 35, 406–413.

144. Maizels, N. 2006 Dynamic roles for G4 DNA in the biology of eukaryotic cells. *Nat. Struct. Mol. Biol.*, 13, 1055–1059.

145. Handa, V. and Jeltsch, A. 2005 Profound flanking sequence preference of Dnmt3a and Dnmt3b mammalian DNA methyltransferases shape the human epigenome. *J. Mol. Biol.*, 348, 1103–1112.

14 Chromatin Remodeling and Cancer

Sari Pennings and Richard R. Meehan

CONTENTS

14.1 MAINTENANCE OF CELL DIFFERENTIATION BY EPIGENETIC MECHANISMS

The organs and tissues of a body are composed of many cell types with distinct characteristics. Differentiation initiates in early embryogenesis when pluripotent cells become restricted to different developmental pathways [1,2]. Specification of each cell lineage involves successive stages of signal induction followed by changes in gene expression profiles, as the cell induction signal is transduced to a nuclear transcription signal. The cell function, morphology, and motility properties that these differentiating cell types acquire are defined by the transcriptome, the set of genes transcribed from the genome [3]. This selective expression of the genetic information is established by gene regulatory networks. Although transcription factor network feedback loops can maintain a form of transcriptional memory [4], developmental signals are typically transitory, and DNA-bound transcription factor associations do not usually persist through DNA replication. The stable transmission of such gene expression patterns through cell division depends on epigenetic mechanisms that can propagate multiple gene expression traits in the absence of changes in the genomic DNA sequence [5]. The resulting clonal variegation of cellular properties creates cell lineages that can further differentiate as tissues develop.

The causal analysis of developmental transitions leading from zygote to adult, from its genetics starting point was termed epigenetics by Waddington [6]. This process cannot normally be reversed, as the developmental potential of somatic cells becomes more restricted as specification proceeds, in spite of their unchanged genetic makeup. In normal development, terminally differentiated cells usually only give rise to the same cell type (fibroblasts) or become post-mitotic (neurons). Germ cells retain the developmental potential of forming a totipotent zygote, which requires the nuclear remodeling of parental haploid genomes upon fertilization. Only in rare instances can

253

somatic cell nuclei be reprogrammed, e.g., by nuclear transfer into donor oocytes, to recapitulate zygotic development for cloning or stem cell generation [7,8]. In experimental models, differentiation-restricted cells can be converted to stem-like cells through the transient expression of pluripotent factors; this is accompanied by changes in the epigenetic signature profile [9,10]. Within normal organs and tissues, any loss of somatic cell type specification coupled with proliferative escape constitutes a major risk of cancer [11]. Alternatively, adult cancers may derive from rare stem or progenitor cells [12], as a result of injury-induced self-renewal escaping normal differentiation control [13]. The accompanying abnormal transcription profiles could promote cancer progression which, using the same epigenetic mechanisms, would be heritably transmitted to form tumor tissue [14].

14.2 CHROMATIN STATES AND THEIR EPIGENETIC MARKS

The stable maintenance of heritable patterns of gene expression relies on chromatin architecture within the cell nucleus. Early microscopic observation distinguished regions of euchromatin and heterochromatin, which approximate active and silenced chromatin compartments, respectively. Genomic DNA is wrapped around octamers of histone proteins to form nucleosomes, the structural repeating unit of chromatin [15]. The core nucleosome structure has been determined to atomic resolution but less is known about the organization of nucleosome arrays into successive higher order chromatin fibers incorporating further structural proteins, such as heterochromatin protein 1 (HP1). This compaction progressively reduces access to DNA and contributes a primary level of gene regulation, as controlled by the active or inactive chromatin states of gene promoter and enhancer regions [16]. These chromatin states are hallmarked by molecular marks on the DNA and histones, which aid their structural organization, as well as leaving an epigenetic signature that can be analyzed [17]. As the term epigenetics has evolved from its developmental origins to a more molecular definition, the focus is now on how these mechanisms operate and how they can be manipulated to control disease.

Best known among epigenetic modifications is the methylation of DNA, which adds a 5-methyl group to cytosines in the context of CpG sequences [18]. This leads to the formation of repressive chromatin structures, mediated by multiple methyl-CpG-binding proteins. The maintenance of this epigenetic mark requires the DNA methyltransferase DNMT1, which recognizes and methylates hemimethylated CpGs after DNA replication. Less well characterized are the mechanisms of de novo DNA methylation involving DNMT3a/b, or DNA demethylation. Concurrent with proposals of the histone code hypothesis, histone modifications have become regarded as epigenetic marks [19,20], although the mechanism of their transmission through replication is not clear [21]. Primarily considered in this context is lysine methylation, as it is the more stable posttranslational modification of histones [22–24]. Most methylated histone lysine marks are associated with gene repression through heterochromatin formation, with the exception of di- and trimethylated H3 lysine 4 and also dimethylated H3 lysine 79. Various histone methyltransferases containing the conserved SET domain have been identified with specificity for particular lysine residues, but some redundancy exists: for example, EuHMTase1 and G9a can dimethylate H3 lysine 9. The recently identified LSD1 and Jumonji histone demethylase family proteins have been found to be responsible for histone demethylation, thus confirming the reversibility of this epigenetic mark [25].

While histone phosphorylation and acetylation are highly dynamic, the active chromatin states with which they are associated are nevertheless directly or indirectly inheritable. Histone acetylation levels represent a balance of rapid acetylation and deacetylation enzymatic activities [26]. Histone acetyl transferases are a large and diverse group, which includes known transcription coactivators such as GCN5 and p300/CBP, whereas histone deacetylase activities can be grouped in three classes based on homology to the yeast histone deacetylase archetypes Rpd3, Hda1, or Sir2 [27].

Acetylation marks are most directly related to gene activity [20], as histone acetylation regulates accessibility to transcription factors in conjunction with chromatin remodeling [28].

14.3   EPIGENETIC SIGNATURES OF CANCER

It was discovered early on that many cancers have a genetic origin, which can stem from translocations, viral integrations, and sporadic or inherited mutations. The genetic defects identified in tumors typically have serious consequences at the level of gene expression or chromosome stability. Often this is the result of the malfunction of a major signaling pathway or a master regulator with many downstream target genes. Established tumors typically show global abnormalities including hundreds of misexpressed genes that are not limited to tumor suppressor and oncogenes [29]. This is typically accompanied by global hypomethylation of DNA contrasted with hypermethylation of promoter CpG islands. This silences large numbers of genes and represents almost a reversal of normal DNA methylation patterns. Detection methods for gene misexpression as well as aberrant CpG methylation profiles at promoters can help to type a cancer, as well as adding prognostic value [29]. Equally, histone modifications are aberrantly distributed in cancer cells, as DNA hypermethylation is coupled with loss of active marks and gain of trimethylation of H3 lysine 27 and lysine 9 [30]. Genome-wide loss of acetylation of H4 lysine 16 and trimethylation of H4 lysine 20 has been linked to oncogenic transformation in an apparent breakdown of the epigenetic mechanism maintaining normal heterochromatin [31].

While the early focus was on the aberrant CpG methylation patterns in cancer, the question was whether these epigenetic modifications were a primary cause or a secondary effect of gene silencing. This cancer-specific issue has its basis in the problem of epigenetic hierarchy in normal development. Recent reports support the view that histone modification might precede DNA methylation, as histone methyl transferases seem required to establish DNA methylation patterns in Neurospora, Arabidopsis, and mouse ES cells [32]. In other systems, however, DNA methylation was shown to be essential for normal histone modification patterns [33,34]. This could mean that different or mutually reinforcing silencing strategies exist. Nevertheless, the mechanism for maintaining an epigenetic memory of histone modifications is not as clear as the one for transmitting the methyl-CpG mark [21,35]. The Polycomb group (PcG) proteins are good candidates based on their role at the developmentally regulated cluster of Hox genes, which are progressively and stably silenced along cell lineages [21]. These conserved regulatory proteins exist in multi-subunit protein complexes, which bind to complex PcG response elements at target genes in *Drosophila*, and less well characterized elements in other organisms [21]. There they form repressive chromatin structures, while they trimethylate histone lysine 27 through their EZ2H subunit [36].

Forced demethylation and activation of silenced CpG island promoters in cancer lines produces unusual juxtapositions of active and persistent inactive marks, including trimethylated H3 lysine 27 [37]. Removal of the demethylating agents results in gene re-silencing, which suggests that the altered transcriptome mediated by aberrant DNA and histone modifications has become epigenetically fixed in a particular cancer type [37]. A bivalent chromatin signature is also observed in embryonic stem cells [38]; in embryonic carcinoma cells it was proposed to leave the DNA vulnerable to DNA hypermethylation during tumor initiation [39] via the connection between DNMTs and PcG complexes or HP1 proteins [40,41].

14.4   EPIGENETIC DEFICIENCIES IN CANCER

Unlike the programmed differentiation of normal development, cancer progression presents an escape from neighboring cell control followed by exposure to an extremely selective environment [11]. This may select for deficiencies in epigenetic mechanisms [42]. Such epigenetic defects could conceivably form either strike in Knudson's "two hit" hypothesis, with many further defects occurring as a consequence of the malignancy [43]. It has also been proposed that progenitor cell disruption and epigenetic variation may precede the onset of tumorigenesis and provide a common basis for epigenetic alterations observed subsequently [44]. This may also promote an environment in which genetic instability leads to an increased mutation rate.

Mutational inactivation of the DNMT1 gene is probably a rare event during human carcinogenesis as it is required to perpetuate novel epigenome profiles [29]. In a limited study, somatic mutations of DNMT1 were observed in unstable colorectal cancers, however, it was not clear whether this was a causative event [45]. Inactivating mutations in histone modifiers are also rare, although they are observed for the histone acetyltransferases CBP, and p300; and deletion in mouse of histone methyltransferases SUV39H, or RIZ1 (PRDM2), predisposes to cancer [29,46,47]. More commonly, the levels of expression of histone modifiers vary significantly across tissue types and cancer types [48], which together make up a characteristic expression profile that might be used as a therapeutic index in drug trials [29,49].

A case in point are the breast cancer primary tumor expression profiles that can also predict disease outcome [14,50]. This set includes components of the Polycomb repressor complexes (PRCs), overexpression of which promotes neoplastic transformation of breast epithelial cells. Depletion of p16INK4a activity in primary human mammary epithelial cells results in upregulation of EZH2 and SUZ12. These are recruited to a PRC2 binding site upstream of HOXA9, a gene expressed during normal breast development but epigenetically silenced in breast cancer [41,51]. This leads to the recruitment of DNA methyltransferases and subsequent hypermethylation, as well as to chromatin remodeling of HOX loci that is dependent on SUZ12 expression. These results reveal the causal role of p16INK4a disruption in modulating DNA and histone modification, and show a dynamic and active process whereby epigenetic modulation of gene expression is an early event in breast tumor progression. In stem cells, Polycomb group targets are up to 12-fold more likely to have cancer-specific promoter DNA hypermethylation than nontargets, supporting a stem cell origin of cancer in which this reversible gene repression is replaced by permanent silencing [39]. Misexpression of the Jumonji proteins GASC1 and PLU-1 has also been implicated in the proliferative capacity of esophageal squamous carcinomas and breast cancer, respectively [52,53].

Covalent modifications to the DNA and histones denote active and inactive chromatin structures while leaving tangible marks that can be analyzed. Recently mutations have also been detected in complexes that solely modify the structure of chromatin. SWI/SNF complexes, which disrupt nucleosomal arrays, have been implicated in cancer, most recently through mutations in the Snf5 subunit [54].

14.5   CHROMATIN-REMODELING COMPLEX FAMILY

The Swi–Snf complex is arguably the archetypal chromatin-remodeling complex since its first isolation from the yeast *Saccharomyces cerevisiae* [55]. While the discoveries of DNA methylation and histone modification had preceded the identification of the enzymes responsible for these modifications by some time, the existence of a family of abundant megadalton multiprotein complexes in the nucleus was unsuspected and might have gone undetected if not for yeast genetics [56]. Classic complementation analyses revealed that a number of gene regulation mutants identified in separate genetic screens for effects on mating type switching and sucrose utilization had mutations in the same group of genes [57,58], with mutations in any one gene of this group producing similar pleiotropic gene activation defects. Confirmation that the gene products acted as a multiprotein complex was provided when this complex was finally biochemically isolated from yeast [59,60]. The identification of orthologous complexes in *Drosophila* and mammals quickly followed [61–64], as well as the discoveries of numerous related complexes [65].

As the SWI/SNF-like protein family expanded, three large subclasses and a subsequent fourth group became evident based on similarity of the primary enzymatic component of the complex, the helicase-like ATPase unit (Table 14.1). Other, more distantly related helicase-like families exist,

TABLE 14.1

Chromatin-Remodeling Complex Subclasses and Defining Features

Grouping	Subfamily	Complexes	Helicase-Like ATPase	Characteristic Domain	Characteristic Functions
Snf2-like	SWI/SNF	Swi-Snf, Rsc, BAF	Snf2, Sth1, BRM, BRG1	Bromodomain	Transcription regulation, binding of acetylated histones, nucleosome mobilization
	Mi-2/CHD	CHD1, Mi-2/NuRD, MeCP1	Mi-2; Chd1	Chromodomain	Transcription elongation, nucleosome spacing
	ISWI	Isw1a/b, Isw2, ACF, RSF, WICH, CHRAC, NURF	Isw1/2, ISWI, hSNF2H/L	SANT domain	Transcription, replication and nucleosome assembly, nucleosome spacing
Swr1-like	INO80/SWR1	Ino80, Swr1, SRCAP, Tip60/TRRAP	Ino80, Swr1, Domino EP400	Spacer in ATPase	Transcription, DNA/chromatin repair Histone exchange

Source: Based on Flaus, A., Martin, D.M., Barton, G.J., and Owen-Hughes, T., *Nucleic Acids Res.*, 34, 2887, 2006; Bao, Y. and Shen, X. *Cell*, 129, 632, 2007 and additional points discussed in the text.

Note: Only the major subfamilies are listed; the Lsh subfamily related to ISWI without a SANT domain, functions in DNA methylation. Mutations in SWI/SNF components SNF5, BRG1; in Lsh; and in CHD5 have so far been associated with cancer. Furthermore, BRM is suppressed in many tumor types. Complexes involved in DNA repair (INO80/SWR1) could potentially contribute. The methyl-binding domain component MBD2 of MeCP1 is linked to abnormal gene silencing in tumors.

which may act less directly on nucleosomes [66]. The SWI/SNF, ISWI, CHD/Mi-2, and INO80/SWR1 subfamilies associate with a variety of protein components in modular complexes [67,68]. Their mechanisms of action are wide ranging [55,69]. SWI/SNF ATPases have been demonstrated to increase accessibility of transcription factors (and nucleases) to nucleosomes [70,71], as well as destabilizing nucleosome arrays [72], and assisting the mobility of nucleosomes in an ATP-dependent way [73] over and above the inherent mobility of these nucleosomes [74]. These properties may reflect a destabilization of histone–DNA contacts, possibly changing the DNA path around the histone octamer and leading to altered nucleosome positioning behavior [75]. By contrast, ISWI ATPases can improve the regular spacing of nucleosome arrays after replication. Intriguingly, on short fragments of DNA, the complexes move nucleosomes in opposite directions [75]. The related Lsh/DDM subclass ATPases have been implicated in maintenance of DNA methylation and aging [76–78]. The chromodomain containing CHD ATPases are involved in gene repression linked to DNA methylation and histone deacetylation (Mi-2) and transcription elongation (CHD1). The yeast INO80 and SWR1 complexes function in the repair of double-strand breaks and histone exchange [66,68,79].

How remodeling complexes can be effectors in cancer is illustrated in the colorectal cancer model. This cancer most frequently results from mutations in the APC tumor suppressor, a component of the β-catenin destruction complex [80]. Inappropriate accumulation of β-catenin increases the nuclear pool of this coactivator of the Wnt signaling pathway, which normally controls body patterning throughout development [81]. Constitutive activation of Wnt target genes, including the oncogene c-Myc, leads to hyperproliferation of intestinal epithelial cells and is closely linked to colorectal tumorigenesis. Wnt signaling is now thought to regulate the expansion of adult stem cell populations, including intestinal crypt stem cells. Nuclear β-catenin binds DNA-bound TCF/LEF-1 and also interacts with various chromatin-remodeling complexes, histone methyltransferases that trimethylate histone H3 lysine 4, and histone acetyltransferases [82]. Thus, the activation mechanism mediated by β-catenin provides all the components to set up active chromatin and remodel its structure.

Nontranscriptional roles can furthermore not be excluded. Signaling pathways typically involve the shuttling of components between different complexes and cellular compartments. β-Catenin binds as a heterodimer with α-catenin to E-cadherin at membrane anchored adherens junctions of epithelial cells [83]. This influences cell adhesion as well as indirectly, cytoskeleton architecture through α-catenin-mediated actin polymerization. Chromatin-remodeling complexes contain β-actin and actin-related proteins [72,84], raising the interesting possibility that their recruitment by β-catenin indicates a structural involvement in nuclear remodeling. Increasing evidence for the involvement of several ATP-dependent chromatin-remodeling complexes in DNA repair and maintenance of genome stability suggests further mechanisms by which these complexes can contribute to cancer progression.

14.6   CHROMATIN REMODELING AND THE EPIGENETIC CONNECTION

The mammalian NuRD (nucleosome remodeling histone deacetylase) complex has been affinity-purified and has seven subunits: Mi-2, MTA2, HDAC1, HDAC2, RbAp48, RbAp46, and MBD3. Mi-2 (CHD4) contains several recognizable sequence motifs, including PHD (plant homeodomain) zinc fingers, chromodomains, and a SWI2/SNF2 helicase/ATPase domain. Recruitment of NuRD to target loci depends on DNA-binding components. The MeCP1 version of NuRD contains the methyl-CpG repressor protein MBD2 and inhibits transcription through preferential binding, remodeling, and deacetylation of methylated nucleosomes [85]. Lsh/DDM subclass ATPases, while not ostensibly associated with other subunits in a complex, are necessary to establish normal DNA methylation patterns and levels. The latter examples show the close connection between chromatin remodeling and DNA methylation [85].

In repressive chromatin-remodeling complexes, the physical association of nucleosome array ordering subunits with methyl-CpG binding and histone deacetylation subunits is suggestive of a three-pronged approach of mutually reinforcing gene silencing mechanisms. By contrast, histone methyltransferases are not found directly associated with ATPase-remodeling complexes. Histone acetyltransferase activity has been identified in the TIP60 complex combining NuA4-like histone acetyltransferase and SWR1-like ATPase functions [68]. More common is the temporal connection of these activities, which is detected upon activation of many genes [86]. In relation to this, the SNF2 ATPases have bromodomains, which bind acetylated histone tails, while the activating mark of trimethylated H3 lysine 4 is recognized by CHD and ISWI in transcription elongation [87]. The need for histone modification as well as chromatin remodeling in order to render a promoter accessible to transcription factors is commonly accepted but variations exist as to the order of events. In one well-documented example, Swi–Snf was recruited first by a transcription factor to the yeast HO gene before histone acetyltransferase, but in many other genes, Swi–Snf may associate with histone tails that are first acetylated [88].

Intriguingly, recent findings in yeast suggest that ATP-dependent chromatin-remodeling enzymes can maintain a form of epigenetic transcriptional memory that may operate independently from the traditional epigenetic marks. In *S. cerevisiae*, a capacity for rapid reinduction of the *GAL1* gene was observed after prior expression of the gene. This acquired expression state persisted after the disassembly of the transcription machinery and survived DNA replication. The heritability was independent of a wide range of histone-modifying enzymes tested. Instead, the ATP-dependent chromatin-remodeling enzyme, Swi–Snf, was found to be essential for the control of transcriptional memory at *GAL1*, by antagonizing the repressive role of ISWI-like chromatin-remodeling enzymes [89].

14.7   CHROMATIN-REMODELING COMPLEXES AND CANCER

Rather than causing epigenetic deficiencies, chromatin-remodeling complexes may affect cancer progression by interpreting the aberrant methylation cues to remodel inappropriate chromatin states.

Tumor promoting effects of methyl-CpG–binding proteins have been attributed to their ability to interpret inappropriate DNA methylation. MBD2, a component of the MeCP1/NuRD complex, is required for silencing of de novo methylated tumor-suppressor genes in human cancer cell lines and it may be required for cancer progression [90]. Another methyl-CpG-binding protein, Kaiso (Zbtb33), is also required for tumor progression in cancer-susceptible Apc(Min/+) mice [91]. One hypothesis is that transcriptional repressors such as Kaiso or Mbd2 mediate the abnormal gene silencing (usually as a result of de novo DNA methylation) that occurs in cancer cells [92]. Loss of Kaiso or Mbd2 function may thus permit the expression of genes that can counter tumor formation.

Additionally, ATPase deficiencies are associated with cell proliferation abnormalities. CHD5 residing on *1p36* shows about 40% identity with Mi-2/CHD4 and in chromosome engineering experiments in mice has been shown to be a tumor suppressor that controls proliferation, apoptosis, and senescence via the $p19^{Arf}/p53$ pathway [93]. Downregulation of CHD5 is also observed in human gliomas [93]. A biochemical characterization of CHD5 has not been reported, but it is likely that it has a role in chromatin remodeling at specific loci.

The Lsh gene locus has been designated SMARCA6 (SWI/SNF2-related, matrix-associated, actin-dependent regulator of chromatin, subfamily A, member 6) [94]. It is highly expressed in human thymus and testis but a specific splice variant that lacks a region known to be critical for the transactivation of genes involved in sucrose fermentation is highly expressed in human leukemia cell lines [94]. It is hypothesized that the leukemia-specific form of Lsh functions in a dominant negative way leading to altered regional specific chromatin remodeling and subsequent gene expression. The region of chromosome 10 where *Lsh* resides is associated with loss of hetero-zygosity in a variety of cancers and it is possible that it may function as a tumor suppressor [94]. *Lsh* mutations have been created in mice, both of which survive to term but one mutant dies immediately after birth whereas the other exhibits developmental growth retardation and a premature aging phenotype [78,95]. *Lsh*-deficient murine embryonal fibroblasts show reduced proliferation with signs of abnormal mitosis and increased levels of p16INK4a levels, although this is independent of changes in promoter methylation [78,96]. Overexpression of an intact Lsh, but not a catalytically dead form, disrupts cell cycle progression in vitro, again suggesting that Lsh (like other Snf2 homologues) functions as a regulator of cell proliferation [97].

14.8   SNF5 AND CANCER

The SWI/SNF complex is targeted by several signaling pathways, which regulate the transcription factors that recruit SWI/SNF, and can also phosphorylate SWI/SNF subunits [98]. In mammals, complexes either contain the ATPase BRG1 or Brm [99]. Further tissue-specific subunits may be involved in recruitment of SWI/SNF to tissue-specific gene subsets [56]. Yeast SWI/SNF regulates 6% of genes expressed in rich media [100]. As expected from single gene studies [101], it was found to act as a transcriptional activator, but also repressed a substantial proportion of genes [100,102]. A similar situation was found with mammalian SWI/SNF. Apart from its role in transcription, SWI/SNF controls cell proliferation through its association with cyclin E, which determines G1/S cell cycle progression. Mutations of BRG1 and SNF5/INI1, and reduced expression of Brm have been found in a variety of cancers [99]. The link is strongest for mutations in the SNF5 subunit, also termed BAF47 or SMARCB1, which is present in all variants of the SWI/SNF complex and which interacts with many transcription regulators and viral proteins. Inactivating Snf5 mutations are prevalent in malignant rhabdoid tumors, an aggressive childhood cancer that can arise in kidney and brain. Snf5 deletion is embryonic lethal in mice. An induced conditional knockout is also lethal, due to hematopoietic failure. When this is experimentally prevented, all mice develop tumors, and this is accelerated upon co-inactivation of p53. This demonstrates that SNF5 is a potent tumor suppressor [54]. SNF5 and BRG1 protect against cancer in different ways, as Brg1 +/− heterozygotes are predisposed to mammary tumors [103]. The function of the SWI/SNF complex itself is not affected by the deletion of Snf5, however. Mutations lead to chromosomal instability due to abrogated chromosome segregation. Furthermore, hSNF5 can regulate Rb activity by either inhibiting the expression of cyclin D1 or by activating the transcription of the p16INK4a tumor suppressor gene [54]. Thus, hSNF5 acts as a dual co-regulator which, depending on the promoter context either activates or represses target genes [104]. One possible consequence of Snf5 deletion is loss of p16INK4a activity resulting in an excess of E2F transcription factors that may, as in breast tissue, upregulate PRC2 components and indirectly contribute to abnormal gene silencing [41]. It has been suggested that PRC2 binding sites in undifferentiated human embryonic stem cells coincide with hyperconserved CpG domains (HCGD). Many HCGD overlap with genes that encode known or putative developmental regulators as well as genes implicated in cancer and differentiation of immune cells [105]. Interestingly, inhibition of p16INK4a activation by siRNA has been shown to block hSNF5-mediated cellular senescence [104].

14.9   PERSPECTIVES

Our understanding of the role of chromatin-remodeling factors in cancer development is still at an early stage. It is expected that global transcriptome, genetic, splicing, and mutational profiling of cancers will provide more clues as to how aberrant regulation of these factors underlies the molecular pathology of a particular cancer or contributes to its progression. The connection between DNA methylation and Ezh2-mediated H3K27 methylation is emerging as an important pathway for de novo silencing of genes in cancer. This raises the question as to what protects CpG islands from silencing by chromatin-modifying activities. In most studied examples, de novo CpG methylation coincides with or is subsequent to transcriptional repression. It is also possible that promoter regions become susceptible to silencing through alterations in DNA replication timing [92]. Alternative models suggest active recruitment of DNMT activity to targeted promoters by components of chromatin-remodeling complexes or that the maintenance methyltransferase Dnmt1 has intrinsic sequence preferences as a de novo methyltransferase [40,106,107]. Whatever the case, it is clear that in cancer a combination of chromatin modifying and remodeling activities contributes either to the initiation or maintenance of the altered expression patterns. Ultimately, we need to identify the best methods for interfering with these processes with minimal effects on normal cells.

14.10   CONCLUSIONS

Stable epigenetic inheritance of a large range of cell type-specific properties in the absence of alterations to the genotype is essential for normal differentiation during development. Following the early discovery that mutagenesis can lead to oncogenesis, epigenetic factors have been shown to exacerbate these genetic defects by propagating malignant expression patterns and by contributing to the destabilization of active and repressive chromatin states. DNA methylation was the first epigenetic mark to be investigated for clues to the abnormal gene expression in cancer and also as a prognostic marker and drug target. After DNA methyltransferases and histone modification enzymes, the latest chromatin modifier to enter the fray is the family of ATP-dependent chromatin remodeling complexes. While their molecular signatures are comparatively difficult to detect, the evidence for their involvement in cancer is emerging from the identification of mutations in components of remodeling complexes. This finding is not surprising, as chromatin remodeling and modification activities go hand in hand in gene activation and both are linked to DNA methylation in gene silencing. Whereas the epigenetic marks ensure the heritability of gene expression patterns in normal development and cancer, chromatin remodeling determines the DNA accessibility of active and repressive chromatin states and is therefore directly connected to normal and malignant gene expression.

ACKNOWLEDGMENTS

Work in SP's laboratory is supported by the BBSRC and by the MRC in RM's laboratory. We would like to thank our laboratory members for useful discussions on chromatin-remodeling complexes and disease.

REFERENCES

1. Loebel, D.A., Watson, C.M., De Young, R.A., and Tam, P.P. Lineage choice and differentiation in mouse embryos and embryonic stem cells. *Dev. Biol.* 264, 1–14 2003.
2. Tam, P.P. and Loebel, D.A. Gene function in mouse embryogenesis: Get set for gastrulation. *Nat. Rev. Genet.* 8, 368–381 2007.
3. Hamatani, T. et al. Global gene expression profiling of preimplantation embryos. *Hum. Cell* 19, 98–117 2006.
4. Acar, M., Becskei, A., and van Oudenaarden, A. Enhancement of cellular memory by reducing stochastic transitions. *Nature* 435, 228–232 2005.
5. Morgan, H.D., Santos, F., Green, K., Dean, W., and Reik, W. Epigenetic reprogramming in mammals. *Hum. Mol. Genet.* 14, Spec No 1, R47–R58 2005.
6. Stern, C.D. and Conrad, H. Waddington's contributions to avian and mammalian development, 1930–1940. *Int. J. Dev. Biol.* 44, 15–22 2000.
7. Hochedlinger, K. et al. Reprogramming of a melanoma genome by nuclear transplantation. *Genes Dev.* 18, 1875–1885 2004.
8. Hochedlinger, K. and Jaenisch, R. Nuclear reprogramming and pluripotency. *Nature* 441, 1061–1067 2006.
9. Takahashi, K. and Yamanaka, S. Induction of pluripotent stem cells from mouse embryonic and adult fibroblast cultures by defined factors. *Cell* 126, 663–676 2006.
10. Wernig, M. et al. In vitro reprogramming of fibroblasts into a pluripotent ES-cell-like state. *Nature* 448, 318–324 2007.
11. Hendrix, M.J. et al. Reprogramming metastatic tumour cells with embryonic microenvironments. *Nat. Rev. Cancer* 7, 246–255 2007.
12. Clarke, M.F. and Fuller, M. Stem cells and cancer: Two faces of eve. *Cell* 124, 1111–1115 2006.
13. Beachy, P.A., Karhadkar, S.S., and Berman, D.M. Tissue repair and stem cell renewal in carcinogenesis. *Nature* 432, 324–331 2004.
14. West, M. et al. Predicting the clinical status of human breast cancer by using gene expression profiles. *Proc. Natl. Acad. Sci. U S A* 98, 11462–11467 2001.
15. Khorasanizadeh, S. The nucleosome: From genomic organization to genomic regulation. *Cell* 116, 259–272 2004.
16. Razin, S.V. et al. Chromatin domains and regulation of transcription. *J. Mol. Biol.* 2007.
17. Meehan, R.R., Dunican, D.S., Ruzov, A., and Pennings, S. Epigenetic silencing in embryogenesis. *Exp. Cell Res.* 309, 241–249 2005.
18. Klose, R.J. and Bird, A.P. Genomic DNA methylation: The mark and its mediators. *Trends Biochem. Sci.* 31, 89–97 2006.
19. Meehan, R.R. DNA methylation in animal development. *Semin. Cell Dev. Biol.* 14, 53–65 2003.
20. Turner, B.M. Defining an epigenetic code. *Nat. Cell Biol.* 9, 2–6 2007 .
21. Ringrose, L. and Paro, R. Epigenetic regulation of cellular memory by the Polycomb and Trithorax group proteins. *Annu. Rev. Genet.* 38, 413–443 2004.
22. Fischle, W., Wang, Y., and Allis, C.D. Histone and chromatin cross-talk. *Curr. Opin. Cell Biol.* 15, 172–183 2003.
23. Goldberg, A.D., Allis, C.D., and Bernstein, E. Epigenetics: A landscape takes shape. *Cell* 128, 635–638 2007.
24. Nightingale, K.P., O'Neill, L.P., and Turner, B.M. Histone modifications: Signalling receptors and potential elements of a heritable epigenetic code. *Curr. Opin. Genet. Dev.* 16, 125–136 2006.
25. Anand, R. and Marmorstein, R. Structure and mechanism of lysine-specific demethylase enzymes. *J. Biol. Chem.* 282, 35425–35429 2007.
26. Clayton, A.L., Hazzalin, C.A., and Mahadevan, L.C. Enhanced histone acetylation and transcription: A dynamic perspective. *Mol. Cell* 23, 289–296 2006.

27. Kurdistani, S.K. and Grunstein, M. Histone acetylation and deacetylation in yeast. *Nat. Rev. Mol. Cell Biol.* 4, 276–284 2003.

28. Millar, C.B. and Grunstein, M. Genome-wide patterns of histone modifications in yeast. *Nat. Rev. Mol. Cell Biol.* 7, 657–666 2006.

29. Esteller, M. Cancer epigenomics: DNA methylomes and histone-modification maps. *Nat. Rev. Genet.* 8, 286–298 2007.

30. Vire, E. et al. The Polycomb group protein EZH2 directly controls DNA methylation. *Nature* 439, 871–874 2006.

31. Fraga, M.F. et al. Loss of acetylation at Lys16 and trimethylation at Lys20 of histone H4 is a common hallmark of human cancer. *Nat. Genet.* 37, 391–400 2005.

32. Ting, A.H., McGarvey, K.M., and Baylin, S.B. The cancer epigenome—components and functional correlates. *Genes Dev.* 20, 3215–3231 2006.

33. Espada, J. et al. Human DNA methyltransferase 1 is required for maintenance of the histone H3 modification pattern. *J. Biol. Chem.* 279, 37175–37184 2004.

34. Tariq, M., Habu, Y., and Paszkowski, J. Depletion of MOM1 in non-dividing cells of Arabidopsis plants releases transcriptional gene silencing. *EMBO Rep.* 3, 951–955 2002 .

35. Ringrose, L. and Paro, R. Polycomb/Trithorax response elements and epigenetic memory of cell identity. *Development* 134, 223–232 2007.

36. Schwartz, Y.B. and Pirrotta, V. Polycomb silencing mechanisms and the management of genomic programmes. *Nat. Rev. Genet.* 8, 9–22 2007.

37. McGarvey, K.M. et al. Silenced tumor suppressor genes reactivated by DNA demethylation do not return to a fully euchromatic chromatin state. *Cancer Res.* 66, 3541–3549 2006.

38. Bernstein, B.E. et al. A bivalent chromatin structure marks key developmental genes in embryonic stem cells. *Cell* 125, 315–326 2006.

39. Ohm, J.E. et al. A stem cell-like chromatin pattern may predispose tumor suppressor genes to DNA hypermethylation and heritable silencing. *Nat. Genet.* 39, 237–242 2007.

40. Smallwood, A., Esteve, P.O., Pradhan, S., and Carey, M. Functional cooperation between HP1 and DNMT1 mediates gene silencing. *Genes Dev.* 21, 1169–1178 2007.

41. Reynolds, P.A. et al. Tumor suppressor p16INK4A regulates polycomb-mediated DNA hypermethylation in human mammary epithelial cells. *J. Biol. Chem.* 281, 24790–24802 2006.

42. Kirschmann, D.A. et al. Down-regulation of HP1Hsalpha expression is associated with the metastatic phenotype in breast cancer. *Cancer Res.* 60, 3359–3363 2000.

43. Knudson, A.G. A personal sixty-year tour of genetics and medicine. *Annu. Rev. Genomics Hum. Genet.* 6, 1–14 2005.

44. Feinberg, A.P., Ohlsson, R., and Henikoff, S. The epigenetic progenitor origin of human cancer. *Nat. Rev. Genet.* 7, 21–33 2006.

45. Kanai, Y., Ushijima, S., Nakanishi, Y., Sakamoto, M., and Hirohashi, S. Mutation of the DNA methyltransferase (DNMT) 1 gene in human colorectal cancers. *Cancer Lett.* 192, 75–82 2003.

46. Iyer, N.G., Ozdag, H., and Caldas, C. p300/CBP and cancer. *Oncogene* 23, 4225–4231 2004.

47. Steele-Perkins, G. et al. Tumor formation and inactivation of RIZ1, an Rb-binding member of a nuclear protein-methyltransferase superfamily. *Genes Dev.* 15, 2250–2262 2001.

48. Ozdag, H. et al. Differential expression of selected histone modifier genes in human solid cancers. *BMC Genomics* 7, 90 2006.

49. Sigalotti, L. et al. Epigenetic drugs as pleiotropic agents in cancer treatment: Biomolecular aspects and clinical applications. *J. Cell Physiol.* 212, 330–344 2007.

50. van't Veer, L.J. et al. Gene expression profiling predicts clinical outcome of breast cancer. *Nature* 415, 530–536 2002.

51. Novak, P. et al. Epigenetic inactivation of the HOXA gene cluster in breast cancer. *Cancer Res.* 66, 10664–10670 2006.

52. Cloos, P.A. et al. The putative oncogene GASC1 demethylates tri- and dimethylated lysine 9 on histone H3. *Nature* 442, 307–311 2006.

53. Yamane, K. et al. PLU-1 is an H3K4 demethylase involved in transcriptional repression and breast cancer cell proliferation. *Mol. Cell* 25, 801–812 2007.

54. Imbalzano, A.N. and Jones, S.N. Snf5 tumor suppressor couples chromatin remodeling, checkpoint control, and chromosomal stability. *Cancer Cell* 7, 294–295 2005.

55. Workman, J.L. and Kingston, R.E. Alteration of nucleosome structure as a mechanism of transcriptional regulation. *Annu. Rev. Biochem.* 67, 545–579 1998.

56. de la Serna, I.L., Ohkawa, Y., and Imbalzano, A.N. Chromatin remodelling in mammalian differentiation: Lessons from ATP-dependent remodellers. *Nat. Rev. Genet.* 7, 461–473 2006.

57. Neigeborn, L. and Carlson, M. Genes affecting the regulation of SUC2 gene expression by glucose repression in Saccharomyces cerevisiae. *Genetics* 108, 845–858 1984.

58. Stern, M., Jensen, R., and Herskowitz, I. Five SWI genes are required for expression of the HO gene in yeast. *J. Mol. Biol.* 178, 853–868 1984.

59. Cairns, B.R., Kim, Y.J., Sayre, M.H., Laurent, B.C., and Kornberg, R.D. A multi-subunit complex containing the SWI1/ADR6, SWI2/SNF2, SWI3, SNF5, and SNF6 gene products isolated from yeast. *Proc. Natl. Acad. Sci. U S A* 91, 1950–1954 1994.

60. Peterson, C.L., Dingwall, A., and Scott, M.P. Five SWI/SNF gene products are components of a large multi-subunit complex required for transcriptional enhancement. *Proc. Natl. Acad. Sci. U S A* 91, 2905–2908 1994.

61. Dingwall, A.K. et al. The *Drosophila* snr1 and brm proteins are related to yeast SWI/SNF proteins and are components of a large protein complex. *Mol. Biol. Cell* 6, 777–791 1995.

62. Khavari, P.A., Peterson, C.L., Tamkun, J.W., Mendel, D.B., and Crabtree, G.R. BRG1 contains a conserved domain of the SWI2/SNF2 family necessary for normal mitotic growth and transcription. *Nature* 366, 170–174 1993.

63. Muchardt, C. and Yaniv, M. A human homologue of *Saccharomyces cerevisiae* SNF2/SWI2 and *Drosophila* BRM genes potentiates transcriptional activation by the glucocorticoid receptor. *EMBO J.* 12, 4279–4290 1993.

64. Tamkun, J.W. et al. Brahma: A regulator of *Drosophila* homeotic genes structurally related to the yeast transcriptional activator SNF2/SWI2. *Cell* 68, 561–572 1992.

65. Kingston, R.E. and Narlikar, G.J. ATP-dependent remodeling and acetylation as regulators of chromatin fluidity. *Genes Dev.* 13, 2339–2352 1999.

66. Flaus, A., Martin, D.M., Barton, G.J., and Owen-Hughes, T. Identification of multiple distinct Snf2 subfamilies with conserved structural motifs. *Nucleic Acids Res.* 34, 2887–2905 2006.

67. Wang, W. et al. Purification and biochemical heterogeneity of the mammalian SWI-SNF complex. *EMBO J.* 15, 5370–5382 1996.

68. Bao, Y. and Shen, X. SnapShot: Chromatin remodeling complexes. *Cell* 129, 632 2007.

69. Mohrmann, L. and Verrijzer, C.P. Composition and functional specificity of SWI2/SNF2 class chromatin remodeling complexes. *Biochim. Biophys. Acta* 1681, 59–73 2005.

70. Côté, J., Quinn, J., Workman, J.L., and Peterson, C.L. Stimulation of GAL4 derivative binding to nucleosomal DNA by the yeast SWI/SNF complex. *Science* 265, 53–60 1994.

71. Kwon, H., Imbalzano, A.N., Khavari, P.A., Kingston, R.E., and Green, M.R. Nucleosome disruption and enhancement of activator binding by a human SW1/SNF complex. *Nature* 370, 477–481 1994.

72. Varga-Weisz, P.D., and Becker, P.B. Regulation of higher-order chromatin structures by nucleosome-remodelling factors. *Curr. Opin. Genet. Dev.* 16, 151–156 2006.

73. Whitehouse, I. et al. Nucleosome mobilization catalysed by the yeast SWI/SNF complex. *Nature* 400, 784–787 1999.

74. Meersseman, G., Pennings, S., and Bradbury, E.M. Mobile nucleosomes—a general behavior. *EMBO J.* 11, 2951–2959 1992.

75. Becker, P.B. and Hörz, W. ATP-dependent nucleosome remodeling. *Annu. Rev. Biochem.* 71, 247–273 2002.

76. Bourc'his, D. and Bestor, T.H. Helicase homologues maintain cytosine methylation in plants and mammals. *Bioessays* 24, 297–299 2002.

77. Meehan, R.R., Pennings, S., and Stancheva, I. Lashings of DNA methylation, forkfuls of chromatin remodeling. *Genes Dev.* 15, 3231–3236 2001.

78. Sun, L.Q. et al. Growth retardation and premature aging phenotypes in mice with disruption of the SNF2-like gene, PASG. *Genes Dev.* 18, 1035–1046 2004.

79. Papamichos-Chronakis, M., Krebs, J.E., and Peterson, C.L. Interplay between Ino80 and Swr1 chromatin remodeling enzymes regulates cell cycle checkpoint adaptation in response to DNA damage. *Genes Dev.* 20, 2437–2449 2006.

80. Clevers, H. Wnt/beta-catenin signaling in development and disease. *Cell* 127, 469–480 2006.

81. Willert, K. and Jones, K.A. Wnt signaling: Is the party in the nucleus? *Genes Dev.* 20, 1394–1404 2006.

82. Barker, N. et al. The chromatin remodelling factor Brg-1 interacts with beta-catenin to promote target gene activation. *EMBO J.* 20, 4935–4943 2001.

83. Nelson, W.J. and Nusse, R. Convergence of Wnt, beta-catenin, and cadherin pathways. *Science* 303, 1483–1487 2004.

84. Miralles, F. and Visa, N. Actin in transcription and transcription regulation. *Curr. Opin. Cell Biol.* 18, 261–266 2006.

85. Feng, Q. and Zhang, Y. The MeCP1 complex represses transcription through preferential binding, remodeling, and deacetylating methylated nucleosomes. *Genes Dev.* 15, 827–832 2001.

86. Robertson, K.D. DNA methylation and chromatin—unraveling the tangled web. *Oncogene* 21, 5361–5379 2002.

87. Flanagan, J.F. et al. Molecular implications of evolutionary differences in CHD double chromodomains. *J. Mol. Biol.* 369, 334–342 2007.

88. Cosma, M.P., Tanaka, T., and Nasmyth, K. Ordered recruitment of transcription and chromatin remodeling factors to a cell cycle- and developmentally regulated promoter. *Cell* 97, 299–311 1999.

89. Kundu, S., Horn, P.J., and Peterson, C.L. SWI/SNF is required for transcriptional memory at the yeast GAL gene cluster. *Genes Dev.* 21, 997–1004 2007.

90. Pulukuri, S.M. and Rao, J.S. CpG island promoter methylation and silencing of 14–3–3sigma gene expression in LNCaP and Tramp-C1 prostate cancer cell lines is associated with methyl-CpG-binding protein MBD2. *Oncogene* 25, 4559–4572 2006.

91. Prokhortchouk, A. et al. Kaiso-deficient mice show resistance to intestinal cancer. *Mol. Cell Biol.* 26, 199–208 2006.

92. Antequera, F. Structure, function and evolution of CpG island promoters. *Cell Mol. Life Sci.* 60, 1647–1658 2003.

93. Bagchi, A. et al. CHD5 is a tumor suppressor at human 1p36. *Cell* 128, 459–475 2007.

94. Lee, D.W. et al. Proliferation-associated SNF2-like gene (PASG): A SNF2 family member altered in leukemia. *Cancer Res.* 60, 3612–3622 2000.

95. Dennis, K., Fan, T., Geiman, T., Yan, Q., and Muegge, K. Lsh, a member of the SNF2 family, is required for genome-wide methylation. *Genes Dev.* 15, 2940–2944 2001.

96. Fan, T. et al. Lsh-deficient murine embryonal fibroblasts show reduced proliferation with signs of abnormal mitosis. *Cancer Res.* 63, 4677–4683 2003.

97. Raabe, E.H., Abdurrahman, L., Behbehani, G., and Arceci, R.J. An SNF2 factor involved in mammalian development and cellular proliferation. *Dev. Dyn.* 221, 92–105 2001.

98. Simone, C. SWI/SNF: The crossroads where extracellular signaling pathways meet chromatin. *J. Cell Physiol* 207, 309–314 2006.

99. Roberts, C.W. and Orkin, S.H. The SWI/SNF complex—chromatin and cancer. *Nat. Rev. Cancer* 4, 133–142 2004.

100. Holstege, F.C. et al. Dissecting the regulatory circuitry of a eukaryotic genome. *Cell* 95, 717–728 1998.

101. Hirschhorn, J.N., Brown, S.A., Clark, C.D., and Winston, F. Evidence that SNF2/SWI2 and SNF5 activate transcription in yeast by altering chromatin structure. *Genes Dev.* 6, 2288–2298 1992.

102. Sudarsanam, P., Iyer, V.R., Brown, P.O., and Winston, F. Whole-genome expression analysis of snf/swi mutants of *Saccharomyces cerevisiae*. *Proc. Natl. Acad. Sci. U S A* 97, 3364–3369 2000.

103. Bultman, S.J. et al. Characterization of mammary tumors from Brg1 heterozygous mice. *Oncogene* 27, 460–468 2008.

104. Oruetxebarria, I. et al. P16INK4a is required for hSNF5 chromatin remodeler-induced cellular senescence in malignant rhabdoid tumor cells. *J. Biol. Chem.* 279, 3807–3816 2004.

105. Tanay, A., O'Donnell, A.H., Damelin, M., and Bestor, T.H. Hyperconserved CpG domains underlie Polycomb-binding sites. *Proc. Natl. Acad. Sci. U S A* 104, 5521–5526 2007.

106. Jair, K.W. et al. De novo CpG island methylation in human cancer cells. *Cancer Res.* 66, 682–692 2006.

107. Esteve, P.O. et al. Direct interaction between DNMT1 and G9a coordinates DNA and histone methylation during replication. *Genes Dev.* 20, 3089–3103 2006.

15 Poly-ADP-Ribosylation in Cancer

Rajeshwar Nath Sharan

CONTENTS

15.1 INTRODUCTION

Poly-ADP-ribosylation (PAR) is a posttranslational modification of proteins. The process was discovered and reported in early 1960s by Mandel and coworkers [1,2]. Since then it has been a subject of extensive research [reviewed in Refs. 3–5]. The interest has endured because of the uniqueness associated with PAR metabolism and its continuously expanding biological involvement, implications, and roles. The ubiquitous, enzyme catalyzed, fully reversible metabolic reaction involves transfer of an ADP-ribose moiety from a metabolic donor, nicotinamide adenine dinucleotide (NAD$^+$), to acceptor amino acid residues of a target protein. The primary biosynthesizing enzyme for PAR reaction is poly-ADP-ribose polymerase (PARP). The commonly modified amino acid residues in eukaryotes are glutamate and aspartate, though occasionally ADP-ribose moiety is found on residues such as arginine, cystine, asparagine, and diphthamide [3]. The target proteins for PAR are mostly nuclear and include, but are not limited to, such diverse array of proteins as histones, endonucleases, DNA pol α and β, DNA ligase I and II, topoisomerase I and II, RNA polymerases, reverse transcriptase, high mobility group (HMG) proteins, p53, Fos, AP endonuclease, Ku70, and the enzyme responsible for biosynthesis of PAR, that is, PARP itself. The metabolic reaction creates a complex, variably sized, and covalently attached homopolymeric, heterogeneous branched or unbranched ADP-ribose polymer adducts on the target protein thereby accomplishing the posttranslational modification of a protein (Figure 15.1). The complex homopolymeric ADP-ribose polymer adducts may contain up to 200 or, occasionally, more of monomers of ADP-ribose in linear or multiple branching architecture. Simultaneously, the main biodegrading enzyme of ADP-ribose polymer, the poly-ADP-ribose glycohydrolase (PARG), acts on the homopolymer attached to a modified protein and rapidly de-poly-ADP-ribosylates it by randomly and sequentially degrading the ADP-ribose monomers from the target protein (Figure 15.1). The two metabolic reactions occur simultaneously but in

265

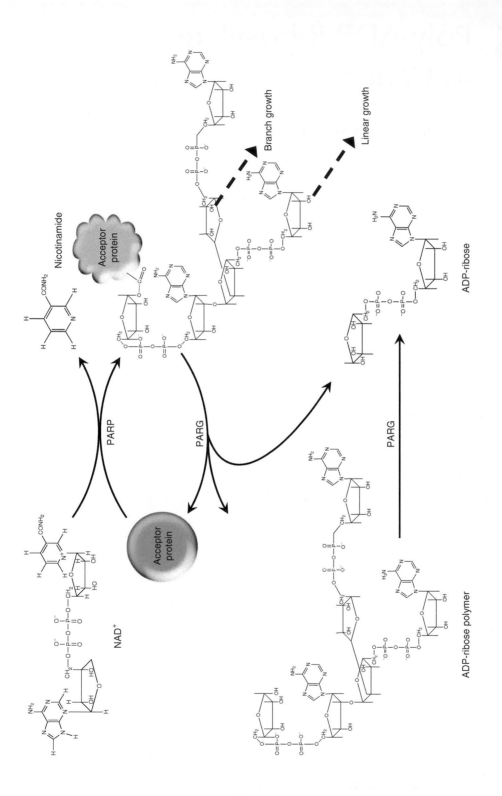

FIGURE 15.1 The biochemical pathway depicting the main features of the PAR metabolism. The ADP-ribose moiety from donor nicotinamide dinucleotide (NAD$^+$) is transferred onto an acceptor protein by PARP enzyme successively in a branched or unbranched complex architecture. The PARG enzyme breaks down the ADP-ribose polymers as ADP-ribose polymer or ADP-ribose monomer.

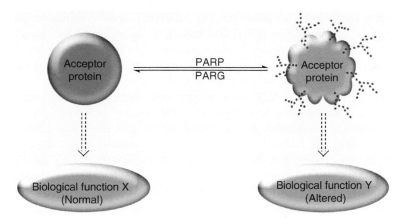

FIGURE 15.2 Schematic representation of the dynamism of the biosynthesizing and biodegrading pathways of PAR metabolism catalyzed by PARP and PARG. The normal biological functionality of an acceptor protein is altered after the modification.

opposite directions (Figure 15.2). Therefore, the net level of ADP-ribose polymer on cellular target proteins is the concerted outcome of these two opposing reactions.

PAR is reported to occur ubiquitously in eukaryotes, including some plants with a possible exception of yeast [6]. The addition of heterogeneous ADP-ribose polymers on a target protein influences the fundamental properties of the protein in several ways. Due to this, the cellular functioning of the modified protein is altered in various ways. Depending on the class of protein (e.g., structural, catalytic, regulatory, etc.) being modified, its biological consequences may vary widely (Figure 15.2). For example, if a structural protein was modified by PAR, structural organization and functions of the protein would be affected. On the other hand, if PAR of a catalytic protein took place the related metabolic pathway would manifest corresponding change. The biological expression of effect is not only dependent on the target proteins but also on the extent of its modification and the role of the modified protein in cellular metabolism. A logical consequence of this would be changes in cellular physiology and metabolism that are dependent on the target protein. This can be a very simple way of looking at the biological implication of PAR of cellular proteins. Therefore, in this review, I shall critically examine the role of PAR of cellular proteins and dwell upon its possible implication in cellular functioning, especially in carcinogenesis. However, before doing so it would be relevant to understand the metabolic process itself including the enzymes that are responsible for the metabolism and its effects on chromatin organization.

PAR is not to be confused with a different but closely related metabolic process called mono-ADP-ribosylation (MAR) in which a singular ADP-ribose moiety is attached to a target protein [reviewed in Ref. 7]. MAR is essentially a phenomenon of prokaryotic (bacterial) system and is associated with cellular signaling involving modification of signal transduction proteins, including G-protein, and expression of bacterial toxicity. The enzyme catalyzing MAR, mono-ADP-ribose transferase, is of considerable medical importance as it is associated with many infectious diseases of humans such as cholera, diphtheria, whooping cough, etc. [8].

15.1.1 Poly-ADP-Ribose Polymerase: The Main Biosynthesizing Enzyme of PAR Metabolism

The primary biosynthesizing enzyme for ADP-ribose polymer on a target protein is PARP [3–5]. Since its discovery about 40 years ago, it is now known to consist of over 17 isoforms in a highly conserved enzyme family in higher eukaryotes [9]. All of them share common properties of having a catalytic domain for recognition of NAD^+ and the ability to perform PAR. They also show high level of amino acid sequence homology between different species. Variable number of orthologues of human PARP isoforms has been found in different species. For instance, 16 out of 17 reported isoforms of human PARP seem to have orthologue genes in mouse, rat, and *T. nigroviridis* (pufferfish). Similarly, human PARP show recognizable orthologues for over 12 isoforms in chicken genome, 2 in *D. melanogaster* (fruit fly), 3 each in few mosquito species, and so on [9]. On the basis of overall similarities in characteristics, the isoforms have been arranged in five subgroups. Nonetheless, the biochemical and other characterization are, at best, partially done for not more than seven isoforms of PARP [10]. Among them, PARP-1 is the most abundant ($\sim 10^6$ molecules per mammalian cell), extensively studied, and well characterized model PARP enzyme accounting for about 90% of total cellular PAR metabolism. PARP-1 has an average molecular weight 113 kDa. It is proposed to have three functional domains. While the 46 kDa N-terminal DNA binding domain contains two zinc finger motifs supposedly mediating strand break recognition and a nuclear location signal, the largest, 54 kDa C-terminal catalytic domain binds to the substrate, NAD^+ [11]. The smallest, 22 kDa central automodification domain is involved with regulation of the catalytic property of the enzyme itself. It is known that PAR of PARP enzyme, appropriately named "automodification," causes loss of catalytic ability of the enzyme to further poly-ADP-ribosylate a target protein (heteromodification) and vice versa. To achieve it, the PARP enzyme functions as a dimer, such that one subunit of the holoenzyme can mutually catalyze addition of ADP-ribose polymer on the other subunit [8].

The significance of existence of multitude of PARP isoforms in eukaryotes is not yet clear and is a subject of extensive study [9,10]. Even though not all isoforms of PARP have been characterized [10], the information available from the best characterized isoforms support the contention that PARP is a conserved family of proteins with consensus amino acid sequences, motifs, intron positions, and domains. Existence of multitude of PARP isoforms in different species and their conservation through evolutionary tree point out to the criticality and importance of PARP enzyme in cellular metabolism. It could be speculated that existence of conserved and multiple isoforms of PARP in a genome is to just ensure continuance of PAR metabolism even under extreme conditions of cellular stress.

PARP was originally recognized as an important DNA damage sensor protein mediating DNA repair, genome integrity, and cell survival on one hand, and cytotoxicity and cell death on the other [12]. It was thought that a damaged DNA, in particular a strand break or nick in DNA, was an essential molecular trigger for PARP activity. It appears that the free-floating cellular PARP responds to a molecular trigger (e.g., single-strand break or SSB), rapidly binds to the damaged or nicked DNA, acquires metabolically active status, and initiates biosynthesis of ADP-ribose polymers on target proteins (heteromodification) as well as itself (automodification). Depending on the rate of this reaction a sudden depletion of endogenous NAD^+ might occur. The automodified PARP correspondingly becomes inactive and the process of PAR of self and other nonself target proteins stops (Figure 15.3). On the other hand, evidences have accumulated over the recent years showing that PARP enzyme may also be metabolically activated by structural status of DNA [13] in absence of a damaged or nicked DNA [14]. These observations suggest that PARP might also participate in a variety of normal metabolic processes of a cell upon appropriate stimulation [15]. In elucidating this aspect, use of a host of inhibitors of PARP has been made thereby paving the way for deeper understanding of biological role of PARP, which appears to be more complicated than what we currently understand [reviewed in Ref. 16]. This line of investigation has opened up a new possibility of therapeutic use of PARP inhibitors after many unresolved issues are settled. That the PARP family of enzymes are involved with diverse cellular functions is also supported by the fact that the enzyme or its isoforms is not only localized in the nucleus but also in extranuclear region of a cell, e.g., in centrosomes, mitochondria, etc. [9,10,15].

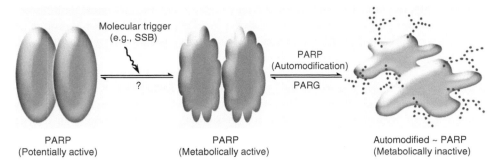

FIGURE 15.3 Schematic depiction of self-regulatory ability of PARP enzyme by automodification. The potentially active dimeric, free-floating form of PARP acquires metabolically active status upon receiving an appropriate molecular trigger. While it mediates heteromodification on other target proteins, the two subunits of PARP mutually modify each other, which inactivate the enzyme. Conformational changes during this cascade are hypothesized.

15.1.2 POLY-ADP-RIBOSE GLYCOHYDROLASE: THE MAIN BIODEGRADING ENZYME OF PAR METABOLISM

The main biodegrading enzyme for the ADP-ribose polymers attached onto a modified protein is PARG [3–5]. The degradation of ADP-ribose units mediated by PARG is random as well as sequential, exhibiting both exo- and endo-glycosidic cleavages, until the last monomer directly attached to amino acid residue of a target protein. In the process, while exoglycosidic cleavage systematically releases monomers of ADP-ribose, the endoglycosidic breakdown randomly creates variable types of ADP-ribose polymers (Figure 15.1). The last, protein proximal ADP-ribose is removed by another enzyme called ADP-ribose protein lyase [7]. In contrast to multiplicity of PARP genes in eukaryotes, there exists only one PARG gene [reviewed in Ref. 15]. The enzyme is found in low abundance and is hypersensitive to proteases [17]. Analysis of organization of PARG genes shows existence of five conserved homologous domains in many higher eukaryotes including mouse, rat, and humans but not necessarily in lower organisms [15]. The amino acid sequence of PARG of *Xenopus laevis* shows complete absence of exon 1 and 2, partial conservation of exon 3, and strong conservation from exon 4 till C-terminal end of the enzyme. This suggests that more critical domains are located downstream and are conserved. Many phylogenetic analysis of PARG gene now point to variable importance of exons 1 to 3 in PARG activity. The enzyme acquires further complexity because the PARG enzyme exists in three distinctly different, alternately spliced, and translated isoforms of molecular weights 111, 102, and 99 kDa with distinct cellular distributions [18]. While the less abundant, full length PARG is a nuclear enzyme, the abundant but truncated isoforms are perinuclear in localization. Research of last few years has indicated that regulation of PARG is very complex and is influenced by specific shuttling of the enzyme under different conditions [19,20]. Existence of two distinctly different species of PARG mRNA, both of which produce the same high molecular weight enzyme isoform, has been reported [21]. Observation of yet another isoform of PARG with a molecular weight of 60 kDa in genetically manipulated cells and in mitochondrion has added another dimension to the already complex matter [22]. Since existence of equivalent mRNA to the 60 kDa isoform has not yet been shown, it has been hypothesized that it might be either a proteolysis product of a normal PARG or a translation product of normal mRNA in which the translation started from a second downstream codon located in exon 4 [12,22]. Both these attractive proposals by which a 60 kDa PARG could be formed in normal cells need further verifications. In this, use of inhibitors of PARG would be useful. Though inhibitors of PARG, e.g., gallotannins, are currently being used in such studies, its cellular impermeability and other pharmacological properties indicate that it may not be a suitable biological inhibitor of PARG [23]. Oka et al. have recently reported existence of a 39 kDa poly-ADP-ribose hydrolase (ARH 3) enzyme in mammalian cells, which exhibited PARG activity despite being structurally different from it [24].

Nonetheless, the polymer of ADP-ribose biosynthesized on a target protein is very rapidly degraded by PARG in vivo; a rough biological half-life of PAR is estimated to be ≤ 1 min in mammalian cell [25]. It appears that the turnover of cellular ADP-ribose polymer is very important in maintenance of normal cellular metabolism, physiology, and development. In absence of a suitable biological inhibitor of PARG, some studies utilized an approach of disruption of PARG gene to understand its role in cellular physiology and metabolism. Partial and complete knockout mouse for PARG showed continued accumulation of ADP-ribose polymers on target proteins that lead to embryonic lethality pointing to strong influence of failure of degradation or turnover of ADP-ribose polymers from proteins on cellular well being [20,26].

15.2 CHARACTERISTIC FEATURES OF ADP-RIBOSE POLYMER

The heterogeneity and bulk of the ADP-ribose polymers as well as the resulting alteration in the net charge of the modified protein make this modification a unique posttranslational modification. There are three special features of this modification that separate it from other posttranslational modifications (e.g., phosphorylation, acetylation, methylation, etc.). In the first, it is to be noted that each moiety of ADP-ribose confers two negative charges emancipating from its two phosphate groups on the modified protein (Figure 15.1). Since this modification is in form of a polymer comprising up to 200 monomer units of ADP-ribose, each target protein being modified is likely to experience a highly significant change in its net charge. No other posttranslational modification of proteins can match the quantum of this change in the net charge of a modified protein. Dramatically altered charge status of a protein would strongly influence its conformational status and, consequently, its functionality. It has been shown that PAR of histone H1 protein lead to significant change in chromatin superstructure even when $\leq 5\%$ of H1 was modified [27]. Secondly, the histone proteins, a preferred target protein family [3–5], are small-sized proteins with molecular weights in the range of approximately 11–21 kDa [28]. The molecular weight of a moiety of ADP-ribose is approximately 550 Da (Figure 15.1). Thus, an ADP-ribose polymer consisting of, for example, 200 monomers of ADP-ribose would be a very bulky adduct of molecular weight of ~ 110 kDa. Thus, in case of PAR of histone proteins, the size and bulk of the modification or ADP-ribose polymer adduct would, on an average, exceed the size and bulk of the target protein itself by 6- to 11-folds. Lastly, it is known that there exists a wide variation in the size, number, and pattern of branching of ADP-ribose polymers conferring a great heterogeneity on the modified cellular proteins [3–5] (Figures 15.1 and 15.3). In different organizational status and conformations, free or protein associated ADP-ribose polymers are likely to influence (1) protein–protein interaction that is very common and critical among histone proteins (e.g., in core histone organization) as well as (2) protein–DNA interaction that is the basis of chromosomal organization [26,29]. All these features are unique to PAR and are not found in other posttranslational modifications of proteins making PAR a unique metabolic process.

15.3 INFLUENCE OF ADP-RIBOSE POLYMERS ON CHROMATIN ORGANIZATION AND CELLULAR PHYSIOLOGY

It is estimated that over 3.2 giga base pairs (Gbp) of nucleotides (NT) constitute a human genomic DNA comprising an estimated 25 to 30 thousand genes [30]. The double helix of mammalian DNA should comprise approximately 25 million nucleosomes at the first level of chromatin organization. This level of structural organization of DNA essentially revolves around an octamer of core histone proteins consisting of two molecules each of H2A, H2B, H3, and H4. The histone protein H1 is associated with the linker, inter-nucleosomal DNA [see Ref. 26]. The resulting polynucleosomal "beads on string" then takes secondary, tertiary, and quaternary levels of supercoiling. In addition, it is estimated that the chromatin forms approximately 50,000 loops per domain in conjunction with nuclear matrix proteins in a mammalian nucleus to complete the chromatin/chromosomal organization [31]. Thus, a host of chromosomal proteins participate in the structural organization of chromatin, foundation of which is laid by the core histone octamer and histone H1 proteins in a nucleosome. Histones, which are highly conserved basic proteins, interact among themselves, with other nonhistone class of chromosomal proteins, and counterions on one hand and with the acidic DNA on the other, by charge and other weak, noncovalent interactions [26,29]. Therefore, any change in the net charge of the histone proteins would naturally greatly influence the charge interaction between these proteins, other nonhistone proteins, and DNA, thereby, affecting the foundation of the structural organization of chromatin [29,32,33].

The biological functions (DNA repair, replication, transcription, etc.) associated with the genetic material or genes on chromatin necessitates complete collapse of the highly compact, positively supercoiled structure of chromatin. To achieve this, the chromatin superstructure must relax locally and transiently, and undergo negative supercoiling to the extent that even the double helix of DNA becomes single stranded. Only in this condition can proteins, enzymes, and other associated factors of DNA replication, transcription, or repair can access the DNA strand and perform their expected biological tasks. Understandably, once the biological task is completed the highly negatively supercoiled DNA should quickly and efficiently go back to its positively super-coiled structural organization. PAR of chromosomal protein is potentially most suitable process to facilitate this winding–unwinding process because it has the capacity to drastically change charge as well as bulk of the histone and other chromosomal proteins. While the charge alteration shall weaken the protein–DNA interaction, the growing bulk of the ADP-ribose polymer adduct on the protein could act as a physical force to further weaken the interaction and vice versa providing a very efficient mechanism of altering chromatin superstructural organization. PAR of catalytic and other regulatory proteins, in general, result in downregulation of activities of the modified proteins. A notable example of this is the automodification of PARP itself, the activity of which is downregulated upon modification (Figure 15.3).

In addition to the influence of PAR of target proteins as discussed above, free ADP-ribose polymer itself also influences the functionality of histones and many other DNA damage response factors by noncovalent interactions [34]. Several recent models based on histone-shuttling, originally proposed by Althaus and coworkers [35], suggest that the high quantum of negative charge from ADP-ribose polymers, the bulk of ADP-ribose polymer adduct, and associated conformational changes of modified proteins cause dissociation of chromosomal proteins from DNA resulting in transient loosening of the chromatin superstructure. This transient influence of PAR of cellular proteins makes it an attractive candidate for association with efficient, quick, and short-term regulation of cellular functions as biological systems are expected to respond quickly to an ever-changing internal and external environs and milieu. Similarly, many new facets of PARG are opening up and it appears that even this enzyme has multifarious but not clearly understood biological roles [see Ref. 15]. Since the level and extent of cellular PAR is a product of two opposing reactions comprising polymer forming PARP and polymer breaking PARG activities (Figures 15.2 and 15.3), monitoring the status and activity of PARP alone is likely to obscure the actual mechanism of influence exerted by ADP-ribose polymer. In that sense it becomes imperative that simultaneous study of PARG is also made to have a complete picture of the influence of this enigmatic metabolic process [36, reviewed in Ref. 37]. Alternatively, one could directly monitor poly-ADP-ribosylated proteins, not withstanding the activities of PARP or PARG, and attempt to correlate it with observed biological influences on various cellular processes. I have targeted my research using the later approach as irrespective of the activities of the involved enzymes, the modification in form of ADP-ribose polymer is the cause of all changes exerted through the modified protein.

The PAR metabolism is known to have extensive involvement in the structural organization and functional status of chromatin as well as in diverse cellular physiology. PAR was initially considered a cellular or metabolic response to DNA damage induced by ionizing radiations and alkylating agents [3–5,12]. Upon induction of DNA damage, particularly SSB or nicks, the biosynthesizing enzyme PARP was rapidly stimulated several folds. Concomitantly ADP-ribose polymer was biosynthesized accompanied by depletion of cellular NAD^+ (Figure 15.1), which facilitated repair of the DNA strand breaks. Rapid NAD^+ depletion in extreme cases resulted in cell death. This suggests that PAR is not only a repair mediating process but also facilitates cell death in heavily damaged cells. Two aspects of the PAR metabolism need special attention. Firstly, stimulated PARP activity alone cannot give precise idea of status of ADP-ribose polymer on a target protein. The activity of the opposite reaction catalyzed by PARG has also to be known simultaneously besides, perhaps, status of the cellular pool of NAD^+, as the status of actual ADP-ribose polymer on a target protein is decided by the interplay of two opposing reactions (Figure 15.2). The interaction of a modified protein with other proteins or DNA, in turn, would be decided by the ADP-ribose polymer adducts rather than PARP or PARG enzyme activities. Secondly, rapid depletion of NAD^+ observed during PARP stimulation would also affect other NAD^+-dependent metabolic processes that might have its own influences on cell response.

Over the years the PAR of cellular proteins has acquired a very broad meaning in the realm of cellular metabolism and is expanding. Its role and implications have penetrated a wide variety of normal cellular physiology and metabolism even in absence of DNA break [13,14,38,39]. At the same time, its original role as a DNA damage sensor remains intact. As a cellular response to DNA damage, over 500-fold induction of ADP-ribose polymers has been reported [40]. The notable inter- and intraspecies sequence/domain homology and widespread intracellular distribution of PARP and its isoforms [9,10] certainly suggest that it may have multifarious metabolic roles to play. Consequently, dysregulation of PAR metabolic reactions might be important in manifestation of disease conditions as well as in signaling pathways [15]. For example, involvement of PAR metabolism has been shown in carcinogenesis [41], maintenance of genome integrity [42] and cellular detoxification, cell division [43], cell cycle progression, strand break repair as a recruiter of associated factors [44], maintenance of long-term cellular memory [45], apoptosis [15] (especially for caspase-independent, apoptosis inducing factor mediated pathway) to name the obvious. To interlink these seemingly very diverse biological responses reported in the literature, I have attempted to at least partially map the metabolic processes and molecules associated with PAR metabolism using PathwayStudio (ver. 4.0) software in Figure 15.4. The map shows a very wide spectrum of possible involvement of PAR in cellular functions with PARP at the center stage (Figure 15.4). Pathway-Studio is a bioinformatics software, which enables in depth analysis of interrelated biological data using pathway reconstruction algorithms [46]. The software finds common regulator molecules, biological entities, and associated pathways from a large collection of databases covering mammalian and other life forms. This analysis gives a new perspective to PAR metabolism in which PARP-1 enzyme exerts influences, either directly or indirectly, on diverse biological functions such as cancer, cell division, proliferation, differentiation, repair pathways, inflammation, senescence, necrosis through apoptosis (Figure 15.4). The pathway map also depicts a large number of metabolites, molecules, and processes that might be involved in expression and regulation of PAR metabolism as well as in manifestation of its biological influence. Even though not every aspect of cellular physiology mapped is clear at this time, it is obvious that a comprehensive and complete understanding of the PAR metabolism is presently far from our reach.

15.4   INVOLVEMENT OF ADP-RIBOSE POLYMER IN CARCINOGENESIS

Carcinogenesis is a complex, multistep process that can be categorized into three clearly defined stages, namely, initiation, promotion, and progression. Conceptually, irreversible molecular events of "initiation" stage lead to monoclonal expansion of the initiated cells during "promotion," which

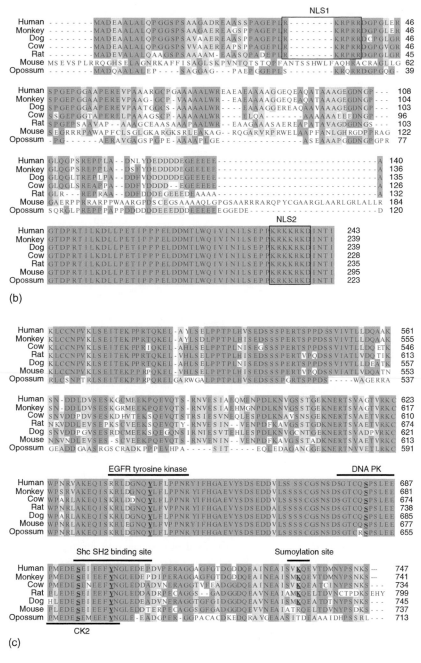

FIGURE 9.3 Members of mammalian SIRT1. (b) Sequence alignment of the N-terminal region of mammalian SIRT1 proteins. The conserved nuclear localization sequences are placed in a square. Perfectly conserved residues are shown in blue and those that are functionally conserved shown in yellow. (c) Sequence alignment of the C-terminal region of mammalian SIRT1 proteins. Perfectly conserved residues are shown in blue and those that are functionally conserved shown in yellow. Conserved residues predicted with Scansite software (http://scansite.mit.edu/motifscan_seq.phtml) at high stringency and Sumoplot (www.abgent.com/sumoplot.html) as probable phosphorylation and sumoylation sites are shown in red. EGFR: epidermal growth factor receptor; DNA-PK: DNA dependent protein kinase; SH2; Src homology 2; CK2: casein kinase 2.

(*continued*)

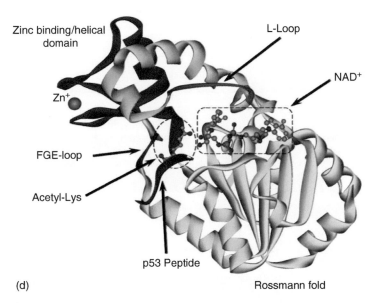

Zinc binding/helical domain

L-Loop

NAD$^+$

Zn$^+$

FGE-loop

Acetyl-Lys

p53 Peptide

(d)

Rossmann fold

FIGURE 9.3 (continued) (d) 3D structure of the catalytic domain of sirtuins. The ribbon structure of SIR2-TM in complex with acetylated p53 peptide and NAD$^+$ is shown. The X-ray data of SIR2-TM [28] was downloaded from Protein Data Bank (PDB) (http://www.rcsb.org) and visualized with Accelrys ViewerLite Version 5.0. The colors for different domains are cyan for the Rossmann fold, green for the helical module, red for the zinc-binding module, blue for the FGE loop that makes contact with acetyl-Lys of the substrate, pink for the K382 acetyl-p53 peptide and brown for the L-1B loop that serves as a ceiling for NAD$^+$ binding.

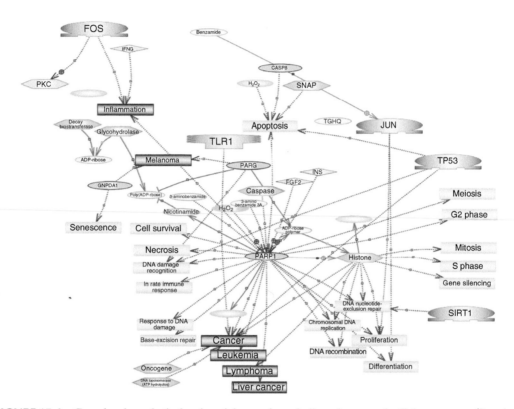

FIGURE 15.4 Complex, hypothetical and partial map of metabolic pathways and cellular processes/functions that are influenced by PAR metabolism and associated molecules with PARP enzyme at the center. The map is based on data retrieval from various data banks using the PathwayStudio software.

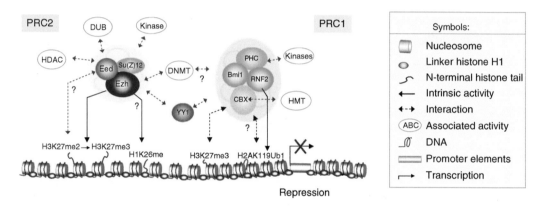

BOX 16.2 Schematic representation of molecular interactions in PcG silencing (see main text for explanations).

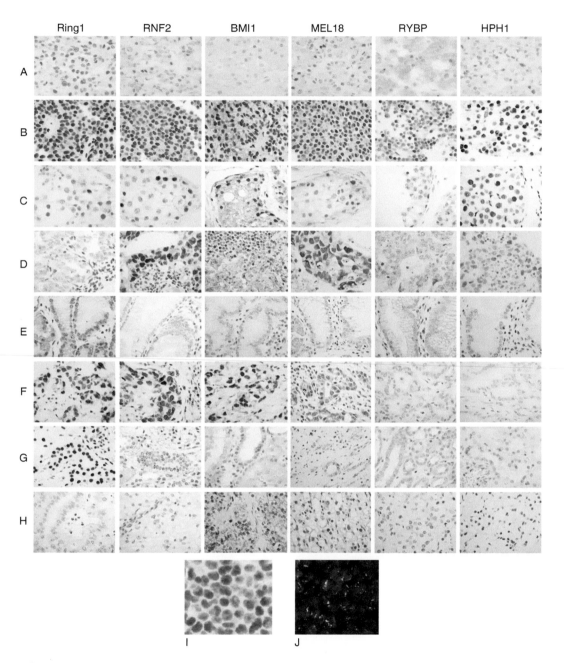

FIGURE 16.1 Examples of PcG expression profiles in normal and tumor tissues: (A) pituitary gland, (B) pituitary adenoma, (C) testis, (D) embryonal carcinoma, (E) gastric surface-epithelial cells, (F) gastric adenocarcinoma, (G) kidney tubules, (H) clear-cell renal cell carcinoma, (I) BMI1 overexpression in a mantle cell lymphoma case, (J) BMI1 amplification detected by FISH in mantle cell lymphoma from (I). The picture shows a group of cells with several copies of BMI1 (green signal) in comparison with two centromeric copies for chromosome 10 (red signal). RING1, RNF2, BMI1, MEL18, and HPH1 are PcG proteins from the PRC1 complex (Box 16.2). RYBP (RING1 and YY1 binding protein) is a PRC1 complex-associated protein [58]. (Photographs reprinted from Sanchez-Beato, M., et al., *Mod. Pathol.*, 19, 684, 2006. With permission.)

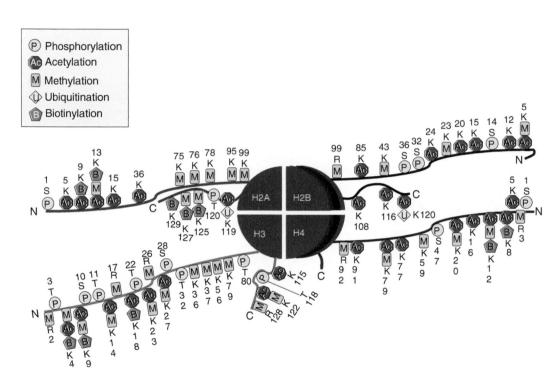

FIGURE 20.1 Histone modifications.

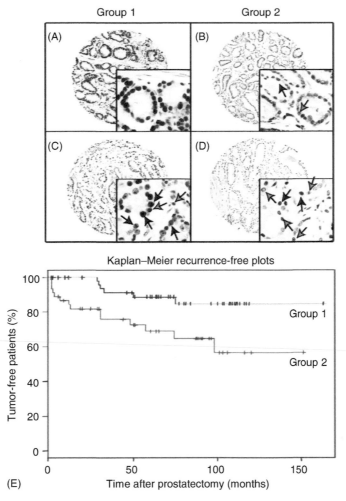

FIGURE 21.2 Cancer tissues exhibit cellular epigenetic heterogeneity. Immunohistochemical (IHC) analysis of histone modifications in malignant prostate glandular epithelial cells from tumors of similar grade and stage reveals heterogeneity in cellular levels of specific modifications. (A, B) H3K18ac (C, D) H4R3me2. Heterogeneity exists within individuals as populations of cells show varying intensities of staining, indicating varying levels of specific modifications within single tumors. (B) Individual cells from a patient's cancer tissue show both relatively high (brown arrow) and low (blue arrow) levels of K18ac. Heterogeneity also exists between patients which show differences in the percentage of cells with high levels of modification. For example, compare (C) and (D) which show IHC results for H4R3me2 staining from two different patients. Cells with high levels of modification stain positively and are colored brown (brown arrow) while those with low level of modification stain negatively and are colored blue (blue arrow). The percentage of cells that stain positively (brown arrow) for this particular modification is much higher in (C) than in (D). Significantly, patterns of histone modifications can be used to group patients of similar grade and stage into two distinct groups that show significantly different outcomes. (E) Kaplan–Meier recurrence-free plots of the two groups (black, group 1; red, group 2) identified among the patients with low-grade tumors ($n = 104$; UCLA) based on the histone modification patterns. Group 1 patients (A, C) generally have higher percentage of cells that stain positively for various modifications than Group 2 patients (B, D). These groups were created using only the histone modification patterns and are independent of all other clinicopathological variables.

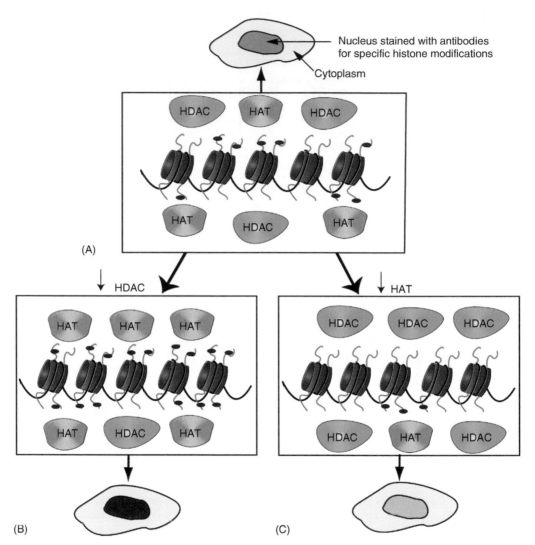

FIGURE 21.3 A hypothetical relationship between cellular levels of histone modifications and molecular activities of histone modifier enzymes. (A) Levels of histone modifications are determined by the actions of opposing groups of enzymes, such as HATs and HDACs for acetylation. IHC staining of whole-cell nuclei can be used to visualize cellular levels of a particular modification as represented by the cells in this picture. While aberrant targeting of histone-modifying enzymes can lead to local changes in levels of particular modifications (Figure 21.1), changes in the activity and expression of these histone-modifying enzymes can account for the epigenetic heterogeneity that is seen in cellular levels of histone modifications in cancer. (B) For example, decreased activity/expression of HDACs can lead to hyperacetylation of histones throughout the genome due to unopposed HAT activity. This results in cells with a high intensity of staining when analyzed by IHC. (C) In a similar fashion, decreased expression of HATs can lead to unbalanced HDAC activity, decreases in cellular levels of acetylation, and cells that stain weakly by IHC.

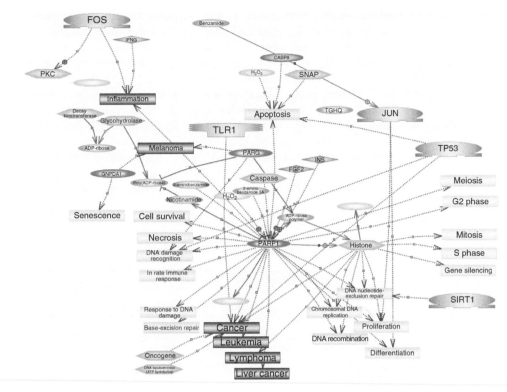

FIGURE 15.4 **(See color insert following page 272.)** Complex, hypothetical and partial map of metabolic pathways and cellular processes/functions that are influenced by PAR metabolism and associated molecules with PARP enzyme at the center. The map is based on data retrieval from various data banks using the PathwayStudio software.

is partially reversible. During the third and last "progression" stage, which is also irreversible, the promoted cells progressively turn malignant characterized by uncontrolled cell division, loss of contact inhibition, and dedifferentiation [reviewed in Ref. 39,47]. During the process multiple genetic alterations occur. This includes mutations of various kinds, chromosomal aberrations, gene rearrangements, gene silencing and shutdown, neogene expression, and gene amplification, to name the obvious. All these molecular events can, in principle, be seriously influenced by PAR metabolism as ADP-ribose polymers on chromosomal proteins potentially alter chromatin super-structure transiently and locally. In a relaxed, negatively supercoiled state of chromatin organization gene expression is permissible. The reverse would lead to shut down, silencing or downregulation of genes, and so on. In general, it is accepted that structural organization of chromatin strongly influences damage and repair of DNA, replication, gene expression, and other genome dependent functions. It essentially means that for the genome to perform its biological functions, the higher order of chromatin superstructure must collapse. Logistically, the entire chromatin superstructure cannot collapse in the confines of a nucleus. Therefore, the collapse has to be transient and local. Once the biological function has been performed, the reverse ought to happen. This dynamism seems to be the hallmark and essence of the highly regulated living process (Figure 15.4). Thus, in a relaxed organizational state, various agents, proteins, biomolecules, regulatory entities, etc., can easily access the genomic DNA. In its condensed state, on the contrary, access is prevented or, at least, not permitted to the same extent [reviewed in Ref. 39]. The genomic functions are known to be drastically altered during carcinogenesis. Because of the accepted role of PAR of chromosomal proteins, especially histone proteins, in the organization of chromosomal superstructure [26], it is only logical to conclude that PAR and carcinogenesis would be intricately involved with each other.

Early observation that PARP enzyme was upregulated in malignant or transformed cell in vitro as compared to normal cells [48], in leukemia and ovarian cancers [49], in biopsies of human hepatocellular carcinomas [50], and many similar observations supported this proposition. The reverse was observed when PARP enzyme was inhibited by chemical inhibitors. For example, pharmacological inhibition of PARP-1 by DPQ delayed 7,12-dimethylbenz(a)anthracene-induced skin carcinogenesis in mice or caused its regression [51]. In this case, essentially some tumor-related and inflammation-associated genes were downregulated. Some inhibitors of PARP have been shown to possess anti-angiogenetic property. While DPQ was found to inhibit skin tumor vascularization [51], other inhibitors like 3-AB, PJ-34, 5-IAQ, and IQD showed angiogenetic properties [52,53]. During the same period inconsistent results (stimulation, inhibition, or no effect) of a variety of chemical inhibitors of PARP on oncogene, chemical, or radiation induced cell transformation were also reported in several studies [reviewed in Ref. 16]. Many specific and newly synthesized inhibitors of PARP-1 and PARP-2 are under preclinical studies to potentially use them as anticancer drugs [54]. Nonetheless, a lot of work and clearer understanding are still required before the full therapeutic potential of PARP inhibitors as anticancer drugs may be exploited [16,52–54].

Even though not everything is absolutely clear about roles of PARP inhibitors, some inconsistent results reported in the literature can be explained. After careful consideration it appears that the observed inconsistency of results related to effects of inhibitors of PARP could be due to more than one reason. In the first, in many of such studies only PARP enzyme activity was monitored. The observation of enhanced or reduced PARP was essentially interpreted as increase or decrease, respectively, in the level of ADP-ribose polymers on chromosomal proteins and correlated to carcinogenesis. Monitoring only PARP activity is unlikely to reveal the metabolic consequence of addition of ADP-ribose polymer onto chromosomal proteins, which actually alters chromatin structural, and, consequently, its functional standing (Figure 15.2). Therefore, inconsistency might have crept into different studies as the activity of PARG, the biodegrading enzyme in PAR metabolism, might have been different under dissimilar experimental conditions. Secondly, it is now known that various isoforms of PARP respond differently to different inhibitors, thereby, exerting variable biological effects [9,10,16]. This may also be a source of inconsistency in observations by different researchers using different inhibitors of PARP. Thirdly, it is possible that cells at different stages of transformation and carcinogenesis responded differently to inhibitors of PARP as different characteristic molecular events occur in three stages of the process. Finally, the inhibitors of PARP, besides inhibiting the enzyme, might also exert some other nonspecific metabolic influence on other enzymes/biomolecules about which we do not know yet. Instead of monitoring PARP or PARG enzymes of PAR metabolism, in some studies the level of ADP-ribose polymers on cellular protein has been directly measured so that correlation between ADP-ribose polymer and DNA damage, repair, and carcinogenesis induced by various factors could be established. High levels of ADP-ribose polymer on cellular proteins, including a 113 kDa protein, was reported in rectal cancer biopsies [55]. In contrast, a series of investigations carried out in my laboratory show that cellular ADP-ribose polymers progressively and statistically significantly reduced during carcinogenesis induced by chemical carcinogens in mice [56–59]. In this series of investigations a novel, sensitive, and specific slot blot and Western blot immunoprobe assay of ADP-ribose polymer has been utilized, which directly assays the polymeric adducts on cellular proteins. The immunoprobe assay utilizes a polyclonal antibody (PAb) raised against natural, heterogeneous ADP-ribose polymers of mouse spleen cells as an antigen. The advantage of this approach is that the PAb directly detects both free as well as protein associated cellular ADP-ribose polymers in a given sample. Thus, the assay gave a true metabolic measure of cellular ADP-ribose polymers [49]. The PAb has been raised against the isolated ADP-ribose polymers, and not against ADP-ribose polymer bound to a particular protein.

Thus, the nature of target protein to which the ADP-ribose polymer is attached does not effect detection. In this way, one could conveniently measure total cellular level of ADP-ribose polymers using slot blot immunoprobe assay [49]. Likewise, when Western blot immunoprobe assay was used, individual poly-ADP-ribosylated proteins could be visualized and quantified. Thus, it became possible to quantify true metabolic levels of either total cellular ADP-ribose polymers of all proteins or of an individual cellular protein. The ADP-ribose polymers of total cellular as well as of histone proteins of mouse liver and spleen cells were found to progressively go down during initiation of carcinogenesis induced by dimethyl- or diethyl-nitrosamine [49,51,52,60,61]. Similarly, aqueous extract of betel nut or arecoline also progressively inhibited ADP-ribose polymer adducts on total, histone or HMG proteins of mice tissues during initiation period [31,53,62,63]. Using a Dalton's lymphoma ascites tumorigenesis model in mice, it was found that a similar negative correlation existed between ADP-ribose polymers on histone proteins and carcinogenesis during latter stages of cancer development [50]. This approach of ADP-ribose polymer assay [49,64] has recently been extended to quantification of total ADP-ribose polymer on peripheral blood lymphocyte proteins isolated from blood samples of patients in advance stages (III/IV) of cancers of breast, cervix, and head and neck. A significant reduction in ADP-ribose polymers on total cellular proteins has been observed in all three cancers [65]. Preliminary calculations convincingly show nearly 45% reduction in the level of total ADP-ribose polymers on blood lymphocyte proteins in these patients as compared to the controls samples obtained from young volunteers with no known history of cancer (R.O. Lakadong and R.N. Sharan, unpublished results). Though more preclinical studies involving early stages of human cancer would need to be done, one may take cue from the mouse model to hypothesize a direct link between ADP-ribose polymers on cellular proteins, particularly histone proteins, and carcinogenesis. In this progressive lowering of metabolic level of ADP-ribose polymer adducts on cellular proteins stands out as a hallmark of carcinogenesis.

Though in these reports ADP-ribose polymer has been directly measured, it has to be noted that the reports of Yalcintepe et al. [48] and that from my laboratory [40–45] essentially show contradictory results. This needs an explanation. I believe, the explanation lies in the methods employed in these studies to assay and quantify ADP-ribose polymers on cellular proteins. ADP-ribose polymer has been assayed in a variety of ways [reviewed in Ref. 66]. Radioisotopic method, a widely used assay of ADP-ribose polymers utilizing radiolabeled-NAD^+ [67], was employed in the investigation of Yalcintepe et al. [48]. We have earlier shown that different methodological interventions associated with the radioisotopic assay per se induced unphysiologically high levels of cellular ADP-ribose polymers, which does not necessarily reflect the in vivo or metabolic level of cellular ADP-ribose polymers [49,57,68]. It is possible that the high ADP-ribose polymer reported by Yalcintepe et al. [48] could be due to such methodological problems. Its assumption derives strength from the same study wherein the authors also report high level of ADP-ribose polymer on a 113 kDa protein. On the basis of molecular weight similarity and the fact the PARP is a preferred target for PAR (automodification) [8], it is likely that this 113 kDa protein is actually PARP enzyme. Since automodified PARP loses its catalytic ability to carry on PAR of other target proteins (heteromodification) [3–5,8] (Figure 15.3), it seems unlikely that metabolic level of ADP-ribose polymers on cellular proteins would actually increase under this condition. The immunoprobe assay employed in the study being carried out in my laboratory [40–45] actually measured the true metabolic level of ADP-ribose polymer adducts on cellular proteins [see Refs. 49,57]. As has been emphasized earlier, downregulation of PARP does not necessarily result in lowering of PAR of histone or other cellular proteins. It would be decided by the level of PARG enzyme activity. With downregulated PARP activity, the level of ADP-ribose polymers on histone or other proteins could still be high provided simultaneously the PARG enzyme was also downregulated and vice versa. Indirect support for this proposition comes from the observation that 3-aminobenzamide (3-AB), a widely used physiological inhibitor of PARP-1, did not completely inhibit PAR signal as revealed by slot- and Western blot immunoprobe assay in some of our studies in mice to quantify ADP-ribose polymers [51,52]. The ADP-ribose polymer signals in both slot blot and Western blots were only subdued at a concentration of 3-AB that is known to fully inhibit PARP-1. This indicates two possibilities. Firstly, all isoforms of PARP may not be inhibited by 3-AB as efficiently as PARP-1. Secondly, it is possible that PARG was either fully or partially inhibited due to which existing ADP-ribose polymer adducts on target proteins were not removed. Much more investigation and deeper insight are required before these and other issues get fully resolved in the realm of PAR metabolism.

In conclusion, it appears that a global understanding of PAR metabolism still eludes us despite the tremendous wealth of scientific information available. The deep involvement of PAR metabolism or its components in diverse normal and pathophysiological processes are yet to be fully uncovered. Till such time no truly conclusive understanding can be arrived at. Full understanding of significance of existence of isoforms of PARP is urgently required. Similarly, importance of PARG enzyme in PAR metabolism needs to be elucidated further. In this, more research will be required to clearly understand contributions of these two enzymes and their isoforms in PAR metabolism and their significance in carcinogenesis. One approach to this could be use of different chemical inhibitors of enzymes PARP and PARG. Particular emphasis needs to be paid to differential influences of different inhibitors on various isoforms of PARP. Another, more direct approach to this could be further elucidation of influence of the heterogeneous ADP-ribose polymer adducts on target proteins itself, and consequent alterations in protein functionality and cellular metabolism. The known heterogeneity of ADP-ribose polymer adducts in terms of their size, branching pattern, and location on target protein suggests differential influence of the polymer on various proteins. This needs to be understood clearly. On the basis of present understanding, it is obvious that PAR metabolism is highly relevant to carcinogenesis and could be potentially exploited in cancer therapeutics and diagnostics. Some of our recent results do suggest potential use of ADP-ribose polymer immunoprobe assay in cancer screening programme [49–52,54,57,59]. More work will be required before the therapeutic as well as diagnostic potential of PAR metabolism in cancer can be fully exploited.

ACKNOWLEDGMENTS

The author acknowledges Ariadne Genomics for kind permission to use PathwayStudio (version 4.0) software. I am grateful to my past and present students and colleagues who have directly or indirectly helped me in research leading to this review. Author is thankful to DST, CSIR, UGC, etc., for various research grants results of which have been used in this review.

REFERENCES

1. Chambon, P., Weill, J.D., and Mandel, P., Nicotinamide mononucleotide activation of new DNA-dependent polyadenylic acid synthesizing nuclear enzyme, *Biochem. Biophys. Res. Commun.*, 11, 39, 1963.
2. Chambon, P. et al., On the formation of a novel adenylytic compound by enzymatic extracts of liver nuclei, *Biochem. Biophys. Res. Commun.*, 25, 638, 1966.
3. Althaus, F.R., Poly-ADP-ribosylation reactions. In: *ADP-Ribosylation of Proteins: Enzymology and Biological Significance*, Althaus, F.R. and Richter, C. (Eds.), Part I, Springer Verlag, Berlin, 1987.
4. Jacobson, M.K. and Jacobson, E.L., *ADP-Ribose Transfer Reactions: Mechanisms and Biological Significance*, Springer, New York, 1989.
5. Poirier, G.G. and Moreau, P., *ADP-Ribosylation Reactions*, Springer, New York, 1992.
6. Park, J.K. et al., Inhibition of topoisomerase I by NAD and enhancement of cytotoxicity of MMS by inhibitors of poly(ADP-ribose) polymerase in *Saccharomyces cerevisiae*, *Cell. Mol. Biol.*, 37, 739, 1991.
7. Richter, C., Mono-ADP-ribosylation reactions. In: *ADP-Ribosylation of Proteins: Enzymology and Biological Significance*, Althaus, F.R. and Richter, C. (Eds.), Part II, Springer Verlag, Berlin, 1987.
8. De Murcia, G., Jacobson, M., and Shall, S., Regulation by ADP-ribosylation, *Trends Cell Biol.*, 5, 78, 1995.
9. Otto, H. et al., In silico characterization of the family PARP-like poly(ADP-ribosyl)transferases (pARTs), *BMC Genomics*, 6, 139, 2005.

10. Ame, J.C., Spenjehauer, C., and De Murcia, G., The PARP superfamily, *BioEssays*, 26, 882, 2004.
11. Scovassi, A.I. and Diedrich, M., Modulation of poly(ADP-ribosylation) in apoptotic cells, *Biochem. Pharmacol.*, 68, 1041, 2004.
12. Virag, L., The expanding universe of poly(ADP-ribosyl)ation, *Cell. Mol. Life Sci.*, 62, 719, 2005.
13. Lonskaya, I. et al., Regulation of poly(ADP-ribose) polymerase-1 by DNA structure-specific binding, *J. Biol. Chem.*, 280, 17076, 2005.
14. Hassa, P.O. et al., The enzymatic and DNA binding activity of PARP-1 are not required for NF-kB coactivator function, *J. Biol. Chem.*, 276, 45588, 2001.
15. Gange, J.-P. et al., The expanding role of poly(ADP-ribose) metabolism: Current challenges and new perspectives, *Curr. Opinion Cell Biol.*, 18, 145, 2006.
16. Plummer, E.R., Inhibitors of poly(ADP-ribose) polymerase in cancer, *Curr. Opin. Pharmacol.*, 6, 364, 2006.
17. Bonicalzi, M.E. et al., Regulation of poly(ADP-ribose) metabolism by poly(ADP-ribose) glycohydrolase: Where and when? *Cell. Mol. Life Sci.*, 62, 739, 2005.
18. Meyer-Ficca, M.L. et al., Human poly(ADP-ribose) glycohydrolase is expressed in alternative splice variants yielding isoforms that localizes to different cell compartments, *Exp. Cell Res.*, 297, 521, 2004.
19. Haince, J.-F. et al., Dynamic relocation of poly(ADP-ribose) glycohydrolase isoforms during radiation-induced DNA damage, *Biochim. Biophys. Acta*, 1763, 226, 2006.
20. Ohashi, S. et al., Subcellular localization of poly(ADP-ribose) glycohydrolase in mammalian cells, *Biochem. Biophys. Res. Commun.*, 307, 915, 2003.
21. Cortes, U. et al., Depletion of the 110 kDa isoform of poly(ADP-ribose) glycohydrolase increases sensitivity to genotoxic and endotoxic stress to mice, *Mol. Cell. Biol.*, 24, 7163, 2004.
22. Lin, W. et al., Isolation and characterization of the cDNA encoding bovine poly(ADP-ribose) glyco-hydrolase, *J. Biol. Chem.*, 272, 11895, 1997.
23. Falsig, J. et al., Poly(ADP-ribose) glycohydrolase as a target for neuroprotective intervention: Assessment of currently available pharmacological tools, *Eur. J. Pharmacol.*, 497, 7, 2004.
24. Oka, S., Kato, J., and Moss, J., Identification and characterization of a mammalian 39 kDa poly(ADP-ribose) glycohydrolase, *J. Biol. Chem.*, 281, 705, 2006.
25. Herceg, Z. and Wang, Z.-Q., Functions of poly(ADP-ribose) polymerase (PARP) in DNA repair, genomic integrity and cell death, *Mutat. Res.*, 477, 97, 2001.
26. Koh, W. et al., Failure to degrade poly(ADP-ribose) causes increased sensitivity to cytotoxicity and early embryonic lethality, *Proc. Natl. Acad. Sci. U S A*, 101, 17699, 2004.
27. Aubin, R.J. et al., Correlation between endogenous nucleosomal hyper(ADP-ribosyl)ation of histone H1 and the induction of chromatin relaxation, *EMBO J.*, 2, 1685, 1983.
28. De Robertis, E.D.P. and De Robertis, E.M.F. Jr., *Cell and Molecular Biology*, Chapter 15, Saunders College, Philadelphia, 1980.
29. Peterman, E., Keil, C., and Oei, L., Importance of poly(ADP-ribose) polymerases in regulation of DNA-dependent processes, *Cell. Mol. Life Sci.*, 62, 731, 2005.
30. Baltimore, D., Our genome unveiled, *Nature*, 409, 814, 2001.
31. Boulikas, T., Nuclear envelope and chromatin structure, *Int. Rev. Cytol. Suppl.*, 17, 493, 1987.
32. Tulin, A. and Spradling, A., Chromatin loosening by poly(ADP)-ribose polymerase (PARP) in *Drosophila* puff loci, *Science*, 299, 560, 2003.
33. Saikia, J.R., Schneeweiss, F.H.A., and Sharan, R.N., Arecoline induced changes of poly-ADP-ribosylation of cellular proteins and its influence on chromatin organization, *Cancer Letts.*, 139, 59, 1999.
34. Pleschke, J.M. et al., Poly(ADP-ribose) binds to specific domains in DNA damage check points, *J. Biol. Chem.*, 275, 40974, 2000.
35. Realini, C.A. and Althaus, F.R., Histone shuttling by poly(ADP-ribosylation), *J. Biol. Chem.*, 267, 18858, 1992.
36. Sharan, R.N., Schneeweiss, F.H.A., and Feinendegen, L.E., Neutrons affect ADP-ribosylation of proteins in human kidney T1-Cells in vitro, *Indian J. Biochem. Biophys.*, 33, 281, 1996.
37. Davidovic, L. et al., Importance of poly(ADP-ribose) glycohydrolase in the control of poly(ADP-ribose) metabolism, *Exp. Cell Res.*, 268, 7, 2001.
38. Althaus, F.R., Poly-ADP ribosylation: A histone shuttle mechanism in DNA excision repair, *J. Cell Sci.*, 102, 663, 1992.

39. Faraone-Mennella, M.R., Chromatin architecture and function: The role(s) of poly(ADP-ribose) polymerase and poly(ADP-ribosyl)ation of nuclear proteins, *Biochem. Cell Biol.*, 83, 396, 2005.

40. D'Amours, D. et al., Poly(ADP-ribosyl)ation reaction in the regulation of nuclear functions, *Biochem. J.*, 342, 249, 1999.

41. Boulikas, T., Relationship between carcinogenesis, chromatin structure and poly(ADP-ribosylation), *Anticancer Res.*, 11, 489, 1991.

42. Menissier de Murcia, J. et al., Functional interaction between PARP-1 and PARP-2 in chromosome stability and embryonic development in mouse, *EMBO J.*, 22, 2255, 2003.

43. Chang, P., Coughlin, M., and Mitchison, T.J., Tankyrase-1 polymerization of poly(ADP-ribose) is required for spindle structure and function, *Natl. Cell Biol.*, 7, 1133, 2005.

44. Leppard, J.B. et al., Physical and functional interaction between DNA ligase IIIα and poly(ADP-ribose) polymerase 1 in DNA single strand break repair, *Mol. Cell. Biol.*, 23, 5919, 2003.

45. Cohen-Armon, M., et al., Long-term memory requires poly-ADP-ribosylation, *Science*, 304, 1820, 2004.

46. www.ariadnegenomics.com/products/pathway-studio/

47. Miwa, M. and Sugimura, T., ADP ribosylation and carcinogenesis. In: *ADP-Ribosylating Toxins and G Proteins: Insights into Signal Transduction*, Moss, J. and Vaughan, M., (Eds.), American Society for Microbiology, Washington DC, pp. 543–560, 1990.

48. Miwa, M. et al., Cell density-dependent increase in chromatin-associated ADP-ribosyltransferase activity in simian virus 40-transformed cells, *Arch. Biochem. Biophys.*, 181, 313, 1977.

49. Singh, N., Enhanced poly ADP-ribosylation in human leukemia lymphocytes and ovarian cancers, *Cancer Letts.*, 58, 131, 1991.

50. Shimizu, S. et al., Expression of poly(ADP-ribose) polymerase in human hepatocellular carcinoma and analysis of biopsy specimens obtained under sonographic guidance, *Oncol. Rep.*, 12, 821, 2004.

51. Martin-Olivia, D. et al., Inhibition of poly(ADP-ribose) polymerase modulates tumor-related gene expression, including hypoxia-inducible factor-1 activation, during skin carcinogenesis, *Cancer Res.*, 66, 5744, 2006.

52. Rajesh, M. et al., Poly(ADP-ribose)polymerase inhibition decreases angiogenesis, *Biochem. Biophys. Res. Commun.*, 350, 1056, 2006.

53. Rajesh, M. et al., Pharmacological inhibition of poly(ADP-ribose) polymerase inhibits angiogenesis, *Biochem. Biophys. Res. Commun.*, 350, 352, 2006.

54. Tentori, L. and Graziani, G., Chemopotentiation by PARP 1 & 2 inhibitors in cancer therapy, *Pharmacol. Res.*, 52, 25, 2005.

55. Yalcintepe, L. et al., Change in NAD/ADP-ribose metabolism in rectal cancer, *Brazilian J. Med. Biol. Res.*, 38, 361, 2005.

56. Sharan, R.N. et al., Detection and quantification of poly-ADP-ribosylated cellular proteins of spleen and liver tissues of mice in vivo by slot and Western blot immunoprobing using polyclonal antibody against mouse ADP-ribose polymer, *Mol. Cell. Biochem.*, 278, 213, 2005.

57. Devi, B.J. and Sharan, R.N., Progressive reduction of poly-ADP-ribosylation of histone proteins during Dalton's lymphoma induced ascites tumorigenesis in mice, *Cancer Letts.*, 238, 135, 2006.

58. Devi, B.J., Schneeweiss, F.H.A., and Sharan, R.N., Negative correlation between poly-ADP-ribosylation of spleen cell histone proteins and initial duration of dimethylnitrosamine exposure to mice in vivo measured by Western blot immunoprobe assay: A possible biomarker for cancer detection, *Cancer Detect. Prev.*, 29, 66, 2005.

59. Kma, L. and Sharan, R.N., In vivo exposure of Swiss Albino mice to chronic low doses of dimethylnitrosamine (DMN) lowers poly-ADP-ribosylation of bone marrow cells and blood lymphocytes, *Mol. Cell. Biochem.*, 288, 143, 2006.

60. Pariat, T. and Sharan, R.N., Qualitative change in mice liver HMG proteins after low dose chronic administration of aqueous extract of betel but and diethylnitrosamine, *Hepatol. Res.*, 12, 177, 1998.

61. Kma, L. and Sharan, R.N., Negative correlation between poly-ADP-ribosylation of mouse blood lymphocyte proteins and dimethylnitrosamine induced initiation of carcinogenesis as revealed by slot- and Western blot immunoassay, *Proc. Natl. Acad. Sci. India*, 73B, 43, 2003.

62. Pariat, T. and Sharan, R.N., Role of mouse spleen cell HMG proteins and their poly-ADP-ribosylation in betel nut induced carcinogenesis, *Indian J. Biochem. Biophys.*, 39, 130, 2002.

63. Sharan, R.N., Association of betel nut with carcinogenesis (review), *Cancer J.*, 9, 13, 1996.

64. Sharan, R.N. et al., Immunodetection of cellular poly-ADP-ribosylation. In: *Trends in Radiation and Cancer Biology*, Sharan, R.N., (Ed.), Forschungszentrum Juelich GmbH, Juelich, Int. Coop. Bilateral Sem. Series Vol. 29, pp. 240–243, 1998.

65. Lakadong, R.O., Kma, L., and Sharan, R.N., Poly-ADP-ribosylation of blood lymphocyte proteins: A potential biomarker of cancer, presented at Natl. Sem. on Adaptation Biochemistry, Shillong, March 22–23, 2007.

66. Shah, G.M. et al., Methods for chemical study of poly(ADP-ribose) metabolism in vitro and in vivo, *Anal. Biochem.*, 227, 1, 1995.

67. Surowy, C.S. and Berger, N.A., Unique acceptors for poly(ADP-ribose) in resting, proliferating and DNA-damaged human lymphocytes, *Biochim. Biophys. Acta*, 740, 8, 1983.

68. Schneeweiss, F.H.A., Sharan, R.N., and Feinendegen, L.E., Change of ADP-ribosylation in human kidney T1 cells by various external stimuli, *Indian J. Biochem. Biophys.*, 32, 119, 1995.

16 Polycomb Group Proteins in Tumorigenesis

Hanneke E.C. Niessen and Jan Willem Voncken

CONTENTS

16.1 SHORT GENERAL INTRODUCTION

The molecular mechanisms underlying cancer development have since long been a subject of study. Traditionally, cancer is thought of as a multistep disease, largely based on accumulation of genetic abnormalities, such as amplification, mutations and deletions, leading to gain-of-function of onco-genes, or loss-of-function of tumor suppressor genes [1]. In recent years it has become evident that not only changes at the genomic level but also at the epigenomic level contribute to cancer development. It is now clear that abnormal epigenetic regulation may act as an alternative mech-anism for, e.g., loss of tumor suppressor function in cancer cells. One of the best-known examples is the inactivation of tumor suppressor genes, such as $p16^{INK4A}$ and BRCA1, by DNA methylation [2]. In this way, DNA-methylation-dependent gene silencing contributes to biallelic loss of tumor suppressor function, as proposed in the "two-hit" model, by Knudson [3]. Processes or proteins that orchestrate epigenetic regulation may affect expression or regulation of tumor suppressor or oncogenes, and in doing so contribute to tumorigenesis. In this chapter, we specifically focus on a group of epigenetic regulators, the Polycomb group (PcG) proteins and their role in normal development and in tumorigenesis.

16.2 PcG INTRODUCTION

Epigenetic regulation of gene expression in essence centers on processes that modify chromatin structure (Box 16.1). Epigenetic chromatin remodeling may take place at any higher-order chro-matin structure level (i.e., ranging from posttranslational changes on nucleosomes to chromatin compaction levels beyond the 30 nm solenoid fiber). As all DNA-templated processes require

BOX 16.1
Epigenetic Mechanisms in Gene Expression

DNA, the macromolecule in which our genetic make-up is hardwired, is tightly packed inside the eukaryotic nucleus. This packing fits 2 m of DNA into a nucleus of approximately 2 μm in diameter. The macromolecular DNA/protein structure in which DNA is packed is called chromatin. The consequence of this packing is that DNA (genes) is not naked and therefore not freely accessible. For transcription factors and DNA-dependent RNA polymerases to access DNA, for example to activate gene expression, chromatin structure has to be remodeled. Thus, chromatin structure regulation is essential for many, if not all, important biological processes, which are DNA templated (e.g., DNA replication, transcription, recombination, repair, chromosome segregation). The direct consequence of chromatin packing is that expression of genes and genetic traits is not solely dependent on the genetic blueprint, as it is stored in the base pair sequence in the DNA. Complex eukaryotic cells have a second, epigenetic, layer of control on top of the genetic blue print; access to DNA (i.e., expression of genetic traits) is controlled by chromatin structure modification: whether a gene is used or not is what matters. Epigenetic regulation explains cellular diversity in our body (despite genetically identical cells). Epigenetics is also what makes genetically identical monozygotic twins different in terms of susceptibility to disease (i.e., neurological, cancer, metabolic disorders). Evidence is accumulating that epigenetic information is also heritable, and that epigenetic inheritance may even contribute to heritable, multifactorial disease.

Epigenetic regulation of gene expression occurs at least at three macromolecular levels: (1) at the DNA level—covalent modification of DNA by coupling methyl-groups to specific nucleotides in DNA (explained elsewhere in this volume), (2) at the RNA level—noncoding RNAs may either target genes or mRNAs, and directly regulate their expression (explained elsewhere in this volume), and (3) at the protein level, for instance by covalent modification of the N-terminal tails of histone proteins. DNA is wrapped almost twice around a nucleosome, an octamer of four different histone proteins. The protruding N-terminal histone tails are accessible to many regulatory proteins. Depending on the modification (i.e., metyl, acetyl, phosphoryl, ubiquityl groups), novel protein binding sites are generated and protein complexes are recruited that either open-up chromatin or compact its structure. Combined, these histone-tail modifications function as an epigenetic register. The epigenetic register is "written" by histone-modifying enzymes. Many proteins have been identified which can "read" the register and thereby become recruited to chromatin to write additional modifications or change its configuration. Many histone-modifying complexes combine writing and reading properties. Recent studies suggest that at least some of these histone-tail-marks recruit nucleosome-remodeling machineries. For a gene to become transcriptionally active, RNA-polymerase has to gain access to the naked DNA. For this to occur, nucleosomes are quite literally moved aside, by ATP-dependent nucleosome-remodeling machineries, to bear the DNA elements required for binding, and transcription initiation. Gene silencing reverses most of these processes.

modification of chromatin structure to allow access to DNA, the epigenetic mechanisms that regulate DNA access and gene expression are pivotal to maintenance of cell identity throughout differentiation and hence normal development. PcG proteins play a key role in the maintenance of gene silencing patterns, whereas their biological counterparts, the trithorax Group (trxG) proteins, generally are or associate with transcription activators [4]. PcG proteins were first identified as repressors of Hox genes in *Drosophila melanogaster*. The descriptive name Polycomb derives from a class of homeotic phenotypes in the fruit fly, which have the common property of determining the number or positioning of a structure called the sex comb, of which male flies normally have only one on each front leg. The phenotypic characteristics of fly extremities (e.g., antennae, legs, wings, halteres) and their positioning (i.e., anteroposterior (A-P) axial patterning) are controlled by the homeotic complex of genes (HOM-C) [5,6]. An oversimplified delineation of HOM-C dependent A-P patterning argues that segment/extremity identity is determined by the combination of HOM-C genes active in any given A-P position. The fly HOM-C genes are structurally conserved throughout evolution. In addition, the organization of HOM-C is conserved from primitive chordates to humans, as is the co-linear regulation of HOM-C gene expression [5]. PcG proteins were identified as a HOM-C repressive activity almost 30 years ago [7]. Whereas initiation of silencing or activation is not dependent on PcG proteins, *Drosophila* PcG mutants typically mis-express specific HOM-C genes at later stages; this, combined with unaltered expression initiation of HOM-C genes in PcG mutants led to the proposition that maintenance of HOM-C expression boundaries (i.e., silencing) was affected in these mutants. Also PcG and trxG proteins appeared conserved throughout evolution and even their antagonistic properties [8,9]. Hence, it would seem likely that their regulatory function and mode of action on the mammalian Hox-gene clusters (counterparts of the fly HOM-C) would also be conserved. Indeed Hox and PcG genes are involved in A-P axial patterning in vertebrates, exemplified by typical skeletal abnormalities in PcG mutant mice, which are accompanied by defective maintenance of Hox-gene expression boundaries [10,11]. Also it is now clear that besides Hox genes, PcG proteins repress many other genes, among which genes that are directly relevant for cell proliferation and survival, and thus oncogenesis. Relevantly, the first mammalian PcG gene, BMI1, was identified as an oncogene in a murine model for oncogene cooperativity in leukemia [12,13]. Besides transcriptional regulation, PcG proteins are implicated in diverse biological processes, such as X-inactivation, genomic imprinting, cell cycle regulation, and stem cell maintenance (Refs. [14–18]). Some of these properties, when deregulated, are likely to contribute to tumorigenesis. Our understanding of the molecular details of PcG repression is still incomplete, however, in recent years our insight into the silencing mechanisms that PcG proteins impinge on has increased significantly.

16.3 PcG BIOLOGY

16.3.1 MECHANISMS IN PcG SILENCING

We now know that PcG proteins function in large multiprotein complexes. PcG biochemistry appears highly conserved among fruit flies and higher eukaryotes, like mice and men. The two most studied Polycomb repressive complexes (PRCs) are PRC1 and PRC2 [19,20]. Reconstitution studies using recombinant proteins identified the core complex of *Drosophila* PRC1, which consists of Polycomb (Pc), Polyhomeotic (Ph), Posterior sex combs (Psc), and dRing/Sex combs extra (Sce) [21–23]. As was the case with evolutionary duplication of HOM-C between *Drosophila* and human (the human genome harbors four *Hox*-gene clusters), mostly single fly PcG genes were also duplicated. Mammalian PRC1 genes comprise multiple orthologues of the *Drosophila* core components: HPC2 and 3, HPH1, 2, and 3, BMI1, and RING1A and RING1B/RNF2, respectively (Box 16.2) [24]. Biochemical purification of *Drosophila* PRC2 members identified Enhancer of zeste (E(z)), Extra sex combs (ESC), and Suppressor of zeste-12 (Su(z)12) as the main core complex members [25]. The mammalian orthologues of PRC2 are EZH2, EED, and SUZ12. As PcG proteins are equipped with multiple protein-interaction domains, and many if not all mammalian homologues are known to homo- or heterodimerize, many different protein–protein binding modalities are expected to contribute to complex composition and diversity. Several observations suggest that also spatio-temporal expression patterns of PcG proteins may not only determine PcG complex composition but also allow composition to change throughout development [20].

Studies of early fly development suggest that the PRC2 complex initiates PcG-mediated repression. The most compelling evidence for this is provided by temporally restricted sequential

BOX 16.2
Polycomb Group Genes and Functions

PcG Complex	*Drosophila*	Human	Protein Domains	Biochemical Activity
PRC2	E(z)	**EZH1**	SET	Histone methyl transferase for H3K27
		EZH2		
	ESC	**EED**	WD40	Cofactor for E(z); methyl-lysine binding motif?
	Su(z)12	**SUZ12**	Zinc-finger	
PRC1	Pc	**CBX2**/HPC1	Chromodomain	HPc2: binding to H3K27me3, RNA binding, Sumo-E3 ligase
		CBX4/HPC2		
		CBX6		
		CBX7		
		CBX8/HPC3		
	Ph	HPH1/EDR1/**PHC1**	SAM	Currently unknown
		HPH2/EDR2/**PHC2**		
		HPH3/EDR3/**PHC3**		
	Sce/dRing	**RING1**/RNF1/RING1A	RING-finger	E3 Ubiquitin ligase for H2AK119
		RNF2/RING1B/RING2		
	Psc	**BMI1**/PCGF4/RNF51	RING-finger	Cofactor for RING proteins
		PCGF2/RNF110/MEL18		
PhoRC	Pho	**YY1**	Zinc-finger	Sequence specific DNA binding

Due to the large number of PcG orthologues, this table is limited to core complex members. Gene names according to HUGO gene nomenclature are shown in bold. BMI, B lymphoma mo-MLV insertion region; CBX, chromobox homologue; EDR, early development regulator; EED, embryonic ectoderm development; ESC, extra sex combs; E(z), enhancer of zeste; HPC, homolog of polycomb; HPH, homolog of polyhomeotic; Pc, polycomb; PCGF, polycomb group ring-finger; Ph, polyhomeotic; PHC, polyhomeotic-like; Pho, pleiohomeotic; Psc, posterior sex combs; RING, really interesting new gene; RNF, ring-finger protein; SAM, self-association motif; Sce, sex combs extra; Su(z), suppressor of zeste; YY, yin–yang transcription factor.

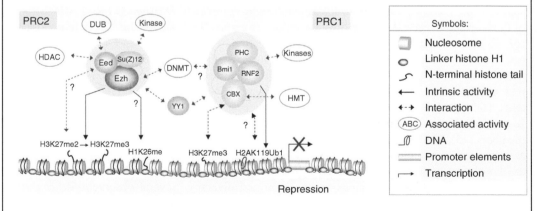

(See color insert following page 272.) Schematic representation of molecular interactions in PcG silencing (see main text for explanations).

expression patterns of the fly PRC2 and PRC1 complexes, respectively [26–28]. One study revealed a transient physical interaction between PRC2 and PRC1 complexes, which was interpreted as perhaps being required to transfer PcG-mediated transcriptional silencing from initiation to maintenance state [29]. Mammalian PcG complex collaboration may not be entirely similar, yet a sequential collaborative action was suggested by recent biochemical studies. EZH2 contains histone methyl transferase (HMT) activity within its SET domain, which mediates methylation of other proteins. *Drosophila* E(z) and its mammalian orthologue EZH2 specifically trimethylate histone H3 on lysine K27 (H3K27me3) and, to a lesser extent, lysine 9 (H3K9me3) [25,30–32]. EZH2 thus writes the epigenetic register (Box 16.1). Other PRC2 proteins read epigenetic markings on other histones. As WD40 domains have been implicated in binding specific histone methyl-lysine marks in other proteins [33–35], EED may likewise read specific methylated histone marks via its WD40 repeat and contribute to sequential recruitment of other histone-modifying enzymes and, as such, provide additional regulation [33]. The H3K27me3 mark is a PcG-associated silencing mark, as it recruits the PRC1 maintenance complex via binding the chromodomain of the *Drosophila* Pc protein [36,37]. Crystallography studies have provided compelling evidence for a differential fit of the H3K9me3 and H3K27me3 in the chromodomains of HP1 and Pc, respectively [36,37]. This represents a functional biochemical bridging mechanism between PRC2 and PRC1 complexes (also see Section 16.3.2). Biochemical interactions and repressive mechanisms in mammalian systems are assumed to function similarly to some extent (based in part on experimental data, in part on inference), in that the SET domain HMTases Ezh1/2 write an epigenetic H3K27me3 mark, to which mammalian Pc orthologs HPc2/3 and possibly other CBXs are recruited [19,38]. Recent studies suggest, however, that the H3K27me3 mark by itself may not be sufficient for HPC2 recruitment, as RNF2 depletion also disrupts HPC2 chromatin binding, whereas the H3K27me3 motif remains [39]. HPC2 itself is a SUMO-E3 ligase, which sumoylates transcription factors and, as such, may provide additional regulation [40,41]. RING-finger domain proteins within PRC1 may all be E3 ubiquitin ligases, yet so far only RNF2 was shown to be responsible for histone H2A mono-ubiquitylation of K119 (H2AK119ub1) [39,42,43]. Although its exact enzymatic substrate is currently not known, BMI1-association somehow stimulates the E3 ligase activity of RNF2 toward H2A [44,45]. In addition, recent studies of X-inactivation revealed that RNF2 is recruited to and marks nucleosomes on the inactivated X-chromosome in female cells, independent of prior PRC2 activity [46]. Clearly, the above studies have only just begun to unravel the complexity and multiple layers of regulation of PcG repressive activity. Finally, PRC complexes also associate with other chromatin-modifying enzymes (Box 16.2). The EED protein interacts with histone deacetylases (HDACs) and deubiquitinases (DUBs), and HPC2 brings HMT SUV39H1 activity to local chromatin, which may help silencing marks spread out over regulatory and into coding sequences [33,47,48]. EZH2 recruits DNA methyltransferases (DNMTs) to specific target genes, suggesting intersection of epigenetic mechanisms at the chromatin level, which may ultimately serve to stabilize repression [49,50]. Two PRC2-like complexes were identified recently, termed PRC3 and PRC4, which incorporate different EED isoforms (RNA splice variants) [51,52]. These isoforms were suggested to differentially target HMT activity toward histone H3K27me or histone H1K26me [51], although other studies found no support for such concept [53]. PcG expression may also influence complex composition through mechanisms other than spatio-temporal presence of complex members: PRC4 is formed when EZH2 is overexpressed in cultured cells and it comprises the NAD^+-dependent histone deacetylase SirT1 and EED-isoform 2. PRC4 appears preferentially expressed in cancer and undifferentiated embryonic stem (ES) cells [52].

Recently, a third PRC complex, Pleiohomeotic (Pho) repressive complex (PhoRC) has been identified in *Drosophila* [54]. It contains Pho and the SFMBT protein (Scm-related gene containing four malignant brain tumor domains). Pho is one of the two PcG proteins, together with pleiohomeotic-like (Phol) that has sequence-specific DNA-binding properties [55–57]. The MBT repeats of SFMBT bind mono- and dimethylated H3-K9 and H4-K20 [54]. So far, the only mammalian orthologue with apparent direct DNA-binding activity is YY1. YY1 interacts with both

PRC2 and PRC1 complexes [58,59]. Whether this dual interaction acts in integrating or bridging silencing by PRC2 and PRC1 is still speculative.

16.3.2 PcG-Mediated Higher-Order Chromatin Structure Remodeling

The PRC1 complex inhibits chromatin remodeling and transcription in vitro [21,23,24,60]. In addition, a reconstituted PRC1 core complex induced in vitro compaction of nucleosomal arrays [61]. Although EEDs' ability to target PRC2-HMT activity to linker histone H1 may point to a possible involvement in higher-order remodeling, there is to date no in vivo data that suggests PcG binding leads to transcriptional silencing by higher-order chromatin compaction.

In *D. melanogaster* elements have been identified in regulatory regions of PcG target genes (i.a. Hox genes) that are necessary for PcG-mediated silencing, termed Polycomb response elements or PREs. PREs are *cis*-acting elements typically several hundred base pairs in size. Although there is apparent variation between PREs in sequence, size, or distance from possible PcG target genes, they share the functional characteristic that they confer PcG-responsiveness on transgenes [62]. Multiple studies have concentrated on the role of PREs in PcG-mediated silencing. Strangely, PREs appear to be nearly void of nucleosomes and hence of PcG-mediated trimethyl marks [62,63]. Recent insights may, however, have uncovered a mechanistic role for PRE in PcG-dependent silencing in the fly: the PhoRC complex is recruited to PRE, via DNA consensus recognition; Pho biochemically interacts with PRC2 and PRC1 [64]. The PRC2 complex mediates trimethylation of H3K27 and H3K9 in upstream regulatory regions, the promoter region and in the coding region of target genes. This in turn would recruit PRC1 complexes via chromodomain binding; via recruitment of yet additional HMTase activity, PcG-dependent silencing would spread over larger chromatin areas [63–66]. The Pho–PRC2 interaction at PREs may provide PREs with the capacity to function as *cis*- or *trans*-acting repressive nucleation sites for adjacent or remote promoters. This mode of action would show resemblance to Sir protein-mediated telomere-proximity silencing in yeast [63,67,68]. Paradoxically, many active promoters in *Drosophila* appear "preoccupied" with Pc, as if on standby, waiting on other local epigenetic events that signal (progressive) silencing; indeed many *Drosophila* PRE and promoters simultaneously bind PcG and trxG proteins, in support of this "bivalent standby" notion [66,69]. Indeed the presence of the trxG protein Ash1 within the promoter and coding regions of a transcriptionally active Ubx locus (one of *Drosophilas*' HOM-C clusters) correlates with loss of H3K27, H3K9, and H4K20 methylation marks and suggests a model in which recruitment of trxG proteins prevents silencing [66]. To date, no mammalian PREs have been identified. However, recent studies have revealed highly conserved noncoding elements (HCNE) in higher eukaryotes, which display specific PcG protein association and may orchestrate PcG-dependence in regulation of developmental programs [70]. The epigenetic indexing on and around these HCNEs is peculiar in that it contains both activating and inactivating histone modifications (e.g., H3K4me3 + H3K27me3), and hence represent "bivalently" marked nuclear elements (see below). These findings are supported by reports that H3K27me3-marking does not necessarily correlate with transcriptional silencing but rather precedes silencing [71,72]. A somewhat anti-quated model suggested PcG-mediated silencing involved sterical blockage of DNA access for transcription factors; recent insights, however, have superseded this view and suggest a role in interference with events that either control pre-initiation complex formation and transcription initiation, despite transcription machinery recruitment and elongation processes [62,66,73]. Recent technological advances have enabled us to directly examine PcG and trxG occupation around and within genes (Section 16.3.3) and interrogate epigenetic marking and promoter occupation, PcG complex composition and other relevant aspects of chromatin mediated silencing. Although PcG-mediated transcriptional silencing already appears a much more complex process than anticipated, ongoing studies will help refine positioning of PcG function in the proper cellular context.

16.3.3 PcG in Stem Cells and Differentiation

Stem cells have two key properties: the ability to self-renew and the ability to differentiate along any given differentiation program. Through asymmetric division, a dividing stem cell will give rise to one new stem cell and one cell that will differentiate into a mature cell of a certain lineage. Two main categories of mammalian stem cells are ES cells and adult stem cells. ES cells are pluripotent, which refers to their capacity to differentiate into any cell type of the adult organism. Adult stem cells, such as hematopoietic stem cells, in contrast, are restricted in their differentiation range. PcG proteins are involved in the maintenance of embryonic and adult stem cells [18,74].

In adult stem cells PcG proteins regulate self-renewal of hematopoietic and neuronal stem cells. BMI1, RAE28/MPH1, and EZH2 all have a role in murine hematopoietic stem cell (HSC) self-renewal. HSCs have the capacity to regenerate fully functional bone marrow, as illustrated by serial transplantation experiments with limited numbers of cells in mice. Both BMI1 and RAE28/MPH1 deficient HSCs are defective in their proliferative and self-renewal capacity, whereas EZH2 over-expression in HSCs prevents hematopoietic stem cell exhaustion [75–79]. BMI1 is also required for the self-renewal of stem cells in the peripheral and central nervous systems [80]. The dependence of self-renewal of these stem cells on BMI1 is believed to contribute to the murine $BMI^{-/-}$ phenotype; besides skeletal transformations, mice show a gradual depletion of progenitor and mature cells in haematopoietic and neural lineages [81].

The observation that EZH2 deficient ES cells cannot be established in vitro, indicates involvement of PcG in ES cell biology [82]. Functional PcG gene ablation studies in mice show that EZH2, EED, SUZ12, and RNF2 are essential for early embryonic development [82–85]. Even though both $EED^{-/-}$ and $SUZ12^{-/-}$ ES cells can be established in tissue culture, they display aberrant activation of differentiation specific genes [86,87].

For a long time we have known that PcG proteins repress Hox clusters. Due to their involvement in diverse biological processes, it was expected that PcG proteins bind to and repress many other genes in addition to Hox genes. A number of studies have recently identified genome-wide binding patterns for PcG proteins in human, mouse, and *Drosophila* cells [70,86,88–92]. Indeed, in all organisms PcG proteins target a large number of loci. Most PcG target genes, as studied in human embryonic fibroblasts and both human and mouse ES cells, constitute developmental regulators [86,88,89]. Among the PcG target genes transcription factors and components of signaling pathways are highly represented. Transcription factor PcG target genes include homeodomain genes of the *Dlx*, *Irx*, *Lhx*, and *Pax* gene families, and *FOX*, *SOX*, and *TBX* gene families, which all control differentiation. Signaling pathways, identified as PcG-regulated, include Wnt, Hedgehog, Notch, TGF-β, FGF, and BMP signaling. Differentiation activates some of these signaling pathways, which corresponds to decreased H3K27me3 and PcG binding at the PcG target genes that regulate the specific developmental process [70,86,88,89,93].

Recent studies have focused on identifying molecular signatures for pluripotency [94]. In ES cells many promoters of developmental regulators are occupied by the transcription factors OCT4, SOX2, and NANOG, all key factors in maintaining ES cell pluripotency [95]. Interestingly, almost all repressed developmental regulators that are occupied by OCT4, SOX2, and NANOG are co-occupied by PcG proteins [89]. Many nontranscribed genes in ES cells have a bivalent chromatin state, harboring both H3K4me3 (catalyzed by trxG proteins) and H3K27me3 epigenetic marks (a PRC2-installed mark), normally associated with gene transcription and gene repression, respectively. Many of these bivalently marked regulatory regions occur in CpG-rich promoters, which in turn are often found in genes expressed during embryogenesis and in "housekeeping" genes [70,71,93]. Housekeeping genes are, however, mostly marked with H3K4me3 only (i.e., void of PRC-association) and their transcriptional status is mostly "on." Bivalent marking is thought to keep developmental control genes (e.g., key developmental factors, morphogens, surface molecules, microRNAs) transcriptionally silent in ES cells, while preserving their potential to become activated upon lineage commitment and during differentiation [70,71,93]. So PcG proteins occupy a set of developmental genes in ES cells that need to be repressed to maintain pluripotency and that are poised for activation during ES cell differentiation.

A model of PcG protein functioning during differentiation, which is gaining acceptance and support by empirical observation, is that in undifferentiated cells PcG proteins activate and maintain gene repression by instating and binding to H3K27me3 marks. Upon lineage commitment or during differentiation, PcG proteins are displaced from critical developmental control genes by as yet unidentified mechanisms. These genes will now become actively transcribed; corresponding regulatory regions will lose their bivalent marking; lineage "nonrelevant" developmental genes will lose their activating marks (i.e., H3K4me3) and will become truly PcG-repressed targets; a third class of high CpG-content promoters loses both me3 marks; these will eventually be shut down transcriptionally. Despite this emerging *modus operandi* many questions remain. Several studies have reported that transcriptional activity is lost upon loss of Polycomb function or that transcription occurs, despite PRC-occupance, suggesting involvement of other spatio-temporal factors or aspects of chromatin status [71,96–98]. In addition, some PcG proteins may harbor as yet unknown biochemical activity which may not be directly linked to epigenetic indexing of genes; in breast cancer cells, Ezh2 appears to transactivate gene expression in estrogen and Wnt signaling pathways in a SET domain independent manner [99]. Combined, these findings clearly stress that PcG-mediated repression is controlled by mechanisms currently far from completely understood. Delineation of chromatin contexts, identification of interaction partners, and associated enzymatic activity in a cell context dependent manner will undoubtedly help further our understanding of the molecular nature of PcG-mediated processes.

16.3.4 PcG Regulation

Signaling molecules in cancer pathways often comprise both oncogenic as well as tumor suppressive activity. Knowing the biochemical relationships between signaling proteins and their regulation is crucial for our understanding of how mutations in these gene products contribute to cancer and how oncogenesis signaling may be averted. Regulation of PcG function is currently poorly understood. Several levels of regulation may be anticipated: transcriptional, posttranscriptional, translational, and posttranslational. Very little is known about transcriptional regulation of PcG genes themselves. A number of reports address possible pathways and transcription factors that control expression of specific PcG genes [98,100–102]. Some of these proteins are known to be involved in cancer as well and may therefore be involved in exposing tumorigenic properties of PcG proteins. Activated JNK signaling in *Drosophila* during *trans*-determination of imaginal discs (larval structures that form extremities) leads to downregulation of PcG genes [103]. Studies in the flies and mammalian systems also indicate that expression of PcG genes is dependent on PcG proteins [88,104–106]. Changes in cellular expression levels may drive differential PcG complex composition, consequently overexpression of particular PcG proteins in cancer may contribute to tumorigenesis. At the posttranscriptional level, i.e., mRNA processing, stability or translation, virtually nothing is known about PcG mRNA regulation. Interestingly though, genetic ablation of Sf3b1, a factor involved in mRNA splicing, evokes a Polycomb phenotype in mice, suggesting a connection with either translation of PcG proteins or their targets [9]. Also TrxG members have been associated with RNA processing: the association of the SWI/SNF-protein Brm with snRNAs and spliceosome factors and the observation that loss of Brm function affects mRNA elongation and inclusion of variant exons suggest a direct role in RNA-splicing and possibly point toward a more fundamental role of PcG and trxG proteins in chromatin structure dependent processes [107]. Regulation of PcG function at the posttranslational level is slowly emerging. As discussed in the preceding paragraphs, PcG complex composition varies between cell types and over time (hematopoiesis). A little over a decade ago, large, functionally distinct complexes, like PcG or trxG complexes, which shared conserved interaction modules and other domains, were tentatively viewed as large bulky multiprotein masses that modified chromatin structure at least in part by binding and physically interfering with accessibility. As we begin to uncover and understand more about the properties of individual PcG proteins, it is now clear that many PcG proteins have specific catalytic or binding activities, relevant to epigenetic regulation. Studies on their posttranslational regulation are slowly emerging, but our knowledge on PcG-regulation is of yet far from complete. Several PcG proteins are regulated by posttranslational modification. Self-ubiquitylation of RNF2 is required to mono-ubiquitylate H2A at lysine 119, which in turn is required for PRC1 repression [39,108]. RNF2 and BMI1 are both ubiquitylated by an as of yet unknown E3 ligase; this results in their proteolytic breakdown [108]. As RNF2 and BMI1 are often found overexpressed in cancers, abnormal ubiquitylation is likely to contribute to tumorigenic processes. EZH2 is phosphorylated by AKT, which controls survival signaling in cells [109]. Although EZH2 overexpression marks malignant progression in some tumors, paradoxically, the tumorigenic action of EZH2 may not solely depend on its chromatin repressive properties, as phosphorylation by AKT inhibits its HMT activity by impeding H3 binding [109]. Two recent mass spectrometric studies identified kinase activity associated with PRC1 complex members CBX8 and RNF2 [110,111]. In addition, a recent study identified the PRC1 complex as targets of canonical MAPK signaling pathways, and identified MAPKAPKs as interaction partners of the PcG complexes [112]. Relevantly, phosphorylation of PRC1 correlates with chromatin dissociation, and activation of repressed tumor suppressors [112]. Although the above observations suggest a link between tumorigenic pathways and regulation of PcG function, clearly much work remains to fully understand the significance of these connections with cancer, and whether or not these regulatory processes provide entries for therapeutic applications.

16.3.5 CROSS TALK BETWEEN PcG AND OTHER EPIGENETIC MECHANISMS

Several levels of epigenetic regulation exist. At its most fundamental level, epigenetic regulation of gene expression occurs at the level of DNA (e.g., DNA methylation), RNA (e.g., RNA interference) and at the protein level (e.g., histone modifications; chromatin structure; Box 16.1). Recent epigenetic studies suggest cross talk between these processes. PcG function most directly impinges on epigenetic regulation at the protein level. However, several reports indicate that epigenetic mechanisms involving DNA and RNA may intersect with PcG-mediated epigenetics. In *Arabidopsis* and *Caenorhabditis elegans* RNA interference is defective in a PcG mutant background [113–115]. In Drosophila PcG-chromatin association and silencing is dependent on components of the RNAi machinery [116]. Chromodomains, which are also used by PcG CBX proteins (Box 16.2), are RNA binding moieties; chromatin association of HP1, a structural relative (chromodomain protein which binds H3K9me3) of the PcG CBX proteins, is dependent on RNA components [117–119]. X-inactivation is dependent on coating of the inactive X-chromosome with a noncoding RNA, Xist, which is at least in part responsible for recruitment of PRC1 [46]. At the DNA-methylation level, several observations suggest a direct link with PcG-mediated silencing: EZH2 binds directly to DNMTs (de novo DNA methyl transferases) and DNA methylation at EZH2 target genes is dependent on EZH2 [50]. Notably, PcG target genes are up to 12 times more likely to have cancer-specific promoter DNA hypermethylation than nontargets [120,121]. RNA, DNA, and protein-mediated epigenetic repressive mechanisms have all been implicated in cancer; therefore these biochemical connections are likely to become increasingly relevant for our understanding of tumorigenic processes and treatment options.

16.4 PcG IN CANCER

Cancer is also an epigenetic disease. It is well established that tumorigenesis is accompanied by global hypomethylation of DNA at CpG dinucleotide sequences, and local hypermethylation at CpG-islands. Loss or mutation of tumor suppressor genes is one of the hallmarks of cancer; tumor suppressors are also targets of aberrant DNA methylation in cancers [1,2,122]. The first indication that PcG proteins are involved in tumorigenesis came from studies that searched for oncogenes that cooperated with c-Myc in lymphomagenesis. In this way the *BMI1* gene (B lymphoma Mo-MLV Insertion region 1) was identified as an oncogene [12,13]. Subsequently, the murine knockout model revealed an inverse correlation between *BMI1* expression and *INK4A/ARF* expression [123]. *INK4A/ARF* encodes two cell cycle inhibitory proteins, $P16^{INK4A}$ and $P14^{ARF}$ ($p19^{Arf}$ in mice) [123], that regulate the PRb and TP53 pathways, respectively (Box 16.3) [124]. $P16^{INK4A}$ is a cyclin-dependent kinase inhibitor (CKI), which inhibits cyclinD-CDK4/6-mediated phosphorylation of PRb, whereas $P14^{ARF}$ inhibits interaction between MDM2 and TP53, thereby stabilizing TP53. BMI1s' ability to downregulate *INK4A/ARF* expression has obvious etiological consequences for tumorigenesis. Indeed this property provided an explanation for its cooperative oncogenic action with c-Myc, as c-Myc-induced $P14^{ARF}$-dependent apoptosis was suppressed by BMI1, effectively favoring proliferation [125]. Although it is believed that overexpression of BMI1 may be functionally equivalent to loss of PRb and TP53, and therefore reflects distinctive tumor etiology, recent reports on simultaneous overexpression of $P16^{INK4A}$ as well as BMI1 in some tumors clearly suggest other relevant players [126,127]. Recent studies revealed an antagonistic action of MEL18 and BMI1 on AKT/PKB-activation, a known tumor promoting signal transduction molecule, suggesting multiple genetic interactions of PRC function and the PI3KAKT/mTOR pathway [109,128]. The intuitive notion that there must be many more oncogenic PcG targets is further supported by the recent identification of several thousand Polycomb targets in human fibroblasts [88].

BOX 16.3
Basic Mechanisms in Cancer

Cumulative scientific evidence over the last decades demonstrates that multiple steps are required for a normal cell, with a limited proliferative life span, to bypass this limitation and eventually become a fully transformed cancerous cell. Proliferative longevity and controlled proliferation of human cells are kept in check by multiple so-called checkpoints. These checkpoints evaluate and integrate input from a myriad of pathways, which collectively provide crucial information on the cell and its relationship with its environment. This input includes replicative age of a cell, the presence of DNA damage, availability of nutrients, metabolic status, environmental cues important for cell cycle entry, arrest or exit, such as growth factor signaling-input, cell–cell, or cell–matrix contacts, etc. Checkpoints provide failsafe mechanisms to protect a cell from genome integrity breaches and unwarranted, potentially malignant proliferation. Inevitably, the outcome of checkpoint activation is cell cycle arrest/quiescence, senescence, or apoptosis. As a cancer becomes increasingly malignant, many of these checkpoints appear to have been compromised. Weinbergs' "Hallmarks of cancer" describe a biologically defined set of common tumorigenic events which ultimately occur in most if not all cancers [1]. Human cells need to lose at least three checkpoints to gain proliferative immortality and require a fourth oncogenic event (e.g., activation of H-RAS) to become fully transformed. A common property of many cancer cells is enhanced telomerase (hTERT) activity, which protects telomere repeats at the ends of chromosomes from further "erosion." Telomere length effectively mirrors replicative age and telomere erosion eventually leads to replicative senescence [178]. Senescence bypass is an important mechanism in the onset and development of cancer. Two tumor-suppressor pathways are pivotal in establishing arrest/quiescence, senescence, or apoptosis in response to many potentially dangerous situations (including nontelomeric signals) when left unchecked. These are the PRb and the TP53 pathways, which control cell cycle entry and progression. Loss or inactivation of PRb and TP53, or mutation of effector or affector molecules in the PRb and TP53 pathways are common to most cancers.

BOX 16.3 (continued)
Basic Mechanisms in Cancer

These same antitumor pathways, which control senescence in normal cells, also limit the life span of stem cells. Increasing scientific evidence supports the presence of cancer stem cells (CSCs) in many different tumors. These CSCs are held responsible for tumor maintenance, growth, and ultimately malignant progression, and because of particular properties inherent to stem cells, are most often difficult to treat with current anticancer therapy. The prototype mammalian PcG-protein BMI1 impinges on several antitumor checkpoints described above, but likely does so in a cell type dependent fashion (see main text).

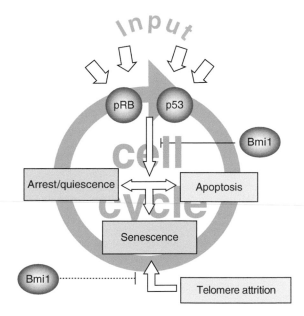

Simplified representation of check-points in cell cycle regulation and cancer; possible role for Polycomb Group protein Bmi1 in these checkpoints.

BMI1 is not the only PcG protein that has been linked to tumorigenesis. Since EZH2 was recognized as a marker for the progression of prostate cancer, it has been linked to a variety of other cancers, such as breast cancer and several lymphomas [129–134] (see Table 16.1). Although BMI1 and EZH2 have received most attention, a general notion of the tumorigenic potential of other PcG proteins is slowly emerging. The use of tissue microarrays (see Figure 16.1), to study PcG expression in normal and tumor tissues will greatly enhance our understanding of PcG in cancer [135]. This particular study analyzed PcG expression immuno-histologically in different normal and tumor tissues; several tumors show very distinctive and sometimes unique PcG expression profiles, pointing to specific etiology. To fully understand the role of PcG in tumorigenesis, it will be crucial to chart all PcG proteins implicated in tumor development, and thoroughly examine their mode of action, *cq* their relative contribution to different types of cancer in years to come.

Not only do PcG proteins appear overexpressed in various cancer types but also expression is often correlated with specific aspects of tumor biology: a more progressive disease state, metastasis, and ultimately poor prognosis. In prostate cancer, EZH2 staining intensity increased from benign, prostatic atrophy, prostatic intraepithelial neoplasia, clinically localized prostate cancer to metastatic prostate cancer [136]. In this study, EZH2 expression was a better predictor of clinical outcome than

TABLE 16.1

Deregulated PcG Expression in Human Cancers

PcG Complex	PcG Protein	Cancer Type	References
PRC2	EZH2	B-cell non-Hodgkin lymphoma	[157]
		Bladder	[158–160]
		Breast	[129–132]
		Colon	[161]
		Endometrium	[129]
		Hodgkin lymphoma	[133]
		Liver	[162]
		Mantle cell lymphoma	[134]
		Melanoma	[129]
		Prostate	[129,136,163–165]
		Stomach	[138,166]
	SUZ12	Breast	[167]
		Colon	[167]
		Liver	[167]
PRC1	CBX7	Follicular lymphoma	[168]
	RING1	Prostate	[165]
	RNF2	Colon	[135]
		Diffuse large B-cell lymphoma	[135]
		Hodgkin lymphoma	[135]
		Stomach	[135]
	BMI1	B-cell non-Hodgkin lymphoma	[157]
		Breast	[169]
		Colon	[170]
		Diffuse large B-cell lymphoma	[135,139]
		Hodgkin lymphoma	[133,135,171]
		Leukemia	[172]
		Mantle cell lymphoma	[135,173]
		Medulloblastoma	[102]
		Neuroblastoma	[174]
		Nonsmall cell lung cancer	[175]
		Oral squamous cell carcinoma	[176]
		Parathyroid adenoma	[135]
		Pituitary adenoma	[135]
		Prostate	[163,165]
	PHC1	Acute lymphoblastic leukemia	[177]

surgical margin status, maximum tumor dimension, Gleason score, and preoperative PSA [136]. Indeed, malignant tumor progression could be mimicked in vitro by overexpression of EZH2. Thus, EZH2 protein expression in prostate specimens may be a valuable prognostic marker. EZH2 overexpression is associated with a poor prognosis in prostate, breast, gastric, and colon cancers as is BMI1 overexpression in different leukaemia's [131,136–139]. Moreover, EZH2 was even upregulated in preneoplastic lesions in breast tissue [140]. In this respect EZH2 may help identify patients at risk for developing breast cancer [141]. Recently, microarray-based gene expression profiling was used to identify a death-from-cancer signature, which is partly based on BMI1-regulated genes and consists of 11 genes [142]. A stem cell-like expression profile of this signature in primary tumors was a powerful predictor of a short interval to disease recurrence, distant metastasis, and death after therapy in 11 different cancer types [142].

Interestingly, EZH2 also embodies a direct bridge between epigenetic mechanisms at the protein level (histone-tail modifications) and the DNA level; recent reports provided experimental evidence

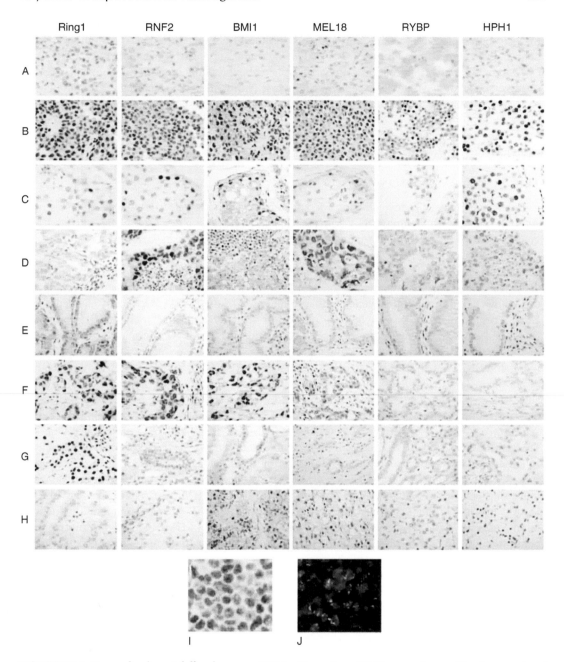

FIGURE 16.1 (See color insert following page 272.) Examples of PcG expression profiles in normal and tumor tissues: (A) pituitary gland, (B) pituitary adenoma, (C) testis, (D) embryonal carcinoma, (E) gastric surface-epithelial cells, (F) gastric adenocarcinoma, (G) kidney tubules, (H) clear-cell renal cell carcinoma, (I) BMI1 overexpression in a mantle cell lymphoma case, (J) BMI1 amplification detected by FISH in mantle cell lymphoma from (I). The picture shows a group of cells with several copies of BMI1 (green signal) in comparison with two centromeric copies for chromosome 10 (red signal). RING1, RNF2, BMI1, MEL18, and HPH1 are PcG proteins from the PRC1 complex (Box 16.2). RYBP (RING1 and YY1 binding protein) is a PRC1 complex-associated protein [58]. (Photographs reprinted from Sanchez-Beato, M., et al., *Mod. Pathol.*, 19, 684, 2006. With permission.)

for a biochemical and functional link between PcG-mediated silencing and DNA methylation at tumor suppressor genes, persuasively arguing that the PRC2 complex may contribute to carcinogenesis through disabling checkpoints by silencing tumor suppressor genes [50,143].

Although sparse information is available on overexpression of YY1 in human cancers, its positive association with DNA repair and inverse correlation with cellular TP53 homeostasis suggests a potential oncogenic role for this PcG and trxG associated factor [144,145]. These observations together imply that screening for altered PcG expression in certain cancer types may in the near future enable physicians to more accurately predict prognosis and personalize patient treatment strategies.

16.4.1 PcG and Senescence

Cellular senescence, an irreversible proliferative arrest, can be induced in vitro by diverse stresses, such as telomere shortening, DNA damage, and strong mitogenic signaling by activated oncogenes [146,147]. As cellular senescence limits the proliferative capacity of cells, bypass of senescence is an important step in tumorigenesis (Box 16.3). Senescence was recently shown to occur in vivo in premalignant or benign conditions [148]. Although diverse stimuli induce senescence, its regulation converges on the TP53 and PRb pathways. Recent studies advanced our knowledge of how PcG proteins regulate expression of the *INK4A/ARF* locus and senescence. Until recently, mechanistic insights into repression of the *INK4A/ARF* locus by PcG proteins as BMI1 and CBX7 were inferred from inverse correlation between PcG and *INK4A/ARF* expression [123,149]. However, chromatin immunoprecipitation (ChIP) experiments identified both PRC2 members, SUZ12 and EZH2, and PRC1 members, BMI1, CBX4, CBX7, CBX8, and RNF2 as direct interactors of the *INK4A/ARF* locus [110,150,151]. In senescing fibroblasts EZH2 expression was downregulated, which coincided with decreased H3K27me3 levels, BMI1 displacement and activation of P16^{INK4A} transcription [150]. Thus, PcG binding to *INK4A/ARF* regulates senescence. Overexpression of the PcG proteins BMI1, CBX7, and CBX8 can overcome senescence [110,123,149]. CBX7 can promote proliferation in the absence of BMI1, whereas the function of CBX8 is dependent on BMI1, indicating that CBX7 might be part of PRC1 complexes that do not include BMI1 [110,149]. BMI1 is also implicated in a different mechanism of bypassing senescence. In human mammary epithelial cells (HMECs) overexpression of BMI1 led to activation of human telomerase reverse transcriptase (hTERT) transcription and induction of telomerase activity [152]. For now this has only been detected in HMECs, and does not appear to contribute to immortalization in human fibroblasts or in normal human oral keratinocytes [153,154], and even so, telomere shortening does occur in BMI1-immortalized HMEC [152]. Clearly, PcG proteins regulate senescence via *INK4A/ARF*-dependent and independent mechanisms and they do so in a cell type dependent fashion, presumably reflecting cell type specific molecular networks and underlying epigenetic profiles and chromatin status. As PcG proteins are upregulated in many different cancers, it is tempting to speculate that PcG-mediated senescence bypass represents an early event in tumorigenesis.

16.4.2 PcG and CSCs

Stem cells have also been detected in cancers, where they can give rise to more CSCs and also nontumorigenic cancer cells [155]. This explains at least in part the heterogeneity often observed in tumors. The identification of CSCs has direct implications for cancer treatment. As CSCs are emerging as the main driving tumorigenic entities, to treat cancer effectively, preferably CSCs should be targeted. Hence, it is crucial to understand the processes that define stem cellness, and the pathways or proteins that are deregulated in CSCs. PcG proteins are also implicated in the self-renewal of CSCs. In a recent study, BMI$^{-/-}$ bone marrow cells were transduced to develop acute myeloid leukemia (AML). Importantly, primary recipients developed AML, whereas BMI$^{-/-}$ AML, in contrast to wild-type derived AML, failed to repopulate secondary recipients [77]. This establishes BMI1 as an important regulator of the proliferative activity of not only normal but also leukemic HSCs. As overexpression of BMI1 is often seen in leukemias, a comprehensive understanding of the regulation of stem cell biology by PcG proteins is a very important aspect of hematologic oncology.

16.4.3 PERSPECTIVES

Recent research position PcG proteins as master regulators of development. Their suppressive action is necessary for proper embryonic development, cell proliferation, and stem cell biology. The initial view that PcG proteins act as stable repressors throughout cell division and differentiation has been adapted to a more dynamic regulation of PcG binding and repression. As with all proteins that regulate normal proliferation, deregulated PcG protein expression levels have been detected in cancer. Putative mechanisms of how PcG proteins contribute to tumorigenesis are bypass of senescence and maintenance of CSCs. Many questions remain. How do PcG proteins exactly function in tumorigenesis? How are PcG binding patterns changed upon differentiation? Which enzymatic activities are associated with PcG complexes and which unknown enzymatic activities do they harbor themselves? How do posttranslational modifications alter PcG function? Answers to these questions will determine whether PcG proteins can become suitable targets for therapeutic intervention. In this respect it is promising that, e.g., loss of BMI1 seems to promote cancer-specific cell death [156]. For now, the use of PcG expression as a biomarker for tumor progression and patient prognosis is a far more attainable concept. First these concepts have to be validated and detection methods require standardization before it can be translated into clinical application. Considering the speed at which the PcG research field is developing, it can be anticipated that many novel aspects of PcG function and regulation will be elucidated in the next few years.

REFERENCES

1. Hanahan, D. and Weinberg, R.A., The hallmarks of cancer, *Cell*, 100, 57, 2000.
2. Esteller, M., Cancer epigenomics: DNA methylomes and histone-modification maps, *Nat Rev Genet*, 8, 286, 2007.
3. Knudson, A.G., Jr., Hereditary cancer, oncogenes, and antioncogenes, *Cancer Res*, 45, 1437, 1985.
4. Francis, N.J. and Kingston, R.E., Mechanisms of transcriptional memory, *Nat Rev Mol Cell Biol*, 2, 409, 2001.
5. Krumlauf, R., Hox genes in vertebrate development, *Cell*, 78, 191, 1994.
6. McGinnis, W. and Krumlauf, R., Homeobox genes and axial patterning, *Cell*, 68, 283, 1992.
7. Lewis, E.B., A gene complex controlling segmentation in *Drosophila*, *Nature*, 276, 565, 1978.
8. Hanson, R.D., et al., Mammalian trithorax and polycomb-group homologues are antagonistic regulators of homeotic development, *Proc Natl Acad Sci U S A*, 96, 14372, 1999.
9. Isono, K., et al., Mammalian polycomb-mediated repression of Hox genes requires the essential spliceosomal protein sf3b1, *Genes Dev*, 19, 536, 2005.
10. Deschamps, J., et al., Initiation, establishment and maintenance of Hox gene expression patterns in the mouse, *Int J Dev Biol*, 43, 635, 1999.
11. van Lohuizen, M., Functional analysis of mouse polycomb group genes, *Cell Mol Life Sci*, 54, 71, 1998.
12. Haupt, Y., et al., Novel zinc finger gene implicated as myc collaborator by retrovirally accelerated lymphomagenesis in E mu-myc transgenic mice, *Cell*, 65, 753, 1991.
13. van Lohuizen, M., et al., Identification of cooperating oncogenes in E mu-myc transgenic mice by provirus tagging, *Cell*, 65, 737, 1991.
14. Delaval, K. and Feil, R., Epigenetic regulation of mammalian genomic imprinting, *Curr Opin Genet Dev*, 14, 188, 2004.
15. Heard, E., Delving into the diversity of facultative heterochromatin: The epigenetics of the inactive x chromosome, *Curr Opin Genet Dev*, 15, 482, 2005.

16. Martinez, A.M. and Cavalli, G., The role of polycomb group proteins in cell cycle regulation during development, *Cell Cycle*, 5, 2006.
17. Sparmann, A. and van Lohuizen, M., Polycomb silencers control cell fate, development and cancer, *Nat Rev Cancer*, 6, 846, 2006.
18. Valk-Lingbeek, M.E., et al., Stem cells and cancer; the polycomb connection, *Cell*, 118, 409, 2004.
19. Lund, A.H. and van Lohuizen, M., Polycomb complexes and silencing mechanisms, *Curr Opin Cell Biol*, 16, 239, 2004.
20. Otte, A.P. and Kwaks, T.H.J., Gene repression by polycomb group protein complexes: A distinct complex for every occasion? *Curr Opin Genet Dev*, 13, 448, 2003.
21. Francis, N.J., et al., Reconstitution of a functional core polycomb repressive complex, *Mol Cell*, 8, 545, 2001.
22. Saurin, A.J., et al., A *Drosophila* polycomb group complex includes zeste and dTAFII proteins, *Nature*, 412, 655, 2001.
23. Shao, Z., et al., Stabilization of chromatin structure by PRC1, a polycomb complex, *Cell*, 98, 37, 1999.
24. Levine, S.S., et al., The core of the polycomb repressive complex is compositionally and functionally conserved in flies and humans, *Mol Cell Biol*, 22, 6070, 2002.
25. Muller, J., et al., Histone methyltransferase activity of a *Drosophila* polycomb group repressor complex, *Cell*, 111, 197, 2002.
26. Bienz, M. and Muller, J., Transcriptional silencing of homeotic genes in *Drosophila*, *Bioessays*, 17, 775, 1995.
27. Kennison, J.A. and Tamkun, J.W., Dosage-dependent modifiers of polycomb and antennapedia mutations in *Drosophila*, *Proc Natl Acad Sci U S A*, 85, 8136, 1988.
28. Paro, R., Imprinting a determined state into the chromatin of *Drosophila*, *Trends Genet*, 6, 416, 1990.
29. Poux, S., et al., Establishment of polycomb silencing requires a transient interaction between pc and ESC, *Genes Dev*, 15, 2509, 2001.
30. Cao, R., et al., Role of histone h3 lysine 27 methylation in polycomb-group silencing, *Science*, 26, 1039, 2002.
31. Czermin, B., et al., Drosophila enhancer of zeste/ESC complexes have a histone h3 methyltransferase activity that marks chromosomal polycomb sites, *Cell*, 111, 185, 2002.
32. Kuzmichev, A., et al., Histone methyltransferase activity associated with a human multiprotein complex containing the enhancer of zeste protein, *Genes Dev*, 16, 2893, 2002.
33. Higa, L.A., et al., Cul4-ddb1 ubiquitin ligase interacts with multiple wd40-repeat proteins and regulates histone methylation, *Nat Cell Biol*, 8, 1277, 2006.
34. Ruthenburg, A.J., et al., Histone h3 recognition and presentation by the wdr5 module of the mll1 complex, *Nat Struct Mol Biol*, 13, 704, 2006.
35. Wysocka, J., et al., Wdr5 associates with histone h3 methylated at k4 and is essential for h3 k4 methylation and vertebrate development, *Cell*, 121, 859, 2005.
36. Fischle, W., et al., Molecular basis for the discrimination of repressive methyl-lysine marks in histone h3 by polycomb and hp1 chromodomains, *Genes Dev*, 17, 1870, 2003.
37. Min, J., et al., Structural basis for specific binding of polycomb chromodomain to histone h3 methylated at lys 27, *Genes Dev*, 17, 1823, 2003.
38. Kim, S.Y., et al., Juxtaposed polycomb complexes co-regulate vertebral identity, *Development*, 133, 4957, 2006.
39. Wang, H., et al., Role of histone h2a ubiquitination in polycomb silencing, *Nature*, 431, 873, 2004.
40. Kagey, M.H., et al., The polycomb protein Pc2 is a sumo e3, *Cell*, 113, 127, 2003.
41. Long, J., et al., Pc2-mediated sumoylation of Smad-interacting protein 1 attenuates transcriptional repression of E-cadherin, *J Biol Chem*, 280, 35477, 2005.
42. Cao, R., et al., Role of Bmi-1 and ring1a in h2a ubiquitylation and Hox gene silencing, *Mol Cell*, 20, 845, 2005.
43. de Napoles, M., et al., Polycomb group proteins ring1a/b link ubiquitylation of histone h2a to heritable gene silencing and X inactivation, *Dev Cell*, 7, 663, 2004.
44. Buchwald, G., et al., Structure and e3-ligase activity of the ring-ring complex of polycomb proteins Bmi1 and ring1b, *Embo J*, 25, 2465, 2006.
45. Li, Z., et al., Structure of a Bmi-1-RING1B polycomb group ubiquitin ligase complex, *J Biol Chem*, 2006.

46. Schoeftner, S., et al., Recruitment of PRC1 function at the initiation of x inactivation independent of PRC2 and silencing, *Embo J*, 2006.

47. Sewalt, R.G., et al., Selective interactions between vertebrate polycomb homologs and the suv39h1 histone lysine methyltransferase suggest that histone h3-k9 methylation contributes to chromosomal targeting of polycomb group proteins, *Mol Cell Biol*, 22, 5539, 2002.

48. van der Vlag, J. and Otte, A.P., Transcriptional repression mediated by the human polycomb-group protein EED involves histone deacetylation, *Nat Genet*, 23, 474, 1999.

49. Reynolds, P.A., et al., Tumor suppressor p16ink4a regulates polycomb-mediated DNA hypermethylation in human mammary epithelial cells, *J Biol Chem*, 2006.

50. Vire, E., et al., The polycomb group protein EZH2 directly controls DNA methylation, *Nature*, 439, 871, 2006.

51. Kuzmichev, A., et al., Different ezh2-containing complexes target methylation of histone h1 or nucleosomal histone h3, *Mol Cell*, 14, 183, 2004.

52. Kuzmichev, A., et al., Composition and histone substrates of polycomb repressive group complexes change during cellular differentiation, *Proc Natl Acad Sci U S A*, 102, 1859, 2005.

53. Martin, C., et al., Substrate preferences of the EZH2 histone methyltransferase complex, *J Biol Chem*, 281, 8365, 2006.

54. Klymenko, T., et al., A polycomb group protein complex with sequence-specific DNA-binding and selective methyl-lysine-binding activities, *Genes Dev*, 20, 1110, 2006.

55. Brown, J.L., et al., The *Drosophila* Pho-like gene encodes a yy1-related DNA binding protein that is redundant with pleiohomeotic in homeotic gene silencing, *Development*, 130, 285, 2003.

56. Brown, J.L., et al., The *Drosophila* polycomb group gene pleiohomeotic encodes a DNA binding protein with homology to the transcription factor yy1, *Mol Cell*, 1, 1057, 1998.

57. Fritsch, C., et al., The DNA-binding polycomb group protein pleiohomeotic mediates silencing of a *Drosophila* homeotic gene, *Development*, 126, 3905, 1999.

58. Garcia, E., et al., Rybp, a new repressor protein that interacts with components of the mammalian polycomb complex, and with the transcription factor YY1, *Embo J*, 18, 3404, 1999.

59. Satijn, D.P., et al., The polycomb group protein EED interacts with YY1, and both proteins induce neural tissue in xenopus embryos, *Mol Cell Biol*, 21, 1360, 2001.

60. King, I.F., et al., Native and recombinant polycomb group complexes establish a selective block to template accessibility to repress transcription in vitro, *Mol Cell Biol*, 22, 7919, 2002.

61. Francis, N.J., et al., Chromatin compaction by a polycomb group protein complex, *Science*, 306, 1574, 2004.

62. Schwartz, Y.B. and Pirrotta, V., Polycomb silencing mechanisms and the management of genomic programmes, *Nat Rev Genet*, 8, 9, 2007.

63. Mohd-Sarip, A., et al., Architecture of a polycomb nucleoprotein complex, *Mol Cell*, 24, 91, 2006.

64. Wang, L., et al., Hierarchical recruitment of polycomb group silencing complexes, *Mol Cell*, 14, 637, 2004.

65. Moehrle, A. and Paro, R., Spreading the silence: Epigenetic transcriptional regulation during *Drosophila* development, *Dev Genet*, 15, 478, 1994.

66. Papp, B. and Muller, J., Histone trimethylation and the maintenance of transcriptional on and off states by trxG and PcG proteins, *Genes Dev*, 20, 2041, 2006.

67. Gartenberg, M.R., The sir proteins of saccharomyces cerevisiae: Mediators of transcriptional silencing and much more, *Curr Opin Microbiol*, 3, 132, 2000.

68. Kahn, T.G., et al., Polycomb complexes and the propagation of the methylation mark at the *Drosophila* ubx gene, *J Biol Chem*, 281, 29064, 2006.

69. Orlando, V., Polycomb, epigenomes, and control of cell identity, *Cell*, 112, 599, 2003.

70. Bernstein, B.E., et al., A bivalent chromatin structure marks key developmental genes in embryonic stem cells, *Cell*, 125, 315, 2006.

71. Azuara, V., et al., Chromatin signatures of pluripotent cell lines, *Nat Cell Biol*, 8, 532, 2006.

72. Kohlmaier, A., et al., A chromosomal memory triggered by Xist regulates histone methylation in x inactivation, *PLoS Biol*, 2, E171, 2004.

73. Dellino, G.I., et al., Polycomb silencing blocks transcription initiation, *Mol Cell*, 13, 887, 2004.

74. Park, I.K., et al., BMI1, stem cells, and senescence regulation, *J Clin Invest*, 113, 175, 2004.

75. Kamminga, L.M., et al., The polycomb group gene EZH2 prevents hematopoietic stem cell exhaustion, *Blood*, 107, 2170, 2006.

76. Kim, J.Y., et al., Defective long-term repopulating ability in hematopoietic stem cells lacking the polycomb-group gene rae28, *Eur J Haematol*, 73, 75, 2004.

77. Lessard, J. and Sauvageau, G., BMI-1 determines the proliferative capacity of normal and leukaemic stem cells, *Nature*, 423, 255, 2003.

78. Ohta, H., et al., Polycomb group gene rae28 is required for sustaining activity of hematopoietic stem cells, *J Exp Med*, 195, 759, 2002.

79. Park, I.K., et al., Bmi-1 is required for maintenance of adult self-renewing haematopoietic stem cells, *Nature*, 423, 302, 2003.

80. Molofsky, A.V., et al., Bmi-1 dependence distinguishes neural stem cell self-renewal from progenitor proliferation, *Nature*, 425, 962, 2003.

81. van der Lugt, N.M., et al., Posterior transformation, neurological abnormalities, and severe hematopoietic defects in mice with a targeted deletion of the Bmi-1 proto-oncogene, *Genes Dev.*, 8, 757, 1994.

82. O'Carroll, D., et al., The polycomb-group gene EZH2 is required for early mouse development, *Mol Cell Biol*, 21, 4330, 2001.

83. Pasini, D., et al., Suz12 is essential for mouse development and for EZH2 histone methyltransferase activity, *Embo J*, 23, 4061, 2004.

84. Schumacher, A., et al., Positional cloning of a global regulator of anterior–posterior patterning in mice, *Nature*, 383, 250, 1996.

85. Voncken, J.W., et al., Rnf2 (ring1b) deficiency causes gastrulation arrest and cell cycle inhibition, *Proc Natl Acad Sci U S A*, 100, 2468, 2003.

86. Boyer, L.A., et al., Polycomb complexes repress developmental regulators in murine embryonic stem cells, *Nature*, 441, 349, 2006.

87. Pasini, D., et al., The polycomb group protein Suz12 is required for embryonic stem cell differentiation, *Mol Cell Biol*, 27, 3769, 2007.

88. Bracken, A.P., et al., Genome-wide mapping of polycomb target genes unravels their roles in cell fate transitions, *Genes and Development*, 20, 1123, 2006.

89. Lee, T.I., et al., Control of developmental regulators by polycomb in human embryonic stem cells, *Cell*, 125, 301, 2006.

90. Negre, N., et al., Chromosomal distribution of PcG proteins during *Drosophila* development, *PLoS Biol*, 4, e170, 2006.

91. Schwartz, Y.B., et al., Genome-wide analysis of polycomb targets in *Drosophila* melanogaster, *Nat Genet*, 38, 700, 2006.

92. Tolhuis, B., et al., Genome-wide profiling of PRC1 and PRC2 polycomb chromatin binding in *Drosophila* melanogaster, *Nat Genet*, 38, 694, 2006.

93. Mikkelsen, T.S., et al., Genome-wide maps of chromatin state in pluripotent and lineage-committed cells, *Nature*, 448, 553, 2007.

94. Spivakov, M. and Fisher, A.G., Epigenetic signatures of stem-cell identity, *Nat Rev Genet*, 8, 263, 2007.

95. Boyer, L.A., et al., Core transcriptional regulatory circuitry in human embryonic stem cells, *Cell*, 122, 947, 2005.

96. LaJeunesse, D. and Shearn, A., E(z): A polycomb group gene or a trithorax group gene? *Development*, 122, 2189, 1996.

97. Kirmizis, A., et al., Silencing of human polycomb target genes is associated with methylation of histone h3 lys 27, *Genes Dev*, 18, 1592, 2004.

98. Bracken, A.P., et al., EZH2 is downstream of the pRB-E2F pathway, essential for proliferation and amplified in cancer, *Embo J*, 22, 5323, 2003.

99. Shi, B., et al., Integration of estrogen and Wnt signaling circuits by the polycomb group protein EZH2 in breast cancer cells, *Mol Cell Biol*, 27, 5105, 2007.

100. Guney, I., et al., Reduced c-myc signaling triggers telomere-independent senescence by regulating Bmi-1 and p16ink4a, *Proc Natl Acad Sci U S A*, 103, 3645, 2006.

101. Kranc, K.R., et al., Transcriptional coactivator cited2 induces Bmi1 and mel18 and controls fibroblast proliferation via INK4a/ARF, *Mol Cell Biol*, 23, 7658, 2003.

102. Leung, C., et al., BMI1 is essential for cerebellar development and is overexpressed in human medulloblastomas, *Nature*, 428, 337, 2004.

103. Lee, N., et al., Suppression of polycomb group proteins by JNK signalling induces transdetermination in *Drosophila* imaginal discs, *Nature*, 438, 234, 2005.

104. Ali, J.Y. and Bender, W., Cross-regulation among the polycomb group genes in *Drosophila melanogaster*, *Mol Cell Biol*, 24, 7737, 2004.

105. Bloyer, S., et al., Identification and characterization of polyhomeotic pres and tres, *Dev Biol*, 261, 426, 2003.

106. Guo, W.-J., et al., Mel-18, a polycomb group protein, regulates cell proliferation and senescence via transcriptional repression of BMI-1 and c-myc oncoproteins, *Mol Biol Cell*, 18, 536, 2007.

107. Batsche, E., et al., The human Swi/Snf subunit BRM is a regulator of alternative splicing, *Nat Struct Mol Biol*, 13, 22, 2006.

108. Ben-Saadon, R., et al., The polycomb protein RING1B generates self atypical mixed ubiquitin chains required for its in vitro histone H2a ligase activity, *Mol Cell*, 24, 701, 2006.

109. Cha, T.L., et al., AKT-mediated phosphorylation of EZH2 suppresses methylation of lysine 27 in histone h3, *Science*, 310, 306, 2005.

110. Dietrich, N., et al., Bypass of senescence by the polycomb group protein cbx8 through direct binding to the INK4A-ARF locus, *Embo J*, 26, 1637, 2007.

111. Sanchez, C., et al., Proteomic analysis of Ring1B/Rnf2 interactors identifies a novel complex with the Fbxl10/ Jhdm1B histone demethylase and the Bcl6 interacting corepressor, *Mol Cell Proteomics*, 6, 820, 2007.

112. Voncken, J.W., et al., Mapkap kinase 3pk phosphorylates and regulates chromatin-association of the polycomb-group protein BMI1, *J Biol Chem*, 280, 5178, 2005.

113. Dudley, N.R., et al., Using RNA interference to identify genes required for RNA interference, *Proc Natl Acad Sci U S A*, 99, 4191, 2002.

114. Kavi, H.H., et al., RNA silencing in *Drosophila*, *FEBS Lett*, 579, 5940, 2005.

115. Kidner, C.A. and Martienssen, R.A., The role of argonaute1 (ago1) in meristem formation and identity, *Dev Biol*, 280, 504, 2005.

116. Grimaud, C., et al., RNAI components are required for nuclear clustering of polycomb group response elements, *Cell*, 124, 957, 2006.

117. Akhtar, A., et al., Chromodomains are protein-RNA interaction modules, *Nature*, 407, 405, 2000.

118. Lachner, M., et al., Methylation of histone H3 lysine 9 creates a binding site for hp1 proteins, *Nature*, 410, 116, 2001.

119. Maison, C., et al., Higher-order structure in pericentric heterochromatin involves a distinct pattern of histone modification and an RNA component, *Nat Genet*, 30, 329, 2002.

120. Schlesinger, Y., et al., Polycomb-mediated methylation on lys27 of histone h3 pre-marks genes for de novo methylation in cancer, *Nat Genet*, 39, 232, 2007.

121. Widschwendter, M., et al., Epigenetic stem cell signature in cancer, *Nat Genet*, 39, 157, 2007.

122. Egger, G., et al., Epigenetics in human disease and prospects for epigenetic therapy, *Nature*, 429, 457, 2004.

123. Jacobs, J.J., et al., The oncogene and polycomb-group gene Bmi-1 regulates cell proliferation and senescence through the INK4A locus, *Nature*, 397, 164, 1999.

124. Sherr, C.J., The INK4A/arf network in tumour suppression, *Nat Rev Mol Cell Biol*, 2, 731, 2001.

125. Jacobs, J.J., et al., Bmi-1 collaborates with c-myc in tumorigenesis by inhibiting c-myc-induced apoptosis via INK4A/ARF, *Genes Dev*, 13, 2678, 1999.

126. Breuer, R.H., et al., Expression of the p16(INK4A) gene product, methylation of the p16(INK4A) promoter region and expression of the polycomb-group gene Bmi-1 in squamous cell lung carcinoma and premalignant endobronchial lesions, *Lung Cancer*, 48, 299, 2005.

127. Raaphorst, F.M., et al., Site-specific expression of polycomb-group genes encoding the hpc–hph/PRC1 complex in clinically defined primary nodal and cutaneous large B-cell lymphomas, *Am J Pathol*, 164, 533, 2004.

128. Guo, W.J., et al., Mel-18 acts as a tumor suppressor by repressing BMI-1 expression and down-regulating AKT activity in breast cancer cells, *Cancer Res*, 67, 5083, 2007.

129. Bachmann, I.M., et al., Ezh2 expression is associated with high proliferation rate and aggressive tumor subgroups in cutaneous melanoma and cancers of the endometrium, prostate, and breast, *J Clin Oncol*, 24, 268, 2006.

130. Collett, K., et al., Expression of enhancer of zeste homologue 2 is significantly associated with increased tumor cell proliferation and is a marker of aggressive breast cancer, *Clin Cancer Res*, 12, 1168, 2006.

131. Kleer, C.G., et al., Ezh2 is a marker of aggressive breast cancer and promotes neoplastic transformation of breast epithelial cells, *Proc Natl Acad Sci U S A*, 100, 11606, 2003.

132. Raaphorst, F.M., et al., Poorly differentiated breast carcinoma is associated with increased expression of the human polycomb group EZH2 gene, *Neoplasia*, 5, 481, 2003.

133. Raaphorst, F.M., et al., Coexpression of bmi-1 and EZH2 polycomb group genes in reed-sternberg cells of Hodgkin's disease, *Am J Pathol*, 157, 709, 2000.

134. Visser, H.P., et al., The polycomb group protein ezh2 is upregulated in proliferating, cultured human mantle cell lymphoma, *Br J Haematol*, 112, 950, 2001.

135. Sanchez-Beato, M., et al., Variability in the expression of polycomb proteins in different normal and tumoral tissues. A pilot study using tissue microarrays, *Mod Pathol*, 19, 684, 2006.

136. Varambally, S., et al., The polycomb group protein EZH2 is involved in progression of prostate cancer, *Nature*, 419, 624, 2002.

137. Chowdhury, M., et al., Expression of polycomb-group (PcG) protein BMI-1 predicts prognosis in patients with acute myeloid leukemia, *Leukemia*, 2007.

138. Matsukawa, Y., et al., Expression of the enhancer of zeste homolog 2 is correlated with poor prognosis in human gastric cancer, *Cancer Sci*, 97, 484, 2006.

139. van Galen, J.C., et al., Expression of the polycomb-group gene BMI1 is related to an unfavourable prognosis in primary nodal dlbcl, *J Clin Pathol*, 60, 167, 2007.

140. Ding, L., et al., Identification of EZH2 as a molecular marker for a precancerous state in morphologically normal breast tissues, *Cancer Res*, 66, 4095, 2006.

141. Ding, L. and Kleer, C.G., Enhancer of zeste 2 as a marker of preneoplastic progression in the breast, *Cancer Res*, 66, 9352, 2006.

142. Glinsky, G.V., et al., Microarray analysis identifies a death-from-cancer signature predicting therapy failure in patients with multiple types of cancer, *J Clin Invest*, 115, 1503, 2005.

143. Beke, L., et al., The gene encoding the prostatic tumor suppressor psp94 is a target for repression by the polycomb group protein EZH2, *Oncogene*, 26, 4590, 2007.

144. Oei, S.L. and Shi, Y., Transcription factor yin yang 1 stimulates poly(adp-ribosyl)ation and DNA repair, *Biochem Biophys Res Commun*, 284, 450, 2001.

145. Sui, G., et al., Yin yang 1 is a negative regulator of p53, *Cell*, 117, 859, 2004.

146. Campisi, J., Senescent cells, tumor suppression, and organismal aging: Good citizens, bad neighbors, *Cell*, 120, 513, 2005.

147. Dimri, G.P., What has senescence got to do with cancer? *Cancer Cell*, 7, 505, 2005.

148. Narita, M. and Lowe, S.W., Senescence comes of age, *Nat Med*, 11, 920, 2005.

149. Gil, J., et al., Polycomb cbx7 has a unifying role in cellular lifespan, *Nat Cell Biol*, 6, 67, 2004.

150. Bracken, A.P., et al., The polycomb group proteins bind throughout the ink4a-arf locus and are disassociated in senescent cells, *Genes Dev*, 21, 525, 2007.

151. Kotake, Y., et al., Prb family proteins are required for h3k27 trimethylation and polycomb repression complexes binding to and silencing p16ink4a tumor suppressor gene, *Genes Dev*, 21, 49, 2007.

152. Dimri, G.P., et al., The Bmi-1 oncogene induces telomerase activity and immortalizes human mammary epithelial cells, *Cancer Res*, 62, 4736, 2002.

153. Itahana, K., et al., Control of the replicative life span of human fibroblasts by p16 and the polycomb protein Bmi-1, *Mol Cell Biol*, 23, 389, 2003.

154. Kim, R.H., et al., Bmi-1 cooperates with human papillomavirus type 16 e6 to immortalize normal human oral keratinocytes, *Exp Cell Res*, 313, 462, 2007.

155. Al-Hajj, M. and Clarke, M.F., Self-renewal and solid tumor stem cells, *Oncogene*, 23, 7274, 2004.

156. Liu, L., et al., Loss of the human polycomb group protein BMI1 promotes cancer-specific cell death, *Oncogene*, 25, 4370, 2006.

157. van Kemenade, F.J., et al., Coexpression of Bmi-1 and EZH2 polycomb-group proteins is associated with cycling cells and degree of malignancy in B-cell non-Hodgkin lymphoma, *Blood*, 97, 3896, 2001.

158. Arisan, S., et al., Increased expression of EZH2, a polycomb group protein, in bladder carcinoma, *Urol Int*, 75, 252, 2005.

159. Raman, J.D., et al., Increased expression of the polycomb group gene, EZH2, in transitional cell carcinoma of the bladder, *Clin Cancer Res*, 11, 8570, 2005.

160. Weikert, S., et al., Expression levels of the EZH2 polycomb transcriptional repressor correlate with aggressiveness and invasive potential of bladder carcinomas, *Int J Mol Med*, 16, 349, 2005.

161. Mimori, K., et al., Clinical significance of enhancer of zeste homolog 2 expression in colorectal cancer cases, *Eur J Surg Oncol*, 31, 376, 2005.

162. Sudo, T., et al., Clinicopathological significance of EZH2 mRNA expression in patients with hepatocellular carcinoma, *Br J Cancer*, 92, 1754, 2005.
163. Berezovska, O.P., et al., Essential role for activation of the polycomb group (PcG) protein chromatin silencing pathway in metastatic prostate cancer, *Cell Cycle*, 5, 1886, 2006.
164. Saramaki, O.R., et al., The gene for polycomb group protein enhancer of zeste homolog 2 (EZH2) is amplified in late-stage prostate cancer, *Genes Chromosomes Cancer*, 45, 639, 2006.
165. van Leenders, G.J., et al., Polycomb-group oncogenes EZH2, BMI1, and RING1 are overexpressed in prostate cancer with adverse pathologic and clinical features, *Eur Urol*, 52, 455, 2007.
166. Mattioli, E., et al., Immunohistochemical analysis of prb2/p130, vegf, EZH2, p53, p16(INK4A), p27(kip1), p21(waf1), ki-67 expression patterns in gastric cancer, *J Cell Physiol*, 210, 183, 2007.
167. Kirmizis, A., et al., Identification of the polycomb group protein su(z)12 as a potential molecular target for human cancer therapy, *Mol Cancer Ther*, 2, 113, 2003.
168. Scott, C.L., et al., Role of the chromobox protein CBX7 in lymphomagenesis, *Proc Natl Acad Sci U S A*, 104, 5389, 2007.
169. Silva, J., et al., Implication of polycomb members Bmi-1, mel-18, and hpc-2 in the regulation of p16INK4a, P14ARF, h-tert, and c-myc expression in primary breast carcinomas, *Clin Cancer Res*, 12, 6929, 2006.
170. Kim, J.H., et al., The BMI-1 oncoprotein is overexpressed in human colorectal cancer and correlates with the reduced p16INK4a/p14ARF proteins, *Cancer Lett*, 203, 217, 2004.
171. Dutton, A., et al., BMI-1 is induced by the Epstein-Barr virus oncogene lmp1 and regulates the expression of viral target genes in Hodgkin lymphoma cells, *Blood*, 109, 2597, 2007.
172. Sawa, M., et al., BMI-1 is highly expressed in M0-subtype acute myeloid leukemia, *Int J Hematol*, 82, 42, 2005.
173. Bea, S., et al., BMI-1 gene amplification and overexpression in hematological malignancies occur mainly in mantle cell lymphomas, *Cancer Res*, 61, 2409, 2001.
174. Nowak, K., et al., BMI1 is a target gene of e2f-1 and is strongly expressed in primary neuroblastomas, *Nucleic Acids Res*, 34, 1745, 2006.
175. Vonlanthen, S., et al., The BMI-1 oncoprotein is differentially expressed in non-small cell lung cancer and correlates with ink4a-arf locus expression, *Br J Cancer*, 84, 1372, 2001.
176. Kang, M.K., et al., Elevated Bmi-1 expression is associated with dysplastic cell transformation during oral carcinogenesis and is required for cancer cell replication and survival, *Br J Cancer*, 96, 126, 2007.
177. Tokimasa, S., et al., Lack of the polycomb-group gene rae28 causes maturation arrest at the early B-cell developmental stage, *Exp Hematol*, 29, 93, 2001.
178. Hayflick, L., The limited in vitro lifetime of human diploid cell strains, *Exp Cell Res*, 37, 614, 1965.

17 Cancer and Aging: The Epigenetic Connection

Craig A. Cooney

CONTENTS

17.1   INTRODUCTION

Epigenetics relies on enzymatic modifications of DNA and histones. The epigenetic modification of DNA in mammals is mainly or entirely enzymatic methylation of some cytosines $5'$ to guanines. Histone modification is more complex and involves several enzymatic covalent additions of which methylation and acetylation are discussed here. The substrates for these modifications come from metabolism and may, to various degrees, respond to environmental influences such as diet and stress-related changes in metabolic regulation. Methylation of DNA nearly always results in gene silencing; however, methylation of histones can promote either gene silencing or activation and thus influences on epigenetics through methyl metabolism may be constrained by these competing influences on histone methylation. Histone acetylation nearly always results in gene activation and thus the effects of acetyl metabolism on epigenetics are potentially stronger because they act in one direction.

17.2   EPIGENETIC MECHANISMS

All eukaryotes studied including *Saccharomyces cerevisiae*, *Caenorhabditis elegans*, *Drosophila melanogaster*, Arabidopsis, mouse, and human use histone modifications as epigenetic mechanisms. Many eukaryotes, including vertebrates, also use DNA methylation to silence genes.

The promoter enhancer regions of many genes contain CpG sequences, which are targets for cytosine methylation by DNA methyltransferases (Dnmts) that use *S*-adenosylmethionine (SAM) as the methyl donating substrate [1]. Regions rich in CpG sequences are called CpG islands. Methylation of DNA nearly always results in gene silencing which is facilitated by the binding of several proteins to DNA and chromatin. These proteins include methylated DNA binding proteins (e.g., MBD proteins and MeCP2), heterochromatin protein 1 (HP1) and specific histone methyltransferases (HMTs) and histone deacetylases (HDACs) [2,3]. This chromatin structure helps maintain an inactive chromatin state and gene silencing.

Methylation of histones can either promote gene silencing (e.g., methylation of histone H3 lysines 9 or 27) or gene activation (e.g., methylation of histone H3 lysines 4 or 79). Methylation of histone H3 lysine 9 is performed by specific histone methyltransferases including Suv39h and ESET that also use SAM as methyl donor and whose action promotes gene silencing. Methylation of these sites is recognized by specific proteins that contribute to an overall inactive chromatin structure [3,4]. On lysine 9 of H3, methylation (which promotes silencing) is in competition with acetylation (which promotes activity) [5]. Methylation of histone H3 lysine 4 leads to the binding of a protein complex that includes a histone acetyltransferase (HAT). This complex uses acetyl-coenzyme A (AcCoA) as acetate donor to acetylate histone H3 and promote transcription [6].

Histone acetylation uses AcCoA and promotes transcription by changing the charge on lysines from positive to uncharged and reducing their interaction with negatively charged DNA. Histone acetylation eliminates H3 lysine 9 as a substrate for methylation and is recognized by binding proteins that promote transcription. Acetylation is recognized by HATs, which continue acetylating and spreading the active chromatin state. This spreading may occur over a chromatin domain but stop at chromatin boundaries [7] leading to "acetylation islands" of active chromatin [8,9]. Histone acetylation is readily reversible by histone deacetylases (HDACs and SIRTs), which promote silencing.

There are additional important histone modifications including phosphorylation, ubiquitination, and sumoylation (reviewed by Kouzarides) [3]. Because in many, if not all cases, gene activity can be switched by affecting DNA methylation or histone acetylation, these are the focus of this chapter.

17.3   AGING EPIGENETICS

Most epigenetic studies with aging have focused on DNA methylation. In nearly all studies, global levels of DNA methylation decline with age in mammalian tissues (reviewed by Cooney) [10]. In a few studies, these levels remain unchanged. Likewise, in most cases methylation of repetitive sequences declines with age. For example, long terminal repeats (LTRs) of endogenous retroviruses (ERVs) often show methylation loss with aging [11,12].

DNA methylation of single copy genes sometimes decreases with age but most often increases with age. A greater degree of DNA methylation is found in several genes including the IGF2 promoter and MYOD1 in normal colon of humans of greater age [13–17]. This is consistent with increased silencing (and a loss of imprinting) in IGF2 with age [15]. IGF2 is further methylated in several human cancers supporting a model of an evolving methylation process from aging to cancer [15]. The estrogen receptor alpha gene is more highly methylated in human colon [14] and right atrium [18] in old than in young subjects. This gene is also methylated in human colon cancer suggesting that its methylation in aging is an early step in cancer development [14]. The estrogen receptor-β gene is also more highly methylated in atherosclerotic compared to normal vascular tissue suggesting epigenetic dysregulation with age [19].

Issa and coworkers [20] showed that a substantially different set of genes is hypermethylated in aging prostate than in aging colon and that these are some of the same genes commonly hypermethylated in prostate cancer (e.g., RASSF1A and GSTP1). The methylation levels on these genes in individual subjects were higher in cancerous prostate than in aged normal prostate sections, suggesting a progressive process. Some genes such as MYOD1 that are silenced in aging colon and in colon cancer [13] were not silenced in prostate suggesting that gene hypermethylation in both cancer and aging are tissue specific. Issa and coworkers [20] suggest that age-related hypermethylation may precede and even predispose subjects to prostate cancer.

In a few cases, there are losses of DNA methylation in specific genes with age (in the Huntington's gene [21], and the ITGAL gene [22]. In each case this demethylation is associated with disease (Huntington's disease and systemic lupus erythematosus, respectively).

Monozygotic twins are genetically identical and yet these twins often differ in various ways including disease (phenotypic discordance). Fraga et al. [23] tested monozygotic twins extensively using assays for global DNA methylation, gene-specific DNA methylation, global histone acetylation, gene expression, and others. They found that younger twins had substantial concordance whereas older twins had significantly less concordance. A separate study of 5 year old monozygotic twins showed substantial differences in DNA methylation of the catechol-O-methyltransferase gene from buccal cells in two sets of twins [24] demonstrating that some young twins already show epigenetic differences (discordance). Thus, epigenetic discordance in monozygotic twins may be the result of aging but may exist from a much younger age or be the result of environmental factors [25,26].

Only a few studies have looked at histone acetylation and aging. Howard and coworkers [27,28] mapped histone acetylation levels in young and old adults over a one megabase region. They found higher acetylation levels at several locations in the young adults.

17.4 CANCER EPIGENETICS

Epigenetic changes with aging may contribute to cancer development, as do age-related mutations and chromosome damage. However, these age-related changes seem modest compared to the radically altered epigenetics and chromosome arrangements found in most cancers [14,15,29–31]. Thus, there is an early evolution in cancer where at least some of these radical changes occur or where the conditions are set up for their subsequent occurrence.

Some cancers clearly do not have an aging component and yet have altered epigenetics [32,33]. These cancers raise the possibility that epigenetic errors or epigenetic patterns left over from earlier development may contribute to cancer development later in life. Similarly, aspects of stem cell epigenetics may contribute to cancer development [34]. Nevertheless, it appears that some age-related gene-specific hypermethylation and global hypomethylation set the stage for cancer development.

All cancers studied show global hypomethylation that is more pronounced than found in the corresponding normal tissue, even when that normal tissue is aged [29,31]. While global methylation changes may be important they have contributed relatively little to understanding or treating cancer. In stark contrast to global loss of methylation, there is hypermethylation, often to an extreme, on the promoter and early transcribed regions of many genes in all cancers studied. These are genes often used by normal cells for cell cycle control, DNA repair, apoptosis, and other functions. Hypermethylation silences these genes that presumably do not serve cancer cell survival and evolution [30].

The number of genes silenced in each tumor is uncertain but a recent estimate puts this at 100–400 [30]. Not all genes have been surveyed and this number may go higher as epigenomic methods are used to screen all genes (or at least all genes with CpG islands). The utility of gene silencing in cancer varies. The silencing of some genes seems necessary for cancer cell survival, proliferation, or metastasis while others may facilitate the evolution of cancer cells and yet others may be silenced by some, as yet undefined, general silencing mechanism but not contribute to cancer survival or progression. Higuchi and coworkers [35] have established a relationship between mitochondrial DNA depletion and nuclear gene silencing in prostate cancer.

In contrast to widespread gene silencing, a relatively few genes show cancer specific hypomethylation. These include synuclein (in several cancers) [36], hexokinase (in a hepatoma cell line) [37], CAGE (in several cancers) [38], and MYOD1 (in rhabdomyosarcoma) [32]. In some cases these genes have apparent functions in cancer and their activity in the face of widespread gene silencing would also indicate that they contribute to cancer survival or progression.

Cancers also show changes in histone modifications compared to normal cells and tissues. A small proportion of prostate cancer patients have fast growing cancer that proves metastatic and fatal. In contrast, most prostate cancer is slow growing, nonmetastatic and a low health risk [39]. Global losses in histone acetylation and histone methylation have been demonstrated by immuno-histochemistry and used for the early identification of these high-risk patients [40]. A wide range of normal human tissues and cancer cells have been studied for histone H4 modifications by Fraga et al. [41]. They studied cancer cell lines as well as fresh tumors and corresponding normal cell lines and normal tissues. They found that cancers were always depleted in monoacetylated H4 and trimethylated H4 and that cancers often had levels of these modifications that were about half of normal. Specifically, they found this loss in H4K20 for trimethylation and primarily in H4K16 for acetylation. As with global losses of DNA methylation, they determined that the bulk of histone modification losses were in repetitive sequences including pericentromeric and subtelomeric sequences. Fraga et al. consider the global loss of these two H4 modifications to be nearly universal epigenetic markers for cancer as are global DNA hypomethylation and CpG island hypermethylation [41].

As with other genome regions, telomeres and subtelomeres are epigenetically modified [42,43]. Telomeres generally grow shorter with aging in many somatic cells due to chromosome duplication without telomere extension or due to telomere damage. The main modifications associated with long telomeres in somatic cells are those characteristic of constitutive heterochromatin. While vertebrate telomere repeats (TTAGGG) do not contain CpG sites they have extensive methylation of histones H3 (trimethylation at lysine 9, K9me3) and H4 (trimethylation at lysine 20, K20me3). Subtelomeres have these same histone modifications as well as DNA methylation. Both regions bind multiple isoforms of HP1 and show hypoacetylation of histones H3 and H4. DNA methylation [44] and H3K9me3 and H4K20me3 are negative regulators of telomere length and telomere recombination [45]. Such closed chromatin conformations are thought to block access to telomere recombination and telomerase and thus block telomere extension.

Short telomeres and their subtelomeres have a more active chromatin structure including acetylation of histones H3 and H4 [46], which suggests that they may be available for elongation by telomerase and by telomere recombination. Such elongation could be a permissive step for the development of cancer. In fact, cancer cells often express telomerase, extend telomeres, and have altered epigenetic control consistent with an active chromatin conformation at telomeres and subtelomeres [42,43].

In gene-specific work, Dahiya and coworkers [47] studied histone modifications of the GSTP1 promoter in human prostate cells in tissue culture (LNCaP cancer cells and virally transformed "normal" adult prostate epithelial cells). They showed that gene-activating H3 and H4 acetylation and H3K4 methylation were present in the GSTP1 gene of the normal cells but were lost in the LNCaP cells. They suggested that loss of H3 and H4 acetylation and H3K4 methylation occurred in the process of prostate carcinogenesis. The silenced GSTP1 promoter of LNCaP cells was heavily methylated on DNA but not on H3K9 or H3K27. They found some similar results with the RASSF1A gene [48]. Using another in vitro prostate cancer model, Clark and coworkers [49] suggested that DNA methylation changes preceded histone acetylation changes in GSTP1 silencing. GSTP1 is expressed and its DNA is hypomethylated in normal prostate tissue. It is silenced and its DNA is hypermethylated in prostate cancer and prostate cancer cell lines [50–53].

Another study shows that RASSF1A gene expression levels and promoter histone acetylation are lost and H3K9 methylation gained before DNA hypermethylation occurs in gene promoters as normal human mammary epithelial cells are passaged [54]. In the MCF7 breast cancer cell line, the RASSF1A gene is silenced and its promoter is DNA hypermethylated. Strunnikova et al. [55] proposed that RASSF1A silencing in senescent cells occurred primarily by histone methylation and deacetylation (and not DNA methylation) [54]. Other studies looking only at DNA methylation show RASSF1A methylation in hyperplastic and cancer tissue but not in normal breast tissue.

It may be that the type of epigenetic modification that changes first in silencing depends on the cellular metabolic state and the signals received by the cells. This has been shown by Meaney and Szyf [56], Pruitt et al. [57], and others where signals acting on either DNA methylation or on histone acetylation can affect the epigenetic state.

17.5 DIET AND OTHER ENVIRONMENTAL FACTORS AFFECT EPIGENETICS

In early development, epigenetics is affected by the diet, metabolism and behavior of the mother. Maternal diet during pregnancy affects epigenetics in mice [58–62]. Other studies clearly show that maternal behavior of rats in the first neonatal week affects the epigenetics of offspring [56]. These studies use specific genes such as agouti or the glucocorticoid receptor. Numerous studies in rats and mice also show an apparent epigenetic effect of maternal glucose intake or maternal acquired diabetes on the glucose levels and diabetes of subsequent generations of offspring [63,64]. However, no particular genes or their epigenetic modifications have been identified to explain these effects of maternal glucose on offspring diabetes.

In adult rats and mice two different groups show that gene silencing and DNA methylation can be promoted in vivo by intraventricular injection of methionine and that gene activation can be promoted by intraventricular injection of HDAC inhibitors (such as trichostatin or valproic acid) [56,65]. These studies show that the silenced or active states can be produced by manipulating metabolism to change DNA methylation or using small molecule enzyme inhibitors to prevent histone deacetylation, respectively. These studies are consistent with models where at least some genes are in a transcriptionally "poised" state where they may have ongoing turnover of histone modifications and where they can be turned on or off based on the availability of metabolites for histone or DNA modification [56,66]. Aside from these studies there are few data about specific metabolic effects on epigenetic regulation of specific genes in adults or with aging [25,67].

17.6 EFFECTS OF METABOLISM ON EPIGENETICS

Methylation of DNA nearly always results in gene silencing; however, methylation of histones can either promote gene silencing (e.g., methylation of histone H3 lysine 9) or gene activation (e.g., methylation of histone H3 lysine 4). Thus, dietary and metabolic influences on epigenetics through methyl metabolism may be constrained by these competing influences of different methylation sites on histones. In contrast, histone acetylation nearly always results in gene activation and thus the effects of acetyl metabolism on epigenetics are potentially stronger.

Epigenetics relies on common metabolic intermediates such as SAM for DNA and histone methylation and AcCoA for histone acetylation and may respond to environmental influences through the availability of SAM, AcCoA, and related metabolic intermediates. Potential influences include diet and stress-related changes in metabolic regulation. For example, SAM and methyl metabolism are dependent on diet because many of the key intermediaries are essential nutrients (e.g., methionine, folate) (Figure 17.1) [4,10,68]. They are also dependent on oxidative stress because cellular requirements for cysteine and glutathione are derived, in part, through thiols derived from methionine [25,69].

There are two main sources of acetate in metabolism. One is from glycolysis by the action of pyruvate dehydrogenase (PDH) (pyruvate and CoA-SH and $NAD+ \rightarrow$ AcCoA and CO_2 and NADH) and the other is from β-oxidation of fatty acids (Figure 17.2). Some AcCoA can also come from catabolism of some amino acids and from the metabolism of ethanol. AcCoA also comes from free acetate and CoA-SH by the actions of AcCoA synthetases (ACSs). Two ACSs have been identified in mammals. One is cytoplasmic (ACS1) and the other is mitochondrial (ACS2) [70,71]. ACS1 is the likely source of AcCoA for histone acetylation (which occurs in the nucleus). ACS1 is regulated by acetylation and is activated by the deacetylase action of

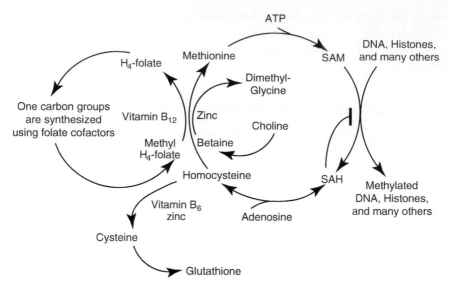

FIGURE 17.1 Methyl metabolism. *S*-adenosylmethionine (SAM) is the methyl donor used for enzymatic methylation of DNA, histones and other molecules. Production of SAM and recycling of *S*-adenosylhomocysteine (SAH) are dependent on methyl metabolism. Many components of this metabolism are essential nutrients or provide alternative metabolic pathways to make SAM and recycle SAH. (Adapted from Cooney, C. A., *Dis. Markers*, 23, 121, 2007.)

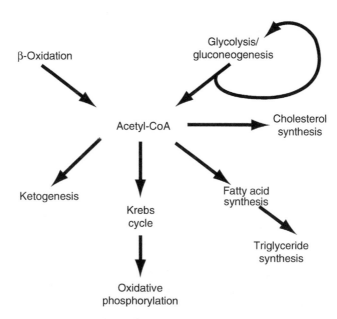

FIGURE 17.2 Acetate metabolism. Acetyl-CoA is the acetyl donor for enzymatic acetylation of histones and other molecules. Glucose and fats can be completely metabolized via acetyl-CoA, the TCA cycle and oxidative phosphorylation. However, when glucose is abundant or mitochondrial function is compromised, cells can rely on glycolysis and avoid production of acetyl-CoA. (Adapted from Kennedy, A.R. et al., *Am. J. Physiol. Endocrinol. Metab.*, 292, E1724, 2007.)

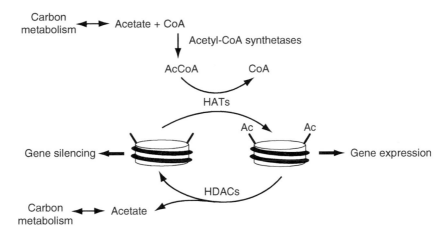

FIGURE 17.3 Acetate is a common currency of the major pathways of carbon and energy metabolism. The availability of acetate may provide a readout of the energy and metabolic state of the cell and may directly affect gene expression via histone acetylation. (Adapted from Takahashi, H. et al., *Mol. Cell*, 23, 207, 2006.)

SIRT1 [72,73]. Takahashi et al. [74] have suggested that ACSs and AcCoA link eukaryotic carbon metabolism and histone acetylation (Figure 17.3).

AcCoA levels are highly dependent on energy metabolism and metabolic regulation. For example glycolysis leading to lactate produces no acetate whereas glycolysis leading to oxidative phosphorylation utilizes AcCoA as does the β-oxidation of fatty acids (Figure 17.2) [75]. Metabolic extremes are seen in cancer cells that are typically glycolytic [76,77] and which extensively silence genes [30]. An emphasis on glycolysis leading to lactate is also found in normal cells under circumstances such as glucose abundance and rapid proliferation [78].

17.6.1 Glucose May Effect Long-Term Changes in Gene Expression in Individuals

Mobbs et al. [79] has proposed that high blood glucose concentrations and induction of gene expression for the glycolytic enzymes results in glucose hysteresis where a memory is developed for glycolytic enzyme expression. This would be a form of transcriptional reprogramming and may be a type of epigenetic reprogramming in the adult; however, this has yet to be determined.

There are few data on the epigenetics of glycolytic genes although epigenetic down regulation of type II hexokinase has been reported for hepatocytes compared to high gene activity in hepatoma [37]. Some studies indicate that the long-term expression of some genes is increased by transient exposure to glucose [80]. Some metabolic states that affect glucose utilization, including caloric restriction, result in substantial changes in gene expression [81–86]. Mobbs and coworkers [87,88] showed that hypoglycemia produces gene expression patterns similar to those produced by caloric restriction where metabolism is shifted away from glycolysis and toward β-oxidation, the pentose pathway and protein turnover.

The coordinated induction of genes for glycolysis or fatty acid oxidation have been described for specific tissues including heart, pancreas [89], liver [90,91], and muscle [92]. Glucose levels and glucose signaling are coupled to the action of transcription factors such as FOXO1 and ChREBP [85,90,93–96].

The amount of glucose taken up, and presumably the amount of glucose signaling, would be much greater if glucose is used just to make lactate and ATP by glycolysis because this process is very energy inefficient. Much less glucose uptake and signaling would be required if glucose is used for the production of pyruvate, AcCoA, and subsequent oxidative phosphorylation to form CO_2 and water. Even less glucose would be required if most energy were generated by fatty acid β-oxidation to AcCoA followed by TCA cycle and oxidative phosphorylation.

The persistent effects proposed by Mobbs et al. [79] could be explained by transcription factors (such as FOXO1 and ChREBP) that continue to have effects once the original signaling has dissipated. Such persistent effects could also be due to epigenetic changes such as histone modification or DNA methylation that maintain active or silent states of genes in the normal cells of animals exposed to high or low glucose levels.

17.6.2 GLUCOSE PRODUCES LONG-TERM CHANGES IN PHENOTYPE OVER ANIMAL GENERATIONS

During pregnancy in rats or mice, high glucose can induce effects in offspring that lead to the inheritance of diabetes or related disorders to one or more subsequent generations. This occurs in strains of rats and mice not considered to have a genetic susceptibility to diabetes.

Diabetes induced in female rats with drugs such as alloxan or streptozotocin can be passed to the offspring even though the offspring never receive the drugs. In some cases hyperglycemia or diabetes is maternally inherited for several generations [97–100]. Similar phenotypes have been induced in mice over multiple generations [101].

Glucose infusions in the last week of pregnancy (third trimester) will also affect rat offspring [64,102]. When adult offspring (F1) were compared with adult offspring from control dams (infused with a glucose-free solution), those from hyperglycemic mothers had mild glucose intolerance and impaired insulin secretion. This worsened with age to basal hyperglycemia and severe glucose intolerance. F2 newborns of F1 hyperglycemic dams were also hyperglycemic, hyperinsulinemic, and demonstrated fetal overgrowth. They later developed basal hyperglycemia and defective glucose tolerance and insulin secretion. These studies show that maternal glucose intake in pregnancy can produce heritable diabetic states in the offspring.

Women with gestational diabetes are significantly more likely to have mothers with type II diabetes than to have fathers with type II diabetes. They are also more likely to have grandmothers with type II diabetes than to have grandfathers with type II diabetes. This data supports a model of maternal transgenerational inheritance of type II diabetes in humans analogous to that in rats and mice [103].

Recent studies provide a model in which maternal behavior affects offspring epigenetics, which affects the behavior of adult offspring. This pattern perpetuates both the epigenetic effects and the behaviors for generations [25,56]. A similar model may work with glucose handling and diabetes where high maternal glucose affects offspring epigenetics, which affects the glucose levels of adult offspring.

17.7   AGING, CALORIC RESTRICTION, AND EXERCISE

Caloric restriction (CR) extends lifespan in *S. cerevisiae, C. elegans, D. melanogaster*, fish, and some mammalian species including rats and mice compared to animals fed ad libitum (AL) [104]. CR acts to reduce or delay many effects of aging including the onset and progression of cancer, and the accumulation of damaged molecules [104,105]. CR maintains lower plasma glucose and insulin levels in rats and mice [106,107].

When glucose levels are high they induce the expression of glycolytic enzymes and suppress the utilization of other fuels such as fatty acids [79]. Aging promotes glycolysis over fatty acid oxidation in AL fed mice but CR preserves fatty acid oxidation [84]. Fetal heart relies on glucose whereas healthy adult heart uses fatty acid oxidation. In human heart failure there is a switch back to reliance on glucose [108–110].

In normal human fibroblasts age-related declines in mitochondrial function necessitate a shift toward glycolysis and away from acetate production and oxidative phosphorylation [111,112]. In mutator mice somatic mitochondrial DNA mutations did not substantially affect reactive oxygen species production. This suggests that the premature aging phenotype of these mice is caused by defects in oxidative phosphorylation [113]. In humans, Melov et al. [114] showed that healthy older adults showed transcriptional profiles indicative of mitochondrial impairment in skeletal muscle compared to younger controls. After 6 months of exercise training, the transcriptional profiles were shifted substantially toward those of young adults.

Exacerbating a reliance on glucose is the almost exclusive use by some cells of glycolysis with conversion of most pyruvate to lactate (and without oxidative metabolism of pyruvate). Brand et al. [78] describe aerobic glycolysis by proliferating normal rat thymocytes where only 14% of ATP is produced by oxidation of glucose to CO_2 and water (most glucose goes to lactate). This is in contrast to 88% of ATP produced by oxidation of glucose to CO_2 and water in resting normal rat thymocytes. Cancer cells commonly use aerobic glycolysis with production of lactate with little oxidative phosphorylation [76].

17.8   HYPOTHESIS

Epigenetic regulation uses covalent modifications of macromolecules to achieve a combination of gene activation and gene silencing in each cell. Acetyl and methyl groups are two of the main covalent modifiers. While modifying enzymes, metabolic enzymes and various enzymes of signaling pathways can influence epigenetics, acetyl and methyl groups and their activated forms, AcCoA and SAM, can also be viewed as signals that represent the state of the cell.

To generate acetate, we need mitochondrial function to convert pyruvate to acetate (as AcCoA) or to perform β-oxidation of fatty acids. Some acetate can also come from amino acid catabolism and from alcohol metabolism. While it might seem that acetate is a ubiquitous metabolic intermediate and not a good candidate for metabolic regulation of epigenetics, as discussed above, acetate metabolism is quite variable among cells in different metabolic states. These metabolic states are, in turn, dependent on diet composition and caloric intake, two environmental variables for nearly all organisms.

It is hypothesized that cellular acetate levels influence the level of histone acetylation and the level of gene expression. These effects are mediated by SIRTs, ACSs, HATs, acetate shuttling pathways etc. Acetate serves as a signal that energy is available and acetate promotes gene expression and growth. Acetate provides the means to keep active genes active through histone acetylation and thus acetate can help preserve the epigenetic status quo. If acetate levels rise they will tend to increase the activity of genes or activate new genes. If acetate levels fall they would tend to lower the activity of genes or silence some currently active genes. A change in acetate levels that persists may change epigenetics by affecting other epigenetic controls. For example, DNA methylation may be decreased through high acetate levels promoting an active chromatin state that is inconsistent with the maintenance of DNA methylation. Alternatively, persistently low acetate levels will not support an active state in many genes and will thus allow silencing mechanisms including DNA methylation to predominate and silence the genes.

When glucose is abundant at least some cells rely mainly on glycolysis terminating in lactate (which is exported). This metabolism blocks fatty acid oxidation and most oxidative phosphorylation. Under these conditions acetate is likely to be in short supply. Persistent high glucose may result in a trend of low histone acetylation and ectopic gene silencing. This may be one facet of the "cumulative toxic effect of glucose" described by Mobbs et al. [79]. Similarly mitochondrial damage with aging may result in little acetate production and little oxidative phosphorylation. This state may be irreversible and also result in low histone acetylation and gene silencing.

Both of these situations occur in cancer cells most of which rely on glycolysis ending in exported lactate and have dysfunctional mitochondria not performing oxidative phosphorylation. In cancer these conditions are persistent. Thus, acetate is likely to be in short supply and there would be little histone acetylation and substantial gene silencing. It is proposed here that a shortage of acetate to maintain gene activity is a factor in maintaining the epigenetic state and phenotype of cancer cells.

It is proposed that cancer occurs, at least in part, because during long-term high glucose exposure or aging, epigenetic control is eroded because acetate levels are too low to maintain gene activity. Eventually ectopic gene silencing contributes to a combination of gene expression that produces cancer cells and allows for their selection and evolution.

ACKNOWLEDGMENTS

I thank Kimberly Cooney for drawing most of the figures. This work is supported by grants P01AG20641 from the NIA/NIH, R01AA016676 from the NIAAA/NIH and a grant from the Arkansas Biosciences Institute (Arkansas Tobacco Settlement Fund).

REFERENCES

1. Goll, M.G. and Bestor, T.H. Eukaryotic cytosine methyltransferases. *Annu. Rev. Biochem. 74*, 481, 2005.
2. Berger, J. and Bird, A. Role of MBD2 in gene regulation and tumorigenesis. *Biochem. Soc. Trans. 33*, (Pt 6), 1537, 2005.
3. Kouzarides, T. Chromatin modifications and their function. *Cell 128*, 693, 2007.
4. Huang, S. Histone methyltransferases, diet nutrients and tumour suppressors. *Nat. Rev. Cancer 2*, 469, 2002.
5. Kondo, Y. et al. Chromatin immunoprecipitation microarrays for identification of genes silenced by histone H3 lysine 9 methylation. *Proc. Natl. Acad. Sci. U S A 101*, 7398, 2004.
6. Taverna, S.D. et al. Yng1 PHD finger binding to H3 trimethylated at K4 promotes NuA3 HAT activity at K14 of H3 and transcription at a subset of targeted ORFs. *Mol. Cell 24*, 785, 2006.
7. Bulger, M. Hyperacetylated chromatin domains: Lessons from heterochromatin. *J. Biol. Chem. 280*, 21689, 2005.
8. Roh, T., Cuddapah, S. and Zhao, K. Active chromatin domains are defined by acetylation islands revealed by genome-wide mapping. *Genes. Dev. 19*, 542, 2005.
9. Roh, T. et al. Genome-wide prediction of conserved and nonconserved enhancers by histone acetylation patterns. *Genome Res. 17*, 74, 2007.
10. Cooney, C.A. Are somatic cells inherently deficient in methylation metabolism? A proposed mechanism for DNA methylation loss, senescence and aging. *Growth Dev. Aging 57*, 261, 1993.
11. Barbot, W. et al. Epigenetic regulation of an IAP retrotransposon in the aging mouse: Progressive demethylation and de-silencing of the element by its repetitive induction. *Nucl. Acids Res. 30*, 2365, 2002.
12. Mays-Hoopes, L. et al. Decreased methylation of the major mouse long interspersed repeated DNA during aging and in myeloma cells. *Dev. Genet. 7*, 65, 1986.
13. Ahuja, N. et al. Aging and DNA methylation in colorectal mucosa and cancer. *Cancer Res. 58*, 5489, 1999.
14. Issa, J.P. et al. Methylation of the oestrogen receptor CpG island links ageing and neoplasia in human colon. *Nat. Genet. 7*, 536, 1994.
15. Issa, J.P. et al. Switch from monoallelic to biallelic human IGF2 promoter methylation during agingáandácarcinogenesis. *PNAS 93*, 11757, 1996.
16. Nakagawa, H. et al. Age-related hypermethylation of the $5'$ region of MLH1 in normal colonic mucosa is associated with microsatellite-unstable colorectal cancer development. *Cancer Res. 61*, 6991, 2001.
17. Yatabe, Y., Tavare, S. and Shibata, D. Investigating stem cells in human colon by using methylation patterns. *Proc. Natl. Acad. Sci. U S A 98*, 10839, 2001.
18. Post, W.S. et al. Methylation of the estrogen receptor gene is associated with aging and atherosclerosis in the cardiovascular system. *Cardiovasc. Res. 43*, 985, 1999.
19. Kim, J. et al. Epigenetic changes in estrogen receptor [beta] gene in atherosclerotic cardiovascular tissues and in-vitro vascular senescence. *Biochim. Biophys. Acta (BBA)—Molecular Basis of Disease 1772*, 72, 2007.
20. Kwabi-Addo, B. et al. Age-related DNA methylation changes in normal human prostate tissues. *Clin. Cancer Res. 13*, 3796, 2007.
21. Reik, W. et al. Age at onset in Huntington's disease and methylation at D4S95. *J. Med. Genet. 30*, 185, 1993.

22. Zhang, Z. et al. Age-dependent DNA methylation changes in the ITGAL (CD11a) promoter. *Mech. Ageing Dev. 123*, 1257, 2002.

23. Fraga, M.F. et al. Epigenetic differences arise during the lifetime of monozygotic twins. *Proc. Natl. Acad. Sci. U S A 102*, 10604, 2005.

24. Mill, J. et al. Evidence for monozygotic twin (MZ) discordance in methylation level at two CpG sites in the promoter region of the catechol-*O*-methyltransferase (COMT) gene. *Am. J. Med. Genet. B Neuropsychiatr. Genet. 141*, 421, 2006.

25. Cooney, C.A. Epigenetics–DNA-based mirror of our environment? *Dis. Markers 23*, 121, 2007.

26. Poulsen, P. et al. The epigenetic basis of twin discordance in age-related diseases. *Pediatr. Res. 61*, 38R, 2007.

27. Russanova, V.R. et al. Mapping development-related and age-related chromatin remodeling by a high throughput ChIP–HPLC approach. *J. Gerontol. A Biol. Sci. Med. Sci. 59*, 1234, 2004.

28. Russanova, V.R. et al. Semirandom sampling to detect differentiation-related and age-related epigenome remodeling. *J. Gerontol. A Biol. Sci. Med. Sci. 59*, 1221, 2004.

29. Ehrlich, M. DNA methylation in cancer: Too much, but also too little. *Oncogene 21*, 5400, 2002.

30. Esteller, M. Cancer epigenomics: DNA methylomes and histone-modification maps. *Nat. Rev. Genet. 8*, 286, 2007.

31. Gama-Sosa, M.A. et al. The 5-methylcytosine content of DNA from human tumors. *Nucleic Acids Res. 11*, 6883, 1983.

32. Chen, B. et al. Methylation alterations of the MyoD1 upstream region are predictive of subclassification of human rhabdomyosarcomas. *Am. J. Pathol. 152*, 1071, 1998.

33. Kurmasheva, R.T. et al. Upstream CpG island methylation of the PAX3 gene in human rhabdomyosarcomas. *Pediatr. Blood Cancer 44*, 328, 2005.

34. Ohm, J.E. et al. A stem cell-like chromatin pattern may predispose tumor suppressor genes to DNA hypermethylation and heritable silencing. *Nat. Genet. 39*, 237, 2007.

35. Xie, C.H. et al. Mitochondrial regulation of cancer associated nuclear DNA methylation. *Biochem. Biophys. Res. Commun. 364*, 656, 2007.

36. Liu, H. et al. Loss of epigenetic control of synuclein-gamma gene as a molecular indicator of metastasis in a wide range of human cancers. *Cancer Res. 65*, 7635, 2005.

37. Goel, A., Mathupala, S.P. and Pedersen, P.L. Glucose metabolism in cancer. Evidence that demethylation events play a role in activating type II hexokinase gene expression. *J. Biol. Chem. 278*, 15333, 2003.

38. Cho, B. et al. Promoter hypomethylation of a novel cancer/testis antigen gene CAGE is correlated with its aberrant expression and is seen in premalignant stage of gastric carcinoma. *Biochem. Biophys. Res. Commun. 307*, 52, 2003.

39. Routh, J.C. and Leibovich, B.C. Adenocarcinoma of the prostate: Epidemiological trends, screening, diagnosis, and surgical management of localized disease. *Mayo Clin. Proc. 80*, 899, 2005.

40. Seligson, D.B. et al. Global histone modification patterns predict risk of prostate cancer recurrence. *Nature 435*, 1262, 2005.

41. Fraga, M.F. et al. Loss of acetylation at Lys16 and trimethylation at Lys20 of histone H4 is a common hallmark of human cancer. *Nat. Genet. 37*, 391, 2005.

42. Blasco, M.A. The epigenetic regulation of mammalian telomeres. *Nat. Rev. Genet. 8*, 299, 2007.

43. Fraga, M.F., Agrelo, R. and Esteller, M. Cross-talk between aging and cancer: The epigenetic language. *Ann. N.Y. Acad. Sci. 1100*, 60, 2007.

44. Gonzalo, S. et al. DNA methyltransferases control telomere length and telomere recombination in mammalian cells. *Nat. Cell Biol. 8*, 416, 2006.

45. Garcia-Cao, M. et al. Epigenetic regulation of telomere length in mammalian cells by the Suv39h1 and Suv39h2 histone methyltransferases. *Nat. Genet. 36*, 94, 2004.

46. Benetti, R., Garcia-Cao, M. and Blasco, M.A. Telomere length regulates the epigenetic status of mammalian telomeres and subtelomeres. *Nat. Genet. 39*, 243, 2007.

47. Okino, S.T. et al. Chromatin changes on the GSTP1 promoter associated with its inactivation in prostate cancer. *Mol. Carcinog. 46*, 839, 2007.

48. Kawamoto, K. et al. Epigenetic modifications of RASSF1A gene through chromatin remodeling in prostate cancer. *Clin. Cancer Res. 13*, 2541, 2007.

49. Stirzaker, C. et al. Transcriptional gene silencing promotes DNA hypermethylation through a sequential change in chromatin modifications in cancer cells. *Cancer Res. 64*, 3871, 2004.

50. Harden, S.V. et al. Quantitative GSTP1 methylation clearly distinguishes benign prostatic tissue and limited prostate adenocarcinoma. *J. Urol. 169*, 1138, 2003.

51. Henrique, R. and Jeronimo, C. Molecular detection of prostate cancer: A role for GSTP1 hypermethylation. *Eur. Urol. 46*, 660, 2004.

52. Li, L.C., Carroll, P.R. and Dahiya, R. Epigenetic changes in prostate cancer: Implication for diagnosis and treatment. *J. Natl. Cancer Inst. 97*, 103, 2005.

53. Nakayama, M. et al. GSTP1 CpG island hypermethylation as a molecular biomarker for prostate cancer. *J. Cell Biochem. 91*, 540, 2004.

54. Strunnikova, M. et al. Chromatin inactivation precedes de novo DNA methylation during the progressive epigenetic silencing of the RASSF1A promoter. *Mol. Cell. Biol. 25*, 3923, 2005.

55. Lehmann, U. et al. Quantitative assessment of promoter hypermethylation during breast cancer development. *Am. J. Pathol. 160*, 605, 2002.

56. Meaney, M.J. and Szyf, M. Maternal care as a model for experience-dependent chromatin plasticity? *Trends Neurosci. 28*, 456, 2005.

57. Pruitt, K. et al. Inhibition of SIRT1 reactivates silenced cancer genes without loss of promoter DNA hypermethylation. *PLoS. Genet. 2*, e40, 2006.

58. Cooney, C.A., Dave, A.A. and Wolff, G.L. Maternal methyl supplements in mice affect epigenetic variation and DNA methylation of offspring. *J. Nutr. 132* (Suppl 8), 2393S, 2002.

59. Cropley, J.E. et al. Germ-line epigenetic modification of the murine Avy allele by nutritional supplementation. *Proc. Natl. Acad. Sci. U S A 103*, 17308, 2006.

60. Dolinoy, D.C. et al. Maternal genistein alters coat color and protects Avy mouse offspring from obesity by modifying the fetal epigenome. *Environ. Health Perspect. 114*, 567, 2006.

61. Waterland, R.A. and Jirtle, R.L. Transposable elements: Targets for early nutritional effects on epigenetic gene regulation. *Mol. Cell Biol. 23*, 5293, 2003.

62. Wolff, G.L. et al. Maternal epigenetics and methyl supplements affect agouti gene expression in Avy/a mice. *FASEB J. 12*, 949, 1998.

63. Aerts, L. and Van Assche, F.A. Animal evidence for the transgenerational development of diabetes mellitus. *Int. J. Biochem. Cell Biol. 38*, 894, 2006.

64. Gauguier, D. et al. Inheritance of diabetes mellitus as consequence of gestational hyperglycemia in rats. *Diabetes 39*, 734, 1990.

65. Dong, E. et al. Histone hyperacetylation induces demethylation of reelin and 67-kDa glutamic acid decarboxylase promoters. *Proc. Natl. Acad. Sci. U S A 104*, 4676, 2007.

66. Clayton, A.L., Hazzalin, C.A. and Mahadevan, L.C. Enhanced histone acetylation and transcription: A dynamic perspective. *Mol. Cell 23*, 289, 2006.

67. Lu, Q. et al. Epigenetics, disease, and therapeutic interventions. *Ageing Res. Rev. 5*, 449, 2006.

68. Ulrey, C.L. et al. The impact of metabolism on DNA methylation. *Hum. Mol. Genet. 14* (Suppl 1), R139, 2005.

69. Banerjee, R. and Zou, C. Redox regulation and reaction mechanism of human cystathionine-[beta]-synthase: A PLP-dependent hemesensor protein. *Arch. Biochem. Biophy. 433*, 144, 2005.

70. Loikkanen, I. et al. Expression of cytosolic acetyl-CoA synthetase gene is developmentally regulated. *Mech. Dev. 115*, 139, 2002.

71. Starai, V.J. and Escalante-Semerena, J.C. Acetyl-coenzyme A synthetase (AMP forming). *Cell Mol. Life Sci. 61*, 2020, 2004.

72. Hallows, W.C., Lee, S. and Denu, J.M. Sirtuins deacetylate and activate mammalian acetyl-CoA synthetases. *Proc. Natl. Acad. Sci. U S A 103*, 10230, 2006.

73. Michan, S. and Sinclair, D. Sirtuins in mammals: Insights into their biological function. *Biochem. J. 404*, 1, 2007.

74. Takahashi, H. et al. Nucleocytosolic acetyl-coenzyme a synthetase is required for histone acetylation and global transcription. *Mol. Cell 23*, 207, 2006.

75. Kennedy, A.R. et al. A high-fat, ketogenic diet induces a unique metabolic state in mice. *Am. J. Physiol. Endocrinol. Metab. 292*, E1724, 2007.

76. Bonnet, S. et al. A mitochondria-K^+ channel axis is suppressed in cancer and its normalization promotes apoptosis and inhibits cancer growth. *Cancer Cell 11*, 37, 2007.

77. Warburg, O. On respiratory impairment in cancer cells. *Science 124*, 269, 1956.

78. Brand, K.A. and Hermfisse, U. Aerobic glycolysis by proliferating cells: A protective strategy against reactive oxygen species. *FASEB J. 11*, 388, 1997.

79. Mobbs, C.V. et al. Secrets of the lac operon. Glucose hysteresis as a mechanism in dietary restriction, aging and disease. *Interdiscip. Top. Gerontol. 35*, 39, 2007.

80. Roy, S. et al. Overexpression of fibronectin induced by diabetes or high glucose: Phenomenon with a memory. *Proc. Natl. Acad. Sci. U S A 87*, 404, 1990.

81. Gonzalez, A.A. et al. Metabolic adaptations to fasting and chronic caloric restriction in heart, muscle, and liver do not include changes in AMPK activity. *Am. J. Physiol. Endocrinol. Metab. 287*, E1032, 2004.

82. Hagopian, K., Ramsey, J.J. and Weindruch, R. Influence of age and caloric restriction on liver glycolytic enzyme activities and metabolite concentrations in mice. *Exp. Gerontol. 38*, 253, 2003.

83. Hagopian, K., Ramsey, J.J. and Weindruch, R. Krebs cycle enzymes from livers of old mice are differentially regulated by caloric restriction. *Exp. Gerontol. 39*, 1145, 2004.

84. Lee, C.K. et al. Transcriptional profiles associated with aging and middle age-onset caloric restriction in mouse hearts. *Proc. Natl. Acad. Sci. U S A 99*, 14988, 2002.

85. Ma, L., Robinson, L.N. and Towle, H.C. ChREBP*Mlx is the principal mediator of glucose-induced gene expression in the liver. *J. Biol. Chem. 281*, 28721, 2006.

86. Weindruch, R. et al. Microarray profiling of gene expression in aging and its alteration by caloric restriction in mice. *J. Nutr. 131*, 918S, 2001.

87. Mastaitis, J.W. et al. Acute induction of gene expression in brain and liver by insulin-induced hypoglycemia. *Diabetes 54*, 952, 2005.

88. Mobbs, C.V. et al. Mining microarrays for metabolic meaning: Nutritional regulation of hypothalamic gene expression. *Neurochem. Res. 29*, 1093, 2004.

89. Roche, E. et al. Induction by glucose of genes coding for glycolytic enzymes in a pancreatic beta-cell line (INS-1). *J. Biol. Chem. 272*, 3091, 1997.

90. Dentin, R. et al. Hepatic glucokinase is required for the synergistic action of ChREBP and SREBP-1c on glycolytic and lipogenic gene expression. *J. Biol. Chem. 279*, 20314, 2004.

91. Pilkis, S.J. and Granner, D.K. Molecular physiology of the regulation of hepatic gluconeogenesis and glycolysis. *Annu. Rev. Physiol. 54*, 885, 1992.

92. Abbot, E.L. et al. Diverging regulation of pyruvate dehydrogenase kinase isoform gene expression in cultured human muscle cells. *FEBS J. 272*, 3004, 2005.

93. Aoyama, H., Daitoku, H. and Fukamizu, A. Nutrient control of phosphorylation and translocation of Foxo1 in C57BL/6 and db/db mice. *Int. J. Mol. Med. 18*, 433, 2006.

94. Southgate, R.J. et al. PGC-1α gene expression is down-regulated by Akt-mediated phosphorylation and nuclear exclusion of FoxO1 in insulin-stimulated skeletal muscle. *FASEB J. 19*, 2072, 2005.

95. Towle, H.C. Glucose as a regulator of eukaryotic gene transcription. *Trends Endocrinol. Metabol. 16*, 489, 2005.

96. Zhang, W. et al. FoxO1 regulates multiple metabolic pathways in the liver: Effects on gluconeogenic, glycolytic, and lipogenic gene expression. *J. Biol. Chem. 281*, 10105, 2006.

97. Baranov, V.G. et al. Development of diabetes mellitus in the progeny of 6 generations of female rats with alloxan diabetes. *Biull. Eksp. Biol. Med. 105*, 13, 1988.

98. Dorner, G. et al. Teratogenetic maternofoetal transmission and prevention of diabetes susceptibility. *Exp. Clin. Endocrinol. 91*, 247, 1988.

99. Spergel, G., Khan, F. and Goldner, M.G. Emergence of overt diabetes in offspring of rats with induced latent diabetes. *Metabolism 24*, 1311, 1975

100. Van Assche, F.A. and Aerts, L. Long-term effect of diabetes and pregnancy in the rat. *Diabetes 34* (Suppl 2), 116, 1985.

101. Shibata, M. and Yasuda, B. New experimental congenital diabetic mice (N.S.Y. mice). *Tohoku J. Exp. Med. 130*, 139, 1980.

102. Gauguier, D. et al. Insulin secretion in adult rats after intrauterine exposure to mild hyperglycemia during late gestation. *Diabetes 40* (Suppl 2), 109, 1991.

103. Harder, T. et al. Maternal and paternal family history of diabetes in women with gestational diabetes or insulin-dependent diabetes mellitus type I. *Gynecol. Obstet. Invest 51*, 160, 2001.

104. Masoro, E.J. Overview of caloric restriction and ageing. *Mech. Ageing Dev. 126*, 913, 2005.

105. Spindler, S.R. Rapid and reversible induction of the longevity, anticancer and genomic effects of caloric restriction. *Mech. Ageing Dev. 126*, 960, 2005.

106. Koizumi, A. et al. Caloric restriction perturbs the pituitary-ovarian axis and inhibits mouse mammary tumor virus production in a high-spontaneous-mammary-tumor-incidence mouse strain (C3H/SHN). *Mech. Ageing Dev. 49*, 93, 1989.
107. Masoro, E.J. et al. Dietary restriction alters characteristics of glucose fuel use. *J. Gerontol. 47*, B202, 1992.
108. Hansson, A. et al. A switch in metabolism precedes increased mitochondrial biogenesis in respiratory chain-deficient mouse hearts. *Proc. Natl. Acad. Sci. U S A 101*, 3136, 2004.
109. Lehman, J.J. and Kelly, D.P. Transcriptional activation of energy metabolic switches in the developing and hypertrophied heart. *Clin. Exp. Pharmacol. Physiol. 29*, 339, 2002.
110. Stanley, W.C. et al. Myocardial substrate metabolism in the normal and failing heart. *Physiol. Rev. 85*, 1093, 2005.
111. Greco, M. et al. Marked aging-related decline in efficiency of oxidative phosphorylation in human skin fibroblasts. *FASEB J. 17*, 1706, 2003.
112. Stockl, P. et al. Sustained inhibition of oxidative phosphorylation impairs cell proliferation and induces premature senescence in human fibroblasts. *Exp. Gerontol. 41*, 674, 2006.
113. Trifunovic, A. et al. Somatic mtDNA mutations cause aging phenotypes without affecting reactive oxygen species production. *Proc. Natl. Acad. Sci. U S A 102*, 17993, 2005.
114. Melov, S. et al. Resistance exercise reverses aging in human skeletal muscle. *PLoS. ONE. 2*, e465, 2007.

Part IV

Epigenetics in the Diagnosis, Prognosis and Therapy of Cancer

18 DNA Methylation in Tumor Diagnosis

Kavitha Ramachandran, Edna Gordian, and Rakesh Singal

CONTENTS

18.1 INTRODUCTION

Cancer is one of the common causes of deaths worldwide. Despite advances in cancer treatment, mortality and morbidity remains high, since most tumors are diagnosed at advanced/incurable stages. Therefore, there is a great need for novel methods that can detect cancer at an early stage. Evaluation of molecular changes in tumor cells in combination with the existing detection methods may help in early detection of cancer. Cancer biomarker is any biological substance that is indicative of the presence of the tumor in the body. It can be derived from the tumor site, remote media or found in circulation. Cancer biomarkers include cell-free tumor derived DNA, RNA, and protein. Elevated levels of DNA have been detected in serum and plasma of cancer patients using quantitative PCR [1–4]. Hence, quantitation of cell-free DNA may be useful in tumor diagnosis. In addition, DNA methylation markers have been used to detect epigenetic changes in cancer. Changes in gene expression contribute to cancer initiation and progression. Reverse transcriptase PCR (RT-PCR) has been used to detect alterations in gene expression patterns. With the advent of microarray technologies, gene expression profiling studies have been carried out using circulating RNA and RNA from remote media [5,6]. Use of RNA as a biomarker is challenging due to decreased stability of the molecule. Proteins secreted by tumor cells into the extracellular environment may also be used as a biomarker in cancer diagnostics [7,8]. However, acquiring proteomic data has been challenged by lack of standardized methodologies, sensitivity, and reproducibility due to tumor heterogeneity.

Owing to their chemical and biological stability, DNA methylation based biomarkers have potential clinical applications in cancer detection, diagnosis, and targeted therapies. Epigenetic changes, associated with loss of gene expression, are common events in cancer. These alterations include hypermethylation of promoter CpG dinucleotides accompanied by global hypomethylation and histone modifications of which DNA hypermethylation is the most studied. Methylation of cytosine within a CpG residue in DNA is catalyzed by the transfer of a methyl group from the donor

S-adenosyl methionine (SAM) by DNA methyl transferase (DNMT) enzymes [9]. The finding that DNA methylation is an early occurrence in cancer coupled with the chemical and biological stability of this alteration makes it a potential diagnostic tool in cancer. Applications for DNA methylation markers include sensitive detection, risk assessment, disease stratification, and prediction of response to treatment or recurrence.

DNA hypermethylation occurs commonly in CpG rich gene promoter sequences. In contrast, cancer related DNA hypomethylation targets repeat DNA sequences such as Alu, LINE 1 elements, and tandem DNA repeats. Some studies have shown that global hypomethylation is linked to specific cancers including advanced or higher-grade ovarian cancers. Moreover, they identified hypomethylation of satellite 2 (Sat 2) in chromosome 1 as a potential predictor of disease prognosis [10]. Frigola et al. showed that DNA hypermethylation and hypomethylation are independent processes that play different roles in colorectal carcinogenesis [11]. Similarly, in Wilms tumors, DNA hypomethylation was found to contribute to tumorigenesis and disease progression independent of aberrant DNA hypermethylation [12].

A large number of studies have focused on use of gene specific DNA hypermethylation as tumor markers. In this chapter, we confine our discussion to the role of aberrant DNA hypermethylation in cancer diagnosis.

18.1.1 IDENTIFICATION OF METHYLATION MARKERS

A number of genes are frequently deregulated by hypermethylation during carcinogenesis. However, the greatest challenge for clinical application of methylation markers is to identify methylation of which specific CpGs among the vast pool of CpG clusters in the entire genome are associated with a particular cancer. For this, it is essential to tap and process the information from screening studies and analyze them critically. For use of methylation markers in clinical diagnosis, it is important to establish the differences in methylation between cancer cases and controls [13]. Many studies have evaluated methylation changes between tumor tissues and adjacent histologically normal tissues. It is also important to conduct methylation analysis on tissue samples from different stages of cancer. This will aid in classification of methylation markers according to the different stages of disease progression. Due to limited availability, very few studies have included tissues from healthy donors for comparison with tumor tissues. Ideally, methylation patterns of DNA obtained from tumor tissue and adjacent "normal" tissue of the same patient must be compared with tissue from a healthy donor. The information from such studies would be vital in identifying early methylation events that occur during oncogenesis. Such markers would then be useful to identify premalignant conditions and in assessing the risk of an individual to develop cancer. Patients who do not show any signs or symptoms of cancer but display a methylation pattern that is associated with the disease can then be periodically followed up to determine if they eventually develop cancers. If methylation changes were able to predict development of cancer, methylation markers would identify high-risk individuals who would be potential candidates for chemoprevention trials. A classic example of early methylation event in cancer is change in glutathione sulfo transferase p1 (GSTP1) methylation during initiation and progression of prostate cancer [14]. GSTP1 encodes the carcinogen detoxification enzyme and hence has been proposed to have a "caretaker" function in the normal prostate. When GSTP1 methylation was assessed at different stages of prostate cancer, gene methylation was detected in 6% of the 64 proliferative inflammatory atrophic (PIA) lesions tested, 69% of the prostatic intraepithelial neoplasia lesions (PIN), and 91% of prostate cancer indicating the increase in GSTP1 methylation during disease progression. Methylation of GSTP1 could not be detected in normal prostatic epithelium suggesting that hypermethylation of the gene was a cancer-specific event occurring early during prostatic carcinogenesis that renders the cells vulnerable to genome damage mediated by environmental carcinogens. Comparison of extent of methylation of 131 CpG sites along the promoter and transcribed regions of GSTP1 gene in prostate cancer versus normal prostate tissue showed that extensive DNA methylation was not confined to CpG sites of the promoter region but occurred extensively across the CpG sites of the exons and the 3' untranslated region (UTR). However, in normal prostate tissues, methylation was absent in the CpG rich promoter region but present in the downstream CpG sites located in the exons and in the 3' UTR of the gene [15]. Increased levels of GSTP1 methylation was found to be associated with higher Gleason grade tumors, indicating that quantitative assessment of GSTP1 methylation could be used to predict clinical outcome of prostate cancer [16].

Another example of caretaker genes that are silenced during cancer is that of O-6 methylguanine methyl transferase (MGMT), which is a DNA repair enzyme that removes O-6 methylguanine adducts from DNA. Methylation mediated silencing of the gene occurs in colon, lung, lymphoid, and other tumors and can cause G to A transition mutations in tumor suppressor genes such as p53 [17] and K-RAS [18] leading to cancer. A third example of early occurrence of DNA methylation during tumor formation is that of methylation of hMLH1 gene in colon cancer, which leads to microsatellite instability [19]. Apart from its role in development of cancer, DNA methylation also correlates with poor prognosis. MGMT promoter methylation increased sensitivity of gliomas to alkylating agents [20], whereas methylation of hMLH1 resulted in increased resistance of colorectal cells to chemotherapy [21].

Another question to be addressed in establishing methylation markers is to determine whether evaluation of methylation of a subset of genes would provide more accurate information than analyzing a single gene. Evaluation of methylation of multiple genes in prostate cancer showed that higher methylation index correlated with stage III disease and Gleason score 7. This study indicates the correlation of methylation of multiple genes with clinicopathological features of disease progression [22]. Other studies have analyzed methylation of a panel of candidate genes in tumors and controls in order to evaluate their efficacy as a tumor marker. Analysis of aberrant promoter methylation profile of a subset of 10 genes in bladder cancer revealed that methylation of 4 genes (CDH1, RASSF1A, APC, and CDH13) correlated with poor prognosis, while methylation of CDH1, FHIT, and a high methylation index was associated with poor survival [23]. Evaluation of promoter methylation status of 10 genes in 101 prostate cancer cases showed that methylation of RASSF1A, GSTP1, RARβ, and CDH13 genes was significantly more in the high Gleason score group compared to that of the low Gleason score group [24]. Comparison of aberrant DNA methylation of lung tumors showed significant differences in methylation profiles between NSCLC and SCLC [25]. These studies indicate that methylation of an individual gene is not observed uniformly in all cancer cases. Since epigenetic alterations affect multiple genes in each cancer type, it would be important to identify groups of methylated genes that could collectively act as a sensitive biomarker in cancer diagnostics.

For successful clinical application, a biomarker should be highly specific. Specificity is assessed by the number of healthy individuals that test negative for the particular biomarker. It is important to keep in mind that methylation markers vary in different cancers; hence methylation of a gene cannot be considered a universal marker for all cancer types. For example, GSTP1 hypermethylation and transcriptional silencing occurs in prostate, breast, and renal carcinomas and is associated with the initiation of the disease. On the contrary, over expression of the gene is observed in gliomas, ovarian, cervical, and colorectal cancers and is associated with drug resistance resulting in poor prognosis. Clinical sensitivity of a biomarker is determined by the ratio of number of cancer cases that test positive to the total number of cases tested. Further, DNA methylation biomarkers can be used for stratifying stage and grade of cancer. Gene methylation can also be used as a marker for predicting response to therapeutic drugs and defining treatment regimens.

In the process of biomarker selection, it is also important to consider other factors that influence gene methylation. Since DNA methylation changes are affected by factors such as age, race, diet, and stage of the disease, it is important to analyze gene methylation data by taking these variables into consideration. A previous study showed age related increase in methylation of N33 and MYOD in normal colon mucosa [26]. Differences in methylation of specific genes in different racial groups can also skew methylation analysis if racial differences are not included in data interpretation. TMS1 methylation was more prevalent in cases than in controls among Whites (Odds Ratio; OR: 7.6) whereas no significant difference in methylation was observed between cases and controls among Blacks (OR: 1.1) [27]. Since incidence of prostate cancer is higher among black population compared to whites, this study indicates that DNA methylation might play the role of a molecular determinant in cancer susceptibility of different race groups. Moreover, effect of diet on DNA methylation and cancer risk has been documented by several studies. Dietary folate has been shown to correlate inversely with DNA methylation and colorectal cancer risk [28]. Methylation profiling studies would be useful in mapping the age, race, and diet related changes in DNA methylation which when analyzed with appropriate statistical tools would become valuable for risk assessment of individuals.

Selection strategy for diagnostic methylation marker involves screening of candidate tumor suppressor genes for methylation in cancer tissue. Another approach is genome wide scanning for identification of novel methylation targets. Genome-wide methylation profiling techniques identify differences in methylation in CpG clusters between cancer cases and normal controls. Once identified, the sequences are then tested for sensitivity and specificity on a number of clinical samples using a simple, yet sensitive laboratory based assay for gene specific methylation analysis (such as MS-PCR or MS-SNuPE) [29]. The principles of the genome wide and gene specific methylation are outlined in the following section.

18.1.2 Techniques for Analysis of DNA Methylation in Tumor

Identification of methylated DNA from tumors relies on highly sensitive methylation detection assays. Global changes in DNA methylation is considered to be a part of carcinogenesis [30]. Genome wide analysis of methylation levels in a DNA pool is carried out by determining the ratio of 5-methyl cytosine to cytosine. This is done by complete enzymatic hydrolysis of DNA followed by high resolution separation using high performance liquid chromatography (HPLC) [31] or by anti-methyl cytosine antibodies [32], reverse phase HPLC [31], thin layer chromatography (TLC) [33], mass spectrometry assays [34] etc. Although, changes in global methylation content offer very little information about gene regulatory sequences that are affected in oncogenesis, it provides preliminary information in risk assessment [35]. Genome wide techniques can be used to identify novel tumor suppressor genes and to identify global changes in DNA methylation between cancer and controls. Identification of candidate genes can be achieved by methylation profiling techniques. Methylation sensitive arbitrarily primed PCR (MS-AP-PCR) screens for methylation changes using a PCR-based fingerprinting method [36]. Restriction landmark genomic scanning (RLGS) determines the methylation status of CpG islands in a genome within a single gel [37]. Methylation specific melting curve analysis (MS-MCA) is based on the principle that the melting temperature of methylated alleles differs from the unmethylated allele after bisulfite treatment [38]. In methylation specific microarray (MSO) bisulfite treated DNA is amplified by bisulfite PCR followed by hybridization of the product to glass slides containing an array of oligonucleotides that discriminate between methylated and unmethylated cytosines at specific CpG sites [39]. However, genome wide techniques require large quantities of DNA. This problem can be overcome by use of whole genome amplification of bisulfite converted DNA using degenerate primers which yields high quantities of the template that can be used for high-throughput methylation analysis of multiple genes [40]. After identification of candidate methylated gene sequences by genome wide methylation profiling, gene/region specific methylation assays must be performed to screen clinical samples. Quantitative or qualitative methylation analysis of individual CpGs provides more definitive information on the methylation pattern of a gene promoter with a high degree of sensitivity and specificity. The techniques for gene specific or region specific methylation analysis are based on methylation sensitive restriction digestion or bisulfite conversion. The former principle has been used in combination with Southern blotting [41], or PCR [42] to obtain measurement of DNA methylation. However, methylation sensitive restriction digestion is limited by the availability of restriction sites in a particular sequence. Bisulfite conversion techniques can be used alone or in combination with other detection methods to reveal important changes in methylation levels and patterns. Bisulfite treatment deaminates unmethylated cytosines to uracil leaving methylated cytosines unchanged [43]. Various methods are available for analyzing the bisulfite converted DNA based on the downstream application. These methods include bisulfite genomic sequencing (BGS) and methylation specific polymerase chain reaction (MS-PCR), among others. Primers for amplification of bisulfite converted DNA can be designed using the available software tools [44–46]. Primer sets spanning the CpGs are used in MS-PCR to discriminate methylated regions from unmethylated sequences. In contrast, primers flanking the CpGs would amplify the bisulfite converted sequence irrespective of the methylation status. This amplified sequence can be further analyzed by methylation specific primers (nested MS-PCR), bisulfite genomic sequencing, or by methylation sensitive single nucleotide primer extension (MS-SNuPE) [47]. Quantitation of gene specific DNA methylation is carried out using the high-throughput MethylLight [48] real time PCR analysis. Combinations of methylation sensitive restriction digestion with bisulfite conversion form the basis of combined bisulfite restriction analysis (COBRA) [49] that enhances the applicability of either individual methylation analysis principle.

After initial selection of the candidate methylated genes by genome wide methylation profiling, MS-PCR technique appears to be an ideal method for routine laboratory analysis. MS-PCR is a sensitive method since it requires low initial amounts of DNA. Quantitative MS-PCR is more advantageous than conventional MS-PCR since it provides information on the relative levels of methylation. Further, qualitative MS-PCR analysis is performed by agarose gel electrophoresis, which may result in contamination by the PCR product leading to spurious data. These can be prevented by performing in silico computation of methylation levels using quantitative MS-PCR methods. Further after careful analysis, it is possible to assign a threshold methylation level to differentiate cancer cases from benign controls.

Advances in epigenetic research and continuous search for markers together with the development of modern technologies for sensitive detection had led to lack of uniform standards for analysis. There is a need for standardized methodologies starting with the DNA extraction procedures, bisulfite treatment protocols to the choice of analysis technique so that results across different laboratories can be cross-validated. Also, the choice of primers for MS-PCR amplification affects the methylation analysis of a single gene. This can be overcome by use of methylation profiling techniques to determine the clinically significant "hot spots" of methylation. Multiple primer sets must then be tested to map the methylation pattern across different CpG clusters and to determine a consensus primer set that is most specific.

18.1.3 DNA METHYLATION IN TUMOR TISSUES

Using these strategies aberrant DNA methylation patterns of candidate genes have been described in tissues of different tumor types. Table 18.1 lists a few examples of potential methylation biomarkers in tumor tissues. The identification of specific hypermethylated genes in tumors confirms that epigenetic silencing in cancer is not a random process but is targeted to genes involved in homeostasis. Using DNA methylation markers not only can yield insight into the complex molecular pathways involved in cancer but also can provide a highly sensitive and specific cancer diagnostic tool. However, methylation patterns in tumor tissue have their limitations. In some cancers, accessibility to tissue is limited. Also, a routine biopsy may exclude the actual tumor site thereby resulting in a false negative result. An attractive alternative to this problem is the use of less invasive and more uniform specimens for determining the methylation status of candidate genes.

18.1.4 METHYLATION MARKERS IN BODY FLUIDS

The finding that tumor DNA could be detected in body fluids came as a great promise for molecular markers as diagnostic tools for early detection of cancer. This paved the way for less invasive

TABLE 18.1

Frequently Methylated Genes in Tumor Tissues

Type of Cancer	Gene	Sensitivity	Specificity	References
Bladder	BCL2	65/125 (52%)	28/35 (80%)	[57]
	RARβ	86/98 (87.8%)	4/7 (57.1%)	[76]
	DAPK	57/98 (58.1%)	7/7 (100%)	[76]
	ECAD	62/98 (63.3%)	7/7 (100%)	[76]
	p16	26/98 (26.5%)	7/7 (100%)	[76]
Brain	RASSF1A	41/44 (93.2%)	5/5 (100%)	[77]
Breast	APC	15/34 (47%)	6/6 (100%)	[71]
	RASSF1A	22/34 (65%)	6/6 (100%)	[71]
	DAPK	17/34 (50%)	6/6 (100%)	[71]
Cervical	DCR1	50/50 (100%)	N/A	[78]
	DCR2	9/50 (18%)	N/A	[78]
	hTERT	38/65 (58.5%)	14/20 (70%)	[79]
	p73	40/103 (38.8%)	120/124 (96.8%)	[80]
Colorectal	MGMT	14/36 (38%)	N/A	[81]
	hMSH2	11/60 (18.3%)	60/60 (100%)	[82]
Endometrial	ERαC	83/88 (94.3%)	46/46 (100%)	[83]
	hMLH1	21/53 (40.4%)	27/27 (100%)	[84]
	APC	11/50 (22%)	27/27 (100%)	[84]
	ECAD	7/50 (14%)	27/27 (100%)	[84]
Lung (SCLC) (NSCLC)	RARβ	63/87 (72%)	57/58 (98.2%)	[85]
	RARβ	43/107 (40.2%)	89/104 (85.6%)	[86]
	TIMP-3	28/107 (26.2%)	96/104 (92.3%)	[86]
	p16	27/107 (25.2%)	104/104 (100%)	[86]
	MGMT	22/107 (20.6%)	104/104 (100%)	[86]
	DAPK	20/107 (18.7%)	98/104 (94.2%)	[86]
	ECAD	19/107 (17.8%)	104/104 (100%)	[86]
	p14	9/107 (8.4%)	99/104 (95.2%)	[86]
	GSTP1	7/107 (6.5%)	104/104 (100%)	[86]
Oral	p16	36/80 (45%)	29/26 (73%)	[87]
	RARβ	68/80 (85%)	6/21 (29%)	[87]
	CYGB	59/80 (74%)	9/21 (43%)	[87]
	CYCA1	56/78 (72%)	8/9 (89%)	[87]
	ECAD	59/71 (83%)	8/18 (44%)	[87]
Prostate	RARβ	54/101 (53.5%)	31/32 (96.9%)	[24]
	RASSF1A	54/101 (53.5%)	27/32 (84.4%)	[24]
	GSTP1	36/101 (35.6%)	31/32 (96.9%)	[24]
	APC	27/101 (26.7%)	30/32 (93.8%)	[24]
	FHIT	15/101 (14.8%)	32/32 (100%)	[24]
	CDH13	31/101 (30.7%)	27/32 (84.4%)	[24]
	CDH1	27/101 (26.7%)	24/32 (75%)	[24]
	p16	3/101 (3%)	32/32 (100%)	[24]

Note: APC, adenomatous polyposis coli; BCL2, B-cell lymphoma gene; DAPK1, death-associated protein kinase 1; GSTP1, glutathione *S*-transferase Pi 1; hTERT, human telomerase reverse transcriptase; hMLH1, Mut L homologue 1; MGMT, O-6 methylguanine-DNA methyltransferase; p16(CDKN2A), cyclin-dependent kinase inhibitor 2A; RARβ, retinoic acid receptor-β RASSF1A, Ras association domain family member 1. Percent sensitivity (value in parenthesis) is determined by the ratio of methylation positive cases to the total number of cases analyzed. Percent specificity (value in parenthesis) is calculated as ratio of controls that test negative for gene methylation to the total number of controls.

methods to screen the presence of cancer and also to monitor the progression, recurrence, and prognosis of the disease. Tumor cells are continuously exfoliated into the body fluids and provide the source of cancer-specific DNA in the remote media. Epigenetic changes in DNA derived from body fluids have been used as a biomarker in detection of several solid tumors. Examples of these include analysis of bronchial aspirates for lung cancer diagnosis [50], ductal lavage for breast cancer [51], pancreatic juice for pancreatic cancer [52,53], and peritoneal fluid for ovarian cancer [53]. Search for noninvasive detection methods led to the use of saliva and sputum for detection of lung cancer [54], saliva for head and neck cancer [55], and urine for kidney [56], bladder [57], and prostate cancer [58] diagnosis. Studies using a highly sensitive nested methylation sensitive PCR (MS-PCR) have validated the use of methylation markers using sputum samples for detection of lung cancer up to 3 years prior to the clinical diagnosis [54]. This enables population based screening assays to identify potential target genes that can be evaluated in chemoprevention trials. Detection of methylated apoptosis related genes in urine sediments of bladder cancer patients but not in normal subjects has emerged as a promising tool for noninvasive and specific diagnostic tool for bladder cancer [57]. GSTP1 hypermethylation in urine can be used as a diagnostic tool to identify groups of suspected prostate cancer patients who are at a higher risk of harboring cancer despite a negative biopsy [58]. MS-PCR assay has been used to detect methylation of p14ARF, APC, and RASSF1A in urine of kidney cancer patients whereas no methylation could be detected in healthy individuals [56]. Examples of methylation biomarkers isolated from remote media are described in Table 18.2.

18.1.5 CIRCULATING DNA

Detection of tumor derived DNA in serum and plasma of patients with cancer suggested its application as a tumor marker in cancer diagnosis. Higher levels of circulating DNA were found in cancer patients compared to those with nonmalignant disease and several studies have focused on quantitation of circulating DNA levels as a screening tool for early detection of cancer [1–3]. The exact mechanism of release of DNA into the blood stream is not clear. In normal individuals circulating DNA originates from lymphocytes or other nucleated cells, whereas in cancer patients a large proportion is derived from tumor cells [4]. It was initially thought that circulating DNA in cancer patients was released due to the lysis of circulating cancer cells or micrometastases shed by the tumor [4]. But this hypothesis does not account for the large amount of DNA as opposed to the limited number of cells in circulation [4,59,60]. Hence cell death by apoptosis resulting in well-fragmented DNA was proposed to be the mechanism of release of DNA into circulation [61]. However, pathological cell death by necrosis, autophagy, or mitotic catastrophe leading to the release of intact DNA cannot be ruled out [62]. It has been postulated that tumors actively release DNA into the intercellular space that is drained by the lymphatics into the blood [62]. Decrease in DNase activity observed in plasma of gastrointestinal cancer patients correlating with increased integrity of DNA in circulation also supports the active DNA release theory [63].

The use of amounts of free circulating DNA as a marker for cancer is not specific since increased amounts of DNA in circulation can occur in benign diseases as well [64]. Hence detection of epigenetic changes in free DNA greatly improves the specificity since changes in methylation patterns of the gene are cancer specific [65]. Extensive studies have been carried out for the development of methylation specific biomarkers for screening and early detection of lung cancer. Using MS-PCR approach, an early study in lung cancer screened 22-non small cell lung cancer for methylation of p16, DAPK, GSTP1, and MGMT in DNA isolated from sera and matched it with tissue methylation patterns [66]. Methylation of any one of the genes was detected in tissue DNA in 68% of the cancer patients but none of the genes were methylated in normal tissue samples. The abnormal methylation pattern in tissues matched with the methylation pattern of the corresponding serum samples in 73% of the cases. Further, none of the above genes were methylated in sera from normal patients. This initial study determined the specificity of serum DNA methylation in relation

TABLE 18.2

Frequently Methylated Genes in Serum/Plasma, Body Fluids

Type of Cancer	Specimen	Gene	Sensitivity	Specificity	References
Bladder	Urine	BCL2	24/37 (64.9%)	N/A	[57]
		TERT	19/37 (51.4%)	N/A	
		DAPK	8/37 (21.6%)	N/A	
	Serum	p16	18/86 (22%)	31/31 (100%)	[88]
Breast	Serum	APC	10/34 (29%)	20/20 (100%)	[71]
		RASSF1A	19/34 (55.9%)	20/20 (100%)	
		DAPK	12/34 (35.3%)	20/20 (100%)	
	Plasma	GSTP1	12/47 (26%)	38/38 (100%)	[89]
		RARβ	12/47 (26%)	35/38 (92%)	
		RASSF1A	15/47 (32%)	36/38 (95%)	
		APC	8/47 (17%)	38/38 (100%)	
Colorectal	Serum	hMLH1	3/19 (15.8%)	N/A	[90,91]
		p16	14/52 (27%)	10/10 (100%)	
Liver	Serum	p16	5/22 (22.7%)	10/10 (100%)	[68]
	Plasma	p16	8/22 (36.4%)	10/10 (100%)	[68]
Lung (SCLC)	Bronchial aspirates	RASSF1A	14/17 (82%)	102/102 (100%)	[50]
Lung (NSCLC)		RARβ	12/16 (75%)	81/102 (79.4%)	
		p16	1/17 (6%)	102/102 (100%)	
	Bronchial aspirates	RASSF1A	5/18 (28%)	102/102 (100%)	[50]
		RARβ	9/18 (50%)	81/102 (79.4%)	
		p16	1/18 (6%)	102/102 (100%)	
	Sputum	p16	8/10 (80%)	N/A	[54]
		MGMT	5/10 (50%)	N/A	
	Serum	p16	3/22 (13.6%)	N/A	[66]
		DAPK	4/22 (18.1%)	N/A	
		MGMT	4/22 (18.1%)	N/A	
	Plasma	p16	12/35 (34.3%)	15/15 (100%)	[92]
Prostate	Urine	GSTP1	7/18 (38.9%)	14/21 (63.6%)	[58]
	Plasma	GSTP1	9/69 (13%)	31/31 (100%)	[93]

Note: APC, adenomatous polyposis coli; BCL2, B-cell lymphoma gene; CDH1, cadherin 1; CDH13, H-cadherin; CYCA1, cyclin A1; CYGB, cytoglobin; DAPK1, death-associated protein kinase 1; DcR1 and DcR2, decoy receptors; ECAD, E-cadherin; ER, estrogen receptor; FHIT, fragile histidine triad; GSTP1, glutathione *S*-transferase Pi 1; hTERT, human telomerase reverse transcriptase; hMLH1, Mut L homologue 1; MGMT, O-6 methylguanine-DNA methyltransferase; p14ARF, alternate reading frame product of the INK4 gene; p16(CDKN2A), cyclin-dependent kinase inhibitor 2A; RAR-β, retinoic acid receptor-β; RASSF1A, Ras association domain family member 1; TIMP-3, tissue inhibitor of metalloproteinase-3. Percent sensitivity (value in parenthesis) is determined by the ratio of methylation positive cases to the total number of cases analyzed. Percent specificity (value in parenthesis) is calculated as ratio of controls that test negative for gene methylation to the total number of controls.

to tissue DNA methylation patterns and established the use of aberrant promoter methylation analysis in screening and detection of cancer. This was followed by a report that p16 hypermethylation in plasma DNA corresponded to the methylation pattern in tissue and could be used as a molecular marker for early detection of lung cancer [67]. Similar studies were also conducted in hepatocellular carcinoma. Aberrant methylation of p16 gene was detected in plasma/serum samples as well as the corresponding tumor tissues in 81% of the cases, whereas no methylated p16 sequences were detected in peripheral plasma or serum of patients with chronic hepatitis/cirrhosis or in healthy control subjects tested. These results pointed out the specificity of the assay and its potential application in screening a wide variety of other cancers as well [68]. Methylation status of

CALCA, hTERT, MYOD1, PGR, and TIMP3 was studied in serum samples from cervical tumor patients [69]. Methylation of at least one of these genes was observed in 87% of the serum samples studied, hence validating the use of serological detection of gene methylation as a diagnostic tool in cervical cancer. More recently, use of serum DNA methylation markers in diagnosing thyroid cancer was studied [70]. Results of quantitative analysis of methylation of five genes, CLAC, CDH1, TIMP3, DAPK, and RARβ2 revealed that the assay had a sensitivity of 68%, specificity of 95%, and diagnostic accuracy of 77%. In patients with indeterminate thyroid nodules, methylation analysis could diagnose 73% of cancers and 100% of benign tumors with a diagnostic accuracy of 80%. Hypermethylation of APC, RASSF1A, and DAPK genes was detected with 100% specificity in serum of patients who showed tissue methylation of the corresponding genes indicating that methylation analysis of serum could be used as a screening tool for early diagnosis of breast cancer as well [71]. Apart from free DNA, DNA in circulation also exists in a cell bound form [72]. Nucleic acids bind to cell surface DNA binding proteins and phospholipids [73]. Recent studies have shown that quantitative MS-PCR analysis on cell bound DNA in combination with analysis on total circulating DNA in plasma provides a sensitive assay to differentiate patients with breast cancer and those with benign breast tumors [72].

18.1.6 ADVANTAGES AND PITFALLS OF FREE CIRCULATING METHYLATED DNA ANALYSIS

Methylated free circulating DNA promises to be a new generation of diagnostic marker with the advantages of high specificity and sensitivity. Analysis of tumor specific epigenetically altered DNA from serum or plasma could emerge as a potential noninvasive tool for cancer diagnosis, thus enabling population based screening for early detection of cancer. It may allow for follow up of suspected cancer patients by eliminating the need for repeated biopsies.

The presence of gene promoter hypermethylation is not homogenous within a tumor, thus increasing the possibility of false negative results. This can be overcome using a more homogenous serum or plasma sample for analysis.

18.1.7 CHALLENGES IN CLINICAL IMPLEMENTATION OF METHYLATION BIOMARKERS

Despite the above advantages and extensive examples of use of aberrantly methylated DNA from body fluids as a diagnostic marker in several cancers, there are several limitations that need to be overcome before the application of these tumor markers in the clinic. As opposed to tissue specimens, low amounts of DNA in body fluids result in reduced sensitivity. Further, lack of standardized procedures for sample collection and processing introduces variability in results [2]. Change of anticoagulants for plasma collection can result in variability. Further, it is important to eliminate cells from plasma in order to obtain cell-free DNA. It has been found that a single centrifugation step may not be sufficient to obtain cell-free plasma [74]. Differences in nucleic acid extraction procedures from plasma/serum are yet another factor that results in varying quantities of DNA that skews the analysis. Another concern is the quality of DNA that is extracted from serum/plasma. Small fragments constitute the majority of circulating DNA, which gets further degraded by bisulfite treatment. To overcome this problem, whole genome amplification using degenerate primers has been developed [40]. Also DNA methylation analysis of a particular gene of interest is affected by the choice of primers used in MS-PCR. Recently methylation hotspots have been described by using CpG island (CGI) microarray for some candidate genes in acute lympho-blastic leukemia [75]. Similar studies are required to establish a consensus for methylation analysis of specific target sequence in each cancer type. Further, lack of consensus in the methylation detection assay introduces variability in results across different laboratories performing similar studies. Also, an important factor that determines the uniformity between studies is definition of appropriate controls. Lack of uniform set of controls introduces disparity in results between clinical studies.

18.2 CONCLUSIONS

Early detection of tumors before they spread and progress to an incurable stage has been one of the greatest challenges in oncology. The finding that changes in DNA methylation can be detected in body fluids and in circulation by sensitive methylation detection assays holds great promise for effective cancer detection, diagnosis, and prediction of prognosis. However, this powerful tool is still far from being applied in clinical cancer diagnosis. Lack of standard methodologies affects all steps of the study beginning from sample collection to final data analysis. Rigorous standardization of each method is required to ensure test reproducibility, to eliminate intra- and inter-assay variability, and to increase sensitivity, specificity, and efficacy of methylation markers. Finally, prospective clinical studies are required to validate the application of these biomarkers in clinical cancer diagnosis.

REFERENCES

1. Leon, S.A., Shapiro, B., Sklaroff, D.M., and Yaros, M.J., Free DNA in the serum of cancer patients and the effect of therapy, *Cancer Res* 37(3), 646–650, 1977.
2. Ziegler, A., Zangemeister-Wittke, U., and Stahel, R.A., Circulating DNA: A new diagnostic gold mine? *Cancer Treat Rev* 28(5), 255–271, 2002.
3. Tong, Y.K. and Lo, Y.M., Diagnostic developments involving cell-free (circulating) nucleic acids, *Clin Chim Acta* 363(1–2), 187–196, 2006.
4. Pathak, A.K., Bhutani, M., Kumar, S., Mohan, A., and Guleria, R., Circulating cell-free DNA in plasma/serum of lung cancer patients as a potential screening and prognostic tool, *Clin Chem* 52(10), 1833–1842, 2006.
5. Li, Y., St John, M.A., Zhou, X., Kim, Y., Sinha, U., Jordan, R.C., Eisele, D., Abemayor, E., Elashoff, D., Park, N.H., and Wong, D.T., Salivary transcriptome diagnostics for oral cancer detection, *Clin Cancer Res* 10(24), 8442–8450, 2004.
6. Li, Y., Zhou, X., St John, M.A., and Wong, D.T., RNA profiling of cell-free saliva using microarray technology, *J Dent Res* 83(3), 199–203, 2004.
7. Zhang, Z., Bast, R.C., Jr., Yu, Y., Li, J., Sokoll, L.J., Rai, A.J., Rosenzweig, J.M., Cameron, B., Wang, Y.Y., Meng, X.Y., Berchuck, A., Van Haaften-Day, C., Hacker, N.F., de Bruijn, H.W., van der Zee, A.G., Jacobs, I.J., Fung, E.T., and Chan, D.W., Three biomarkers identified from serum proteomic analysis for the detection of early stage ovarian cancer, *Cancer Res* 64(16), 5882–5890, 2004.
8. Kozak, K.R., Su, F., Whitelegge, J.P., Faull, K., Reddy, S., and Farias-Eisner, R., Characterization of serum biomarkers for detection of early stage ovarian cancer, *Proteomics* 5(17), 4589–4596, 2005.
9. Das, P.M. and Singal, R., DNA methylation and cancer, *J Clin Oncol* 22(22), 4632–4642, 2004.
10. Widschwendter, M., Jiang, G., Woods, C., Muller, H.M., Fiegl, H., Goebel, G., Marth, C., Muller-Holzner, E., Zeimet, A.G., Laird, P.W., and Ehrlich, M., DNA hypomethylation and ovarian cancer biology, *Cancer Res* 64 (13), 4472–4480, 2004.
11. Frigola, J., Sole, X., Paz, M.F., Moreno, V., Esteller, M., Capella, G., and Peinado, M.A., Differential DNA hypermethylation and hypomethylation signatures in colorectal cancer, *Hum Mol Genet* 14(2), 319–326, 2005.
12. Ehrlich, M., Jiang, G., Fiala, E., Dome, J.S., Yu, M.C., Long, T.I., Youn, B., Sohn, O.S., Widschwendter, M., Tomlinson, G.E., Chintagumpala, M., Champagne, M., Parham, D., Liang, G., Malik, K., and Laird, P.W., Hypomethylation and hypermethylation of DNA in Wilms tumors, *Oncogene* 21(43), 6694–6702, 2002.
13. Laird, P.W., The power and the promise of DNA methylation markers, *Nat Rev Cancer* 3(4), 253–266, 2003.
14. Nakayama, M., Gonzalgo, M.L., Yegnasubramanian, S., Lin, X., De Marzo, A.M., and Nelson, W.G., GSTP1 CpG island hypermethylation as a molecular biomarker for prostate cancer, *J Cell Biochem* 91(3), 540–552, 2004.
15. Millar, D.S., Ow, K.K., Paul, C.L., Russell, P.J., Molloy, P.L., and Clark, S.J., Detailed methylation analysis of the glutathione S-transferase pi (GSTP1) gene in prostate cancer, *Oncogene* 18(6), 1313–1324, 1999.
16. Zhou, M., Tokumaru, Y., Sidransky, D., and Epstein, J.I., Quantitative GSTP1 methylation levels correlate with Gleason grade and tumor volume in prostate needle biopsies, *J Urol* 171(6 Pt 1), 2195–2198, 2004.

17. Esteller, M., Risques, R.A., Toyota, M., Capella, G., Moreno, V., Peinado, M.A., Baylin, S.B., and Herman, J.G., Promoter hypermethylation of the DNA repair gene O(6)-methylguanine-DNA methyltransferase is associated with the presence of G:C to A:T transition mutations in p53 in human colorectal tumorigenesis, *Cancer Res* 61(12), 4689–4692, 2001.

18. Esteller, M., Toyota, M., Sanchez-Cespedes, M., Capella, G., Peinado, M.A., Watkins, D.N., Issa, J.P., Sidransky, D., Baylin, S.B., and Herman, J.G., Inactivation of the DNA repair gene O6-methylguanine-DNA methyltransferase by promoter hypermethylation is associated with G to A mutations in K-ras in colorectal tumorigenesis, *Cancer Res* 60(9), 2368–2371, 2000.

19. Nakagawa, H., Nuovo, G.J., Zervos, E.E., Martin, E.W., Jr., Salovaara, R., Aaltonen, L.A., and de la Chapelle, A., Age-related hypermethylation of the 5' region of MLH1 in normal colonic mucosa is associated with microsatellite-unstable colorectal cancer development, *Cancer Res* 61(19), 6991–6995, 2001.

20. Esteller, M., Garcia-Foncillas, J., Andion, E., Goodman, S.N., Hidalgo, O.F., Vanaclocha, V., Baylin, S.B., and Herman, J.G., Inactivation of the DNA-repair gene MGMT and the clinical response of gliomas to alkylating agents, *N Engl J Med* 343(19), 1350–1354, 2000.

21. Arnold, C.N., Goel, A., and Boland, C.R., Role of hMLH1 promoter hypermethylation in drug resistance to 5-fluorouracil in colorectal cancer cell lines, *Int J Cancer* 106(1), 66–73, 2003.

22. Singal, R., Ferdinand, L., Reis, I.M., and Schlesselman, J.J., Methylation of multiple genes in prostate cancer and the relationship with clinicopathological features of disease, *Oncol Rep* 12(3), 631–637, 2004.

23. Maruyama, R., Toyooka, S., Toyooka, K.O., Harada, K., Virmani, A.K., Zochbauer-Muller, S., Farinas, A.J., Vakar-Lopez, F., Minna, J.D., Sagalowsky, A., Czerniak, B., and Gazdar, A.F., Aberrant promoter methylation profile of bladder cancer and its relationship to clinicopathological features, *Cancer Res* 61(24), 8659–8663, 2001.

24. Maruyama, R., Toyooka, S., Toyooka, K.O., Virmani, A.K., Zochbauer-Muller, S., Farinas, A.J., Minna, J.D., McConnell, J., Frenkel, E.P., and Gazdar, A.F., Aberrant promoter methylation profile of prostate cancers and its relationship to clinicopathological features, *Clin Cancer Res* 8(2), 514–519, 2002.

25. Toyooka, S., Toyooka, K.O., Maruyama, R., Virmani, A.K., Girard, L., Miyajima, K., Harada, K., Ariyoshi, Y., Takahashi, T., Sugio, K., Brambilla, E., Gilcrease, M., Minna, J.D., and Gazdar, A.F., DNA methylation profiles of lung tumors, *Mol Cancer Ther* 1(1), 61–67, 2001.

26. Ahuja, N., Li, Q., Mohan, A.L., Baylin, S.B., and Issa, J.P., Aging and DNA methylation in colorectal mucosa and cancer, *Cancer Res* 58(23), 5489–5494, 1998.

27. Das, P.M., Ramachandran, K., Vanwert, J., Ferdinand, L., Gopisetty, G., Reis, I.M., and Singal, R., Methylation mediated silencing of TMS1/ASC gene in prostate cancer, *Mol Cancer* 5, 28, 2006.

28. Kim, Y.I., Folate and DNA methylation: A mechanistic link between folate deficiency and colorectal cancer? *Cancer Epidemiol Biomarkers Prev* 13(4), 511–519, 2004.

29. Paluszczak, J. and Baer-Dubowska, W., Epigenetic diagnostics of cancer—the application of DNA methylation markers, *J Appl Genet* 47(4), 365–375, 2006.

30. Gama-Sosa, M.A., Slagel, V.A., Trewyn, R.W., Oxenhandler, R., Kuo, K.C., Gehrke, C.W., and Ehrlich, M., The 5-methylcytosine content of DNA from human tumors, *Nucleic Acids Res* 11(19), 6883–6894, 1983.

31. Kuo, K.C., McCune, R.A., Gehrke, C.W., Midgett, R., and Ehrlich, M., Quantitative reversed-phase high performance liquid chromatographic determination of major and modified deoxyribonucleosides in DNA, *Nucleic Acids Res* 8(20), 4763–4776, 1980.

32. Oakeley, E.J., Schmitt, F., and Jost, J.P., Quantification of 5-methylcytosine in DNA by the chloroacetaldehyde reaction, *Biotechniques* 27(4), 744–746, 748–750, 752, 1999.

33. Wilson, V.L., Smith, R.A., Autrup, H., Krokan, H., Musci, D.E., Le, N.N., Longoria, J., Ziska, D., and Harris, C.C., Genomic 5-methylcytosine determination by 32P-postlabeling analysis, *Anal Biochem* 152(2), 275–284, 1986.

34. Annan, R.S., Kresbach, G.M., Giese, R.W., and Vouros, P., Trace detection of modified DNA bases via moving-belt liquid chromatography–mass spectrometry using electrophoric derivatization and negative chemical ionization, *J Chromatogr* 465(2), 285–296, 1989.

35. Ting Hsiung, D., Marsit, C.J., Houseman, E.A., Eddy, K., Furniss, C.S., McClean, M.D., and Kelsey, K.T., Global DNA methylation level in whole blood as a biomarker in head and neck squamous cell carcinoma, *Cancer Epidemiol Biomarkers Prev* 16(1), 108–114, 2007.

36. Gonzalgo, M.L., Liang, G., Spruck, C.H., 3rd, Zingg, J.M., Rideout, W.M., 3rd, and Jones, P.A., Identification and characterization of differentially methylated regions of genomic DNA by methylation-sensitive arbitrarily primed PCR, *Cancer Res* 57(4), 594–599, 1997.

37. Costello, J.F., Fruhwald, M.C., Smiraglia, D.J., Rush, L.J., Robertson, G.P., Gao, X., Wright, F.A., Feramisco, J.D., Peltomaki, P., Lang, J.C., Schuller, D.E., Yu, L., Bloomfield, C.D., Caligiuri, M.A., Yates, A., Nishikawa, R., Su Huang, H., Petrelli, N.J., Zhang, X., O'Dorisio, M.S., Held, W.A., Cavenee, W.K., and Plass, C., Aberrant CpG-island methylation has non-random and tumour-type-specific patterns, *Nat Genet* 24(2), 132–138, 2000.

38. Worm, J., Aggerholm, A., and Guldberg, P., In-tube DNA methylation profiling by fluorescence melting curve analysis, *Clin Chem* 47(7), 1183–1189, 2001.

39. Gitan, R.S., Shi, H., Chen, C.M., Yan, P.S., and Huang, T.H., Methylation-specific oligonucleotide microarray: A new potential for high-throughput methylation analysis, *Genome Res* 12(1), 158–164, 2002.

40. Mill, J., Yazdanpanah, S., Guckel, E., Ziegler, S., Kaminsky, Z., and Petronis, A., Whole genome amplification of sodium bisulfite-treated DNA allows the accurate estimate of methylated cytosine density in limited DNA resources, *Biotechniques* 41(5), 603–607, 2006.

41. Southern, E.M., Detection of specific sequences among DNA fragments separated by gel electrophoresis, *J Mol Biol* 98(3), 503–517, 1975.

42. Singer-Sam, J., LeBon, J.M., Tanguay, R.L., and Riggs, A.D., A quantitative HpaII-PCR assay to measure methylation of DNA from a small number of cells, *Nucleic Acids Res* 18(3), 687, 1990.

43. Frommer, M., McDonald, L.E., Millar, D.S., Collis, C.M., Watt, F., Grigg, G.W., Molloy, P.L., and Paul, C.L., A genomic sequencing protocol that yields a positive display of 5-methylcytosine residues in individual DNA strands, *Proc Natl Acad Sci U S A* 89(5), 1827–1831, 1992.

44. Singal, R. and Grimes, S.R., Microsoft Word macro for analysis of cytosine methylation by the bisulfite deamination reaction, *Biotechniques* 30(1), 116–120, 2001.

45. Li, L.C. and Dahiya, R., MethPrimer: Designing primers for methylation PCRs, *Bioinformatics* 18(11), 1427–1431, 2002.

46. Anbazhagan, R., Herman, J.G., Enika, K., and Gabrielson, E., Spreadsheet-based program for the analysis of DNA methylation, *Biotechniques* 30(1), 110–114, 2001.

47. Gonzalgo, M.L. and Jones, P.A., Rapid quantitation of methylation differences at specific sites using methylation-sensitive single nucleotide primer extension (Ms-SNuPE), *Nucleic Acids Res* 25(12), 2529–2531, 1997.

48. Eads, C.A., Danenberg, K.D., Kawakami, K., Saltz, L.B., Blake, C., Shibata, D., Danenberg, P.V., and Laird, P.W., MethyLight: A high-throughput assay to measure DNA methylation, *Nucleic Acids Res* 28(8), E32, 2000.

49. Xiong, Z. and Laird, P.W., COBRA: A sensitive and quantitative DNA methylation assay, *Nucleic Acids Res* 25(12), 2532–2534, 1997.

50. Schmiemann, V., Bocking, A., Kazimirek, M., Onofre, A.S., Gabbert, H.E., Kappes, R., Gerharz, C.D., and Grote, H.J., Methylation assay for the diagnosis of lung cancer on bronchial aspirates: A cohort study, *Clin Cancer Res* 11(21), 7728–7734, 2005.

51. Fackler, M.J., Malone, K., Zhang, Z., Schilling, E., Garrett-Mayer, E., Swift-Scanlan, T., Lange, J., Nayar, R., Davidson, N.E., Khan, S.A., and Sukumar, S., Quantitative multiplex methylation-specific PCR analysis doubles detection of tumor cells in breast ductal fluid, *Clin Cancer Res* 12(11 Pt 1), 3306–3310, 2006.

52. Jiang, P., Watanabe, H., Okada, G., Ohtsubo, K., Mouri, H., Tsuchiyama, T., Yao, F., and Sawabu, N., Diagnostic utility of aberrant methylation of tissue factor pathway inhibitor 2 in pure pancreatic juice for pancreatic carcinoma, *Cancer Sci* 97(11), 1267–1273, 2006.

53. Ibanez de Caceres, I., Battagli, C., Esteller, M., Herman, J.G., Dulaimi, E., Edelson, M.I., Bergman, C., Ehya, H., Eisenberg, B.L., and Cairns, P., Tumor cell-specific BRCA1 and RASSF1A hypermethylation in serum, plasma, and peritoneal fluid from ovarian cancer patients, *Cancer Res* 64(18), 6476–6481, 2004.

54. Palmisano, W.A., Divine, K.K., Saccomanno, G., Gilliland, F.D., Baylin, S.B., Herman, J.G., and Belinsky, S.A., Predicting lung cancer by detecting aberrant promoter methylation in sputum, *Cancer Res* 60(21), 5954–5958, 2000.

55. Righini, C.A., de Fraipont, F., Timsit, J.F., Faure, C., Brambilla, E., Reyt, E., and Favrot, M.C., Tumor-specific methylation in saliva: A promising biomarker for early detection of head and neck cancer recurrence, *Clin Cancer Res* 13(4), 1179–1185, 2007.

56. Battagli, C., Uzzo, R.G., Dulaimi, E., Ibanez de Caceres, I., Krassenstein, R., Al-Saleem, T., Greenberg, R.E., and Cairns, P., Promoter hypermethylation of tumor suppressor genes in urine from kidney cancer patients, *Cancer Res* 63(24), 8695–8699, 2003.

57. Friedrich, M.G., Weisenberger, D.J., Cheng, J.C., Chandrasoma, S., Siegmund, K.D., Gonzalgo, M.L., Toma, M.I., Huland, H., Yoo, C., Tsai, Y.C., Nichols, P.W., Bochner, B.H., Jones, P.A., and Liang, G., Detection of methylated apoptosis-associated genes in urine sediments of bladder cancer patients, *Clin Cancer Res* 10(22), 7457–7465, 2004.
58. Gonzalgo, M.L., Pavlovich, C.P., Lee, S.M., and Nelson, W.G., Prostate cancer detection by GSTP1 methylation analysis of postbiopsy urine specimens, *Clin Cancer Res* 9(7), 2673–2677, 2003.
59. Stroun, M., Anker, P., Lyautey, J., Lederrey, C., and Maurice, P.A., Isolation and characterization of DNA from the plasma of cancer patients, *Eur J Cancer Clin Oncol* 23(6), 707–712, 1987.
60. Stroun, M., Lyautey, J., Lederrey, C., Olson-Sand, A., and Anker, P., About the possible origin and mechanism of circulating DNA apoptosis and active DNA release, *Clin Chim Acta* 313(1–2), 139–142, 2001.
61. Giacona, M.B., Ruben, G.C., Iczkowski, K.A., Roos, T.B., Porter, D.M., and Sorenson, G.D., Cell-free DNA in human blood plasma: Length measurements in patients with pancreatic cancer and healthy controls, *Pancreas* 17(1), 89–97, 1998.
62. Widschwendter, M. and Menon, U., Circulating methylated DNA: A new generation of tumor markers, *Clin Cancer Res* 12(24), 7205–7208, 2006.
63. Tamkovich, S.N., Laktionov, P.P., Rykova, E.Y., and Vlassov, V.V., Simple and rapid procedure suitable for quantitative isolation of low and high molecular weight extracellular nucleic acids, *Nucleosides Nucleotides Nucleic Acids* 23(6–7), 873–877, 2004.
64. Shapiro, B., Chakrabarty, M., Cohn, E.M., and Leon, S.A., Determination of circulating DNA levels in patients with benign or malignant gastrointestinal disease, *Cancer* 51(11), 2116–2120, 1983.
65. Wallner, M., Herbst, A., Behrens, A., Crispin, A., Stieber, P., Goke, B., Lamerz, R., and Kolligs, F.T., Methylation of serum DNA is an independent prognostic marker in colorectal cancer, *Clin Cancer Res* 12(24), 7347–7352, 2006.
66. Esteller, M., Sanchez-Cespedes, M., Rosell, R., Sidransky, D., Baylin, S.B., and Herman, J.G., Detection of aberrant promoter hypermethylation of tumor suppressor genes in serum DNA from non-small cell lung cancer patients, *Cancer Res* 59(1), 67–70, 1999.
67. An, Q., Liu, Y., Gao, Y., Huang, J., Fong, X., Li, L., Zhang, D., and Cheng, S., Detection of p16 hypermethylation in circulating plasma DNA of non-small cell lung cancer patients, *Cancer Lett* 188(1–2), 109–114, 2002.
68. Wong, I.H., Lo, Y.M., Zhang, J., Liew, C.T., Ng, M.H., Wong, N., Lai, P.B., Lau, W.Y., Hjelm, N.M., and Johnson, P.J., Detection of aberrant p16 methylation in the plasma and serum of liver cancer patients, *Cancer Res* 59(1), 71–73, 1999.
69. Widschwendter, A., Muller, H.M., Fiegl, H., Ivarsson, L., Wiedemair, A., Muller-Holzner, E., Goebel, G., Marth, C., and Widschwendter, M., DNA methylation in serum and tumors of cervical cancer patients, *Clin Cancer Res* 10(2), 565–571, 2004.
70. Hu, S., Ewertz, M., Tufano, R.P., Brait, M., Carvalho, A.L., Liu, D., Tufaro, A.P., Basaria, S., Cooper, D.S., Sidransky, D., Ladenson, P.W., and Xing, M., Detection of serum deoxyribonucleic acid methylation markers: A novel diagnostic tool for thyroid cancer, *J Clin Endocrinol Metab* 91(1), 98–104, 2006.
71. Dulaimi, E., Hillinck, J., Ibanez de Caceres, I., Al-Saleem, T., and Cairns, P., Tumor suppressor gene promoter hypermethylation in serum of breast cancer patients, *Clin Cancer Res* 10(18 Pt 1), 6189–6193, 2004.
72. Skvortsova, T.E., Rykova, E.Y., Tamkovich, S.N., Bryzgunova, O.E., Starikov, A.V., Kuznetsova, N.P., Vlassov, V.V., and Laktionov, P.P., Cell-free and cell-bound circulating DNA in breast tumours: DNA quantification and analysis of tumour-related gene methylation, *Br J Cancer* 94(10), 1492–1495, 2006.
73. Kuroi, K., Tanaka, C., and Toi, M., Plasma nucleosome levels in node-negative breast cancer patients, *Breast Cancer* 6(4), 361–364, 1999.
74. Chiu, R.W., Poon, L.L., Lau, T.K., Leung, T.N., Wong, E.M., and Lo, Y.M., Effects of blood-processing protocols on fetal and total DNA quantification in maternal plasma, *Clin Chem* 47(9), 1607–1613, 2001.
75. Taylor, K.H., Pena-Hernandez, K.E., Davis, J.W., Arthur, G.L., Duff, D.J., Shi, H., Rahmatpanah, F.B., Sjahputera, O., and Caldwell, C.W., Large-scale CpG methylation analysis identifies novel candidate genes and reveals methylation hotspots in acute lymphoblastic leukemia, *Cancer Res* 67(6), 2617–2625, 2007.
76. Chan, M.W., Chan, L.W., Tang, N.L., Tong, J.H., Lo, K.W., Lee, T.L., Cheung, H.Y., Wong, W.S., Chan, P.S., Lai, F.M., and To, K.F., Hypermethylation of multiple genes in tumor tissues and voided urine in urinary bladder cancer patients, *Clin Cancer Res* 8(2), 464–470, 2002.
77. Lindsey, J.C., Lusher, M.E., Anderton, J.A., Bailey, S., Gilbertson, R.J., Pearson, A.D., Ellison, D.W., and Clifford, S.C., Identification of tumour-specific epigenetic events in medulloblastoma development by hypermethylation profiling, *Carcinogenesis* 25(5), 661–668, 2004.

78. Shivapurkar, N., Toyooka, S., Toyooka, K.O., Reddy, J., Miyajima, K., Suzuki, M., Shigematsu, H., Takahashi, T., Parikh, G., Pass, H.I., Chaudhary, P.M., and Gazdar, A.F., Aberrant methylation of trail decoy receptor genes is frequent in multiple tumor types, *Int J Cancer* 109(5), 786–792, 2004.

79. Widschwendter, A., Muller, H.M., Hubalek, M.M., Wiedemair, A., Fiegl, H., Goebel, G., Mueller-Holzner, E., Marth, C., and Widschwendter, M., Methylation status and expression of human telomerase reverse transcriptase in ovarian and cervical cancer, *Gynecol Oncol* 93(2), 407–416, 2004.

80. Liu, S.S., Leung, R.C., Chan, K.Y., Chiu, P.M., Cheung, A.N., Tam, K.F., Ng, T.Y., Wong, L.C., and Ngan, H.Y., p73 expression is associated with the cellular radiosensitivity in cervical cancer after radiotherapy, *Clin Cancer Res* 10(10), 3309–3316, 2004.

81. Esteller, M., Hamilton, S.R., Burger, P.C., Baylin, S.B., and Herman, J.G., Inactivation of the DNA repair gene O6-methylguanine-DNA methyltransferase by promoter hypermethylation is a common event in primary human neoplasia, *Cancer Res* 59(4), 793–797, 1999.

82. Zhang, H., Fu, W.L., and Huang, Q., Mapping of the methylation pattern of the hMSH2 promoter in colon cancer, using bisulfite genomic sequencing, *J Carcinog* 5, 22, 2006.

83. Sasaki, M., Kotcherguina, L., Dharia, A., Fujimoto, S., and Dahiya, R., Cytosine-phosphoguanine methylation of estrogen receptors in endometrial cancer, *Cancer Res* 61(8), 3262–3266, 2001.

84. Banno, K., Yanokura, M., Susumu, N., Kawaguchi, M., Hirao, N., Hirasawa, A., Tsukazaki, K., and Aoki, D., Relationship of the aberrant DNA hypermethylation of cancer-related genes with carcinogenesis of endometrial cancer, *Oncol Rep* 16(6), 1189–1196, 2006.

85. Virmani, A.K., Rathi, A., Zochbauer-Muller, S., Sacchi, N., Fukuyama, Y., Bryant, D., Maitra, A., Heda, S., Fong, K.M., Thunnissen, F., Minna, J.D., and Gazdar, A.F., Promoter methylation and silencing of the retinoic acid receptor-beta gene in lung carcinomas, *J Natl Cancer Inst* 92(16), 1303–1307, 2000.

86. Zochbauer-Muller, S., Fong, K.M., Virmani, A.K., Geradts, J., Gazdar, A.F., and Minna, J.D., Aberrant promoter methylation of multiple genes in non-small cell lung cancers, *Cancer Res* 61(1), 249–255, 2001.

87. Shaw, R.J., Liloglou, T., Rogers, S.N., Brown, J.S., Vaughan, E.D., Lowe, D., Field, J.K., and Risk, J.M., Promoter methylation of P16, RARbeta, E-cadherin, cyclin A1 and cytoglobin in oral cancer: Quantitative evaluation using pyrosequencing, *Br J Cancer* 94(4), 561–568, 2006.

88. Valenzuela, M.T., Galisteo, R., Zuluaga, A., Villalobos, M., Nunez, M.I., Oliver, F.J., and Ruiz de Almodovar, J.M., Assessing the use of p16(INK4a) promoter gene methylation in serum for detection of bladder cancer, *Eur Urol* 42(6), 622–628; discussion 628–630, 2002.

89. Hoque, M.O., Feng, Q., Toure, P., Dem, A., Critchlow, C.W., Hawes, S.E., Wood, T., Jeronimo, C., Rosenbaum, E., Stern, J., Yu, M., Trink, B., Kiviat, N.B., and Sidransky, D., Detection of aberrant methylation of four genes in plasma DNA for the detection of breast cancer, *J Clin Oncol* 24(26), 4262–4269, 2006.

90. Grady, W.M., Rajput, A., Lutterbaugh, J.D., and Markowitz, S.D., Detection of aberrantly methylated hMLH1 promoter DNA in the serum of patients with microsatellite unstable colon cancer, *Cancer Res* 61(3), 900–902, 2001.

91. Zou, H.Z., Yu, B.M., Wang, Z.W., Sun, J.Y., Cang, H., Gao, F., Li, D.H., Zhao, R., Feng, G.G., and Yi, J., Detection of aberrant p16 methylation in the serum of colorectal cancer patients, *Clin Cancer Res* 8(1), 188–191, 2002.

92. Bearzatto, A., Conte, D., Frattini, M., Zaffaroni, N., Andriani, F., Balestra, D., Tavecchio, L., Daidone, M.G., and Sozzi, G., p16(INK4A) hypermethylation detected by fluorescent methylation-specific PCR in plasmas from non-small cell lung cancer, *Clin Cancer Res* 8(12), 3782–3787, 2002.

93. Jeronimo, C., Usadel, H., Henrique, R., Silva, C., Oliveira, J., Lopes, C., and Sidransky, D., Quantitative GSTP1 hypermethylation in bodily fluids of patients with prostate cancer, *Urology* 60(6), 1131–1135, 2002.

19 DNA Methylation Profiles as Prognostic Markers for Cancer

Sam Thiagalingam and Panagiotis Papageorgis

CONTENTS

19.1 PROGNOSIS OF CANCER

Prognosis of cancer is broadly defined as the estimation of the likely course and future outcome of a cancer patient's disease and therefore, it is an attempt to predict whether the patient will recover or have a recurrence of the tumor. Many factors affect a cancer patient's prognosis and overall survival rates. Survival rates are a measure of the percentage of patients with a certain type and stage of cancer who survive the disease for a specific period after the diagnosis. Some of the most important factors that determine the survival rates are the type and location of the cancer as well as the stage and grade. Other features that may also affect the prognosis include the patient's age, health history, and response to treatment.

It is widely accepted that besides traditional approaches (i.e., histopathological methods) for determining the prognosis of patients with different cancer types, more molecular methods are urgently needed for rapid, painless, and accurate determination of prognosis. Novel molecular methods and markers have begun to emerge during the past few years in order to facilitate and strengthen the prognostic procedures.

One of the major criteria for selecting an ideal molecular prognostic marker is the ability to utilize it with high level of sensitivity that allows the determination of the status in clinical samples that consist of a small number of affected cells in a background of a large number of normal cells. Molecular tests usually begin with the preparation of a protein or RNA/DNA extract from a clinical sample or certain type of body fluid. Even within the same types of samples, the ratio of neoplastic to normal cells varies considerably from one sample to another. Often, it is difficult to specifically isolate neoplastic cells for analysis from the available clinical samples as these are generally composed primarily of cellular debris and cell free substrate along with a small number of the affected target cells. As a result, the extracts from clinical samples frequently consist of a hetero-geneous mixture of normal and cancer cell DNA/RNA and protein requiring sensitive tests that selectively allow the evaluation of a single desired parameter.

Therefore, the use of molecular approaches with a high level of sensitivity will be necessary for more accurate and sensitive detection of the cancer prognostic markers if there were only a few affected cells against a large background of normal unaffected cells. Since the DNA molecule exhibits significantly higher stability compared to RNA during preparations from clinical samples, it remains as the preferred substrate for molecular evaluations in clinical practice. More importantly, because of the availability of various methods to assess the promoter DNA hypermethylation of specific genes even when there were only a few affected cells in the clinical sample of choice, it has rapidly become a very attractive approach for assessing the prognosis of cancer patients [1] as discussed in detail in this chapter.

19.2 DNA METHYLATION IS A HALLMARK OF CANCER

DNA methylation plays crucial roles in the control of gene expression and nuclear architecture. It is the most widely studied epigenetic modification in humans [2]. In general, the CpG dinucleotides that can potentially be methylated are not randomly distributed within the human genome; instead, CpG-rich regions, also called as "CpG islands," are the targets for this epigenetic modification and exist usually in an unmethylated state in normal cells [3]. Typically, the CpG islands are a contiguous region of 200 bp or longer-DNA with a high density of cytosine–phosphoguanine (CpG) nucleotides, which are often located in the promoter region or the first exons of any given gene. The GC content in these islands is >0.5 and the ratio of observed CpG frequency (CpG_{obs}/CpG_{exp}, observed to expected ratio based on GC content) usually exceeds 0.6 [4]. The unmethylated status of the CpG-island-containing genes corresponds with the ability to be tran-scribed into mRNA in the presence of the necessary transcriptional activators. In cancer cells, loss of function due to the transcriptional silencing of tumor-suppressor genes by CpG-island-promoter hypermethylation is a key event for the initiation and progression of the tumorigenic process [5], contributing to all of the typical hallmarks of a cancer cell that result from the inactivation of a tumor-suppressor gene [6]. On the other hand, repetitive genomic sequences are also highly methylated. There is evidence to suggest that the maintenance of DNA methylation plays an important role in the protection of chromosomal integrity by silencing transposons that are often localized to the repetitive sequences and may cause chromosomal instability due to the transposi-tions that mediate translocations and insertions leading to gene disruptions [7]. Even though it is not well studied and understood as DNA hypermethylation events, the global DNA hypomethylation that is characteristic of the cancer cells is also likely to contribute to both hyperexpression of specific genes as well as the large-scale genomic stability changes that are central features of tumorigenesis [8]. On the basis of this phenomenon, it is likely that in the future DNA hypomethylated genes or DNA regions might as well emerge as potentially useful prognostic markers. Some examples include hypomethylation of urokinase plasminogen activator (uPA) [9], P-Cadherin (CDH3) [10] as well as hypomethylation of repetitive genomic sequences such as the long interspersed nuclear element-1 (LINE-1), short interspersed elements (SINE) of repeated sequence originally characterized by the action of the Alu restriction endonuclease (Alu) and satellite 2 (Sat2) [11].

Until recently, cancer epigenetic studies have primarily relied upon investigating DNA methylation at promoter regions of specific candidate genes (Table 19.1). Despite much insight has been gained into the role of epigenetic defects in cancer, more studies will be required to assemble overall patterns of DNA methylation in cancer cells with the use of unbiased genome-wide approaches to further expand this knowledge base to cancer diagnostics/prognostics in the near future. Furthermore, gene silencing in cancer mediated due to DNA hypermethylation should be more accurately considered in the context of other epigenetic modifications. It is associated with the formation of more "compact,"

TABLE 19.1

Selected Genes That Are Hypermethylated in Cancer and Are Associated with Poor Prognosis

Gene Name	Function	Cancer Types	Methylation Detection in Body Fluids and Surface Epithelial Linings	References
CCND2	Cell cycle	Breast, ovarian	Nipple aspirate	[55]
p14/ARF	Cell cycle	Bladder, breast	Urine, nipple aspirate	[20,28,56]
CDKN2B (Ink4B/p15)	Cell cycle	Leukemias liver	Blood lymphocytes	[57–59]
CDKN2A (Ink4A/p16)	Cell cycle	Colon, breast, lung, head and neck, pancreas, prostate, melanoma, neuroblastoma, esophageal, bladder	Serum, sputum, saliva, urine, nipple aspirate, stool, blood, bronchial brushings	[20,24,28, 56,60–65]
p73	Cell cycle	Leukemia	Blood lymphocytes	[66]
Rb	Cell cycle	Retinoblastoma	N/A	[67]
BRCA1	DNA repair	Breast, ovarian, esophageal	Serum, peritoneal fluid	[68–70]
MGMT	DNA repair	Colon, lung, head and neck, pancreas, lymphomas, esophageal, bladder, glioma/glioblastoma	Serum, sputum, saliva, urine, stool, blood, bronchial brushings	[24–26,28, 60–62,64, 71–74]
MLH1	DNA repair	Colon, ovarian, gastric	Serum, stool	[24,27,60]
DAPK	Apoptosis	Colon, lung, head and neck, esophageal, breast, cervical	Serum, saliva, sputum, nipple aspirate, blood, bronchial brushings	[25,26,61, 64,65,75]
FHIT	Apoptosis	Bladder	N/T	[76]
RIL	Apoptosis	Leukemia, colon	N/T	[77]
RUNX3	Apoptosis	Prostate, gastric, lung, colon, breast	N/T	[78–82]
TMS1	Apoptosis	Neuroblastoma, prostate, lung	Sputum	[83–85]
APC	Wnt signaling	Colon, bladder, lung, esophageal	Serum, urine, stool	[23,28,31, 32,60–62]
RASSF1A	Ras signaling	Bladder, breast, ovarian, cervical	Serum, peritoneal fluid, urine, nipple aspirate	[20,28,70,86]
SOCS1	JAK-STAT pathway	Liver, myeloma	N/T	[87]
SOCS3	JAK-STAT pathway	Lung	N/T	[88]
EDNRB	Endothelin receptor	Prostate	N/T	[89]
SYK	Tyrosine kinase	Liver	N/T	[90]
ER	Growth factor response	Breast, melanoma	Serum	[91,92]
PR	Growth factor response	Breast	N/T	[91]
RARβ2	Growth factor response	Breast, bladder, cervical	Nipple aspirate, urine	[20,28,75]
IGFBP3	Growth factor signaling	Lung, skin	N/T	[93]
HOXA9	Transcription factor	Neuroblastoma	N/T	[83]

(*continued*)

TABLE 19.1 (continued)
Selected Genes That Are Hypermethylated in Cancer
and Are Associated with Poor Prognosis

Gene Name	Function	Cancer Types	Methylation Detection in Body Fluids and Surface Epithelial Linings	References
ID4	Transcription factor	Colon	N/T	[94]
IRF8	Transcription factor	Colon	N/T	[95]
MYOD1	Transcription factor/differentiation	Cervical, colon	Serum	[86,96]
CDH1/E-cadherin	Cell adhesion	Breast, bladder, prostate, esophageal, lung, cervical	Urine, blood, bronchial brushings	[28,61,62,64, 86,97,98]
CDH13	Cell adhesion	Cervical, lung, breast, colon	Serum	[75,86, 97–101]
PCDH20	Cell adhesion	Lung	N/T	[102]
CD44	Cell–cell, cell–matrix interactions	Prostate	Serum	[103,104]
Lamin A/C	Nuclear structure	Lymphoma, leukemia	N/T	[105]
LKB1	Tumor invasion/architecture	Colon, breast, lung	N/T	[106]
TIMP3	Metalloproteinase inhibitor	Bladder, colon, esophageal, gastric	Serum, urine	[28,61,107]
TPEF/HPP1	Transmembrane protein	Colon, bladder	Serum, stool	[60,108]
TSP-1	Anti-angiogenic	Glioblastoma	N/T	[109]
COX2	Inflammation	Colon, stomach	N/T	[110]
GSTP1	Detoxification	Prostate, kidney, breast, bladder	Serum, urine, nipple aspirate	[28,29,111–113]
VHL	Hypoxic response	Renal	N/T	[114]

Note: N/A, not applicable; N/T, not tested.

nuclease-resistant chromatin formed due to the coordinated effects of the methyl-CpG binding proteins, DNA methyltransferases (DNMTs), histone deacetylases, and histone methyltransferases [12,13]. The DNA methyltransferases that have been more extensively studied and shown to regulate the methylation status of the DNA are DNMT1 (involved in the maintenance of DNA methylation) and DNMT3a and DNMT3b (responsible for de novo DNA methylation). Interestingly, there have been some reports suggesting overexpression of these enzymes in various types of cancer. However, this has not been established as a universal phenomenon correlating to the cancer development [14]. To obtain an accurate understanding of the epigenome of various cancers, more studies in the future will be necessary to establish a cause and effect relationship between differential expression of DNA methylases/demethylases and cancer as well as the deciphering of the intricacies underlying the differential and coordinated activities of the various other chromatin modifier enzymes.

19.3   SILENCING OF GENES DUE TO PROMOTER DNA HYPERMETHYLATION AND POOR PROGNOSIS

For a molecular prognostic marker to be clinically applicable, it must be specific, sensitive, and ideally should be detectable in specimens obtained through minimally invasive procedures.

Promising results have already been obtained with the detection of aberrantly methylated CpG islands in the promoter region of several genes in DNA samples derived from nipple aspirates, saliva, serum, sputum, stool, and urine of cancer patients (Table 19.1). Gene silencing associated downregulation of expression or loss of function of a number of these genes has significantly been associated with poor prognosis of cancer patients. Therefore, detection of promoter DNA hyper-methylation of a specific set of genes could be directly or indirectly correlated with and it could provide useful insight about the prognostic outcome of these patients. It should also be noted that changes in DNA methylation also occur in normal epithelia [15,16]. However, in the past few years, it has become increasingly compelling that one of the most encouraging types of prognostic marker would be to detect DNA hypermethylation of CpG islands localized to the promoter regions of specific set of cancer-related genes that impair normal gene expression [17,18]. A high degree of cytosine methylation in a gene's promoter can lead to complete block in transcription of the affected genes [17]. Multiple types of cancers have been found to exhibit inactivation of tumor-suppressor genes via this mechanism. Because of the promise for differential DNA hypermethylation of specific genes as prognostic markers, there is extensive research currently underway to identify tumor-specific DNA methylation events that afford enough sensitivity and specificity to be utilized as biomarkers to predict prognostic outcome of cancer patients. One of the major obstacles to overcome is the fact that tumor DNA is present in minimal amounts in tumor specimens or biological fluids. Thus, exquisitely sensitive techniques need to be utilized to detect and analyze DNA-methylation patterns in tumor-derived DNA.

PCR-based assays, most common of which is methylation-specific PCR (MSP), have been developed to rapidly and accurately evaluate methylation status of regions of DNA [19] and are discussed in detail later in the chapter. The sensitivity of MSP is approximately one cancer cell among 1,000 normal cells, which is sensitive enough to detect tumor DNA in most body fluids. This is a very important parameter for the application of these assays for the prognosis of cancer. The collection of biological fluids, the isolation of cancer cells from them, and reliable detection of hypermethylated genes can greatly enhance the prognostic process in terms of time, cost, sensitivity, and accuracy. This approach has been successfully used to detect cancer cells in the nipple aspirate from women with breast cancer [20], in the saliva of patients with oral cancer [21], in the sputum and bronchoalveolar lavage fluid of lung cancer patients [22,23], in the stool of colon cancer patients [24], in the serum of patients with lung [25], head and neck [26], and colorectal cancers [27] and in the urine of bladder [28] and prostate cancer patients [29]. In this manner, groups of hypermethylated genes in different combinations can be associated with different tumor types as well as different stages of cancer progression [30].

Approaches that allow the simultaneous analysis of multiple DNA methylation markers are critically important as combinations of several markers are likely to provide a more accurate and reliable prognosis than analysis of one marker at a time. For example, methylation of cell cycle regulation genes that encode *INK4A* (also known as *p16*; a cyclin-dependent kinase inhibitor that is encoded by *CDKN2A*), *INK4B* (*p15*), *p14/ARF*, *CCND2* (Cyclin D2), retinoblastoma protein (Rb), and p73 (p53 homologue), DNA repair genes like *BRCA1* (breast cancer associated-1), *MLH1* (mutL homologue 1) and *MGMT* (O6-methylguanine-DNA methyltransferase), cell adhesion like *CDH1* (E-cadherin), *CDH13* (H-cadherin), and *PCDH20* (protocadherin 20) as well as apoptosis related genes like *DAPK* (death-associated protein kinase) and *RUNX3*, have been associated with poor prognosis of patients with numerous types of cancers Table 19.1. Some of these genes were found to be methylated in serum samples of over 50% of lung and head and neck cancer patients [25,26]. Moreover, methylation of *APC* (adenomatous polyposis coli), which is involved in Wnt signaling, was observed in the serum and plasma DNA of early-stage lung cancer and esophageal cancer patients [31,32]. Finally, promoter hypermethylation of the detoxification gene *GSTP1* (Glutathione *S*-transferase protein 1) has been detected in the serum and urine of prostate cancer patients [29]. The long list of cancer-associated methylated genes has continued to expand as methylated targets have been catalogued in many primary cancers [33]. Because of the variability in the levels of methylation affecting specific CpG sites and the possibility that a range in the degree of methylation could also play an equally important role in tumorigenesis, the real-time MSP-based assays could be used to measure the degree of methylation of the affected genes as quantitative prognostic markers. A representative summary of the various DNA methylation associated prognostic markers are listed in Table 19.1.

19.4 TECHNOLOGICAL APPROACHES USED IN CANCER METHYLOMICS FOR PROGNOSIS: WHAT IS AN IDEAL ASSAY FOR DNA METHYLATION DETECTION? (SPECIFICITY VERSUS SENSITIVITY)

The most important parameters that have to be considered when one determines the efficacy and success of a molecular marker to be utilized in an assay are the degree of sensitivity (the least amount of the substrate that can be detected by the assay) and specificity (the percentage of assays that will accurately distinguish between normal versus tumor tissue derived samples). When an assay is chosen for application, one is always faced with striking a balance between sensitivity and specificity. Often, high levels of sensitivity might reduce specificity leading to the apparent detection when the intended substrate is not actually present in the sample (false-positive results). On the other hand, a highly specific assay might not always be very sensitive to detect substrate (false-negative results). Therefore, before large scale applications are pursued, the sensitivity and specificity should be clearly established well in advance using the control samples that contain already known amounts and types of substrates to be tested. The validity of these assays should then be formally assessed in clinical trials by comparing these results to previously established test results. The overall conclusion from several studies is that, in general, the most sensitive tests are not always the better ones and vice versa.

19.5 METHODS FOR EVALUATING DNA METHYLATION STATUS OF GENES

Since the early 1980s, when the first oncogene was discovered, genetics has taken the leading role in cancer research. At the beginning, epigenetics did not attract much of the scientists' attention. Interestingly, the fact is that the first major findings in the fields of cancer genetics and epigenetics occurred around the same period [3,34]. One of the main reasons for the delay in the attention to the epigenetics has been the technical limitations. In general, the first generation of the techniques that were developed for studying DNA methylation such as bisulfite treatment, methylation-sensitive restriction enzymes, and quantitative PCR-based methods has been largely limited to candidate genes.

Although techniques such as high-performance liquid chromatography (HPLC) and high-performance capillary electrophoresis (HPCE) have been available for many years in order to accurately quantify the total amount of 5-methylcytosine [35], the whole-genome study of DNA methylation of particular sequences was initially almost entirely based on the use of enzymes that can differentially distinguish between methylated and unmethylated recognition sites at the gene promoter (for example, MspI and HpaII or SmaI and XmaI). This approach has many disadvantages, including incomplete digestion with the restriction enzymes and limitations with the number and type of regions that can be sampled. It also involves Southern blotting which requires significant amounts of DNA, a parameter which can be particularly challenging when screening primary tumors or body fluid samples from only which limited quantities of DNA material can be obtained.

19.5.1 METHYLATION-SPECIFIC PCR AND BISULFITE SEQUENCING

The predominant and the most widely used technique which has become the gold standard for the determination of the methylation status of DNA is the methylation-specific PCR (MSP). The discovery and establishment of this methodology was an important advance in cancer epigenetics [19,36].

The molecular basis of this assay is that it uses sodium bisulfite modified genomic DNA to distinguish methylated over unmethylated CpG regions of DNA. Sodium bisulfite modification converts cytosine to uracil at almost 100% efficiency; however, if the cytosine residue is methylated, it will be resistant to conversion. PCR products can be generated to determine the methylated status of specific genes with the use of primer pairs specific to unmethylated or methylated template DNA. Therefore, the major advantage of this method is its great sensitivity. The MSP assay allows detection of gene-specific promoter DNA hypermethylation in small amounts of DNA template (a few ng) recovered from primary tumors or biological fluids of cancer patients. Because the two pairs of primers are designed with each pair being able to recognize specifically either the methylated or unmethylated alleles, the contaminating normal tissue will not interfere with detecting DNA methylation. Moreover, unlike mutation screens that require sequencing using several primers covering long coding DNA sequences to detect gene inactivation or dysfunction, the MSP approach uses one set of two primer pairs to assay a common genomic region where the detection of DNA hypermethylation is associated with loss of gene expression.

Furthermore, it is also very important that the MSP assay is amenable to and can also be performed directly on tissue sections by in situ MSP [37]. This approach can be very useful in cases where tumor cells cannot be detected in any type of biological fluids. Upon surgical removal of the tumor, such technique can be applied solely or in combination with histopathological approaches to more accurately predict the prognosis of the patient. Recent innovations have improved the sensitivity of the MSP procedure to detect one methylated allele among the $\sim$50,000 unmethylated alleles by incorporating a nested PCR approach [23]. For the nested MSP assay, the first amplification primers are placed outside the CpG islands of the gene that is being evaluated without preference for methylated or unmethylated alleles. A portion of the first round PCR product is then used in a second round PCR with nested primers specific to methylated or unmethylated alleles. This approach improves the sensitivity of MSP for examining biological fluids which are highly contaminated with normal cells, or formalin-fixed tissues where the DNA might be highly degraded. Furthermore, the standardization of the MSP assay with respect to primers, amount of DNA template, and PCR conditions are important parameters for minimization of assay variability between different laboratories. Because these assays rely on PCR of CG-rich DNA, the use of positive and negative control samples for methylated DNA (for example, in vitro methylated DNA) and unmethylated DNA (normal tissue, usually placental DNA) is always required when one uses the MSP approach.

The combination of these methods with genomic sequencing [38] has enabled many research groups to evaluate DNA methylation, even when limited amounts of material are available. It is important to remember that the PCR-based methods interrogate the methylation status only at CpG sites that are complementary to the primers that are directed to specific genomic sequences. Therefore, the predominant methylation profile in a sample will not necessarily be reflected in the results of such experiments, and therefore bisulfite treatment followed by genomic sequencing is necessary to obtain a complete picture of the heterogeneous DNA-methylation patterns that exist within the probed sequences of a tumor sample.

During the past few years, promoter hypermethylation detection has been greatly enhanced by the development of quantitative assays. Real-time PCR-based approaches, which involve amplification by DNA polymerases with monitoring of fluorescent signals during the actual amplification process, such as SYBR green or Taqman methods can be used to quantify the number of methylated alleles (in a single region) in DNA samples from normal or affected individuals. Other recent innovations that have been introduced to the quantitative PCR-based methods include bisulfite treatment in combination with MethyLight [39] or pyrosequencing [40], which have greatly improved the ability to detect minimal amounts of aberrant DNA methylation in samples with questionable data obtained using other methods. When MethyLight method is used, bisulfite-converted genomic DNA is amplified using gene locus and methylation-specific PCR primers flanking an oligonucleotide probe with a $5'$ fluorescent reporter dye and a $3'$ quencher dye.

The 5′ to 3′ nuclease activity of Taq DNA polymerase releases the reporter dye from the oligonu-cleotide probe, whose fluorescence can be quantified by the laser detector for the cycle number at which the fluorescent signal crosses a threshold in the exponential phase of the PCR reaction. On the other hand, pyrosequencing is based on designating a methylation-depended DNA sequence as a kind of single nucleotide polymorphism (methylSNP) of the C/T type. On the basis of this concept, DNA is treated with sodium bisulfite and subjected to SNP typing using PyroMeth sequencing in a quantitative manner. A range of novel approaches have recently been developed for assessing patterns of DNA methylation in normal and cancer cells at the level of the genome. In Section 19.5.2, we provide an overview of these methods, focusing on recent technological advancements that have allowed genome-wide approaches to be taken advantage of with respect to their applications to cancer prognosis.

19.5.2 Assessing Global DNA Hypermethylation Patterns as Prognostic Markers and Screening for Novel Targets in Cancer

When the concepts were developed to evaluate the DNA methylation status of individual genes, it became clear that simultaneous screening of multiple genes using genomic approaches may provide accurate tools for prognostic purposes. Restriction landmark genomic scanning (RLGS) was one of the first methods that was adapted for genome-wide DNA methylation analysis [41]. With this technique, DNA is first digested using methylation-sensitive restriction enzymes, radioactively labeled at unmethylated sites, and size-fractionated in one dimension using gel electrophoresis. The resulting products are digested again with a second restriction enzyme that is specific for high-frequency targets, and the fragments are separated in a second dimension to generate a number of scattered hot spots of DNA methylation. When normal and tumor samples are compared, the relative positions and intensity of the various spots could be used to reveal their location and the frequency of the corresponding restriction site, respectively [42]. However, this technique has some disad-vantages including its dependence on specific restriction sites that are not present in all CpG islands and also the fact that not all of the resulting DNA fragments can be resolved by the two-dimensional electrophoresis. Even under these circumstances, the global analysis of the methylation status of approximately 1,000 unselected CpG islands can be achieved using this method [41], which is highly satisfactory for most research/prognostic purposes.

Other useful approaches for detecting genome-wide differential DNA methylation profiles involve variations of arbitrarily primed PCR [43]. Some examples include methylation-sensitive arbitrary primed PCR [44], methylated CpG-island amplification (MCA) [45], and amplification of intermethylated sites (AIMS) [46]. The usefulness of these methods lies in the fact that arbitrary primed PCR is carried out using DNA templates that have been enriched for methyl sequences as the test samples to engineer preferential amplification of CpG islands and gene-rich regions [46]. However, the potentially methylated regions that are identified using these techniques require further validation by bisulfite genomic sequencing and one should be aware of the background due to PCR "noise" from repetitive sequences.

One of the recently developed successful approaches for studying genome-wide CpG-island methylation patterns involves the screening for CpG-island-containing promoters using promoter microarrays. The major advantage of this technique over arbitrarily primed PCR methods is that it avoids further cloning and sequencing. An example of such an approach is differential methylation hybridization (DMH), which allows the simultaneous detection of methylation in a large number of CpG-island loci [47]. The CpG-island-containing library of DNA fragments are spotted on high-density arrays and the genomic DNA from the various tissues to be tested is digested using methylation-sensitive enzymes. The digestion products are then used as templates for PCR upon ligation to linkers in order to generate targets for screening for the hypermethylated sequences within the CpG-island library to identify sequences that are hypermethylated in cancer patients but not in normal DNA samples. A related method is the HELP assay (HpaII tiny fragment enrichment by ligation-mediated PCR), which is based on digestion of the DNA samples with a methylation-sensitive restriction enzyme or its methylation-insensitive isoschizomer followed by co-hybridization to a genomic DNA microarray [48]. This assay has been successfully used to identify a large number of tissue-specific, differentially methylated regions [47], and can therefore be used in prognostic applications by comparing normal and cancer cells.

High-throughput DNA methylation profiling has also been performed using universal bead arrays [49], which is based on designing two pairs of probes in order to interrogate either the top or the bottom DNA strand. The first pair of probes consists of an allele-specific oligonucleotide (ASO) and a locus-specific oligonucleotide (LSO) for the methylated state of the CpG site whereas the other probe consists of the corresponding ASO-LSO pair for the unmethylated state, with the 3′-portion of the ASO determining the specificity to each state. ASOs and LSOs incorporate a universal PCR primer sequence at the 5′ and 3′ end, respectively, and upon hybridization of pooled oligonucleotides to bisulfite-treated, biotinylated genomic DNA immobilized on paramagnetic beads, an allele-specific extension step is performed followed by ligation in order to generate amplifiable templates. After a PCR reaction using fluorescently labeled universal PCR primers, the product is hybridized to a microarray containing complementary sequences for 1536 CpG sites. This method has been successfully applied for identification of a human embryonic stem cell epigenetic signature [50] and it has the potential for adaptation for cancer prognostic screenings.

ChIP(chromatin immunoprecipitation)-on-chip (arrays) based approaches have also provided another important advance in the epigenomic profiling of tumors. In a recent study, DNA that was immunoprecipitated from breast cancer cells using antibodies against methyl-CpG-binding domain proteins (MBDs), which have high-binding affinity for methylated cytosines [51], has been used for hybridization to genomic DNA microarray platforms (chip) in order to identify novel hypermethylated genes involved in mammary tumorigenesis [52]. A related promising approach for nontargeted analysis of the DNA methylome comes from the important finding that DNA that is immunoprecipitated with an antibody against 5-methylcytosine (methyl-DIP) can be used as a target for hybridization to genomic DNA microarray platforms [53], allowing the rapid identification of multiple novel methylated CpG sites. An important parameter that limits efficient use of these approaches is that although several promoter CpG-island and tiling microarrays are available from different companies, the entire human genome is not yet represented in any microarray. Moreover, the fact that immunoprecipitated DNA needs to be amplified prior to hybridization to the microarrays can sometimes introduce PCR bias, and should be taken into account when using the methyl-DIP related assays.

Finally, one of the most logical approaches for studying genome-wide DNA-methylation patterns is exploiting the gene-expression profiling data obtained using expression microarrays. Since it has now become a widely used method, one can take advantage of the data already available from several previous studies and generated in targeted studies to identify genes that are potentially regulated by differential DNA methylation by comparative evaluation of the transcriptomes of normal versus cancer cells. The candidate genes identified in these comparative evaluations of the expression data for consistent hypo-expression are legitimate candidates bisulfite DNA sequencing evaluations of the likely regulatory regions at the level of the genome.

19.6 FUTURE PROSPECTS

It has recently become increasingly clear that complex differential gene-expression patterns that define the functionality of the various genes depend not only on the genomic sequence, but also on the epigenome. Therefore, comprehensive genome evaluation methodologies to survey and identify the various epigenetic DNA methylation changes will also undoubtedly play a key role in this feat. There have been concerted efforts to obtain the epigenetic signatures to understand the molecular basis of cancer progression in the form of Human Epigenome Projects (HEP) in Europe and the United States. While we are poised to succeed in identifying the various epigenetic alterations through these efforts, it will be critical for us to obtain a "big picture" of molecular alterations that occur during cancer progression to effectively utilize them as markers for diagnosis, prognosis, and therapy. One of the most important aspects for consideration during these analyses is the observed heterogeneity in the genetic and epigenetic alterations that have consistently posed a challenge to the comprehensive diagnostic as well as prognostic evaluation of the various cancers. Recently, we proposed a multi-modular molecular network (MMMN) cancer progression model to address this issue by advocating the consideration that the various observed alterations could potentially affect alternate target genes in modules of a cascade of events [54]. On the basis of these facts and challenges, it is apparent that the ultimate goal of finding the optimal pool of genes affected by DNA methylation changes to increase the accuracy of cancer prognosis will depend on both the expansion of the pool of all of the possible alterations specific to any particular type of cancer as well as the understanding of the functional significance of these changes in the context of MMMN of events.

REFERENCES

1. Laird, P.W., The power and the promise of DNA methylation markers, *Nat Rev Cancer*, 3, 253, 2003.
2. Herman, J.G. and Baylin, S.B., Gene silencing in cancer in association with promoter hypermethylation, *N Engl J Med*, 349, 2042, 2003.
3. Feinberg, A.P. and Tycko, B., The history of cancer epigenetics, *Nat Rev Cancer*, 4, 143, 2004.
4. Gardiner-Garden, M. and Frommer, M., CpG islands in vertebrate genomes, *J Mol Biol*, 196, 261, 1987.
5. Baylin, S.B. and Ohm, J.E., Epigenetic gene silencing in cancer—a mechanism for early oncogenic pathway addiction? *Nat Rev Cancer*, 6, 107, 2006.
6. Hanahan, D. and Weinberg, R.A., The hallmarks of cancer, *Cell*, 100, 57, 2000.
7. Walsh, C.P., Chaillet, J.R., and Bestor, T.H., Transcription of IAP endogenous retroviruses is constrained by cytosine methylation, *Nat Genet*, 20, 116, 1998.
8. Gaudet, F., et al., Induction of tumors in mice by genomic hypomethylation, *Science*, 300, 489, 2003.
9. Pakneshan, P., Szyf, M., and Rabbani, S.A., Hypomethylation of urokinase (uPA) promoter in breast and prostate cancer: Prognostic and therapeutic implications, *Curr Cancer Drug Targets*, 5, 471, 2005.
10. Paredes, J., et al., P-cadherin overexpression is an indicator of clinical outcome in invasive breast carcinomas and is associated with CDH3 promoter hypomethylation, *Clin Cancer Res*, 11, 5869, 2005.
11. Tangkijvanich, P., et al., Serum LINE-1 hypomethylation as a potential prognostic marker for hepatocellular carcinoma, *Clin Chim Acta*, 379, 127, 2007.
12. Esteller, M., Cancer epigenomics: DNA methylomes and histone-modification maps, *Nat Rev Genet*, 8, 286, 2007.
13. Thiagalingam, S., et al., Histone deacetylases: Unique players in shaping the epigenetic histone code, *Ann N Y Acad Sci*, 983, 84, 2003.
14. Girault, I., et al., Expression analysis of DNA methyltransferases 1, 3A, and 3B in sporadic breast carcinomas, *Clin Cancer Res*, 9, 4415, 2003.
15. Strichman-Almashanu, L.Z., et al., A genome-wide screen for normally methylated human CpG islands that can identify novel imprinted genes, *Genome Res*, 12, 543, 2002.
16. Song, F., et al., Association of tissue-specific differentially methylated regions (TDMs) with differential gene expression, *Proc Natl Acad Sci U S A*, 102, 3336, 2005.
17. Baylin, S.B. and Herman, J.G., DNA hypermethylation in tumorigenesis: Epigenetics joins genetics, *Trends Genet*, 16, 168, 2000.
18. Sidransky, D., Emerging molecular markers of cancer, *Nat Rev Cancer*, 2, 210, 2002.
19. Herman, J.G., et al., Methylation-specific PCR: A novel PCR assay for methylation status of CpG islands, *Proc Natl Acad Sci U S A*, 93, 9821, 1996.
20. Krassenstein, R., et al., Detection of breast cancer in nipple aspirate fluid by CpG island hypermethylation, *Clin Cancer Res*, 10, 28, 2004.
21. Rosas, S.L., et al., Promoter hypermethylation patterns of p16, O6-methylguanine-DNA-methyltransferase, and death-associated protein kinase in tumors and saliva of head and neck cancer patients, *Cancer Res*, 61, 939, 2001.
22. Ahrendt, S.A., et al., Molecular detection of tumor cells in bronchoalveolar lavage fluid from patients with early stage lung cancer, *J Natl Cancer Inst*, 91, 332, 1999.

23. Palmisano, W.A., et al., Predicting lung cancer by detecting aberrant promoter methylation in sputum, *Cancer Res*, 60, 5954, 2000.
24. Petko, Z., et al., Aberrantly methylated CDKN2A, MGMT, and MLH1 in colon polyps and in fecal DNA from patients with colorectal polyps, *Clin Cancer Res*, 11, 1203, 2005.
25. Esteller, M., et al., Detection of aberrant promoter hypermethylation of tumor suppressor genes in serum DNA from non-small cell lung cancer patients, *Cancer Res*, 59, 67, 1999.
26. Sanchez-Cespedes, M., et al., Gene promoter hypermethylation in tumors and serum of head and neck cancer patients, *Cancer Res*, 60, 892, 2000.
27. Grady, W.M., et al., Detection of aberrantly methylated hMLH1 promoter DNA in the serum of patients with microsatellite unstable colon cancer, *Cancer Res*, 61, 900, 2001.
28. Hoque, M.O., et al., Quantitation of promoter methylation of multiple genes in urine DNA and bladder cancer detection, *J Natl Cancer Inst*, 98, 996, 2006.
29. Cairns, P., et al., Molecular detection of prostate cancer in urine by GSTP1 hypermethylation, *Clin Cancer Res*, 7, 2727, 2001.
30. Silva, J.M., et al., Presence of tumor DNA in plasma of breast cancer patients: Clinicopathological correlations, *Cancer Res*, 59, 3251, 1999.
31. Kawakami, K., et al., Hypermethylated APC DNA in plasma and prognosis of patients with esophageal adenocarcinoma, *J Natl Cancer Inst*, 92, 1805, 2000.
32. Usadel, H., et al., Quantitative adenomatous polyposis coli promoter methylation analysis in tumor tissue, serum, and plasma DNA of patients with lung cancer, *Cancer Res*, 62, 371, 2002.
33. Esteller, M., et al., A gene hypermethylation profile of human cancer, *Cancer Res*, 61, 3225, 2001.
34. Barbacid, M., ras genes, *Annu Rev Biochem*, 56, 779, 1987.
35. Fraga, M.F. and Esteller, M., DNA methylation: A profile of methods and applications, *Biotechniques*, 33, 632, 2002.
36. Merlo, A., et al., 5′ CpG island methylation is associated with transcriptional silencing of the tumour suppressor p16/CDKN2/MTS1 in human cancers, *Nat Med*, 1, 686, 1995.
37. Nuovo, G.J., et al., In situ detection of the hypermethylation-induced inactivation of the p16 gene as an early event in oncogenesis, *Proc Natl Acad Sci U S A*, 96, 12754, 1999.
38. Clark, S.J., et al., High sensitivity mapping of methylated cytosines, *Nucleic Acids Res*, 22, 2990, 1994.
39. Eads, C.A., et al., MethyLight: A high-throughput assay to measure DNA methylation, *Nucleic Acids Res*, 28, E32, 2000.
40. Uhlmann, K., et al., Evaluation of a potential epigenetic biomarker by quantitative methyl-single nucleotide polymorphism analysis, *Electrophoresis*, 23, 4072, 2002.
41. Costello, J.F., et al., Aberrant CpG-island methylation has non-random and tumour-type-specific patterns, *Nat Genet*, 24, 132, 2000.
42. Zardo, G., et al., Integrated genomic and epigenomic analyses pinpoint biallelic gene inactivation in tumors, *Nat Genet*, 32, 453, 2002.
43. Welsh, J. and McClelland, M., Fingerprinting genomes using PCR with arbitrary primers, *Nucleic Acids Res*, 18, 7213, 1990.
44. Gonzalgo, M.L., et al., Identification and characterization of differentially methylated regions of genomic DNA by methylation-sensitive arbitrarily primed PCR, *Cancer Res*, 57, 594, 1997.
45. Toyota, M., et al., Identification of differentially methylated sequences in colorectal cancer by methylated CpG island amplification, *Cancer Res*, 59, 2307, 1999.
46. Frigola, J., et al., Methylome profiling of cancer cells by amplification of inter-methylated sites (AIMS), *Nucleic Acids Res*, 30, e28, 2002.
47. Huang, T.H., Perry, M.R., and Laux, D.E., Methylation profiling of CpG islands in human breast cancer cells, *Hum Mol Genet*, 8, 459, 1999.
48. Khulan, B., et al., Comparative isoschizomer profiling of cytosine methylation: The HELP assay, *Genome Res*, 16, 1046, 2006.
49. Bibikova, M., et al., High-throughput DNA methylation profiling using universal bead arrays, *Genome Res*, 16, 383, 2006.
50. Bibikova, M., et al., Human embryonic stem cells have a unique epigenetic signature, *Genome Res*, 16, 1075, 2006.
51. Lopez-Serra, L., et al., A profile of methyl-CpG binding domain protein occupancy of hypermethylated promoter CpG islands of tumor suppressor genes in human cancer, *Cancer Res*, 66, 8342, 2006.

52. Ballestar, E., et al., Methyl-CpG binding proteins identify novel sites of epigenetic inactivation in human cancer, *Embo J*, 22, 6335, 2003.

53. Weber, M., et al., Chromosome-wide and promoter-specific analyses identify sites of differential DNA methylation in normal and transformed human cells, *Nat Genet*, 37, 853, 2005.

54. Thiagalingam, S., A cascade of modules of a network defines cancer progression, *Cancer Res*, 66, 7379, 2006.

55. Sakuma, M., et al., Promoter methylation status of the Cyclin D2 gene is associated with poor prognosis in human epithelial ovarian cancer, *Cancer Sci*, 98, 380, 2007.

56. Kawamoto, K., et al., p16INK4a and p14ARF methylation as a potential biomarker for human bladder cancer, *Biochem Biophys Res Commun*, 339, 790, 2006.

57. Issa, J.P., Baylin, S.B., and Herman, J.G., DNA methylation changes in hematologic malignancies: Biologic and clinical implications, *Leukemia*, 11 Suppl 1, S7, 1997.

58. Wong, I.H., et al., Frequent p15 promoter methylation in tumor and peripheral blood from hepatocellular carcinoma patients, *Clin Cancer Res*, 6, 3516, 2000.

59. Wong, I.H., et al., Aberrant p15 promoter methylation in adult and childhood acute leukemias of nearly all morphologic subtypes: Potential prognostic implications, *Blood*, 95, 1942, 2000.

60. Belshaw, N.J., et al., Use of DNA from human stools to detect aberrant CpG island methylation of genes implicated in colorectal cancer, *Cancer Epidemiol Biomarkers Prev*, 13, 1495, 2004.

61. Brock, M.V., et al., Prognostic importance of promoter hypermethylation of multiple genes in esophageal adenocarcinoma, *Clin Cancer Res*, 9, 2912, 2003.

62. Chan, M.W., et al., Hypermethylation of multiple genes in tumor tissues and voided urine in urinary bladder cancer patients, *Clin Cancer Res*, 8, 464, 2002.

63. Herman, J.G., et al., Inactivation of the CDKN2/p16/MTS1 gene is frequently associated with aberrant DNA methylation in all common human cancers, *Cancer Res*, 55, 4525, 1995.

64. Russo, A.L., et al., Differential DNA hypermethylation of critical genes mediates the stage-specific tobacco smoke-induced neoplastic progression of lung cancer, *Clin Cancer Res*, 11, 2466, 2005.

65. Takita, J., et al., The p16 (CDKN2A) gene is involved in the growth of neuroblastoma cells and its expression is associated with prognosis of neuroblastoma patients, *Oncogene*, 17, 3137, 1998.

66. Kawano, S., et al., Loss of p73 gene expression in leukemias/lymphomas due to hypermethylation, *Blood*, 94, 1113, 1999.

67. Stirzaker, C., et al., Extensive DNA methylation spanning the Rb promoter in retinoblastoma tumors, *Cancer Res*, 57, 2229, 1997.

68. Catteau, A., et al., Methylation of the BRCA1 promoter region in sporadic breast and ovarian cancer: Correlation with disease characteristics, *Oncogene*, 18, 1957, 1999.

69. Dobrovic, A. and Simpfendorfer, D., Methylation of the BRCA1 gene in sporadic breast cancer, *Cancer Res*, 57, 3347, 1997.

70. Ibanez de Caceres, I., et al., Tumor cell-specific BRCA1 and RASSF1A hypermethylation in serum, plasma, and peritoneal fluid from ovarian cancer patients, *Cancer Res*, 64, 6476, 2004.

71. Esteller, M., et al., Hypermethylation of the DNA repair gene O(6)-methylguanine DNA methyltransferase and survival of patients with diffuse large B-cell lymphoma, *J Natl Cancer Inst*, 94, 26, 2002.

72. Esteller, M., et al., Inactivation of the DNA-repair gene MGMT and the clinical response of gliomas to alkylating agents, *N Engl J Med*, 343, 1350, 2000.

73. Esteller, M., et al., Inactivation of the DNA repair gene O6-methylguanine-DNA methyltransferase by promoter hypermethylation is a common event in primary human neoplasia, *Cancer Res*, 59, 793, 1999.

74. Hegi, M.E., et al., MGMT gene silencing and benefit from temozolomide in glioblastoma, *N Engl J Med*, 352, 997, 2005.

75. Feng, Q., et al., Detection of hypermethylated genes in women with and without cervical neoplasia, *J Natl Cancer Inst*, 97, 273, 2005.

76. Maruyama, R., et al., Hypermethylation of FHIT as a prognostic marker in nonsmall cell lung carcinoma, *Cancer*, 100, 1472, 2004.

77. Boumber, Y.A., et al., RIL, a LIM gene on 5q31, is silenced by methylation in cancer and sensitizes cancer cells to apoptosis, *Cancer Res*, 67, 1997, 2007.

78. Lau, Q.C., et al., RUNX3 is frequently inactivated by dual mechanisms of protein mislocalization and promoter hypermethylation in breast cancer, *Cancer Res*, 66, 6512, 2006.

79. Ku, J.L., et al., Promoter hypermethylation downregulates RUNX3 gene expression in colorectal cancer cell lines, *Oncogene*, 23, 6736, 2004.

80. Kang, G.H., et al., Aberrant CpG island hypermethylation of multiple genes in prostate cancer and prostatic intraepithelial neoplasia, *J Pathol*, 202, 233, 2004.

81. Sakakura, C., et al., Possible involvement of RUNX3 silencing in the peritoneal metastases of gastric cancers, *Clin Cancer Res*, 11, 6479, 2005.

82. Yanagawa, N., et al., Promoter hypermethylation of RASSF1A and RUNX3 genes as an independent prognostic prediction marker in surgically resected non-small cell lung cancers, *Lung Cancer*, 58, 131, 2007.

83. Alaminos, M., et al., Clustering of gene hypermethylation associated with clinical risk groups in neuroblastoma, *J Natl Cancer Inst*, 96, 1208, 2004.

84. Machida, E.O., et al., Hypermethylation of ASC/TMS1 is a sputum marker for late-stage lung cancer, *Cancer Res*, 66, 6210, 2006.

85. Suzuki, M., et al., Methylation of apoptosis related genes in the pathogenesis and prognosis of prostate cancer, *Cancer Lett*, 242, 222, 2006.

86. Muller, H.M., et al., Prognostic DNA methylation marker in serum of cancer patients, *Ann N Y Acad Sci*, 1022, 44, 2004.

87. Yoshikawa, H., et al., SOCS-1, a negative regulator of the JAK/STAT pathway, is silenced by methylation in human hepatocellular carcinoma and shows growth-suppression activity, *Nat Genet*, 28, 29, 2001.

88. He, B., et al., SOCS-3 is frequently silenced by hypermethylation and suppresses cell growth in human lung cancer, *Proc Natl Acad Sci U S A*, 100, 14133, 2003.

89. Jeronimo, C., et al., Endothelin B receptor gene hypermethylation in prostate adenocarcinoma, *J Clin Pathol*, 56, 52, 2003.

90. Yuan, Y., et al., Frequent epigenetic inactivation of spleen tyrosine kinase gene in human hepatocellular carcinoma, *Clin Cancer Res*, 12, 6687, 2006.

91. Lapidus, R.G., et al., Methylation of estrogen and progesterone receptor gene 5' CpG islands correlates with lack of estrogen and progesterone receptor gene expression in breast tumors, *Clin Cancer Res*, 2, 805, 1996.

92. Mori, T., et al., Estrogen receptor-alpha methylation predicts melanoma progression, *Cancer Res*, 66, 6692, 2006.

93. Chang, Y.S., et al., Correlation between insulin-like growth factor-binding protein-3 promoter methylation and prognosis of patients with stage I non-small cell lung cancer, *Clin Cancer Res*, 8, 3669, 2002.

94. Umetani, N., et al., Epigenetic inactivation of ID4 in colorectal carcinomas correlates with poor differentiation and unfavorable prognosis, *Clin Cancer Res*, 10, 7475, 2004.

95. Yang, D., et al., Repression of IFN regulatory factor 8 by DNA methylation is a molecular determinant of apoptotic resistance and metastatic phenotype in metastatic tumor cells, *Cancer Res*, 67, 3301, 2007.

96. Hiranuma, C., et al., Hypermethylation of the MYOD1 gene is a novel prognostic factor in patients with colorectal cancer, *Int J Mol Med*, 13, 413, 2004.

97. Widschwendter, A., et al., CDH1 and CDH13 methylation in serum is an independent prognostic marker in cervical cancer patients, *Int J Cancer*, 109, 163, 2004.

98. Widschwendter, A., et al., DNA methylation in serum and tumors of cervical cancer patients, *Clin Cancer Res*, 10, 565, 2004.

99. Sato, M., et al., The H-cadherin (CDH13) gene is inactivated in human lung cancer, *Hum Genet*, 103, 96, 1998.

100. Toyooka, K.O., et al., Loss of expression and aberrant methylation of the CDH13 (H-cadherin) gene in breast and lung carcinomas, *Cancer Res*, 61, 4556, 2001.

101. Toyooka, S., et al., Aberrant methylation of the CDH13 (H-cadherin) promoter region in colorectal cancers and adenomas, *Cancer Res*, 62, 3382, 2002.

102. Imoto, I., et al., Frequent silencing of the candidate tumor suppressor PCDH20 by epigenetic mechanism in non-small-cell lung cancers, *Cancer Res*, 66, 4617, 2006.

103. Lou, W., et al., Methylation of the CD44 metastasis suppressor gene in human prostate cancer, *Cancer Res*, 59, 2329, 1999.

104. Vis, A.N., et al., Feasibility of assessment of promoter methylation of the CD44 gene in serum of prostate cancer patients, *Mol Urol*, 5, 199, 2001.

105. Agrelo, R., et al., Inactivation of the lamin A/C gene by CpG island promoter hypermethylation in hematologic malignancies, and its association with poor survival in nodal diffuse large B-cell lymphoma, *J Clin Oncol*, 23, 3940, 2005.

106. Esteller, M., et al., Epigenetic inactivation of LKB1 in primary tumors associated with the Peutz-Jeghers syndrome, *Oncogene*, 19, 164, 2000.
107. Leung, W.K., et al., Potential diagnostic and prognostic values of detecting promoter hypermethylation in the serum of patients with gastric cancer, *Br J Cancer*, 92, 2190, 2005.
108. Wallner, M., et al., Methylation of serum DNA is an independent prognostic marker in colorectal cancer, *Clin Cancer Res*, 12, 7347, 2006.
109. Li, Q., et al., Methylation and silencing of the Thrombospondin-1 promoter in human cancer, *Oncogene*, 18, 3284, 1999.
110. Toyota, M., et al., Aberrant methylation of the Cyclooxygenase 2 CpG island in colorectal tumors, *Cancer Res*, 60, 4044, 2000.
111. Jeronimo, C., et al., Quantitation of GSTP1 methylation in non-neoplastic prostatic tissue and organ-confined prostate adenocarcinoma, *J Natl Cancer Inst*, 93, 1747, 2001.
112. Jeronimo, C., et al., Quantitative GSTP1 hypermethylation in bodily fluids of patients with prostate cancer, *Urology*, 60, 1131, 2002.
113. Lee, W.H., et al., Cytidine methylation of regulatory sequences near the pi-class glutathione *S*-transferase gene accompanies human prostatic carcinogenesis, *Proc Natl Acad Sci U S A*, 91, 11733, 1994.
114. Herman, J.G., et al., Silencing of the VHL tumor-suppressor gene by DNA methylation in renal carcinoma, *Proc Natl Acad Sci U S A*, 91, 9700, 1994.

20 Diagnosing Cancer Using Histone Modification Analysis

Mukesh Verma and Deepak Kumar

CONTENTS

20.1 BACKGROUND

Cancer is a genetic and epigenetic disease [1–5]. Cancer cells are characterized by epigenetic dysregulation, including global genome hypomethylation, regional hypo- and hypermethylation, altered histone modifications, and disturbed genomic imprinting [6–11]. Histones have four basic subunits, and in the native state they exist as an octamer. DNA winds around this octamer. Acidic histones neutralize DNA charge and maintain chromatin stability. The octamer histone binding is independent of surrounding DNA sequence and the N-terminal region of histones is subject to phosphorylation, acetylation, and other modifications. Multiple histone modifications can take place within a short stretch of amino acids of histone tails [12–14]. These modifications regulate transcription, DNA replication, and DNA repair, and thus are part of the transformation process. Modification of histones may occur in large regions of chromatin, including coding and nonpromoter sequences, termed global histone modifications. Global histone modifications have been suggested to play a role in cancer recurrence [13,15–18].

The histone acetyltransferases (HATs) add the acetyl group to histones whereas the histone deacetylases (HDACs) remove the acetyl group. The opposing activities of HATs and HDACs tightly regulate gene expression through chromatin modifications. HATs activity results in chromatin relaxation, whereas HDACs make chromatin compact. Recently, histones have been investigated for use in cancer diagnosis and follow up of treatments [19]. In this article, we have described the current status and challenges in this relatively new field of biomedical research.

20.2   HISTONE MODIFICATIONS

A number of modifications in the tail region of histones have been reported, such as acetylation, deacetylation, phosphorylation, poly-ADP ribosylation, methylation, ubiquitination, sumoylation, carbonylation, citrullination, and glycosylation (Figure 20.1) [20–22]. It is important to understand the significance of these modifications as they are not only important in disease diagnosis but also to understand histone–DNA interaction, nucleosome–nucleosome interaction, and interaction of non-histone proteins with chromatin [23,24]. Histone modifications may produce two opposite effects at the level of transcription by either increasing or decreasing transcription of genes [25–31]. It has been proposed that a histone modification cassette exists which controls histone activities. In the following sections, we have described different types of histone modifications with relevant clinical information.

20.2.1   HISTONE ACETYLATION

Acetylation is mediated by HATs where transfer of an acetyl group occurs from acetyl-coenzyme A to the epsilon-amino group of specific lysine side chain [29,32]. The most common free amino

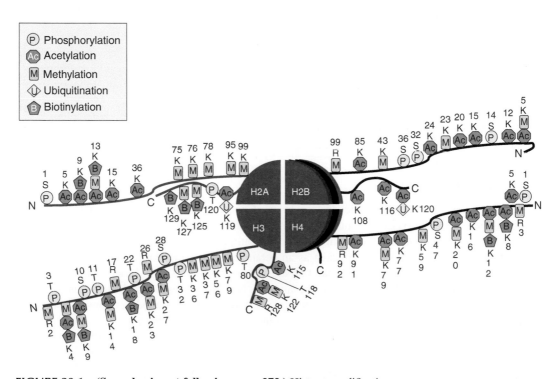

FIGURE 20.1   (See color insert following page 272.) Histone modifications.

terminal ends of lysine that undergo acetylation are lysine 9, 14, 18, and 23 in histone H3, and 5, 8, 12, and 16 of lysine in histone H4 [25,33–35]. The negatively charged DNA, which generally interacts with histones is neutralized because acetylation interacts with the epsilon-amino group of lysine. Furthermore, acetylated histones provide binding sites for different transcription factors. Modifications of lysine residues provide an antigenic substrate for generating specific antibodies, which can be used for diagnosis of cancer. Five HATs have been identified to date.

HDACs have been categorized in three major classes: class I (HDACs 1,2, 3, and 18), class II (HDACs 4, 5, 6, and 9), and class III (SIR2 family of NADP dependent HDACs) [36–38]. HDACs deacetylase all four histones and sometimes act on nonhistone proteins as well [39].

20.2.2 HISTONE METHYLATION

The enzyme histone methyl transferase transfers the methyl group from *S*-adenosylmethionine to lysine or arginine of histones H3 and H4 [25,31,40–42]. Most of the time lysine 4, 9, and 27 of histone H3 and lysine 20 of histone H4 get methylated. There are three types of methyltransferases: arginine methyl transferase (involved in maintenance of chromatin modifications), SET domain containing methyl transferase (induces ectopic heterochromatin), and Dot 1-like methyl transferase (involved in cell proliferation) [43–49].

20.2.3 HISTONE PHOSPHORYLATION

Many transcription factors get activated through phosphorylation. Aurora kinase is involved in histone H3 phosphorylation at serine 10 and 28 positions. Different forms of aurora kinase have different efficiencies of phosphorylation [50–55]. Two other kinases that phosphorylate histones are Mst1 and RSK2 kinases. Mst1 has a preference for histone H2B, whereas RSK2 does not show specific preferences [56].

20.2.4 HISTONE UBIQUITINATION

Histone H2B gets ubiquitinated by Rad6, which is an ubiquitinin conjugating enzyme involved in DNA damage repair and meiosis [57,58].

20.2.5 HISTONE SUMOYLATION

For transcription activity sumoylation of histones occur. The role of Ubc9 in sumoylation affecting normal mitosis and recovery from DNA damage has been proposed [59].

20.3 HISTONE MODIFICATIONS AND CANCER DIAGNOSIS

Although not much research has been conducted and validated in this area, findings published to date indicate that histone modifications can play a role in cancer diagnosis, including cancers which are difficult to diagnose by histopathology alone, such as pediatric cancers [60–63].

As shown in Figure 20.1, major cancers where histone modifications have been reported are breast and colorectal cancer, glioblastoma, hepatocellular carcinoma, leukemia, non-Hodgkin's lymphoma, lung carcinoma, osteosarcoma, melanoma, and prostate cancer. In monozygotic twins, the role of histone modifications has been very well demonstrated [61].

20.3.1 BLADDER CANCER

Histone H3 methylation at lysine 4 and 9 has been studied in bladder cancer to investigate the effect of this modification on cancer-related gene expression [64]. However, there are so few relevant studies on bladder cancer that it is difficult to judge the clinical utility of this modification in bladder cancer.

20.3.2   Breast Cancer

In a recent study, epigenetic modifications were followed in three human breast cancer cell lines MCF-7, MDA-MB-231, and MDA-MB-231(S30) representing different stages of human breast cancer. Decreased trimethylation of lysine 20 of histone H4 and hyperacetylation of histone H4, was observed in these cell lines. This dramatic decrease in trimethylation of lysine 20 of histone H4, especially in MDA-MB-231 cells, was accompanied by diminished expression of Suv4-20h2 histone methyltransferase. Thus, differential expression of histone modifications could be interpreted in cell lines as representing the aggressiveness of the tumor or different stages of tumor development. Such epigenetic dysregulation may contribute to and may be indicative of the formation of a more aggressive tumor phenotype during tumor progression. In another study, one of the HDACs, namely HDAC6, was found to contribute to responsiveness to estrogen treatment [7,8,65,66]. Apart form a downregulation of Suv4-20h2, the loss of 3MeK20-H4 in cancer can also be explained through the deregulation of the tumor suppressor RB [7,8,14,67] or other histone-modifying enzymes. In another study, microarray analysis of samples from breast cancer patients was performed to follow histone expression and a correlation between tumor stages and grade was established [68]. Zhang et al. reported quantitative expression of HDAC1 and its correlation with a patient's age, negative lymph-node status, small tumor size, her2/neu negative status, and ER/PR status [69].

20.3.3   Cervical Cancer

Dysplasia and neoplasia are difficult to distinguish by histopathological analyses. However, nuclear phosphatase staining of histone H3 distinguishes these two stages of cancer development [70]. Cytological smears from case and control squamous epithelial cells were found positive with phosphorylated H3 antibodies.

Success in the early detection and treatment of intraepithelial neoplasia results in protecting patients from developing invasive cervical cancer. Recently, the mortality due to cervical cancer decreased due to excellent screening programs using early marker detection [71].

20.3.4   Colon Cancer

Colon cancer diagnosis involves determining the ratio of methylation versus acetylation at the lysine 9 position of histone H3 [72]. This ratio also helps in distinguishing normal versus cancer cells. A high-throughput Chip-on-Chip assay has been developed for clinical application of histone modifications in cancer diagnosis. Hypoacetylation of histone H3 is shown to be useful, along with methylation markers in diagnosis of colorectal cancer [73].

20.3.5   Esophageal Cancer

Histone H4 acetylation and hypermethylation of the fragile histidine triad (FHIT) gene are useful markers in following esophageal cancer development [74]. H4 acetylation has been shown to be inversely proportional to the depth of cancer invasion and pathological stage [75]. These markers have also been utilized for prognosis.

20.3.6   Gastric Cancer

Like other cancers, methylation and histone modifications also have been reported in gastric cancer. In one study, hypoacetylation of histones H3 and H4 was observed in the *p21* promoter region of the gene in approximately 50% of the samples examined [76]. Acetylation was correlated with advanced tumor stage, tumor invasion, and lymph-node metastasis.

20.3.7 Head and Neck Cancer

In head and neck cancer, increased expression of *RARbeta* was observed. This was not due to methylation changes, but to increased acetylation of histone H3 at lysine 9. The expression of the receptor showed a correlation with cell density [77].

20.3.8 Leukemia

In acute myeloblastoid leukemia (AML), CpG islands of the *p15* gene are surrounded by acetylated histones and the trimethylated histone H3 [78]. In another study, demethylation of histone H3 at lysine 36 has been observed in leukemia [79]. In myeloid leukemia, H3 methylation was shown to be correlated with differentiation and incomplete chromatin condensation [80].

20.3.9 Lung Cancer

Compared to other modifications, ubiquitination seems to dominate in lung cancer [81]. Ubiquitination of histone H2A and histone H2B at different amino acid sites was studied in lung cancer cells exposed to nickel. In another study, class II HDACs were found to be associated with poor prognosis of lung cancer [39].

20.3.10 Ovarian Cancer

Hypoacetylation of histone H3 and histone H4 and loss of trimethylation at lysine 4 of histone H3 was observed in five ovarian epithelial and carcinoma cell lines (human "immortalized" ovarian surface epithelium (HIO)-117, HIO-114, A2780, SKOV3, and ES2). Altered expression of GATA factors plays a crucial role in ovarian carcinogenesis [82]. In the same set of experiments, lysine K9 di-methylation of histone H3 and HP1-γ association was not observed, excluding reorganization of GATA genes into heterochromatic structures. Results were confirmed using appropriate inhibitors of acetylation and methylation. Considering the short survival time of ovarian cancer patients after diagnosis, the observations reported in these studies have great potential for improving ovarian cancer outcomes.

20.3.11 Pediatric Adrenocortical Tumors

These are rare tumors in which adenomas and carcinomas are difficult to distinguish by histopathology analysis. However, when expression of histone LI was followed and compared with *Ki67* expression, histone expression was fivefold higher in the S phase than in the G1, G2, and S1 phases, whereas *Ki67* expression was uniform in all phases [83]. This suggested that histone expression could be used as a differentiating marker in different stages of cancer.

20.3.12 Prostate Cancer

Epigenetic changes, including both methylation of promoter and modification of histone, occur quite early during cancer development [84]. In prostate cancer diagnosis and recurrence histone modifications have been utilized [13,84–87]. Acetylation and methylation of different histones result in different outcomes and has potential in prediction of clinical outcome and disease recurrence [13]. Kahl et al. correlated histone demethylase I expression with androgen receptor level, Gleason score, Gleason grade, and *p53* levels and suggested histone demethylase I as a predictive marker for relapse [88]. Studying the histone modification profile can also help us understand different clinical outcomes [13,85].

20.3.13 SKIN CANCER

Loss of monoacetylated and trimethylated forms of histone H4 in skin cancer has been observed in early tumors and accumulates during cancer progression [61,62,89–91]. As previously mentioned, loss of histone acetylation and trimethylation has been also found in other types of cancer [9–11,75,82,92–98], which suggests that these alterations can be common in tumorogenesis [62]. Although validation has to be done by different groups, it seems that the global loss of mono-acetylation and trimethylation of histone H4 is a common hallmark of human tumor cells.

20.4 CHALLENGES AND POTENTIAL SOLUTIONS

Chromatin is a highly dynamic environment playing critical roles in the regulation of gene expression. The posttranslational modification of the core histones is critical to the regulation of chromatin structure. Generally it is much easier to study DNA modifications than protein modifications [1,4,99–101]. We now describe current technologies to detect histone modifications, advantages, and disadvantages associated with these techniques and potential problems and their possible solutions. Studying histone modifications may provide an advantage in diagnosing cancer early [5]. Antibody-based assays to detect histone modifications have been developed [102,103]. Recently, quantitative proteomic assays have been developed to check profiling of histone modifications and it is possible that in the coming years these assays will be useable in clinics for hundreds of samples [12,23]. A hydrophilic interaction liquid chromatographic separation method in combination with mass spectrometric analysis also has been used to follow the alterations in histone H4 methylation/acetylation status and the interplay between H4 methylation and acetylation during differentiation of erythroleukemia cells and study modifications which affect the chromatin structure [104].

Traditional methods utilize immunoassay techniques to determine the extent and site of post-translational modification to detect histone modification. Although these methods are sensitive, they require site-specific antibodies and since a large number of modifications occur in histones, several antibodies would need to be available. Another approach, which seems promising, involves the application of reverse-phase high-pressure liquid chromatography and mass spectrometry (LC–MS) to analyze global modification levels of core histones. Relative to other methods, this method is fast, sensitive, and easily automated [105]. Furthermore, the LC–MS approach provides the global patterns of modification for all four-core histones in a single experiment. The characterization of changes in histone modification in acute myeloid leukemia (AML) has been accomplished using LC–MS [105]. Determination of dose-dependent changes in the distribution of modified core histones has been observed and results have been validated in primary leukemia cells from patients with refractory or relapsed AML or chronic lymphocytic leukemia (CLL). For example, clinically significant higher levels of histone H4 were detected in patients at intervals of 4 and 24 h after treatment with inhibitors.

Attempts should be made to establish a panel of antibodies against histone modifications that could discriminate different stages of tumor development. This would help in screening patient sera on microarrays with a combination of antibodies. Ultimately, samples from populations with cancer or populations at high risk of developing cancer would be identified utilizing approaches described above. It would also help in the identification of associated etiological factors, which contribute to the initiation and progression of cancer.

One area that has not yet been explored extensively is the identification of chromatin images in normal and disease stages and their correlation with cancer initiation and development [5]. In vivo images should also be captured when chromatin starts compacting followed by relaxing during gene expression in disease stages. As an extension to this work, studies should be conducted to correlate biochemical changes in histones with the conformational changes in chromatin during the progression of the disease.

For several histone modifiers, we do not know the mechanism of action. Technological advancements in microarray technologies and proteomics may shed some light on this aspect. Genome-wide mapping approaches of histones provide new opportunities to decipher histone code. Integration of datasets and improved technologies will provide the future directions. The major challenge in the area of cancer diagnosis is validation of markers that have been studied by one or few groups. As always, this will be a very big project in which the collaborative efforts of biochemists, proteomic and genomic experts, clinicians, and molecular biologists will be needed.

REFERENCES

1. Verma, M., Maruvada, P., and Srivastava, S. 2004 Epigenetics and cancer. *Crit Rev Clin Lab Sci*, 41, 585–607
2. Jones, P.A. and Baylin, S.B. 2007 The epigenomics of cancer. *Cell*, 128, 683–692.
3. Ting, A.H., McGarvey, K.M., and Baylin, S.B. 2006 The cancer epigenome—components and functional correlates. *Genes Dev*, 20, 3215–3231.
4. Verma, M. and Manne, U. 2006 Genetic and epigenetic biomarkers in cancer diagnosis and identifying high risk populations. *Crit Rev Oncol Hematol*, 60, 9–18.
5. Verma, M., Seminara, D., Arena, F.J., John, C., Iwamoto, K., and Hartmuller, V. 2006 Genetic and epigenetic biomarkers in cancer: Improving diagnosis, risk assessment, and disease stratification. *Mol Diagn & Ther*, 10, 1–15.
6. Bernstein, B.E., Meissner, A., and Lander, E.S. 2007 The mammalian epigenome. *Cell*, 128, 669–681.
7. Gonzalo, S. and Blasco, M.A. 2005 Role of Rb family in the epigenetic definition of chromatin. *Cell Cycle (Georgetown, Texas)*, 4, 752–755.
8. Gonzalo, S., Garcia-Cao, M., Fraga, M.F., Schotta, G., Peters, A.H., Cotter, S.E., Eguia, R., Dean, D.C., Esteller, M., Jenuwein, T., and Blasco, M.A. 2005 Role of the RB1 family in stabilizing histone methylation at constitutive heterochromatin. *Nat Cell Biol*, 7, 420–428.
9. Pogribny, I.P., Ross, S.A., Tryndyak, V.P., Pogribna, M., Poirier, L.A., and Karpinets, T.V. 2006 Histone H3 lysine 9 and H4 lysine 20 trimethylation and the expression of Suv4–20h2 and Suv-39h1 histone methyltransferases in hepatocarcinogenesis induced by methyl deficiency in rats. *Carcinogenesis*, 27, 1180–1186.
10. Tryndyak, V.P., Kovalchuk, O., and Pogribny, I.P. 2006 Loss of DNA methylation and histone H4 lysine 20 trimethylation in human breast cancer cells is associated with aberrant expression of DNA methyl-transferase 1, Suv4–20h2 histone methyltransferase and methyl-binding proteins. *Cancer Biol Ther*, 5, 65–70.
11. Tryndyak, V.P., Muskhelishvili, L., Kovalchuk, O., Rodriguez-Juarez, R., Montgomery, B., Churchwell, M.I., Ross, S.A., Beland, F.A., and Pogribny, I.P. 2006 Effect of long-term tamoxifen exposure on genotoxic and epigenetic changes in rat liver: Implications for tamoxifen-induced hepatocarcinogenesis. *Carcinogenesis*, 27, 1713–1720.
12. Beck, H.C., Nielsen, E.C., Matthiesen, R., Jensen, L.H., Sehested, M., Finn, P., Grauslund, M., Hansen, A.M., and Jensen, O.N. 2006 Quantitative proteomic analysis of post-translational modifications of human histones. *Mol Cell Proteomics*, 5, 1314–1325.
13. Seligson, D.B., Horvath, S., Shi, T., Yu, H., Tze, S., Grunstein, M., and Kurdistani, S.K. 2005 Global histone modification patterns predict risk of prostate cancer recurrence. *Nature*, 435, 1262–1266.
14. Isaac, C.E., Francis, S.M., Martens, A.L., Julian, L.M., Seifried, L.A., Erdmann, N., Binne, U.K., Harrington, L., Sicinski, P., Berube, N.G., Dyson, N.J., and Dick, F.A. 2006 The retinoblastoma protein regulates pericentric heterochromatin. *Mol Cell Biol*, 26, 3659–3671.
15. Verdone, L., Wu, J., van Riper, K., Kacherovsky, N., Vogelauer, M., Young, E.T., Grunstein, M., Di Mauro, E., and Caserta, M. 2002 Hyperacetylation of chromatin at the ADH2 promoter allows Adr1 to bind in repressed conditions. *EMBO J*, 21, 1101–1111.
16. Vogelauer, M., Wu, J., Suka, N., and Grunstein, M. 2000 Global histone acetylation and deacetylation in yeast. *Nature*, 408, 495–498.
17. Seligson, D., Horvath, S., Huerta-Yepez, S., Hanna, S., Garban, H., Roberts, A., Shi, T., Liu, X., Chia, D., Goodglick, L., and Bonavida, B. 2005 Expression of transcription factor Yin Yang 1 in prostate cancer. *Int J Oncol*, 27, 131–141.

18. Kogai, T., Hershman, J.M., Motomura, K., Endo, T., Onaya, T., and Brent, G.A. 2001 Differential regulation of the human sodium/iodide symporter gene promoter in papillary thyroid carcinoma cell lines and normal thyroid cells. *Endocrinology*, 142, 3369–3379.

19. Polo, S.E. and Almouzni, G. 2005 Histone metabolic pathways and chromatin assembly factors as proliferation markers. *Cancer Lett*, 220, 1–9.

20. Bolden, J.E., Peart, M.J., and Johnstone, R.W. 2006 Anticancer activities of histone deacetylase inhibitors. *Nature Reviews*, 5, 769–784.

21. Nightingale, K.P., O'Neill, L.P., and Turner, B.M. 2006 Histone modifications: Signalling receptors and potential elements of a heritable epigenetic code. *Curr Opin Genet Dev*, 16, 125–136.

22. Nguyen, T.T., Cho, K., Stratton, S.A., and Barton, M.C. 2005 Transcription factor interactions and chromatin modifications associated with p53-mediated, developmental repression of the alpha-fetoprotein gene. *Mol Cell Biol*, 25, 2147–2157.

23. Rando, O.J. 2007 Global patterns of histone modifications. *Curr Opin Gen Dev*, 17, 94–99.

24. Marte, B. 2005 Cancer: A changing global view. *Nature*, 435, 1172.

25. An, W. 2007 Histone acetylation and methylation: Combinatorial players for transcriptional regulation. *SubCell Biochem*, 41, 351–369.

26. He, S., Dunn, K.L., Espino, P.S., Drobic, B., Li, L., Yu, J., Sun, J.M., Chen, H.Y., Pritchard, S., and Davie, J.R. 2007 Chromatin organization and nuclear microenvironments in cancer cells. *J Cell Biochem*, 104, 2004–2015.

27. Kondo, Y., Shen, L., Suzuki, S., Kurokawa, T., Masuko, K., Tanaka, Y., Kato, H., Mizuno, Y., Yokoe, M., Sugauchi, F., Hirashima, N., Orito, E., Osada, H., Ueda, R., Guo, Y., Chen, X., Issa, J.P., and Sekido, Y. 2007 Alterations of DNA methylation and histone modifications contribute to gene silencing in hepatocellular carcinomas. *Hepatol Res*, 37, 974–983.

28. Kurdistani, S.K. 2007 Histone modifications as markers of cancer prognosis: A cellular view. *Br J Cancer*, 97, 1–5.

29. Orr, J.A. and Hamilton, P.W. 2007 Histone acetylation and chromatin pattern in cancer. *A review. Analytical and Quantitative Cytology and Histology/the International Academy of Cytology [and] American Society of Cytology*, 29, 17–31.

30. Provenzano, M.J. and Domann, F.E. 2007 A role for epigenetics in hearing: Establishment and maintenance of auditory specific gene expression patterns. *Hear Res*, 233, 1–13.

31. Trtkova, K., Bouchal, J., and Kolar, Z. 2007 Histone acetylation and methylation in the signaling of steroid hormone receptors. *Cell Mol Biol (Noisy-le-Grand, France)*, 53 Suppl, OL930–OL942.

32. Wischnewski, F., Pantel, K., and Schwarzenbach, H. 2006 Promoter demethylation and histone acetylation mediate gene expression of MAGE-A1, -A2, -A3, and -A12 in human cancer cells. *Mol Cancer Res*, 4, 339–349.

33. Thomas, N.R. 2006 Histone post-translational modification: From discovery to the clinic. *IDrugs*, 9, 398–401.

34. Verdone, L., Agricola, E., Caserta, M., and Di Mauro, E. 2006 Histone acetylation in gene regulation. *Brief Funct Genomic Proteomic*, 5, 209–221.

35. Wu, J., Wang, S.H., Potter, D., Liu, J.C., Smith, L.T., Wu, Y.Z., Huang, T.H., and Plass, C. 2007 Diverse histone modifications on histone 3 lysine 9 and their relation to DNA methylation in specifying gene silencing. *BMC Genomics*, 8, 131.

36. Santini, V., Gozzini, A., and Ferrari, G. 2007 Histone deacetylase inhibitors: Molecular and biological activity as a premise to clinical application. *Curr Drug Metab*, 8, 383–393.

37. Broide, R.S., Redwine, J.M., Aftahi, N., Young, W., Bloom, F.E., and Winrow, C.J. 2007 Distribution of histone deacetylases 1–11 in the rat brain. *J Mol Neurosci*, 31, 47–58.

38. Qian, D.Z., Kachhap, S.K., Collis, S.J., Verheul, H.M., Carducci, M.A., Atadja, P., and Pili, R. 2006 Class II histone deacetylases are associated with VHL-independent regulation of hypoxia-inducible factor 1{alpha}. *Cancer Res*, 66, 8814–8821.

39. Osada, H., Tatematsu, Y., Saito, H., Yatabe, Y., Mitsudomi, T., and Takahashi, T. 2004 Reduced expression of class II histone deacetylase genes is associated with poor prognosis in lung cancer patients. *Int J Cancer*, 112, 26–32.

40. Barski, A., Cuddapah, S., Cui, K., Roh, T.Y., Schones, D.E., Wang, Z., Wei, G., Chepelev, I., and Zhao, K. 2007 High-resolution profiling of histone methylations in the human genome. *Cell*, 129, 823–837.

41. Osada, S. 2007 Histone modification enzymes induced during chemical hepatocarcinogenesis. *Yakugaku Zasshi*, 127, 469–479.

42. Zhu, X., Lee, K., Asa, S.L., and Ezzat, S. 2007 Epigenetic silencing through DNA and histone methylation of fibroblast growth factor receptor 2 in neoplastic pituitary cells. *Am J Pathol*, 170, 1618–1628.

43. Brueckner, B., Kuck, D., and Lyko, F. 2007 DNA methyltransferase inhibitors for cancer therapy. *Cancer J*, 13, 17–22.

44. Eoli, M., Menghi, F., Bruzzone, M.G., De Simone, T., Valletta, L., Pollo, B., Bissola, L., Silvani, A., Bianchessi, D., D'Incerti, L., Filippini, G., Broggi, G., Boiardi, A., and Finocchiaro, G. 2007 Methylation of O6-methylguanine DNA methyltransferase and loss of heterozygosity on 19q and/or 17p are overlapping features of secondary glioblastomas with prolonged survival. *Clin Cancer Res*, 13, 2606–2613.

45. Fanelli, M., Caprodossi, S., Ricci-Vitiani, L., Porcellini, A., Tomassoni-Ardori, F., Amatori, S., Andreoni, F., Magnani, M., De Maria, R., Santoni, A., Minucci, S., and Pelicci, P.G. 2008 Loss of pericentromeric DNA methylation pattern in human glioblastoma is associated with altered DNA methyltransferases expression and involves the stem cell compartment. *Oncogene*, 27, 358–365.

46. Ghoshal, K. and Bai, S. 2007 DNA methyltransferases as targets for cancer therapy. *Drugs Today (Barc)*, 43, 395–422.

47. Menigatti, M., Pedroni, M., Verrone, A.M., Borghi, F., Scarselli, A., Benatti, P., Losi, L., Di Gregorio, C., Schar, P., Marra, G., Ponz de Leon, M., and Roncucci, L. 2007 O6-methylguanine-DNA methyltransferase promoter hypermethylation in colorectal carcinogenesis. *Oncol Rep*, 17, 1421–1427.

48. Murakami, J., Lee, Y.J., Kokeguchi, S., Tsujigiwa, H., Asaumi, J., Nagatsuka, H., Fukui, K., Kuroda, M., Tanaka, N., and Matsubara, N. 2007 Depletion of O6-methylguanine-DNA methyltransferase by O6-benzylguanine enhances 5-FU cytotoxicity in colon and oral cancer cell lines. *Oncol Rep*, 17, 1461–1467.

49. Ouerhani, S., Oliveira, E., Marrakchi, R., Ben Slama, M.R., Sfaxi, M., Ayed, M., Chebil, M., Amorim, A., El Gaaied, A.B., and Prata, M.J. 2007 Methylenetetrahydrofolate reductase and methionine synthase polymorphisms and risk of bladder cancer in a Tunisian population. *Cancer Genet Cytogenet*, 176, 48–53.

50. Hirota, T., Lipp, J.J., Toh, B.H., and Peters, J.M. 2005 Histone H3 serine 10 phosphorylation by Aurora B causes HP1 dissociation from heterochromatin. *Nature*, 438, 1176–1180.

51. Lee, E.C., Frolov, A., Li, R., Ayala, G., and Greenberg, N.M. 2006 Targeting Aurora kinases for the treatment of prostate cancer. *Cancer Res*, 66, 4996–5002.

52. Li, Y., Kao, G.D., Garcia, B.A., Shabanowitz, J., Hunt, D.F., Qin, J., Phelan, C., and Lazar, M.A. 2006 A novel histone deacetylase pathway regulates mitosis by modulating Aurora B kinase activity. *Genes Dev*, 20, 2566–2579.

53. Monier, K., Mouradian, S., and Sullivan, K.F. 2007 DNA methylation promotes Aurora-B-driven phosphorylation of histone H3 in chromosomal subdomains. *J Cell Sci*, 120, 101–114.

54. Sistayanarain, A., Tsuneyama, K., Zheng, H., Takahashi, H., Nomoto, K., Cheng, C., Murai, Y., Tanaka, A., and Takano, Y. 2006 Expression of Aurora-B kinase and phosphorylated histone H3 in hepatocellular carcinoma. *Anticancer Res*, 26, 3585–3593.

55. Wang, Q., Wang, C.M., Ai, J.S., Xiong, B., Yin, S., Hou, Y., Chen, D.Y., Schatten, H., and Sun, Q.Y. 2006 Histone phosphorylation and pericentromeric histone modifications in oocyte meiosis. *Cell Cycle*, 5, 1974–1982.

56. Ura, S., Nishina, H., Gotoh, Y., and Katada, T. 2007 Activation of the c-Jun N-terminal kinase pathway by MST1 is essential and sufficient for the induction of chromatin condensation during apoptosis. *Mol Cell Biol*, 27, 5514–5522.

57. Giannattasio, M., Lazzaro, F., Plevani, P., and Muzi-Falconi, M. 2005 The DNA damage checkpoint response requires histone H2B ubiquitination by Rad6-Bre1 and H3 methylation by Dot1. *J Biol Chem*, 280, 9879–9886.

58. Zhang, X., Kolaczkowska, A., Devaux, F., Panwar, S.L., Hallstrom, T.C., Jacq, C., and Moye-Rowley, W.S. 2005 Transcriptional regulation by Lge1p requires a function independent of its role in histone H2B ubiquitination. *J Biol Chem*, 280, 2759–2770.

59. Gayther, S.A., Batley, S.J., Linger, L., Bannister, A., Thorpe, K., Chin, S.F., Daigo, Y., Russell, P., Wilson, A., Sowter, H.M., Delhanty, J.D., Ponder, B.A., Kouzarides, T., and Caldas, C. 2000 Mutations truncating the EP300 acetylase in human cancers. *Nat Genet*, 24, 300–303.

60. Boix-Chornet, M., Fraga, M.F., Villar-Garea, A., Caballero, R., Espada, J., Nunez, A., Casado, J., Largo, C., Casal, J.I., Cigudosa, J.C., Franco, L., Esteller, M., and Ballestar, E. 2006 Release of hypoacetylated and trimethylated histone H4 is an epigenetic marker of early apoptosis. *J Biol Chem*, 281, 13540–13547.

61. Fraga, M.F., Ballestar, E., Paz, M.F., Ropero, S., Setien, F., Ballestar, M.L., Heine-Suner, D., Cigudosa, J.C., Urioste, M., Benitez, J., Boix-Chornet, M., Sanchez-Aguilera, A., Ling, C., Carlsson, E., Poulsen, P.,

Vaag, A., Stephan, Z., Spector, T.D., Wu, Y.Z., Plass, C., and Esteller, M. 2005 Epigenetic differences arise during the lifetime of monozygotic twins. *Proc Natl Acad Sci U S A*, 102, 10604–10609.

62. Fraga, M.F., Ballestar, E., Villar-Garea, A., Boix-Chornet, M., Espada, J., Schotta, G., Bonaldi, T., Haydon, C., Ropero, S., Petrie, K., Iyer, N.G., Perez-Rosado, A., Calvo, E., Lopez, J.A., Cano, A., Calasanz, M.J., Colomer, D., Piris, M.A., Ahn, N., Imhof, A., Caldas, C., Jenuwein, T., and Esteller, M. 2005 Loss of acetylation at Lys16 and trimethylation at Lys20 of histone H4 is a common hallmark of human cancer. *Nat Genet*, 37, 391–400.

63. Idikio, H.A. 2006 Spindle checkpoint protein hMad2 and histone H3 phosphoserine 10 mitosis marker in pediatric solid tumors. *Anticancer Res*, 26, 4687–4694.

64. Nguyen, C.T., Weisenberger, D.J., Velicescu, M., Gonzales, F.A., Lin, J.C., Liang, G., and Jones, P.A. 2002 Histone H3-lysine 9 methylation is associated with aberrant gene silencing in cancer cells and is rapidly reversed by 5-aza-2′-deoxycytidine. *Cancer Res*, 62, 6456–6461.

65. Zhang, Z., Yamashita, H., Toyama, T., Sugiura, H., Omoto, Y., Ando, Y., Mita, K., Hamaguchi, M., Hayashi, S., and Iwase, H. 2004 HDAC6 expression is correlated with better survival in breast cancer. *Clin Cancer Res*, 10, 6962–6968.

66. Saji, S., Kawakami, M., Hayashi, S., Yoshida, N., Hirose, M., Horiguchi, S., Itoh, A., Funata, N., Schreiber, S.L., Yoshida, M., and Toi, M. 2005 Significance of HDAC6 regulation via estrogen signaling for cell motility and prognosis in estrogen receptor-positive breast cancer. *Oncogene*, 24, 4531–4539.

67. Siddiqui, H., Fox, S.R., Gunawardena, R.W., and Knudsen, E.S. 2007 Loss of RB compromises specific heterochromatin modifications and modulates HP1alpha dynamics. *J Cell Physiol*, 211, 131–137.

68. Krusche, C.A., Wulfing, P., Kersting, C., Vloet, A., Bocker, W., Kiesel, L., Beier, H.M., and Alfer, J. 2005 Histone deacetylase-1 and -3 protein expression in human breast cancer: A tissue microarray analysis. *Breast Cancer Res Treat*, 90, 15–23.

69. Zhang, Z., Yamashita, H., Toyama, T., Sugiura, H., Ando, Y., Mita, K., Hamaguchi, M., Hara, Y., Kobayashi, S., and Iwase, H. 2005 Quantitation of HDAC1 mRNA expression in invasive carcinoma of the breast. *Breast Cancer Res Treat*, 94, 11–16.

70. Anton, M., Horky, M., Kuchtickova, S., Vojtesek, B., and Blaha, O. 2004 Immunohistochemical detection of acetylation and phosphorylation of histone H3 in cervical smears. *Ceska gynekologie/Ceska lekarska spolecnost J Ev*, 69, 3–6.

71. Duenas-Gonzalez, A., Lizano, M., Candelaria, M., Cetina, L., Arce, C., and Cervera, E. 2005 Epigenetics of cervical cancer. An overview and therapeutic perspectives. *Mol Cancer*, 4, 38.

72. Kondo, Y., Shen, L., Yan, P.S., Huang, T.H., and Issa, J.P. 2004 Chromatin immunoprecipitation microarrays for identification of genes silenced by histone H3 lysine 9 methylation. *Proc Natl Acad Sci U S A*, 101, 7398–7403.

73. Kondo, Y. and Issa, J.P. 2004 Epigenetic changes in colorectal cancer. *Cancer Metastasis Rev*, 23, 29–39.

74. Tzao, C., Sun, G.H., Tung, H.J., Hsu, H.S., Hsu, W.H., Wang, Y.C., Cheng, Y.L., and Lee, S.C. 2006 Reduced acetylated histone H4 is associated with promoter methylation of the fragile histidine triad gene in resected esophageal squamous cell carcinoma. *Ann Thorac Sur*, 82, 396–401; discussion 401.

75. Toh, Y., Ohga, T., Endo, K., Adachi, E., Kusumoto, H., Haraguchi, M., Okamura, T., and Nicolson, G.L. 2004 Expression of the metastasis-associated MTA1 protein and its relationship to deacetylation of the histone H4 in esophageal squamous cell carcinomas. *Int J Cancer*, 110, 362–367.

76. Yasui, W., Sentani, K., Motoshita, J., and Nakayama, H. 2006 Molecular pathobiology of gastric cancer. *Scand J Surg*, 95, 225–231.

77. Youssef, E.M., Issa, J.P., and Lotan, R. 2004 Regulation of RARbeta1 expression in head and neck cancer cells by cell density-dependent chromatin remodeling. *Cancer Biol Ther*, 3, 1002–1006.

78. Ogawa, M., Sakashita, K., Zhao, X.Y., Hayakawa, A., Kubota, T., and Koike, K. 2007 Analysis of histone modification around the CpG island region of the p15 gene in acute myeloblastic leukemia. *Leuk Res*, 31, 611–621.

79. Linggi, B.E., Brandt, S.J., Sun, Z.W., and Hiebert, S.W. 2005 Translating the histone code into leukemia. *J Cell Bioch*, 96, 938–950.

80. Lukasova, E., Koristek, Z., Falk, M., Kozubek, S., Grigoryev, S., Kozubek, M., Ondrej, V., and Kroupova, I. 2005 Methylation of histones in myeloid leukemias as a potential marker of granulocyte abnormalities. *J Leukoc Biol*, 77, 100–111.

81. Karaczyn, A.A., Golebiowski, F., and Kasprzak, K.S. 2006 Ni(II) affects ubiquitination of core histones H2B and H2A. *Exp Cell Res*, 312, 3252–3259.

82. Caslini, C., Capo-chichi, C.D., Roland, I.H., Nicolas, E., Yeung, A.T., and Xu, X.X. 2006 Histone modifications silence the GATA transcription factor genes in ovarian cancer. *Oncogene*, 25, 5446–5461.

83. Orhan, D., Kale, G., Caglar, M., Gogus, S., and Karaagaoglu, E. 2006 Histone mRNA in situ hybridization and Ki 67 immunohistochemistry in pediatric adrenocortical tumors. *Virchows Arch*, 448, 591–596.

84. Manoharan, M., Ramachandran, K., Soloway, M.S., and Singal, R. 2007 Epigenetic targets in the diagnosis and treatment of prostate cancer. *Int Braz J Urol*, 33, 11–18.

85. Dobosy, J.R., Roberts, J.L., Fu, V.X., and Jarrard, D.F. 2007 The expanding role of epigenetics in the development, diagnosis and treatment of prostate cancer and benign prostatic hyperplasia. *J Urol*, 177, 822–831.

86. Schulz, W.A. and Hatina, J. 2006 Epigenetics of prostate cancer: Beyond DNA methylation. *J Cell Mol Med*, 10, 100–125.

87. Li, L.C., Carroll, P.R., and Dahiya, R. 2005 Epigenetic changes in prostate cancer: Implication for diagnosis and treatment. *J Natl Cancer Inst*, 97, 103–115.

88. Kahl, P., Gullotti, L., Heukamp, L.C., Wolf, S., Friedrichs, N., Vorreuther, R., Solleder, G., Bastian, P.J., Ellinger, J., Metzger, E., Schule, R., and Buettner, R. 2006 Androgen receptor coactivators lysine-specific histone demethylase 1 and four and a half LIM domain protein 2 predict risk of prostate cancer recurrence. *Cancer Res*, 66, 11341–11347.

89. Fraga, M.F., Agrelo, R., and Esteller, M. 2007 Cross-talk between aging and cancer: The epigenetic language. *Ann N Y Acad Sci*, 1100, 60–74.

90. Fraga, M.F. and Esteller, M. 2005 Towards the human cancer epigenome: A first draft of histone modifications. *Cell Cycle*, 4, 1377–1381.

91. Fraga, M.F. and Esteller, M. 2007 Epigenetics and aging: The targets and the marks. *Trends Genet*, 23, 413–418.

92. Olins, D.E. and Olins, A.L. 2005 Granulocyte heterochromatin: Defining the epigenome. *BMC Cell Biol*, 6, 39.

93. Ono, S., Oue, N., Kuniyasu, H., Suzuki, T., Ito, R., Matsusaki, K., Ishikawa, T., Tahara, E., and Yasui, W. 2002 Acetylated histone H4 is reduced in human gastric adenomas and carcinomas. *J Exp Clin Cancer Res*, 21, 377–382.

94. Sarg, B., Koutzamani, E., Helliger, W., Rundquist, I., and Lindner, H.H. 2002 Postsynthetic trimethylation of histone H4 at lysine 20 in mammalian tissues is associated with aging. *J Biol Chem*, 277, 39195–39201.

95. Toh, Y., Yamamoto, M., Endo, K., Ikeda, Y., Baba, H., Kohnoe, S., Yonemasu, H., Hachitanda, Y., Okamura, T., and Sugimachi, K. 2003 Histone H4 acetylation and histone deacetylase 1 expression in esophageal squamous cell carcinoma. *Oncol Rep*, 10, 333–338.

96. Yan, N. and Shi, Y. 2003 Histone H1.2 as a trigger for apoptosis. *Nat Struct Biol*, 10, 983–985.

97. Yan, Q., Huang, J., Fan, T., Zhu, H., and Muegge, K. 2003 Lsh, a modulator of CpG methylation, is crucial for normal histone methylation. *EMBO J*, 22, 5154–5162.

98. Yasui, W., Oue, N., Ono, S., Mitani, Y., Ito, R., and Nakayama, H. 2003 Histone acetylation and gastrointestinal carcinogenesis. *Ann N Y Acad Sci*, 983, 220–231.

99. Verma, M. 2003 Viral genes and methylation. *Ann N Y Acad Sci*, 983, 170–180.

100. Verma, M., Dunn, B.K., Ross, S., Jain, P., Wang, W., Hayes, R., and Umar, A. 2003 Early detection and risk assessment: Proceedings and recommendations from the Workshop on Epigenetics in Cancer Prevention. *Ann N Y Acad Sci*, 983, 298–319.

101. Verma, M. and Srivastava, S. 2002 Epigenetics in cancer: Implications for early detection and prevention. *Lancet Oncol*, 3, 755–763.

102. Ronzoni, S., Faretta, M., Ballarini, M., Pelicci, P., and Minucci, S. 2005 New method to detect histone acetylation levels by flow cytometry. *Cytometry A*, 66, 52–61.

103. Ho, S.M. and Tang, W.Y. 2007 Techniques used in studies of epigenome dysregulation due to aberrant DNA methylation: An emphasis on fetal-based adult diseases. *Reprod Toxicol*, 23, 267–282.

104. Sarg, B., Helliger, W., Talasz, H., Koutzamani, E., and Lindner, H.H. 2004 Histone H4 hyperacetylation precludes histone H4 lysine 20 trimethylation. *J Biol Chem*, 279, 53458–53464.

105. Zhang, L., Freitas, M.A., Wickham, J., Parthun, M.R., Klisovic, M.I., Marcucci, G., and Byrd, J.C. 2004 Differential expression of histone post-translational modifications in acute myeloid and chronic lymphocytic leukemia determined by high-pressure liquid chromatography and mass spectrometry. *J Am Soc Mass Spectrom*, 15, 77–86.

21 Histone Modifications in Cancer Biology and Prognosis

Matthew A. McBrian, David B. Seligson, and Siavash K. Kurdistani

CONTENTS

21.1 INTRODUCTION

The DNA of eukaryotic cells is packaged into nucleosomes, the basic repeating unit of chromatin. The nucleosome can present an efficient barrier to all DNA-templated processes, including transcription, replication, and DNA repair, necessitating a dynamic chromatin structure that includes highly condensed heterochromatin and the more relaxed and accessible euchromatin. Nucleosomes are comprised of 147 base pairs of DNA wrapped 1.65 times around the globular core of the histone octamer which is assembled from two copies each of histones H2A, H2B, H3, and H4 [1]. Importantly, the N-terminal tail of each histone protrudes out from the cylindrically shaped nucleosomal core in an unstructured fashion and is accessible to nuclear proteins. These tails are the site of numerous, well-characterized covalent modifications of specific amino acid residues which, along with ATP-dependent chromosome remodeling complexes, function to regulate chromatin architecture and accessibility of the underlying DNA [2]. Mass spectrometry has revealed modification of histones in the globular region as well, and emerging evidence indicates a role for these modifications in the regulation of chromatin structure [3–7]. In addition, chromatin structure is affected by incorporation of histone variants that have specific effects on various DNA-based processes.

Histones, especially their N-terminal tails, are covalently modified in numerous ways, including methylation, acetylation (ac), phosphorylation, ubiquitination, sumoylation, and ADP-ribosylation [8].

These posttranslational modifications (PTMs) affect chromatin structure by modulating the contacts between DNA and the histones and by serving as binding sites for a variety of effector protein complexes [9]. The former function is largely mediated by acetylation of lysine (K) residues that neutralizes the positively charged amino acid, thereby weakening its association with the negatively charged polyphosphate backbone of the DNA. In addition to controlling chromatin structure at the level of specific genes, acetylation may also serve to modulate higher order chromatin structure as witnessed by the inhibitory effects of histone H4 lysine 16 acetylation (H4K16ac) on the formation of the 30 nM chromatin fiber [10]. Because the histone PTMs are dynamic, they can effectively act as on/off switches to regulate DNA-templated processes [2].

21.2 HISTONE ACETYLATION

Histone acetylation levels are controlled by the relative activity of two groups of modifying enzymes, histone acetyltransferases (HATs) and histone deacetylases (HDACs) [11]. Some of the residues that are acetylated include H2A at K5; H2B at K12 and 15; H3 at K9, 14, 18, 23, 27, and 56; and H4 at K5, 8, 12, and 16. The first nuclear HAT was cloned in 1996 and was termed HAT A [12]. It showed remarkable homology to the yeast Gcn5p (general control nonderepressible-5 protein), a protein known to be essential for the robust activity of a number of transcriptional activators. Gcn5 was subsequently shown to possess HAT activity, thereby providing a link between histone acetylation and transcriptional activation. Many HATs have since been identified and grouped into the GNAT (Gcn5-related N-acetyltransferase), MYST (MOZ, Ybf2/Sas3, Sas2, and Tip60), and CBP/p300 families based on conserved regions of homology. The best character-ized family is the GNAT family which includes Gcn5 and PCAF (p300/CBP-associated factor) in humans. These enzymes robustly acetylate H3 and show some activity on H4 as well. The MYST family includes Tip60 (Tat-interactive protein 60), Sas2 (something about silencing), MOZ (monocytic leukemia zinc finger), MORF (MOZ-related factor), and HBO1 (histone acetyl-transferase bound to ORC), with Tip60 being the first MYST HAT identified in humans. These enzymes show preferential activity toward lysine residues in H4 and also acetylate H3 residues to a lesser extent. CBP (CREB-binding protein) and p300 are considered structural and functional homologs but they do show subtle differences. Both are able to acetylate all four histones and show little specificity in vitro. The importance of p300 in transcription is exemplified by in vitro transcription assays on chromatin templates which require addition of p300 to allow transcription to occur. p300 interacts with sequence-specific transcription factors in vivo, thus demonstrating its relevance to cellular transcription [13]. In addition to the HATs listed above, nuclear receptors, such as SRC-1 (steroid receptor coactivator-1), and components of the general transcription machinery, such as TAF1 (TATA-binding protein-associated factor 1), also exhibit HAT activity [11]. Notably, the in vivo activity of many HATs differs from their in vitro activity. This likely reflects the fact that HATs are found as part of large, multiprotein complexes in vivo, the subunits of which are able to alter the activity and specificity of the associated HAT [14]. The difficult nature of determining the in vivo specificity of HATs is also due to the functional redundancy exhibited by this class of enzymes.

HDACs oppose the function of HATs in removing the acetyl moiety from lysine residues [15]. Like HATs, these enzymes are typically found as subunits of large protein complexes. They can be divided into three classes based on their similarity to yeast proteins [16]. Class I HDACs (HDAC1, 2, 3, and 8) closely resemble the yeast transcriptional regulator Rpd3 (reduced potassium deficiency 3), while class II HDACs (HDAC4, 5, 6, 7, 9, and 10) more closely resemble the yeast deacetylase HDAC1 (histone deacetylase 1). Class I HDACs are generally expressed in all tissues whereas class II HDACs exhibit more tissue-specific expression. Both classes are sensitive to a variety of HDAC inhibitors, including trichostatin A (TSA) and sodium butyrate. Class III HDACs exhibit pronounced differences from class I and class II enzymes. Class III HDACs are related to the yeast Sir2 (silent information regulator 2) protein and are nicotinamide adenine dinucleotide (NAD)-dependent for their enzymatic activity.

This class is not inhibited by molecules that inhibit class I and class II enzymes but can be inhibited by NAD analogues such as nicotinamide.

21.3   HISTONE METHYLATION

Methylation (me) of lysine and arginine (R) residues by lysine and arginine histone methyltransferases (HMTs) has also been well characterized. Lysines can be mono-, di-, or trimethylated by HMTs that contain a conserved catalytic SET domain. Arginines are mono- or dimethylated, either symmetrically (both methyl groups on the same nitrogen atom) or, more commonly, asymmetrically (one methyl group on each nitrogen atom) by HMTs that contain a conserved catalytic domain that differs from the SET domain [17]. HMTs show much more specificity for individual histone residues than HATs, as many HMTs modify only a single residue. Methylation occurs on H3 at R2, 8, 17, and 26 and K4, 9, 18, 27, 36, and 79 and also on H4 at R3 and K20. Methylation has also been detected on H2A at R3 and K17 and 238 [18,19] and on H2B at K47, 57, and 108 [6]. Early findings on HMTs came from studies on position effect variegation (PEV). PEV is a phenomenon described in yeast and *Drosophila* in which expression levels of a gene depend on the location of the gene within the genome. *Drosophila* screens engineered to identify modifiers of PEV revealed the presence of suppressors, Su(var), and enhancers, E(var), of variegation [20]. Subsequent characterization of some of these modifiers identified Su(var)3–9 as HMTs. The first lysine-specific human HMT was found to be a homolog of these proteins and was named SUV39H1 [21]. SUV39H1 exhibits specificity for lysine 9 of histone H3. H3K9 can also be acetylated and the authors of this study showed that acetylation of lysine 9, but not other nearby lysines such as K14, precluded methylation of lysine 9. Additional suppressors of variegation were identified as HMTs, including the Polycomb group (PcG) protein EZH2 (enhancer of zeste human 2) and the trithorax group protein trithorax (Trx) [22]. Each HMT was found to exhibit much more residue selectivity than is found in HATs. Certain HMTs display a preference for euchromatic or heterochromatic regions of chromatin, thereby generating an additional level of specificity to histone methylation [23]. Many other HMTs have been discovered, including the lysine HMTs G9a, MLL (mixed lineage leukemia), SMYD3 (SET and MYND domain-containing 3), and NSD1 (nuclear receptor-binding SET domain-containing 1), and the arginine HMTs CARM1 (coactivator-associated arginine methyltransferase 1) (the first HMT identified [24]), PRMT4 (protein arginine methyltransferase 4), and PRMT5.

An interesting feature of histone methylation is that it imposes regulation of transcription at two levels. At the level of individual genes, histone methylation is associated with both activation and repression of transcription, depending on the residue that has been modified as well as the other modifications that occur at the same genetic loci. In Tetrahymena, methylation of H3K4 is associated with transcriptionally active macronuclei whereas inactive micronuclei were depleted of this modification [25]. Paradoxically, K4 trimethylation has more recently been linked to active gene repression [26]. It appears that the context within which the modification occurs, i.e., the specific gene and the array of modifications that exist there, is important in determining functional output. However, certain methylation events may be less ambiguous in their functional significance. Methylation of H3K9 and H3K27 has been observed at the promoters of repressed but not active genes. At the genome-wide level, bulk chromatin is described as highly condensed heterochromatin and less condensed euchromatin. Heterochromatin is characterized by di- and trimethylation of H3K9, serving as a mark for the binding of the heterochromatin-associated protein (HP1, see Section 21.4) while euchromatin is marked by methylation of H3K4 and H3K79 [22].

Efficient silencing of gene expression depends not only on PTMs of histone residues but on DNA methylation as well. Accordingly, molecular processes that establish histone modifications and DNA methylation patterns are mutually reinforcing. For instance, methylated cytosine residues are recognized by proteins containing a methyl-CpG-binding domain (MBD). These MBD-containing proteins have been found to associate with repressive complexes that contain

HDACs. The deacetylation of lysines may be necessary to provide a substrate for HMTs that cannot methylate acetylated residues. SETDB1 (SET domain bifurcated 1) is a human HMT that selectively methylates H3K9, a mark associated with transcriptional repression. SETDB1 contains an MBD, thus providing a potentially direct link between DNA methylation and histone methylation [27]. In addition, DNA methylation of EZH2 target promoters is dependent on the presence of EZH2 with an intact SET domain, indicating a mechanistic link between histone and DNA methylation [28]. Recent studies have further elucidated the cooperative effects on silencing between histone and DNA methylation [29]. The authors find that G9a methylates H3K9 which leads to binding of HP1. G9a also stimulates the recruitment of Dnmt1 (DNA-methyltransferase 1), and HP1 binding is stabilized by the presence of Dnmt1. In return, HP1 stimulates the activity of Dnmt1, forming a reinforcing loop between histone and DNA methylation. This system employs three different types of proteins: a histone methyltransferase, a DNA methyltransferase, and a chromodomain-containing protein (see below) to silence euchromatic genes [29]. These results argue that the presence of all three types of proteins is necessary for efficient silencing and that HP1 provides a link between histone and DNA methylation.

While histone methylation was previously thought to be an irreversible epigenetic mark, the past several years have ushered in the arrival of histone demethylases (HDMs). LSD1 (lysine-specific demethylase 1) was the first enzyme found to possess demethylase activity [30]. LSD1 is part of several corepressor complexes, and its demethylating activity was suggested based on its homology with flavin adenine dinucleotide (FAD)-dependent amine oxidases that had been proposed to function as HDMs [31]. The mechanism of this class of enzymes requires a protonated nitrogen atom in the substrate, and thus amine oxidases can demethylate mono- and dimethylated lysines but not trimethylated lysines [32]. Subsequent to the discovery of the amine oxidase HDMs, several groups uncovered the existence of an entire family of Fe(II) and α-ketoglutarate-dependent dioxygenases with histone demethylase activity [33]. The common feature of this family is the presence of the catalytic Jumonji C (JmjC) domain that does not require protonated nitrogen for its activity and is therefore capable of reversing the trimethyl mark on lysines. Indeed, the JMJD2 subfamily of JmjC-containing HDMs demethylates trimethylated residues. The specificity of HDMs is important for proper gene regulation as demonstrated for androgen receptor (AR) target genes. LSD1 and JMJD2C are found complexed with AR and participate in the expression of androgen target genes [34]. JMJD2C demethylates trimethylated H3K9 generating dimethylated K9 which serves as a substrate for LSD1 leading to removal of the remaining two methyl groups, thereby relieving repression of transcription imposed by the methylation of K9. In addition to its role in gene activation, LSD1 also has a role in transcriptional repression by interacting with the Co-REST (corepressor to the RE1 silencing transcription factor) corepressor [35]. Thus, it appears that the effects of HDMs such as LSD1 on transcriptional output depend partly on their interactions with other proteins [36]. Proper targeting of HDMs is crucial for maintaining correct expression programs as shown by the repression of the developmentally important Hox genes in embryonic stem (ES) cells by a JmjC domain-containing HDM, RBP2 (retinoblastoma-binding protein 2) [37,38]. Depletion of the RPB2 homolog in *C. elegans* by RNAi led to increases in H3K4me3 and developmental defects. In ES cells, RBP2 is displaced from the promoters of the Hox genes upon differentiation correlating with an increase in H3K4 methylation and expression of the Hox genes.

21.4 INTERPRETING HISTONE MODIFICATIONS

The covalent modification of histones serves not only to weaken the DNA-histone interactions but also to provide binding surfaces for proteins that interact with chromatin. The first domain identified to interact with modified histones was the bromodomain, which recognizes and binds to acetylated lysine residues [39]. The bromodomain is found in a variety of transcriptionally related proteins such as SNF2 (sucrose nonfermenting 2), TAF1, GCN5, and PCAF. The crystal structure of the bromodomain of GCN5 complexed with a peptide corresponding to acetylated H4K16 suggests that the bromodomain may distinguish between different acetylated lysines depending on the context in which the lysine is displayed [40]. Another chromatin-binding domain, the chromodomain is able to recognize and bind to methylated lysine residues. Chromodomains are found in transcriptionally related proteins such as Polycomb, HP1, and MOF (males absent on the first). Much like the bromodomain, the chromodomain of different proteins has the ability to recognize specific methylated residues, as shown for the HP1 protein which binds methylated H3K9 [41,42]. The domains differ in their affinity for their respective binding sites with the chromodomain showing higher affinity for methylated residues than the bromodomain for acetylated residues. The dissociation constant (K_d) for the HP1 chromodomain binding to methylated H3K9 is 4 μM [43] whereas the K_d for the single bromodomain of PCAF binding to acetylated histone H4 is 346 μM [39]. The higher K_d of the bromodomain relative to the chromodomain may explain why bromodomains tend to show less site-specificity than chromodomains. However, the presence of a second bromodomain, as occurs in TAF1, can increase the affinity, lowering the K_d to levels similar to that for the chromodomain, suggesting that efficient binding requires a precise combination of appropriately spaced modifications [44]. The PHD finger domain is another domain that reads the methyl mark of lysine residues. The PHD finger of the NURF (nucleosome remodeling factor) complex was specifically associated with tri- but not mono- or dimethylated H3K4, and this association was abrogated by depletion of the K4 trimethyl mark [45]. The recognition of specific PTMs of histones by specific domains has led to the proposal of the histone code [9]. This hypothesis states that the pattern of PTMs of histones at a particular locus allows for the binding of specific effector protein complexes leading to distinct transcriptional outputs.

Although histone modifications generally tend to recruit proteins to chromatin, they can also serve to exclude the binding of proteins. This was first shown in yeast where acetylation of H4K16 excludes the binding of Sir3p at telomeres [46,47]. Another study in yeast demonstrated the inhibitory effect of methylation of H3K4 on binding of the NuRD (nucleosome remodeling and deacetylase) complex to histone peptides [48]. Evidence of an exclusion function for histone modifications was later shown in mammals where phosphoacetylation of histone H3 (lysine 9 acetylation and serine 10 phosphorylation) leads to eviction of HP1 from chromatin [49]. Therefore, either presence or absence of histone modifications may regulate interactions of protein complexes with chromatin.

21.5   ATP-DEPENDENT CHROMATIN REMODELING

In addition to PTMs of histones, chromatin remodeling is accomplished by a group of related ATP-dependent remodeling complexes that function by sliding or relocating nucleosomes to expose previously inaccessible DNA [50]. Thus, remodelers regulate both the distribution of nucleosomes and their relative mobility on the chromatin fiber. Nucleosome density is heterogeneous throughout the genome, and patterns of distribution likely reflect patterns of gene expression [51]. For example, the promoters of highly expressed ribosomal genes have a low density of nucleosomes in yeast that are growing exponentially. These genes are repressed upon heat shock, and this occurs concomitantly with an increase in nucleosomal density at these same promoters [52]. The remodeling complexes all possess a related ATPase catalytic subunit but they have different subunits with various chromatin-recognizing domains [53]. These distinct domains found in chromatin remodelers have been used to group these enzymes into four families found in mammalian cells. These include the SWI/SNF (switching defective/sucrose nonfermenting) family whose members possess a bromodomain in the catalytic subunit; the ISWI (imitation SWI) family whose members have a SANT (SWI3, ADA2, NCOR, and TFIIIB DNA-binding) and SLIDE (SANT-like ISWI) domain which directs the complex to unmodified histone tails; the NuRD/Mi-2/CHD (nucleosome remodeling and deacetylation/chromodomain, helicase, DNA binding) family in which the CHD subunits possess tandem chromodomains; and the INO80 (inositol requiring 80) family which have subunits containing a split ATPase domain. The different chromatin-related domains may allow for different chromatin-remodeling complexes to be targeted to distinct regions of the genome [54]. The SWI/SNF remodeler is large (~2 MDa), and its representative complexes all contain a catalytic subunit that resembles the yeast SWI2/SNF2 family of ATPases. These complexes are primarily involved in the destabilization of nucleosomes which promotes gene activation. The ISWI family includes the NURF, ACF (ATP-dependent chromatin assembly and remodeling factor), and CHRAC (chromatin-accessibility complex) complexes. They all share the same ATPase subunit that also resembles SWI2/SNF2 [55]. CHRAC and ACF generally repress transcription by aiding in the assembly of nucleosomes. The INO80 and NuRD/Mi-2/CHD family complexes support both active and repressed gene expression. The INO80 complex is also recruited to sites of DNA double-strand breaks (DSBs) in yeast via an interaction with the H2AX histone variant (see below) [56].

21.6   COOPERATIVE CHROMATIN REMODELING

Histone-modifying enzymes and chromatin remodelers work cooperatively to regulate gene expression. For example, the NURF complex component BPTF (bromodomain and PHD finger transcription factor) possesses a PHD finger which binds to H3K4me3, leading to remodeling of loci containing the trimethyl mark, thus linking histone methylation to chromatin remodeling and gene activation [45]. Distinct histone modifications also cooperate with one another to regulate transcription. The promoters of actively transcribed genes are enriched for H3K4me3 and H3 acetylation. The PHD finger of Yng1, a subunit of the NuA3 (nucleosome acetyltransferase of histone H3) HAT complex, recognizes and binds the trimethylated K4. Chromatin immunoprecipitation (ChIP) experiments demonstrated that H3 hyperacetylation depended on Yng1 binding to H3K4me3, thereby providing a direct link between histone methylation and acetylation [57]. The fact that histone modifications cooperate with one another implies that distinct patterns of modifications may be necessary for proper gene regulation and suggests a significant amount of "crosstalk" between modifications on the same and adjacent tails. These ideas are consistent with the histone code hypothesis and several studies have revealed an ordered series of events leading to transcriptional activation [58,59]. Of particular relevance is that there is not a single ordered series of events regulating transcription at every locus. Accordingly, individual histone modifications may have different transcriptional effects based on the specific genomic loci, presence of other histone modification, and the collection of other proteins [60]. However, groups of genes that are functionally related to each other and have similar expression patterns, also share similar patterns of acetylated and deacetylated residues at their promoters or coding regions [61]. The existence of consistent and combinatorial patterns of histone acetylation supports the histone code hypothesis and may contribute to coordinate expression of biologically related genes.

21.7   HISTONE VARIANTS

The deposition of histone variants, with the assistance of specific chaperones, at precise genomic locations also has effects on chromatin architecture [62]. Histone variants have been identified for each of the histones H2A, H2B, and H3 but not for H4. Differences between variants can be as small as a single amino acid, yet this can have significant functional effects, especially if the altered amino acid is one that can be covalently modified. For example, the H2AX variant is very similar to the canonical H2A histone but differs at the C-terminus by the presence of a serine, glutamine (SQ) motif. The serine of the SQ motif is phosphorylated in response to DNA damage. Phospho-H2AX (also known as γH2AX) permits the accumulation of a variety of DNA damage response proteins. Notably, mice deficient for H2AX exhibit increased genomic instability and develop T- and B-cell lymphomas in a p53 null background [63]. Another histone variant, H3.3, is found to coexist at sites enriched for methylated H3K4 and RNA Pol II and thus marks actively transcribed regions of chromatin [64].

21.8 COORDINATED GENE REGULATION

Taken altogether, the proper regulation of gene expression depends on the coordinated effects of histone modifications, chromatin remodeling, DNA methylation, and deposition of histone variants. The enzymes that mediate these effects show significant functional interactions amongst each other in creating a dynamic chromatin structure that is essential for all DNA-based processes. The appropriate chromatin conformation may be achieved by various combinations of events at different genetic loci. For example, histone acetylation may provide binding sites for nucleosome remodeling complexes that contain bromodomains during activation of some genes. Alternatively, chromosome remodelers may be necessary to create a more relaxed chromatin conformation to allow access of HATs at other loci. Detailed studies of gene regulation at individual loci will likely provide numerous variations in the order of events leading to transcriptional activation or repression. However, a unifying theme has emerged in which the concerted actions of multiple chromatin-affecting proteins lead to precise control over DNA-templated processes that are critical for proper cell function.

21.9 EPIGENETICS AND CANCER

Historically, cancer has been viewed as a disease owing to the accumulation of genetic mutations in oncogenes and tumor suppressors. Activating mutations in oncogenes and inactivating mutations in tumor suppressor genes have been extensively described. More recently, the role of epigenetics in the initiation and progression of cancer has come to share the spotlight with studies of genetic mutations [65–69].

Epigenetics is the study of the inheritance of phenotypes that occur without an alteration in DNA sequence. The original use of the term "epigenetics" is credited to the biologist C.H. Waddington who used it in 1942 to describe how genotypes give rise to phenotypes, especially in determination of cell fate during development. Although unknown at that time, the molecular mechanisms of epigenetic inheritance as it relates to chromatin include the inter-related processes of DNA methylation and histone modifications. Through small chemical moieties that covalently attach to DNA or histones, the epigenetic processes can increase the capacity of the genome to store and transmit biological information beyond the DNA sequence. Despite its heritability, the epigenetic information is dynamic and maintained through reversible and regulated molecular pathways that are essential for normal development and function of cells. In cancer, the epigenetic program can be subverted by cancer cells to gain biological advantage over their normal counterpart.

It has been proposed that epigenetic alterations not only contribute to cancer progression but may also make cells susceptible to genetic mutations for initiation of cancer [70,71]. Although far from conclusive, some studies have provided evidence that epigenetic alterations may in fact precede tumor initiation. For example, loss of imprinting (LOI), loss of parent-of-origin DNA methylation pattern, of the IGF2 (insulin-like growth factor 2) gene is seen in many patients with colorectal cancer. Importantly, biallelic expression of IGF2 is found in the normal-appearing colonic epithelium of colorectal cancer patients [72]. Elevated levels of IGF2 in normal cells may confer a growth advantage that is necessary for tumor initiation. In support of this idea, mice engineered to express IGF2 from both alleles are predisposed to tumor formation [73]. These mice have a structurally altered colonic epithelium marked by lengthening of crypt foci, the site of epithelial stem cell renewal, and increases in the levels of progenitor cell markers. Whether LOI of IGF2 and the expansion of the progenitor pool plays a direct causal role in tumor initiation has yet to be determined.

Alterations in chromatin-regulating proteins lead to aberrations in chromatin structure and gene expression which play important roles in tumorigenesis and cancer progression. Mechanistically, general hypoacetylation of histone lysine residues along with the methylation of specific lysines and arginines contributes to repression of tumor suppressors (Figure 21.1A). For example, ChIP

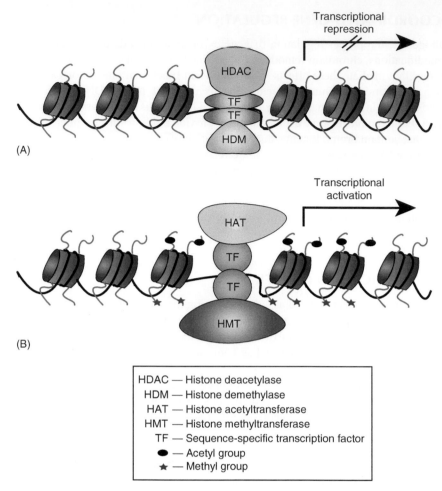

FIGURE 21.1 Transcription factors recruit a variety of proteins to chromatin to regulate gene expression. Aberrant targeting of histone-modifying enzymes to specific loci leads to altered expression of genes involved in cancer development and progression. (A) Tumor suppressors can be silenced by the recruitment of histone deacetylases to remove acetyl groups from lysines—such as H3K9, H3K18, H4K8, and H4K12—and demethylases to remove methyl groups from lysines—such as H3K4 and H3K36—to reverse histone modifications that are associated with transcriptional activation. (B) Alternatively, the aberrant recruitment of histone acetyltransferases and methyltransferases can result in increased acetylation of lysines—such as H3K9, H3K18, H4K8, and H4K12—and increased methylation of lysines—such as H3K4 and K3K36—to upregulate the expression of oncogenes due to changes in these histone modifications.

experiments show that silencing of the CDKN2A locus—which encodes the p16^{INK4A} and p14^{ARF} tumor suppressors—is associated with hypermethylation of H3K9 and hypomethylation of H3K4 [74]. Interestingly, the epigenetic silencing of this locus is seen in lung cancer as early as preneoplastic lesions [75]. Conversely, modifications associated with gene activation can lead to the upregulation of oncogene expression (Figure 21.1B). While this mechanism has been less well documented than epigenetic silencing of tumor suppressors, examples have been described. For instance, the H3K4 HMT, MLL, is translocated in certain forms of leukemia [76], leading to the upregulation of the Hox genes, a potentially necessary step for leukemic transformation [77].

Deregulation of histone-modifying enzymes has also been described in a multitude of cancer types (Table 21.1) [78–82]. Significantly, the differential expression of a select group

TABLE 21.1
List of Epigenetic Genes Disrupted in Human Cancer

Gene	Alteration	Cancer Type
Histone Acetyltransferases		
p300 [85]	Mutations, translocations	Multiple epithelial cancers, hematological malignancy
CBP [84]	Mutations, deletions	Colon, stomach, lung, Rubinstein–Taybi syndrome
pCAF	Mutations	Colon
MOZ	Translocations	Hematological malignancy
MORF	Translocations	Hematological malignancy
MOF	Translocations	Hematological malignancy
Histone Deacetylases		
HDAC1 [83]	Misexpression, mistargeting	Hematological malignancy, colon, endometrial
HDAC2 [93,94]	Misexpression, mistargeting, mutation	Hematological malignancy, colon, endometrial
SIRT1	Overexpression	Breast, colon, prostate
Histone Methyltransferases		
RIZ1	Mutation, silenced expression	Multiple
CARM1 [103]	Overexpression	Prostate
MLL1 [76,77]	Translocations	Hematological malignancy
SMYD3 [99,100]	Overexpression	Colorectal and hepatocellular carcinoma
EZH2 [109,110]	Overexpression, amplification	Prostate, breast
Histone Demethylases		
PLU-1 [117]	Overexpression	Breast
GASC1 [118,119]	Amplification, overexpression	Esophageal squamous carcinoma
LSD1 [121]	Overexpression	Prostate
ATP-Dependent Chromatin-Remodeling Components		
BRG1 [134]	Misexpression	Prostate
BRM [134]	Misexpression, mutation, deletion	Prostate, lung
SNF5 [50]	Mutation	Lymphoma, rhabdoid tumors

Other Epigenetic Players

Gene	Function	Alteration	Cancer Type
BMI1 [108]	PcG protein	Overexpression, amplification	Lymphoma, brain, hematological malignancy
EED2 [133]	PcG protein	Misexpression	Multiple
DNMT1	DNA methyltransferase	Overexpression	Multiple
MBD1–4	Methyl-CpG-binding protein	Overexpression, mutation	Multiple

of histone-modifying genes was shown to distinguish cancer cells from their normal counterparts and have the ability to specify cancer type [83]. Representatives of each type of histone modifier can be found to be misexpressed or have altered activity in tumors leading to alterations in the levels of histone acetylation and methylation at specific loci and on a global scale as well.

In addition to abnormalities in histone modifications, alterations in chromatin structure—as influenced by chromatin-remodeling complexes and histone variants—have been indicated to play a role in tumor progression. Sections 21.10 and 21.11 give an overview of the enzymes that have been found to be deregulated in various cancer types and highlights a few well-characterized examples that give detailed insight into the mechanism of how epigenetic abnormalities can serve as surrogates for genetic mutations in the development of cancer.

21.10   HISTONE ACETYLATION IN CANCER

The acetylation status of histones is determined by the balancing activities of HATs and HDACs. Altered patterns of histone acetylation have been observed in many cancer types, partly owing to altered activity of HATs. For example, mutations that abrogate the activity of CBP are associated with Rubinstein–Taybi syndrome (RTS), and individuals with this disease are predisposed to developing cancer. Forty percent of transgenic mice that are heterozygous for CBP ($CBP^{+/-}$) develop tumors or have tumorigenic cells, and in all cases tumor cells were found to have lost the second allele [84]. Several inactivating mutations have been found in p300 in cancers of epithelial origin, thus providing evidence for a tumor suppressor role for this HAT [85]. Additional studies have supported this notion, as cell-cycle progression of HeLa cells driven by adenoviral E1A oncoprotein is dependent on its association with p300 and is inhibited by overexpression of p300 and pCAF [86]. CBP, p300, MOZ, and MORF are found as parts of fusion proteins in certain cases of leukemia due to translocation events.

HDACs have also been shown to be altered in cancer and play a role in the silencing of tumor suppressors as evidenced by the ability of HDAC inhibitors to retard growth of cancer cell lines [87–89]. A role for the HDACs in the derepression of tumor suppressors has been more directly established by investigation of the $p21^{WAF1}$ tumor suppressor in bladder carcinoma cells and breast cancer cells [90,91]. In these studies, the HDAC inhibitor SAHA (suberoylanilide hydroxamic acid) was found to have antiproliferative effects that correlated with increased levels of $p21^{WAF1}$. This effect is at least partly a transcriptional one, as SAHA induced hyperacetylation of H3 and H4 at the CDKN1A locus from which $p21^{WAF1}$ is transcribed. The increase in histone acetylation exhibited some selectivity for the CDKN1A locus, as treatment with SAHA did not induce changes in acetylation levels at the p27 or actin genes. This is in agreement with studies that have shown HDAC inhibitors to affect the expression of only a small subset of genes [92]. In addition to silencing tumor suppressors, HDACs are also associated with known tumor suppressor pathways. For example, HDAC1 associates with the RB (retinoblastoma) tumor suppressor to silence the E2F transcription factor target genes [81]. Loss of the APC (adenomatosis polyposis coli) tumor suppressor in colon cancer leads to overexpression of HDAC2 [93]. In addition, a truncating mutation of HDAC2 was discovered in colorectal and endometrial cancer cell lines [94]. Altered activity of HDACs can lead to the aberrant expression of select genes. This has been shown in the case of the AML-ETO (acute myelogenous leukemia-eight twenty-one) fusion protein in acute myeloid leukemia [95,96]. This fusion protein associates with HDACs through contact with ETO, resulting in histone hypoacetylation and diminished transcription at AML target genes. These genes are important for the differentiation of hematopoietic cells and loss of their expression leads to a differentiation block and an increase in the fraction of proliferative progenitor cells in the population.

21.11   HISTONE METHYLATION IN CANCER

Just as histone acetylation levels are determined by the opposing roles of HATs and HDACs, so too are methylation levels set by the activity of opposing classes of enzymes, the HMTs and HDMs. The activity of both classes of enzymes has been found to be altered in cancer. RIZ1 (retinoblastoma interacting zinc finger) was the first HMT to be identified to have a role in tumor suppression [97]. Since then, many translocations found in leukemia have been shown to involve proteins known to be HMTs, such as MLL. Mice deficient for the SUV39H1 homolog, which is important in establishing heterochromatic domains, exhibit increased chromosomal instability and are predisposed to tumor formation [98]. Gene expression profiling of colorectal carcinoma (CRC) and hepatocellular carcinoma (HCC) has revealed the overexpression of SMYD3, a H3K4 specific HMT, in these types of cancer [99]. Transfection of SMYD3 into an SMYD3 nonexpressing cell line, NIH3T3, enhanced cell growth, an effect that was negated by siRNA-mediated knockdown of SMYD3. This activity was shown to be dependent on the catalytic SET domain. Examination of the SMYD3 promoter region identified a polymorphism in the number of tandem repeats of a CCGCC motif that was identified as an E2F-1 binding sequence [100]. Reporter assays confirmed that an increased number of repeats resulted in increased transcription from the SMYD3 promoter. Importantly, the frequency of the expanded number of repeats was significantly higher in individuals with CRC, HCC, and breast cancer than in healthy individuals. Knockdown of SMYD3 by siRNA leads to growth inhibition in breast cancer cell lines as well [101]. A recent report has identified hTERT (human telomerase reverse transcriptase), a gene essential for immortalization and transformation, as a direct target of SMYD3, thus giving some insight into the mechanism of the role of SMYD3 in cancer [102]. The arginine HMT, CARM1, has also been implicated in the progression of prostate cancer. CARM1 is a transcriptional coactivator that enhances transcription from nuclear receptor target genes. Levels of CARM1 were found to correlate with progression of the disease [103]. ChIP experiments showed the recruitment of CARM1 to the androgen responsive enhancer of the PSA (prostate specific antigen) and hK2 (human kallikrein 2) genes with a concomitant increase in H3R17 methylation and transcription of these genes [104]. CARM1 knockdown abrogated the effect of androgen in stimulating the expression of these genes.

PcG proteins are also important in establishing methylation patterns, and many recent studies have begun to elucidate their role in tumor initiation and progression [105]. PcG proteins were originally identified in *Drosophila* as repressors of the developmentally important Hox genes. Most PcG proteins are classified according to the complex in which they participate. The human version of Polycomb repressive complex 1 (PRC1) consists of a core comprised of RING1A, HPC1–3, HPH1–3, BMI1, and SCHMH1–2, while human PRC2 has a core of EZH2, EED, SUZ12, and RbAp46/48 [106]. In humans, the catalytic subunit of PRC2 is EZH2 which has the ability to methylate H3K27. Methylated K27 is recognized by the chromodomain of the PC subunit of PRC1 leading to transcriptional silencing [107]. Silencing is thought to be effected by preventing access to chromatin remodelers and components of the transcriptional machinery. In addition, EZH2 was shown to physically interact with DNA methyltransferases (Dnmts), and methylation of cytosine nucleotides also contributes to silencing. Association of Dnmts with Polycomb-silenced targets is dependent on an intact SET domain in EZH2 [28].

Both the PRC1 and PRC2 complexes are associated with oncogenes: BMI1 in PRC1 and EZH2 in PRC2 [78]. BMI1 is overexpressed in tumors and has been shown to downregulate the expression of p16^{INK4A} and p19^{Arf} from the CDKN2A locus in cooperation with MYC in the formation of B- and T-cell lymphomas [108]. EZH2 overexpression has also been noted in prostate and breast cancers, and elevated levels correlate with poor prognosis [109,110]. Another study found an inverse correlation between EZH2 levels and DAB2IP (DAB interacting protein 2) levels [111]. DAB2IP is a GTPase-activating protein that exerts its growth inhibitory effects by modulating Ras-mediated signaling. A series of ChIP experiments confirmed the presence of PRC components at the DAB2IP promoter along with increased levels of di- and trimethylated K27 and the presence of HDAC1 and low levels of histone acetylation. Significantly, knockdown of EZH2 resulted in increases in histone acetylation, dissociation of HDAC1 and PRC components, and re-expression of DAB2IP.

PcG proteins are also important in maintaining the identity of pluripotent cells through the repression of a set of developmental regulators [112]. Several recent reports have suggested that the ES cell PcG targets may be marked for silencing in cancer cells by promoter hypermethylation, thus imparting a stem cell-like phenotype to cancer cells. One study examined the relationship between PRC2 components and K27me3 in ES cells with the DNA methylation status of ~200 genes in normal and cancer cells [113]. This analysis revealed that genes that are hypermethylated specifically in cancer are much more likely to be silenced in ES cells by PcG proteins than are genes that are constitutively methylated or not methylated at all. Another study compared the DNA methylation status of genes in adult cancer, normal ES cells, and embryonal carcinoma (EC) cells [114]. The authors found that the genes frequently hypermethylated in adult cancer are typically not methylated in ES and EC cells but that these genes are commonly marked in ES cells by a bivalent domain. Bivalent chromatin domains are characterized by the presence of histone modifications that are associated with both active (H3K4me3) and repressed (H3K27me3) gene transcription, thereby allowing for plasticity of expression [115]. Methylation of H3K27 in bivalent domains is established by PcG proteins. Interestingly, treatment of adult cancer cells with DNA-demethylating reagents results in a chromatin signature that closely resembles EC cells, leading to the suggestion that EC cells may represent a transition state in chromatin architecture from ES cells to adult cancer cells. These results were confirmed by ChIP experiments that showed the presence of K27me3 in ES cells at genes hypermethylated in cancer but not at genes that are unmethylated in cancer [116]. Together, these reports suggest that silencing of genes that are also silenced in ES cells may impart a stem cell-like phenotype to cancer cells. This is particularly interesting in light of studies indicating that blocking differentiation leads to expansion of progenitor cell pools and predisposition to tumor development [65,71].

As methylation status is also dependent on the activity of histone demethylases, it follows that several HDMs have been observed to be deregulated in cancer. Deregulation of HDMs can both activate and repress transcriptional programs as shown for LSD1 [34,35]. PLU-1 is a JmjC domain-containing HDM that specifically demethylates H3K4me3. It was originally identified by its over-expression in breast cancer. Knockdown of PLU-1 in breast cancer cells resulted in growth inhibition due to an inability to progress through the G1 phase of the cell cycle [117]. Consistent with its demethylation activity towards H3K4me3, PLU-1 acts as a transcriptional repressor. PLU-1 knockdown cells showed upregulation of several genes associated with growth inhibition. ChIP experiments showed a concomitant occupancy of PLU-1 at several of these genes with low levels of K4me3 in normal cells, while the knockdown cells showed increases in levels of K4me3. Another HDM, GASC1 (gene amplified in squamous cell carcinoma), was originally identified as a gene amplified in esophageal squamous carcinoma (ESC) [118]. Consistent with its role in cancer, knockdown of GASC1 in ESC cell lines resulted in decreased proliferation [119]. GASC1 ectopic expression led to decreased levels of K9 di- and trimethylation, delocalization of HP1, and reduction in levels of heterochromatin, likely contributing to genomic instability. HDMs also play a role in transcriptional regulation by hormone receptors such as the androgen receptor [34,120]. This may be important for androgen dependent prostate cancer as LSD1 has been postulated as a biomarker for this disease [121]. In addition, knockdown of either of the AR interacting HDMs, LSD1, or JMJD2C, has antiproliferative effects in prostate cancer cell lines [34,122].

21.12 CELLULAR LEVELS OF HISTONE MODIFICATIONS AND CANCER PROGNOSIS

The overwhelming majority of studies investigating cancer epigenetics have looked at the level of individual genes that may play a role in tumor biology. One study examining hematological malignancies and colorectal adenocarcinoma found decreased levels of H4K16ac and K20me3 in large regions of the genome associated with DNA repetitive elements. Interestingly, loss of both marks occurred early in the development of a mouse model of multistage skin carcinogenesis [123]. In addition to being targeted to specific genomic loci through sequence-specific transcription factors, histone-modifying enzymes can also affect histones throughout the genome in an untargeted, global manner [60]. Both targeted and global histone modifications affect gene activity [124]. While studies of localized histone modifications continue to identify novel targets for therapy, none of them have been related to clinical outcome of cancer patients.

Predictions of clinical outcome are generally based on grade, the degree of tumor differentiation, and stage, a measure of tumor size, and spread beyond the primary site. However, heterogeneity exists within tumors of equivalent grade and stage, with subtypes of patients experiencing different clinical outcomes. The use of biomarkers including gene expression array analysis has been particularly useful in distinguishing between patient subtypes in multiple cancers such as lymphomas and breast cancer [125–128]. As epigenetics is increasingly realized to have an important role in cancer, it becomes more apparent that epigenetic patterns may constitute important biomarkers to maximize the efficacy of cancer therapy.

Studies of histone modifications have been, by and large, focused on gene-to-gene differences in occurrences of various modifications. However, when examined at the level of single nuclei (i.e., cellular or global levels), individual cells exhibit dissimilar levels of specific histone modifications, generating cellular diversity in epigenetic patterns in cell populations (Figure 21.2A through D). Unlike gene-specific histone modifications, the biological significance of cell-to-cell variability in histone modifications, and the underlying mechanisms are very poorly understood. Remarkably, recent investigations by our lab have revealed that cellular levels of various histone modifications are predictive of clinical outcome in low-grade prostate cancer (Figure 21.2E) [129]. Cellular levels of histone modifications in primary cancer tissues were assessed by immunohistochemical (IHC) staining with antibodies specific for a particular modification. Interestingly, IHC reveals a great deal of heterogeneity in the percentage of cells that are positively stained and in the intensity of staining, not only between patients but within a single patient as well. As epigenetics are increasingly being recognized to play an integral role in cellular identity it becomes tempting to speculate that this epigenetic heterogeneity may partly underlie the cellular heterogeneity that is seen within individual tumors [64,105].

Heterogeneity was seen for all modifications examined, including H3K4me2, K9ac, K18ac, H4R3me2, and K12ac. While no single modification was predictive on its own, analysis of the entire dataset defined two groups of patients with significantly different risks of tumor recurrence. Patients with lower levels of histone modifications (i.e., decreased percent cell staining) had poorer prognosis. This predictive power was independent of all clinicopathological variables tested. This is somewhat surprising given that all the examined modifications are associated with activated gene expression. Tumors rely on the expression of genes that allow for the cancer phenotype [130]. Silencing of these genes due to the loss of activating histone modifications would seemingly result in loss of the cancer phenotype and reduction in tumor growth. However, the low cellular levels of histone modifications do not likely represent the complete absence of these marks at the molecular level from the genome. A possible scenario is the redistribution of these modifications to select locations including specific promoters within the genome, with an overall decrease in percent of histones that are modified. This redistribution may lead to a gene expression program that confers a more aggressive phenotype to the tumor cells.

Alterations in cellular levels of histone modifications could be due to the altered expression and activity of histone-modifying enzymes (Figure 21.3). It is conceivable that alterations in the activity of one or more histone-modifying enzymes may serve to rewire the entire gene expression program, leading to altered cellular identity with oncogenic potential. For example, loss of p300 activity due to mutation has been observed in human cancers, and loss of p300 in cultured cancer cell lines leads to a more aggressive phenotype [85,131]. However, aberrations in cellular levels of histone modifications do not necessarily require changes in the expression or sequence of a histone-modifying enzyme. Rather, changes in activity can be brought about by posttranslational modifications of the enzymes themselves. For example, EZH2 is phosphorylated by Akt, resulting in a reduced affinity of EZH2 for histone H3 with subsequent loss of methylation at H3K27 [132]. Alternatively, the activity of histone-modifying enzymes can be altered by changes in the subunit composition of the multiprotein complexes in which the enzymes reside. This phenomenon has also been described for EZH2. The presence of a cancer-specific PRC complex, PRC4, has been shown to contain the EED2 isoform of EED which is only expressed in pluripotent cells and in cancer cells [133]. PRC4 shows altered

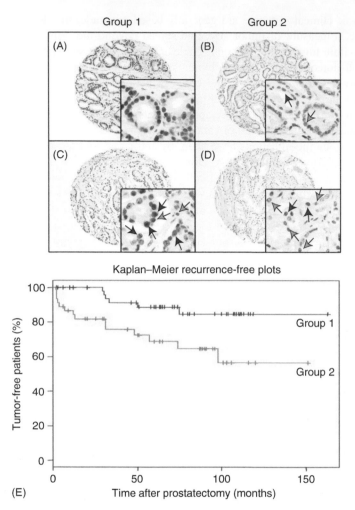

FIGURE 21.2 **(See color insert following page 272.)** Cancer tissues exhibit cellular epigenetic heterogeneity. IHC analysis of histone modifications in malignant prostate glandular epithelial cells from tumors of similar grade and stage reveals heterogeneity in cellular levels of specific modifications. (A, B) H3K18ac (C, D) H4R3me2. Heterogeneity exists within individuals as populations of cells show varying intensities of staining, indicating varying levels of specific modifications within single tumors. (B) Individual cells from a patient's cancer tissue show both relatively high (brown arrow) and low (blue arrow) levels of K18ac. Heterogeneity also exists between patients which show differences in the percentage of cells with high levels of modification. For example, compare (C) and (D) which show IHC results for H4R3me2 staining from two different patients. Cells with high levels of modification stain positively and are colored brown (brown arrow) while those with low level of modification stain negatively and are colored blue (blue arrow). The percentage of cells that stain positively (brown arrow) for this particular modification is much higher in (C) than in (D). Significantly, patterns of histone modifications can be used to group patients of similar grade and stage into two distinct groups that show significantly different outcomes. (E) Kaplan–Meier recurrence-free plots of the two groups (black, group 1; red, group 2) identified among the patients with low-grade tumors ($n = 104$; UCLA) based on the histone modification patterns. Group 1 patients (A, C) generally have higher percentage of cells that stain positively for various modifications than Group 2 patients (B, D). These groups were created using only the histone modification patterns and are independent of all other clinicopathological variables.

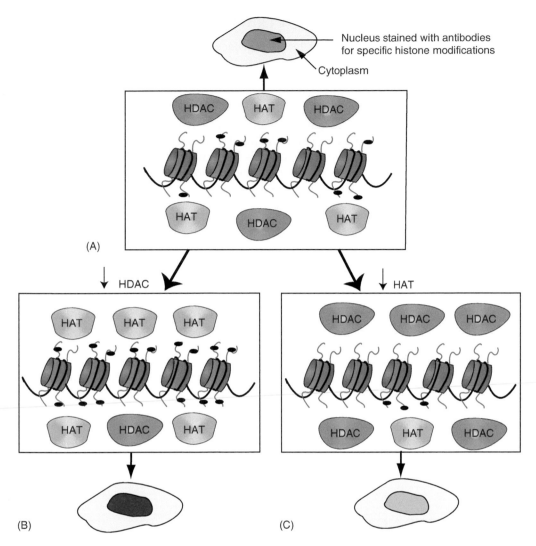

FIGURE 21.3 **(See color insert following page 272.)** A hypothetical relationship between cellular levels of histone modifications and molecular activities of histone modifying enzymes. (A) Levels of histone modifications are determined by the actions of opposing groups of enzymes, such as HATs and HDACs for acetylation. IHC staining of whole-cell nuclei can be used to visualize cellular levels of a particular modification as represented by the cells in this picture. While aberrant targeting of histone-modifying enzymes can lead to local changes in levels of particular modifications (Figure 21.1), changes in the activity and expression of these histone-modifying enzymes can account for the epigenetic heterogeneity that is seen in cellular levels of histone modifications in cancer. (B) For example, decreased activity/expression of HDACs can lead to hyperacetylation of histones throughout the genome due to unopposed HAT activity. This results in cells with a high intensity of staining when analyzed by IHC. (C) In a similar fashion, decreased expression of HATs can lead to unbalanced HDAC activity, decreases in cellular levels of acetylation, and cells that stain weakly by IHC.

substrate specificity, methylating linker histone H1 in addition to H3K27. So, while mistargeting of histone-modifying enzymes may lead to changes in patterns of histone modifications at individual loci, the alterations in the expression or activity of the enzymes that mediate these PTMs provide a mechanism to regulate levels of these modifications throughout the genome. Studies addressing the regulation of histone-modifying enzymes should begin to unravel the molecular mechanisms whereby altered cellular patterns of histone modifications are created.

21.13   ATP-DEPENDENT CHROMATIN REMODELING IN CANCER

In addition to enzymes that covalently modify histones, deregulation of chromatin-remodeling enzymes also contributes to cancer. Aberrant expression of the SWI/SNF catalytic subunits BRG1 and BRM were found in invasive prostate cancer, and overexpression led to enhanced invasion in an in vitro assay [134]. The Snf5 subunit of SWI/SNF has also been implicated in cancer. While Snf5 null mice are embryonic lethal, heterozygotes develop tumors 15%–30% of the time with an average age of onset of 12 months, with tumor cells exhibiting loss of heterozygosity. Nearly 100% of adult mice that are conditionally null for Snf5 develop mature T-cell lymphomas or rhabdoid tumors with an onset of only 11 weeks. Inactivating mutations further solidify the role of Snf5 as a tumor suppressor [50]. The tumor suppressor role of Snf5 is due to its control of E2F target genes. Snf5 inactivation leads to the upregulation of E2F targets, thereby stimulating cell-cycle progression. When crossed with p53 null mice, Snf5 mutations accelerated tumor formation [135,136]. Several SWI/SNF subunits, including Snf5, interact with various proteins involved in growth regulation including β-catenin, p53, MYC, BRCA1, and RB [137]. In general, aberrant expression of chromatin remodelers is not sufficient to induce transformation but may work in concert with other genetic and epigenetic aberrations in the process of tumor initiation [66].

21.14   HISTONE VARIANTS IN CANCER

Histone variants have also been observed to have roles in cancer. High levels of phosphorylated H2AX (γH2AX) designates sites of double-strand break repair, and large γH2AX foci are seen in untreated tumor cells. High numbers of γH2AX foci could be a result of high levels of genomic instability as cell lines with greater numbers of chromosomal rearrangements tended to show higher levels of γH2AX [138]. Accordingly, H2AX deficient mice exhibit increased genomic instability and an accelerated rate of tumor formation in a p53 null background [139]. In addition, CENP-A, which substitutes for H3 at centromeres, is overexpressed and mislocalized in colorectal cancers [140].

21.15   CONCLUDING REMARKS

Accumulating evidence indicates that epigenetic events contribute to both tumor initiation and progression. Epigenetic studies of individual gene promoters have provided much information about tumor biology but have not typically been correlated to clinical outcome. Analysis of cellular levels of histone modifications has revealed a novel cellular heterogeneity that exists within tumors that can be used for cancer prognosis. Much work remains to determine the molecular mechanisms responsible for generating this cellular epigenetic heterogeneity. It will also be interesting to see whether modification patterns change during different stages of tumor progression and whether cells with high or low levels of histone modifications derive from a clonal origin. How do levels of modifications confer a poor prognosis? Can the observed epigenetic heterogeneity explain the diversity of cells seen within a single tumor? Answering these questions remains an exciting challenge that will likely bear significance on the way we think of cancer.

REFERENCES

1. Luger, K., et al., Crystal structure of the nucleosome core particle at 2.8 Å resolution. *Nature*, 1997. 389 (6648): 251–260.
2. Cheung, P., C.D. Allis, and P. Sassone-Corsi, Signaling to chromatin through histone modifications. *Cell*, 2000. 103(2): 263–271.

3. Xu, F., K. Zhang, and M. Grunstein, Acetylation in histone H3 globular domain regulates gene expression in yeast. *Cell*, 2005. 121(3): 375–385.

4. Ng, H.H., et al., Lysine-79 of histone H3 is hypomethylated at silenced loci in yeast and mammalian cells: A potential mechanism for position-effect variegation. *Proc Natl Acad Sci U S A*, 2003. 100(4): 1820–1825.

5. van Leeuwen, F., P.R. Gafken, and D.E. Gottschling, Dot1p modulates silencing in yeast by methylation of the nucleosome core. *Cell*, 2002. 109(6): 745–756.

6. Beck, H.C., et al., Quantitative proteomic analysis of post-translational modifications of human histones. *Mol Cell Proteomics*, 2006. 5(7): 1314–1325.

7. Ye, J., et al., Histone H4 lysine 91 acetylation a core domain modification associated with chromatin assembly. *Mol Cell*, 2005. 18(1): 123–130.

8. Kouzarides, T., Chromatin modifications and their function. *Cell*, 2007. 128(4): 693–705.

9. Jenuwein, T. and C.D. Allis, Translating the histone code. *Science*, 2001. 293(5532): 1074–1080.

10. Shogren-Knaak, M., et al., Histone H4-K16 acetylation controls chromatin structure and protein interactions. *Science*, 2006. 311(5762): 844–847.

11. Sterner, D.E. and S.L. Berger, Acetylation of histones and transcription-related factors. *Microbiol Mol Biol Rev*, 2000. 64(2): 435–459.

12. Brownell, J.E., et al., Tetrahymena histone acetyltransferase A: A homolog to yeast Gcn5p linking histone acetylation to gene activation. *Cell*, 1996. 84(6): 843–851.

13. Black, J.C., et al., A mechanism for coordinating chromatin modification and preinitiation complex assembly. *Mol Cell*, 2006. 23(6): 809–818.

14. Turner, B.M., Histone acetylation and an epigenetic code. *Bioessays*, 2000. 22(9): 836–845.

15. Ekwall, K., Genome-wide analysis of HDAC function. *Trends Genet*, 2005. 21(11): 608–615.

16. de Ruijter, A.J., et al., Histone deacetylases (HDACs): Characterization of the classical HDAC family. *Biochem J*, 2003. 370(Pt 3): 737–749.

17. Stallcup, M.R., Role of protein methylation in chromatin remodeling and transcriptional regulation. *Oncogene*, 2001. 20(24): 3014–3020.

18. Ancelin, K., et al., Blimp1 associates with Prmt5 and directs histone arginine methylation in mouse germ cells. *Nat Cell Biol*, 2006. 8(6): 623–630.

19. Chu, F., et al., Mapping post-translational modifications of the histone variant MacroH2A1 using tandem mass spectrometry. *Mol Cell Proteomics*, 2006. 5(1): 194–203.

20. Reuter, G. and P. Spierer, Position effect variegation and chromatin proteins. *Bioessays*, 1992. 14(9): 605–612.

21. Rea, S., et al., Regulation of chromatin structure by site-specific histone H3 methyltransferases. *Nature*, 2000. 406(6796): 593–599.

22. Sims, R.J., 3rd, K. Nishioka, and D. Reinberg, Histone lysine methylation: A signature for chromatin function. *Trends Genet*, 2003. 19(11): 629–639.

23. Tachibana, M., et al., Set domain-containing protein, G9a, is a novel lysine-preferring mammalian histone methyltransferase with hyperactivity and specific selectivity to lysines 9 and 27 of histone H3. *J Biol Chem*, 2001. 276(27): 25309–25317.

24. Chen, D., et al., Regulation of transcription by a protein methyltransferase. *Science*, 1999. 284(5423): 2174–2177.

25. Strahl, B.D., et al., Methylation of histone H3 at lysine 4 is highly conserved and correlates with transcriptionally active nuclei in Tetrahymena. *Proc Natl Acad Sci U S A*, 1999. 96(26): 14967–14972.

26. Shi, X., et al., ING2 PHD domain links histone H3 lysine 4 methylation to active gene repression. *Nature*, 2006. 442(7098): 96–99.

27. Schultz, D.C., et al., SETDB1: A novel KAP-1-associated histone H3, lysine 9-specific methyltransferase that contributes to HP1-mediated silencing of euchromatic genes by KRAB zinc-finger proteins. *Genes Dev*, 2002. 16(8): 919–932.

28. Vire, E., et al., The Polycomb group protein EZH2 directly controls DNA methylation. *Nature*, 2006. 439 (7078): 871–874.

29. Smallwood, A., et al., Functional cooperation between HP1 and DNMT1 mediates gene silencing. *Genes Dev*, 2007. 21(10): 1169–1178.

30. Shi, Y., et al., Histone demethylation mediated by the nuclear amine oxidase homolog LSD1. *Cell*, 2004. 119(7): 941–953.

31. Bannister, A.J., R. Schneider, and T. Kouzarides, Histone methylation: Dynamic or static? *Cell*, 2002. 109(7): 801–806.

32. Shi, Y. and J.R. Whetstine, Dynamic regulation of histone lysine methylation by demethylases. *Mol Cell*, 2007. 25(1): 1–14.

33. Tsukada, Y., et al., Histone demethylation by a family of JmjC domain-containing proteins. *Nature*, 2006. 439(7078): 811–816.

34. Wissmann, M., et al., Cooperative demethylation by JMJD2C and LSD1 promotes androgen receptor-dependent gene expression. *Nat Cell Biol*, 2007. 9(3): 347–353.

35. Shi, Y.J., et al., Regulation of LSD1 histone demethylase activity by its associated factors. *Mol Cell*, 2005. 19(6): 857–864.

36. Wang, J., et al., Opposing LSD1 complexes function in developmental gene activation and repression programmes. *Nature*, 2007. 446(7138): 882–887.

37. Christensen, J., et al., RBP2 belongs to a family of demethylases, specific for tri- and dimethylated lysine 4 on histone 3. *Cell*, 2007. 128(6): 1063–1076.

38. Seward, D.J., et al., Demethylation of trimethylated histone H3 Lys4 in vivo by JARID1 JmjC proteins. *Nat Struct Mol Biol*, 2007. 14(3): 240–242.

39. Dhalluin, C., et al., Structure and ligand of a histone acetyltransferase bromodomain. *Nature*, 1999. 399(6735): 491–496.

40. Owen, D.J., et al., The structural basis for the recognition of acetylated histone H4 by the bromodomain of histone acetyltransferase gcn5p. *EMBO J*, 2000. 19(22): 6141–6149.

41. Bannister, A.J., et al., Selective recognition of methylated lysine 9 on histone H3 by the HP1 chromo domain. *Nature*, 2001. 410(6824): 120–124.

42. Lachner, M., et al., Methylation of histone H3 lysine 9 creates a binding site for HP1 proteins. *Nature*, 2001. 410(6824): 116–120.

43. Jacobs, S.A. and S. Khorasanizadeh, Structure of HP1 chromodomain bound to a lysine 9-methylated histone H3 tail. *Science*, 2002. 295(5562): 2080–2083.

44. Jacobson, R.H., et al., Structure and function of a human TAFII250 double bromodomain module. *Science*, 2000. 288(5470): 1422–1425.

45. Wysocka, J., et al., A PHD finger of NURF couples histone H3 lysine 4 trimethylation with chromatin remodelling. *Nature*, 2006. 442(7098): 86–90.

46. Carmen, A.A., L. Milne, and M. Grunstein, Acetylation of the yeast histone H4 N terminus regulates its binding to heterochromatin protein SIR3. *J Biol Chem*, 2002. 277(7): 4778–4781.

47. Suka, N., K. Luo, and M. Grunstein, Sir2p and Sas2p opposingly regulate acetylation of yeast histone H4 lysine16 and spreading of heterochromatin. *Nat Genet*, 2002. 32(3): 378–383.

48. Zegerman, P., et al., Histone H3 lysine 4 methylation disrupts binding of nucleosome remodeling and deacetylase (NuRD) repressor complex. *J Biol Chem*, 2002. 277(14): 11621–11624.

49. Mateescu, B., et al., Tethering of HP1 proteins to chromatin is relieved by phosphoacetylation of histone H3. *EMBO Rep*, 2004. 5(5): 490–496.

50. Roberts, C.W. and S.H. Orkin, The SWI/SNF complex—chromatin and cancer. *Nat Rev Cancer*, 2004. 4(2): 133–142.

51. Ercan, S., M.J. Carrozza, and J.L. Workman, Global nucleosome distribution and the regulation of transcription in yeast. *Genome Biol*, 2004. 5(10): 243.

52. Lee, C.K., et al., Evidence for nucleosome depletion at active regulatory regions genome-wide. *Nat Genet*, 2004. 36(8): 900–905.

53. Cairns, B.R., Chromatin remodeling complexes: Strength in diversity, precision through specialization. *Curr Opin Genet Dev*, 2005. 15(2): 185–190.

54. Wang, G.G., C.D. Allis, and P. Chi, Chromatin remodeling and cancer, part II: ATP-dependent chromatin remodeling. *Trends Mol Med*, 2007. 13(9): 373–380.

55. Johnson, C.N., N.L. Adkins, and P. Georgel, Chromatin remodeling complexes: ATP-dependent machines in action. *Biochem Cell Biol*, 2005. 83(4): 405–417.

56. Morrison, A.J., et al., INO80 and gamma-H2AX interaction links ATP-dependent chromatin remodeling to DNA damage repair. *Cell*, 2004. 119(6): 767–775.

57. Taverna, S.D., et al., Yng1 PHD finger binding to H3 trimethylated at K4 promotes NuA3 HAT activity at K14 of H3 and transcription at a subset of targeted ORFs. *Mol Cell*, 2006. 24(5): 785–796.

58. Agalioti, T., G. Chen, and D. Thanos, Deciphering the transcriptional histone acetylation code for a human gene. *Cell*, 2002. 111(3): 381–392.
59. Cosma, M.P., T. Tanaka, and K. Nasmyth, Ordered recruitment of transcription and chromatin remodeling factors to a cell cycle- and developmentally regulated promoter. *Cell*, 1999. 97(3): 299–311.
60. Kurdistani, S.K. and M. Grunstein, Histone acetylation and deacetylation in yeast. *Nat Rev Mol Cell Biol*, 2003. 4(4): 276–284.
61. Kurdistani, S.K., S. Tavazoie, and M. Grunstein, Mapping global histone acetylation patterns to gene expression. *Cell*, 2004. 117(6): 721–733.
62. Wolffe, A.P. and D. Pruss, Deviant nucleosomes: The functional specialization of chromatin. *Trends Genet*, 1996. 12(2): 58–62.
63. Bernstein, E. and S.B. Hake, The nucleosome: A little variation goes a long way. *Biochem Cell Biol*, 2006. 84(4): 505–517.
64. Cavalli, G., Chromatin and epigenetics in development: Blending cellular memory with cell fate plasticity. *Development*, 2006. 133(11): 2089–2094.
65. Baylin, S.B. and J.E. Ohm, Epigenetic gene silencing in cancer—a mechanism for early oncogenic pathway addiction? *Nat Rev Cancer*, 2006. 6(2): 107–116.
66. Ducasse, M. and M.A. Brown, Epigenetic aberrations and cancer. *Mol Cancer*, 2006. 5: 60.
67. Feinberg, A.P. and B. Tycko, The history of cancer epigenetics. *Nat Rev Cancer*, 2004. 4(2): 143–153.
68. Jones, P.A. and S.B. Baylin, The epigenomics of cancer. *Cell*, 2007. 128(4): 683–692.
69. Hake, S.B., A. Xiao, and C.D. Allis, Linking the epigenetic language of covalent histone modifications to cancer. *Br J Cancer*, 2004. 90(4): 761–769.
70. Feinberg, A.P., The epigenetics of cancer etiology. *Semin Cancer Biol*, 2004. 14(6): 427–432.
71. Feinberg, A.P., R. Ohlsson, and S. Henikoff, The epigenetic progenitor origin of human cancer. *Nat Rev Genet*, 2006. 7(1): 21–33.
72. Ting, A.H., K.M. McGarvey, and S.B. Baylin, The cancer epigenome—components and functional correlates. *Genes Dev*, 2006. 20(23): 3215–3231.
73. Sakatani, T., et al., Loss of imprinting of Igf2 alters intestinal maturation and tumorigenesis in mice. *Science*, 2005. 307(5717): 1976–1978.
74. Nguyen, C.T., et al., Histone H3-lysine 9 methylation is associated with aberrant gene silencing in cancer cells and is rapidly reversed by 5-aza-2′-deoxycytidine. *Cancer Res*, 2002. 62(22): 6456–6461.
75. Belinsky, S.A., et al., Aberrant methylation of p16 (INK4a) is an early event in lung cancer and a potential biomarker for early diagnosis. *Proc Natl Acad Sci U S A*, 1998. 95(20): 11891–11896.
76. Ayton, P.M. and M.L. Cleary, Molecular mechanisms of leukemogenesis mediated by MLL fusion proteins. *Oncogene*, 2001. 20(40): 5695–5707.
77. Ayton, P.M. and M.L. Cleary, Transformation of myeloid progenitors by MLL oncoproteins is dependent on Hoxa7 and Hoxa9. *Genes Dev*, 2003. 17(18): 2298–2307.
78. Esteller, M., Epigenetics provides a new generation of oncogenes and tumour-suppressor genes. *Br J Cancer*, 2006. 94(2): 179–183.
79. Gibbons, R.J., Histone modifying and chromatin remodelling enzymes in cancer and dysplastic syndromes. *Hum Mol Genet*, 2005. 14 : : Spec No 1: R85–R92.
80. Klochendler-Yeivin, A. and M. Yaniv, Chromatin modifiers and tumor suppression. *Biochim Biophys Acta*, 2001. 1551(1): M1–M10.
81. Lund, A.H. and M. van Lohuizen, Epigenetics and cancer. *Genes Dev*, 2004. 18(19): 2315–2335.
82. Aniello, F., et al., Expression of four histone lysine-methyltransferases in parotid gland tumors. *Anticancer Res*, 2006. 26(3A): 2063–2967.
83. Ozdag, H., et al., Differential expression of selected histone modifier genes in human solid cancers. *BMC Genomics*, 2006. 7: 90.
84. Kung, A.L., et al., Gene dose-dependent control of hematopoiesis and hematologic tumor suppression by CBP. *Genes Dev*, 2000. 14(3): 272–277.
85. Gayther, S.A., et al., Mutations truncating the EP300 acetylase in human cancers. *Nat Genet*, 2000. 24(3): 300–303.
86. Yang, X.J., et al., A p300/CBP-associated factor that competes with the adenoviral oncoprotein E1A. *Nature*, 1996. 382(6589): 319–324.
87. Espino, P.S., et al., Histone modifications as a platform for cancer therapy. *J Cell Biochem*, 2005. 94(6): 1088–1102.

88. Savickiene, J., et al., The novel histone deacetylase inhibitor BML-210 exerts growth inhibitory, proapoptotic and differentiation stimulating effects on the human leukemia cell lines. *Eur J Pharmacol*, 2006. 549(1–3): 9–18.

89. Xia, Q., et al., Chronic administration of valproic acid inhibits prostate cancer cell growth in vitro and in vivo. *Cancer Res*, 2006. 66(14): 7237–7244.

90. Huang, L., et al., Activation of the p21WAF1/CIP1 promoter independent of p53 by the histone deacetylase inhibitor suberoylanilide hydroxamic acid (SAHA) through the Sp1 sites. *Oncogene*, 2000. 19(50): 5712–5719.

91. Richon, V.M., et al., Histone deacetylase inhibitor selectively induces p21WAF1 expression and gene-associated histone acetylation. *Proc Natl Acad Sci U S A*, 2000. 97(18): 10014–10019.

92. Van Lint, C., S. Emiliani, and E. Verdin, The expression of a small fraction of cellular genes is changed in response to histone hyperacetylation. *Gene Expr*, 1996. 5(4–5): 245–253.

93. Zhu, P., et al., Induction of HDAC2 expression upon loss of APC in colorectal tumorigenesis. *Cancer Cell*, 2004. 5(5): 455–463.

94. Ropero, S., et al., A truncating mutation of HDAC2 in human cancers confers resistance to histone deacetylase inhibition. *Nat Genet*, 2006. 38(5): 566–569.

95. Nimer, S.D. and M.A. Moore, Effects of the leukemia-associated AML1-ETO protein on hematopoietic stem and progenitor cells. *Oncogene*, 2004. 23(24): 4249–4254.

96. Peterson, L.F. and D.E. Zhang, The 8;21 translocation in leukemogenesis. *Oncogene*, 2004. 23(24): 4255–4262.

97. Steele-Perkins, G., et al., Tumor formation and inactivation of RIZ1, an Rb-binding member of a nuclear protein-methyltransferase superfamily. *Genes Dev*, 2001. 15(17): 2250–2262.

98. Peters, A.H., et al., Loss of the Suv39h histone methyltransferases impairs mammalian heterochromatin and genome stability. *Cell*, 2001. 107(3): 323–337.

99. Hamamoto, R., et al., SMYD3 encodes a histone methyltransferase involved in the proliferation of cancer cells. *Nat Cell Biol*, 2004. 6(8): 731–740.

100. Tsuge, M., et al., A variable number of tandem repeats polymorphism in an E2F-1 binding element in the 5′ flanking region of SMYD3 is a risk factor for human cancers. *Nat Genet*, 2005. 37(10): 1104–1107.

101. Hamamoto, R., et al., Enhanced SMYD3 expression is essential for the growth of breast cancer cells. *Cancer Sci*, 2006. 97(2): 113–118.

102. Liu, C., et al., The telomerase reverse transcriptase (hTERT) gene is a direct target of the histone methyltransferase SMYD3. *Cancer Res*, 2007. 67(6): 2626–2631.

103. Hong, H., et al., Aberrant expression of CARM1, a transcriptional coactivator of androgen receptor, in the development of prostate carcinoma and androgen-independent status. *Cancer*, 2004. 101(1): 83–89.

104. Majumder, S., et al., Involvement of arginine methyltransferase CARM1 in androgen receptor function and prostate cancer cell viability. *Prostate*, 2006. 66(12): 1292–1301.

105. Sparmann, A. and M. van Lohuizen, Polycomb silencers control cell fate, development and cancer. *Nat Rev Cancer*, 2006. 6(11): 846–856.

106. Schuettengruber, B., et al., Genome regulation by polycomb and trithorax proteins. *Cell*, 2007. 128(4): 735–745.

107. Cao, R., et al., Role of histone H3 lysine 27 methylation in Polycomb-group silencing. *Science*, 2002. 298 (5595): 1039–1043.

108. Jacobs, J.J., et al., The oncogene and Polycomb-group gene bmi-1 regulates cell proliferation and senescence through the ink4a locus. *Nature*, 1999. 397(6715): 164–168.

109. Kleer, C.G., et al., EZH2 is a marker of aggressive breast cancer and promotes neoplastic transformation of breast epithelial cells. *Proc Natl Acad Sci U S A*, 2003. 100(20): 11606–11611.

110. Varambally, S., et al., The Polycomb group protein EZH2 is involved in progression of prostate cancer. *Nature*, 2002. 419(6907): 624–629.

111. Chen, H., S.W. Tu, and J.T. Hsieh, Down-regulation of human DAB2IP gene expression mediated by polycomb Ezh2 complex and histone deacetylase in prostate cancer. *J Biol Chem*, 2005. 280(23): 22437–22444.

112. Lee, T.I., et al., Control of developmental regulators by Polycomb in human embryonic stem cells. *Cell*, 2006. 125(2): 301–313.

113. Widschwendter, M., et al., Epigenetic stem cell signature in cancer. *Nat Genet*, 2007. 39(2): 157–158.

114. Ohm, J.E., et al., A stem cell-like chromatin pattern may predispose tumor suppressor genes to DNA hypermethylation and heritable silencing. *Nat Genet*, 2007. 39(2): 237–242.

115. Bernstein, B.E., et al., A bivalent chromatin structure marks key developmental genes in embryonic stem cells. *Cell*, 2006. 125(2): 315–326.

116. Schlesinger, Y., et al., Polycomb-mediated methylation on Lys27 of histone H3 pre-marks genes for de novo methylation in cancer. *Nat Genet*, 2007. 39(2): 232–236.

117. Yamane, K., et al., PLU-1 Is an H3K4 demethylase involved in transcriptional repression and breast cancer cell proliferation. *Mol Cell*, 2007. 25(6): 801–812.

118. Yang, Z.Q., et al., Identification of a novel gene, GASC1, within an amplicon at 9p23–24 frequently detected in esophageal cancer cell lines. *Cancer Res*, 2000. 60(17): 4735–4739.

119. Cloos, P.A., et al., The putative oncogene GASC1 demethylates tri- and dimethylated lysine 9 on histone H3. *Nature*, 2006. 442(7100): 307–311.

120. Yamane, K., et al., JHDM2A, a JmjC-containing H3K9 demethylase, facilitates transcription activation by androgen receptor. *Cell*, 2006. 125(3): 483–495.

121. Kahl, P., et al., Androgen receptor coactivators lysine-specific histone demethylase 1 and four and a half LIM domain protein 2 predict risk of prostate cancer recurrence. *Cancer Res*, 2006. 66(23): 11341–11347.

122. Metzger, E., et al., LSD1 demethylates repressive histone marks to promote androgen-receptor-dependent transcription. *Nature*, 2005. 437(7057): 436–439.

123. Fraga, M.F., et al., Loss of acetylation at Lys16 and trimethylation at Lys20 of histone H4 is a common hallmark of human cancer. *Nat Genet*, 2005. 37(4): 391–400.

124. Vogelauer, M., et al., Global histone acetylation and deacetylation in yeast. *Nature*, 2000. 408(6811): 495–498.

125. Alizadeh, A.A., et al., Distinct types of diffuse large B-cell lymphoma identified by gene expression profiling. *Nature*, 2000. 403(6769): 503–511.

126. Perou, C.M., et al., Molecular portraits of human breast tumours. *Nature*, 2000. 406(6797): 747–752.

127. Sorlie, T., et al., Gene expression patterns of breast carcinomas distinguish tumor subclasses with clinical implications. *Proc Natl Acad Sci U S A*, 2001. 98(19): 10869–10874.

128. Sotiriou, C., et al., Breast cancer classification and prognosis based on gene expression profiles from a population-based study. *Proc Natl Acad Sci U S A*, 2003. 100(18): 10393–10398.

129. Seligson, D.B., et al., Global histone modification patterns predict risk of prostate cancer recurrence. *Nature*, 2005. 435(7046): 1262–1266.

130. Hanahan, D. and R.A. Weinberg, The hallmarks of cancer. *Cell*, 2000. 100(1): 57–70.

131. Krubasik, D., et al., Absence of p300 induces cellular phenotypic changes characteristic of epithelial to mesenchyme transition. *Br J Cancer*, 2006. 94(9): 1326–1332.

132. Cha, T.L., et al., Akt-mediated phosphorylation of EZH2 suppresses methylation of lysine 27 in histone H3. *Science*, 2005. 310(5746): 306–310.

133. Kuzmichev, A., et al., Composition and histone substrates of polycomb repressive group complexes change during cellular differentiation. *Proc Natl Acad Sci U S A*, 2005. 102(6): 1859–1864.

134. Sun, A., et al., Aberrant expression of SWI/SNF catalytic subunits BRG1/BRM is associated with tumor development and increased invasiveness in prostate cancers. *Prostate*, 2007. 67(2): 203–213.

135. Isakoff, M.S., et al., Inactivation of the Snf5 tumor suppressor stimulates cell cycle progression and cooperates with p53 loss in oncogenic transformation. *Proc Natl Acad Sci U S A*, 2005. 102(49): 17745–17750.

136. Klochendler-Yeivin, A., E. Picarsky, and M. Yaniv, Increased DNA damage sensitivity and apoptosis in cells lacking the Snf5/Ini1 subunit of the SWI/SNF chromatin remodeling complex. *Mol Cell Biol*, 2006. 26(7): 2661–2674.

137. Guidi, C.J., et al., Functional interaction of the retinoblastoma and ini1/snf5 tumor suppressors in cell growth and pituitary tumorigenesis. *Cancer Res*, 2006. 66(16): 8076–8082.

138. Yu, T., et al., Endogenous expression of phosphorylated histone H2AX in tumors in relation to DNA double-strand breaks and genomic instability. *DNA Repair (Amst)*, 2006. 5(8): 935–946.

139. Celeste, A., et al., H2AX haploinsufficiency modifies genomic stability and tumor susceptibility. *Cell*, 2003. 114(3): 371–383.

140. Tomonaga, T., et al., Overexpression and mistargeting of centromere protein-A in human primary colorectal cancer. *Cancer Res*, 2003. 63(13): 3511–3516.

22 Histone Deacetylase Inhibitors in Cancer Therapy

Aaron M. Sargeant, Samuel K. Kulp, Yen-Shen Lu, Ann-Lii Cheng, and Ching-Shih Chen

CONTENTS

22.1 INTRODUCTION

The transcriptional capacity of genes is intricately controlled in part by the opposing actions of histone acetyltransferases (HATs) and histone deacetylases (HDACs), which contribute to the structural remodeling of chromatin by adding or removing, respectively, acetyl groups from the amino-terminal lysine residues of histones. Acetylation is one of the many posttranslational histone modifications and promotes transcription by locally expanding chromatin to permit the access of regulatory proteins to DNA, whereas removal of acetyl groups leads to transcriptional repression via chromatin condensation. These modifications of histones by acetylation are among the epigenetic mechanisms controlling gene expression that is suggested to contribute to malignant transformation [1–4]. HDACs have been implicated in the actions of well-known cellular oncogenes and tumor suppressor genes [5–8], and a tumor suppressor role for HATs has been described [9,10]. Moreover, associations have been reported between the expression levels of HDAC enzymes and various features of human malignancies, such as steroid hormone receptor status and clinical outcome in breast cancer [11–13], prognosis in lung cancer [14], and response to therapy in acute myeloblastic leukemia [15]. Collectively, these studies suggest a role for histone acetylation in the development and progression of cancer.

Epigenetic regulators of gene transcription such as HDACs are attractive anticancer targets in light of the influence of chromatin structure on cancer development, and considering the reversible nature of epigenetic aberrations. Accordingly, the synthesis of small-molecule inhibitors of HDACs (HDACi) has been an active focus in the field of anticancer drug discovery in recent years. To this end, vorinostat has been approved by the Food and Drug Administration (FDA) for the treatment of refractory cutaneous T-cell lymphoma [16], and many others follow in various stages of preclinical and clinical development in pursuit of regulatory sanction for use against an array of solid and hematopoietic malignancies [17–19].

A noteworthy feature of HDACi is the selectivity of their antiproliferative effects for malignant versus normal cells [20]. Proposed mechanisms for this selective toxicity include the differential upregulation of thioredoxin, a protective antioxidant protein, in normal cells in response to HDACi treatment [21]. The activation of cell cycle checkpoints in nonmalignant cells that are dysfunctional in tumor cells has also been proposed as a mechanism for the selectivity of HDACi. Targeting of these checkpoints, which include the HDACi-sensitive G2-phase checkpoint and the drug-dependent bypass of the mitotic spindle checkpoint, has been shown to confer the selective cytotoxicity against tumor cells which is a desirable feature of effective anticancer agents [22,23].

The re-expression of genes inducing cell death, cell cycle arrest, or differentiation by HDACi underscores the applicability of these agents to the field of cancer therapy and is the conventional rationale for their development [24]. Importantly however, in addition to histone-mediated remodeling of chromatin, HDACs cause a plethora of histone-independent biological effects by modifying the acetylation state of a diverse set of protein substrates including transcription factors that influence gene expression, and mediators of signal transduction that induce apoptosis independent of gene expression [17,18,25]. HDACi consequently convey a broad and manifold arsenal of anticancer activities via these nonhistone mechanisms not only in cancer cells but also in the tumor vasculature and host immune system [26–28]. These histone-independent anticancer activities of HDACi reveal molecular targets more diverse than perhaps originally predicted and are a focus of this chapter.

22.2 PHARMACOPHORE MODEL OF HDACi

Natural product-derived and synthetic deacetylase inhibitors are classified into structurally distinct categories including short-chain fatty acids, hydroxamic acids, benzamide derivatives, cyclic peptides, and epoxyketones [17,18] (Figure 22.1B). Most of these compounds demonstrate activity against class I, II, and IV HDACs, and all function by binding to various portions of the enzyme catalytic domains. The propensity of HDACi to bind zinc cations within the active catalytic sites of HDACs directly correlates with their inhibitory activity. Many are synthesized on the basis of a three-component working model consisting of a cap group, linker, and zinc-chelating motif [20], as demonstrated for trichostatin A (TSA) in Figure 22.1A. Specifically, the zinc-binding domain is separated by a linker domain from a surface domain which makes contact with the rim of the HDAC [18]. This mode of protein–ligand interaction serves as a framework for the design and synthesis of HDACi, including the production of trithiocarbonate analogs [29] and our own development of a novel class of phenylbutyrate-derived HDACi, the latter of which is summarized below [30–32].

22.2.1 Case in Point: Preclinical Development of Phenylbutyrate-Based HDACi

On the basis of the pharmacophore model of HDAC inhibition, a series of phenylbutyrate analogs were tethered to a zinc-chelating motif through a hydrophobic linker resulting in the synthesis of hydroxamate-tethered phenylbutyrate (HTPB), which exhibited an IC_{50} against in vitro HDAC activity of 44 nM compared to 0.4 mM for the parent phenylbutyrate molecule. These studies suggest that the active-site pocket of HDAC accommodates cap groups of diverse stereoelectronic properties, considering the variability in cap group structures among HTPB, TSA, vorinostat, and MS-275 [30]. Lead optimization strategies with HTPB culminated in the synthesis of

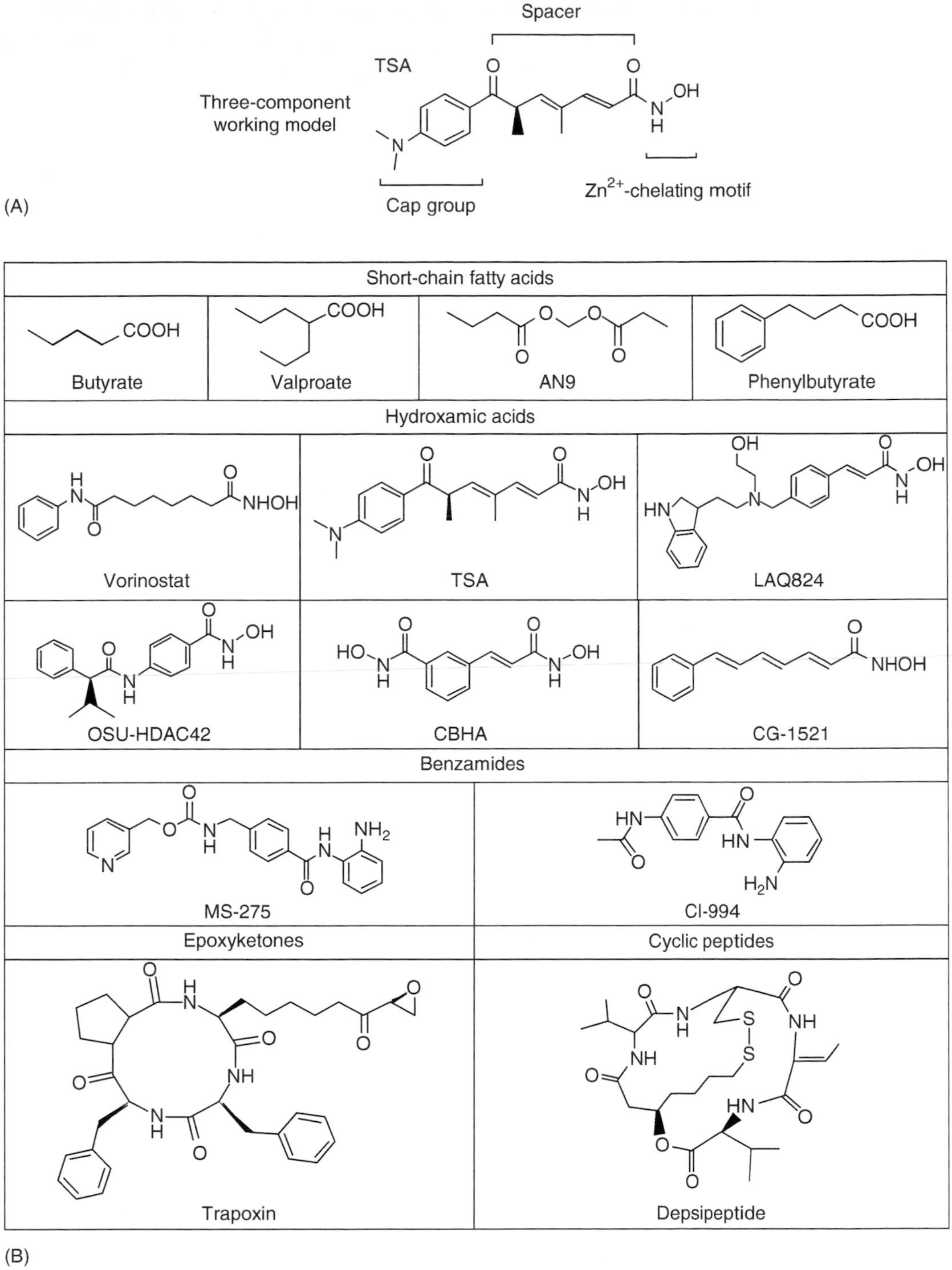

FIGURE 22.1 Pharmacophore model (A) and chemical structures (B) of HDACi.

OSU-HDAC42 (Figure 22.2), previously designated as (*S*)-HDAC-42, which inhibited HDAC with even greater potency (IC$_{50}$ 16 nM versus 44 nM for HTPB) [31].

As described in the following sections on the anticancer mechanisms of these agents, HDACi typically exhibit pleiotropic antitumor activities that encompass histone acetylation-dependent

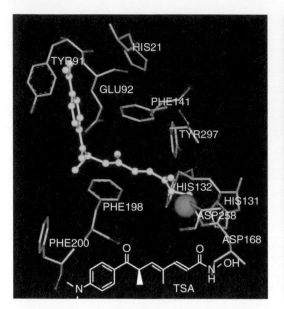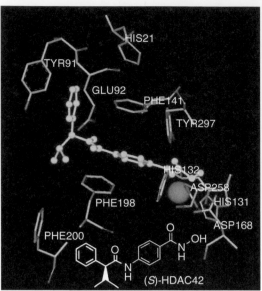

FIGURE 22.2 Modeled docking of TSA (left) and (S)-HDAC42 (right) into the active site of histone deacetylase-like protein (HDLP). The small molecules are in the center of the images.

and -independent mechanisms. We have demonstrated this wide range of activity, for instance, with OSU-HDAC42 in models of human cancers in vitro and in vivo. This broad spectrum of activity includes not only the potent inhibition of HDAC, as indicated by hallmark features of HDAC inhibition such as histone hyperacetylation and upregulated p21 expression, but also of targets regulating multiple aspects of cancer cell survival including Akt signaling, mitochondrial integrity, and caspase activity. Our evaluation of the antitumor efficacy of OSU-HDAC42 in in vitro and in vivo models of prostate cancer [33,34] and hepatocellular carcinoma (HCC) [35] reveals noteworthy features of OSU-HDAC42 that suggest its potential clinical value. First, the assessment of drug effects in nonmalignant cells, specifically prostate epithelial cells and normal hepatocytes, showed them to be approximately 8- to 12-fold less sensitive to the antiproliferative effects of OSU-HDAC42 than panels of prostate cancer and HCC cells [33,35]. Second, in addition to expected effects on HDAC-related biomarkers (p21 expression, histone acetylation), OSU-HDAC42 induced significant reductions in the levels of phospho-Akt, Bcl-xL, and the inhibitor of apoptosis protein (IAP) family members, survivin, cIAP-1, and cIAP-2 [33]. Third, the oral administration of OSU-HDAC significantly suppressed the growth of subcutaneous (prostate cancer and HCC) and orthotopic (HCC) tumor xenografts in association with intratumoral changes in the biomarkers of drug activity described above, i.e., phospho-Akt, Bcl-xL, and IAPs, and in the absence of limiting toxicity [33,35]. This broad spectrum of activity exhibited by OSU-HDAC42 is common to many HDACi in preclinical and clinical development, and suggests their viability as part of therapeutic strategies for various human cancers.

22.3 ANTICANCER MECHANISMS OF HDACi

The conventional paradigm for the use of HDACi in the treatment of cancer is the activation of growth–arrest genes secondary to histone acetylation [17,36]. Accordingly, although insufficient to molecularly characterize an observed clinical effect, histone acetylation assays are routinely used to assess the activity and bioavailability of HDACi. This measurement may be accomplished on tumor tissue itself or more commonly on peripheral white blood cells which are easily accessible [16,17]. Genes and gene products that negatively impact cancer cell survival and growth that have been upregulated subsequent to treatment with HDACi include the tumor suppressor genes p53, gelsolin, and mapsin, and other regulators such as $p21^{WAF1/Cip1}$, Bcl2, Fas/Fas ligand, and caspase 3 among others [17,33,37–43].

Given the roles of HDACs in chromatin remodeling and transcriptional regulation, as well as in the modification of diverse transcription factors, hypothetically HDAC inhibition might be expected to have a profound global genomic effect. While the percentage of all genes transcriptionally affected by HDAC inhibition is not precisely known, genetic profiling studies estimate that only a small fraction of genes is regulated [18,24,44]. In fact, Glaser et al. evaluated the effects of three HDACi on gene expression in human urinary bladder and breast cancer cells and showed that less than 10% of 6800 genes evaluated were up- or downregulated from which a core set of thirteen genes regulated by the three HDACi in three different cell lines was identified. Moreover, the number of genes upregulated secondary to HDACi treatment was reported to be approximately equal to that downregulated [18,24]. The suppression of gene expression by HDACi treatment may result, in part, from the increased transcription of genes whose products have negative effects on gene expression. Additionally, the acetylation-dependent alteration of transcription factor function may also contribute to downregulated gene expression. For example, acetylation of Stat1 by HDACi treatment results in its interaction with nuclear factor-κB (NF-κB) and subsequent suppression of NF-κB nuclear localization and of target gene expression [45].

While some HDAC inhibitors exhibit isoform-specific properties [46,47], the majority of these structurally diverse molecules are comparably nonselective with respect to inhibition of class I and II enzymes. The jury is still out regarding the clinical significance of improving HDAC isoform selectivity [16,25]. In support of developing discriminative HDACi, growing evidence indicates that individual HDACs play different physiological roles [48], and even that some enzymes show strong correlation with disease. Enzyme knockdown via siRNA for example, supports the relationship of class I enzymes with cancer [49] and suggests that inhibitors of these HDACs are more likely to achieve desirable clinical outcomes [25,50]. Moreover, antagonizing certain HDACs within class I such as HDAC1 and HDAC3 may be more desirable in cancer therapy than targeting the entire class [25], and considering that inhibiting HDAC2 activity may be detrimental in patients with chronic obstructive pulmonary disease [51]. In contrast, similar experiments with class II HDACs failed to significantly impact cancer cells [49,50]. However, it should be noted that, while expression data for certain HDACs in primary tumor tissues have been reported [52,53], much of the data implicating different roles of class I and class II enzymes, and therefore their validity as anticancer targets, was generated by using siRNA-mediated knockdown of expression which requires careful interpretation of its clinical relevance. Until the biological functions of individual HDACs are more fully characterized, the potential value of broad-spectrum versus narrow-spectrum HDACi in the context of cancer therapy will remain unclear. While adverse reactions may be theoretically circumvented by streamlining activity to cancer-relevant HDACs [25], broad-spectrum HDACi may provide advantages by targeting multiple signaling pathways associated with cancer cell survival and proliferation, which could be envisioned to induce more efficient cell death or delay/prevent the acquisition of resistance [54].

Although a variety of HDACi have demonstrated tumor-specific anticancer selectivity and success in preclinical and clinical studies, there are no reliable methods to predict the sensitivity of a certain tumor to any class of HDAC inhibitor, and the molecular groundwork for this activity remains largely unknown [17]. Continued mechanistic research is needed considering that the use of HDACi actually pre-dated the characterization of HDACs themselves, and that HDACi have been our most valuable means of defining the roles of these enzymes in normal and neoplastic biology [18]. Recent advances in proteomics illustrate one discipline by which acetylation events may be more thoroughly defined. These methodologies have permitted the characterization of specific acetylation patterns of histones in response to HDACi treatment [55,56], which may help to account for differences in effects induced by these agents. For example, the specific posttranslational modification sites on histone H3 and H4 have been determined by mass spectrometry following separation by acetic acid urea-polyacrylamide gel electrophoresis (AU-PAGE). Peptide mass mapping of histone H4 using this technique revealed an identical order of acetylation in mouse lymphosarcoma cells treated with TSA and depsipeptide [55]. These and other strategies such as genetic profiling of HDACi-treated cells and tumors will further our understanding of this complex group of molecules and perhaps extend their usefulness in preclinical and clinical settings.

22.4   EFFECTS OF HDACi ON NONHISTONE PROTEINS

Although histones are by bulk mass the most abundant HDAC substrate, the list of other identified key substrates (Table 22.1) is ever-growing, as is an appreciation for their potentially immense influence on the anticancer activities of HDACi. The role of these enzymes as protein deacetylases should not be underestimated [57]. Since some HDACs at least partially reside in extranuclear cytoplasmic compartments [58,59] and considering their extensive phylogenetic characterization in prokaryotes as well as eukaryotes, it is believed that HDACs evolved in the absence of histone proteins [18,60]. It is not surprising then that alteration of the histone code alone is insufficient to account for the antitumor success of these agents [20]. The Sections 22.4.1 through 22.4.6 summarize some of the nonhistone anticancer mechanisms of HDACi.

22.4.1   HDACi-Mediated Acetylation of Ku70 Induces Caspase-Dependent Apoptosis

The acetylation-dependent, Bax-binding activity of Ku70 represents another nonhistone target of HDACi that may be exploited for anticancer use. Ku70 is commonly upregulated in neoplastic cells where it prevents cell death by binding to and sequestering the proapoptotic protein Bax [61]. By inhibiting Bax function, Ku70 acquires an antiapoptotic character and even increases the resistance of tumors to chemotherapy. The Bax-binding ability of Ku70 has recently been characterized as acetylation dependent, at least in neuroblastoma cells, and its acetylation status is tightly controlled by multiple HDACs [61,62]. When acetylated Ku70 can no longer bind to Bax, allowing Bax to translocate to the mitochondria where it causes cytochrome c release and caspase 9-mediated apoptosis. Treatment of neuroblastoma cells with TSA caused hyperacetylation of Ku70 and subsequent cell death. This study by Subramanian et al. demonstrates the in vitro success of single agent treatment with HDACi and suggests that Ku70 can serve as a target for proapoptotic regulation by HDACi [62].

TABLE 22.1
Representative Nonhistone Substrates of HDACs

Substrate	Intracellular Function	Deacetylases	References
Ku70	Antiapoptosis	SIRT1, TSA-sensitive HDACs	[61,62,144,145]
FOXO1	Transcription factor	SIRT1	[146]
p300	Transcription factor	SIRT1	[147]
p53	Tumor suppressor	HDAC1, SIRT1	[69–71,148–150]
Androgen receptor	Hormone receptor	HDAC1, SIRT1	[151–154]
Smad7	Signal transducer of TGF-β	HDACs 1, 3, and 6	[155]
Stat3	Signal transducer of cytokines	HDACs 1, 2, and 3	[89,91]
NF-κB (RelA)	Nuclear transcription factor	HDAC3, SIRT1	[96,156–158]
SRY	Y-chromosome-encoded DNA-binding protein	HDAC3	[159]
α-Tubulin	Microtubule component	HDAC6, SIRT2	[58,93,160–162]
Hsp90	Molecular chaperone	HDAC6	[80,87,163]

In addition to sequestration of Bax, Ku70 mediates a second cytoprotective function through its critical role in the repair of DNA double-strand breaks (DSB). We have shown that TSA, OSU-HDAC42, and MS-275, but not vorinostat, potently induced Ku70 acetylation leading to reductions in the binding affinity of Ku70 to broken DNA ends. Moreover, this inhibitory effect on Ku70's DNA DSB repair activity could be exploited to sensitize prostate cancer cells to cell killing by agents that produce DNA DSB, such as bleomycin, doxorubicin, and VP-19 [34].

22.4.2 HDACi Stabilizes p53 and Causes Cell Cycle Arrest and Apoptosis

The dysregulation of the p53 tumor suppressor gene in cancer is well established [63]. Evidence shows that posttranslational modifications including acetylation play an important role in its regulation [64–66]. The HAT p300/CBP commonly acetylates p53 in response to p53-activating agents [67], and degradation of p53 is dependent on HDAC1-associated deacetylation [68]. Interestingly, Tip60-dependent acetylation of p53 at K120 is crucial for its apoptotic, but not growth arrest, activity [66].

Treatment of prostate cancer cells with HDACi demonstrated profound effects of posttranslational modification on p53 function [69–73]. CG-1521, an HDACi with structural similarity to vorinostat, was shown to modify p53 acetylation in a site-specific manner in LNCaP cells. Specifically, treatment with CG-1521 stabilized the acetylation of p53 at Lys373, which induced cell cycle arrest by increasing p21 expression, and caused apoptosis via translocation of Bax. TSA, in comparison, stabilized acetylation of p53 at Lys382 but failed to change steady-state levels of p21 mRNA and had no apoptotic effect [73]. Hyperacetylation of H3 and H4 occurred in the cells much earlier than morphologic evidence of apoptosis or cell cycle arrest, suggesting that these traditional targets may have contributed to cell death by re-expression of antiproliferative genes [72]. However, the effect on chromatin remodeling by CG-1521 may be secondary to changes in cell function induced by other, nonhistone targets and is likely not the key anticancer mediator. Importantly, other studies have shown that p53 expression appears to not be requisite for the antitumor activity of HDACi [74–76].

22.4.3 Hsp90 Loses Chaperone Function Subsequent to HDACi-Induced Acetylation

In addition to the well-known stress-induced effects of heat-shock proteins (Hsp), Hsp90 is an important modulator of normal and neoplastic cell signaling and is a potential target in cancer therapy [77,78]. Hsp90 is a molecular chaperone that, with assistance from cochaperones, facilitates the assembly and maturation of client proteins including glucocorticoid receptor (GR) [79,80]. Its activity is acetylation dependent, with deacetylation mediated specifically by HDAC6 [80]. Kovacs et al. demonstrated that siRNA-mediated repression of HDAC6 caused hyperacetylation of Hsp90 and loss of its chaperone activity by dissociation from p23, an essential cochaperone, and subsequent degradation of GR [80]. Similar degradation of Hsp90 client proteins including Akt, Her2/neu, BCR-ABL, ERBB1, and ERBB2 has been observed upon pharmacological inhibition of HDAC, presumably HDAC6 [81–86]. The HDACi depsipeptide likewise diminished the level of many oncoproteins normally stabilized by Hsp90 in nonsmall-cell lung cancer cells [87]. While the isozyme-specific regulation of Hsp90 predicts a positive antitumor response to pharmacological inhibition of HDAC6, the necessity of HDAC6 inhibition for the induction of cell death by HDACi is not certain since several HDACi with potent antitumor properties are not capable of inhibiting this enzyme [19].

22.4.4 Stat3 Acetylation Mediates Its Nuclear Translocation
for Transcriptional Regulation

Upon activation, signal transducers and activators of transcription (Stat) proteins dimerize and translocate to the nucleus to regulate transcription [88]. Recent studies demonstrate that Stat3 can be modified by acetylation both in vitro and in vivo which stimulates its sequence-specific, DNA-binding ability [89–91]. Its coactivator p300/CREB-binding protein, a member of an important class of HATs, is responsible for its acetylation and capable of regulating many cellular processes. Treatment of cells with TSA mimicked intensified activity of this protein through increased Stat3 nuclear localization owing to its hyperacetylation [90]. This study defined a novel, alternative mechanism for Stat3 activation by acetylation, the functional consequences of which are yet to be determined. The activity of another Stat protein has been shown to be governed by acetylation [45]. Kramer et al. showed that HDACi-induced acetylation of Stat1 promotes its interaction with NF-κB. The nuclear localization of NF-κB, and associated expression of antiapoptotic target genes, is suppressed subsequent to this interaction [45].

22.4.5 α-TUBULIN ACETYLATION BY HDAC6 INHIBITION ABATES CELL MOTILITY

Hypoacetylation of microtubules by HDACs including TSA led to the discovery of "tubacin," a small-molecule agent designed to inhibit the deacetylation of α-tubulin in mammalian cells [47]. Tubacin specifically inhibits HDAC6, the enzyme responsible for removing acetyl groups from tubulin. The specificity of this novel compound negates other cellular effects such as cell cycle arrest and apoptosis induced by pleiotropic HDACi such as TSA. This specificity was determined by the comparison of genome-wide transcriptional profiles of mouse embryonic stem cells treated with either TSA or tubacin. Whereas 232 genes were increased above a 1.3-fold threshold value after TSA treatment, gene expression in cells treated with tubacin was relatively unaltered and similar to DMSO-treated control cells [47]. Altering the acetylation status of tubulin affects microtubules regulating cell motility and has important potential therapeutic applications in modulating cell movement necessary for metastasis and angiogenesis [92,93].

22.4.6 OTHER NONHISTONE EFFECTS OF HDACI

The phosphorylated, active form of Akt is overexpressed in many cancer types where it promotes cell survival and exerts antiapoptotic effects [94]. Some HDACs (HDAC1, 6, and 10) form complexes containing protein serine/threonine phosphatases in the cytosol, and inactivation of HDAC6 by mutagenesis or HDACi has been shown to disrupt the HDAC6/protein phosphatase 1 (PP1) complex [95]. Chen et al. demonstrated that the reshuffling of PP1 complexes is one mechanism by which HDACi inactivate Akt [32]. In this study, the HDACi TSA and OSU-HDAC42 caused dose-dependent inactivation of Akt in PTEN-null glioblastoma (U87MG) and prostate (PC3) cancer cells exhibiting constitutively active Akt. Co-immunoprecipitation experiments showed that HDAC1 and 6 form complexes with PP1. Upon treatment with HDACi, PP1 dissociates from HDAC and reshuffles to phospho-Akt leading to its inactivating dephosphorylation, and ultimately caspase 9-mediated apoptosis [32]. Similar results obtained by repressing HDAC1 and 6 with isozyme-specific siRNA validate the role of these HDACs in decreasing Akt phosphorylation and activity [20,32].

Numerous other HDAC effects have been described with less mechanistic certainty. For example, through an unknown mechanism, HDACi can inhibit DNA repair responses in certain cell lines, potentially increasing the sensitivity of these cells to chemo- and radiotherapy. The DNA-binding affinity of the transcription factor NF-κB and subsequent transcriptional activation is regulated by acetylation, as is the activity of κB-α (IκBα), the regulator of NF-κB [96]. NF-κB activity is additionally regulated by the acetylation of Stat1 [45]. Also, acetylation of transcription factors of several members of the death receptor pathway is thought to promote apoptosis; these members include Fas ligand, death receptor 5 (DR5), and TRAIL (tumor-necrosis factor-related apoptosis-inducing ligand), the ligand of DR5 [18].

22.5 CLINICAL DEVELOPMENT OF HDACi

Nearly a dozen small-molecule HDAC inhibitors are currently in clinical phases of testing in pursuit of regulatory approval, all of which have exhibited some degree of antitumor activity to warrant continued development. These are structurally diverse molecules used as monotherapies or in combination with other agents to treat leukemias, lymphomas, lung cancer, cervical cancer, and other advanced or refractory solid tumors [18,19]. In cell culture and in vivo, these inhibitors are consistently found to induce differentiation, cell cycle arrest, or apoptosis [17], with the latter being the most typical mode of cell death and therefore the most thoroughly investigated [19]. Of note, most of the HDACi used in these studies degrade quickly following administration and are associated with a rapid rise and fall of histone hyperacetylation in tumor cells [17].

The efficiency of enzyme activity inhibition and clinical effectiveness vary greatly among these structurally heterogeneous compounds [19]. Short-chain fatty acids including butyrate and valproic acid are effective only at high therapeutic doses (IC_{50} in low millimolar range) and, while seemingly well-tolerated in patients in clinical trials, have very short plasma half-lives. Hydroxamic acids have lower IC_{50} values (high nanomolar) against class I and II enzymes. Of these, TSA, a natural product, has high potency against cancer cells but adverse toxic effects preclude its use in the clinic [20]. Synthetic suberoylanilide hydroxamic acid (SAHA) or vorinostat (marketed as Zolinza; Merck & Co. Inc., NJ, USA) is a pleiotropic hydroxamate-derived compound that was recently approved by the FDA for use in refractory cutaneous T-cell lymphoma, and demonstrates potent antitumor activity in a variety of solid and hematopoietic tumors [16].

PXD101 (belinostat) and LBH589 are two other hydroxamate-based HDACi that have shown promising preclinical results, with the latter exhibiting effective pharmacodynamic modulation in a phase I trial in patients with hematopoietic neoplasms [97]. LBH589 increases apoptosis of multiple myeloma cells, including those resistant to conventional chemotherapy [98], and targets angiogenesis in vitro [26]. PXD101 exerts antitumor effects in preclinical studies of ovarian cancer in the absence of limiting toxicity [99,100]. Numerous clinical trials with these agents are currently recruiting patients for their evaluation in solid and hematopoietic tumors (www.clinicaltrials.gov).

Benzamides including MS-275 and CI-994 (acetyldinaline) are in phase I, II, and III clinical trials and inhibit HDAC activity at micromolar concentrations [19]. MS-275 may be regarded as the most advanced selective class I inhibitor [25] and targets HDAC 1 and 3 more potently than HDAC8 (IC_{50} 0.3, 8, and >100 μM, respectively) [46]. Cyclic peptides are structurally complex inhibitors with IC_{50}'s in the nanomolar range. Of these, trapoxin A, apicidin, and CHAPs inhibit class I HDACs with some degree of specificity, but have no or limited use in the clinic [101–103]. Depsipeptide, a cyclic peptolide, is generally classified as a broad-spectrum inhibitor [25], but has been shown to inhibit class I enzymes with greater potency than class II enzymes [104]. This compound acts as a prodrug with a reduced sulfhydryl group liberated only intracellularly [104].

Several phase I/II trials have been completed with acceptable toxicity profiles; however, only limited conclusions can be drawn from these preliminary studies [18]. A common adverse event is the rapid lowering of white blood cell and thrombocyte counts in the blood [105,106]. Interestingly, the time course of this neutropenia and thrombocytopenia in HDACi-treated patients is distinct and different from the myelosuppression that occurs with conventional chemotherapeutic drugs. It is likely that HDACi affect mature myeloid cells, rather than early myeloid precursors, since the occurrence of neutropenia and thrombocytopenia is rapid (nadirs at 5 and 10 days, respectively) and transient (return to pretreatment levels within 10 days) [106]. The hydroxamate vorinostat and cyclic peptide depsipeptide share similar toxicities including neutropenia, thrombocytopenia, fatigue, and dehydration, all of which appear to be reversible following withdrawal of the drug [105–108].

More serious adverse events have been reported for some agents such as depsipeptide, TSA, and valproic acid derivatives, and include cardiotoxicity and teratogenicity [109–111]. While a phase II clinical trial with depsipeptide was terminated prematurely due to potentially serious adverse cardiac events [20,109], other clinical studies with depsipeptide have shown no association with myocardial damage or impaired cardiac function [112,113], and fourteen clinical studies are listed as recruiting patients at the time of this review (see www.clinicaltrials.gov). In many instances toxicity may be attributed to nonspecific, HDACi-independent side effects since some of these compounds are only efficacious at high concentrations, especially the short-chain fatty acids [20]. The relationship between HDACi pharmacodynamic and kinetic properties and toxicity is yet to be determined [18,20].

Importantly, the key targets for HDACi in the context of cancer therapy are unknown as is the prediction of which patients will respond to treatment. Much work in this area, such as gene profiling of tumor samples from responders compared to nonresponders, is critical for optimization of HDACi therapy. And since acetylation is a reversible process, pharmacokinetic considerations are critical because HDACi effects may be diminished soon after treatment and vary significantly between enteral and parenteral administration [18,107,108]. The clinical development of vorinostat, for example, switched from intravenous to oral formulation in early phases of clinical testing due to differences in pharmacokinetic profiles and toxicity [114].

Indeed the limited pharmacokinetic data available for these agents to date suggest a need for continuous drug exposure to achieve significant in vivo antitumor activity [17]. Accordingly, it is logical to predict that continuous inhibition of HDAC activity will lead to a more positive long-term response [18]. To this end, a recent review article discusses the chemopreventive potential of HDACi in the diet [115]. Dashwood et al. propose that continued, subtle regulation of gene expression by dietary weak ligand HDACi may prevent cancer by affecting cell growth and apoptosis. Examples of dietary HDAC inhibitors with weak ligand properties include butyrate, diallyl sulfide, and sulforaphane [115].

22.6   HDACi IN COMBINATION WITH OTHER AGENTS

Additive or synergistic activity, with or without a preconceived molecular basis, has been observed with HDACi in combination with a variety of functionally diverse chemical compounds [19] (Table 22.2). The described broad-spectrum anticancer activity and relative safety in normal cells make HDACi an attractive choice for combinatorial use. This promising activity, which may be

TABLE 22.2
Combination Strategies of HDACi and Other Anticancer Agents

HDACi	Other Stimulus	Effect	References
TSA, depsipeptide, vorinostat, sodium butyrate	5-aza-2′-deoxycytidine	Enhanced apoptosis; re-expression of hypermethylated genes	[123–127]
CI-994, LAQ824, vorinostat	Gemcitabine, docetaxel	Enhanced cell cycle arrest and apoptosis	[19,85,128,130]
Magnesium valproate	Neoadjuvant doxorubicin, cyclophosphamide	HDAC inhibition and gene reactivation	[131]
OSU-HDAC42, TSA, vorinostat, MS-275	DNA-damaging agents bleomycin, doxorubicin, etoposide	Enhanced cell killing	[34,132]
LAQ824, MS-275, sodium butyrate	TRAIL	Enhanced apoptosis; apoptosis in TRAIL-resistant cells	[119,133,134]
MS-275, vorinostat, sodium butyrate	LY294002	Enhanced apoptosis	[135]
Vorinostat	PD184352	Enhanced apoptosis	[136]
Vorinostat, depsipeptide, apicidin	Imatinib (STI571)	Enhanced apoptosis; apoptosis in imatinib-resistant cells	[121,122,137,138]
Vorinostat, LBH589, sodium butyrate	17-AAG	Enhanced apoptosis and degradation of Hsp90 client proteins	[17,82,118]
Vorinostat, TSA, depsipeptide, MS-275, M344	Radiation	Enhanced radiosensivity; cell cycle arrest and inhibition of DNA synthesis	[120,139–141]

the greatest therapeutic potential of HDACi, has been demonstrated in preclinical and clinical studies in combinations with standard chemotherapeutics, signal transduction inhibitors, demethylating agents, nuclear receptor ligands, Hsp90 antagonists, and proteasome inhibitors [17,82,116–118]. While enhanced apoptosis is the most common synergistic effect, HDACi have also been shown to overcome resistance to specific targeted therapy [98,119–122].

Given that epigenetic control of gene expression involves more than acetylation, potent targeting of transcription in neoplasia may require a combination of chromatin-modifying agents including inhibitors of histone methylase, DNA methyltransferase (DNMT), and HDAC, and specific activators of transcription factors [123]. Since hypomethylation of DNA induces the reactivation of methylation-silenced tumor suppressor genes, there is currently much interest in the therapeutic strategy of combining DNMT inhibitors with HDACi [124]. The demethylating agent 5-aza-2′-deoxycytidine (5-aza-CdR) used in combination with TSA, for example, profoundly induced the re-expression of genes silenced by hypermethylated promoters, including MLH1, TIMP3 (TIMP3), CDKN2B (INK4B, p15), and CDKN2A (INK4, p16) [125]. Synergistic effects on the induction of apoptosis, differentiation, and cell growth arrest have also been observed in human breast, thoracic, lung, leukemia, and colon cancer cell lines treated with combinations of HDACi and inhibitors of DNMT [124]. In its limited use in the clinic to date, this combination strategy has shown promising results in the treatment of solid and hematopoietic tumors [126,127].

HDACi in combination with standard chemotherapeutics have demonstrated enhanced anticancer activity. CI-994 in combination with the antimetabolite gemcitabine or the antimicrotubule agent docetaxel caused synergistic effects on cell cycle arrest and apoptosis of nonsmall-cell lung cancer cell lines [128]. While limited conclusions on antitumor activity could be drawn from a phase I study of CI-994 and gemcitabine in patients with advanced cancer [129], a phase II trial with pancreatic cancer showed that CI-994 confers no advantage over gemcitabine alone [130]. Clinical trials of gemcitabine and CI-994 in lung cancer are in progress (see www.clinicaltrials.gov) [19]. Gemcitabine and docetaxel have also demonstrated enhanced killing of breast cancer cell lines in combination with either of the hydroxamates LAQ824 or SAHA [85]. A clinical trial of the HDACi magnesium valproate added to neoadjuvant doxorubicin and cyclophosphamide in locally advanced breast cancer caused HDAC inhibition and gene reactivation in primary tumors [131]; a randomized phase III study involving this combination strategy is ongoing.

Synergistic effects have been observed with many other combination strategies using HDACi with or without a strong molecular basis. As discussed above, HDACi combined with DNA-damaging agents appear to be a promising therapeutic strategy for prostate and other cancers [34,132]. Death-receptor signaling is enhanced by the HDACi-mediated hypersensitization of malignant cells to apoptosis. This potentiated apoptotic effect has been reported with different HDACi combined with activators of the TRAIL and Fas pathways, and TRAIL-resistant tumor cells have been sensitized to TRAIL-mediated apoptosis by low doses of HDACi [119,133,134]. HDACi also are capable of potentiating the effects of kinase inhibitors [121,122,135–138] and radiation [139–141] on cancer cell death, and synergistic activity from these combinations are commonly reported.

22.7 SUMMARY AND PERSPECTIVES

In summary, acetylation is an important posttranslational modification that regulates gene expression through pleiotropic effects on chromatin structure, protein stability, protein–protein interactions, protein localization, and DNA binding [142,143]. The balance of histone acetylation is regulated by the cooperation of HDACs and HATs. Beyond chromatin remodeling, the HDAC enzyme complex has been linked with numerous regulatory pathways for cellular growth and differentiation [72]. Given the array of nonhistone substrates, additional targets such as polyamines and metabolic intermediates should be expected [60]. The growing characterization of nontranscriptional mechanisms suggests that HDACi be considered as partially epigenetic therapy [18] and their targets be extended to protein deacetylases. Although the selectivity of most deacetylase inhibitors is poor, the development of tubacin [47] exemplifies the feasibility of developing HDACi-specific agents. With growing enthusiasm and resources devoted to the discovery, preclinical, and clinical evaluation of this class of compounds, HDACi will undoubtedly remain an important consideration in the development of novel cancer therapeutic strategies. The present intense interest in these compounds by the scientific community predicts an improved understanding of important mechanistic, pharmacodynamic, and pharmacokinetic properties that will extend the usefulness of HDACi in the clinic and broaden their future scope in the treatment of cancer.

REFERENCES

1. Stephen, J.K., Vaught, L.E., Chen, K.M. et al., Epigenetic events underlie the pathogenesis of sinonasal papillomas, *Mod Pathol* 20 (10), 1019–1027, 2007.
2. Farwell, D.G., Shera, K.A., Koop, J.I. et al., Genetic and epigenetic changes in human epithelial cells immortalized by telomerase, *Am J Pathol* 156 (5), 1537–1547, 2000.
3. Esteller, M. and Herman, J.G., Cancer as an epigenetic disease: DNA methylation and chromatin alterations in human tumours, *J Pathol* 196 (1), 1–7, 2002.
4. Deng, G., Bell, I., Crawley, S. et al., Braf mutation is frequently present in sporadic colorectal cancer with methylated hmlh1, but not in hereditary nonpolyposis colorectal cancer, *Clin Cancer Res* 10 (1 Pt 1), 191–195, 2004.
5. Feng, W., Lu, Z., Luo, R.Z. et al., Multiple histone deacetylases repress tumor suppressor gene ARHI in breast cancer, *Int J Cancer* 120 (8), 1664–1668, 2007.
6. Yu, J., Angelin-Duclos, C., Greenwood, J. et al., Transcriptional repression by Blimp-1 (PRDI-BF1) involves recruitment of histone deacetylase, *Mol Cell Biol* 20 (7), 2592–2603, 2000.
7. Brehm, A., Miska, E.A., McCance, D.J. et al., Retinoblastoma protein recruits histone deacetylase to repress transcription, *Nature* 391 (6667), 597–601, 1998.
8. Magnaghi-Jaulin, L., Groisman, R., Naguibneva, I. et al., Retinoblastoma protein represses transcription by recruiting a histone deacetylase, *Nature* 391 (6667), 601–605, 1998.
9. Gayther, S.A., Batley, S.J., Linger, L. et al., Mutations truncating the EP300 acetylase in human cancers, *Nat Genet* 24 (3), 300–303, 2000.
10. Giles, R.H., Peters, D.J., and Breuning, M.H., Conjunction dysfunction: Cbp/p300 in human disease, *Trends Genet* 14 (5), 178–183, 1998.
11. Zhang, Z., Yamashita, H., Toyama, T. et al., Quantitation of HDAC1 mRNA expression in invasive carcinoma of the breast*, *Breast Cancer Res Treat* 94 (1), 11–16, 2005.
12. Krusche, C.A., Wulfing, P., Kersting, C. et al., Histone deacetylase-1 and -3 protein expression in human breast cancer: A tissue microarray analysis, *Breast Cancer Res Treat* 90 (1), 15–23, 2005.
13. Saji, S., Kawakami, M., Hayashi, S. et al., Significance of HDAC6 regulation via estrogen signaling for cell motility and prognosis in estrogen receptor-positive breast cancer, *Oncogene* 24 (28), 4531–4539, 2005.
14. Osada, H., Tatematsu, Y., Saito, H. et al., Reduced expression of class II histone deacetylase genes is associated with poor prognosis in lung cancer patients, *Int J Cancer* 112 (1), 26–32, 2004.
15. Bradbury, C.A., Khanim, F.L., Hayden, R. et al., Histone deacetylases in acute myeloid leukaemia show a distinctive pattern of expression that changes selectively in response to deacetylase inhibitors, *Leukemia* 19 (10), 1751–1759, 2005.
16. Marks, P.A. and Breslow, R., Dimethyl sulfoxide to vorinostat: Development of this histone deacetylase inhibitor as an anticancer drug, *Nat Biotechnol* 25 (1), 84–90, 2007.
17. Drummond, D.C., Noble, C.O., Kirpotin, D.B. et al., Clinical development of histone deacetylase inhibitors as anticancer agents, *Annu Rev Pharmacol Toxicol* 45, 495–528, 2005.
18. Minucci, S. and Pelicci, P.G., Histone deacetylase inhibitors and the promise of epigenetic (and more) treatments for cancer, *Nat Rev Cancer* 6 (1), 38–51, 2006.
19. Bolden, J.E., Peart, M.J., and Johnstone, R.W., Anticancer activities of histone deacetylase inhibitors, *Nat Rev Drug Discov* 5 (9), 769–784, 2006.
20. Lin, H.Y., Chen, C.S., Lin, S.P. et al., Targeting histone deacetylase in cancer therapy, *Med Res Rev* 26 (4), 397–413, 2006.

21. Ungerstedt, J.S., Sowa, Y., Xu, W.S. et al., Role of thioredoxin in the response of normal and transformed cells to histone deacetylase inhibitors, *Proc Natl Acad Sci U S A* 102 (3), 673–678, 2005.

22. Qiu, L., Burgess, A., Fairlie, D.P. et al., Histone deacetylase inhibitors trigger a G2 checkpoint in normal cells that is defective in tumor cells, *Mol Biol Cell* 11 (6), 2069–2083, 2000.

23. Warrener, R., Beamish, H., Burgess, A. et al., Tumor cell-selective cytotoxicity by targeting cell cycle checkpoints, *FASEB J* 17 (11), 1550–1552, 2003.

24. Glaser, K.B., Staver, M.J., Waring, J.F. et al., Gene expression profiling of multiple histone deacetylase (HDAC) inhibitors: Defining a common gene set produced by HDAC inhibition in T24 and MDA carcinoma cell lines, *Mol Cancer Ther* 2 (2), 151–163, 2003.

25. Karagiannis, T.C. and El-Osta, A., Will broad-spectrum histone deacetylase inhibitors be superseded by more specific compounds? *Leukemia* 21 (1), 61–65, 2007.

26. Qian, D.Z., Kato, Y., Shabbeer, S. et al., Targeting tumor angiogenesis with histone deacetylase inhibitors: The hydroxamic acid derivative LBH589, *Clin Cancer Res* 12 (2), 634–642, 2006.

27. Bonaldi, T., Talamo, F., Scaffidi, P. et al., Monocytic cells hyperacetylate chromatin protein HMGB1 to redirect it towards secretion, *EMBO J* 22 (20), 5551–5560, 2003.

28. Wang, X.Q., Alfaro, M.L., Evans, G.F. et al., Histone deacetylase inhibition results in decreased macrophage CD9 expression, *Biochem Biophys Res Commun* 294 (3), 660–666, 2002.

29. Dehmel, F., Ciossek, T., Maier, T. et al., Trithiocarbonates-exploration of a new head group for HDAC inhibitors, *Bioorg Med Chem Lett* 17 (17), 4746–4752, 2007.

30. Lu, Q., Yang, Y.T., Chen, C.S. et al., Zn^{2+} -chelating motif-tethered short-chain fatty acids as a novel class of histone deacetylase inhibitors, *J Med Chem* 47 (2), 467–474, 2004.

31. Lu, Q., Wang, D.S., Chen, C.S. et al., Structure-based optimization of phenylbutyrate-derived histone deacetylase inhibitors, *J Med Chem* 48 (17), 5530–5535, 2005.

32. Chen, C.S., Weng, S.C., Tseng, P.H. et al., Histone acetylation-independent effect of histone deacetylase inhibitors on Akt through the reshuffling of protein phosphatase 1 complexes, *J Biol Chem* 280 (46), 38879–38887, 2005.

33. Kulp, S.K., Chen, C.S., Wang, D.S. et al., Antitumor effects of a novel phenylbutyrate-based histone deacetylase inhibitor, (s)-HDAC-42, in prostate cancer, *Clin Cancer Res* 12 (17), 5199–5206, 2006.

34. Chen, C.S., Wang, Y.C., Yang, H.C. et al., Histone deacetylase inhibitors sensitize prostate cancer cells to agents that produce DNA double-strand breaks by targeting Ku70 acetylation, *Cancer Res* 67 (11), 5318–5327, 2007.

35. Lu, Y.S., Kashida, Y., Kulp, S.K. et al., Efficacy of a novel histone deacetylase inhibitor in murine models of hepatocellular carcinoma, *Hepatology* 46 (4), 1119–1130, 2007.

36. Rosato, R.R., Almenara, J.A., Yu, C. et al., Evidence of a functional role for p21WAF1/CIP1 down-regulation in synergistic antileukemic interactions between the histone deacetylase inhibitor sodium butyrate and flavopiridol, *Mol Pharmacol* 65 (3), 571–581, 2004.

37. Hoshikawa, Y., Kwon, H.J., Yoshida, M. et al., Trichostatin A induces morphological changes and gelsolin expression by inhibiting histone deacetylase in human carcinoma cell lines, *Exp Cell Res* 214 (1), 189–197, 1994.

38. Kim, J.S., Lee, S., Lee, T. et al., Transcriptional activation of p21(WAF1/CIP1) by apicidin, a novel histone deacetylase inhibitor, *Biochem Biophys Res Commun* 281 (4), 866–871, 2001.

39. Kim, M.S., Kwon, H.J., Lee, Y.M. et al., Histone deacetylases induce angiogenesis by negative regulation of tumor suppressor genes, *Nat Med* 7 (4), 437–443, 2001.

40. Kwon, S.H., Ahn, S.H., Kim, Y.K. et al., Apicidin, a histone deacetylase inhibitor, induces apoptosis and Fas/Fas ligand expression in human acute promyelocytic leukemia cells, *J Biol Chem* 277 (3), 2073–2080, 2002.

41. Han, J.W., Ahn, S.H., Park, S.H. et al., Apicidin, a histone deacetylase inhibitor, inhibits proliferation of tumor cells via induction of p21WAF1/CIP1 and gelsolin, *Cancer Res* 60 (21), 6068–6074, 2000.

42. Almenara, J., Rosato, R., and Grant, S., Synergistic induction of mitochondrial damage and apoptosis in human leukemia cells by flavopiridol and the histone deacetylase inhibitor suberoylanilide hydroxamic acid (SAHA), *Leukemia* 16 (7), 1331–1343, 2002.

43. Sasakawa, Y., Naoe, Y., Inoue, T. et al., Effects of FK228, a novel histone deacetylase inhibitor, on tumor growth and expression of p21 and c-myc genes in vivo, *Cancer Lett* 195 (2), 161–168, 2003.

44. Van Lint, C., Emiliani, S., and Verdin, E., The expression of a small fraction of cellular genes is changed in response to histone hyperacetylation, *Gene Expr* 5 (4–5), 245–253, 1996.

45. Kramer, O.H., Baus, D., Knauer, S.K. et al., Acetylation of Stat1 modulates NF-κB activity, *Genes Dev* 20 (4), 473–485, 2006.

46. Hu, E., Dul, E., Sung, C.M. et al., Identification of novel isoform-selective inhibitors within class I histone deacetylases, *J Pharmacol Exp Ther* 307 (2), 720–728, 2003.

47. Haggarty, S.J., Koeller, K.M., Wong, J.C. et al., Domain-selective small-molecule inhibitor of histone deacetylase 6 (HDAC6)-mediated tubulin deacetylation, *Proc Natl Acad Sci U S A* 100 (8), 4389–4394, 2003.

48. Cho, H.H., Park, H.T., Kim, Y.J. et al., Induction of osteogenic differentiation of human mesenchymal stem cells by histone deacetylase inhibitors, *J Cell Biochem* 96 (3), 533–542, 2005.

49. Glaser, K.B., Li, J., Staver, M.J. et al., Role of class I and class II histone deacetylases in carcinoma cells using siRNA, *Biochem Biophys Res Commun* 310 (2), 529–536, 2003.

50. Inoue, S., Mai, A., Dyer, M.J. et al., Inhibition of histone deacetylase class I but not class II is critical for the sensitization of leukemic cells to tumor necrosis factor-related apoptosis-inducing ligand-induced apoptosis, *Cancer Res* 66 (13), 6785–6792, 2006.

51. Barnes, P.J., Reduced histone deacetylase in COPD: Clinical implications, *Chest* 129 (1), 151–155, 2006.

52. Wilson, A.J., Byun, D.S., Popova, N. et al., Histone deacetylase 3 (HDAC3) and other class I HDACs regulate colon cell maturation and p21 expression and are deregulated in human colon cancer, *J Biol Chem* 281 (19), 13548–13558, 2006.

53. Zhu, P., Martin, E., Mengwasser, J. et al., Induction of HDAC2 expression upon loss of APC in colorectal tumorigenesis, *Cancer Cell* 5 (5), 455–463, 2004.

54. Peart, M.J., Smyth, G.K., van Laar, R.K. et al., Identification and functional significance of genes regulated by structurally different histone deacetylase inhibitors, *Proc Natl Acad Sci U S A* 102 (10), 3697–3702, 2005.

55. Ren, C., Zhang, L., Freitas, M.A. et al., Peptide mass mapping of acetylated isoforms of histone h4 from mouse lymphosarcoma cells treated with histone deacetylase (HDACs) inhibitors, *J Am Soc Mass Spectrom* 16 (10), 1641–1653, 2005.

56. Zhang, L., Su, X., Liu, S. et al., Histone H4 N-terminal acetylation in Kasumi-1 cells treated with depsipeptide determined by acetic acid-urea polyacrylamide gel electrophoresis, amino acid coded mass tagging, and mass spectrometry, *J Proteome Res* 6 (1), 81–88, 2007.

57. Yoshida, M., Matsuyama, A., Komatsu, Y. et al., From discovery to the coming generation of histone deacetylase inhibitors, *Curr Med Chem* 10 (22), 2351–2358, 2003.

58. Hubbert, C., Guardiola, A., Shao, R. et al., HDAC6 is a microtubule-associated deacetylase, *Nature* 417 (6887), 455–458, 2002.

59. Fischle, W., Dequiedt, F., Fillion, M. et al., Human HDAC7 histone deacetylase activity is associated with HDAC3 in vivo, *J Biol Chem* 276 (38), 35826–35835, 2001.

60. Gregoretti, I.V., Lee, Y.M., and Goodson, H.V., Molecular evolution of the histone deacetylase family: Functional implications of phylogenetic analysis, *J Mol Biol* 338 (1), 17–31, 2004.

61. Subramanian, C., Opipari, A.W., Jr., Castle, V.P. et al., Histone deacetylase inhibition induces apoptosis in neuroblastoma, *Cell Cycle* 4 (12), 1741–1743, 2005.

62. Subramanian, C., Opipari, A.W., Jr., Bian, X. et al., Ku70 acetylation mediates neuroblastoma cell death induced by histone deacetylase inhibitors, *Proc Natl Acad Sci U S A* 102 (13), 4842–4847, 2005.

63. Levine, A.J., P53, the cellular gatekeeper for growth and division, *Cell* 88 (3), 323–331, 1997.

64. Gu, W. and Roeder, R.G., Activation of p53 sequence-specific DNA binding by acetylation of the p53 C-terminal domain, *Cell* 90 (4), 595–606, 1997.

65. Vogelstein, B., Lane, D., and Levine, A.J., Surfing the p53 network, *Nature* 408 (6810), 307–310, 2000.

66. Tang, Y., Luo, J., Zhang, W. et al., Tip60-dependent acetylation of p53 modulates the decision between cell-cycle arrest and apoptosis, *Mol Cell* 24 (6), 827–839, 2006.

67. Ito, A., Lai, C.H., Zhao, X. et al., p300/CBP-mediated p53 acetylation is commonly induced by p53-activating agents and is inhibited by MDM2, *EMBO J* 20, 1331–1340, 2001.

68. Ito, A., Kawaguchi, Y., Lai, C.H. et al., MDM2-HDAC1-mediated deacetylation of p53 is required for its degradation, *EMBO J* 21, 6236–6245, 2002.

69. Juan, L.J., Shia, W.J., Chen, M.H. et al., Histone deacetylases specifically down-regulate p53-dependent gene activation, *J Biol Chem* 275 (27), 20436–20443, 2000.

70. Luo, J., Su, F., Chen, D. et al., Deacetylation of p53 modulates its effect on cell growth and apoptosis, *Nature* 408 (6810), 377–381, 2000.

71. Gu, W., Luo, J., Brooks, C.L. et al., Dynamics of the p53 acetylation pathway, *Novartis Found Symp* 259, 197–205; discussion 205–207, 223–225, 2004.

72. Roy, S., Packman, K., Jeffrey, R. et al., Histone deacetylase inhibitors differentially stabilize acetylated p53 and induce cell cycle arrest or apoptosis in prostate cancer cells, *Cell Death Differ* 12 (5), 482–491, 2005.

73. Roy, S. and Tenniswood, M., Site-specific acetylation of p53 directs selective transcription complex assembly, *J Biol Chem* 282 (7), 4765–4771, 2007.

74. Greenberg, V.L., Williams, J.M., Cogswell, J.P. et al., Histone deacetylase inhibitors promote apoptosis and differential cell cycle arrest in anaplastic thyroid cancer cells, *Thyroid* 11 (4), 315–325, 2001.

75. Hague, A., Manning, A.M., Hanlon, K.A. et al., Sodium butyrate induces apoptosis in human colonic tumour cell lines in a p53-independent pathway: Implications for the possible role of dietary fibre in the prevention of large-bowel cancer, *Int J Cancer* 55 (3), 498–505, 1993.

76. Yamamoto, H., Fujimoto, J., Okamoto, E. et al., Suppression of growth of hepatocellular carcinoma by sodium butyrate in vitro and in vivo, *Int J Cancer* 76 (6), 897–902, 1998.

77. Drysdale, M.J., Brough, P.A., Massey, A. et al., Targeting Hsp90 for the treatment of cancer, *Curr Opin Drug Discov Devel* 9 (4), 483–495, 2006.

78. Whitesell, L. and Lindquist, S.L., Hsp90 and the chaperoning of cancer, *Nat Rev Cancer* 5 (10), 761–772, 2005.

79. Kovacs, J.J., Cohen, T.J., and Yao, T.P., Chaperoning steroid hormone signaling via reversible acetylation, *Nucl Recept Signal* 3, e004, 2005.

80. Kovacs, J.J., Murphy, P.J., Gaillard, S. et al., HDAC6 regulates Hsp90 acetylation and chaperone-dependent activation of glucocorticoid receptor, *Mol Cell* 18 (5), 601–607, 2005.

81. Rahmani, M., Reese, E., Dai, Y. et al., Coadministration of histone deacetylase inhibitors and perifosine synergistically induces apoptosis in human leukemia cells through Akt and erk1/2 inactivation and the generation of ceramide and reactive oxygen species, *Cancer Res* 65 (6), 2422–2432, 2005.

82. George, P., Bali, P., Annavarapu, S. et al., Combination of the histone deacetylase inhibitor LBH589 and the Hsp90 inhibitor 17-AAG is highly active against human CML-BC cells and aml cells with activating mutation of FLT-3, *Blood* 105 (4), 1768–1776, 2005.

83. Bali, P., Pranpat, M., Swaby, R. et al., Activity of suberoylanilide hydroxamic acid against human breast cancer cells with amplification of Her-2, *Clin Cancer Res* 11 (17), 6382–6389, 2005.

84. Chen, L., Meng, S., Wang, H. et al., Chemical ablation of androgen receptor in prostate cancer cells by the histone deacetylase inhibitor LAQ824, *Mol Cancer Ther* 4 (9), 1311–1319, 2005.

85. Fuino, L., Bali, P., Wittmann, S. et al., Histone deacetylase inhibitor LAQ824 down-regulates her-2 and sensitizes human breast cancer cells to trastuzumab, taxotere, gemcitabine, and epothilone B, *Mol Cancer Ther* 2 (10), 971–984, 2003.

86. Nimmanapalli, R., Fuino, L., Bali, P. et al., Histone deacetylase inhibitor LAQ824 both lowers expression and promotes proteasomal degradation of Bcr-Abl and induces apoptosis of imatinib mesylate-sensitive or -refractory chronic myelogenous leukemia-blast crisis cells, *Cancer Res* 63 (16), 5126–5135, 2003.

87. Yu, X., Guo, Z.S., Marcu, M.G. et al., Modulation of p53, ErbB1, ErbB2, and Raf-1 expression in lung cancer cells by depsipeptide FR901228, *J Natl Cancer Inst* 94 (7), 504–513, 2002.

88. Darnell, J.E., Jr., Stats and gene regulation, *Science* 277 (5332), 1630–1635, 1997.

89. O'Shea, J.J., Kanno, Y., Chen, X. et al., Cell signaling. Stat acetylation—a key facet of cytokine signaling? *Science* 307 (5707), 217–218, 2005.

90. Wang, R., Cherukuri, P., and Luo, J., Activation of Stat3 sequence-specific DNA binding and transcription by p300/creb-binding protein-mediated acetylation, *J Biol Chem* 280 (12), 11528–11534, 2005.

91. Yuan, Z.L., Guan, Y.J., Chatterjee, D. et al., Stat3 dimerization regulated by reversible acetylation of a single lysine residue, *Science* 307 (5707), 269–273, 2005.

92. Palazzo, A., Ackerman, B., and Gundersen, G.G., Cell biology: Tubulin acetylation and cell motility, *Nature* 421 (6920), 230, 2003.

93. Tran, A.D., Marmo, T.P., Salam, A.A. et al., HDAC6 deacetylation of tubulin modulates dynamics of cellular adhesions, *J Cell Sci* 120 (Pt 8), 1469–1479, 2007.

94. Hennessy, B.T., Smith, D.L., Ram, P.T. et al., Exploiting the pi3k/Akt pathway for cancer drug discovery, *Nat Rev Drug Discov* 4 (12), 988–1004, 2005.

95. Brush, M.H., Guardiola, A., Connor, J.H. et al., Deacetylase inhibitors disrupt cellular complexes containing protein phosphatases and deacetylases, *J Biol Chem* 279 (9), 7685–7691, 2004.

96. Chen, L.F. and Greene, W.C., Shaping the nuclear action of NF-κB, *Nat Rev Mol Cell Biol* 5 (5), 392–401, 2004.

97. Giles, F., Fischer, T., Cortes, J. et al., A phase I study of intravenous LBH589, a novel cinnamic hydroxamic acid analogue histone deacetylase inhibitor, in patients with refractory hematologic malignancies, *Clin Cancer Res* 12 (15), 4628–4635, 2006.

98. Maiso, P., Carvajal-Vergara, X., Ocio, E.M. et al., The histone deacetylase inhibitor LBH589 is a potent antimyeloma agent that overcomes drug resistance, *Cancer Res* 66 (11), 5781–5789, 2006.

99. Plumb, J.A., Finn, P.W., Williams, R.J. et al., Pharmacodynamic response and inhibition of growth of human tumor xenografts by the novel histone deacetylase inhibitor PXD101, *Mol Cancer Ther* 2 (8), 721–728, 2003.

100. Qian, X., LaRochelle, W.J., Ara, G. et al., Activity of PXD101, a histone deacetylase inhibitor, in preclinical ovarian cancer studies, *Mol Cancer Ther* 5 (8), 2086–2095, 2006.

101. Bhalla, K.N., Epigenetic and chromatin modifiers as targeted therapy of hematologic malignancies, *J Clin Oncol* 23 (17), 3971–3993, 2005.

102. Vannini, A., Volpari, C., Filocamo, G. et al., Crystal structure of a eukaryotic zinc-dependent histone deacetylase, human HDAC8, complexed with a hydroxamic acid inhibitor, *Proc Natl Acad Sci U S A* 101 (42), 15064–15069, 2004.

103. Furumai, R., Komatsu, Y., Nishino, N. et al., Potent histone deacetylase inhibitors built from trichostatin A and cyclic tetrapeptide antibiotics including trapoxin, *Proc Natl Acad Sci U S A* 98 (1), 87–92, 2001.

104. Furumai, R., Matsuyama, A., Kobashi, N. et al., FK228 (depsipeptide) as a natural prodrug that inhibits class I histone deacetylases, *Cancer Res* 62 (17), 4916–4921, 2002.

105. Rubin, E.H., Agrawal, N.G., Friedman, E.J. et al., A study to determine the effects of food and multiple dosing on the pharmacokinetics of vorinostat given orally to patients with advanced cancer, *Clin Cancer Res* 12 (23), 7039–7045, 2006.

106. Sandor, V., Bakke, S., Robey, R.W. et al., Phase I trial of the histone deacetylase inhibitor, depsipeptide (fr901228, nsc 630176), in patients with refractory neoplasms, *Clin Cancer Res* 8 (3), 718–728, 2002.

107. Kelly, W.K., O'Connor, O.A., Krug, L.M. et al., Phase I study of an oral histone deacetylase inhibitor, suberoylanilide hydroxamic acid, in patients with advanced cancer, *J Clin Oncol* 23 (17), 3923–3931, 2005.

108. Kelly, W.K., Richon, V.M., O'Connor, O. et al., Phase I clinical trial of histone deacetylase inhibitor: Suberoylanilide hydroxamic acid administered intravenously, *Clin Cancer Res* 9 (10 Pt 1), 3578–3588, 2003.

109. Shah, M.H., Binkley, P., Chan, K. et al., Cardiotoxicity of histone deacetylase inhibitor depsipeptide in patients with metastatic neuroendocrine tumors, *Clin Cancer Res* 12 (13), 3997–4003, 2006.

110. Menegola, E., Di Renzo, F., Broccia, M.L. et al., Inhibition of histone deacetylase activity on specific embryonic tissues as a new mechanism for teratogenicity, *Birth Defects Res B Dev Reprod Toxicol* 74 (5), 392–398, 2005.

111. Eikel, D., Hoffmann, K., Zoll, K. et al., S-2-pentyl-4-pentynoic hydroxamic acid and its metabolite s-2-pentyl-4-pentynoic acid in the nmri-exencephaly-mouse model: Pharmacokinetic profiles, teratogenic effects, and histone deacetylase inhibition abilities of further valproic acid hydroxamates and amides, *Drug Metab Dispos* 34 (4), 612–620, 2006.

112. Piekarz, R.L., Frye, A.R., Wright, J.J. et al., Cardiac studies in patients treated with depsipeptide, FK228, in a phase II trial for T-cell lymphoma, *Clin Cancer Res* 12 (12), 3762–3773, 2006.

113. Molife, R., Fong, P., Scurr, M. et al., HDAC inhibitors and cardiac safety, *Clin Cancer Res* 13 (3), 1068; author reply 1068–1069, 2007.

114. O'Connor, O.A., Heaney, M.L., Schwartz, L. et al., Clinical experience with intravenous and oral formulations of the novel histone deacetylase inhibitor suberoylanilide hydroxamic acid in patients with advanced hematologic malignancies, *J Clin Oncol* 24 (1), 166–173, 2006.

115. Dashwood, R.H., Myzak, M.C., and Ho, E., Dietary HDAC inhibitors: Time to rethink weak ligands in cancer chemoprevention? *Carcinogenesis* 27 (2), 344–349, 2006.

116. Catley, L., Weisberg, E., Kiziltepe, T. et al., Aggresome induction by proteasome inhibitor bortezomib and α-tubulin hyperacetylation by tubulin deacetylase (TDAC) inhibitor LBH589 are synergistic in myeloma cells, *Blood* 108 (10), 3441–3449, 2006.

117. Duan, J., Friedman, J., Nottingham, L. et al., Nuclear factor-kappaB p65 small interfering RNA or proteasome inhibitor bortezomib sensitizes head and neck squamous cell carcinomas to classic histone deacetylase inhibitors and novel histone deacetylase inhibitor PXD101, *Mol Cancer Ther* 6 (1), 37–50, 2007.

118. Rahmani, M., Yu, C., Dai, Y. et al., Coadministration of the heat shock protein 90 antagonist 17-allylamino-17-demethoxygeldanamycin with suberoylanilide hydroxamic acid or sodium butyrate synergistically induces apoptosis in human leukemia cells, *Cancer Res* 63 (23), 8420–8427, 2003.

119. Rosato, R.R., Almenara, J.A., Dai, Y. et al., Simultaneous activation of the intrinsic and extrinsic pathways by histone deacetylase (HDAC) inhibitors and tumor necrosis factor-related apoptosis-inducing ligand (trail) synergistically induces mitochondrial damage and apoptosis in human leukemia cells, *Mol Cancer Ther* 2 (12), 1273–1284, 2003.

120. Kim, M.S., Baek, J.H., Chakravarty, D. et al., Sensitization to UV-induced apoptosis by the histone deacetylase inhibitor trichostatin A (TSA), *Exp Cell Res* 306 (1), 94–102, 2005.

121. Yu, C., Rahmani, M., Almenara, J. et al., Histone deacetylase inhibitors promote sti571-mediated apoptosis in sti571-sensitive and -resistant Bcr/Abl + human myeloid leukemia cells, *Cancer Res* 63 (9), 2118–2126, 2003.

122. Nimmanapalli, R., Fuino, L., Stobaugh, C. et al., Cotreatment with the histone deacetylase inhibitor suberoylanilide hydroxamic acid (SAHA) enhances imatinib-induced apoptosis of Bcr-Abl-positive human acute leukemia cells, *Blood* 101 (8), 3236–3239, 2003.

123. Johnstone, R.W. and Licht, J.D., Histone deacetylase inhibitors in cancer therapy: Is transcription the primary target? *Cancer Cell* 4 (1), 13–18, 2003.

124. Zhu, W.G. and Otterson, G.A., The interaction of histone deacetylase inhibitors and DNA methyltransferase inhibitors in the treatment of human cancer cells, *Curr Med Chem Anticancer Agents* 3 (3), 187–199, 2003.

125. Cameron, E.E., Bachman, K.E., Myohanen, S. et al., Synergy of demethylation and histone deacetylase inhibition in the re-expression of genes silenced in cancer, *Nat Genet* 21 (1), 103–107, 1999.

126. Gore, S.D., Baylin, S., Sugar, E. et al., Combined DNA methyltransferase and histone deacetylase inhibition in the treatment of myeloid neoplasms, *Cancer Res* 66 (12), 6361–6369, 2006.

127. Garcia-Manero, G., Kantarjian, H.M., Sanchez-Gonzalez, B. et al., Phase 1/2 study of the combination of 5-aza-2'-deoxycytidine with valproic acid in patients with leukemia, *Blood* 108 (10), 3271–3279, 2006.

128. Loprevite, M., Tiseo, M., Grossi, F. et al., In vitro study of CI-994, a histone deacetylase inhibitor, in non-small cell lung cancer cell lines, *Oncol Res* 15 (1), 39–48, 2005.

129. Nemunaitis, J.J., Orr, D., Eager, R. et al., Phase I study of oral CI-994 in combination with gemcitabine in treatment of patients with advanced cancer, *Cancer J* 9 (1), 58–66, 2003.

130. Richards, D.A., Boehm, K.A., Waterhouse, D.M. et al., Gemcitabine plus CI-994 offers no advantage over gemcitabine alone in the treatment of patients with advanced pancreatic cancer: Results of a phase II randomized, double-blind, placebo-controlled, multicenter study, *Ann Oncol* 17 (7), 1096–1102, 2006.

131. Arce, C., Perez-Plasencia, C., Gonzalez-Fierro, A. et al., A proof-of-principle study of epigenetic therapy added to neoadjuvant doxorubicin cyclophosphamide for locally advanced breast cancer, *PLoS ONE* 1, e98, 2006.

132. Kim, M.S., Blake, M., Baek, J.H. et al., Inhibition of histone deacetylase increases cytotoxicity to anticancer drugs targeting DNA, *Cancer Res* 63 (21), 7291–7300, 2003.

133. Guo, F., Sigua, C., Tao, J. et al., Cotreatment with histone deacetylase inhibitor LAQ824 enhances Apo-2l/tumor necrosis factor-related apoptosis inducing ligand-induced death inducing signaling complex activity and apoptosis of human acute leukemia cells, *Cancer Res* 64 (7), 2580–2589, 2004.

134. Kim, E.H., Kim, H.S., Kim, S.U. et al., Sodium butyrate sensitizes human glioma cells to trail-mediated apoptosis through inhibition of Cdc2 and the subsequent downregulation of survivin and XIAP, *Oncogene* 24 (46), 6877–6889, 2005.

135. Rahmani, M., Yu, C., Reese, E. et al., Inhibition of PI-3 kinase sensitizes human leukemic cells to histone deacetylase inhibitor-mediated apoptosis through p44/42 map kinase inactivation and abrogation of p21(CIP1/WAF1) induction rather than Akt inhibition, *Oncogene* 22 (40), 6231–6242, 2003.

136. Yu, C., Dasmahapatra, G., Dent, P. et al., Synergistic interactions between mek1/2 and histone deacetylase inhibitors in Bcr/Abl + human leukemia cells, *Leukemia* 19 (9), 1579–1589, 2005.

137. Kim, J.S., Jeung, H.K., Cheong, J.W. et al., Apicidin potentiates the imatinib-induced apoptosis of Bcr-Abl-positive human leukaemia cells by enhancing the activation of mitochondria-dependent caspase cascades, *Br J Haematol* 124 (2), 166–178, 2004.

138. Kawano, T., Horiguchi-Yamada, J., Iwase, S. et al., Depsipeptide enhances imatinib mesylate-induced apoptosis of Bcr-Abl-positive cells and ectopic expression of cyclin D1, c-Myc or active MEK abrogates this effect, *Anticancer Res* 24 (5A), 2705–2712, 2004.

139. Zhang, Y., Jung, M., and Dritschilo, A., Enhancement of radiation sensitivity of human squamous carcinoma cells by histone deacetylase inhibitors, *Radiat Res* 161 (6), 667–674, 2004.
140. Nome, R.V., Bratland, A., Harman, G. et al., Cell cycle checkpoint signaling involved in histone deacetylase inhibition and radiation-induced cell death, *Mol Cancer Ther* 4 (8), 1231–1238, 2005.
141. Camphausen, K., Scott, T., Sproull, M. et al., Enhancement of xenograft tumor radiosensitivity by the histone deacetylase inhibitor ms-275 and correlation with histone hyperacetylation, *Clin Cancer Res* 10 (18 Pt 1), 6066–6071, 2004.
142. Yang, W.M., Tsai, S.C., Wen, Y.D. et al., Functional domains of histone deacetylase-3, *J Biol Chem* 277 (11), 9447–9454, 2002.
143. Kouzarides, T., Acetylation: A regulatory modification to rival phosphorylation? *EMBO J* 19 (6), 1176–1179, 2000.
144. Cohen, H.Y., Lavu, S., Bitterman, K.J. et al., Acetylation of the C terminus of Ku70 by CBP and PCAF controls Bax-mediated apoptosis, *Mol Cell* 13 (5), 627–638, 2004.
145. Cohen, H.Y., Miller, C., Bitterman, K.J. et al., Calorie restriction promotes mammalian cell survival by inducing the SIRT1 deacetylase, *Science* 305 (5682), 390–392, 2004.
146. Yang, Y., Hou, H., Haller, E.M. et al., Suppression of FOXO1 activity by FHL2 through SIRT1-mediated deacetylation, *EMBO J* 24 (5), 1021–1032, 2005.
147. Bouras, T., Fu, M., Sauve, A.A. et al., SIRT1 deacetylation and repression of p300 involves lysine residues 1020/1024 within the cell cycle regulatory domain 1, *J Biol Chem* 280 (11), 10264–10276, 2005.
148. Vaziri, H., Dessain, S.K., Ng Eaton, E. et al., Hsir2(SIRT1) functions as an NAD-dependent p53 deacetylase, *Cell* 107 (2), 149–159, 2001.
149. Langley, E., Pearson, M., Faretta, M. et al., Human SIR2 deacetylates p53 and antagonizes PML/p53-induced cellular senescence, *EMBO J* 21 (10), 2383–2396, 2002.
150. Luo, J., Nikolaev, A.Y., Imai, S. et al., Negative control of p53 by sir2alpha promotes cell survival under stress, *Cell* 107 (2), 137–148, 2001.
151. Fu, M., Rao, M., Wang, C. et al., Acetylation of androgen receptor enhances coactivator binding and promotes prostate cancer cell growth, *Mol Cell Biol* 23 (23), 8563–8575, 2003.
152. Fu, M., Liu, M., Sauve, A.A. et al., Hormonal control of androgen receptor function through SIRT1, *Mol Cell Biol* 26 (21), 8122–8135, 2006.
153. Gaughan, L., Logan, I.R., Cook, S. et al., Tip60 and histone deacetylase 1 regulate androgen receptor activity through changes to the acetylation status of the receptor, *J Biol Chem* 277 (29), 25904–25913, 2002.
154. Gaughan, L., Logan, I.R., Neal, D.E. et al., Regulation of androgen receptor and histone deacetylase 1 by MDM2-mediated ubiquitylation, *Nucleic Acids Res* 33 (1), 13–26, 2005.
155. Simonini, M.V., Camargo, L.M., Dong, E. et al., The benzamide ms-275 is a potent, long-lasting brain region-selective inhibitor of histone deacetylases, *Proc Natl Acad Sci U S A* 103 (5), 1587–1592, 2006.
156. Quivy, V. and Van Lint, C., Regulation at multiple levels of NF-κB-mediated transactivation by protein acetylation, *Biochem Pharmacol* 68 (6), 1221–1229, 2004.
157. Greene, W.C. and Chen, L.F., Regulation of NF-κb action by reversible acetylation, *Novartis Found Symp* 259, 208–217; discussion 218–225, 2004.
158. Yeung, F., Hoberg, J.E., Ramsey, C.S. et al., Modulation of NF-κB-dependent transcription and cell survival by the SIRT1 deacetylase, *EMBO J* 23 (12), 2369–2380, 2004.
159. Thevenet, L., Mejean, C., Moniot, B. et al., Regulation of human SRY subcellular distribution by its acetylation/deacetylation, *EMBO J* 23 (16), 3336–3345, 2004.
160. Matsuyama, A., Shimazu, T., Sumida, Y. et al., In vivo destabilization of dynamic microtubules by HDAC6-mediated deacetylation, *EMBO J* 21 (24), 6820–6831, 2002.
161. Zhang, Y., Li, N., Caron, C. et al., HDAC-6 interacts with and deacetylates tubulin and microtubules in vivo, *EMBO J* 22 (5), 1168–1179, 2003.
162. Canettieri, G., Morantte, I., Guzman, E. et al., Attenuation of a phosphorylation-dependent activator by an HDAC-pp1 complex, *Nat Struct Biol* 10 (3), 175–181, 2003.
163. Atadja, P., Hsu, M., Kwon, P. et al., Molecular and cellular basis for the anti-proliferative effects of the HDAC inhibitor LAQ824, *Novartis Found Symp* 259, 249–266; discussion 266–268, 285–288, 2004.

23 Methylation Changes in Neoplasia: Diagnostic and Therapeutic Implications

Eleftheria Hatzimichael, Justin Stebbing, and Tim Crook

CONTENTS

23.1   INTRODUCTION

Many genetic and environmental factors contribute to development of cancer. Traditional approaches to identification of cancer suppressor genes focused on structural changes in genes, including point mutations, gene deletions, and rearrangements and it is now well recognized that acquired mutations in key genes, for example p53 in multiple cancer types and B-raf, most strikingly in malignant melanoma, are important genetic changes in tumorigenesis in many cancer types. Furthermore, the unequivocal association of high penetrance germ-line alleles in genes such as BRCA1 (breast cancer) and p53 (Li Fraumeni syndrome) with cancer clearly confirm the importance of genetic changes in contributing to neoplasia. However, one of the major emerging themes of the last 10 years has been the recognition of the importance in carcinogenesis of downregulation of gene expression through transcriptional silencing. A large volume of experimental evidence now suggests that transcriptional silencing, via epigenetic mechanisms, is a mode of gene inactivation at least as common as the disruption of classic tumor-suppressor genes in human cancer by mutation, and possible more so [1]. For example, although germ-line mutations in BRCA1 are strongly associated with increased predisposition to familial breast cancer, somatically acquired mutations in the gene are exceedingly rare in sporadic breast and ovarian cancers, and the gene is more frequently inactivated by transcriptional silencing.

As described elsewhere in this volume, a simple definition of epigenetics is of changes in gene expression that occur in the absence of alterations in gene sequence or structure. The changes in gene expression are stable and are passed from the parental to daughter cells, i.e., they are heritable but, importantly, potentially reversible. The heritable nature of epigenetic changes implies that the phenotypes they control will be subject to selective pressures in the same way as genetic changes, and affords a mechanistic explanation for their clonal selection in neoplasia. Epigenetic modifications of DNA that influence gene expression include acetylation and phosphorylation of histones and methylation at CpG dinucleotides. Alterations in the patterns of these modifications may result in changes in gene expression. It is interesting and instructive to speculate why epigenetic silencing is a more common mechanism of inactivation than mutation. Insight into this question can be gained by consideration of p53, which is rarely subject to direct epigenetic inactivation in human cancers. Human tumor-associated mutants frequently have a gain of function, thus mutation not only inactivates the wild-type function of the gene but also confers additional "gain-of-function" properties that provide a clear selective advantage to cells expressing the mutant protein and an explanation as to why mutation is so frequent in human cancers. For the majority of other genes that are putative suppressors of cancer, there is no evidence of gain of function mutants and epigenetic silencing is a more frequent mechanism of inactivation than somatic mutation. A second consideration is that epigenetic mechanisms underlying changes in gene expression are, as we have stated above, potentially reversible. This may have important implications in allowing both upregulation of previously silenced genes and downregulation of initially expressed genes, allowing cells to rapidly respond to environmental selective pressures such as exposure to chemotherapeutic agents. Consistent with these observations, there is good experimental evidence that inactivation by epigenetic modification occurs at a far higher rate than mutational inactivation [2].

It is well known that the risk of cancer increases with age. Evidence is also accumulating regarding age-dependent methylation changes that are involved in carcinogenesis in elderly people. It has been proposed that increased global hypermethylation in intestinal crypts may be an early and predisposing event, accounting for the increased risk of colon cancer with advancing age. A pattern of age-related methylation for several genes, including ER, IGF2, N33, and MyoD, has also been shown to progress to full methylation in adenomas and neoplasms [3]. The concept that age-related methylation is a predisposing factor for neoplasia implies that it may serve as a diagnostic risk marker in cancer, and as a novel target for chemoprevention.

Methylation of DNA occurs via addition of methyl groups to C5 of cytosine. This occurs, with very few exceptions, at cytosine residues that precede guanines (CpG), the p simply indicating the existence of a phosphodiester bond between the cytosine and guanine residues. Assuming a random distribution of nucleotides, the probability of a cytosine occurring adjacent to a guanine is 1/16. However, CpG sites are much less common than this (appearing on average only once per 80 dinucleotides) in eukaryotic genomes because of the action of DNA methyltransferase (DNMT). DNMTs are enzymes that recognize CpG sites and methylate them thus generating 5-methylcytosine. 5-Methylcytosine is subject to spontaneous deamination, converting it to thymine. These events have two sequelae. First, the cytosine in isolated CpG dinucleotides is usually methylated. Second, isolated CpG dinucleotides are relatively rare in the genome. There are, however, regions of the genome in which CpG dinucleotides occur at higher frequency and these are called CpG islands. CpG islands are associated with genes, most commonly housekeeping genes, and typically (but not invariably) occur close to the transcriptional start site of the gene. Cytosine residues in CpG islands are usually, but not always, unmethylated in normal cells. In cancer cells, aberrant methylation is associated with transcriptional silencing and this relationship has afforded abundant opportunities for investigators to identify novel cancer suppressor genes.

23.2 IDENTIFICATION AND ANALYSIS OF METHYLATED GENES

The free availability of the human genome and the development of bioinformatics have facilitated and prompted investigators to identify genes potentially subject to methylation-dependent silencing. More than 50% of human genes are associated with CpG islands. Identification of CpG islands that are aberrantly methylated in cancer thus necessitates a more targeted approach than simply randomly analyzing each island and a number of experimental approaches have been described to identify genes regulated by methylation. These are described elsewhere in this volume. The most popular analytical methods rely on the use of bisulfite treatment followed by polymerase chain reaction (PCR). Bisulfite converts unmethylated cytosine residues into uracil but methylated cytosine residues are protected and remain as cytosines. In methylation specific PCR (MSP), the sequence differences generated by bisulfite treatment are sufficient to allow PCR primers to be designed which differentiate between the unmethylated and methylated forms [4]. It is highly sensitive, allowing the detection of <1% methylated alleles and does not require expensive infrastructure. It is also readily amenable to automation [5]. The gold standard technique for methylation analysis remains bisulfite sequencing. Primers are designed to amplify a region of bisulfite modified DNA irrespective of its methylation status and the amplified fragment is then either sequenced directly or cloned and the sequence of individual clones determined. Each individual CpG dinucleotide within the amplified fragment is thereby sampled and even single methylated CpG dinucleotides can be easily detected using this technique.

23.3 METHYLATION-DEPENDENT SILENCING TARGETS GENES FROM MULTIPLE FUNCTIONAL GROUPS

The demonstration of a strong relationship between aberrant CpG methylation in specific transcriptional regulatory elements and absence of expression, together with increasingly user friendly and robust analytical techniques, has encouraged numerous studies of methylation, and methylation-dependent silencing of a large number of genes has been reported in a wide variety of cancer types. A survey of the literature shows that genes silenced in cancer come from many functional classes involved in pathways of cancer development. In Table 23.1, we show a selection of genes in which methylation-dependent silencing may have a clinical phenotype and we now explore in more detail some of the clinical implications and applications of epigenetics in cancer medicine.

23.4 METHYLATION PROFILING IDENTIFIES CLINICAL SUBTYPES OF COMMON CANCERS

On the basis of the simple hypothesis that differences in gene expression must underlie the emergence of different clinical phenotypes, it is not unreasonable to predict that profiling of specific patterns of gene methylation might also help to define distinct clinical subtypes of cancers. Several groups are currently attempting to define the DNA methylation signature (methylotype) of each type of human cancer. For example in T-cell acute lymphoblastic leukemia, a clear prognostic difference was established between patient groups with and without a CpG island methylator phenotype [6]. As we discuss below, analysis of specific genes reveals correlations between aberrant hypermethylation and prognosis in myelodysplastic syndrome. Moreover, the strong association between hypermethylation of specific genes and poor prognosis may provide, at least in part, a mechanistic explanation for the clinical activity of demethylating agents in this disease (see Section 23.9).

TABLE 23.1

Some of the Cancer Phenotypes Affected by Gene Silencing in Human Tumours

Gene	Main Cancer Affected	Reported Effect When Gene Silenced	References
Apaf1	Melanoma	Cytotoxic drug resistance	[71]
ASPP1	Acute leukemia	Loss of pro-apoptotic p53 signaling	[72]
ASS	Mesothelioma	Arginine auxotrophy	[73]
Chfr	Endometrium	Taxane sensitivity	[74]
Dab-1	Pancreas	Metastasis	[75]
E-cadherin	Breast	Metastasis	[76]
ER-β	Breast	Tamoxifen sensitivity	[77]
FANCF	Ovary	Sensitivity to cross-linking agents	[14]
IRF8	Colon	Metastasis	[78]
MASPIN	Breast	Metastasis	[79]
MGMT	Glioma	Temozolomide sensitivity	[14]
hMLH1	Ovarian	Platinum resistance	[10]
Reelin	Pancreatic	Metastasis	[75]
14-3-3σ	Lung	Response to cisplatin/gemcitabine	[80]
hSulf-1	Ovarian	Platinum resistance	[81]
PSAT1	Breast	Endocrine therapy response	[82]
RAR-β	Uterine cervix	Cancer detection	[83]
RASSF1A	Melanoma	Response to biochemotherapy	[84]
RECK	Lung	Metastasis	[85]
WRN	Colorectal	Response to Irinotecan	[86]

23.5 METHYLATION AND CHEMOTHERAPY

The correlation of specific methylation profiles with clinically and phenotypically distinct subtypes of hematological disorders further implies that it may be possible to correlate specific properties of individual cancers with distinct methylation profiles or even methylation within a single gene. Clearly, the ability to predict phenotypes such as metastatic potential from methylation analysis of diagnostic biopsies obtained during routine diagnostic workup would be enormously helpful in clinically triaging patients for management. However, for many oncologists the most exciting prospect is that methylation profiling will have utility in predicting the sensitivity of cancers to specific anticancer agents and thus contribute to individualization of chemotherapy. It is likely that hundreds (at least) of genes are subject to aberrant methylation in any individual cancer. As such, attempting to predict phenotypes (and the effect of therapeutic attempts at methylation reversal) in the face of multiple coexistent epigenetic events is likely to be a complex task. The situation is further complicated because individual epigenetically silenced genes may have opposing effects on phenotypes such as sensitivity/resistance to chemotherapeutic agents and epigenetic changes in these genes may occur simultaneously in a given cancer. There is a significant body of evidence that both germ-line and somatically acquired genetic variants in genes such as p53 mutation may have utility in prediction of de novo response to treatment, i.e., as determinants of innate tumor sensitivity to specific chemotherapeutic agents [7]. However, there is little to support the hypothesis that acquired resistance to cytotoxic chemotherapy results from acquisition of novel mutations during treatment. Rather, the available evidence strongly favors changes in gene expression, associated with dynamic changes in CpG methylation, as an important mechanism by which acquired resistance to anticancer chemotherapy occurs [8]. In vitro studies support this possibility. Changes in DNA methylation within the CpG islands of specific genes are detectable very rapidly after exposure to cytotoxic agents [9]. This observation supports a model in which exposure to cytotoxic drugs causes seeding of methylation and thus acts as an initiating epigenetic event in the evolution of drug resistance. In vivo evidence to support this comes from studies of the DNA mismatch repair gene hMLH1. Inactivation of hMLH1 is associated with increased resistance to cisplatin in vitro. By analysis of methylated DNA in the serum of individuals receiving platinum-based adjuvant chemotherapy for epithelial ovarian cancer, it was shown that levels of methylation changed during chemotherapy. Further, acquired methylation during treatment was associated with poor survival [10]. More recent studies, profiling global changes in the cancer cell methylome, confirm the importance of dynamic epigenetic changes in the evolution of multidrug resistance in cancer cells [11]. A general model of acquired drug resistance would therefore envisage opposing and ongoing hypo- and hypermethylation in multiple genes, with complementary phenotypic effects, as a major mechanism driving the process of drug resistance [11,12]. In other words, natural selection operates via epigenetic changes to confer a survival advantage to cancer cells exposed to chemotherapy. A major task now, therefore, is to identify the critical genes and verify their importance in well-characterized clinical series. To this end, investigators have already identified a number of genes whose epigenetic inactivation is associated with drug resistance and others in which transcriptional silencing results in increased sensitivity to anticancer drugs (Table 23.1). To illustrate this, we now briefly discuss examples of both classes of genes. One gene whose downregulation is associated with drug resistance is a p53 homologue, called p73. p73 is a transcription factor and a tumor-suppressor gene regulating cell cycle and apoptosis. There is good evidence that DNA damage induced signaling through p73 is a key mechanism by which many agents such as the anthracyclines, the platinum compounds, etoposide and taxanes exert cytotoxicity [7]. Silencing of p73, which occurs in a number of (predominantly but not exclusively) hematological neoplasms, particularly acute leukemias and non-Hodgkin's lymphomas, would therefore be predicted to decrease cellular sensitivity to these agents. The DNA mismatch repair gene hMLH1 is a second example of a gene whose methylation-dependent silencing results in relative drug resistance as discussed above [10]. In contrast, methylation-dependent silencing of other genes results in increased sensitivity to anticancer agents. Such genes are often cell cycle checkpoint genes or involved in DNA repair pathways. It is a very attractive model that loss of expression of such genes, with abrogation of their function in maintaining genomic fidelity and stability, occurs as important events in generation of genetic diversity during tumor development. However, because these genes have critical functions in pathways of DNA damage response, their loss sensitizes cells to DNA damage. One interesting example is the mitotic checkpoint Chfr (checkpoint with forkhead and ring finger domain). This gene is subject to methylation-dependent silencing in a number of human cancers and cells lacking Chfr have increased sensitivity to microtubule acting agents such as the taxanes [13]. Another example of how methylation of certain genes can predict response to treatment is in patients with gliomas. In this case, the methylation-associated silencing of the DNA repair gene MGMT indicates which patients are going to be sensitive to carmustine (BCNU). The relationship between clinical response and epigenetic change is due to the fact that MGMT protein is responsible for the removal of alkyl groups from, which is the preferred point of DNA attack of several alkylating agents. Therefore, tumors, which lack MGMT function, are more sensitive to the action of these drugs since the alkylation damage will persist and the cancer cell will die [14,15]. The Fanconi anemia/BRCA pathway is critical in repair of lesions induced in DNA by cross-linking agents. Methylation-dependent silencing of one component gene in the pathway, FANCF, was reported in cisplatin-sensitive ovarian carcinoma cell lines [16]. Pharmacological reversal of methylation increased resistance to cisplatin in these cell lines. Despite these caveats of complexity and opposing effects, there is emerging evidence that some epigenetic events have such a strong effect that single gene phenotype: epigenotype correlations can be established (for example the hMLH1 gene described above) in ovarian cancer [10]. We now extend our discussion to encompass situations in which understanding of epigenetics is already having a direct impact on clinical practice.

23.6 CLINICAL IMPACT OF EPIGENETICS

Thus far, we have considered the basic science of epigenetics together with some examples of how epigenetic changes might contribute to clinically important phenotypes. We now consider how an understanding of epigenetic changes has led to its use in bench to bedside diagnostic and therapeutic applications.

23.6.1 DIAGNOSTICS

As we have discussed above, there is now evidence that different subtypes of some tumors have distinct methylation patterns or profiles. This has been successfully applied in lung cancer subtypes [17], leukemias [18], and lymphomas [19] and in predicting the response to endocrine therapy in patients with recurrent breast cancer [20]. Epigenomics (www.epigenomics.com) have collaborated with Roche in an attempt to take these tests further with a number of prospective randomized studies, designed to investigate the utility of methylation testing.

In hormone receptor positive, node-negative breast cancer, many patients receive chemo-endocrine therapy although endocrine therapy with tamoxifen alone would have been sufficient in view of their excellent prognosis. In a microarray study, PITX2 methylation correlated strongly with the risk of recurrence after adjuvant tamoxifen in a small and larger cohort (Figure 23.1). PITX2, paired-like homeodomain transcription factor 2, is involved in the transcription of pituitary-specific genes and hypermethylation has only been previously described in one study in AML [21].

A large multicenter study initiated to validate PITX2 methylation as an outcome predictor for adjuvant tamoxifen using paraffin-embedded tumor tissue using a real-time PCR assay was developed to test PITX2 methylation in paraffin-embedded tissue. Matched frozen and embedded samples ($n = 89$) were analyzed and compared with paraffin-embedded tumors of 422 node-negative patients from nine clinical centers treated with tamoxifen alone. In the independent cohort, PITX2 methylation was strongly correlated with outcome (Cox proportional hazard model, $p = 0.025$). In the group with low PITX2 methylation (45% of the cohort), 98% of the patients were metastasis-free after 10 years, compared to only 85% in the group with high PITX2 methylation. In a multivariate model, PITX2 methylation added significant information to conventional factors such as tumor size, grade, and age [20].

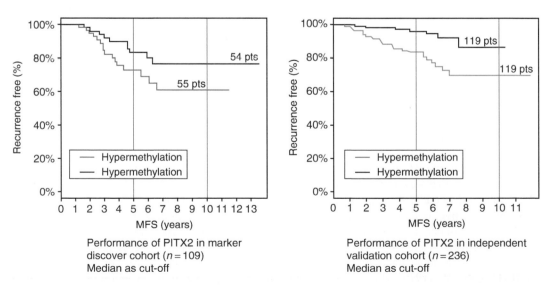

FIGURE 23.1 Performance of PITX2 in predicting recurrence from breast cancer in two cohorts using a 117 gene microarray.

However, testing in a prospective randomized manner on a larger sample of paraffin-embedded tissue from Berlin failed to identify patients who were more likely to respond. The use of candidate methylation sequences has also failed to precisely identify early prostate specific antigen relapse post prostatectomy. These data are yet to be presented.

More encouragingly, recent data from 1,400 individuals indicated that using a standard blood draw and at a set specificity of 95%, the lead undisclosed marker for colon cancer demonstrated sensitivity values of 51%, 65%, and 50%, respectively in three independent clinical studies. Importantly, early-stage cancers were identified with the same sensitivity as later stage cancers, proving the utility of this approach for general population screening. Results were obtained measuring only a single DNA methylation "anchor" marker. In addition, the test was able to detect colorectal cancers regardless of their location, addressing a critical medical need and shortfall of the existing fecal occult blood tests (FOBT), currently the most widely used screening test.

23.7   HYPERMETHYLATION AND DETECTION OF MINIMAL RESIDUAL DISEASE

Hypermethylation may also serve as a molecular disease marker for the detection of minimal residual disease (MRD) in hematological malignancy. By MSP, the sensitivity of methylated MSP for the detection of aberrant promoter methylation ranges from 10^{-3} to 10^{-5} [22–24] depending on the tumor burden in the initial sample. As a test for MRD in leukemia, the sensitivity of MSP is intermediate between two other strategies, based on the PCR of rearranged immunoglobulin/T-cell receptor gene [25–27] and RT-PCR for leukemia-specific fusion transcripts [28]. As a result, hypermethylation may serve as a disease marker, especially in diseases without reciprocal translocation/fusion genes available for RT-PCR. Moreover, as methylated DNA can be amplified from the patients' serum [29], monitoring MRD could be conveniently achieved by quantitative MSP of serum.

23.8   HYPERMETHYLATION AND PROGNOSIS

It is well known that cytogenetic abnormalities are one of the most important prognostic factors in hematological malignancies. However, karyotyping is relatively time-consuming and tedious, and has been difficult to incorporate into prospective clinical trials. As methylation affects genes that are critical either to the initiation or progression of cancer, it might be of prognostic significance.

For instance, $p15^{INK4B}$ gene methylation has been shown to mark disease progression in myelodysplastic syndromes (MDS) [30] and acute myelogenous leukemia (AML) [31,32] and has also been associated with high-grade lymphoma and an unfavorable prognosis [33]. Silencing by methylation of cadherin-13 has been found in patients with CML and was found to correlate with pretreatment risk profile and cytogenetic response to interferon [34]. One study showed a trend of superior survival (OS) in both AML and acute lymphoblastic leukemia (ALL) patients without $p15^{INK4B}$ methylation [33]. In patients with acute promyelocytic leukemia (APL), Chim et al. [25] proposed that hypermethylation of the $p15^{INK4B}$ gene is a new molecular marker for disease-free survival (DFS). Therefore, $p15^{INK4B}$ methylation at diagnosis might indicate a poor prognosis in APL, in which prognostic factors are less well-defined than other AML subtypes [35]. Another study showed that methylation of the ER gene in AML was associated with superior overall survival [36]. In a cohort of childhood and adult ALL patients, methylation of the calcitonin gene was found to be an independent prognostic factor in predicting DFS [37]. In addition, aberrant gene methylation of putative tumor-suppressor gene (TSG) have been shown to be associated with invasive disease in melanoma [38] and confer poor prognosis to soft tissue sarcoma [39]. Large prospective studies, of course, are needed in order to further validate these results. If validated though, gene promoter hypermethylation may be incorporated as a prognostic factor to optimize patient management.

23.9 THERAPEUTICS

To overcome cytosine arabinoside (ara-C) resistance, a series of cytidine analogues resistant to deamination were synthesized that did not require activation by deoxycytidine kinase [40]. Cytidine analogues, i.e., 5-azacytidine, 5-aza-29-deoxycytidine (decitabine), pseudoisocytidine, and 59-fluoro-29-deoxycytidine, are potent inhibitors of DNA methylation. Of these substances, decitabine has the most potent hypomethylating effect [41]; it can demethylate 55% of DNA at concentrations 10-fold lower than those of 5-azacytidine. Dihydro-5-azacytidine [42] has been introduced into clinical trials. Arabinofuranosyl-5-azacytosine (fazarabine) showed potent antitumor activity in preclinical models but not in clinical trials [43]. The hypomethylating effect of cytidine analogues appears to depend primarily on the presence of an altered C5 position; other cytidine analogues, such as ara-C, 6-azacytidine, and gemcitabine, do not possess this property.

It is important to note that much work remains to determine whether the beneficial effects of 5-azacytidine and decitabine, as antineoplastic agents, are being mediated through DNA demethylation alone or cytotoxicity as well. Cytotoxicity results from incorporation into RNA and DNA and DNA hypomethylation through inhibition of DNMT, as mentioned above. Induction of DNA hypomethylation though appears to require lower azacytidine doses than does cytotoxicity, as the concentration of azacytidine required for maximum inhibition of DNA methylation in vitro does not suppress DNA synthesis [44]. In addition, in one trial that MDS patients responding to decitabine appeared to have reversal of p15^{INK4b} hypermethylation during treatment and reestablishment of the protein expression in the bone marrow, suggesting that the clinical efficacy may in fact be by the reversal of gene silencing [45].

Azacytidine (Vidaza, Pharmion, Boulder, Colorado) and decitadine (Dacogentrade; SuperGen Inc, Dublin, California and MGI Pharma Inc, Bloomington, Minnesota), have had a significant impact on the treatment paradigm of MDS [46–48], previously managed mainly by supportive care and hematopoietic-stem-cell transplantation. The use of DNA methylation technologies has also been studied as a pharmacodiagnostic product (Epigenomics, Berlin), comparing patients' cells versus healthy and sick reference samples to enable earlier diagnosis of disease, and accurate prognostic and predictive information [49].

Irreversible hypomethylating agents have been successfully trialed in MDS, a heterogeneous hematopoietic disorder which is characterized by refractory multilineage cytopenias that result in dependence on transfusion, excessive risk of infection or hemorrhage, and heightened potential for transformation to AML. Recent evidence demonstrates that epigenetic silencing of genes is associated with myelodysplasia, that CpG island methylation is a poor prognostic factor and that a worse prognosis correlates with hypermethylation of the cyclin-dependent kinase inhibitor p15^{INK4b}. The disease is characterized by ineffective hematopoiesis, which is due to a complex interaction between hematopoietic progenitors and the microenvironment, resulting in premature apoptotic death of progenitors and their maturing progeny. Refractory anemia (RA) with or without ringed sideroblasts can persist for years, but RA with excess blasts (RAEBs) or RAEBs in transformation to leukemia (RAEB-T) exhibit an accelerated course. In 35%–40% of patients, transformation to acute leukemia occurs, which is particularly refractory to present therapy [50,51]. Four previous randomized studies in MDS, involving cis-retinoic acid, low-dose cytaradine, Filgastrim and Sargramostim have shown evidence of hematologic responses but no differences in time to progression, frequency, or transformation to leukemia.

23.9.1 Azacytidine

5-Azacytidine is currently approved for the treatment of patients in low and high-risk subtypes of MDS following results of the Cancer and Leukemia Group B (9221) study [52]. Here, a randomized controlled trial was undertaken in 191 patients with MDS to compare azacytidine (75 mg/m^2/d subcutaneously for 7 days every 28 days) with supportive care. MDS was defined by FAB criteria.

Although new rigorous response criteria were applied, there were many criticisms of the trial including the fact that patients in the supportive care arm whose disease worsened were permitted to cross over to azacytidine [53].

In the azacytidine arm, responses occurred in 60% of patients (7% complete response, 16% partial response, 37% improved) compared with 5% (improved) receiving supportive care ($p < 0.001$). Median time to leukemic transformation or death was 21 months for azacytidine versus 13 months for supportive care ($p = 0.007$). Transformation to AML occurred as the first event in 15% of patients on the drug arm and in 38% receiving supportive care ($p = 0.001$). Eliminating the confounding effect of early crossover to azacytidine, a landmark analysis after 6 months showed median survival of an additional 18 months for azacytidine and 11 months for supportive care ($p = 0.03$), though on an intent to treat basis, overall survival was not statistically significant. Quality-of-life assessment found significant major advantages in physical function, symptoms, and psychological state for patients initially randomized to the hypomethylating agent.

Most patients demonstrate a response beginning in the third of fourth month, consistent with possible dual mechanisms of action: a low-dose cytotoxic effect and the drug acting as a biological response modifier. The hypomethylating effect is S-phase dependent, and two or more cycles of DNA synthesis are required to alter gene transcription and expression [54,55]. Thus, repetitive exposure on the present low-dose intermittent schedule may have affected small numbers of cells during each treatment, requiring three to four cycles before the effects became clinically apparent. Recent data has demonstrated that alterations in methylation in the CpG island of the p15^{INK4b} gene are implicated in the transformation of MDS to AML, emphasizing the existence of dynamic changes in the methylation status of specific genes during disease progression. These changes can be modulated by azacytidine, thus reducing risk of leukemic transformation [56]. Specifically, azacytidine changes myeloid progenitor cell expansion and differentiation by upregulating p15^{INK4b} protein expression, a gene in which promoter hypermethylation identifies a subgroup of early MDS with a particularly poor prognosis [57,58]. In solid tumors, few clinical trials with small patient numbers have been performed so far and results are not that encouraging. In a trial involving of 58 patients with metastatic breast cancer, 2/27 patients receiving azacytidine and 3/27 patients receiving lomustine responded to treatment [59]. Azacytidine was also given at a dose of 1.6 mg/kg per day for 10 days in 170 patients with solid tumors and antitumor activity was reported in 17% of patients with breast cancer, while in 21% of patients with lymphoma "clinical improvement" was noted [60]. Common side effects were nausea, vomiting, diarrhea, and myelosuppression.

The above mentioned results are particularly exciting because they show that in certain malignancies mainly hematological, epigenetic changes associated with disease progression can be pharmacologically reversed, using existing agents, with sufficient specificity to derive useful therapeutic effects.

23.9.2 DECITABINE

First made in 1964, 5-aza-2′-deoxycytidine (decitabine) is currently awaiting approval by the U.S. Food and Drug Administration after submission in 2004 of a new drug application [61]. This is a prodrug that requires activation via phosphorylation by deoxycytidine kinase. Like azacytidine, it is an S-phase specific agent and the nucleotide analogue is incorporated into DNA, where it produces an irreversible inactivation of DNMT. The drug is given intravenously, not subcutaneously where problems with azacytidine's solubility have led to painful injection site issues. The use of decitabine in MDS, leukemias—both AML and CML, stem cell transplants, sickle cell anemia, and thalassemia appears promising with an epigenetic dose that is lower than a cytotoxic dose.

During the last 10 years, three European phase II studies were performed to investigate the treatment of elderly patients with MDS with low-dose decitabine [62]. All these European trial data were reviewed on the basis of the International Prognostic Scoring System (IPSS) risk criteria and the response criteria as recently published by an international working group. To investigate the results in a larger cohort of patients and to determine risk factors, all data were pooled with some observations from the PCH 95–06 U.S. phase II study. The response rate in the 177 patients evaluated (median age 70 years) was 49%. The median response duration was 36 weeks and the median survival was 15 months. Analysis of the data according to sex, age, FAB classification, percentage of blasts in the bone marrow, IPSS risk group, lactate dehydrogenase, and cytogenetics did not reveal any factor predictive of response. Overall, 69% of patients benefited including those with stable disease during therapy. Response duration was significantly shorter with increasing risk (according to the IPSS classification). Hemoglobin level and neutrophil count showed an inverse correlation to the IPSS classification. Univariate analysis showed a significantly inferior survival for elderly patients (>75 years of age) and for those with high levels of serum lactate dehydrogenase (LDH) (more than two times the normal values). Patients with high-risk cytogenetic abnormalities according to the IPSS risk criteria showed better overall survival than those with intermediate-risk abnormalities. When analyzed according to the IPSS risk classification, high-risk patients had worse survival prospects following decitabine therapy than those with intermediate risk; however, compared to the originally reported IPPS outcomes for high-risk patients, they probably showed better survival. During the treatment period, 18% of the patients progressed toward acute leukemia.

A phase III study has yet to be published though interim results of a study designed to compare three dosing regimens were presented at the American Society of Hematology meeting December 2005 (Oral abstracts 371, 408, 495, 525, and 790). In this study, patients with intermediate-1, intermediate-2, and high-risk MDS were randomized to receive one of the three decitabine regimens every 4 weeks: (1) 20 mg/m^2 intravenous one hour infusion once per day for 5 days; (2) 10 mg/m^2 1 h intravenous infusion once per day for 10 days; or (3) 10 mg/m^2 subcutaneous injection twice per day for 5 days. Randomization of patients to each of the three dosing regimens was equal for the first 50 patients enrolled in this study. After the 50th patient was enrolled, a Bayesian randomization was implemented based on complete response rates, and all additional study participants were treated using the 5 day 20 mg/m^2 intravenous infusion regimen.

For 96 evaluable patients, the overall response rate was 47%, including a 42% complete response rate and a 5% partial response rate. In the 65 patients treated with a 20 mg/m^2 intravenous decitabine infusion once per day for 5 days, the complete response rate was 49%.

As for azacytidine, the most frequently observed adverse events were primarily a result of myelosuppression and included fever (5%) and infection (10%). On the basis of standard CALGB criteria, grade 3 or 4 leukopenia occurred in approximately 60%, granulocytopenia in 80% and thrombocytopenia in 70% of patients receiving these hypomethylating agents. The mean number of red blood cell transfusions appears to increase during the first month of treatment and decline thereafter.

These trials highlight whether improvement in survival should always be the ultimate treatment end point, or whether other surrogate end points for patient benefit can be used. These may include time to transformation; high-quality responses which are consistently associated with improved survival, such as major cytogenetic response in chronic myelogenous leukemia or molecular complete responses in APL or other tumors; significant and durable responses from treatments with negligible side effects or extremely low mortality rates; or significant objective improvements in quality-of-life measures. These surrogate end points have been accepted by the FDA for the approval of several drugs. Future surrogate end points may include modulation of cancer target signals for particular target-specific strategies.

23.9.3 ZEBULARINE

Zebularine [1-(beta-D-ribofuranosyl)-1,2-dihydro-pyrimidin-2-one] is another cytidine analogue that has recently been developed. The most interesting feature of this drug, compared to azacytidine and decitabine, is that it is chemically stable and of low toxicity and also it is the first demethylating agent given orally. Although initially it was designed as a cytidine deaminase inhibitor, it has been shown that it induces reactivation of hypermethylated genes in yeast models, of p16 in bladder cancer cells [63], and also that is effective against the development of a murine T-cell lymphoma [64].

An obvious caveat to the use of all the current demethylating agents is the lack of specificity of these drugs, with the implication that these agents cause global hypomethylation and cannot reactivate specific tumor-suppressor genes. Therefore, it is not known whether treatment with demethylating agents disrupts essential methylation at certain sites, causing inappropriate inactivation of genes in normal cells. Although important, this might not be an insurmountable problem, and this is because in cases where methylation is important, such as X-chromosome inactivation in mammals, the existence of many levels of gene silencing [65] suggests that these genes cannot be easily activated by DNA methylation inhibitors.

Mouse-model studies have been inconclusive: in one study DNA hypomethylation was shown to promote chromosomal instability [66], whereas earlier studies showed that DNA hypomethylation had an adenoma protective effect in mice prone to colon cancer [67]. These latter findings, however, have not been observed in humans [68,69] in whom a completely different pattern is observed. Although these concerns need to be addressed, patients who have received demethylating agents for nonmalignant and malignant diseases have not shown massive toxicity due to inappropriate gene activation.

The aim of reverting to a methylator phenotype remains unchanged though and thus far, unidentified hypermethylated tumor-suppressor genes may also be the targets for demethylating agents. Moreover, it is worth remembering that DNA methylation is one of the several levels involved in epigenetic silencing. Histone deacetylation is another mechanism of gene silencing and it has been shown that the combination of demethylating agents with histone deacetylase inhibitors causes a synergism in the reactivation of hypermethylated tumor-suppressor genes [70]. These in vitro observations have prompted the development of clinical trial testing their efficacy in vivo.

23.10   CONCLUSION

The human epigenome project and an understanding of methylation at the clinical level increases our ability to prospectively identify patients at risk for severe toxicity, or those likely to benefit from a particular therapy and thus promises to help us move toward the often talked about goal of individualized treatment. Thus far, few genetic markers are used routinely to predict clinical efficacy and toxicity despite the fact that physicians and their patients are consistently confronted with this balance. As one of the goals of methylation-based pharmaco-epigenomics is to identify individuals and target populations that may have adverse outcomes, pharmaceutical companies have been reluctant to embrace a strategy that may select patients who are not eligible for a particular treatment. This environment is changing, however, with targeted drug discovery programmes and treatments of specific pathways, with the consequent improvements in surrogate and survival end-points. A "methylome map" will likely have significant utility in this area, identifying new drugs and their targets, as well as enabling earlier diagnosis, decreasing systemic toxicity, and increasing efficacy. While this clearly offers the potential of developing DNA-based tests to help maximize drug efficacy and enhance drug safety, it is prerequisite on the predictive ability of such tests to be established by trials with prespecified end-points. This has particular implications in the current health care environment, where cost containment and evidence-based initiatives are having a significant influence on patient care.

REFERENCES

1. Jones PA and Baylin SB. The fundamental role of epigenetic events in cancer. *Nature Rev Genet* 2002; 3: 415–428.
2. Bhattacharya NP, Skandalis A, Ganesh A et al. Mutator phenotype in human colorectal carcinoma cell lines. *Proc Natl Acad Sci U S A* 1994; 91: 6319–6323.

3. Rushid A and Issa JP. CpG island methylation in gastroenterologic neoplasia: A maturing field. *Gastroenterology* 2004; 127: 1578–1588

4. Herman JG, Graff JR, Myohanen S et al. Methylation-specific PCR: A novel PCR assay for methylation status of CpG islands. *Proc Natl Acad Sci U S A* 1996; 93: 9821–9826.

5. Eads CA, Danenberg KD, Kawakami K et al. MethyLight: A high-throughput assay to measure DNA methylation. *Nucl Acid Res* 2000; 28: E32.

6. Roman-Gomez J, Jimenez-Velasco A, Agirre X et al. Lack of CpG island methylator phenotype defines a clinical subtype of T-cell acute lymphoblastic leukaemia associated with good prognosis. *J Clin Onc* 2005; 23: 7043–7049.

7. Bergamaschi D, Gasco M, Hiller L et al. p53 polymorphism influences response in cancer chemotherapy via modulation of p73-dependent apoptosis. *Cancer Cell* 2003; 3: 387–402.

8. Teodoridis JM, Strathdee G, and Brown R. Epigenetic silencing mediated by CpG island methylation: Potential as a therapeutic target and as a biomarker. *Drug Resist Update* 2004; 7: 267–278.

9. Bredberg A and Bodmer W. Cytostatic drug treatment causes seeding of gene promoter methylation. *Eur J Cancer.* 2007; 43: 947–954.

10. Gifford G, Paul J, Vasey PA et al. The acquisition of hMLH1 methylation in plasma DNA after chemotherapy predicts poor survival for ovarian cancer patients. *Clin Cancer Res* 2004; 10: 4420–4426.

11. Chekhun VF, Lukyanova NY, Kovalchuk O et al. Epigenetic profiling of multi-drug resistant MCF-7 breast adenocarcinomas cells reveals novel hyper-and hypomethylated targets. *Mol Cancer Ther* 2007; 6: 1089–1098.

12. Glasspool RM, Teodoridis JM, and Brown R. Epigenetics as a mechanism driving polygenic clinical drug resistance. *Br J Cancer* 2006; 94: 1087–1092.

13. Scolnick DM and Halazonetis TD. Chfr defines a mitotic stress checkpoint that delays entry into metaphase. *Nature* 2000; 406: 430–445.

14. Esteller M, Garcia-Foncillas J, Andion E, et al. Activity of the DNA repair gene MGMT and the clinical response of gliomas to alkylating agents. *New Engl J Med* 2000; 343: 1350–1354.

15. Esteller M, Hamilton SR, Burger PC, et al. Inactivation of the DNA repair gene O^6-methylguanine DNA methyltransferase by promoter hypermethylation is a common event in primary human neoplasia. *Cancer Res* 1999; 59: 793–797.

16. Taniguchi T, Tischkowitz M, Ameziane N et al. Disruption of the Fanconi anemia-BRCA pathway in cisplatin-sensitive ovarian tumors. *Nat Med* 2003; 9: 568–574.

17. Field JK, Liloglou T, Warrak S et al. Methylation discriminators in NSCLC identified by a microarray based approach. *Int J Oncol* 2005; 27: 105–111.

18. Scholz C, Nimmrich I, Burger M et al. Distinction of acute lymphoblastic leukemia from acute myeloid leukemia through microarray-based DNA methylation analysis. *Ann Hematol* 2005; 84: 236–244.

19. Guo J, Burger M, Nimmrich I, Maier S et al. Differential DNA methylation of gene promoters in small B-cell lymphomas. *Am J Clin Pathol* 2005; 124: 430–439.

20. Harbeck N, Bohlmann I, Ross J et al. Multicenter study validates PITX2 DNA methylation for risk prediction in tamoxifen-treated node negative breast cancer using paraffin-embedded tumor tissue. ASCO Annual Meeting Proceedings. *J Clin Oncol.* 2005; 23(16S): 505.

21. Toyota M, Kopecky KJ, Toyota MO et al. Methylation profiling in acute myeloid leukemia. *Blood* 2001; 97: 2823–2829.

22. Herman JG, Civin CI, Issa JP et al. Distinct patterns of inactivation of p15INK4B and p16INK4A characterize the major types of hematological malignancies. *Cancer Res* 1997; 57: 837–841.

23. Chim CS, Liang R, Tam CYY et al. Methylation of p15 and p16 genes in acute promyelocytic leukemia: Diagnostic and prognostic significance. *J Clin Oncol* 2001; 19: 2033–2040.

24. Chim CS, Tam CYY, Liang R et al. Methylation of p15 and p16 genes in adult acute leukemia: Lack of prognostic significance. *Cancer* 2001; 91: 2222–2229.

25. Chim JC, Coyle LA, Yaxley JC et al. Investigation of minimal residual disease by the use of a semiquantitative multi-step test combination in ALL. *Br J Haematol* 1996; 92: 104–115.

26. Foroni L, Coyle LA, Papioannou M et al. Molecular detection of minimal residual disease in adult and childhood ALL reveals differences in treatment responses. *Leukemia* 1997; 11: 1732–1741.

27. Campana D and Pui CH. Detection of minimal residual disease in acute leukemia: Methodological advances and clinical significance. *Blood* 1995; 85: 1416–1434.

28. Wong IHN, Ng MHL, Huang DP et al. Aberrant p15 promoter methylation in adult and childhood acute leukemias of nearly all morphologic subtypes: Potential prognostic implications. *Blood* 2000; 95: 1942–1949.

29. Sanchez-Beato M, Sanchez-Aguilera A, and Piris MA. Cell cycle deregulation in B-cell lymphomas. *Blood* 2003; 101: 1220–1235.

30. Tien HF, Tang JH, Tsay W et al. Methylation of the p15(INK4B) gene in myelodysplastic syndrome: It can be detected early at diagnosis or during disease progression and is highly associated with leukaemic transformation. *Br J Haematol* 2001; 112: 148–154.

31. Shimamoto T, Ohyashiki JH, and Obyashiki K. Methylation of p15(INK4b) and E-cadherin genes is independently correlated with poor prognosis in acute myeloid leukemia. *Leuk Res* 2005; 29: 653–659.

32. Christiansen DH, Andersen MK, and Pedersen-Bjergaard J. Methylation of p15INK4B is common, is associated with deletion of genes on chromosome arm 7q and predicts poor prognosis in therapy-related myelodysplasia and acute myeloid leukemia. *Leukemia* 2003; 17: 1813–1819.

33. Herman JG, Civin CI, Issa JP et al. Distinct patterns of inactivation of p15INK4B and p16INK4A characterize the major types of hematological malignancies. *Cancer Res* 1997; 57: 837–841.

34. Roman-Gomez J, Castillejo JA, Jimenez A et al. Cadherin-13, a mediator of calcium-dependent cell-cell adhesion, is silenced by methylation in chronic myeloid leukemia and correlates with pretreatment risk profile and cytogenetic response to interferon alfa. *J Clin Oncol* 2003; 21: 1472–1479.

35. Asou N, Adachi K, Tamura J et al. Analysis of prognostic factors in newly diagnosed acute promyelocytic leukemia treated with all-trans retinoic acid and chemotherapy. Japan adult leukemia study group. *J Clin Oncol* 1998; 16: 78–85.

36. Li Q, Kopecky KJ, Mohan A et al. Estrogen receptor methylation is associated with improved survival in adult acute myeloid leukemia. *Clin Cancer Res* 1999; 5: 1077–1084.

37. Roman J, Castillejo JA, Jimenez A et al. Hypermethylation of the calcitonin gene in acute lymphoblastic leukaemia is associated with unfavourable clinical outcome. *Br J Haematol* 2001; 113: 329–338.

38. Reed JA, Loganzo F, Shea CR et al. Loss of expression of the p16/cyclin-dependent kinase inhibitor 2 tumor suppressor gene in melanocytic lesions correlates with invasive stage of tumor progression. *Cancer Res* 1995; 55: 2713–2718.

39. Orlow I, Drobnjak M, Zhang ZF et al. Alterations of INK4A and INK4B genes in adult soft tissue sarcomas: Effect on survival. *J Natl Cancer Inst* 1999; 91: 73–79.

40. Momparler RL. 5-Aza-29-deoxycytidine: An overview. Momparler PL and de Vos D, (Eds.) 5- In: *Aza-29-Deoxycytidine: Preclinical and Clinical Studies*. Haarlem, the Netherlands: PCH Publications; 1990: 9–15.

41. Creusot F, Acs G, and Christman JK. Inhibition of DNA methyltransferase and induction of Friend erythroleukemia cell differentiation by 5-azacytidine and 5-aza-29-deoxycytidine. *J Biol Chem* 1982; 257: 2041–2048.

42. Powell WC and Avramis VI. Biochemical pharmacology of 5,6-dihydro-5-azacytidine (DHAC) and DNA hypomethylation in tumor (L1210)-bearing mice. *Cancer Chemother Pharmacol.* 1988; 21: 117–121.

43. Goldberg RM, Reid JM, Ames MM et al. Phase I and pharmacological trial of fazarabine (Ara-AC) with granulocyte colony-stimulating factor. *Clin Cancer Res* 1997; 3: 2363–2370.

44. Glover AB and Leyland-Jones B. Biochemistry of azacytidine: A review. *Cancer Treat Rep* 1987; 71: 959–964.

45. Daskalakis M, Nguyen TT, Nguyen C, Guldbery P, Kohler G, Wijermaus P et al. Demethylation of a hypermethylated p15/INK4B gene in patients with myelodysplastic syndrome by 5-Aza-2'-deoxycitidine (decitabine) treatment. *Blood* 2002; 100: 2957–2969.

46. Gore SD. Combination therapy with DNA methyltransferase inhibitors in hematologic malignancies. *Nat Clin Pract Oncol* 2005; 2 Suppl 1: S30–35.

47. Baylin SB. DNA methylation and gene silencing in cancer. *Nat Clin Pract Oncol* 2005; 2 Suppl 1: S4–11.

48. Fenaux P. Inhibitors of DNA methylation: Beyond myelodysplastic syndromes. *Nat Clin Pract Oncol* 2005; 2 Suppl 1: S36–44.

49. Novik KL, Nimmrich I, Genc B et al. Epigenomics: Genome-wide study of methylation phenomena. *Curr Issues Mol Biol* 2002; 4: 111–128.

50. Economopoulos T, Stathakis N, Foudoulakis A et al. Myelodysplastic syndromes: Analysis of 131 cases according to the FAB classification. *Eur J Haematol* 1987; 38: 338–344.

51. Kerkhofs H, Hermans J, Haak HL et al. Utility of the FAB classification for myelodysplastic syndromes: Investigation of prognostic factors in 237 cases. *Br J Haematol* 1987; 65: 73–81.

52. Silverman LR, Demakos EP, Peterson BL et al. Randomized controlled trial of azacytidine in patients with the myelodysplastic syndrome: A study of the cancer and leukemia group B. *J Clin Oncol* 2002; 20: 2429–2440.

53. Kantarjian HM. Treatment of myelodysplastic syndrome: Questions raised by the azacitidine experience. *J Clin Oncol* 2002; 20: 2415–2416.

54. Christman JK, Mendelsohn N, Herzog D et al. Effect of 5-azacytidine on differentiation and DNA methylation in human promyelocytic leukemia cells (HL-60). *Cancer Res* 1983; 43: 763–769.

55. Razin A and Riggs AD. DNA methylation and gene function. *Science* 1980; 210: 604–610.

56. Aggerholm A, Holm MS, Guldberg P et al. Promoter hypermethylation of p15, HIC1, CDH1, and ER is frequent in myelodysplastic syndrome and predicts poor prognosis in early-stage patients. *Eur J Haematol* 2006; 76: 23–32.

57. Hackanson B, Guo Y, and Lubbert M. The silence of the genes: Epigenetic disturbances in haematopoietic malignancies. *Expert Opin Ther Targets* 2005; 9: 45–61.

58. Guo Y, Engelhardt M, Wider D et al. Effects of 5-aza-2′-deoxycytidine on proliferation, differentiation and p15/INK4b regulation of human hematopoietic progenitor cells. *Leukemia* 2006; 20: 115–121.

59. Cunningham TJ, Nemoto T, Rosner D et al. Comparison of 5-azacytidine (NSC-102816) with CCNU (NSC- 79037) in the treatment of patients with breast cancer and evaluation of the subsequent use of cyclophosphamide (NSC-26271). *Cancer Chemother Rep* 1974; 58: 677–681.

60. Weiss AJ, Metter GE, Nealon TF et al. Phase II study of 5-azacytidine in solid tumors. *Cancer Treat Rep* 1977; 61: 55–58.

61. de Vos D. Epigenetic drugs: A longstanding story. *Semin Oncol* 2005; 32: 437–442.

62. Wijermans PW, Lubbert M, Verhoef G et al. An epigenetic approach to the treatment of advanced MDS; the experience with the DNA demethylating agent 5-aza-2′-deoxycytidine (decitabine) in 177 patients. *Ann Hematol* 2005; 84 Suppl 13: 9–17.

63. Cheng JC, Matsen CB, Gonzales FA et al. Inhibition of DNA methylation and reactivation of silenced genes by zebularine. *J Natl Cancer Inst* 2003; 95: 399–409.

64. Herranz M, Martin-Caballero J, Fraga MF et al. The novel DNA methylation inhibitor zebularine is effective against the development of murine T-cell lymphoma. *Blood* 2006; 107: 1174–1177.

65. Bird A. DNA methylation patterns and epigenetic memory. *Genes Dev* 2002; 16: 6–21.

66. Eden A.Gaudet F, Waghmare et al. Chromosomal instability and tumors promoted by DNA hypomethylation. *Science* 2003; 300: 455.

67. Laird PW, Jackson-Grusby L, Fazeli A et al. Suppression of intestinal neoplasia by DNA hypomethylation. *Cell* 1995; 81: 197–205.

68. Yang AS and Estecio MR, Garcia-Manero M et al. Comment on chromosomal instability and tumors promoted by DNA hypomethylation and induction of tumors in mice by genomic hypomethylation. *Science* 2003; 302: 1153.

69. Lubbert M, Wijermans P, Kunzmann R et al. Cytogenetic responses in high risk myelodysplastic syndrome following low-dose treatment with the DNA methylation inhibitor 5-aza-2′-deoxycitidine. *Br J Haematol:* 2001; 114: 349–357.

70. Cameron EE, Bachman KE, Myohanen S et al. Synergy of demethylation and histone deacetylase inhibition in the re-expression of genes silenced in cancer. *Nat Genet* 1999; 21: 103–107.

71. Soengas MS, Capodieci P, Polsky D et al. Inactivation of the apoptosis effector Apaf-1 in malignant melanoma. *Nature* 2001; 409: 207–211.

72. Agirre X, Roman-Gomez J, Jimenez-Velasco A et al. ASPP1, a common activator of TP53, is inactivated by aberrant methylation of its promoter in acute lymphoblastic leukemia. *Oncogene* 2006; 25: 1862–1870.

73. Szlosarek PW, Klabatsa A, Pallaska A et al. In vivo loss of expression of argininosuccinate synthetase in malignant pleural mesothelioma is a biomarker for susceptibility to arginine depletion. *Clin Cancer Res* 2006; 12: 7126–7131.

74. Yanokura M, Banno K, Kawaguchi M et al. Relationship of aberrant DNA hypermethylation of CHFR with sensitivity to taxanes in endometrial cancer. *Oncol Rep* 2007; 17: 41–48.

75. Sato N, Fukushima N, Chang R et al. Differential and epigenetic gene expression profiling identifies frequent disruption of the RELN pathway in pancreatic cancers. *Gastroenterology* 2006; 130: 548–565.

76. Graff JR, Herman JG, Lapidus RG et al. E-cadherin expression is silenced by DNA hypermethylation in human breast and prostate carcinomas. *Cancer Res* 1995; 55: 5195–5199.

77. Zhao C, Lam EW, Sunters A et al. Expression of estrogen receptor beta isoforms in normal breast epithelial cells and breast cancer: Regulation by methylation. *Oncogene* 2003; 22: 7600–7606.
78. Yang D, Thangaraju M, Greeneltch K et al. Repression of IFN regulatory factor 8 by DNA methylation is a molecular determinant of apoptotic resistance and metastatic phenotype in metastatic tumor cells. *Cancer Res*. 2007; 67: 3301–3309.
79. Domann FE, Rice JC, Hendrix MJ et al. Epigenetic silencing of maspin gene expression in human breast cancers. *Int J Cancer* 2000; 85: 805–810.
80. Ramirez JL, Rosell R, Taron M et al. 14–3–3sigma methylation in pretreatment serum circulating DNA of cisplatin-plus-gemcitabine-treated advanced non-small-cell lung cancer patients predicts survival: The Spanish lung cancer group. *J Clin Oncol* 2005; 23: 9105–9112.
81. Staub J, Chien J, Pan Y et al. Epigenetic silencing of HSulf-1 in ovarian cancer: Implications in chemoresistance. *Oncogene* 2007; 26(34), 4969–4978.
82. Martens JW, Nimmrich I, Koenig T et al. Association of DNA methylation of phosphoserine aminotransferase with response to endocrine therapy in patients with recurrent breast cancer. *Cancer Res* 2005; 65: 4101–4117.
83. Wisman GB, Nijhuis ER, Hoque MO et al. Assessment of gene promoter hypermethylation for detection of cervical neoplasia. *Int J Cancer* 2006; 119: 1908–1914.
84. Mori T, O'Day SJ, Umetani N et al. Predictive utility of circulating methylated DNA in serum of melanoma patients receiving biochemotherapy. *J Clin Oncol* 2005; 23: 9351–9358.
85. Chang HC, Cho CY, Hung WC. Downregulation of RECK by promoter methylation correlates with lymph node metastasis in non-small cell lung cancer. *Cancer Sci* 2007; 98: 169–173.
86. Agrelo R, Cheng WH, Setien F et al. Epigenetic inactivation of the premature aging Werner syndrome gene in human cancer. *Proc Natl Acad Sci U S A* 2006; 103: 8822–8827.

24 Clinical Trials and Approved Drugs for Epigenetic Cancer Therapy

Debby M.E.I. Hellebrekers and Manon van Engeland

CONTENTS

24.1   EPIGENETIC THERAPY IN CANCER

Epigenetic changes play a major role in the initiation and progression of human cancer, in addition to genetic alterations. Overall, the genome of tumor cells is characterized by global CpG dinucleotide hypomethylation and reductions of specific histone modifications, i.e., loss of mono-acetylation at lysine 16 and trimethylation at lysine 20 of histone H4 [1,2]. These global epigenetic alterations are thought to contribute to carcinogenesis through harmful expression of inserted viral sequences, oncogene activation, loss of imprinting and X chromosome inactivation, and genomic instability through hypomethylation of structural elements. On the other hand, promoter hyper-methylation and deacetylation of CpG islands of tumor suppressor genes results in aberrant transcriptional silencing [3]. Many genes, located across all chromosome locations, are epigenetically silenced in cancer cells, of which some methylated genes are shared and others are tumor-type-specific. Examples are genes involved in cell cycle regulation and apoptosis (p14ARF, p15INK4b, p16INK4a, APC, RASSF1A, HIC1), DNA repair genes (hMLH1, GSTP1, MGMT, BRCA1), and genes related to metastasis and invasion (CDH1, TIMP-3, DAPK, p73, maspin, TSP1, VHL) [4].

The importance of epigenetic alterations in tumor growth and development creates novel therapeutic targets. In contrast to genetic alterations, epigenetic changes in cancer are potentially reversible, resulting in the development of pharmacologic inhibitors of DNA methylation and histone deacetylation, called epigenetic therapy. DNA methyltransferase (DNMT) inhibitors and histone deacetylase (HDAC) inhibitors can reverse epigenetic silencing of tumor suppressor genes by inducing DNA demethylation and histone acetylation, resulting in reactivation of these genes in tumor cells and restoring of crucial cellular pathways.

24.2 CLINICAL APPLICATION OF EPIGENETIC THERAPY

24.2.1 DNMT INHIBITORS AS ANTICANCER THERAPEUTICS

5-Azacytidine (Vidaza, Pharmion Corp., Boulder, Colorado) and 5-aza-2′-deoxycytidine (Decitabine, Dacogen, SuperGen, Inc., Dublin, California, and MGI Pharma, Inc., Bloomington, Minnesota) are the most widely studied DNMT inhibitors (Table 24.1). 5-Azacytidine and 5-aza-2′-deoxycytidine are nucleoside analogs that are incorporated into the DNA in place of the natural base cytosine during DNA replication, and are therefore only active during S phase. By forming a complex with active sites of DNMTs, these drugs covalently trap these enzymes, resulting in demethylation of DNA after several cell divisions. 5-Azacytidine is partly incorporated into RNA, thereby interfering with protein translation, while 5-aza-2′-deoxycytidine is incorporated only into DNA, causing more efficient inhibition of DNMTs [5,6]. 5-Azacytidine and 5-aza-2′-deoxycytidine reactivate silenced tumor suppressor genes in cultured tumor cells by demethylation of their hypermethylated promoter, thereby restoring their normal function [7]. Reactivation of dormant tumor suppressor genes is suggested to be the mechanism by which these compounds suppress growth and induce differentiation and apoptosis of human tumor cell lines.

5-Azacytidine (Vidaza) and 5-aza-2′-deoxycytidine (decitabine) were both synthesized by Sorm and coworkers in 1964 [8]. There has been a shift in the clinical use of these drugs from chemotherapeutic to demethylating agents. These drugs were initially developed as chemotherapeutic

TABLE 24.1
DNMT and HDAL Inhibitors and Their Clinical Status

Target	Class	Drug	Clinical Trial Status
DNA methylation	Nucleoside analogs	5-Azacytidine (Vidaza)	FDA approved (2004)
		5-Aza-2′-deoxycytidine (Decitabine, Dacogen)	FDA approved (2006)
		5,6-Dihydro-5-azacytidine	Phase I/II
		5-Fluoro-2′-deoxycytidine	Phase I/II
		Zebularine	
	Non-nucleoside	Procainamide	
		Procaine	
		Hydralazine	
		Epigallocatechin	
		RG108	
	DNMT1 antisense	MG98	Phase II
Histone deacetylation	Short-chain fatty acids	Phenylbutyrate	Phase II
		Valproic acid	Phase III
	Hydroxamic acids	Trichostatin A (TSA)	
		Suberoylanilide hydroxamic acid (SAHA)	FDA approved (2006)
		Pyroxamide	Phase I
		PXD101	Phase I
		NVP-LAQ824	Phase I
		LBH589	Phase I
	Cyclic tetrapeptide	Depsipeptide (FK-228)	Phase II
		Apicidin	
		Trapoxin	
	Benzamides	MS-275	Phase II
		CI-994	Phase II

agents and were used at high, often quite toxic doses for the treatment of leukemia. In the early 1980s, in vitro studies showed a different mechanism of action of these drugs at high versus low doses. High concentrations inhibit DNA replication and induce cell cycle arrest, leading to cytotoxicity, whereas low doses are incorporated into DNA inducing DNMT inhibition and DNA hypomethylation with minimal cytotoxicity [9]. This resulted in the use of much lower doses of 5-azacytidine and 5-aza-2′-deoxycytidine in the clinic since the 1990s, especially in the field of hematological malignancies. Although the hypomethylating activity of 5-azacytidine is less than that of 5-aza-2′-deoxycytidine [9], clinical efficacy of both drugs in myelodysplastic syndromes (MDS) seems to be comparable [10]. In contrast with chemotherapy agents, both drugs require repeated therapy courses to achieve a clinical response (delayed response).

The use of DNMT inhibitors in the treatment of MDS results from the knowledge that epigenetic gene silencing of, in particular, p15INK4b is present in poor-risk MDS subtypes and often predicts transformation to acute myeloid leukemia (AML) [11]. Clinical trials of low-dose 5-azacytidine (75 mg/m^2 daily) in patients with MDS have been encouraging. In a phase III multicenter, randomized trial in 191 patients with all MDS subtypes, 5-azacytidine (75 mg/m^2/day in 7-day cycles beginning on days 1, 29, 57, and 85) was compared with supportive care only. Significantly higher response rates were reported in the 5-azacytidine group (7% complete response, 16% partial response, 37% improved) as compared with the group receiving supportive care only (5% improved) [12]. Furthermore, quality of life was significantly improved in the 5-azacytidine group of the same study population. These results led to the FDA approval of 5-azacytidine (Vidaza, Pharmion Corp., Boulder, Colorado) in 2004 for treatment of all MDS subtypes, including refractory anemia, refractory anemia with ringed sideroblasts, refractory anemia with excess blasts, refractory anemia with excess blasts in transformation, and chronic myelomonocytic leukemia.

In several phase I/II/III studies, decitabine (5-aza-2′-deoxycytidine) has also shown promising data in patients with MDS and AML [13,14]. In a phase II study of high-risk MDS in elderly patients, decitabine administered at 45 mg/m^2/day for 3 days every 6 weeks showed encouraging results; the overall response rate was 49%, the median response duration was 31 weeks, and the median survival time from the start of therapy was 15 months. In addition, decitabine treatment showed a clinically significant, often long lasting, effect on the platelet count in a substantial number of high-risk MDS patients, already occurring after one cycle of decitabine therapy in the majority of responding patients [15]. In a randomized phase III trial in which MDS patients were treated with either decitabine (intravenously, 15 mg/m^2 over a 3 h period every 8 h for 3 consecutive days, repeating this treatment cycle every 6 weeks) or supportive care only (blood and blood product transfusions, prophylactic antibiotics, and hematopoietic growth factors), the overall-response rate was 17% in the decitabine-treated group as compared with 0% in the supportive care group [16]. In addition, quality of life was improved in the decitabine group. On May 2, 2006, the FDA approved decitabine (Dacogen, SuperGen, Inc., Dublin, California, and MGI Pharma, Inc., Bloomington, Minnesota) for the treatment of patients with MDS including previously treated and untreated, de novo and secondary MDS of all French–American–British subtypes (refractory anemia, refractory anemia with ringed sideroblasts, refractory anemia with excess blasts, refractory anemia with excess blasts in transformation, and chronic myelomonocytic leukemia) and intermediate-1, intermediate-2, and high-risk International Prognostic Scoring System groups.

Despite the promising effects of trials of 5-azacytidine and 5-aza-2′-deoxycytidine in hematologic malignancies, clinical experience with these drugs in solid tumor malignancies has been less successful [17]. This might be due to differences in drug penetration, drug conversion, or cell cycle duration. Ongoing clinical trials are exploring the use of low doses of DNMT inhibitors for the treatment of solid tumors including melanoma, and cancers of the breast, renal cell, colon, and bladder.

The instability of 5-azacytidine and 5-aza- 2′-deoxycytidine in neutral aqueous solution resulted in the development of more stable cytidine analogs, such as 5,6-dihydro-5-azacytidine and 5-fluoro-2′-deoxycytidine [18]. Due to inconsistent data on the efficacy of 5,6-dihydro-5-azacytidine in phase I and II trials, clinical studies on this drug were not continued. 5-Fluoro-2′-deoxycytidine is currently undergoing phase I/II studies, but generates 5-fluorodeoxyuridine and its metabolites, which may be toxic.

Zebularine is a novel DNMT inhibitor which is very stable at acidic and neutral pHs and in aqueous solution, and has been shown to be minimally toxic both in vitro and in vivo, enabling oral administration of the drug [19]. In addition, this cytidine analogue preferentially targets tumor cells. Although these properties make zebularine a promising candidate for cancer treatment [20], the requirement of higher concentrations (up to 1 g/kg body weight in mice) in comparison with 5-aza-2′-deoxycytidine has kept this drug from entering clinical trials.

The toxicity associated with incorporation of nucleoside analogs into DNA resulted in the search for non-nucleoside DNMT inhibitors [21]. Examples are procainamide and procaine, which inhibit DNMTs by perturbing interactions between the protein and its target sites [18]. Furthermore, the antihypertensive compound hydralazine inhibits DNA methylation by the interaction between its nitrogen atoms and the DNMT active site [22]. Epigallocathechin-3-gallate is a natural product derived from green tea, and has shown to inhibit DNMT activity by binding to and blocking the active site of human DNMT1 [23]. The novel small molecule RG108 blocks the DNMT active site and, intriguingly, causes demethylation and reactivation of tumor suppressor genes, but does not affect methylation of centromeric satellite sequences [24]. These characteristics make RG108 useful for development of new DNMT inhibitors. In contrast with most demethylating agents, which are not specific for a specific DNMT, MG98 is an antisense oligonucleotide that is developed to specifically inhibit DNMT1 mRNA, which might reduce unfavorable toxicity [25]. MG98 is currently being tested in phase II clinical trials.

Despite the exciting results from trials of DNMT inhibitors, the clinical application of these drugs has a number of drawbacks. Several DNMT inhibitors have been associated with serious side effects [26]. The toxicity of nucleoside demethylating agents might be caused by the formation of covalent adducts between DNA and trapped DNMTs [27]. Furthermore, unfavorable effects can be due to the fact that many of the DNMT inhibitors are not specific for a particular DNMT or gene. Toxicity, a central problem in interpretation of clinical data, might be reversed by optimization of treatment schedules, e.g., giving lower doses over longer periods, thereby exposing more cells during S phase. Another approach to minimize side effects might be development of non-nucleoside inhibitors which are not incorporated into DNA. Also, compounds that specifically target a particular DNMT, such as MG98, might reduce nonspecific effects. Another important aspect that should be taken into account in the clinical use of demethylating agents is induction of global hypomethylation, which might induce tumorigenesis by activation of oncogenes, induction of chromosomal instability, and mutagenesis.

Despite the therapeutic effects of decitabine and Vidaza in patients with hematologic malignancies, the connection between the clinical activity of these drugs and their demethylating activity is not clear [10]. Analysis of patterns and levels of DNA methylation in patients may be important surrogate endpoints for investigating the in vivo effects of demethylating drugs in patients, thereby helping the interpretation of clinical responses. Laboratory studies of clinical trails have shown that decitabine induces global hypomethylation as well as promoter-specific hypomethylation and gene activation at doses that are associated with clinical responses. Paradoxically, studies have suggested that patients with p15INK4b promoter hypermethylation at baseline have a lower chance of response than patients without hypermethylation of this gene [10].

An important issue of the clinical use of DNMT inhibitors is the re-methylation problem. Present treatment schedules do not lead to the continuous demethylation of a patient's DNA [28,29] and so may not result in stable epigenetic reprogramming in tumor cells. Therefore, an alternative approach might be to use the demethylation period (before remethylation can occur) as a window of epigenetic sensitization for combination therapies. For example, demethylation may increase sensitivity of tumors to existing cytostatic therapies or to HDAC inhibitors.

24.2.2   HDAC Inhibitors as Anticancer Therapeutics

HDACs are involved in the remodeling of chromatin, and have a key role in the epigenetic regulation of gene expression. By inhibiting histone deacetylation, HDAC inhibitors cause increased histone acetylation, thereby inducing an open chromatin conformation and transcription of previously dormant genes. Several classes of HDAC inhibitors have been found to have potent and specific anticancer activities in preclinical studies [30] and very recently, the first of these, suberoylanilide hydroxamic acid or SAHA (Zolinza, vorinostat, Merck & Co., Inc., the United States), has been approved for cancer treatment. Both naturally existing and synthetic HDAC inhibitors have been characterized (Table 24.1) [31]. Phenylbutyrate and valproic acid are short-chain fatty acids that have been used for nononcological purposes and were recently shown to have activity as HDAC inhibitors. These drugs act as HDAC inhibitors at relatively high concentrations, because their acyl group which contacts the catalytic HDAC zinc ion cannot make significant contact with the catalytic pocket due to their very short side chains. Phenylbutyrate was the first HDAC inhibitor to be tested in patients, and currently phase I and II trails have been performed. Phenylbutyrate has been evaluated in AML and MDS [32], as well as solid tumor malignancies [33]. The short-chain fatty acid HDAC inhibitor valproic acid has been used for decades as an anti-epileptic drug. Phase I and II clinical trials for evaluation as an antitumor agent have recently been reported [34], and currently phase III trials are ongoing.

The hydroxamic acids bind more strongly to the HDAC catalytic site, and are very potent but reversible HDAC inhibitors. Among these compounds is TSA, originally developed as an antifungal agent, which is active at nanomolar concentrations [35]. SAHA is another hydroxamic acid HDAC inhibitor [36]. The discovery of SAHA as anticancer agent resulted from the observation that dimethylsulfoxide (DMSO) induces differentiation of cancer cells. The polar group of DMSO was required for these effects and simple polar amides were more potent than DMSO, leading to the discovery of hexamethylene bisacetamide (HMBA). Although HMBA inhibits growth and induces differentiation of cancer cells, clinical trials found that the doses required for clinical effects were not well tolerated by cancer patients. SAHA was one of a series of synthesized bishydroxamic acids that were much more potent than HMBA in decreasing growth of cancer cells. It was proved to be effective and selective in inhibiting growth and inducing differentiation of a broad variety of transformed cells at low micromolar concentrations that have relatively little toxicity. More potent analogs of SAHA have shown unacceptable toxicity. On the basis of the similarity of the structure of SAHA to that of TSA, its anticancer activity was explained by inhibition of both class I and II HDACs through binding in the catalytic site of the enzyme [36]. SAHA is one of the HDAC inhibitors most advanced in development. In phase I and II clinical trials, SAHA administered intravenously demonstrated significant antitumor activity in both hematologic and solid tumors, was well tolerated, and its biological activity was reflected by accumulation of acetylated histones in tumor and normal tissues [37]. In a phase I study of orally administered SAHA in patients with advanced cancer, a broad range of antitumor activity was found in both hematologic and solid tumor malignancies. Major dose-limiting toxicities were anorexia, dehydration, diarrhea, and fatigue [38]. Currently, SAHA is used in many different clinical trials both as monotherapy and in combination with other agents in different hematologic and solid tumor malignancies. On October 6, 2006, the FDA granted approval to SAHA (Zolinza, vorinostat, Merck & Co., Inc.) for the treatment of cutaneous manifestations of cutaneous T-cell lymphoma (CTCL) in patients with progressive, persistent, or recurrent disease on or following two systemic therapies. The major trial supporting approval was a phase IIB study in CTCL patients who had failed prior therapy. In this trial a significant response rate was shown, including objective recession of lesions symptomatic relief of pruritus [39].

Pyroxamide, PXD101, NVP-LAQ824, and LBH589 are hydroxamic acid HDAC inhibitors in phase I clinical trials [30]. The cyclic hydroxamic acid–containing peptide (CHAP) compounds are built from TSA and cyclic tetrapeptides and inhibit HDACs at nanomolar concentrations [40].

Class III HDAC inhibitors are the cyclic tetrapeptides, including depsipeptide (FK-228, FR901228), apicidin, and trapoxin. Depsipeptide is a prodrug that is activated by reduction upon cellular uptake and inhibits class I HDACs, and is tested in phase I/II clinical trials [41]. The benzamides are a structurally diverse fourth class of HDAC inhibitors, which bind the active zinc in the HDAC catalytic site. Among this class of HDAC inhibitors are the synthetic HDAC inhibitor MS-275, which inhibits HDAC at micromolar concentrations, and CI-994 (*N*-acetyl dinaline), a relatively weak HDAC inhibitor, both of which are tested in phase I/II clinical trials [37].

The anticancer activities of HDAC inhibitors possibly result from their ability to regulate the expression of specific proliferative and apoptotic genes, by inducing histone acetylation and re-expression of genes that have become epigenetically silenced in cancer cells. Different HDAC inhibitors were found to regulate a highly overlapping gene set within distinct apoptosis and cell cycle pathways [42]. Examples of genes consistently induced by HDAC inhibitors are p21(CDKN1A), p16 (INK4A), cyclin E, and thioredoxin-binding protein 2 (TBP2). Besides inducing histone acetylation, HDAC inhibitors also cause increased acetylation and thereby altered activity of nonhistone proteins, including proteins involved in cell proliferation (e.g., Rb [43]), apoptosis (e.g., p53 [44]), cell motility (e.g., tubulin [45]), and angiogenesis (e.g., HIF-1α [46]). The fact that most cancers are characterized by dysregulation in these processes might cause (partly) the efficacy of HDAC inhibitors as anticancer therapeutics. The targeting of multiple histone- and nonhistone proteins by these drugs may provide an advantage of HDAC inhibitors over other chemotherapeutic agents because suppression of a single pathway may not confer resistance to these agents.

Another important effect of HDAC inhibitors besides acetylation of histone and nonhistone proteins is demethylation of DNA. Valproic acid and TSA have been shown to trigger replication-independent active demethylation of DNA [47,48]. Furthermore, VPA and MS-275 have been shown to counteract L-methionine-induced methylation of the reelin/GAD67 promoters in vivo [49]. The induction of DNA demethylation by HDAC inhibitors is a very interesting issue and might, at least partly, contribute to the anticancer activity of these drugs.

Instead of pan-HDAC inhibitors such as SAHA or depsipeptide, the discovery of new HDAC inhibitors selectively targeting one or another HDAC might render more useful anticancer drugs. The potential to alter the expression of a more focused, disease-related subset of genes and to limit adverse effects has prompted the development of isoform-specific HDAC inhibitors [50–53].

Preclinical studies have reported HDAC inhibitors to be additive or synergistic with a number of anticancer agents [21]. Therefore, the clinical use of HDAC inhibitors in combination with other therapeutics might increase treatment efficacy.

24.2.3 COMBINATION THERAPY OF DNMT- AND HDAC INHIBITORS

DNA hypermethylation and histone deacetylation are dynamically linked in gene silencing. This is reflected by the synergy between 5-aza-2′-deoxycytidine and TSA in reactivation of epigenetically silenced tumor suppressor genes [54] and by the finding that cotreatment of tumor cells with DNMT- and HDAC inhibitors produces stronger antineoplastic effects than by either compound alone [55]. Therefore, clinical trials of combining DNMT- and HDAC inhibitors have started and have shown promising efficacy in patients with AML and MDS [56,57]. One of these studies suggests that 5-azacytidine in combination with phenylbutyrate yields a potential enrichment in major hematologic responses in patients with MDS or AML who received aza-CR at 50 mg/m^2/day for 10 days followed by sodium phenylbutyrate. Furthermore, it was demonstrated that molecular mechanisms responsible for these responses may include reversal of aberrant epigenetic gene silencing [56]. In a phase I/II study of the combination of decitabine and valproic acid in patients with advanced leukemia this combination of epigenetic therapy in leukemia was safe and active, and was associated with transient reversal of aberrant epigenetic marks [57].

An important aspect of the clinical use of epigenetic drugs, especially with regard to central nervous system neoplasias, is whether they are able to penetrate the blood-brain barrier.

5-Aza- 2′-deoxycytidine has been shown to be able to cross the BBB effectively [58]. For the HDAC inhibitors, in vivo murine experiments strongly suggest that SAHA and MS-275 can pass the BBB [59,60]. These compounds increase acetylated histone H3 levels in mouse brain tissue, making them promising candidate drugs for the treatment of malignant brain tumors.

24.3 MULTITARGETING OF CANCER BY EPIGENETIC THERAPY

Activation of hypermethylated tumor suppressor genes by promoter hypomethylation and histone acetylation has been proposed to be the major mechanism behind the anticancer activity of epigenetic drugs. However, upcoming evidences indicate that the therapeutic effects of DNMT- and HDAC inhibitors can be extended to immune modulatory- and antiangiogenesis activities. Epigenetic alterations seem to play a prominent role in tumor escape from the host's immune recognition, by downregulating the expression of several genes required for an efficient recognition of cancer cells, including cancer testis antigens (CTAs), HLA antigens, as well as accessory/ co-stimulatory molecules. This suggests that demethylating agents and HDAC inhibitors might effect immune response mechanisms by induction of immunologically active cell surface proteins like CTAs, HLA class I and II antigens, and accessory/co-stimulatory molecules. 5-Aza-2′-deoxycytidine was shown to induce expression of CTAs in solid and hematopoietic tumors of different histotype, thereby increasing their immunological recognition by CTA-specific cytotoxic T lymphocytes [61–63]. Mouse studies demonstrate long lasting de novo expression of cancer testis antigens in melanoma xenografts from 5-aza-2′-deoxycytidine-treated mice. Serological and biochemical analyses identified a de novo expression of NY-ESO-1 protein and a concomitant and persistent upregulation of HLA class I antigens and of HLA-A1 and -A2 alleles. Furthermore, immunization of BALB/c mice with 5-aza-2′-deoxycytidine-treated melanoma cells generated high titer circulating anti-NY-ESO-1 antibodies [64]. 5-Aza-2′-deoxycytidine also upregulated the expression of intercellular adhesion molecule-1 (ICAM-1) on different types of tumor cells [62] and of leukocyte function-associated antigen-3 (LFA-3) on melanoma cells [65]. The HDAC inhibitor TSA was demonstrated to upregulated MAGE-A expression in different human tumor cell lines [66]. TSA and sodium butyrate also effectively induced HLA class I antigen expression in a human neuroblastoma cell line [67]. Furthermore, different HDAC inhibitors have been shown to be able to upregulate the expression of the accessory/co-stimulatory molecules CD40, CD86, and ICAM-1 in tumor cell lines and leukemia blasts of AML [67,68]. These findings are opening the road to clinical trials evaluating the immunomodulatory activities of epigenetic drugs, utilized alone or in combination therapies.

Among the epigenetically silenced tumor suppressor genes in tumor cells are genes with angiogenesis-inhibiting properties. By re-expression of these genes in tumor cells, DNMT- and HDAC inhibitors might indirectly, via the tumor cells, exhibit angiostatic effects in vivo [69]. Examples of angiogenesis-inhibiting tumor suppressor genes that are silenced by epigenetic promoter modifications in tumor cells are p16INK4a and p73, which downregulate expression of the angiogenic factor VEGF (vascular endothelial cell growth factor), maspin, TIMP-2/3, thrombospondin-1, and ADAMTS-8. In addition, we and others have shown that DNMT- and HDAC inhibitors directly inhibit endothelial cell growth and angiogenesis in vitro and in vivo [70,71]. Therefore, angiostatic activities of epigenetic drugs include both indirect anti-angiogenesis effects, through modulation of angiogenesis-regulating genes in tumor cells, and direct inhibition of endothelial cells lining the tumor vasculature.

In conclusion, the reactivation of multiple epigenetically silenced tumor suppressor genes in tumor cells by epigenetic drugs influences virtually all pathways suggested by Hanahan and Weinberg to be involved in the cancer process, decreasing the development of resistance to these compounds. In addition, the effects of DNMT- and HDAC inhibitors can be extended beyond tumor cells alone, since growth and sprouting of the tumor vasculature is also (in)directly decreased by these drugs. Epigenetic therapy affects the uncontrolled growth of neoplastic cells, the escape of

TABLE 24.2

Multitargeting of Epigenetic Anticancer Therapy

Tumor Characteristic	Effect of DNMT- and HDAC Inhibitors	Anticancer Effect
Epigenetic silencing of tumor suppressor genes	Reactivation of tumor suppressor genes by demethylation and histone acetylation	Inhibition of tumor cell growth/migration/invasion Induction of tumor cell apoptosis
Escape from immune response	Reactivation of epigenetically silenced cancer testis antigens, HLA antigens, and accessory/co-stimulatory molecules by demethylation and histone acetylation	Stimulation of antitumor immune recognition and response
Generation of tumor vasculature (tumor angiogenesis)	Reactivation of angiogenesis-inhibiting tumor suppressor genes in tumor cells by demethylation and histone acetylation	Indirect inhibition of tumor angiogenesis
	Reactivation of genes in tumor endothelial cells	Direct inhibition of endothelial cell growth and tumor angiogenesis

tumors from the host's immune response, as well as the generation of tumor vasculature, three processes pivotal in cancer biology (Table 24.2). Clearly, the multitargeting of epigenetic drugs makes them attractive antitumor therapeutics and encourages the development of improved treatment schedules to reach maximal clinical success.

REFERENCES

1. Fraga MF, Ballestar E, Villar-Garea A, et al. Loss of acetylation at Lys16 and trimethylation at Lys20 of histone H4 is a common hallmark of human cancer. *Nat Genet* 2005;37:391–400.
2. Feinberg AP and Vogelstein B. Hypomethylation distinguishes genes of some human cancers from their normal counterparts. *Nature* 1983;301:89–92.
3. Herman JG and Baylin SB. Gene silencing in cancer in association with promoter hypermethylation. *N Engl J Med* 2003;349:2042–2054.
4. Esteller M, Corn PG, Baylin SB, and Herman JG. A gene hypermethylation profile of human cancer. *Cancer Res* 2001;61:3225–3229.
5. Bouchard J and Momparler RL. Incorporation of 5-Aza-2′-deoxycytidine-5′-triphosphate into DNA. Interactions with mammalian DNA polymerase alpha and DNA methylase. *Mol Pharmacol* 1983;24:109–114.
6. Santi DV, Garrett CE, and Barr PJ. On the mechanism of inhibition of DNA-cytosine methyltransferases by cytosine analogs. *Cell* 1983;33:9–10.
7. Gilbert J, Gore SD, Herman JG, and Carducci MA. The clinical application of targeting cancer through histone acetylation and hypomethylation. *Clin Cancer Res* 2004;10:4589–4596.
8. Sorm F and Vesely J. The activity of a new antimetabolite, 5-azacytidine, against lymphoid leukaemia in Ak mice. *Neoplasma* 1964;11:123–130.
9. Jones PA and Taylor SM. Cellular differentiation, cytidine analogs and DNA methylation. *Cell* 1980;20:85–93.
10. Oki Y, Aoki E, and Issa JP. Decitabine—bedside to bench. *Crit Rev Oncol Hematol* 2007;61:140–152.
11. Quesnel B, Guillerm G, Vereecque R, et al. Methylation of the p15(INK4b) gene in myelodysplastic syndromes is frequent and acquired during disease progression. *Blood* 1998;91:2985–2990.
12. Silverman LR, Demakos EP, Peterson BL, et al. Randomized controlled trial of azacitidine in patients with the myelodysplastic syndrome: A study of the cancer and leukemia group B. *J Clin Oncol* 2002;20:2429–2440.

13. Kantarjian H, Oki Y, Garcia-Manero G, et al. Results of a randomized study of 3 schedules of low-dose decitabine in higher-risk myelodysplastic syndrome and chronic myelomonocytic leukemia. *Blood* 2007;109:52–57.
14. Issa JP, Garcia-Manero G, Giles FJ, et al. Phase 1 study of low-dose prolonged exposure schedules of the hypomethylating agent 5-aza-2′-deoxycytidine (decitabine) in hematopoietic malignancies. *Blood* 2004;103:1635–1640.
15. Wijermans P, Lubbert M, Verhoef G, et al. Low-dose 5-aza-2′-deoxycytidine, a DNA hypomethylating agent, for the treatment of high-risk myelodysplastic syndrome: A multicenter phase II study in elderly patients. *J Clin Oncol* 2000;18:956–962.
16. Kantarjian H, Issa JP, Rosenfeld CS, et al. Decitabine improves patient outcomes in myelodysplastic syndromes: Results of a phase III randomized study. *Cancer* 2006;106:1794–1803.
17. Aparicio A and Weber JS. Review of the clinical experience with 5-azacytidine and 5-aza-2′-deoxycytidine in solid tumors. *Curr Opin Investig Drugs* 2002;3:627–633.
18. Villar-Garea A and Esteller M. DNA demethylating agents and chromatin-remodelling drugs: Which, how and why? *Curr Drug Metab* 2003;4:11–31.
19. Cheng JC, Matsen CB, Gonzales FA, et al. Inhibition of DNA methylation and reactivation of silenced genes by zebularine. *J Natl Cancer Inst* 2003;95:399–409.
20. Herranz M, Martin-Caballero J, Fraga MF, et al. The novel DNA methylation inhibitor zebularine is effective against the development of murine T-cell lymphoma. *Blood* 2006;107:1174–1177.
21. Yoo CB and Jones PA. Epigenetic therapy of cancer: Past, present and future. *Nat Rev Drug Discov* 2006;5:37–50.
22. Segura-Pacheco B, Trejo-Becerril C, Perez-Cardenas E, et al. Reactivation of tumor suppressor genes by the cardiovascular drugs hydralazine and procainamide and their potential use in cancer therapy. *Clin Cancer Res* 2003;9:1596–1603.
23. Fang MZ, Wang Y, Ai N, et al. Tea polyphenol (−)-epigallocatechin-3-gallate inhibits DNA methyltransferase and reactivates methylation-silenced genes in cancer cell lines. *Cancer Res* 2003;63:7563–7570.
24. Brueckner B, Boy RG, Siedlecki P, et al. Epigenetic reactivation of tumor suppressor genes by a novel small-molecule inhibitor of human DNA methyltransferases. *Cancer Res* 2005;65:6305–6311.
25. Yan L, Nass SJ, Smith D, Nelson WG, Herman JG, and Davidson NE. Specific inhibition of DNMT1 by antisense oligonucleotides induces re-expression of estrogen receptor-alpha (ER) in ER-negative human breast cancer cell lines. *Cancer Biol Ther* 2003;2:552–556.
26. Silverman LR, Holland JF, Weinberg RS, et al. Effects of treatment with 5-azacytidine on the in vivo and in vitro hematopoiesis in patients with myelodysplastic syndromes. *Leukemia* 1993;7 Suppl 1:21–29.
27. Juttermann R, Li E, and Jaenisch R. Toxicity of 5-aza-2′-deoxycytidine to mammalian cells is mediated primarily by covalent trapping of DNA methyltransferase rather than DNA demethylation. *Proc Natl Acad Sci U S A* 1994;91:11797–11801.
28. Issa JP, Gharibyan V, Cortes J, et al. Phase II study of low-dose decitabine in patients with chronic myelogenous leukemia resistant to imatinib mesylate. *J Clin Oncol* 2005;23:3948–3956.
29. Mund C, Hackanson B, Stresemann C, Lubbert M, and Lyko F. Characterization of DNA demethylation effects induced by 5-aza-2′-deoxycytidine in patients with myelodysplastic syndrome. *Cancer Res* 2005;65:7086–7090.
30. Minucci S and Pelicci PG. Histone deacetylase inhibitors and the promise of epigenetic (and more) treatments for cancer. *Nat Rev Cancer* 2006;6:38–51.
31. Marks P, Rifkind RA, Richon VM, Breslow R, Miller T, and Kelly WK. Histone deacetylases and cancer: Causes and therapies. *Nat Rev Cancer* 2001;1:194–202.
32. Gore SD, Weng LJ, Figg WD, et al. Impact of prolonged infusions of the putative differentiating agent sodium phenylbutyrate on myelodysplastic syndromes and acute myeloid leukemia. *Clin Cancer Res* 2002;8:963–970.
33. Carducci MA, Gilbert J, Bowling MK, et al. A phase I clinical and pharmacological evaluation of sodium phenylbutyrate on an 120-h infusion schedule. *Clin Cancer Res* 2001;7:3047–3055.
34. Pilatrino C, Cilloni D, Messa E, et al. Increase in platelet count in older, poor-risk patients with acute myeloid leukemia or myelodysplastic syndrome treated with valproic acid and all-trans retinoic acid. *Cancer* 2005;104:101–109.
35. Yoshida M, Kijima M, Akita M, and Beppu T. Potent and specific inhibition of mammalian histone deacetylase both in vivo and in vitro by trichostatin A. *J Biol Chem* 1990;265:17174–17179.

36. Marks PA. Discovery and development of SAHA as an anticancer agent. *Oncogene* 2007;26:1351–1356.
37. Kelly WK and Marks PA. Drug insight: Histone deacetylase inhibitors—development of the new targeted anticancer agent suberoylanilide hydroxamic acid. *Nat Clin Pract Oncol* 2005;2:150–157.
38. Kelly WK, O'Connor OA, Krug LM, et al. Phase I study of an oral histone deacetylase inhibitor, suberoylanilide hydroxamic acid, in patients with advanced cancer. *J Clin Oncol* 2005;23:3923–3931.
39. Olsen EO, Kim Y, Kuzel T, Pacheco T, Foss F, and Parker S. Vorinostat (suberoylanilide hydroxamic acid, SAHA) is clinically active in advanced cutaneous T-cell lymphoma (CTCL): Results of a phase IIb trial. *Journal of Clinical Oncology*, 2006 ASCO Annual Meeting Proceedings Part I. Vol 24, No. 18S (June 20 Supplement), 2006: 7500.
40. Furumai R, Komatsu Y, Nishino N, Khochbin S, Yoshida M, and Horinouchi S. Potent histone deacetylase inhibitors built from trichostatin A and cyclic tetrapeptide antibiotics including trapoxin. *Proc Natl Acad Sci U S A* 2001;98:87–92.
41. Piekarz R and Bates S.A review of depsipeptide and other histone deacetylase inhibitors in clinical trials. *Curr Pharm Des* 2004;10:2289–2298.
42. Peart MJ, Smyth GK, van Laar RK, et al. Identification and functional significance of genes regulated by structurally different histone deacetylase inhibitors. *Proc Natl Acad Sci U S A* 2005;102:3697–3702.
43. Chan HM, Krstic-Demonacos M, Smith L, Demonacos C, and La Thangue NB. Acetylation control of the retinoblastoma tumour-suppressor protein. *Nat Cell Biol* 2001;3:667–674.
44. Sakaguchi K, Herrera JE, Saito S, et al. DNA damage activates p53 through a phosphorylation-acetylation cascade. *Genes Dev* 1998;12:2831–2841.
45. Palazzo A, Ackerman B, and Gundersen GG. Cell biology: Tubulin acetylation and cell motility. *Nature* 2003;421:230.
46. Jeong JW, Bae MK, Ahn MY, et al. Regulation and destabilization of HIF-1alpha by ARD1-mediated acetylation. *Cell* 2002;111:709–720.
47. Detich N, Bovenzi V, and Szyf M. Valproate induces replication-independent active DNA demethylation. *J Biol Chem* 2003;278:27586–27592.
48. Milutinovic S, D'Alessio AC, Detich N, and Szyf M. Valproate induces widespread epigenetic reprogramming which involves demethylation of specific genes. *Carcinogenesis* 2007;28:560–571.
49. Dong E, Guidotti A, Grayson DR, and Costa E. Histone hyperacetylation induces demethylation of reelin and 67-kDa glutamic acid decarboxylase promoters. *Proc Natl Acad Sci U S A* 2007;104:4676–4681.
50. Park JH, Jung Y, Kim TY, et al. Class I histone deacetylase-selective novel synthetic inhibitors potently inhibit human tumor proliferation. *Clin Cancer Res* 2004;10:5271–5281.
51. Mai A, Massa S, Pezzi R, Rotili D, Loidl P, and Brosch G. Discovery of (aryloxopropenyl)pyrrolyl hydroxyamides as selective inhibitors of class IIa histone deacetylase homologue HD1-A. *J Med Chem* 2003;46:4826–4829.
52. Heltweg B, Dequiedt F, Marshall BL, et al. Subtype selective substrates for histone deacetylases. *J Med Chem* 2004;47:5235–5243.
53. Haggarty SJ, Koeller KM, Wong JC, Grozinger CM, and Schreiber SL. Domain-selective small-molecule inhibitor of histone deacetylase 6 (HDAC6)-mediated tubulin deacetylation. *Proc Natl Acad Sci U S A* 2003;100:4389–4394.
54. Cameron EE, Bachman KE, Myohanen S, Herman JG, and Baylin SB. Synergy of demethylation and histone deacetylase inhibition in the re-expression of genes silenced in cancer. *Nat Genet* 1999;21:103–107.
55. Belinsky SA, Klinge DM, Stidley CA, et al. Inhibition of DNA methylation and histone deacetylation prevents murine lung cancer. *Cancer Res* 2003;63:7089–7093.
56. Gore SD, Baylin S, Sugar E, et al. Combined DNA methyltransferase and histone deacetylase inhibition in the treatment of myeloid neoplasms. *Cancer Res* 2006;66:6361–6369.
57. Garcia-Manero G, Kantarjian HM, Sanchez-Gonzalez B, et al. Phase 1/2 study of the combination of 5-aza-2'-deoxycytidine with valproic acid in patients with leukemia. *Blood* 2006;108:3271–3279.
58. Chabot GG, Rivard GE, and Momparler RL. Plasma and cerebrospinal fluid pharmacokinetics of 5-Aza-2'-deoxycytidine in rabbits and dogs. *Cancer Res* 1983;43:592–597.
59. Eyupoglu IY, Hahnen E, Trankle C, et al. Experimental therapy of malignant gliomas using the inhibitor of histone deacetylase MS-275. *Mol Cancer Ther* 2006;5:1248–1255.
60. Yin D, Ong JM, Hu J, et al. Suberoylanilide hydroxamic acid, a histone deacetylase inhibitor: Effects on gene expression and growth of glioma cells in vitro and in vivo. *Clin Cancer Res* 2007;13:1045–1052.

61. Gattei V, Fonsatti E, Sigalotti L, et al. Epigenetic immunomodulation of hematopoietic malignancies. *Semin Oncol* 2005;32:503–510.
62. Sigalotti L, Coral S, Fratta E, et al. Epigenetic modulation of solid tumors as a novel approach for cancer immunotherapy. *Semin Oncol* 2005;32:473–478.
63. Weber J, Salgaller M, Samid D, et al. Expression of the MAGE-1 tumor antigen is up-regulated by the demethylating agent 5-aza-2′-deoxycytidine. *Cancer Res* 1994;54:1766–1771.
64. Coral S, Sigalotti L, Colizzi F, et al. Phenotypic and functional changes of human melanoma xenografts induced by DNA hypomethylation: Immunotherapeutic implications. *J Cell Physiol* 2006;207:58–66.
65. Coral S, Sigalotti L, Gasparollo A, et al. Prolonged upregulation of the expression of HLA class I antigens and costimulatory molecules on melanoma cells treated with 5-aza-2′-deoxycytidine (5-AZA-CdR). *J Immunother* (*1997*) 1999;22:16–24.
66. Wischnewski F, Pantel K, and Schwarzenbach H. Promoter demethylation and histone acetylation mediate gene expression of MAGE-A1, -A2, -A3, and -A12 in human cancer cells. *Mol Cancer Res* 2006;4: 339–349.
67. Magner WJ, Kazim AL, Stewart C, et al. Activation of MHC class I, II, and CD40 gene expression by histone deacetylase inhibitors. *J Immunol* 2000;165:7017–7024.
68. Maeda T, Towatari M, Kosugi H, and Saito H. Up-regulation of costimulatory/adhesion molecules by histone deacetylase inhibitors in acute myeloid leukemia cells. *Blood* 2000;96:3847–3856.
69. Hellebrekers DM, Griffioen AW, and van Engeland M. Dual targeting of epigenetic therapy in cancer. *Biochim Biophys Acta* 2007;1775:76–91.
70. Kim MS, Kwon HJ, Lee YM, et al. Histone deacetylases induce angiogenesis by negative regulation of tumor suppressor genes. *Nat Med* 2001;7:437–443.
71. Hellebrekers DM, Jair KW, Vire E, et al. Angiostatic activity of DNA methyltransferase inhibitors. *Mol Cancer Ther* 2006;5:467–475.

Part V

Future Directions

25 Future Directions in Epigenetic Cancer Research

Jacob Peedicayil

CONTENTS

25.1 INTRODUCTION

Epigenetics, the study of mitotically and meiotically heritable changes in gene expression not involving changes in DNA sequence, involves three interacting molecular mechanisms: DNA methylation, histone modification, and RNA-mediated gene silencing (Figure 25.1) [1]. Epigenetic mechanisms are known to show stability as well as flexibility in regulating gene expression during mammalian development [2]. Although each of the three epigenetic mechanisms is physiologically important, it is thought that of the three mechanisms, DNA methylation is predominant in controlling gene expression [3] and the most stable in modulating the transcriptional plasticity of mammalian genomes [4]. A disruption in one of the three epigenetic mechanisms can lead to inappropriate expression or silencing of genes, resulting in disease. It has been predicted that epigenetics, although a new field, will have an enormous impact on medicine, offering new opportunities for the diagnosis and treatment of complex clinical disorders [5]. Previous chapters have discussed the roles and applications of each of the three molecular epigenetic mechanisms in relation to cancer. The field of cancer epigenetics is evolving rapidly on several fronts. This chapter discusses possible future directions in research on cancer epigenetics.

25.1.1 NEW EPIGENETIC TARGETS FOR ANTICANCER DRUGS

Most work on epigenetic drugs for the treatment of cancer to date has focused on the development of drugs that inhibit the enzymes DNA methyltransferase (DNMT) and histone deacetylase (HDAC) [6] as well as on the technology of RNA interference [7,8]. DNMT inhibitors, also called DNA demethylating drugs, comprise two major drug groups: nucleoside analogue inhibitors and non-nucleoside analogue inhibitors [6]. These drugs inhibit methylation of DNA by inhibiting DNMT enzymes. HDAC inhibitors, which inhibit HDAC enzymes, comprise several chemically different drug groups [6]. Extensive preclinical and clinical trials on DNMT and HDAC inhibitors are being conducted and they are gaining approval for clinical use for the management of patients with

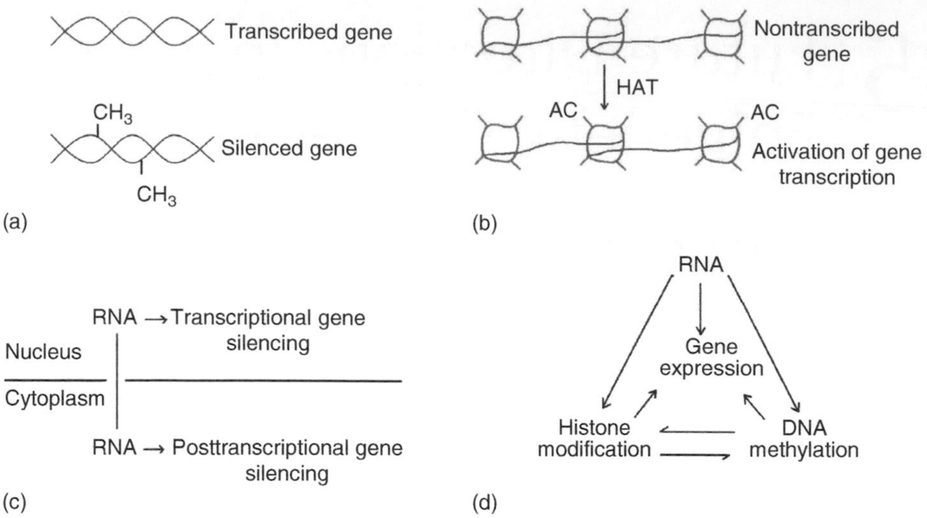

FIGURE 25.1 Schematic representation of the three epigenetic mechanisms. (A) Methylation of DNA leading to silencing of a gene. (B) Acetylation of histones in chromatin catalyzed by HAT leading to activation of gene transcription. (C) RNA-mediated transcriptional and posttranscriptional gene silencing. (D) Inter-relationships of the three epigenetic mechanisms. (Modified and reprinted by permission from Peedicayil, J., *Indian J. Med. Res.*, 123, 17, 2006. © *The Indian Journal of Medical Research.*)

cancer [9]. However, several issues regarding the clinical use of these drugs remain to be addressed and much more research is necessary for their appropriate use [9,10].

There are several other approaches that can be taken for the development of new epigenetic drugs for treating cancer (Table 25.1). One such approach is inhibition of histone acetyltransferase (HAT) enzymes. HAT enzymes are involved in the posttranslational acetylation of core histones, primarily acetylating the N-terminus and at least five families of HAT enzymes have been identified. Histone acetylation is a fundamental process that strongly affects the regulation of gene transcription. It also affects other processes like cell cycle progression, chromosome dynamics, DNA recombination as well as DNA repair and apoptosis [11]. Genes that encode HAT enzymes have been found

TABLE 25.1
New Epigenetic Targets for Anticancer Drugs

Target	Potential Drugs	References
Histone-Modifying Enzymes		
Histone acetyltransferases	Garcinol	[13]
(HAT enzymes)	Curcumin	[14]
	Anacardic acid	[15]
	γ-butyrolactones	[16]
	Isothiazolones	[17]
Histone methyltransferases	—	—
(HMT enzymes)		
Histone demethylases	Pargyline	[20]
	Tranylcypromine	[21]
Methyl-CpG-Binding Proteins		
MBD2	MBD2 antisense oligonucleotide	[23]

to be translocated, amplified, overexpressed, and mutated in both solid and hematological malignancies and there is evidence that HAT enzymes may promote or support malignancy [12]. HAT inhibitors, like HDAC inhibitors, may have therapeutic potential and several naturally occurring HAT inhibitors obtained from plant sources such as garcinol, curcumin, and anacardic acid are being studied for their anticancer properties [13–15]. Small molecule inhibitors of HAT enzymes such as γ-butyrolactones [16] and isothiazolones [17] are also being investigated and may also be useful as anticancer drugs.

Histone methylation regulates many processes like heterochromatin formation, X-chromosome inactivation, genomic imprinting, transcriptional regulation, and DNA repair [18]. Histones may be methylated on either lysine (K) or arginine (R) residues. Histone arginine methylation generally correlates with transcriptional activation while histone lysine methylation leads to either activation or repression, depending upon the particular lysine residue [18]. Since histone methyltransferases (HMT enzymes) such as EZH2 and SMYD3 are overexpressed in many cancers, they also represent inviting targets for inhibition by drugs [19]. At present no pharmacological inhibitors of HMT enzymes have been identified.

Histone methylation is a reversible process, and histone demethylase, which catalyzes the removal of methyl groups on lysine or arginine residues of histones, has been identified. Two types of histone lysine demethylases are known and include lysine-specific demethylase1 (LSD1) and Jumonji C (JmjC) domain family proteins. Imbalance of histone methylation has been associated with the pathogenesis of cancer and histone demethylases may also be useful targets for drugs used to treat cancer [18]. LSD1 has close homology to the enzyme monoamine oxidase (MAO) and two MAO inhibitors, pargyline [20], and tranylcypromine [21] have been shown to inhibit LSD1.

Methyl-CpG-binding proteins (MBDs) constitute a family of proteins that share a methyl-CpG-binding domain which comprises about 70 amino acid residues [22]. Five MBDs named MeCP2, MBD1, MBD2, MBD3, and MBD4 have been identified in vertebrates. MBDs recognize methylated DNA and are responsible for creating, maintaining, and interacting with the epigenome. One of the key functions of MBDs is to repress gene transcription [22]. MBDs are known to play a role in silencing of tumor suppressor genes and are potential targets for anticancer drugs [19]. For example, it has been shown that inhibition of MBD2 by sequence-specific MBD2 antisense oligonucleotides suppresses tumor growth in both lung cancer cell lines in vitro and in vivo when the cell lines are implanted into nude mice [23].

25.1.2 EPIGENETICS OF CANCER STEM CELLS

Stem cells are generally defined as clonogenic cells capable of self-renewal and differentiating into multiple cell lineages [24,25]. One way of classifying stem cells is into three kinds [26]: embryonal stem cells, which are derived from the first five or six divisions of the fertilized egg; germinal stem cells in the adult, which produce eggs and sperm and which are responsible for reproduction; and somatic or adult stem cells, which are considered more limited in their potential and which produce cells that differentiate into mature functioning cells responsible for normal tissue renewal. Progenitor cells are derived from stem cells and have a limited capacity for self-renewal [27].

It is now known that stem cells have a critical role not only in the generation of complex multicellular organisms but also in the development of tumors [28]. The contribution of a stem cell state is integral to current thinking in cancer biology [9]. It is now believed that cancer stem cells constitute the population that is ultimately responsible for perpetuating cancers. These cells have many properties common to normal stem cells but their exact origins are unclear [9]. One possible mode of origin of cancer stem cells is by epigenetic aberrations in physiological progenitor cells [29]. Recent findings suggest that eradication of cancer stem cells may be essential for achieving stable, long-lasting remission, and cure from cancer [30]. However, there are many challenges in developing treatments that target cancer stem cells such as differentiating cancer stem cells from somatic stem cells and selectively targeting the former without affecting the latter [28]. There is now evidence that epigenetic changes are involved in the normal physiology of stem cells [31] and are also associated with cancer stem cells in the pathogenesis of cancer [32,33] and it has been suggested that in the future the epigenetic changes that affect the differentiation potential of cancer stem/progenitor cells should be investigated [27].

Strategies for the targeting of cancer stem cells include the use of RNA interference [34], the chronic administration of epigenetic drugs like DNMT inhibitors to target cancer stem cells after debulking the cancer by standard chemotherapy [9], and the use of differentiation therapy [26]. Differentiation therapy is based on the principle that if the malignant cells of cancers are cancer stem cells, then it should be possible to treat cancers by inducing differentiation of the cancer stem cells [26].

25.1.3 GENOMIC APPROACHES IN CANCER EPIGENETICS

25.1.3.1 RNAi Consortium and Cancer Epigenetics

Although only 1.2% of the human genome encodes protein, a large part of it is transcribed [35]. About 98% of the transcriptional output of the human genome comprises nonprotein-coding RNAs from the introns of protein-coding genes and the exons and introns of nonprotein-coding genes, including many that are antisense to or overlapping protein-coding genes [35]. There are two large families of noncoding RNAs [36] (1) the small nucleolar RNA family (snoRNAs) and (2) the microRNA (miRNA) and small interfering RNA (siRNA) family.

miRNAs and siRNAs are involved in RNA interference (RNAi). RNAi refers to the phenomenon in which a wide variety of eukaryotic organisms, when exposed to foreign genetic material (RNA or DNA), mount a highly specific response to silence the genetic material before it can integrate into the host genome or subvert cellular processes [37,38]. RNAi is mediated by miRNAs and siRNAs that are produced from long double-stranded RNAs by an endonuclease of the ribonuclease-III type called Dicer. This results in small RNA duplexes that are about 21 to 23 nucleotides long, which are incorporated into a nuclease complex, the RNA-induced silencing complex (RISC), which then targets and cleaves mRNA that is complementary to the small RNA [37]. RNAi is thought to have evolved as an ancient cellular antiviral mechanism for silencing genes. RNAi is already being widely used as a research tool to artificially inhibit any gene of interest and holds promise in treating any disease that is linked to elevated expression of an identified gene [39]. RNAi considerably bolsters functional genomics to help identify novel genes involved in disease processes and is helping in the identification of genes and pathways involved in cancer [40,41].

The great power of RNAi as a research tool has stimulated the formation of the RNAi consortium (TRC), which is using RNAi to systematically uncover the function of genes in the human and mouse genomes [42,43]. The goal of TRC is to enable large-scale loss-of-function screens through the development of genome-scale RNAi libraries and methodologies for their use. TRC aims to produce genome-wide mammalian RNAi libraries that achieve stable and specific gene knockdown in a wide variety of cell types and that are practical for routine use in a high-throughput screening context. In particular, TRC is focusing on developing an RNAi library that reproducibly suppresses gene expression in a wide range of dividing and nondividing cells, is stable during propagation in bacterial hosts, targets each gene at multiple sites within its sequence, and is suitable for both arrayed and pooled screening applications [43]. RNAi libraries have already begun to prove their worth in the hunt for genes underlying cancer in humans [44]. For example, RNAi libraries have identified a tumor suppressor role for REST, a transcriptional repressor of neuronal gene expression [45], and identified that the inhibition of PITX1 induces the RAS pathway and tumorigenicity [46]. It is thought that TRC will help identify which genes play an important part in the pathogenesis of cancer and other diseases [42].

25.1.3.2   Cancer Epigenomics

The Human Epigenome Project (HEP) was launched by the Wellcome Trust Sanger Institute (Hinxton, United Kingdom) and Epigenomics AG (Berlin, Germany) in the wake of the Human Genome Project as an ambitious international project aimed at cataloguing and describing genome-wide DNA methylation patterns in all major tissues [47]. The Pilot Study of the HEP identified the methylation profile of the human major histocompatibility complex [48] and DNA methylation profiles of human chromosomes 6, 20, and 22 have been published [4]. Subsequent to the European initiative, The U.S. Human Epigenome Project evolved in meetings sponsored by the American Association for Cancer Research and the National Cancer Institute and was envisaged as a more comprehensive approach to the human epigenome, including the cataloguing of changes in chromatin in addition to that of DNA methylation patterns [49,50].

It is thought that the HEP will help answer many questions in biology and medicine such as [51]: how many tumor suppressor genes undergo CpG island promoter hypermethylation in transformation; how our epigenome changes with the aging process; the impact of the environment in modulating epigenetic marks and gene function; the contribution of DNA methylation and histone modifications to cell and tissue-specific differentiation; and the epigenetic environment of a stem cell. The HEP will provide a "reference epigenome" by resequencing different tissues and adding 5-methylcytosine to the DNA sequence [52]. This information will support the creation of epigenome projects of disease genomes. Human cancers are thought to be characterized by tissue-specific DNA methylation [53] and hence the creation of a "Cancer Epigenome Project" based on the information obtained from the HEP is thought to be a viable option in future [52]. In fact, there already is preliminary data on the epigenomic profile of cancer. Paz and colleagues [54] analyzed 70 widely used human cancer cell lines of 12 different tumor types for CpG island promoter hypermethylation of 15 tumor suppressor genes, global 5-methylcytosine genomic content, chemical response to 5-azacytidine and their genetic haplotype for methyl-group metabolism genes. Esteller et al. [55] analyzed promoter hypermethylation changes in 12 genes in DNA obtained from more than 600 primary tumors representing 15 major tumor types.

Recent advances in technology such as chromatin immunoprecipitation coupled with whole genome DNA microarrays (ChIP-chip) [56,57] have made it possible to also assess histone modifications at genome-scale in mammalian cells. A first draft of a global view of how the histone landscape is distorted in cancer cells has been published [58] in which a global view of histone modifications in cancer cells demonstrated the association between DNA methylation and histone modifications. This draft suggests novel pathways by which enzymes involved in the modification of histones such as HDACs, HATs, and HMTs may play roles as oncogenes and tumor suppressor genes.

Plass and Smiraglia [59] have stressed the importance of defining potential target sequences among DNA methylation patterns since DNA methylation is used as a therapeutic target. They developed a model to define target sequences that could be used as diagnostic and therapeutic markers. Advances in cancer epigenomics are likely to help identify epimutations [6,60], epigenetic polymorphisms [6], and epialleles [61] that underlie various types of cancers, developments which are likely to help in the diagnosis, treatment, and prevention of cancers.

25.1.3.3   Human Cancer Genome Project

The National Institutes of Health in the United States [62,63] and the Institute for Cancer Research, and the Wellcome Trust in the United Kingdom [64] have launched the Human Cancer Genome Project (HCGP), also called the Cancer Genome Atlas [65], whose overall goals are to identify all genomic alterations significantly associated with all major cancer types. The project has an epigenetic component since one of the goals of the project is to study aberrant DNA methylation patterns associated with cancers [62]. The Working Group on Biomedical Technology in February 2005 [62] described the specific goals of the project as being to (1) create a large collection of appropriate, clinically annotated samples from all major types of cancer and (2) completely characterize each sample in terms of: all regions of genomic loss or amplification; all mutations in the coding regions of all human genes; all chromosomal rearrangements; all regions of aberrant methylation; and complete gene expression profile, as well as other appropriate technologies. The HCGP is a natural extension of the Human Genome Project which in fact was initially proposed as the best way of moving forward in order to vanquish cancer [66]. The HCGP plans to use high-throughput methods for the analysis of genome-wide analysis of DNA methylation patterns. The project also plans to analyze tumors for genome-wide RNA expression.

It is thought that the information obtained from the HCGP will provide a permanent foundation for all future cancer research, including the understanding of epigenetic changes that may lead to malignancy, and have major implications for basic, clinical, and commercial efforts to understand, prevent, and treat cancer. Moreover, the HCGP is likely to reveal the subtypes of cancers and systematically reveal the cellular pathways that are deranged in each subtype. This would increase the effectiveness of research to understand tumor initiation, progression, and susceptibility to carcinogenesis. The project is also expected to help early diagnosis of cancer, the development of anticancer drugs, and the design of clinical trials [62].

The HCGP has had its opponents. They have offered objections to the project such as the possibilities that the project may not achieve its goals, that the expenditure incurred may decrease funding for investigator-initiated projects and that the funds could be better used to support other work such as genetic screens for factors required for the growth and survival of cancer cells [67–69]. However, these objections have had their rebuttals [70,71].

25.2 OUTLOOK

Epigenetics is a frontline area in cancer research. Since epigenetics plays a major role in the pathogenesis of cancer and since epigenetic events affect virtually every step in cancer progression [72], advances in epigenetic cancer research are likely to have a major impact in the future on the diagnosis, prevention, and treatment of cancer. Due to these advances one can expect that in the future many previously fatal cancers will become manageable.

REFERENCES

1. Jones, P.A., Overview of cancer epigenetics. *Semin. Hematol.*, 42, S3, 2005.
2. Reik, W., Stability and flexibility of epigenetic gene regulation in mammalian development. *Nature*, 447, 425, 2007.
3. Laird, P.W., Cancer epigenetics. *Hum. Mol. Genet.*, 14, R65, 2005.
4. Eckhardt, F. et al., DNA methylation profiling of human chromosomes 6, 20 and 22. *Nat. Genet.*, 38, 1378, 2006.
5. Rodenhiser, D. and Mann, M., Epigenetics and human disease: Translating basic biology into clinical applications. *CMAJ*, 174, 341, 2006.
6. Peedicayil, J., Epigenetic therapy—A new development in pharmacology. *Indian J. Med. Res.*, 123, 17, 2006.
7. Putral, L.N., Gu, W., and McMillan, N.A.J., RNA interference for the treatment of cancer. *Drug. News. Perspect.*, 19, 317, 2006.
8. Takeshita, F. and Ochiya, T., Therapeutic potential of RNA interference against cancer. *Cancer Sci.*, 97, 689, 2006.
9. Jones, P.A. and Baylin, S.B., The epigenomics of cancer. *Cell*, 128, 683, 2007.
10. Egger, G. et al., Epigenetics in human disease and prospects for epigenetic therapy. *Nature*, 429, 457, 2004.
11. Roth, S.Y., Denu, J.M., and Allis, C.D., Histone acetyltransferases. *Annu. Rev. Biochem.*, 70, 81, 2001.
12. Kristeleit, R. et al., Histone modification enzymes: Novel targets for cancer drugs. *Expert Opin. Emerg. Drugs*, 9, 135, 2004.

13. Balasubramanyam, K. et al., Polyisoprenylated benzophenone, garcinol, a natural histone acetyltransferase inhibitor, represses chromatin transcription and alters global gene expression. *J. Biol. Chem.*, 279, 33716, 2004.

14. Marcu, M.G. et al., Curcumin is an inhibitor of p300 acetyltransferase. *Med. Chem.*, 2, 169, 2006.

15. Sun, Y. et al., Inhibition of histone acetyltransferase activity by anacardic acid sensitizes tumor cells to ionizing radiation. *FEBS. Lett.*, 580, 4353, 2006.

16. Biel, M. et al., Design, synthesis, and biological evaluation of a small-molecule inhibitor of the histone acetyltransferase Gcn5. *Angew Chem. Int. Ed.*, 43, 3974, 2004.

17. Stimson, L. et al., Isothiazolones as inhibitors of PCAF and p300 histone acetyltransferase activity. *Mol. Cancer Ther.*, 4, 1521, 2005.

18. Tian, X. and Fang, J., Current perspectives on histone demethylases. *Acta Biochim. Biophys. Sin.*, 39, 81, 2007.

19. Balch, C. et al., New anti-cancer strategies: Epigenetic therapies and biomarkers. *Front. Biosci.*, 10, 1897, 2005.

20. Metzger, E. et al., LSD1 demethylates repressive histone marks to promote androgen-receptor-dependent transcription. *Nature*, 437, 436, 2005.

21. Schmidt, D.M.Z. and McCafferty, D.G., *trans*-2-Phenylcyclopropylamine is a mechanism-based inactivator of the histone demethylase LSD1. *Biochemistry*, 46, 4408, 2007.

22. Ballestar, E. and Wolffe, A.P., Methyl-CpG-binding proteins. Targeting specific gene expression. *Eur. J. Biochem.*, 268, 1, 2001.

23. Campbell, P.M., Bovenzi, V., and Szyf, M., Methylated DNA-binding protein 2 antisense inhibitors suppress tumourigenesis of human cancer cell lines in vitro and in vivo. *Carcinogenesis*, 25, 499, 2004.

24. Weissman, I.L., Stem cells: Units of development, units of regeneration, and units in evolution. *Cell*, 100, 157, 2000.

25. Körbling, M. and Estrov, Z., Adult stem cells for tissue repair—A new therapeutic concept? *N. Engl. J. Med.*, 349, 570, 2003.

26. Sell, S., Stem cell origin of cancer and differentiation therapy. *Crit. Rev. Oncol. Hematol.*, 51, 1, 2004.

27. Feinberg, A.P., Ohlsson, R., and Henikoff, S., The epigenetic progenitor origin of human cancer. *Nat. Rev. Genet.*, 7, 21, 2006.

28. Jordan, C.T., Guzman, M.L., and Noble, M., Cancer stem cells. *N. Engl. J. Med.*, 355, 1253, 2006.

29. Baylin, S.B. and Ohm, J.E., Epigenetic gene silencing in cancer—A mechanism for early oncogenic pathway addiction? *Nat. Rev. Cancer*, 6, 107, 2006.

30. Clarke, M.F. and Fuller, M., Stem cells and cancer: Two faces of Eve. *Cell*, 124, 1111, 2006.

31. Spivakov, M. and Fisher, A.G., Epigenetic signatures of stem-cell identity. *Nat. Rev. Genet.*, 8, 263, 2007.

32. Ohm, J.E. et al., A stem cell-like chromatin pattern may predispose tumor suppressor genes to DNA hypermethylation and heritable silencing. *Nat. Genet.*, 39, 237, 2007.

33. Widschwendter, M. et al., Epigenetic stem cell signature in cancer. *Nat. Genet.*, 39, 157, 2007.

34. Sell, S., Potential gene therapy strategies for cancer stem cells. *Curr. Gene Ther.*, 6, 579, 2006.

35. Mattick, J.S. and Makunin, I.V., Small regulatory RNAs in mammals. *Hum. Mol. Genet.*, 14, R121, 2005.

36. Hüttenhofer, A., Schattner, P., and Polacek, N., Non-coding RNAs: Hope or hype? *Trends. Genet.*, 21, 289, 2005.

37. Wall, N.R. and Shi, Y., Small RNA: Can RNA interference be exploited for therapy? *Lancet*, 362, 1401, 2003.

38. Mello, C.C. and Conte, D., Revealing the world of RNA interference. *Nature*, 431, 338, 2004.

39. Downward, J., RNA interference. *BMJ.*, 328, 1245, 2004.

40. Gillies, J.K. and Lorimer, A.J., Regulation of $p27^{Kip1}$ by miRNA 221/222 in glioblastoma. *Cell Cycle*, 6, 2005, 2007.

41. Jin, K. et al., The survival kinase Mirk/Dyrk1B is a downstream effector of oncogenic K-ras in pancreatic cancer. *Cancer Res.*, 67, 7247, 2007.

42. Frankish, H., Consortium uses RNAi to uncover genes' function. *Lancet*, 361, 584, 2003.

43. Root, D.E. et al., Genome-scale loss-of-function screening with a lentiviral RNAi library. *Nat. Methods.*, 3, 715, 2006.

44. Downward, J., RNA interference libraries prove their worth in hunt for tumor suppressor genes. *Cell*, 121, 813, 2005.

45. Westbrook, T.F. et al., A genetic screen for candidate tumor suppressors identifies REST. *Cell*, 121, 837, 2005.

46. Kolfschoten, I.G.M. et al., A genetic screen identifies PITX1 as a suppressor of RAS activity and tumorigenicity. *Cell*, 121, 849, 2005.

47. Bradbury, J., Human epigenome project—Up and running. *PloS Biol.*, 1, 316, 2003.

48. Rakyan, V.K. et al., DNA methylation profiling of the human major histocompatibility complex: A pilot study for the human epigenome project. *PloS Biol.*, 2, 2170, 2004.

49. Jones, P.A. and Martienssen, R., A blueprint for a human epigenome project: The AACR human epigenome workshop. *Cancer Res.*, 65, 11241, 2005.

50. Garber, K., Momentum building for human epigenome project. *J. Natl. Cancer Inst.*, 98, 84, 2006.

51. Esteller, M., The necessity of a human epigenome project. *Carcinogenesis*, 27, 1121, 2006.

52. Brena, R.M. et al., Toward a human epigenome. *Nat. Genet.*, 38, 1359, 2006.

53. Costello, J.F. et al., Aberrant CpG-island methylation has non-random and tumor-type-specific patterns. *Nat. Genet.*, 25, 132, 2000.

54. Paz, M.F. et al., A systematic profile of DNA methylation in human cancer cell lines. *Cancer Res.*, 63, 1114, 2003.

55. Esteller, M. et al., A gene hypermethylation profile of human cancer. *Cancer Res.*, 61, 3225, 2001.

56. van Steensel, B. and Henikoff, S., Epigenomic profiling using microarrays. *BioTechniques*, 35, 346, 2003.

57. Buck, M.J. and Lieb, J.D., ChIP-chip: Considerations for the design, analysis, and application of genome-wide chromatin immunoprecipitation experiments. *Genomics*, 83, 349, 2004.

58. Fraga, M.F. and Esteller, M., Towards the human cancer epigenome. A first draft of histone modifications. *Cell Cycle*, 4, 1377, 2005.

59. Plass, C. and Smiraglia, D.J., Genome-wide analysis of DNA methylation changes in human malignancies. *Curr. Top. Microbiol. Immunol.*, 310, 179, 2006.

60. Horsthemke, B., Epimutations in human disease. *Curr. Top. Microbiol. Immunol.*, 310, 45, 2006.

61. Peedicayil, J., Epialleles and common disease. *Med. Hypotheses*, 64, 215, 2005.

62. Recommendation for a Human Cancer Genome Project. Report of Working Group on Biomedical Technology, February 2005 (available at www.genome.gov/Pages/About/NACHGR/May2005NACHGR Agenda/ ReportoftheWorkingGrouponBiomedicalTechnology.pdf.).

63. Garber, K., Human cancer genome project moving forward despite some doubts in community. *J. Natl. Cancer Inst.*, 97, 1322, 2005.

64. Bonetta, L., Going on a cancer gene hunt. *Cell*, 123, 735, 2005.

65. Check, E., Cancer atlas maps out sample worries. *Nature*, 447, 1036, 2007.

66. Vastag, B., NIH Institutes launch joint venture to map cancer genome. *J. Natl. Cancer Inst.*, 98, 162, 2006.

67. Elledge, S.J. and Hannon, G.J., An open letter to cancer researchers. *Science*, 310, 439, 2005.

68. Gabor Miklos, G.L., The human cancer genome project—One more misstep in the war on cancer. *Nat. Biotechnol.*, 23, 535, 2005.

69. Chng, W.J., Limits to the human cancer genome project? *Science*, 315, 762, 2007.

70. Varmus, H. and Stillman, B., Support for the human cancer genome project. *Science*, 310, 1615, 2005.

71. Sjöblom, T. et al., Response. *Science*, 315, 764, 2007.

72. Jones, P.A. and Baylin, S.B., The fundamental role of epigenetic events in cancer. *Nat. Rev. Cancer*, 3, 415, 2002.

Index

A

3-AB, *see* 3-Aminobenzamide
AcCoA synthetases, 307
Acetate metabolism, 308
Acetylcoenzyme A (AcCoA), 304
ACSs, *see* AcCoA synthetases
Acute myelogenous leukemia
    HAT/HDAC in, 183
    HMTs in, 186
    NSD1 protein, 124
    NUP98 gene in, 124
    superior survival in, 405
Acute myeloid leukemia, 294
    bone marrow cells, transduction of, 294
    HAT/HDAC in, 183
    histone modification in, 352
    transformation of MDS to, 417
Acute promyelocytic leukemia, 405
Adenomatous polyposis coli (*APC*)
    methylation of, 337
    Wnt signaling pathway, 83, 335
ADP-ribose polymer adducts, 265, 275–276
ADP-ribose protein lyase, 269
ADP-ribosyltransferase, 152
Aerobic glycolysis, 311
Age-dependent methylation changes
    and carcinogenesis, 400
Aging
    caloric restriction and exercise, 310–311
    epigenetics, 304–305
    and glycolysis, 310
    involvement of sirtuins in, 159–162, 164,
        170–171
    SAH levels and, 127
AKT, phosphorylation by, 289
Allele-specific oligonucleotide, 341
American Association for Cancer Research, 433
3-Aminobenzamide, 275–276
AML, *see* Acute myelogenous leukemia;
    Acute myeloid leukemia
Androgen receptor (AR), 182
Aneuploidy, 134
Antibody-based assays, 352
Anticancer drugs, epigenetic targets for, 429–431
Anti-NY-ESO-1 antibodies, 421
Antisense RNA mediated chromatin alterations,
    53–55
APL, *see* Acute promyelocytic leukemia
Apoptosis
    caspase-dependent, 386–387
    dietary ingredients, due to, 201
    DNA circulation, 325
    HDACi, 388, 390
    miR-21 overexpression, 225
    modulators, 43
    PAR metabolism, 272–273

p53, *BNIP3*, and *caspase-8* gene, 43, 46
    p21 protein and, 104
    p53 tumor suppressor gene and, 387
    Sphk2 role in, 220
    suppression by SIRT1, 166–167, 171–172
    TNFα-induced, 168
ARH 3 enzyme, 270
ART, *see* ADP-ribosyltransferase
ASO, *see* Allele-specific oligonucleotide
Aurora kinase, 349
Azacytidine, 406–407
5-Azacytidine, 416
5-Aza-2′-deoxycytidine, *see* Decitabine

B

Bacteria
    genetic diversity of, 135, 137
    ncRNA gene expression, 218
Bcl6 acetylation, 168
Beckwith–Wiedemann syndrome, 56
Benzamides, 420
Biotination of histone, 190
Biotinylation of histone, dietary influence
    of, 204
Bisulfite conversion techniques, 322–323
Bisulfite genomic sequencing, 339–340
    for methylation analysis, 401
Bivalent chromatin domains, 370
Bivalent marking, 287
Bladder cancer, 349
BMI1, 283
    self-renewal of stem cells, 287
    self-ubiquitylation of, 289
BMI$^{-/-}$ bone marrow cells, 294
BMI1 gene, 290
BNIP3 gene silencing, 43
Body fluids, 323–325
BORIS expression and CT antigen genes, 11
Brain cytoplasmic RNA 1 (RNAs BC1), 223
Breast cancer, 350, 369–371
    CDH1 expression in primary and metastatic, 88
    HDAC inhibition, 391
    histone acetylation in, 101
    histone trimethylation, 350
    hypomethylation of Sat2 sequences, 14
    methyl acceptance capacity, 13
    PCR overexpression, 256
    PITX2 methylation and, 404
    Plu-1/JARID1B gene and, 126
    SMYD3 overexpression in, 186
Breast cancer 1 (*BRCA1*) gene, early onset,
    germline mutations in, 85
Breast cancer cell, *see* MCF7 cell
Bromodomain, 362–363
BWS, *see* Beckwith–Wiedemann syndrome

437

T - #0237 - 251019 - C472 - 254/178/21 - PB - 9780367386863